THE CHEMISTRY OF RADICAL POLYMERIZATION

SECOND FULLY REVISED EDITION

GRAEME MOAD
CSIRO Molecular and Health Technologies
Bayview Ave, Clayton, Victoria 3168, AUSTRALIA

and

DAVID H. SOLOMON
Department of Chemical and Biomolecular Engineering
University of Melbourne, Victoria 3010, AUSTRALIA

2006

ELSEVIER

Amsterdam – Boston – Heidelberg – London – New York – Oxford – Paris – San Diego
San Francisco – Singapore – Sydney – Tokyo

ELSEVIER B.V.	ELSEVIER Inc.	**ELSEVIER Ltd**	ELSEVIER Ltd
Radarweg 29	525 B Street, Suite 1900	**The Boulevard, Langford Lane**	84 Theobalds Road
P.O. Box 211, 1000 AE Amsterdam	San Diego, CA 92101-4495	**Kidlington, Oxford OX5 1GB**	London WC1X 8RR
The Netherlands	USA	**UK**	UK

© 2006 Elsevier Ltd. All rights reserved.

This work is protected under copyright by Elsevier Ltd, and the following terms and conditions apply to its use:

Photocopying
Single photocopies of single chapters may be made for personal use as allowed by national copyright laws. Permission of the Publisher and payment of a fee is required for all other photocopying, including multiple or systematic copying, copying for advertising or promotional purposes, resale, and all forms of document delivery. Special rates are available for educational institutions that wish to make photocopies for non-profit educational classroom use.

Permissions may be sought directly from Elsevier's Rights Department in Oxford, UK: phone (+44) 1865 843830, fax (+44) 1865 853333, e-mail: permissions@elsevier.com. Requests may also be completed on-line via the Elsevier homepage (http://www.elsevier.com/locate/permissions).

In the USA, users may clear permissions and make payments through the Copyright Clearance Center, Inc., 222 Rosewood Drive, Danvers, MA 01923, USA; phone: (+1) (978) 7508400, fax: (+1) (978) 7504744, and in the UK through the Copyright Licensing Agency Rapid Clearance Service (CLARCS), 90 Tottenham Court Road, London W1P 0LP, UK; phone: (+44) 20 7631 5555; fax: (+44) 20 7631 5500. Other countries may have a local reprographic rights agency for payments.

Derivative Works
Tables of contents may be reproduced for internal circulation, but permission of the Publisher is required for external resale or distribution of such material. Permission of the Publisher is required for all other derivative works, including compilations and translations.

Electronic Storage or Usage
Permission of the Publisher is required to store or use electronically any material contained in this work, including any chapter or part of a chapter.

Except as outlined above, no part of this work may be reproduced, stored in a retrieval system or transmitted in any form or by any means, electronic, mechanical, photocopying, recording or otherwise, without prior written permission of the Publisher.
Address permissions requests to: Elsevier's Rights Department, at the fax and e-mail addresses noted above.

Notice
No responsibility is assumed by the Publisher for any injury and/or damage to persons or property as a matter of products liability, negligence or otherwise, or from any use or operation of any methods, products, instructions or ideas contained in the material herein. Because of rapid advances in the medical sciences, in particular, independent verification of diagnoses and drug dosages should be made.

First edition 2006

Library of Congress Cataloging in Publication Data
A catalog record is available from the Library of Congress.

British Library Cataloguing in Publication Data
A catalogue record is available from the British Library.

ISBN-10: 0-08-044288-9 (Hardbound)
ISBN-13: 978-0-08-044288-4 (Hardbound)
ISBN-10: 0-08-044286-2 (Paperback)
ISBN-13: 978-0-08-044286-0 (Paperback)

∞ The paper used in this publication meets the requirements of ANSI/NISO Z39.48-1992 (Permanence of Paper).
Printed in The Netherlands.

Working together to grow
libraries in developing countries

www.elsevier.com | www.bookaid.org | www.sabre.org

ELSEVIER   BOOK AID International   Sabre Foundation

THE CHEMISTRY OF RADICAL POLYMERIZATION

SECOND FULLY REVISED EDITION

Contact Details

Dr Graeme Moad
CSIRO Molecular and Health Technologies
Bayview Ave, Clayton, Victoria 3168
AUSTRALIA
Email: graeme.moad@csiro.au

Prof David H. Solomon
Department of Chemical and Biomolecular Engineering
University of Melbourne
Victoria 3010
AUSTRALIA
Email: davids@unimelb.edu.au

Contents

Contents v
Index to Tables xvi
Index to Figures xx
Preface to the First Edition xxiii
Preface to the Second Edition xxv
Acknowledgments xxvi
1   INTRODUCTION 1
1.1   References 8
2   RADICAL REACTIONS 11
2.1   Introduction 11
2.2   Properties of Radicals 12
    2.2.1   Structures of Radicals 12
    2.2.2   Stabilities of Radicals 14
    2.2.3   Detection of Radicals 14
2.3   Addition to Carbon-Carbon Double Bonds 16
    2.3.1   Steric Factors 19
    2.3.2   Polar Factors 21
    2.3.3   Bond Strengths 22
    2.3.4   Stereoelectronic Factors 23
    2.3.5   Entropic Considerations 24
    2.3.6   Reaction Conditions 24
        2.3.6.1   Temperature 24
        2.3.6.2   Solvent 25
    2.3.7   Theoretical Treatments 26
    2.3.8   Summary 28
2.4   Hydrogen Atom Transfer 29
    2.4.1   Bond Dissociation Energies 30
    2.4.2   Steric Factors 30
    2.4.3   Polar Factors 31
    2.4.4   Stereoelectronic Factors 32
    2.4.5   Reaction Conditions 33
    2.4.6   Abstraction *vs* Addition 34
    2.4.7   Summary 36

Contents

2.5 Radical-Radical Reactions 36
  2.5.1 Pathways for Combination 37
  2.5.2 Pathways for Disproportionation 38
  2.5.3 Combination *vs* Disproportionation 39
    2.5.3.1 Statistical factors 39
    2.5.3.2 Steric factors 40
    2.5.3.3 Polar factors 41
    2.5.3.4 Stereoelectronic and other factors 41
    2.5.3.5 Reaction conditions 42
  2.5.4 Summary 43
2.6 References 44

3 INITIATION 49
3.1 Introduction 49
3.2 The Initiation Process 50
  3.2.1 Reaction with Monomer 51
  3.2.2 Fragmentation 54
  3.2.3 Reaction with Solvents, Additives, or Impurities 55
  3.2.4 Effects of Temperature and Reaction Medium on Radical Reactivity 55
  3.2.5 Reaction with Oxygen 56
  3.2.6 Initiator Efficiency in Thermal Initiation 57
  3.2.7 Photoinitiation 58
  3.2.8 Cage Reaction and Initiator-Derived Byproducts 60
  3.2.9 Primary Radical Termination 61
  3.2.10 Transfer to Initiator 62
  3.2.11 Initiation in Heterogeneous Polymerization 63
3.3 The Initiators 64
  3.3.1 Azo-Compounds 68
    3.3.1.1 Dialkyldiazenes 68
      3.3.1.1.1 Thermal decomposition 72
      3.3.1.1.2 Photochemical decomposition 74
      3.3.1.1.3 Initiator efficiency 74
      3.3.1.1.4 Transfer to initiator 77
    3.3.1.2 Hyponitrites 78
  3.3.2 Peroxides 79
    3.3.2.1 Diacyl or diaroyl peroxides 82
      3.3.2.1.1 Thermal decomposition 82
      3.3.2.1.2 Photochemical decomposition 83
      3.3.2.1.3 Initiator efficiency 84
      3.3.2.1.4 Transfer to initiator and induced decomposition 85
      3.3.2.1.5 Redox reactions 85
    3.3.2.2 Dialkyl peroxydicarbonates 87
    3.3.2.3 Peroxyesters 88

Contents

- 3.3.2.3.1 Thermal decomposition ... 88
- 3.3.2.3.2 Photochemical decomposition ... 90
- 3.3.2.4 Dialkyl peroxides ... 90
- 3.3.2.5 Alkyl hydroperoxides ... 92
- 3.3.2.6 Inorganic peroxides ... 93
  - 3.3.2.6.1 Persulfate ... 94
  - 3.3.2.6.2 Hydrogen peroxide ... 96
- 3.3.3 Multifunctional Initiators ... 96
  - 3.3.3.1 Concerted decomposition ... 97
  - 3.3.3.2 Non-concerted decomposition ... 97
- 3.3.4 Photochemical Initiators ... 98
  - 3.3.4.1 Aromatic carbonyl compounds ... 98
    - 3.3.4.1.1 Benzoin and related compounds ... 99
    - 3.3.4.1.2 Carbonyl compound-tertiary amine systems ... 102
  - 3.3.4.2 Sulfur compounds ... 103
- 3.3.5 Redox Initiators ... 104
  - 3.3.5.1 Metal complex-organic halide redox systems ... 104
  - 3.3.5.2 Ceric ion systems ... 105
- 3.3.6 Thermal Initiation ... 106
  - 3.3.6.1 Styrene homopolymerization ... 107
  - 3.3.6.2 Acrylate homopolymerization ... 109
  - 3.3.6.3 Copolymerization ... 110
- **3.4 The Radicals ... 111**
  - 3.4.1 Carbon-Centered Radicals ... 112
    - 3.4.1.1 Alkyl radicals ... 112
      - 3.4.1.1.1 α-Cyanoalkyl radicals ... 113
    - 3.4.1.2 Aryl radicals ... 117
    - 3.4.1.3 Acyl radicals ... 117
  - 3.4.2 Oxygen-Centered Radicals ... 118
    - 3.4.2.1 Alkoxy radical ... 118
      - 3.4.2.1.1 t-Butoxy radicals ... 119
      - 3.4.2.1.2 Other t-alkoxy radicals ... 124
      - 3.4.2.1.3 Primary and secondary alkoxy radical ... 125
    - 3.4.2.2 Acyloxy and alkoxycarbonyloxy radicals ... 125
      - 3.4.2.2.1 Benzoyloxy radicals ... 126
      - 3.4.2.2.2 Alkoxycarbonyloxy radicals ... 127
    - 3.4.2.3 Hydroxy radicals ... 128
    - 3.4.2.4 Sulfate radical anion ... 129
    - 3.4.2.5 Alkylperoxy radicals ... 130
  - 3.4.3 Other Heteroatom-Centered Radicals ... 131
    - 3.4.3.1 Silicon-centered radicals ... 131
    - 3.4.3.2 Sulfur- and selenium-centered radicals ... 132
    - 3.4.3.3 Phosphorus-centered radicals ... 132

3.5 Techniques 133
3.5.1 Kinetic Studies 133
3.5.2 Radical Trapping 133
3.5.2.1 Spin traps 134
3.5.2.2 Transition metal salts 136
3.5.2.3 Metal hydrides 137
3.5.2.4 Nitroxides 138
3.5.2.5 α-Methystyrene dimer 140
3.5.3 Direct Detection of End Groups 141
3.5.3.1 Infra-red and UV-visible spectroscopy 141
3.5.3.2 Nuclear magnetic resonance spectroscopy 142
3.5.3.3 Electron paramagnetic resonance spectroscopy 143
3.5.3.4 Mass spectrometry 143
3.5.3.5 Chemical methods 144
3.5.4 Labeling Techniques 145
3.5.4.1 Radiolabeling 145
3.5.4.2 Stable isotopes and nuclear magnetic resonance 146
3.6 References 149

4 PROPAGATION 167
4.1 Introduction 167
4.2 Stereosequence Isomerism - Tacticity 168
4.2.1 Terminology and Mechanisms 168
4.2.2 Experimental Methods for Determining Tacticity 173
4.2.3 Tacticities of Polymers 173
4.3 Regiosequence Isomerism - Head vs Tail Addition 176
4.3.1 Monoene Polymers 176
4.3.1.1 Poly(vinyl acetate) 178
4.3.1.2 Poly(vinyl chloride) 179
4.3.1.3 Fluoro-olefin polymers 180
4.3.1.4 Allyl polymers 181
4.3.1.5 Acrylic polymers 182
4.3.2 Conjugated Diene Polymers 182
4.3.2.1 Polybutadiene 184
4.3.2.2 Polychloroprene, polyisoprene 184
4.4 Structural Isomerism - Rearrangement 185
4.4.1 Cyclopolymerization 185
4.4.1.1 1,6-Dienes 186
4.4.1.2 Triene monomers 191
4.4.1.3 1,4- and 1,5-dienes 192
4.4.1.4 1,7- and higher 1,n-dienes 193
4.4.1.5 Cyclo-copolymerization 194
4.4.2 Ring-Opening Polymerization 194

        4.4.2.1   Vinyl substituted cyclic compounds 196
        4.4.2.2   Methylene substituted cyclic compounds.. 199
        4.4.2.3   Double ring-opening polymerization 205
    4.4.3   Intramolecular Atom Transfer 208
        4.4.3.1   Polyethylene and copolymers 208
        4.4.3.2   Vinyl polymers... 211
        4.4.3.3   Acrylate esters and other monosubstituted monomers 211
        4.4.3.4   Addition-abstraction polymerization 212

4.5   Propagation Kinetics and Thermodynamics 213
    4.5.1   Polymerization Thermodynamics... 213
    4.5.2   Measurement of Propagation Rate Constants 216
    4.5.3   Dependence of Propagation Rate Constant on Monomer Structure 218
    4.5.4   Chain Length Dependence of Propagation Rate Constants.......................... 220

4.6   References 221

5   TERMINATION.. 233

5.1   Introduction 233

5.2   Radical-Radical Termination 234
    5.2.1   Termination Kinetics.. 235
        5.2.1.1   Classical kinetics 235
        5.2.1.2   Molecular weights and molecular weight averages 238
        5.2.1.3   Molecular weight distributions 240
        5.2.1.4   Diffusion controlled termination 242
             5.2.1.4.1 Termination at low conversion 244
             5.2.1.4.2 Termination at medium to high conversions 248
        5.2.1.5   Termination in heterogeneous polymerization 249
        5.2.1.6   Termination during living radical polymerization 250
    5.2.2   Disproportionation *vs* Combination 251
        5.2.2.1   Model studies 252
             5.2.2.1.1 Polystyrene and derivatives 253
             5.2.2.1.2 Poly(alkyl methacrylates)... 255
             5.2.2.1.3 Poly(methacrylonitrile)... 256
             5.2.2.1.4 Polyethylene.. 258
        5.2.2.2   Polymerization 258
             5.2.2.2.1 Polystyrene.. 260
             5.2.2.2.2 Poly(alkyl methacrylates)... 261
             5.2.2.2.3 Poly(methacrylonitrile)... 262
             5.2.2.2.4 Poly(alkyl acrylates) 262
             5.2.2.2.5 Poly(acrylonitrile) 262
             5.2.2.2.6 Poly(vinyl acetate) 263
             5.2.2.2.7 Poly(vinyl chloride) 263
        5.2.2.3   Summary.. 263

5.3   Inhibition and Retardation... 264

	5.3.1 'Stable' Radicals	267
	5.3.2 Oxygen	268
	5.3.3 Monomers	269
	5.3.4 Phenols	270
	5.3.5 Quinones	271
	5.3.6 Phenothiazine	272
	5.3.7 Nitrones, Nitro- and Nitroso-Compounds	272
	5.3.8 Transition Metal Salts	273
5.4	References	273
5.4	References	279
6	CHAIN TRANSFER	279
6.1	Introduction	279
6.2	Chain Transfer	279
	6.2.1 Measurement of Transfer Constants	283
	6.2.1.1 Addition-fragmentation	287
	6.2.1.2 Reversible chain transfer	288
	6.2.2 Homolytic Substitution Chain Transfer Agents	289
	6.2.2.1 Thiols	290
	6.2.2.2 Disulfides	291
	6.2.2.3 Monosulfides	292
	6.2.2.4 Halocarbons	293
	6.2.2.5 Solvents and other reagents	294
	6.2.3 Addition-Fragmentation Chain Transfer Agents	296
	6.2.3.1 Vinyl ethers	298
	6.2.3.2 Allyl sulfides, sulfonates, halides, phosphonates, silanes	299
	6.2.3.3 Allyl peroxides	303
	6.2.3.4 Macromonomers	305
	6.2.3.5 Thionoester and related transfer agents	308
	6.2.4 Abstraction-Fragmentation Chain Transfer	309
	6.2.5 Catalytic Chain Transfer	310
	6.2.5.1 Mechanism	310
	6.2.5.2 Catalysts	313
	6.2.5.2.1 Cobalt porphyrin and related complexes	313
	6.2.5.2.2 Cobalt (II) cobaloximes	313
	6.2.5.2.3 Cobalt (III) cobaloximes	314
	6.2.5.2.4 Other catalysts	315
	6.2.5.3 Reaction conditions	315
	6.2.6 Transfer to Monomer	316
	6.2.6.1 Styrene	317
	6.2.6.2 Vinyl acetate	318
	6.2.6.3 Vinyl chloride	318
	6.2.6.4 Allyl monomers	319

Contents

6.2.7	Transfer to Polymer	320
6.2.7.1	Polyethylene	321
6.2.7.2	Poly(alkyl methacrylates)	321
6.2.7.3	Poly(alkyl acrylates)	322
6.2.7.4	Poly(vinyl acetate)	323
6.2.7.5	Poly(vinyl chloride)	325
6.2.7.6	Poly(vinyl fluoride)	325
6.2.8	Transfer to Initiator	325
6.3	**References**	**325**
7	**COPOLYMERIZATION**	**333**
7.1	**Introduction**	**333**
7.2	**Copolymer Depiction**	**335**
7.3	**Propagation in Statistical Copolymerization**	**335**
7.3.1	Propagation Mechanisms in Copolymerization	337
7.3.1.1	Terminal model	337
7.3.1.2	Penultimate model	342
7.3.1.2.1	Model description	342
7.3.1.2.2	Remote substituent effects on radical addition	344
7.3.1.2.3	MMA-S copolymerization	347
7.3.1.2.4	Other copolymerizations	348
7.3.1.2.5	Origin of penultimate unit effects	349
7.3.1.3	Models involving monomer complexes	350
7.3.1.4	Copolymerization with depropagation	353
7.3.2	Chain Statistics	354
7.3.2.1	Binary copolymerization according to the terminal model	354
7.3.2.2	Binary copolymerization according to the penultimate model	355
7.3.2.3	Binary copolymerization according to other models	356
7.3.2.4	Terpolymerization	357
7.3.3	Estimation of Reactivity Ratios	359
7.3.3.1	Composition data	360
7.3.3.2	Monomer sequence distribution	362
7.3.4	Prediction of Reactivity Ratios	363
7.3.4.1	Q-e scheme	363
7.3.4.2	Patterns of reactivity scheme	365
7.4	**Termination in Statistical Copolymerization**	**366**
7.4.1	Chemical Control Model	366
7.4.2	Diffusion Control Models	368
7.4.3	Combination and Disproportionation during Copolymerization	370
7.4.3.1	Poly(methyl methacrylate-co-styrene)	371
7.4.3.2	Poly(methacrylonitrile-co-styrene)	373
7.4.3.3	Poly(butyl methacrylate-co-methacrylonitrile)	374
7.4.3.4	Poly(butyl methacrylate-co-methyl methacrylate)	374

| 7.4.3.5 Poly(ethylene-co-methacrylonitrile) 374
7.5   Functional and End-Functional Polymers 374
   7.5.1   Functional Initiators 375
   7.5.2   Functional Transfer Agents 377
   7.5.3   Thiol-ene Polymerization 378
   7.5.4   Functional Monomers 379
   7.5.5   Functional Inhibitors 381
   7.5.6   Compositional Heterogeneity in Functional Copolymers 381
7.6   Block & Graft Copolymerization 384
   7.6.1   Polymeric and Multifunctional Initiators 385
   7.6.2   Transformation Reactions 387
   7.6.3   Radiation-Induced Grafting Processes 389
   7.6.4   Radical-Induced Grafting Processes 390
      7.6.4.1   Maleic anhydride graft polyolefins 392
      7.6.4.2   Maleate ester and maleimide graft polyolefins 396
      7.6.4.3   (Meth)acrylate graft polyolefins 397
      7.6.4.4   Styrenic graft polyolefins 399
      7.6.4.5   Vinylsilane graft polyolefins 399
      7.6.4.6   Vinyl oxazoline graft polyolefins 400
   7.6.5   Polymerization and Copolymerization of Macromonomers 400
7.7   References 401

8   CONTROLLING POLYMERIZATION 413
8.1   Introduction 413
8.2   Controlling Structural Irregularities 414
   8.2.1   "Defect Structures" in Polystyrene 414
   8.2.2   "Defect Structures" in Poly(methyl methacrylate) 417
   8.2.3   "Defect Structures" in Poly(vinyl chloride) 420
8.3   Controlling Propagation 421
   8.3.1   Organic Solvents and Water 425
      8.3.1.1   Homopolymerization 426
      8.3.1.2   Copolymerization 429
   8.3.2   Supercritical Carbon Dioxide 432
   8.3.3   Ionic liquids 432
   8.3.4   Lewis Acids and Inorganics 433
      8.3.4.1   Homopolymerization 433
      8.3.4.2   Copolymerization 435
   8.3.5   Template Polymerization 437
      8.3.5.1   Non-covalently bonded templates 437
      8.3.5.2   Covalently bonded templates 438
   8.3.6   Enzyme Mediated Polymerization 440
   8.3.7   Topological Radical Polymerization 441

| 8.4 | References | 443 |

9 LIVING RADICAL POLYMERIZATION ... 451
9.1 Introduction ... 451
9.1.1 Living? Controlled? Mediated? ... 451
9.1.2 Tests for Living (Radical) Polymerization ... 452
9.2 Agents Providing Reversible Deactivation ... 454
9.3 Deactivation by Reversible Coupling and Unimolecular Activation .. 457
9.3.1 Kinetics and Mechanism ... 457
    9.3.1.1 Initiators, iniferters, initers ... 457
    9.3.1.2 Molecular weights and distributions ... 458
    9.3.1.3 Polymerization kinetics ... 460
9.3.2 Sulfur-Centered Radical Mediated Polymerization ... 461
    9.3.2.1 Disulfide initiators ... 461
    9.3.2.2 Monosulfide initiators ... 463
    9.3.2.3 Monomers, mechanism, side reactions ... 465
9.3.3 Selenium-Centered Radical Mediated Polymerization ... 466
9.3.4 Carbon-Centered Radical Mediated Polymerization ... 467
    9.3.4.1 Monomers, mechanism, side reactions ... 469
9.3.5 Reversible Addition-Fragmentation ... 470
9.3.6 Nitroxide Mediated Polymerization ... 471
    9.3.6.1 Nitroxides ... 472
    9.3.6.2 Initiation ... 475
    9.3.6.3 Side reactions ... 478
    9.3.6.4 Rate enhancement ... 479
    9.3.6.5 Monomers ... 480
        9.3.6.5.1 Styrene, vinyl aromatics ... 480
        9.3.6.5.2 Acrylates ... 480
        9.3.6.5.3 Methacrylates ... 481
        9.3.6.5.4 Diene monomers ... 481
    9.3.6.6 Heterogeneous polymerization ... 481
9.3.7 Other Oxygen-Centered Radical Mediated Polymerization ... 483
9.3.8 Nitrogen-Centered Radical Mediated Polymerization ... 483
9.3.9 Metal Complex-Mediated Radical Polymerization ... 484
9.4 Atom Transfer Radical Polymerization ... 486
9.4.1 Initiators ... 488
    9.4.1.1 Molecular weights and distributions ... 490
    9.4.1.2 Reverse ATRP ... 491
    9.4.1.3 Initiator activity ... 492
9.4.2 Catalysts ... 492
    9.4.2.1 Copper complexes ... 493
    9.4.2.2 Ruthenium complexes ... 495
    9.4.2.3 Iron complexes ... 496

        9.4.2.4   Nickel complexes...496
    9.4.3   Monomers and Reaction Conditions 497
        9.4.3.1   Solution polymerization..497
        9.4.3.2   Heterogeneous polymerization 497

9.5   Reversible Chain Transfer 498
    9.5.1   Molecular Weights and Distributions...499
    9.5.2   Macromonomer RAFT...501
    9.5.3   Thiocarbonylthio RAFT 502
        9.5.3.1   Mechanism 503
        9.5.3.2   RAFT agents 505
        9.5.3.3   RAFT agent synthesis 515
        9.5.3.4   Side Reactions..517
        9.5.3.5   Reaction conditions 518
        9.5.3.6   Heterogeneous polymerization 520
    9.5.4   Iodine-Transfer Polymerization..521
    9.5.5   Telluride-Mediated Polymerization 522
    9.5.6   Stibine-Mediated Polymerization 524

9.6   Living Radical Copolymerization 525
    9.6.1   Reactivity Ratios...525
    9.6.2   Gradient Copolymers 526
    9.6.3   NMP 527
    9.6.4   ATRP...528
    9.6.5   RAFT...529

9.7   End-Functional Polymers 531
    9.7.1   NMP 531
        9.7.1.1   ω-Functionalization...531
        9.7.1.2   α-Functionalization 533
    9.7.2   ATRP...533
        9.7.2.1   ω-Functionalization...533
        9.7.2.2   α-Functionalization 536
    9.7.3   RAFT...538
        9.7.3.1   ω-Functionalization...538
        9.7.3.2   α-Functionalization 539

9.8   Block Copolymers..540
    9.8.1   Direct Diblock Synthesis 541
        9.8.1.1   NMP 541
        9.8.1.2   ATRP 541
        9.8.1.3   RAFT...543
    9.8.2   Transformation Reactions...544
        9.8.2.1   Second step NMP 545
        9.8.2.2   Second step ATRP..545
        9.8.2.3   Second step RAFT..546

| 9.8.3 Triblock Copolymers 546
| 9.8.4 Segmented Block Copolymers 547
| **9.9** **Star Polymers** 548
| 9.9.1 Core-first Star Synthesis 549
| 9.9.2 Arm-first Star Synthesis 554
| 9.9.3 Hyperbranched Polymers 555
| 9.9.3.1 Self-condensing vinyl polymerization 555
| 9.9.3.2 Dendritic polymers 556
| **9.10** **Graft Copolymers/Polymer Brushes** 558
| 9.10.1 Grafting Through - Copolymerization of Macromonomers 558
| 9.10.2 Grafting From - Surface Initiated Polymerization 560
| 9.10.2.1 Grafting from polymer surfaces 560
| 9.10.2.2 Grafting from inorganic surfaces 562
| 9.10.3 Grafting To - Use of End-Functional Polymers 563
| **9.11** **Outlook for Living Radical Polymerization** 563
| **9.12** **References** 564
| **ABBREVIATIONS** 587
| **SUBJECT INDEX** 591

Index to Tables

Table 2.1 Carbon-Hydrogen and Heteroatom-Hydrogen Bond Dissociation Energies (D in kJ mol^{-1}) 15
Table 2.2 Relative Rate Constants and Regiospecificities for Addition of Radicals to Halo-Olefins 17
Table 2.3 Relative Rate Constants for Reactions of Radicals with Alkyl-Substituted Acrylate Esters 18
Table 2.4 Hammett ρ and ρ^+ Parameters for Reactions of Radicals 22
Table 2.5 Specificity of Intramolecular Hydrogen Abstraction 32
Table 2.6 Bond Dissociation Energies 34
Table 2.7 Values of k_{td}/k_{tc} for the Cross-Reaction between Fluoromethyl and Ethyl Radicals at 25 °C 41
Table 2.8 Values of k_{td}/k_{tc} for t-Butyl Radicals at 25 °C 43
Table 3.1 Guide to Properties of Polymerization Initiators 66
Table 3.2 Selected Kinetic Data for Decomposition of Azo-Compounds 70
Table 3.3 Solvent Dependence of Rate Constants for AIBMe Decomposition 73
Table 3.4 Zero-Conversion Initiator Efficiency for AIBMe under Various Reaction Conditions 76
Table 3.5 Selected Kinetic Data for Decomposition of Peroxides 80
Table 3.6 Kinetic Data for Reactions of Carbon-Centered Radicals 114
Table 3.7 Selected Rate Data for Reactions of Oxygen-Centered Radicals 119
Table 3.8 Specificity Observed in the Reactions of Oxygen-Centered Radicals with Various Monomers at 60 °C 120
Table 3.9 Kinetic Data for Reactions of t-Butoxy Radicals in Various Solvents 124
Table 3.10 Selected Rate Data for Reactions of Heteroatom-Centered Radicals 131
Table 3.11 Radical Trapping Agents for Studying Initiation 134
Table 3.12 Application of MALDI-TOF or ESI Mass Spectrometry to Polymers Prepared by Radical Polymerization 144
Table 3.13 Radical Polymerizations Performed with Initiators Labeled with Stable Isotopes 147
Table 4.1 Tacticities of Selected Homopolymers 175
Table 4.2 Temperature Dependence of Head vs Tail Addition for Fluoro-olefin Monomers 181
Table 4.3 Microstructure of Poly(chloroprene) vs Temperature 185

Index to Tables

Table 4.4 Ring Sizes Formed in Cyclopolymerization of Symmetrical 1,6-Diene Monomers 190
Table 4.5 Extent of Ring-opening During Polymerizations of 2-Methylene-1,3-dioxolane and Related Species 200
Table 4.6 Extent of Ring-Opening During Polymerizations of 4-Methylene-1,3-dioxolane and 2-Methylene-1,4-dioxane Derivatives 203
Table 4.7 Extent of Ring-Opening During Polymerizations of 2-Methylenetetrahydrofuran and Related Compounds 204
Table 4.8 Extent of Double Ring-Opening During Polymerization of Polycyclic Monomers 207
Table 4.9 Structures Formed by Backbiting in Ethylene Copolymerization 210
Table 4.10 Heats of Polymerization for Selected Monomers 215
Table 4.11 Kinetic Parameters for Propagation in Selected Radical Polymerizations in Bulk Monomer 219
Table 4.12 Rate Constants and Arrhenius Parameters for Propagation of Monomers Compared with Rate Constants for Addition of Small Radicals 221
Table 5.1 Parameters Characterizing Chain Length Dependence of Termination Rate Coefficients in Radical Polymerization of Common Monomers 247
Table 5.2 Values of k_{td}/k_{tc} for Polystyryl Radical Model Systems 254
Table 5.3 Values of k_{td}/k_{tc} for Methacrylate Ester Model Systems 255
Table 5.4 Values of k_{td}/k_{tc} for Reactions involving Cyanoisopropyl Radicals 257
Table 5.5 Determinations of k_{td}/k_{tc} for MMA Polymerization 261
Table 5.6 Kinetic Data for Various Inhibitors with Some Common Monomers 265
Table 5.7 Absolute Rate Constants for the Reaction of Carbon-Centered Radicals with Some Common Inhibitors 266
Table 6.1 Chain Length Dependence of Transfer Constants (C_n) 283
Table 6.2 Transfer Constants (60 °C, bulk) for Thiols (RSH) with Various Monomers 290
Table 6.3 Transfer Constants for Disulfides (R-S-S-R) With Various Monomers 292
Table 6.4 Transfer Constants (80 °C, bulk) for Halocarbons with Various Monomers 293
Table 6.5 Transfer Constants (60 °C, bulk) for Selected Solvents and Additives with Various Monomers 295
Table 6.6 Transfer Constants for Vinyl Ethers at 60 °C 299
Table 6.7 Transfer Constants for Allyl Sulfides at 60 °C 300
Table 6.8 Transfer Constants for Allyl Sulfonates and Sulfoxides at 60 °C 302
Table 6.9 Transfer Constants for Allyl Halides, Phosphonates, Silanes and Stannanes at 60 °C 303

Index to Tables

Table 6.10 Transfer Constants for Allyl Peroxide and Related Transfer Agents at 60 °C 304
Table 6.11 Transfer Constants for Macromonomers 307
Table 6.12 Transfer Constants for Thionoester and Related Transfer Agents at 60 °C 309
Table 6.13 Transfer Constants for Cobalt Complexes..316
Table 6.14 Transfer Constants to Monomer...317
Table 6.15 Transfer Constants to Polymer...320
Table 7.1 Reactivity Ratios for Some Common Monomer Pairs.............................339
Table 7.2 Relative Rates for Addition of Substituted Propyl Radicals to AN and S..345
Table 7.3 Relative Rates for Addition of Substituted Methyl Radicals ($R^3R^2R^1C\bullet$) to MMA and S at ~25 °C..346
Table 7.4 Rate Constants (295 K) for Addition of Substituted Propyl Radicals to (Meth)acrylate Esters..347
Table 7.5 Implicit Penultimate Model Reactivity Ratios 348
Table 7.6 List of Donor and Acceptor Monomers 351
Table 7.7 Q-e and Patterns Parameters for Some Common Monomers................365
Table 7.8 Identity of Chain End Units Involved in Radical-Radical Termination in MMA-S Copolymerization...372
Table 8.1 Solvent Effect on Homopropagation Rate Constants for VAc at 30°C 427
Table 8.2 Effect of Solvent on Tacticity of Poly(alkyl methacrylate) at -40 °C..428
Table 8.3 Effect of Amines on Tacticity of Poly(methacrylic acid) at 60 °C......429
Table 8.4 Solvent Dependence of Reactivity Ratios for MMA-MAA Copolymerization at 70°C..430
Table 8.5 Solvent Dependence of Penultimate Model Reactivity Ratios for S-AN Copolymerization at 60°C..430
Table 8.6 Effect of Lewis Acids on Tacticity of Polymers Formed in High Conversion Radical Polymerizations at 60 °C..................................435
Table 9.1 Five-Membered Ring Nitroxides for NMP...473
Table 9.2 Six-Membered Ring Nitroxides for NMP..474
Table 9.3 Open-Chain Nitroxides for NMP..475
Table 9.4 Seven-Membered Ring Nitroxides for NMP..475
Table 9.5 Structures of Ligands for Copper Based ATRP Catalysts.....................494
Table 9.6 Structures of Ruthenium Complexes Used as ATRP Catalysts............495
Table 9.7 Structures of Iron Complexes Used as ATRP Catalysts.......................496
Table 9.8 Structures of Nickel Complexes Used as ATRP Catalysts...................496
Table 9.9 Block Copolymers Prepared by Macromonomer RAFT Polymerization 502
Table 9.10 Tertiary Dithiobenzoate RAFT Agents..508
Table 9.11 Other Aromatic Dithioester RAFT Agents 509

Index to Tables

Table 9.12 Primary and Secondary Dithiobenzoate RAFT Agents 510
Table 9.13 Bis-RAFT Agents 511
Table 9.14 Dithioacetate and Dithiophenylacetate RAFT Agents 511
Table 9.15 Symmetrical Trithiocarbonate RAFT Agents 512
Table 9.16 Non-Symmetrical Trithiocarbonate RAFT Agents 512
Table 9.17 Xanthate RAFT Agents 513
Table 9.18 Dithiocarbamate RAFT Agents 514
Table 9.19 Initiators for Telluride-Mediated Polymerization 524
Table 9.20 Statistical/Gradient Copolymers Synthesized by NMP 528
Table 9.21 Statistical/Gradient Copolymers Synthesized by ATRP 529
Table 9.22 Statistical/Gradient Copolymers Synthesized by RAFT
Polymerization 529
Table 9.23 Methods for End Group Transformation of Polymers Formed
by NMP 532
Table 9.24 Methods for End Group Transformation of Polymers Formed
by ATRP by Addition or Addition-Fragmentation 534
Table 9.25 End Group Transformations for Polymers Formed by ATRP 535
Table 9.26 Methods for End Group Removal from Polymers Formed by
RAFT Polymerization 539
Table 9.27 Diblock Copolymers Prepared by ATRP 543
Table 9.28 Diblock Copolymers Prepared by RAFT Polymerization 543
Table 9.29 Star Precursors for NMP 550
Table 9.30 Star Precursors for ATRP 550
Table 9.31 Star Precursors for RAFT Polymerization 551

Index to Figures

Figure 1.1 Publication rate of papers on radical polymerization and on living, controlled or mediated radical polymerization for period 1975-2002 based on SciFinder™ search. 7

Figure 2.1 Transition state for methyl radical addition to ethylene. Geometric parameters are from *ab initio* calculation with QCISD(T)/6-31GT(d) basis set. 20

Figure 2.2 Effect of polar factors on regiospecificity of radical addition 22

Figure 2.3 Relative rate constants for addition of alkyl radicals to fumarodinitrile (k_1) and methyl α-chloroacrylate (k_2) as a function of temperature 25

Figure 2.4 SOMO-HOMO and SOMO-LUMO orbital interaction diagrams 27

Figure 2.5 Schematic state correlation diagram for free radical addition to a carbon-carbon double bond showing configuration energies as a function of the reaction coordinate 28

Figure 2.6 Transition state for hydrogen atom abstraction 29

Figure 2.7 Predicted order of reactivity of X-H compounds 30

Figure 2.8 Preferred site of attack in hydrogen abstraction by various radicals 32

Figure 2.9 Relative reactivity per hydrogen atom of indicated site towards *t*-butoxy radicals 33

Figure 2.10 Dependence of abstraction:addition ratio on nucleophilicity for oxygen-centered radicals 35

Figure 2.11 Dependence of abstraction:addition ratio on nucleophilicity for carbon-centered radicals 35

Figure 2.12 Trend in k_{td}/k_{tc} for radicals $(CH_3)_2C(\bullet)$-X 42

Figure 2.13 Temperature dependence of k_{td}/k_{tc} values for *t*-butyl radicals with dodecane or 3-methyl-3-pentanol as solvent 43

Figure 3.1 Temperature dependence of rate constants for reactions of cumyloxy radicals (a) β-scission to methyl radicals (b) abstraction from cumene and (c) addition to styrene. Data are an extrapolation based on literature Arrhenius parameters 56

Figure 3.2 Jablonski diagram describing photoexcitation process 59

Figure 3.3 Cumulative and instantaneous initiator efficiency (*f*) of AIBN as initiator in S polymerization as a function of monomer conversion 76

Figure 3.4 Relative reactivity of indicated site towards *t*-butoxy radicals for allyl methacrylate and allyl acrylate 122

Index to Figures

Figure 3.5 Relative reactivity of indicated site towards t-butoxy radicals for BMA..................123

Figure 3.6 Relative rate constants for β-scission of t-alkoxy radicals at 60 °C..................124

Figure 4.1 Representation of meso (m) and racemic (r) dyads with polymer chains..................169

Figure 4.2 Representation of meso (m) and racemic (r) diastereoisomers of low molecular weight compounds..................170

Figure 4.3 Representation of $mrrrmr$ heptad..................170

Figure 4.4 Dependence of K_{eq} on temperature for selected monomers..................214

Figure 4.5 Experimental molecular weight distribution obtained by GPC and its first derivative with respect to chain length for PS prepared by PLP..................218

Figure 5.1 (a) Number and (b) GPC distributions for two polymers both with $\bar{X}_n=100$..................241

Figure 5.2 Dispersity(D) as a function of $\bar{X}_n$ for polymers formed by (a) disproportionation or chain transfer and (b) combination..................242

Figure 5.3 Conversion-time profile for bulk MMA polymerization at 50 °C with AIBN initiator illustrating the three conversion regimes..................243

Figure 5.4 Chain length dependence of $k_t^{i,j}$ predicted by (a) the geometric mean or (b) the harmonic mean approximation..................246

Figure 5.5 Chain length dependence of $k_t^{i,j}$ predicted by the Smoluchowski mean with $\alpha=0.5$ and $k_{to}=10^9$ and the geometric mean with $\alpha=0.2$ and $k_{to}=10^8$; i and j are the lengths of the reacting chains..................248

Figure 6.1 'Mayo plots' in which the calculated limiting slopes, 'last 10% slopes' and 'top 20% slopes' are graphed as a function of [T]/[M]..................285

Figure 7.1 Plot of the instantaneous copolymer composition (F_A) vs monomer feed composition (f_A) for the situation where (a) $r_A=r_B=1.0$, (b) $r_A=r_B=0.5$, (c) $r_A=r_B=0.01$, (d) $r_A=0.5$, $r_B=2.0$..................340

Figure 7.2 Chain end terminology..................344

Figure 7.3 Molecular weight distributions for HEA:BA:S copolymer prepared with butanethiol chain transfer agent..................382

Figure 7.4 Molecular weight distributions for HEA:BA:S copolymer prepared with butanethiol chain transfer agent..................383

Figure 9.1 Predicted evolution of molecular weight with monomer conversion for a conventional radical polymerization with constant rate of initiation and a living polymerization..................453

Figure 9.2 (a) Number and (b) GPC distributions for three polymers each with $\bar{X}_n=100$..................454

Figure 9.3 General description of macromonomer and thiocarbonylthio RAFT agents...................501

Figure 9.4 Effect of Z substituent on effectiveness of RAFT agents..................505

Figure 9.5 Canonical forms of thiocarbonylthio compounds..................506

Index to Figures

Figure 9.6 Effect of R substituent on effectiveness of RAFT agents. 507
Figure 9.7 Evolution of molecular weight and dispersity with conversion for MMA polymerizations in the presence of RAFT agent. 519
Figure 9.8 Comparison of molecular weight distributions for a conventional and RAFT polymerization 520
Figure 9.9 General description of organochalcogenide transfer agents 524
Figure 9.10 General description of organostibine transfer agents 525
Figure 9.11 Star Architectures 548
Figure 9.12 GPC distributions obtained during bulk thermal polymerization of styrene at 110 °C with tetrafunctional RAFT agents 553

Preface to the First Edition

In recent years, the study of radical polymerization has gone through something of a renaissance. This has seen significant changes in our understanding of the area and has led to major advances in our ability to control and predict the outcome of polymerization processes. Two major factors may be judged responsible for bringing this about and for spurring an intensified interest in all aspects of radical chemistry:

Firstly, the classical theories on radical reactivity and polymerization mechanism do not adequately explain the rate and specificity of simple radical reactions. As a consequence, they can not be used to predict the manner in which polymerization rate parameters and details of polymer microstructure depend on reaction conditions, conversion and molecular weight distribution.

Secondly, new techniques have been developed which allow a more detailed characterization of both polymer microstructures and the kinetics and mechanism of polymerizations. This has allowed mechanism-structure-property relationships to be more rigorously established.

The new knowledge and understanding of radical processes has resulted in new polymer structures and in new routes to established materials; many with commercial significance. For example, radical polymerization is now used in the production of block copolymers, narrow polydispersity homopolymers, and other materials of controlled architecture that were previously available only by more demanding routes. These commercial developments have added to the resurgence of studies on radical polymerization.

We believe it is now timely to review the recent developments in radical polymerization placing particular emphasis on the organic and physical-organic chemistry of the polymerization process. In this book we critically evaluate the findings of the last few years, where necessary reinterpreting earlier work in the light of these ideas, and point to the areas where current and future research is being directed. The overall aim is to provide a framework for further extending our understanding of free radical polymerization and create a definable link between synthesis conditions and polymer structure and properties. The end result should be polymers with predictable and reproducible properties.

The book commences with a general introduction outlining the basic concepts. This is followed by a chapter on radical reactions that is intended to lay the theoretical ground-work for the succeeding chapters on initiation, propagation, and termination. Because of its importance, radical copolymerization is treated in a separate chapter. We then consider some of the implications of these chapters by

Preface to the First Edition

discussing the prospects for controlling the polymerization process and structure-property relationships. In each chapter we describe some of the techniques that have been employed to characterize polymers and polymerizations and which have led to breakthroughs in our understanding of radical polymerization. Emphasis is placed on recent developments.

This book will be of major interest to researchers in industry and in academic institutions as a reference source on the factors which control radical polymerization and as an aid in designing polymer syntheses. It is also intended to serve as a text for graduate students in the broad area of polymer chemistry. The book places an emphasis on reaction mechanisms and the organic chemistry of polymerization. It also ties in developments in polymerization kinetics and physical chemistry of the systems to provide a complete picture of this most important subject.

Graeme Moad
David H Solomon

Preface to the Second Edition

In the ten years since the first edition appeared, the *renaissance* in Radical Polymerization has continued and gained momentum. The period has seen the literature with respect to controlled and, in particular, living radical polymerization expand dramatically. The end of 1995, saw the first reports on atom transfer radical polymerization (ATRP) and in 1998 polymerization with reversible addition fragmentation chain transfer (RAFT) was introduced. The period has also seen substantial development in nitroxide-mediated polymerization (NMP) first reported in 1987 and discussed in the first edition. A new generation of control agents has added greater versatility and new applications. The area of living radical polymerization is now responsible for a very substantial fraction of the papers in the field. In this edition, we devote a new chapter to living radical polymerization.

The initial thrust of work in the area of living radical polymerization was aimed at capitalizing on the versatility of radical polymerization with respect to reaction conditions and the greater range of suitable monomers as compared to anionic systems. Anionic polymerizations were seen as the standard. This has now changed, and living radical polymerizations are now seen as offering polymers with unique compositions and properties not achievable with other methodologies. Living radical polymerization has also been combined with other processes and mechanisms to give structures and architectures that were not previously thought possible. The developments have many applications particularly in the emerging areas of electronics, biotechnology and nanotechnology.

A small change has been made to the title and the text of this edition to reflect the current IUPAC recommendation that radicals are no longer 'free'. Of the classical steps of a radical polymerization, while there remains some room for improvement, it can be stated that we now have methodologies that give control over the termination and initiation steps to the extent that specific structures, molecular weight distributions, and architectures can be confidently obtained. The remaining 'holy grail' in the field of radical polymerization is control over the stereochemistry and regiospecificity in the propagation step. Although some small steps have been taken towards achieving this goal, much remains to be done.

The last ten years have also seen significant advances in other areas of radical polymerization. Chapters one through eight have been updated and many new references added to reflect these developments.

Graeme Moad
David H Solomon

Acknowledgments

We gratefully acknowledge the contribution of the following for their assistance in the preparation and proof reading of the manuscript.

Dr Agnes Ho
Dr Catherine L. Moad
Dr Almar Postma
Dr Greg Qiao
Dr Tiziana Russo

In addition, we thank again those who contributed to the production of the first edition.

We also thank Max McMaster of McMaster Indexing for his efforts in producing the index for this volume.

1
Introduction

From an industrial stand-point, a major virtue of radical polymerizations is that they can often be carried out under relatively undemanding conditions. In marked contrast to ionic or coordination polymerizations, they exhibit a tolerance of trace impurities. A consequence of this is that high molecular weight polymers can often be produced without removal of the stabilizers present in commercial monomers, in the presence of trace amounts of oxygen, or in solvents that have not been rigorously dried or purified. Indeed, radical polymerizations are remarkable amongst chain polymerization processes in that they can be conveniently conducted in aqueous media.

It is this apparent simplicity of radical polymerization that has led to the technique being widely adopted for both industrial and laboratory scale polymer syntheses. Today, a vast amount of commercial polymer production involves radical chemistry during some stage of the synthesis, or during subsequent processing steps. These factors have, in turn, provided the driving force for extensive research efforts directed towards more precisely defining the kinetics and mechanisms of radical polymerizations. The aim of these studies has been to define the parameters necessary for predictable and reproducible polymer syntheses and to give better understanding of the properties of the polymeric materials produced. With understanding comes control. Most recently, we have seen radical polymerization move into new fields of endeavor where control and precision are paramount requirements. Indeed, these aspects now dominate the literature.

The history of polymers, including the beginnings of addition and of radical polymerization, is recounted by Morawetz.[1] The repeat unit structure (**1**) of many common polymers, including PS, PVC and PVAc, was established in the latter half of the 19th century. However, the concept that these were materials of high molecular weight took longer to be accepted. Staudinger was one of the earliest and most strident proponents of the notion that synthetic polymers were high molecular weight compounds with a chain structure and he did much to dispel the then prevalent belief that polymers were composed of small molecules held together by colloidal forces.[2] Staudinger and his colleagues are also often credited with coming up with the concept of a chain polymerization. In an early paper in 1920, he proposed that polymer chains might retain unsatisfied valencies at the chain ends (**2**).[3] In 1929, it was suggested that the monomer units might be connected by covalent linkages in large cyclic structures (**3**) to solve the chain end problem.[4] In 1910, Pickles[5] had proposed such a structure for natural rubber. However, by 1935 it was recognized that polymers have discrete functional groups at the chain ends formed by initiation and termination reactions.[6]

1a*   **1b**   **2**   **3**

In the period 1910-1950 many contributed to the development of free-radical polymerization.[1] The basic mechanism as we know it today (Scheme 1.1), was laid out in the 1940s and 50s.[7-9] The essential features of this mechanism are initiation and propagation steps, which involve radicals adding to the less substituted end of the double bond ("tail addition"), and a termination step, which involves disproportionation or combination between two growing chains.

In this early work, both initiation and termination were seen to lead to formation of structural units different from those that make up the bulk of the chain. However, the quantity of these groups, when expressed as a weight fraction of the total material, appeared insignificant. In a polymer of molecular weight 100,000 they represent only *ca* 0.2% of units.[†] Thus, polymers formed by radical polymerization came to be represented by, and their physical properties and chemistry interpreted in terms of, the simple formula **1**.

However, it is now quite apparent that the representation **1** while convenient, and useful as a starting point for discussion, has serious limitations when it comes to understanding the detailed chemistry of polymeric materials. For example, how can we rationalize the finding that two polymers with nominally the same chemical and physical composition have markedly different thermal stability? PMMA (**1**, X=CH$_3$, Y=CO$_2$CH$_3$) prepared by anionic polymerization has been reported to be more stable by some 50 °C than that prepared by a radical process.[10] The simplified representation, (**1**), also provides no ready explanation for the discrepancy in chemical properties between low molecular weight model compounds and polymers even though both can be represented ostensibly by the same structure (**1**). Consideration of the properties of simple models indicates that the onset of thermal degradation of PVC (**1**, X=H, Y=Cl) should occur at a temperature 100 °C higher than is actually found.[11]

---

*   IUPAC recommendations suggest that polymers derived from 1,1-disubstituted monomers CXY=CH$_2$ (or CH$_2$=CXY) be drawn as **1b** rather than as **1a**. However, formula **1a** follows logically from the traditional way of writing the mechanism of radical addition (*e.g.* Scheme 1.1). Because of our focus on mechanism, the style **1a** has been adopted throughout this book.

†   Based on a monomer molecular weight of 100.

Introduction

Such problems have led to a recognition of the importance of defect groups[*] or structural irregularities.[12-16] If we are to achieve an understanding of radical polymerization, and the ability to produce polymers with optimal, or at least predictable, properties, a much more detailed knowledge of the mechanism of the polymerization and of the chemical microstructure of the polymers formed is required.[16]

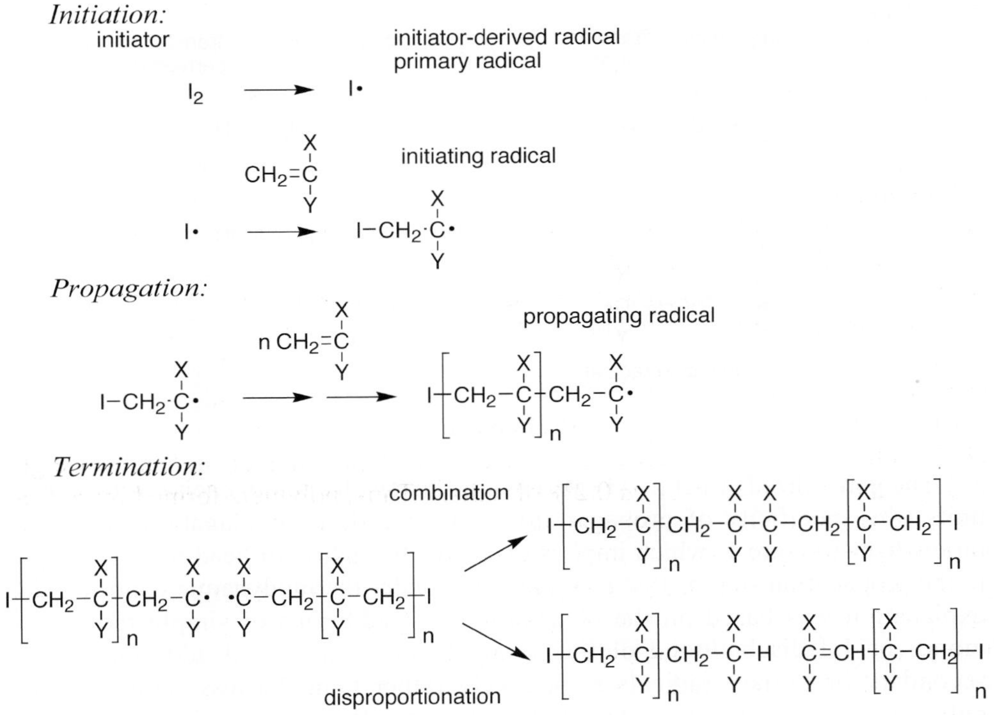

Initiation:

initiator → initiator-derived radical / primary radical

$I_2 \longrightarrow I\cdot$

$I\cdot + CH_2=C(X)(Y) \longrightarrow I-CH_2-\overset{X}{\underset{Y}{C}}\cdot$ initiating radical

Propagation:

$I-CH_2-\overset{X}{\underset{Y}{C}}\cdot \ + \ n\ CH_2=C(X)(Y) \longrightarrow I-[CH_2-\overset{X}{\underset{Y}{C}}]_n-CH_2-\overset{X}{\underset{Y}{C}}\cdot$ propagating radical

Termination:

$I-[CH_2-\overset{X}{\underset{Y}{C}}]_n-CH_2-\overset{X}{\underset{Y}{C}}\cdot \ \ \cdot\overset{X}{\underset{Y}{C}}-CH_2-[\overset{X}{\underset{Y}{C}}-CH_2]_n-I$

combination → $I-[CH_2-\overset{X}{\underset{Y}{C}}]_n-CH_2-\overset{X}{\underset{Y}{C}}-\overset{X}{\underset{Y}{C}}-CH_2-[\overset{X}{\underset{Y}{C}}-CH_2]_n-I$

disproportionation → $I-[CH_2-\overset{X}{\underset{Y}{C}}]_n-CH_2-\overset{X}{\underset{Y}{C}}-H \ \ \ \overset{X}{\underset{Y}{C}}=CH-[\overset{X}{\underset{Y}{C}}-CH_2]_n-I$

Scheme 1.1

Structural irregularities are introduced into the chain during each stage of the polymerization and we must always question whether it is appropriate to use the generalized formula (**1**) for representing the polymer structure. Obvious examples of defect structures are the groups formed by chain initiation and termination. Initiating radicals[†] are not only formed directly from initiator decomposition (Scheme 1.1) but also indirectly by transfer to monomer, solvent, transfer agent, or impurities (Scheme 1.2).

---

[*] 'Defect groups' or 'structural irregularities' need not impair polymer properties, they are simply units that differ from those described by the generalized formula **1**
[†] Initiating radicals are formed from those initiator- or transfer agent-derived radicals that add monomer so as to form propagating radicals (see 3.1).

The Chemistry of Free Radical Polymerization

In termination, unsaturated and saturated ends are formed when the propagating species undergo disproportionation, head-to-head linkages when they combine, and other functional groups may be introduced by reactions with inhibitors or transfer agents (Scheme 1.2). In-chain defect structures (within the polymer molecule) can also arise by copolymerization of the unsaturated byproducts of initiation or termination.

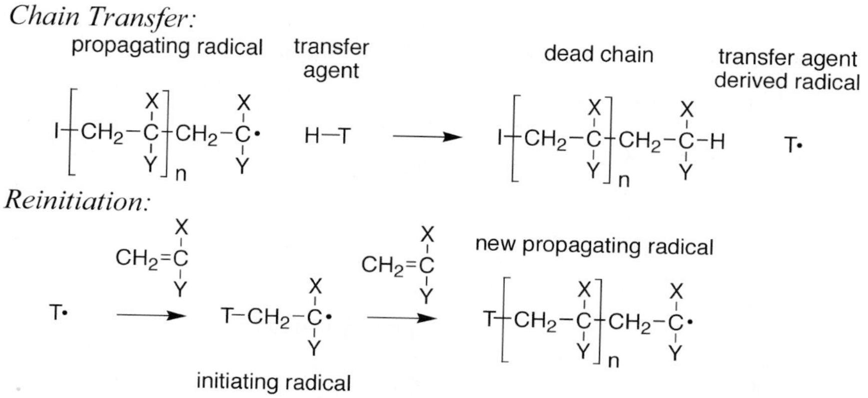

Chain Transfer:

propagating radical   transfer agent   dead chain   transfer agent derived radical

$$I\text{-}[CH_2\text{-}C(X)(Y)\text{-}CH_2\text{-}C(X)(Y)]_n\cdot + H\text{-}T \longrightarrow I\text{-}[CH_2\text{-}C(X)(Y)\text{-}CH_2\text{-}C(X)(Y)\text{-}H]_n + T\cdot$$

Reinitiation:

$$T\cdot + CH_2\text{=}C(X)(Y) \longrightarrow T\text{-}CH_2\text{-}C(X)(Y)\cdot \xrightarrow{CH_2=C(X)(Y)} T\text{-}[CH_2\text{-}C(X)(Y)]_n\text{-}CH_2\text{-}C(X)(Y)\cdot$$

initiating radical → new propagating radical

Scheme 1.2

The generalized structure (**1**) also overestimates the homogeneity of the repeat units (the specificity of propagation). The traditional explanation offered to rationalize structure **1**, which implies exclusive formation of head-to-tail linkages in the propagation step, is that the reaction is under thermodynamic control. This explanation was based on the observation that additions of simple radicals to mono- or 1,1-disubstituted olefins typically proceed by tail addition to give secondary or tertiary radicals respectively rather than the less stable primary radical (Scheme 1.3) and by analogy with findings for ionic reactions where such thermodynamic considerations are of demonstrable importance.

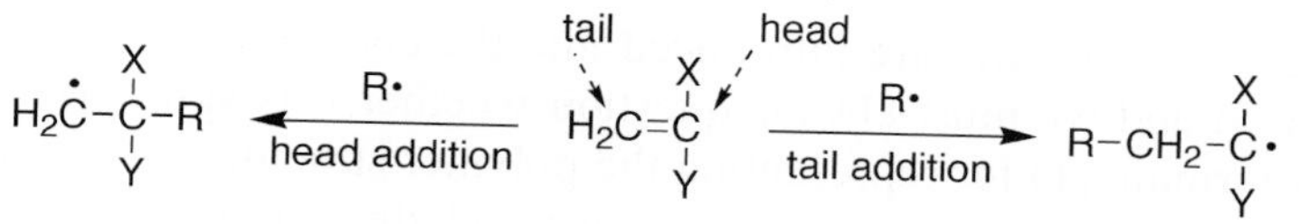

$$H_2\dot{C}\text{-}C(X)(Y)\text{-}R \xleftarrow{R\cdot,\text{ head addition}} H_2C\text{=}C(X)(Y) \xrightarrow{R\cdot,\text{ tail addition}} R\text{-}CH_2\text{-}\dot{C}(X)(Y)$$

Scheme 1.3

Until the early 1970s, the absence of suitable techniques for probing the detailed microstructure of polymers or for examining the selectivity and rates of radical reactions prevented the traditional view from being seriously questioned. In more recent times, it has been established that radical reactions, more often than not, are under kinetic rather than thermodynamic control and the preponderance of

Introduction

head-to-tail linkages in polymers is determined largely by steric and polar influences (see 2.2).[17]

It is now known that a proportion of "head" addition occurs during the initiation and propagation stages of many polymerizations (see 4.3). For example, poly(vinyl fluoride) chains contain in excess of 10% head-to-head linkages.[18] Benzoyloxy radicals give *ca* 5% head addition with styrene (see 3.4.2.2).[19,20] However, one of the first clear-cut examples demonstrating that thermodynamic control is not of overriding importance in determining the outcome of radical reactions is the cyclopolymerization of diallyl compounds (see 4.4.1).[21-24]

Monomers containing multiple double bonds might be anticipated to initially yield polymers with pendant unsaturation and ultimately crosslinked structures. The pioneering studies of Butler and coworkers[23,24] established that diallyl compounds, of general structure (**4**), undergo radical polymerization to give linear saturated polymers. They proposed that the propagation involved a series of inter- and intramolecular addition reactions. The presence of cyclic units in the polymer structure was rigorously established by chemical analysis.[25] Addition of a radical to the diallyl monomer (**4**) could conceivably lead to the formation of 5-, 6- or even 7-membered rings as shown in Scheme 1.4. However, application of the then generally accepted hypothesis, that product radical stability was the most important factor determining the course of radical addition, indicated that the intermolecular step should proceed by tail addition (to give **5**) and that the intramolecular step should afford a 6-membered ring and a secondary radical (**7**). On the basis of this theory, it was proposed that the cyclopolymer was composed of 6-membered rings (**9**) rather than 5-membered rings (**8**).

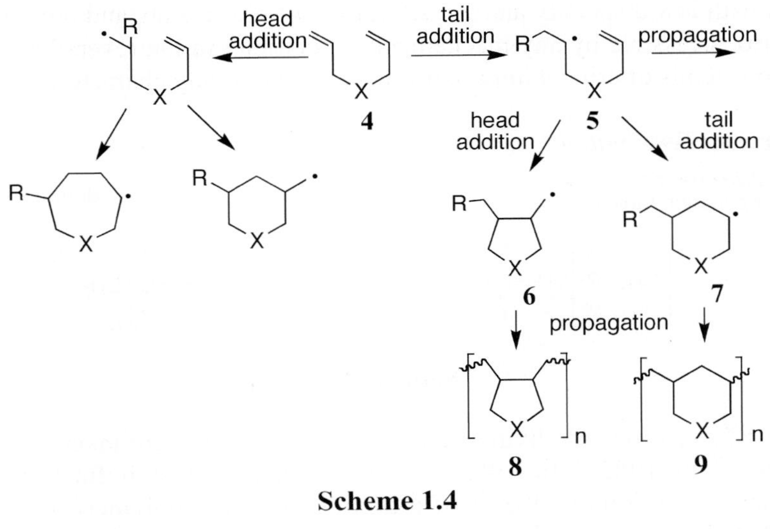

Scheme 1.4

It was established in the early 1960s that hexenyl radicals and simple derivatives gave 1,5- rather than 1,6-ring closure under conditions of kinetic control.[26] However, it was not until 1976 that the structures of cyclopolymers formed from 1,6-dienes (**4**) were experimentally determined and Hawthorne *et al.*[27] showed that the intramolecular cyclization step gives preferentially the less stable radical (**6**) (5- *vs* 6-membered ring, primary *vs* secondary radical) - *i.e.* ≥99% head addition. Over the last two decades, many other examples of radical reactions which preferentially afford the thermodynamically less stable product have come to light. A discussion of various factors important in determining the course and rate of radical additions will be found in Chapter 2.

The examples described in this chapter serve to illustrate two well-recognized, though often overlooked, principles, which lie at the heart of polymer, and, indeed, all forms of chemistry. These are:

(a) The dependence of a reaction (polymerization, polymer degradation, etc.) on experimental variables cannot be understood until the reaction mechanism is established.
(b) The reaction mechanism cannot be fully defined, when the reaction products are unknown.

The recent development of radical polymerizations that show the attributes of living polymerization is a prime example of where the quest for knowledge on polymerization mechanism can take us (Chapter 9). Living radical polymerization relies on the introduction of a reagent that undergoes reversible termination with the propagating radicals thereby converting them to a dormant form (Scheme 1.5). This enables control of the active species concentration allowing conditions to be chosen such that all chains are able to grow at a similar rate (if not simultaneously) throughout the polymerization. This has, in turn, enabled the synthesis of polymers with low dispersity and a wide variety of block, stars and other structures not hitherto accessible by any mechanism. Specificity in the reversible initiation-termination step is of critical importance in achieving living characteristics.

Reversible Termination:

active species
propagating radical   dormant species

$$\left[\text{I}\!-\!\text{CH}_2\!-\!\underset{\underset{Y}{|}}{\overset{\overset{X}{|}}{C}}\right]_n\!\text{CH}_2\!-\!\underset{\underset{Y}{|}}{\overset{\overset{X}{|}}{C}}\bullet \quad \bullet Z \quad \rightleftharpoons \quad \left[\text{I}\!-\!\text{CH}_2\!-\!\underset{\underset{Y}{|}}{\overset{\overset{X}{|}}{C}}\right]_n\!\text{CH}_2\!-\!\underset{\underset{Y}{|}}{\overset{\overset{X}{|}}{C}}\!-\!Z$$

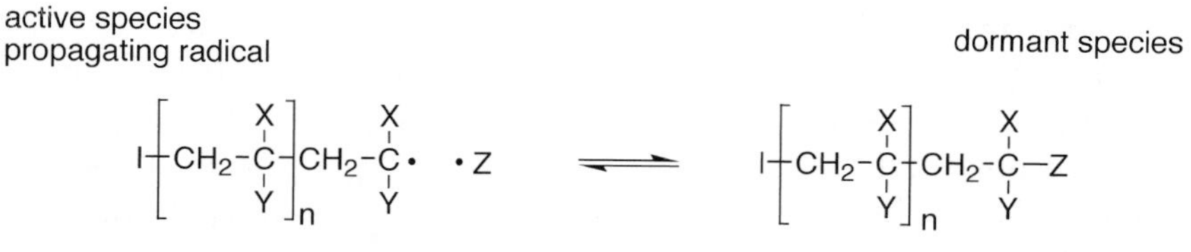

Scheme 1.5

The first steps towards living radical polymerization were taken by Otsu and colleagues[28,29] in 1982. In 1985, this was taken one step further with the development by Solomon *et al.*[30] of nitroxide-mediated polymerization (NMP). This work was first reported in the patent literature[30] and in conference papers but was not widely recognized until 1993 when Georges *et al.*[31] applied the method in the synthesis of narrow polydispersity polystyrene. NMP was described in detail in a small section in the first edition of this book. Since that time the area has expanded dramatically. The scope of NMP has been greatly extended[32] and new, more versatile, methods have appeared. The most notable are atom transfer radical polymerization (ATRP)[33,34] and polymerization with reversible addition fragmentation (RAFT).[35,36] From small beginnings pre-1995, this area now accounts for a third of all papers in the field of radical polymerization. Moreover, the growth in the field since 1995 is almost totally attributable to developments in this area (Figure 1.1).

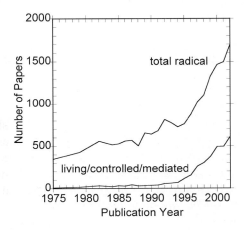

Figure 1.1 Publication rate of journal papers on radical polymerization and on living, controlled or mediated radical polymerization* for period 1975-2002 based on SciFinder™ search (as of Mar 2005).

In the succeeding chapters we detail the current state of knowledge of the chemistry of each stage of polymerization. We consider the details of the mechanisms, the specificity of the reactions, the nature of the group or groups incorporated in the polymer chain, and any byproducts. The intention is to create an awareness of the factors that must be borne in mind in selecting the conditions for a given polymerization and provide the background necessary for a more thorough understanding of polymerizations and polymer properties. In the final chapters, we examine the current status of efforts to control polymerization using either conventional technology or using the various approaches to living radical polymerization.

---

* Does not distinguish forms of controlled radical polymerization. Includes most papers on ATRP, RAFT and NMP and would also include conventional (non-living) but controlled radical polymerizations. It would not include papers, which do not mention the terms 'living', 'controlled' or 'mediated'.

1.1 References

1. Morawetz, H. *Polymers. The Origins and Growth of a Science*; Dover: New York, 1995.
2. Furukawa, Y. *Inventing Polymer Science: Staudinger, Carothers, and the Emergence of Macromolecular Chemistry*; University of Pennsylvania Press: Philadelphia, 1998.
3. Staudinger, H. *Chem. Ber.* **1920**, *53*, 1073.
4. Staudinger, H.; Signer, R.; Johner, H.; Lüthy, M.; Kern, W.; Russidis, D.; Schwetzer, O. *Ann.* **1929**, *474*, 145.
5. Pickles, S.S. *J. Chem. Soc.* **1910**, *97*, 1085.
6. Staudinger, H.; Steinhofer, A. *Ann.* **1935**, *517*, 35.
7. Flory, P.J. *Principles of Polymer Chemistry*; Cornell University Press: Ithaca, New York, 1953.
8. Walling, C. *Free Radicals in Solution*; Wiley: New York, 1957.
9. Bamford, C.H.; Barb, W.G.; Jenkins, A.D.; Onyon, P.F. *The Kinetics of Vinyl Polymerization by Radical Mechanisms*; Butterworths: London, 1958.
10. McNeill, I.C. *Eur. Polym. J.* **1968**, *4*, 21.
11. Mayer, Z. *J. Macromol. Sci., Rev. Macromol. Chem.* **1974**, *C10*, 263.
12. Solomon, D.H.; Cacioli, P.; Moad, G. *Pure Appl. Chem.* **1985**, *57*, 985.
13. Hwang, E.F.J.; Pearce, E.M. *Polym. Eng. Rev.* **1983**, *2*, 319.
14. Mita, I. In *Aspects of Degradation and Stabilization of Polymers*; Jellineck, H.H.G., Ed.; Elsevier: Amsterdam, 1978; p 247.
15. Solomon, D.H. *J. Macromol. Sci., Chem.* **1982**, *A17*, 337.
16. Moad, G.; Solomon, D.H. *Aust. J. Chem.* **1990**, *43*, 215.
17. Tedder, J.M. *Angew. Chem., Int. Ed. Engl.* **1982**, *21*, 401.
18. Cais, R.E.; Kometani, J.M. *ACS Symp. Ser.* **1984**, *247*, 153.
19. Moad, G.; Rizzardo, E.; Solomon, D.H. *Macromolecules* **1982**, *15*, 909.
20. Moad, G.; Rizzardo, E.; Solomon, D.H.; Johns, S.R.; Willing, R.I. *Makromol. Chem., Rapid Commun.* **1984**, *5*, 793.
21. Butler, G.B. *Acc. Chem. Res.* **1982**, *15*, 370.
22. Solomon, D.H.; Hawthorne, D.G. *J. Macromol. Sci., Rev. Macromol. Chem.* **1976**, *C15*, 143.
23. Butler, G.B. In *Encyclopedia of Polymer Science and Engineering*, 2nd ed.; Mark, H.F.; Bikales, N.M.; Overberger, C.G.; Menges, G., Eds.; Wiley: New York, 1986; Vol. 4, p 543.
24. Butler, G.B. In *Comprehensive Polymer Science*; Eastmond, G.C.; Ledwith, A.; Russo, S.; Sigwalt, P., Eds.; Pergamon: Oxford, 1989; Vol. 4, p 423.
25. Butler, G.B.; Crawshaw, A.; Miller, W.L. *J. Am. Chem. Soc.* **1958**, *80*, 3615.
26. Beckwith, A.L.J.; Ingold, K.U. In *Rearrangements in Ground and Excited States*; de Mayo, P., Ed.; Academic Press: New York, 1980; Vol. 1, p 162.
27. Hawthorne, D.G.; Johns, S.R.; Solomon, D.H.; Willing, R.I. *Aust. J. Chem.* **1976**, *29*, 1955.
28. Otsu, T.; Yoshida, M. *Makromol. Chem., Rapid Commun.* **1982**, *3*, 127.
29. Otsu, T.; Yoshida, M.; Tazaki, T. *Makromol. Chem., Rapid Commun.* **1982**, *3*, 133.
30. Solomon, D.H.; Rizzardo, E.; Cacioli, P. US 4581429, 1986 (*Chem. Abstr.* **1985**, *102*, 221335q).
31. Georges, M.K.; Veregin, R.P.N.; Kazmaier, P.M.; Hamer, G.K. *Macromolecules* **1993**, *26*, 2987.
32. Hawker, C.J.; Bosman, A.W.; Harth, E. *Chem. Rev.* **2001**, *101*, 3661.

33. Matyjaszewski, K.; Xia, J. *Chem. Rev.* **2001**, *101*, 2921.
34. Kamigaito, M.; Ando, T.; Sawamoto, M. *Chem. Rev.* **2001**, *101*, 3689.
35. Chiefari, J.; Chong, Y.K.; Ercole, F.; Krstina, J.; Jeffery, J.; Le, T.P.T.; Mayadunne, R.T.A.; Meijs, G.F.; Moad, C.L.; Moad, G.; Rizzardo, E.; Thang, S.H. *Macromolecules* **1998**, *31*, 5559.
36. Moad, G.; Rizzardo, E.; Thang, S. *Aust. J. Chem.* **2005**, *58*, 379.

1
Radical Reactions

1.1 Introduction

The intention of this chapter is to discuss in some detail the factors that determine the rate and course of radical reactions. Emphasis is placed on those reactions most frequently encountered in radical polymerization:

(a) Addition to carbon-carbon double bonds (*e.g.* initiation - Chapter 3, propagation - Chapter 4).

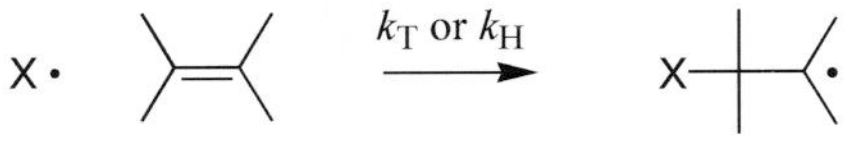

$$X\cdot + \text{C=C} \xrightarrow{k_T \text{ or } k_H} X-C-C\cdot$$

(a) The self-reaction of carbon-centered radicals (*e.g.* termination - Chapter 5).

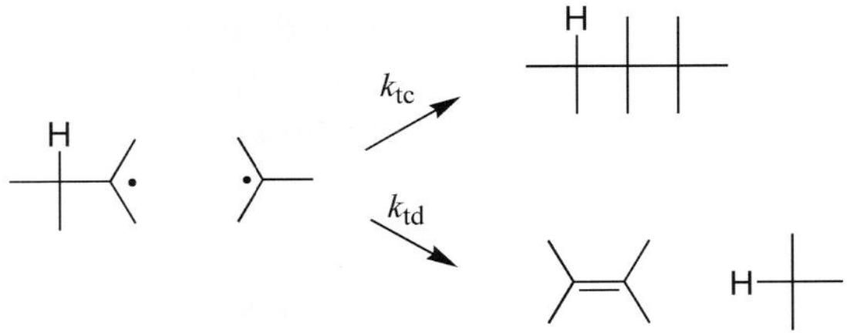

$$\cdot C-H + \cdot C- \xrightarrow{k_{tc}} -C-C- $$
$$\xrightarrow{k_{td}} \text{C=C} + H-C-$$

(a) Hydrogen atom transfer (*e.g.* chain transfer - Chapter 6).

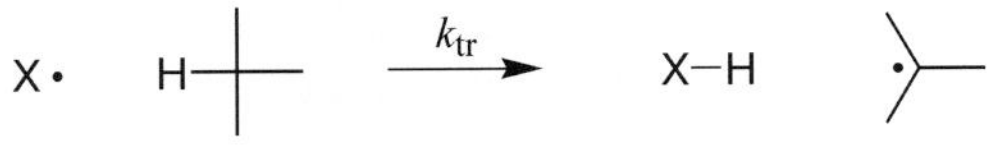

$$X\cdot + H-C- \xrightarrow{k_{tr}} X-H + \cdot C-$$

Other radical reactions not covered in this chapter are mentioned in the chapters that follow. These include additions to systems other than carbon-carbon double bonds [*e.g.* additions to aromatic systems (Section 3.4.2.2.1) and strained ring systems (Section 4.4.2)], transfer of heteroatoms [*e.g.* chain transfer to disulfides (Section 6.2.2.2) and halocarbons (Section 6.2.2.4)] or groups of atoms [*e.g.* in RAFT polymerization (Section 9.5.3)], and radical-radical reactions involving heteroatom-centered radicals or metal complexes [*e.g.* in inhibition (Sections 3.5.2 and 5.3), NMP (Section 9.3.6) and ATRP (Section 9.4)].

Until the early 1970s, views of radical reactions were dominated by two seemingly contradictory beliefs: (a) that radical reactions, in that they involve highly reactive species, should not be expected to show any particular selectivity, and (b) that (as is often possible with ionic reactions) the outcome could be predicted purely on the basis of the relative thermochemical stability of the product radicals. For condition (a) to apply, a reaction should have an early reactant-like transition state and near-zero activation energy. For condition (b) to apply the transition state should be late (or product-like) or the reaction leading to products must be under thermodynamic control by virtue of being rapidly reversible. While either of the above conditions may apply in specific cases, for radical reactions in general, neither need apply.

It is now recognized that radical reactions are, more often than not, under kinetic rather than thermodynamic control. The reactions can nonetheless show a high degree of specificity which is imposed by steric (non-bonded interactions), polar (relative electronegativities), stereoelectronic (requirement for overlap of frontier orbitals), bond-strength (relative strengths of bonds formed and broken) and perhaps other constraints.[1-4] In the following sections we discuss these factors, consider their relative importance in specific reactions and suggest guidelines for predicting the outcome of radical reactions.

1.1 Properties of Radicals

Radicals are chemical species that possess an unpaired electron sometimes called a free spin. The adjective "free", often used to designate radicals, relates to the state of the unpaired electron; it is not intended to indicate whether the compound bearing the free spin is complexed or uncomplexed. In this section we provide a brief overview of the structure, energetics and detection of radicals.

1.1.1 Structures of Radicals

Most radicals located on saturated bonds are π-radicals with a planar configuration and may be depicted with the free spin located in a p-orbital (**1**). Because such radical centers are achiral, stereochemical integrity is lost during radical formation. A new configuration will be assumed (or a previous configuration resumed) only upon reaction. Stereoselectivity in radical reactions is therefore dependent on the environment and on remote substituents.

      1   **2**   **3**

Radicals with very polar substituents (*e.g.* trifluoromethyl radical **2**), and radicals that are part of strained ring systems (*e.g.* cyclopropyl radical **3**) are σ-radicals. They have a pyramidal structure and are depicted with the free spin resident in an sp^3 hybrid orbital. σ-Radicals with appropriate substitution are potentially chiral, however, barriers to inversion are typically low with respect to the activation energy for reaction.

Radical Reactions

Most radicals located on double bonds (*e.g.* **4, 5**) or aromatic systems (*e.g.* **6**) are σ-radicals. The free spin is located in an orbital orthogonal to the π-bond system and it is not delocalized. The orbital of the vinyl radical (**4**) containing the free spin can be *cis*- or *trans*- with respect to substituents on the double bond. The barrier for isomerization of vinyl radicals can be significant with respect to the rate of reaction.

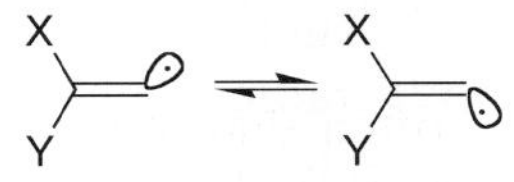

4a   **4b** 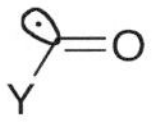   **5** 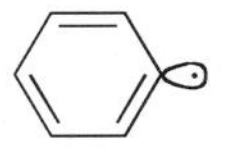   **6**

Radicals with adjacent π-bonds [*e.g.* allyl radicals (**7**), cyclohexadienyl radicals (**8**), acyl radicals (**9**) and cyanoalkyl radicals (**10**)] have a delocalized structure. They may be depicted as a hybrid of several resonance forms. In a chemical reaction they may, in principle, react through any of the sites on which the spin can be located. The preferred site of reaction is dictated by spin density, steric, polar and perhaps other factors. Maximum orbital overlap requires that the atoms contained in the delocalized system are coplanar.

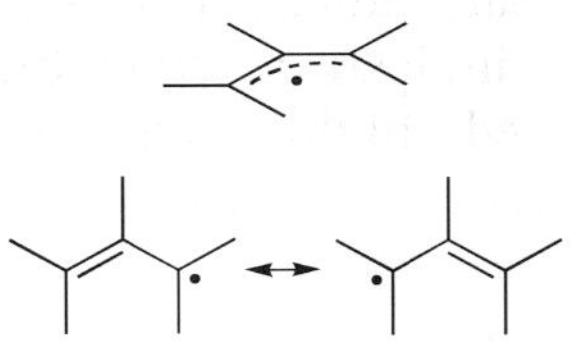

7 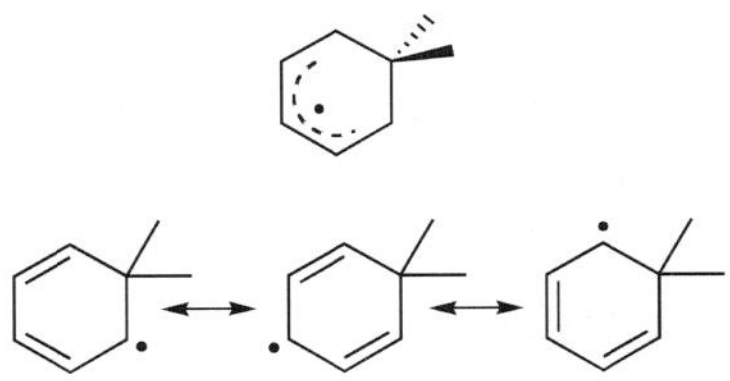   **8** 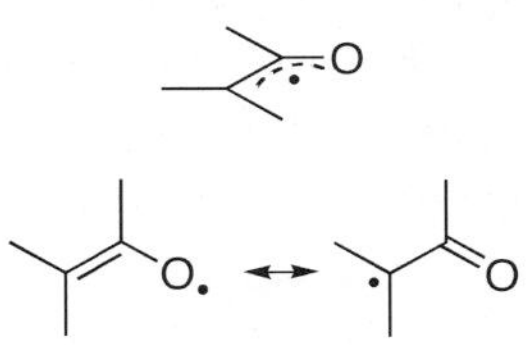   **9**

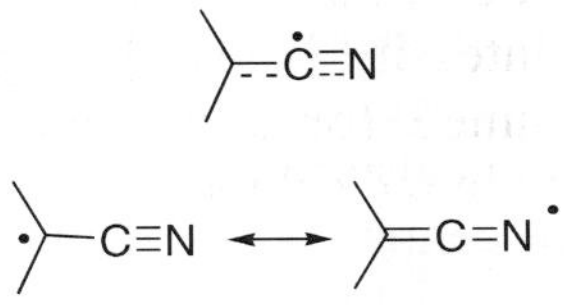

10 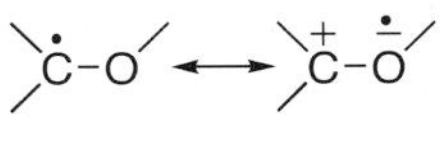   **11** 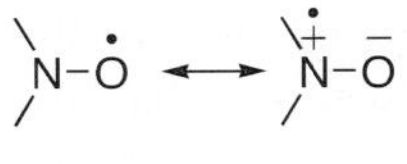   **12**

Radicals with adjacent heteroatoms bearing lone pairs (N, O, Cl, *etc.*), *e.g.* **11**, **12** can also be depicted as a resonance hybrid involving charged structures. The free spin may also be delocalized into adjacent C-H and C-C single bonds through a phenomenon known as hyperconjugation. Maximal hyperconjugative interaction requires coplanarity of the *p*-orbital containing the unpaired electron and the C-H and C-C bonds. Hyperconjugation is used to rationalize the relative stability and the nucleophilicity of alkyl radicals (tertiary > secondary > primary).

1.1.2 Stabilities of Radicals

Most radicals are transient species. They (*e.g.* **1-10**) decay by self-reaction with rates at or close to the diffusion-controlled limit (Section 1.4). This situation also pertains in conventional radical polymerization. Certain radicals, however, have thermodynamic stability, kinetic stability (persistence) or both that is conferred by appropriate substitution. Some well-known examples of stable radicals are diphenylpicrylhydrazyl (DPPH), nitroxides such as 2,2,6,6-tetramethylpiperidin-*N*-oxyl (TEMPO), triphenylmethyl radical (**13**) and galvinoxyl (**14**). Some examples of carbon-centered radicals which are persistent but which do not have intrinsic thermodynamic stability are shown in Section 1.4.3.2. These radicals (DPPH, TEMPO, **13**, **14**) are comparatively stable in isolation as solids or in solution and either do not react or react very slowly with compounds usually thought of as substrates for radical reactions. They may, nonetheless, react with less stable radicals at close to diffusion controlled rates. In polymer synthesis these species find use as inhibitors (to stabilize monomers against polymerization or to quench radical reactions - Section 5.3.1) and as reversible termination agents (in living radical polymerization - Section 9.3).

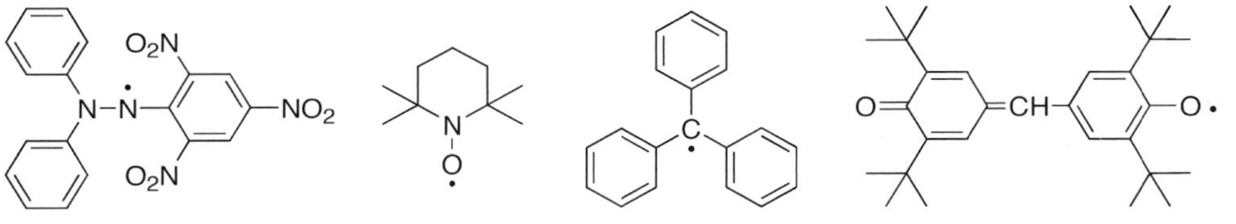

DPPH   **TEMPO**   **13**   **14**

Hydrogen-other atom/group bond dissociation energies are often used as an indication of radical stability. Substitution at a radical center almost invariably increases stability as indicated by a reduced bond dissociation energy. Thus, for alkyl radicals, stability increases in the order primary<secondary<tertiary. Fluorine substitution provides the exception to this rule. Radicals are inductively destabilized by fluorine substituents α- or β- to the radical center.[5] The greatest stabilizing effect is observed with substituents that are able to delocalize the free spin (Ph, CN, C=C). Experimental gas phase bond dissociation energies are tabulated in Table 1.1.[6] Bond dissociation energies can often be estimated with reasonable accuracy using group additivity rules.[7]

While it is desirable and important to have some knowledge of radical stabilities, the following sections will show that this is only one, and often not the major, factor in determining the outcome of radical reactions.

1.1.3 Detection of Radicals

In radical polymerization and in most radical reactions the radical species are present only in low concentrations (total concentration ~ 10^{-8}-10^{-7} M). Radicals are either generated in a chain reaction in which the radical species attain a low steady state concentration or they are generated reversibly and their concentration is controlled by an equilibrium process.

Largely for these reasons, radicals are most often characterized indirectly by examining the products of their reaction. Many of the methods used to study radical reactions have been applied to study initiation of polymerization. Some of these techniques are detailed in Section 3.5.

Table 1.1 Carbon-Hydrogen and Heteroatom-Hydrogen Bond Dissociation Energies (D in kJ mol^{-1})[a,6]

C-H Bond	D	X-H Bond	D
CF_3-H	450	HO-H	497
CH_3-H	439	$CH_3C(=O)O$-H	442
C_2H_5-H	423	CH_3O-H	436
i-C_3H_7-H	409	$(CH_3)_3CO$-H	440
t-C_4H_9-H	404	$(CH_3)_3COO$-H	374
$HOCH_2$-H	402	HOO-H	369
$H(C=O)CH_2$-H	394	PhO-H	362
$CH_2(CN)$-H	393		
CCl_3-H	393	CH_3S-H	365
$PhCH_2$-H	376	PhS-H	349
$CH_2=CHCH_2$-H	362	PhSe-H	326
$(CH_3)_2C(CN)$-H	362		
$CH_3CH(Ph)$-H	357	NH_2-H	453
$(CH_3)_2C(Ph)$-H	353	CH_3NH-H	418
		PhNH-H	368
CH≡C-H	556	NH_2NH-H	366
Ph-H	473		
$CH_2=CH$-H	465	$(CH_3)_3Si$-H	378
cC_3H_5-H	445	$(CH_3)_3Ge$-H	339
O=CH-H	369	$(C_4H_9)_3Sn$-H	308

a All values rounded to the nearest integer.

Electron paramagnetic resonance spectroscopy (EPR), also called electron spin resonance spectroscopy (ESR), may be used for direct detection and conformational and structural characterization of paramagnetic species. Good introductions to EPR have been provided by Fischer[8] and Leffler[9] and most books on radical chemistry have a section on EPR. EPR detection limits are dependent on radical structure and the signal complexity. However, with modern instrumentation, radical concentrations $>10^{-9}$ M can be detected and concentrations $>10^{-7}$ M can be reliably quantified.

UV-visible spectrophotometry and fluorescence spectrophotometry are also used for the direct observation of radical species and their reactions in some

16 The Chemistry of Radical Polymerization

circumstances. Radical species typically absorb at significantly higher wavelengths than similar saturated compounds (bathochromic shift).

Molecular orbital calculations (*ab initio* or semiempirical methods) are also often used to provide a description of radical species and their reactions. High levels of theory are required to provide reliable data. However, rapid advances in computer power and computational methods are seeing these methods more widely used and with greater success (for leading references on the application of theory to describe radical addition reactions, see Section 1.2.7).

1.2 Addition to Carbon-Carbon Double Bonds

With few exceptions, radicals are observed to add preferentially to the less highly substituted end of unsymmetrically substituted olefins (*i.e.* give predominantly tail addition[*] - Scheme 1.1).

For a long time, this finding was correlated with the observation that substituents at a radical center tend to enhance its stability (Section 1.1.2). This in turn led to the belief that the degree of stabilization conferred on the product radical by the substituents was the prime factor determining the orientation and rate of radical addition to olefins. That steric, polar, or other factors might favor the same outcome was either considered to be of secondary importance or simply ignored.[†]

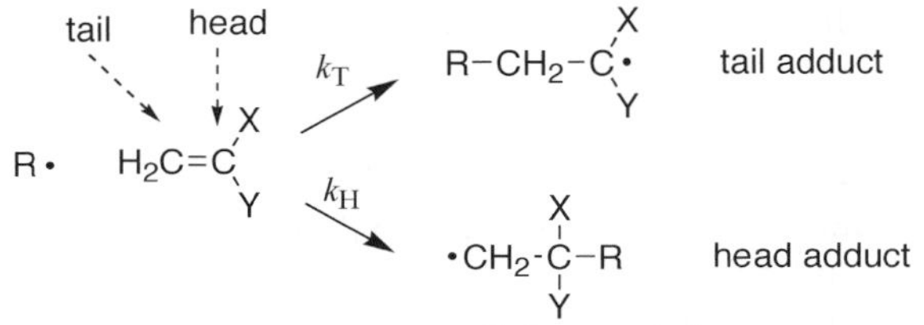

$$R\cdot + H_2C=C\begin{smallmatrix}X\\Y\end{smallmatrix} \quad \xrightarrow{k_T} \quad R-CH_2-C\begin{smallmatrix}X\\\cdot\\Y\end{smallmatrix} \quad \text{tail adduct}$$

$$\xrightarrow{k_H} \quad \cdot CH_2-\underset{Y}{\overset{X}{C}}-R \quad \text{head adduct}$$

Scheme 1.1

Indeed, while alternative hypotheses were entertained by some,[10] there was no serious questioning of the dominant role of thermochemistry in the wider community until the 1970s. Many factors were important in bringing about this change in thinking. Three of the more significant were:

(a) A few isolated examples appeared where "wrong way" addition (formation of the less thermodynamically stable radical) was a significant, or even the major, pathway. Notable examples are predominantly head addition in the intramolecular step of cyclopolymerization of 1,6-dienes (Scheme 1.2)[11] and in the reaction of *t*-butoxy radicals with difluoroethylene (Scheme 1.3).[12]

---

[*] The term tail addition is used to refer to addition to the less highly substituted end of the double bond.

[†] To this day some texts put forward product stability as the sole explanation for preferential tail addition.

Scheme 1.2

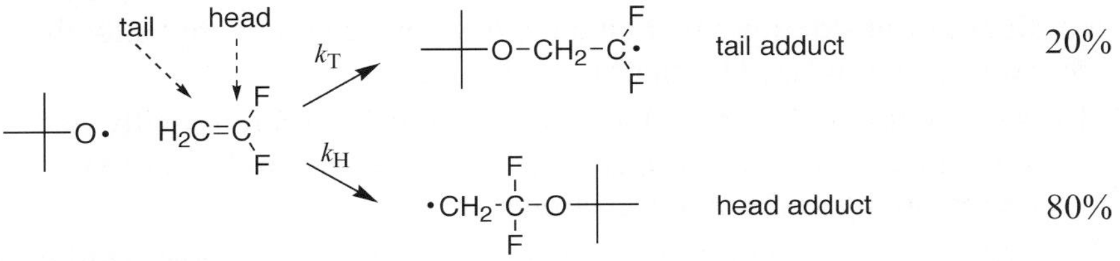

Scheme 1.3

tail adduct — 20%
head adduct — 80%

(b) Dependable measurements of rate constants for radical reactions became available which allowed structure-reactivity relationships to be reliably assessed.[13]

(c) Data on bond dissociation energies were evaluated to demonstrate that the amount of stabilization provided to a radical center by adjacent alkyl substituents is small. The relative stability of primary *vs* secondary *vs* tertiary radicals, even if fully reflected in the transition state, is not sufficient to account for the degree of regioselectivity observed in additions to alkenes.[14]

It is now established that product radical stability is a consideration in determining the outcome of radical addition reactions only where a substituent provides substantial delocalization of the free spin into a π-system. Even then, because these reactions are generally irreversible and exothermic (and consequently have early transition states), resonance stabilization of the incipient radical center may play only a minor role in determining reaction rate and specificity.[2,15-19] Thermodynamic factors will be the dominant influence only when polar and steric effects are more or less evenly balanced.[20,21]

The importance of the various factors determining the rate and regiospecificity of addition is illustrated by the data shown in Table 1.2 and Table 1.3.

Table 1.2 Relative Rate Constants and Regiospecificities for Addition of Radicals to Halo-Olefins[a]

Olefin	$(CH_3)_3CO\cdot$[b]		$CH_3\cdot$[c]		$CF_3\cdot$[c]		$CCl_3\cdot$[c]	
	k_{rel}	k_H/k_T	k_{rel}	k_H/k_T	k_{rel}	k_H/k_T	k_{rel}	k_H/k_T
$CH_2=CH_2$	1.0	-	1.0	-	1.0	-	1.0	-
$CH_2=CHF$	0.7	0.35	1.1	0.2	0.5	0.12	0.62	0.11
$CH_2=CF_2$	1.1	4.0	-	1	0.2	0.04	0.25	0.016
$CHF=CF_2$	6.6	4.5	5.8	2.1	0.05	0.55	0.29	0.32

a k_{rel} is overall rate constant for addition (k_H+k_T) relative to that for addition to ethylene (=1.0). All values have been rounded to 2 significant figures.   b At 60 °C.[22]   c At 164 °C.[13]

Relative rate constants for reaction of methyl, trifluoromethyl, trichloromethyl,[13] and *t*-butoxy radicals[22,23] with the fluoro-olefins are summarized in Table 1.2. Note the following points:

(a) Overall rates of addition for methyl and *t*-butoxy radicals are accelerated by fluorine substitution. In contrast, rates for trifluoromethyl and trichloromethyl radicals are reduced by fluorine substitution.
(b) Trifluoromethyl and trichloromethyl radicals preferentially add to the less substituted end of trifluoroethylene. Methyl and *t*-butoxy radicals add preferentially to the more substituted end.
(c) Trifluoromethyl and trichloromethyl radicals give predominantly tail addition to vinylidene fluoride, methyl radicals give both tail and head addition, *t*-butoxy radicals give predominantly head addition.

The overall trend of reactivities for *t*-butoxy radicals with the fluoro-olefins more closely parallels that for methyl radicals than that for the electrophilic trifluoromethyl or trichloromethyl radicals.

Table 1.3 Relative Rate Constants for Reactions of Radicals with Alkyl-Substituted Acrylate Esters $CHR^1=CR^2CO_2CH_3$ [a]

Monomer	R^1	R^2	$PhCO_2\cdot$ [b]		$Ph\cdot$ [b]		$(CH_3)_3CO\cdot$ [b]		$c\text{-}C_6H_{11}\cdot$ [c]	
			k_H	k_T	k_H	k_T	k_H	k_T	k_H	k_T
MA	H	H	0.2	1.0	0.03	1.0	0.02	1.0	0.002	1.0
MMA	H	CH_3	0.35	4.5	≤0.01	1.6	0	2.9	≤0.001	0.71
MC [d]	CH_3	H	1.6	1.3	0.07	0.12	≤0.03	0.3	0.001	0.011

a Rate constants relative to that for tail addition to MA (=1.0). All data have been rounded to 2 significant figures. b At 60 °C.[24] c At 20 °C.[25] d Methyl *trans*-2-butenoate (methyl crotonate).

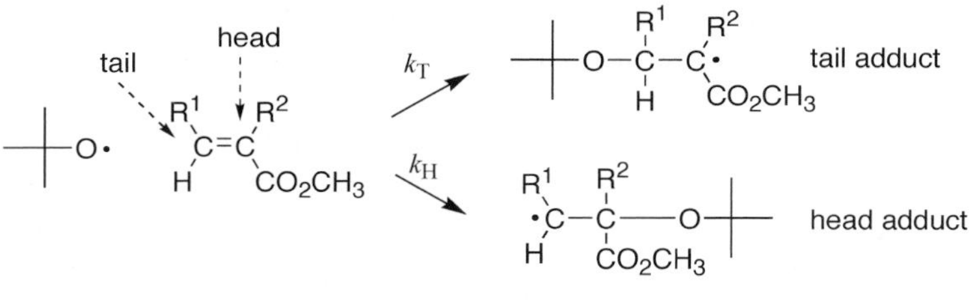

Scheme 1.4

Outcomes from the reactions of radicals with substituted acrylate esters depend on the attacking radical (refer Table 1.3 and Scheme 1.4). The results may be summarized as follows (the methyl substituent is usually considered to be electron donating – Section 1.2.2):

(a) Irrespective of the attacking radical, there is preferential addition to the tail of the double bond (to the end remote from the carbomethoxy group).
(b) For the nucleophilic cyclohexyl radicals (c-C_6H_{11}•), the rate of addition to the unsubstituted end of the double bond is slightly retarded by alkyl substitution (*ca* 30% for MMA *vs* MA). The rate of addition to the substituted end of the double bond is dramatically retarded by alkyl substitution (*ca* 90-fold for MC *vs* MA).[25]
(c) For the slightly electrophilic phenyl and *t*-butoxy radicals [Ph•, $(CH_3)_3CO$•]: the rate of addition to the unsubstituted end of the double bond is enhanced (2-3-fold) by alkyl substitution; the rate of addition to the substituted end of the double bond is retarded (>3-fold for MC *vs* MA) by alkyl substitution.[24,26]
(d) For the electrophilic benzoyloxy radicals ($PhCO_2$•): the rate of addition to the unsubstituted (tail) end of the double bond is enhanced (4.5-fold for MMA *vs* MA) by alkyl substitution; the rate of addition to the substituted (head) end of the double bond is slightly enhanced (75% for MMA *vs* MA) by alkyl substitution.[24]

The data of Table 1.2 and Table 1.3 clearly cannot be rationalized purely in terms of the relative stabilities of the product radicals. Rather, "a complex interplay of polar, steric, and bond strength terms" must be invoked.[13] In the following sections, each of these factors will be examined separately to illustrate their role in determining the outcome of radical addition.

1.2.1 Steric Factors

A clear demonstration of the relative importance of steric and resonance factors in radical additions to carbon-carbon double bonds can be found by considering the effect of (non-polar) substituents on the rate of attack of (non-polar) radicals. Substituents on the double bond strongly retard addition at the substituted carbon while leaving the rate of addition to the other end essentially unaffected (for example, Table 1.3). This is in keeping with expectation if steric factors determine the regiospecificity of addition, but contrary to expectation if resonance factors are dominant.

It is possible to resolve steric factors into several terms:
(a) B-strain engendered by the change from sp^2 towards sp^3 hybridization at the site of attack.[2,14] B-strain is a consequence of the substituents on the (planar) α-carbon of the double bond being brought closer together on moving towards a tetrahedral disposition (Figure 1.1). This term is important in all radical additions and is thought to be the main factor responsible for preferential attack at the less substituted end of the double bond.
(b) Steric hindrance to approach of the attacking radical to the site of attack on the olefin. This term is usually only a minor factor except where substituents on the radical or on the olefin are very bulky.[14,27]

(c) Steric hindrance to adoption of the required transition state geometry. This is not usually a determining factor in intermolecular addition of small radicals, but is extremely important in intramolecular addition where the approach of the reacting centers is constrained by the molecular geometry (Section 1.2.4).[28]

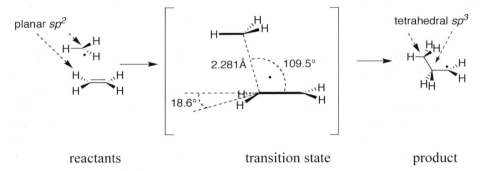

Figure 1.1 Transition state for methyl radical addition to ethylene. Geometric parameters are from *ab initio* calculation with QCISD(T)/6-31GT(d) basis set.[29]

Radical additions are typically highly exothermic and activation energies are small for carbon[30,31] and oxygen-centered[32,33] radicals of the types most often encountered in radical polymerization. Thus, according to the Hammond postulate,* these reactions are expected to have early reactant-like transition states in which there is little localization of the free spin on C_β. However, for steric factors to be important at all, there must be significant bond deformation and movement towards sp^3 hybridization at C_α.

Various *ab initio* and semi-empirical molecular orbital calculations have been carried out on the reaction of radicals with simple alkenes with the aim of defining the nature of the transition state (Section 1.2.7).[29,35,36] These calculations all predict an unsymmetrical transition state for radical addition (*i.e.* Figure 1.1) though they differ in other aspects. Most calculations also indicate a degree of charge development in the transition state.

The rate of radical addition is most dramatically affected by substituents either at the site of attack or at the radical center. Remote substituents generally have only a small influence on the stereochemistry and regiospecificity of addition unless these groups are very bulky or the geometry of the molecules is constrained (*e.g.* intramolecular addition – Section 1.2.4).

It is a common assumption that the influence of steric factors will be manifested mainly as a higher activation energy. In fact, there is good evidence[37] to show that steric factors are mainly reflected in a less favorable entropy of activation or Arrhenius frequency factor. This is due to the degrees of freedom

---

* A highly exothermic (low activation energy) reaction will generally have a transition state that resembles the reactants[34]

that are lost as the radical center approaches the terminus of the double bond and the α-substituents on the double bond are brought closer together on rehybridization.

1.2.2 Polar Factors

The rates of addition to the unsubstituted terminus of monosubstituted and 1,1-disubstituted olefins (this includes most polymerizable monomers) are thought to be determined largely by polar factors.[2,16] Polymer chemists were amongst the first to realize that polar factors were an important influence in determining the rate of addition. Such factors can account for the well-known tendency for monomer alternation in many radical copolymerizations and provide the basis for the Q-e, the Patterns of Reactivity, and many other schemes for estimating monomer reactivity ratios (Section 7.3.4).

The traditional means of assessment of the sensitivity of radical reactions to polar factors and establishing the electrophilicity or nucleophilicity of radicals is by way of a Hammett $\sigma\rho$ correlation. Thus, the reactions of radicals with substituted styrene derivatives have been examined to demonstrate that simple alkyl radicals have nucleophilic character[38,39] while haloalkyl radicals[40] and oxygen-centered radicals[23] have electrophilic character (Table 1.4). It is anticipated that electron-withdrawing substituents (*e.g.* Cl, F, CO$_2$R, CN) will enhance overall reactivity towards nucleophilic radicals and reduce reactivity towards electrophilic radicals. Electron-donating substituents (alkyl) will have the opposite effect.

Many researchers have applied similar approaches to develop or apply linear free energy relationships, when the substituent is directly attached to the double bond, with some success. Two of the more notable examples can be found in the Patterns of Reactivity Scheme (Section 7.3.4) and the works of Giese and coworkers.[16,19]

While steric terms may be the most significant factor in determining that tail addition is the predominant pathway in radical addition, polar factors affect the overall reactivity and have a significant influence on the degree of regiospecificity. In the reaction of benzoyloxy radicals with MMA, even though there is still a marked preference for tail addition, the methyl substituent enhances the rate constants for attack at both head and tail positions over those seen for MA (Table 1.3). With cyclohexyl radicals the opposite behavior is seen. Relative rate constants are reduced and the preference for tail addition is reinforced. For olefins substituted with electron-donor substituents, nucleophilic radicals give the greatest tail *vs* head specificity. The converse generally also applies.

In the reactions of the fluoro-olefins, steric factors are of lesser importance because of the relatively small size of the fluoro-substituent.[5] Fluorine and hydrogen are of similar bulk. In these circumstances, it should be expected that polar factors could play a role in determining regiospecificity. Application of the usual rules to vinylidene fluoride leads to a prediction that, for nucleophilic radicals, the rate of head addition will be enhanced. Similarly, for electrophilic radicals, the rate of tail addition will be enhanced (Figure 1.2).

Table 1.4 Hammett ρ and ρ^+ Parameters for Reactions of Radicals

	Addition to styrenes		H Abstraction from toluenes		
radical	ρ^+	ρ	ρ^+	ρ	ρ^d
(CH$_3$)$_3$C•	1.1^{a,38}	-	0.49^{b,41}	-	-
c-C$_6$H$_{11}$•	0.68^{a,38}	-	-	-	-
n-C$_6$H$_{13}$•	-	-	0.45^{a,38}	-	-
n-C$_{11}$H$_{23}$•	-	-	-	0.45^{a,42}	-
CH$_3$•	-	-	-0.1^{c,43}	-0.12^{c,43}	-0.21
(CH$_3$)$_3$CO•	-0.27^{e,23}	-0.31^{e,23}	-0.32^{f,44}	-0.36^{f,44}	-0.36
(CH$_3$)$_3$COO•	-	-	-0.56^{g,45}	-0.78^{g,45}	-0.73
(CH$_3$)$_2$N•	-	-	-1.08^{h,46}	-1.66^{h,46}	-0.96
CCl$_3$•	-0.42^{i,40}	-0.43^{i,40}	-1.46^{j,47}	-1.46^{j,47}	-1.67
n-C$_8$F$_{17}$•	-	-0.53^{48}	-	-	-

(nucleophilicity increases upward)

a 42 °C.   b 80 °C.   c 100 °C.   d ρ values recalculated by Pryor et al.[49] based on m-substituted derivatives only.   e 60 °C, benzene.   f 45 °C, chlorobenzene. Value shows solvent dependence.   g 40 °C.   h 136 °C.   i 70 °C.   j 50 °C.

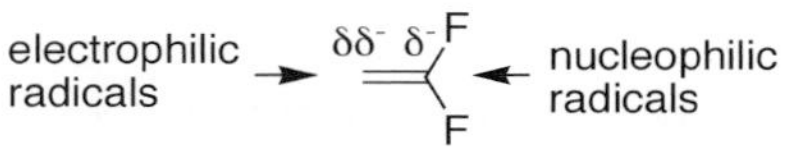

electrophilic radicals → $\delta\delta^-$ δ^-F / F ← nucleophilic radicals

Figure 1.2 Effect of polar factors on regiospecificity of radical addition.

The behavior of methyl and halomethyl radicals in their reactions with the fluoro-olefins (Table 1.2), can thus be rationalized in terms of a more dominant role of polar factors and the nucleophilic or electrophilic character of the radicals involved.[2] Methyl radicals are usually considered to be slightly nucleophilic, trifluoromethyl and trichloromethyl radicals are electrophilic (Table 1.4).

However, consideration of polar factors in the traditional sense does not provide a ready explanation for the regiospecificity shown by the t-butoxy radicals (which are electrophilic, Table 1.3) in their reactions with the fluoro-olefins (Table 1.2).[22,23] Apparent ambiphilicity has been reported[21] for other "not very electrophilic radicals" in their reactions with olefins and has been attributed to the polarizability of the radical.

1.2.3 Bond Strengths

The overriding importance of polar factors in determining rates of addition has recently been questioned by Fischer and Radom[4] who argue that reaction enthalpy should be considered the dominant factor in determining the rate of tail addition. Tedder and Walton[13] have stated: "If an experimentalist requires a simple qualitative theory, he should seek to estimate the strength of the new bond formed during the initial addition step...". Historically, a perceptual problem has been that the bond strength or reaction enthalpy term cannot be separated rigorously from the polar and steric factors discussed above since the latter both play an important role in determining the strength of the new bond. Fischer and Radom's[4] rationale is discussed below (1.2.7).

Just as steric factors may in some cases retard addition, factors that favor bond formation should be anticipated to facilitate addition. A pertinent example is the influence of α-fluorine substitution on C-X bond strength.[50] The C-C bond in $CH_3\text{-}CF_3$ is 46 kJ mol^{-1} stronger than that in $CH_3\text{-}CH_3$. Further fluorine substitution leads to a progressive strengthening of the bond. The effect is even greater for C-O bonds. The C–O bond dissociation energies in $CF_3\text{–}O\text{–}CF_3$ and $CF_3\text{–}OH$ are greater by 92 and 75 kJ mol^{-1}, respectively, than those in $CH_3\text{–}O\text{–}CH_3$ and $CH_3\text{–}OH$. This effect offers an explanation for the differing specificity shown by oxygen- and carbon-centered radicals in their reactions with the fluoro-olefins (Table 1.2).[23,51-53] The finding, that t-butoxy radicals give predominantly head addition with vinylidene fluoride (Scheme 1.3), can therefore be understood in terms of the relative strengths of the $CF_2\text{–}O$ and $CH_2\text{–}O$ bonds.[23]

1.2.4 Stereoelectronic Factors

A stereoelectronic requirement in radical addition to carbon-carbon double bonds first became apparent from studies on radical cyclization and the reverse (fragmentation) reactions.[54-56] It provides a rationalization for the preferential formation of the less thermodynamically stable *exo*-product (*i.e.* head addition) from the cyclization of ω-alkenyl radicals (**16** - Scheme 1.5).[18,57-64]

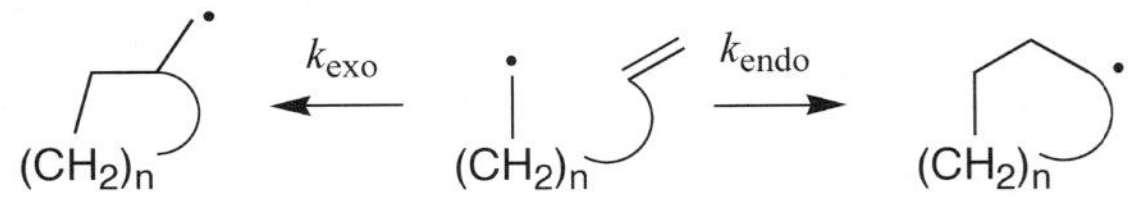

$$15 \quad \xleftarrow{k_{exo}} \quad 16 \quad \xrightarrow{k_{endo}} \quad 17$$

Scheme 1.5

It was proposed that the transition state requires approach of the radical directly above the site of attack and perpendicular to the plane containing the carbon-carbon double bond. An examination of molecular models shows that for the 3-butenyl and 4-pentenyl radicals (**16**, n=1,2) such a transition state can only be reasonably achieved in *exo*-cyclization (*i.e.* **16**→**15**). With the 5-hexenyl and 6-heptenyl radicals (**16**, n=3,4), the transition state for *exo*-cyclization (**16**→**15**) is more easily achieved than that for *endo*-cyclization (*i.e.* **16**→**17**).

The mode and rate of cyclization can be modified substantially by the presence of substituents at the radical center, on the double bond, and at positions on the connecting chain. As with intermolecular addition, substituents at the site of attack on the double bond strongly retard addition. For the 5-hexenyl system (**16**, *n*=3) the magnitude of the effect is such that methyl substitution at the 5-position causes *endo*-cyclization to be favored. For the 5,6-disubstituted radical the rates for both *exo*- and *endo*-addition are slowed and *exo*-cyclization again dominates. A full discussion of substituent effects on intramolecular addition can be found in the reviews cited above.

Stereoelectronic factors may also become important in polymerization when bulky substituents may hinder adoption of the required transition state. They may help explain why rate constants for addition of monomeric radicals may be very different from those for addition of dimeric or higher radicals.[4]

1.2.5 Entropic Considerations

The Arrhenius frequency factors [$\log(A/M^{-1}s^{-1})$] for addition of carbon centered radicals to the unsubstituted terminus of monosubstituted or 1,1-disubstituted olefins cover a limited range (6.0-9.0), depend primarily on the steric demand of the attacking radical and are generally unaffected by remote alkene substituents. Typical values of $\log(A/M^{-1}s^{-1})$ are *ca* 6.5 for tertiary polymeric (*e.g.* PMMA•), *ca* 7.0 for secondary polymeric (PS•, PMA•), and *ca* 7.5, 8.0 and 8.5 for small tertiary (*e.g. t*-C_4H_9•), secondary (*i*-C_3H_7•) and primary (CH_3•, C_2H_5•) radicals respectively (Section 4.5.4).[4] For 1,2,2-trisubstituted alkenes the frequency factors are about an order of magnitude lower.[4] The trend in values is consistent with expectation based on theoretical calculations.

Frequency factors are often determined from data obtained within a narrow temperature window. For this reason, it has been recommended[4] that when extrapolating rate constants less error might be introduced by adopting the standard values for frequency factors (above) than by using experimentally measured values. The standard values may also be used to estimate activation energies from rate constants measured at a single temperature.

1.2.6 Reaction Conditions

There is ample evidence to show that the outcome of radical addition is dependent on reaction conditions and, in particular, the reaction temperature and the reaction medium.

1.2.6.1 Temperature

Radical additions to double bonds are, in general, highly exothermic processes and rates increase with increasing temperature. The regiospecificity of addition to double bonds and the relative reactivity of various olefins towards radicals are also temperature dependent. Typically, specificity decreases with increasing temperature (the Reactivity-Selectivity Principle applies). However, a number of exceptions to this general rule have been reported.[38,65]

Giese and Feix[65] examined the temperature dependence of the relative reactivity of fumarodinitrile and methyl α-chloroacrylate towards a series of alkyl radicals (Scheme 1.6). The temperature dependence was such that they predicted that the order of reactivity of the radicals would be reversed for temperatures above 280 K (the isoselective temperature - Figure 1.3). This finding clearly indicates the need for care when comparing relative reactivity data.[66]

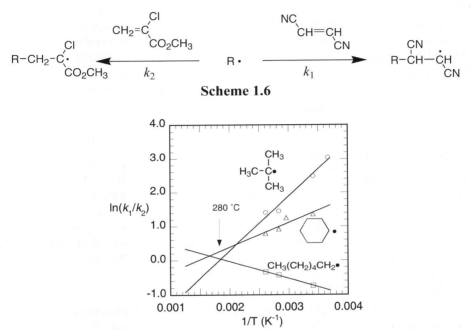

Scheme 1.6

Figure 1.3 Relative rate constants for addition of alkyl radicals to fumarodinitrile (k_1) and methyl α-chloroacrylate (k_2) as a function of temperature (Scheme 1.6).[65]

1.2.6.2 Solvent

It is established that rates of propagation in radical polymerization and reactivity ratios in copolymerization can show significant variation according to the solvent employed (Section 8.3.1).[67-71] For polymerizations of ethylene and vinyl acetate, effects on low conversion values of k_p in excess of an order of magnitude have been reported.[68,72] Smaller though measurable solvent effects on k_p are seen for other monomers. However, conventional wisdom has it that, except for those reactions involving charged intermediates, solvent effects on the rate and regioselectivity of radical addition to olefins are small and, consequently, they have not been widely studied. Nonetheless, reports of measurable solvent effects continue to appear.

Gas phase rate constants are typically an order of magnitude higher than solution phase rate constants. Fischer and Radom[4] have postulated that gas phase frequency factors should be similar to liquid phase numbers and the higher rate constants should therefore be largely attributed to lower activation energies (by *ca* 6.5 kJ mol^{-1}).

Giese and Kretzschmar[73] found the rate of addition of hexenyl radicals to methyl acrylate increased 2-fold between aqueous tetrahydrofuran and aqueous ethanol. Salikhov and Fischer[74] reported that the rate constant for *t*-butyl radical addition to acrylonitrile increased 3.6-fold between tetradecane and acetonitrile. Bednarek *et al.*[75] found that the relative reactivity of S *vs* MMA towards phenyl radicals was *ca* 20% greater in ketone solvents than it was in aromatic solvents.

More pronounced solvent effects have been observed in special cases where substrates or products possess ionic character. Ito and Matsuda[76] found a 35-fold reduction in the rate of addition of the arenethiyl radical **18** to α-methylstyrene when the solvent was changed from dimethylsulfoxide to cyclohexane. Rates for addition of other arenethiyl radicals do not show such a marked solvent dependence. The different behavior was attributed to the radical **18** existing partly in a zwitterionic quinonoid form (Scheme 1.7).[77]

Scheme 1.7

1.2.7 Theoretical Treatments

There have been many theoretical studies of radical addition reactions using *ab initio* methods,[4,35,36,53,78-88] semi-empirical calculations,[89,90] molecular mechanics[54,55] and other procedures. While geometries do not vary substantially with the level of theory, to obtain meaningful activation parameters with *ab initio* methods, a very high level of theory is required.[4,36] Such calculations are, at this stage, only practicable for small systems. However, computational power and method efficiency have improved substantially over the past few years and there is no evidence that this trend is leveling off. Heuts *et al.*[82,84] have argued that reliable Arrhenius *A* factors may be available using lower levels of theory.

The calculations using semi-empirical and low level *ab initio* methods do not give good values of reaction enthalpies or activation parameters and appear to fail dismally in some circumstances.[4] However, they have been shown to be useful in predicting relative energies for structurally similar systems and can give useful insights into mechanism. Methods for obtaining estimates of relative activation energies by molecular mechanics have also been devised.[55]

Various empirical schemes have also been proposed as predictive tools with respect to the outcome of radical addition reactions.[91,92] Two-parameter schemes, including the *Q-e* scheme (Section 7.3.4.1), Patterns of Reactivity (Section 7.3.4.2) and another developed by Ito and Matsuda[93] have been used with some success. Bakken and Jurs have used an approach based on multiple regression and neural networks and tested it by predicting rate constants for addition of methyl[92] and hydroxyl radicals[91] to various substrates. Denisov[94] proposed the "parabolic model" which involves eight descriptors.

Frontier Molecular Orbital (FMO) theory[95] may also be applied to provide qualitative understanding.[16,19] The frontier orbital of the radical is that bearing the free spin (the SOMO) and during radical addition this will interact with both the π* antibonding orbital (the LUMO) and the π-orbital (the HOMO) of the olefin. Both the SOMO-HOMO and the SOMO-LUMO interactions lead to a net drop in energy [i.e. $2(E_2)-E_3$ or E_1 respectively - Figure 1.4]. The dominant interaction and the reaction rates depend on the relative energies of these orbitals. Most radicals have high energy SOMO's and the SOMO-LUMO interaction is likely to be the most important. However, with highly electron deficient radicals, the SOMO may be of sufficiently low energy for the SOMO-HOMO interaction to be dominant.

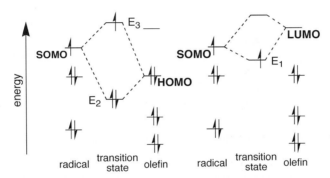

Figure 1.4 SOMO-HOMO and SOMO-LUMO orbital interaction diagrams.

For olefins with π-substituents, whether electron-withdrawing or electron-donating, both the HOMO and LUMO have the higher coefficient on the carbon atom remote from the substituent. A predominance of tail addition is expected as a consequence. However, for non-conjugated substituents, or those with lone pairs (e.g. the halo-olefins), the HOMO and LUMO are polarized in opposite directions. This may result in head addition being preferred in the case of a nucleophilic radical interacting with such an olefin. Thus, the data for attack of alkyl and fluoroalkyl radicals on the fluoro-olefins (Table 1.2) have been rationalized in terms of FMO theory.[16] Where the radical and olefin both have near "neutral" philicity, the situation is less clear.[21]

The State Correlation Diagram (SCD) approach introduced by Shaik and Pross[96] appears similar in some respects. However, the LUMO, HOMO and the first two excited states are considered. (refer Figure 1.5)[4,53] Thus, if we consider the interaction of the radical with the olefin in its ground (singlet) state (R• + C=C[1]) and excited (triplet) state (R• + C=C[3]) and two charge transfer configurations (R^+ + $C=C^-$) and (R^- + $C=C^+$), the energy of ground state configuration increases while those of the excited state configurations decrease as the reactants approach,. In the transition state, the various configurations mix according to their relative energies. A lucid description of the application of this approach to rationalize rate constants to addition of carbon centered radicals to olefins has recently been provided by Fischer and Radom.[4] Guided by the SCD analysis, they devised a scheme to predict absolute rate constants of radical addition based on knowledge of the reaction enthalpy, the singlet-triplet energy gap, the ionization potential and electron affinity of the olefin and the radical and the Coulomb interaction energy.

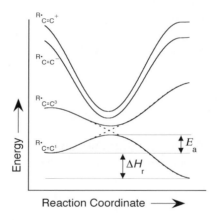

Figure 1.5 Schematic state correlation diagram for radical addition to a carbon-carbon double bond showing configuration energies as a function of the reaction coordinate.

1.2.8 Summary

No single factor can be identified as determining the outcome of radical addition. Nonetheless, there is a requirement for a set of simple guidelines to allow qualitative prediction. This need was recognized by Tedder and Walton,[2,17] Beckwith et al.,[59] Giese,[16] and, most recently, Fischer and Radom.[4] With the current state of knowledge, any such rules must be partly empirical and, therefore, it is to be expected that they may have to be revised from time to time as more results become available and further theoretical studies are carried out. However, this does not diminish their usefulness.

The following set of guidelines is a refinement of those suggested by Tedder:[2]

(a) For mono- or 1,1-disubstituted olefins, there is usually preferential addition to the unsubstituted (tail) end of the double bond. This selectivity can be largely attributed to the degree of steric compression associated with the formation of the new bond which usually overrides other influences on the regioselectivity.

(b) Substituents with π-orbitals (*e.g.* -CH=CH$_2$, -Ph) that can overlap with the half-filled atomic orbital of the incipient radical center may enhance the rate of addition at the remote end of the double bond. However, substituents with non-bonding pairs of electrons (*e.g.* -F, -Cl, -OR) have only a very small resonance effect. Most radical additions are exothermic and have early transition states and delocalization of the unpaired electron in the adduct radical is of small importance.

(c) Polarity can have a major effect on the overall rate of addition. Electron withdrawing substituents facilitate the addition of nucleophilic radicals while electron donating substituents enhance the addition of electrophilic radicals.

(d) The regioselectivity of addition to polysubstituted olefins is primarily controlled by the degree of steric compression associated with forming the new bond. However, if steric effects are small or mutually opposed, polarity can be the deciding factor.

(e) Even though the regioselectivity of addition to polysubstituted olefins is governed mainly by steric compression, polarity can influence the magnitude of the regioselectivity, making it larger or smaller depending on the relative electronegativity of the radical and the substituents on the olefin. The net result may be that the more reactive radical is the more selective.

1.3 Hydrogen Atom Transfer

Atom or radical transfer reactions generally proceed by a S$_H$2 mechanism (substitution, homolytic, bimolecular) that can be depicted as shown in Figure 1.6. This area has been the subject of a number of reviews.[1-3,27,97-99] The present discussion is limited, in the main, to hydrogen atom abstraction from aliphatic substrates and the factors which influence rate and specificity of this reaction.

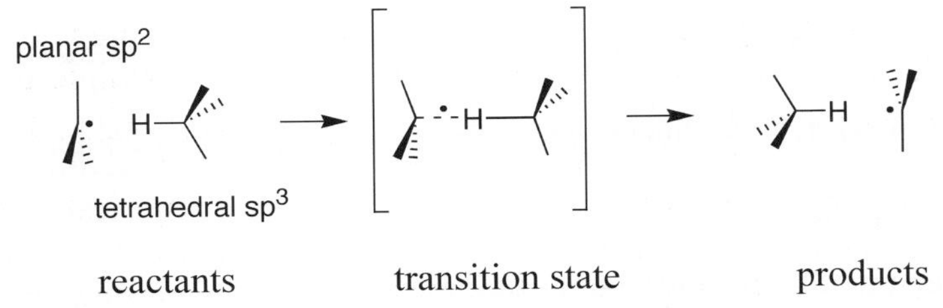

Figure 1.6 Transition state for hydrogen atom abstraction.

1.3.1 Bond Dissociation Energies

Simple thermochemical criteria can often be used to predict the relative facility of hydrogen atom transfer reactions. Evans and Polanyi[100] recognized this and suggested the following relationship (the Evans-Polanyi equation, eq. 1) between the activation energy for hydrogen atom abstraction (E_a) and the difference between the bond dissociation energies for the bonds being formed and broken ($\Delta H°$):

$$E_a = \alpha \Delta H° + \beta \quad (1)$$

where α and β are constants. It follows that for hydrogen abstraction by a given radical from a compound X-H, since the strength of the bond being formed is a constant, there should be a straight line relationship between the activation energy and the strength of the bond being broken [$D(X-H)$] (eq. 2):

$$E_a = \alpha' [D(X-H)] + \beta' \quad (2)$$

where α' and β' are constants. Examples of the application of the Evans-Polanyi equation can be found in reviews by Russell[97] and Tedder.[2,3] In the absence of severe steric constraints, straight line correlations between the relative reactivity of substrates towards a given radical can be found for systems: (a) where there is little polarity in the transition state, or (b) when the transition states are of like polarity. Tedder[2,3] has also stressed that, in these reactions, it is important to take note of the strength of the bond being formed. If there is no polarity in the transition state, the more exothermic reaction will generally be the less selective.

Bond dissociation energies qualitatively predict the order of reactivity of X-H bonds shown in Figure 1.7 (for examples see Table 1.1). However, as will become apparent, a variety of factors may perturb this order.

$R_2N-H \sim RO-H < R_3C-H < R_3Si-H < ROO-H < R_2P-H < RS-H < RSe-H$

Figure 1.7 Predicted order of reactivity of X-H compounds.

1.3.2 Steric Factors

Steric factors fall into four main categories:[27]

(a) The release or occurrence of steric compression due to rehybridization in the transition state where the attacking radical and site of attack are each undergoing rehybridization (from $sp^2 \rightarrow sp^3$ and $sp^3 \rightarrow sp^2$ respectively for aliphatic carbons – refer Figure 1.6). As a consequence, substituents on the attacking radical are brought closer together while those at the site of attack move apart. Thus, depending on the nature of the substituents at these centers, steric retardation or acceleration may accompany rehybridization.

(b) Steric hindrance of the approach of the attacking radical to the point of reaction in the substrate. This is important for the attack of very bulky radicals on hindered substrates.

(c) Steric inhibition of resonance - important in conformationally constrained molecules (Section 1.3.4).

(d) Steric hindrance to adoption of the required co-linear arrangement of atoms in the transition state. This is important in intramolecular reactions (Section 1.3.4).

The first term is of importance in all atom abstraction reactions, however, since the reactions are often highly exothermic with consequent early transition states, the effect may be small.

1.3.3 Polar Factors

Polar factors can play an extremely important role in determining the overall reactivity and specificity of homolytic substitution.[97] Theoretical studies on atom abstraction reactions support this view by showing that the transition state has a degree of charge separation.[101,102]

The traditional method of assessing the polarity of reactive intermediates is to examine the effect of substituents on rates and establish a linear free energy relationship (*e.g.* the Hammett relationship). The reactions of numerous radicals with substituted toluenes have been examined in this context. The value of the Hammett ρ parameter provides an indication of the sensitivity of the reaction to polar factors and gives a measure of the electrophilic or nucleophilic character of the attacking radical. For example, methyl radicals, usually considered to be slightly nucleophilic, have a slightly negative ρ value with respect to abstraction of benzylic hydrogens (Table 1.4).[43] Other simple alkyl radicals typically have positive ρ values.[41,42,103,104] Heteroatom-centered radicals (*e.g.* R$_2$N•, RO•, Cl•) generally have negative ρ values.[44,46,105,106] However, care must be taken in interpreting the results purely in terms of polar effects since electron withdrawing substituents typically also increase bond dissociation energies.[41,49,102,105]

The basic Hammett scheme often does not offer a perfect correlation and a number of variants on this scheme have been proposed to better explain reactivities in radical reactions.[23] However, none of these has achieved widespread acceptance. It should also be noted that linear free energy relationships are the basis of the *Q-e* and Patterns of Reactivity schemes for understanding reactivities of propagating species in chain transfer and copolymerization.

A striking illustration of the influence of polar factors in hydrogen abstraction reactions can be seen in the following examples (Figure 1.8) where different sites on the molecule are attacked preferentially according to the nature of the attacking radical.[97]

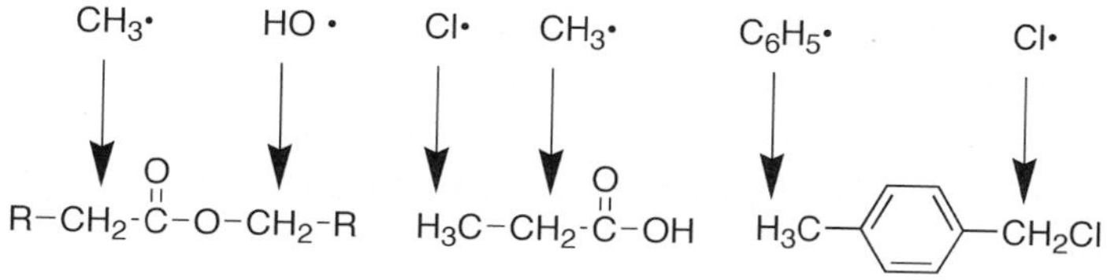

Figure 1.8 Preferred site of attack in hydrogen abstraction by various radicals.

1.3.4 Stereoelectronic Factors

There is a demonstrated requirement for a near co-linear arrangement of the orbital bearing the unpaired electron and the breaking C-H bond in the transition state for hydrogen atom transfer.[28,60,107,108] This becomes of particular importance for intramolecular atom transfer and accounts for the well-known preference for these reactions to occur by way of a six-membered transition state. The adoption of the chair conformation in the transition state for 1,5-atom transfer allows the requisite arrangement of atoms to be adopted readily. Such a transition state cannot be as readily achieved in smaller rings without significant strain being incurred, or in larger rings due to the severe non-bonded interactions and/or a less favorable entropy of activation.[107-110]

Thus, for radicals **19**, there is a strong preference for 1,5-hydrogen atom transfer (Table 1.5).[111] Although 1,6-transfer is also observed, the preference for 1,5-hydrogen atom transfer over 1,6-transfer is substantial even where the latter pathway would afford a resonance stabilized benzylic radical.[111,112] No sign of 1,2-, 1,3-, 1,4-, or 1,7-transfer is seen in these cases. Similar requirements for a co-linear transition state for homolytic substitution on sulfur and oxygen have been postulated.[18,60]

Table 1.5 Specificity of Intramolecular Hydrogen Abstraction[111]

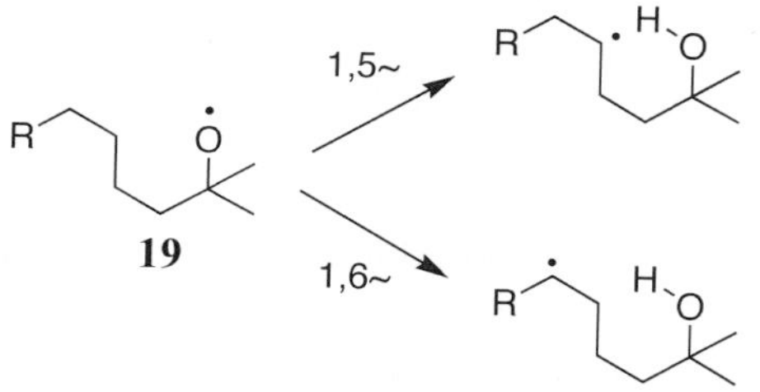

	R=CH$_3$	R=Ph
1,5~	94%	91%
1,6~	6%	9%

It is expected from simple thermochemical considerations that adjacent π-, σ- or lone pair orbitals should have a significant influence over the facility of atom transfer reactions. Thus, the finding that *t*-butoxy radicals show a marked preference for abstracting hydrogens α to ether oxygens (Figure 1.9) is not surprising. The reduced reactivity of the hydrogens β to oxygen in these compounds is attributed to polar influences.[113,114]

The most direct evidence that stereoelectronic effects are also important in these reactions follows from the specificity observed in hydrogen atom abstraction from conformationally constrained compounds.[18,60] C-H bonds adjacent to oxygen[113-118] or nitrogen[119] and which subtend a small dihedral angle with a lone pair orbital (<30°) are considerably activated in relation to those where the dihedral angle is or approaches 90°. Thus, the equatorial H in **20** is reported to be 12 times more reactive towards *t*-butoxy radicals than the axial H in **21**.[115]

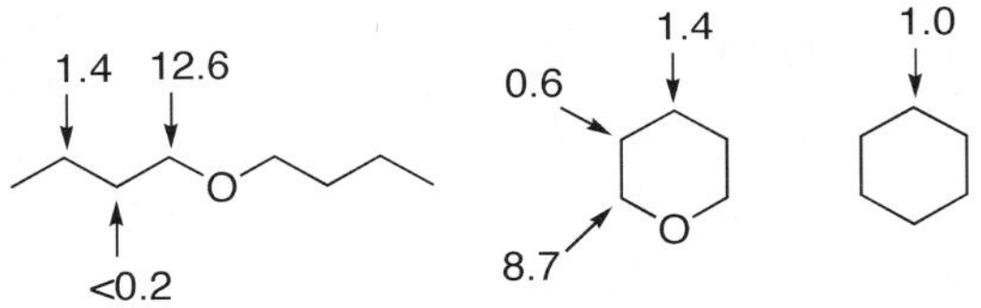

Figure 1.9 Relative reactivity per hydrogen atom of indicated site towards *t*-butoxy radicals.[113,114]

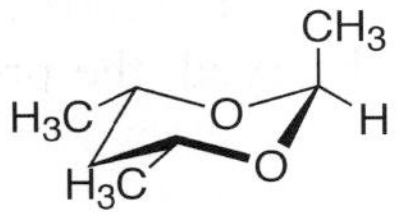

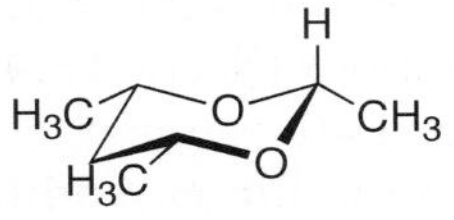

A further example of the importance of this type of stereoelectronic effect is seen in the reactions of *t*-butoxy radicals with spiro[2,n]alkanes (**22**) where it is found that hydrogens from the position α- to the cyclopropyl ring are specifically abstracted. This can be attributed to the favorable overlap of the breaking C-H bond with the cyclopropyl σ bonds.[120,121] No such specificity is seen with bicyclo[n,1,0]alkanes (**23**) where geometric constraints prevent overlap.

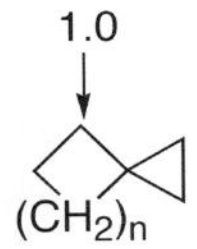

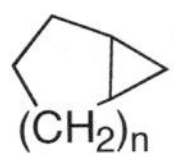

1.3.5 Reaction Conditions

Even though dissociation energies for X-H bonds appear insensitive to solvent changes,[122,123] the nature of the reaction medium[70,71,124] and the reaction temperature[66] can significantly affect the specificity and rate of atom abstraction reactions. One of the more controversial cases concerns the effect of aromatic solvents on hydrogen abstraction by atomic chlorine.

It has been proposed that aromatic solvents, carbon disulfide, and sulfur dioxide form a complex with atomic chlorine and that this substantially modifies both its overall reactivity and the specificity of its reactions.[125] For example, in reactions of Cl• with aliphatic hydrocarbons, there is a dramatic increase in the specificity for abstraction of tertiary or secondary over primary hydrogens in benzene as opposed to aliphatic solvents. At the same time, the overall rate constant for abstraction is reduced by up to two orders of magnitude in the aromatic solvent.[126] The exact nature of the complex responsible for this effect, whether a π-complex (**24**) or a chlorocyclohexadienyl radical (**25**), is not yet resolved.[126-132]

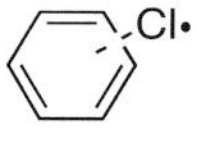

24 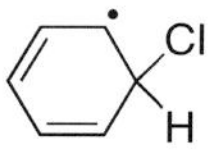   **25**

Significant, though smaller, solvent effects have also been reported for alkoxy radical reactions (Section 3.4.2.1).[133-137]

1.3.6 Abstraction vs Addition

The relative propensity of radicals to abstract hydrogen or add to double bonds is extremely important. In radical polymerization, this factor determines the significance of transfer to monomer, solvent, *etc.* and hence the molecular weight and end group functionality (Chapter 6). It also provides one basis for initiator selection (Section 3.2.1).

Table 1.6 Bond Dissociation Energies (D in kJ mol^{-1})[a,7]

Bond	D	Bond	D	Bond	D	Bond	D
(a) C-R bonds							
C_2H_5-C_2H_5	343	i-C_3H_7-C_2H_5	335	t-C_4H_9-C_2H_5	326	allyl-C_2H_5	299
C_2H_5-H	410	i-C_3H_7-H	395	t-C_4H_9-H	384	allyl-H	364
Δ^b	67		60		58		65
(b) X-R bonds							
H_2N-C_2H_5	351	HO-C_2H_5	381	C_2H_5O-C_2H_5	339	Cl-C_2H_5	339
H_2N-H	460	HO-H	498	C_2H_5O-H	435	Cl-H	431
Δ^b	109		117		96		92

a Values rounded to nearest kJ mol^{-1}. b Difference between $D(C$-$C_2H_5)$ and $D(C$-H) c Difference between $D(X$-$C_2H_5)$ and $D(X$-H).

The hydrogen abstraction:addition ratio is generally greater in reactions of heteroatom-centered radicals than it is with carbon-centered radicals. One factor is the relative strengths of the bonds being formed and broken in the two reactions (Table 1.6). The difference in exothermicity (Δ) between abstraction and addition reactions is much greater for heteroatom-centered radicals than it is for carbon-centered radicals. For example, for an alkoxy as opposed to an alkyl radical, abstraction is favored over addition by ca 30 kJ mol^{-1}. The extent to which this is reflected in the rates of addition and abstraction will, however, depend on the particular substrate and the other influences discussed above.

A number of studies have found that increasing nucleophilicity of the attacking radical favors abstraction over addition to an unsaturated system (benzene ring or double bond).[41,138,139] Bertrand and Surzur[139] surveyed the literature on the reactions of oxygen-centered radicals and observed that the ratio of abstraction to addition increased as shown in Figure 1.10.

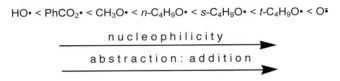

HO• < PhCO$_2$• < CH$_3$O• < n-C$_4$H$_9$O• < s-C$_4$H$_9$O• < t-C$_4$H$_9$O• < O$^{\bullet-}$

nucleophilicity ⟶

abstraction : addition ⟶

Figure 1.10 Dependence of abstraction:addition ratio on nucleophilicity for oxygen-centered radicals.

They, and later Houk,[140] attempted to establish a theoretical basis for this trend in terms of FMO theory. Pryor et al.[41] have found a similar trend for a series of aryl and alkyl radicals (Figure 1.11).

p-NO$_2$Ph• < p-BrPh• < CH$_3$• ~ Ph• < i-C$_3$H$_7$• < t-C$_4$H$_9$• ~ Ph$_3$C•

nucleophilicity ⟶

abstraction : addition ⟶

Figure 1.11 Dependence of abstraction:addition ratio on nucleophilicity for carbon-centered radicals.

However, the situation is not as clear-cut as it might at first seem since a variety of other factors may also contribute to the above-mentioned trend. Abuin et al.[141] pointed out that the transition state for addition is sterically more demanding than that for hydrogen-atom abstraction. Within a given series (alkyl or alkoxy), the more nucleophilic radicals are generally the more bulky (i.e. steric factors favor the same trends). It can also be seen from Table 1.6 that, for alkyl radicals, the values of D decrease in the series primary>secondary>tertiary (i.e. relative bond strengths favor the same trend).

1.3.7 Summary

A simple unifying theory to explain rate and specificity in atom abstraction reactions has yet to be developed. However, as with addition reactions, it is possible to devise a set of guidelines to predict qualitatively the rate and outcome of radical transfer processes. The following are based on those suggested by Tedder:[2]

(a) When there is little polarity in the transition state (or where the polarity is constant in a reaction series), the relative rates of atom transfer by a particular radical (selectivity) will correlate with the strengths of the bonds being broken.

(b) The strength of the bond being formed will be important in determining the absolute rate and the degree of selectivity.

(c) Steric strain relieved or incurred with formation of the new radical center may be important particularly for endothermic or near thermoneutral reactions.

(d) Nucleophilic radicals will prefer to attack electron rich sites. Electrophilic radicals will prefer to attack electron poor sites. If ΔH is small, polar factors may override thermodynamic considerations.

1.4 Radical-Radical Reactions

The last comprehensive review of reactions between carbon-centered radicals appeared in 1973.[142] Rate constants for radical-radical reactions in the liquid phase have been tabulated by Griller.[143] The area has also been reviewed by Alfassi[144] and Moad and Solomon.[145] Radical-radical reactions are, in general, very exothermic and activation barriers are extremely small even for highly resonance-stabilized radicals. As a consequence, reaction rate constants often approach the diffusion-controlled limit (typically ~10^9 M^{-1} s^{-1}).

The reaction may take several pathways:

(a) Combination, which usually but not invariably (Section 1.4.1), takes place by a simple head-to-head coupling of radicals.

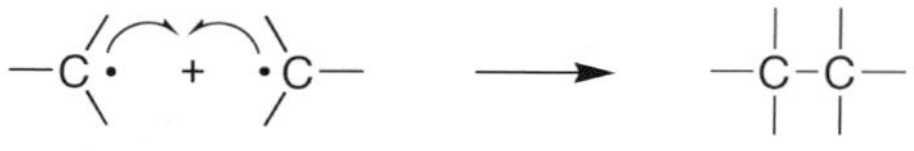

(b) Disproportionation, which involves the transfer of a β-hydrogen from one radical of the pair to the other (Section 1.4.2).

(c) Electron transfer, in which the product is an ion pair.

1.4.1 Pathways for Combination

The combination of carbon-centered radicals usually involves head-to-head (α,α-) coupling. Exceptions to this general rule occur where the free spin can be delocalized into a π-system. The classic example involves the triphenylmethyl radical (**13**) which combines to give exclusively the α-*para* coupling product (**26**), Scheme 1.8).[27] This chemistry is also seen in cross reactions of **13** with other tertiary radicals.[146]

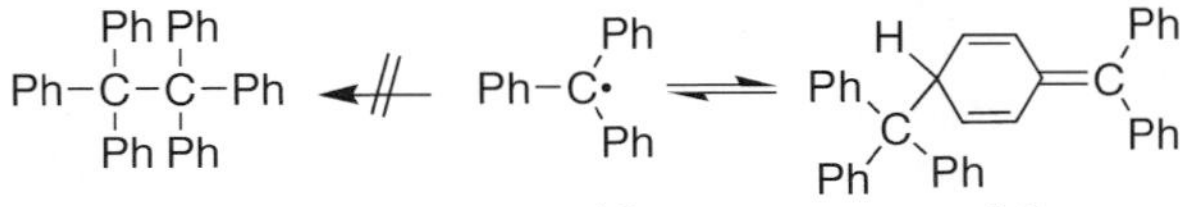

Ph Ph
| |
Ph–C–C–Ph   ≠   Ph–C•   ⇌   Ph   H   Ph
| | | \\ /\\   /
Ph Ph   Ph   C   C
                                     / \\   / \\
                                    Ph Ph Ph
                      13   **26**

Scheme 1.8

Other benzyl radicals, including the parent benzyl radical, give reversible formation of quinonemethide derivatives (typically a mixture of α,*p*- and α,*o*-coupling products) in competition with α,α-coupling (see also Section 5.2.2.1.1).[147-151] The kinetic product distribution appears to be determined by steric factors: α-substitution favors quinonemethide formation; ring substitution favors α,α-coupling. However, since quinonemethide formation is reversible, the only isolable product is often that from α,α-coupling.

For combination processes involving cyanoalkyl radicals, reversible C,N-coupling occurs in competition with C,C-coupling. Steric factors appear to be important in determining the relative amounts of C,C- and C,N-coupling[152] and exclusive C,N-coupling is observed when two bulky radicals combine.[153] For cyanoisopropyl radicals, C,N-coupling is the kinetically preferred pathway (Scheme 1.9).[105-107] However, since the formation of the ketenimine is thermally reversible, the C,C-coupling product is usually the major isolated product (Section 5.2.2.1.3).

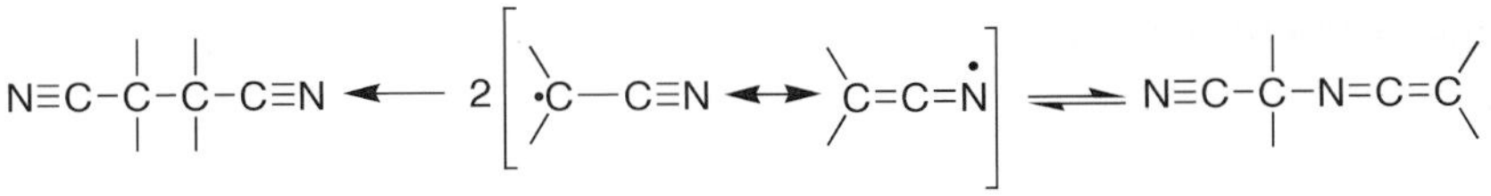

N≡C–C–C–C≡N   ←   2 [•C–C≡N ↔ C=C=N•]   ⇌   N≡C–C–N=C=C

Scheme 1.9

An example of C,O-coupling of α-ketoalkyl radicals with reversible formation of an enol ether has also been reported for a system where C,C-coupling is very hindered (Scheme 1.10).[154] However, this pathway is not observed for simpler species (Section 5.2.2.1.2).

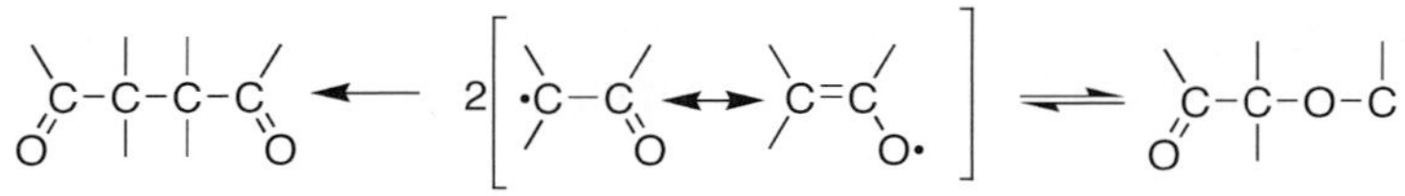

Scheme 1.10

1.4.2 Pathways for Disproportionation

For simple alkyl radicals, the product distribution appears to be predictable using statistical arguments.

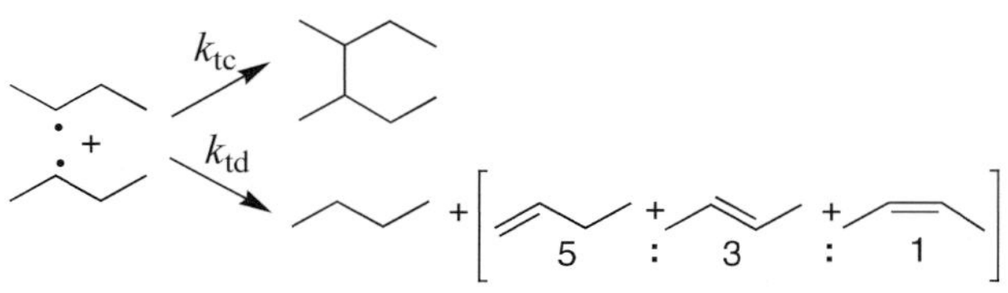

Scheme 1.11

For example, disproportionation of but-2-yl radicals produces a mixture of butenes as shown (Scheme 1.11).[138] Thermodynamic considerations suggest that but-1-ene and but-2-enes should be formed in a ratio of ca 2:98. However, the observed 5:4 ratio of but-1-ene:but-2-enes is little different from the 3:2 ratio that is expected on statistical grounds (*i.e.* ratio of β-hydrogens in the 1- and 3- positions).

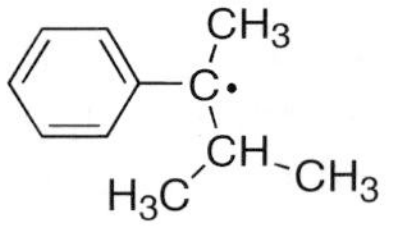

27

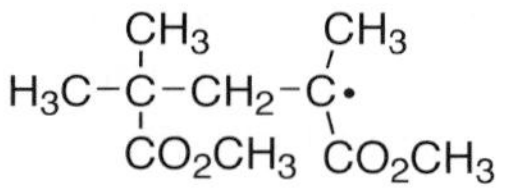

28

For more highly substituted examples, it is clear that other factors are also important. Substitution at the radical center has a profound effect. For example, in disproportionation, radicals **27**[155] and **28**[156] show a marked preference for loss of a hydrogen from the α-methyl substituent.

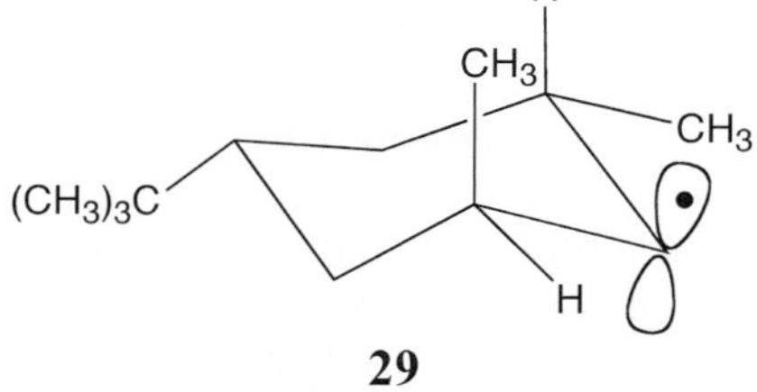

29

With the radical **29**, even though loss of an equatorial hydrogen should be sterically less hindered and is favored thermodynamically (by relief of 1,3 interactions of the axial methyl), there is an 8-fold preference for loss of the axial hydrogen (at 100 °C). The selectivity observed in the disproportionation of this and other substituted cyclohexyl radicals led Beckwith[18] to propose that disproportionation is subject to stereoelectronic control which results in preferential breaking of the C–H bond which has best overlap with the orbital bearing the unpaired spin.

1.4.3 Combination *vs* Disproportionation

Reactions between carbon-centered radicals generally give a mixture of disproportionation and combination. Much effort has been put into establishing the relative importance of these processes. The ratio of disproportionation to combination (k_{td}/k_{tc}) is dependent on the structural features of the radicals involved and generally shows only minor variation with solvent, pressure, temperature, *etc.*

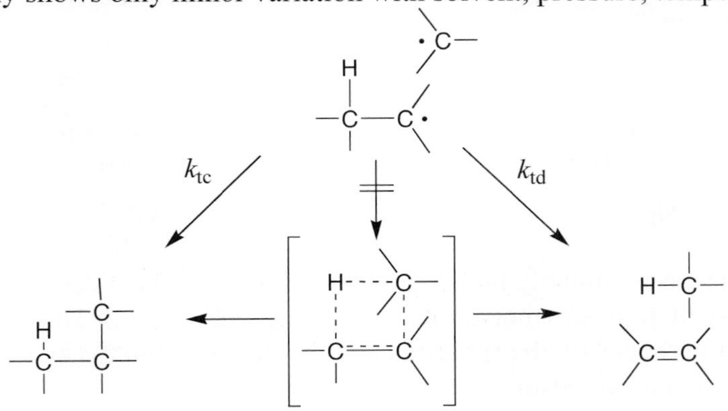

Scheme 1.12

Early workers in the area[157,158] suggested the involvement of a single 4-center transition state or intermediate which could lead to either disproportionation or combination (Scheme 1.12). The hypothesis fell from favor when it was established that k_{td}/k_{tc} showed a small though measurable dependence on temperature and pressure.[142] It is now generally recognized that combination and disproportionation should be considered as two separate reactions with distinct transition states. This view is supported by theoretical studies.[159-163]

1.4.3.1 *Statistical factors*

For a given series of radicals, the ratio k_{td}/k_{tc} increases with the number of β-hydrogen atoms. However, in general, there is no straight-forward relationship between k_{td}/k_{tc} and the number of β-hydrogens and it is clear that other factors are involved.[27,142] It is usually observed that even after allowing for the different number of β-hydrogens, the importance of disproportionation increases with increasing substitution at the radical center. For example, in the self-reaction of simple primary, secondary, and tertiary alkyl radicals, the values of $k_{td}/k_{tc}n$ are ca 0.06, 0.2, and 0.8 respectively, where n is the number of β-hydrogens.[27,142]

1.4.3.2 Steric factors

It has been suggested that the discrepancies between the value of k_{td}/k_{tc} observed and that predicted on the basis of simple statistics may reflect the greater sensitivity of combination to steric factors. Beckhaus and Rüchardt[164] reported a correlation between $\log(k_{td}/k_{tc})$ (after statistical correction) and Taft steric parameters for a series of alkyl radicals.

A graphic demonstration of the importance of steric factors on k_{td}/k_{tc} is provided by the contrasting behavior of radicals **30** and **31**. The self-reaction of cumyl radicals (**30**) affords predominantly combination while the radical **31**, in which an α-methyl is replaced by a neopentyl group, gives predominantly disproportionation.[165]

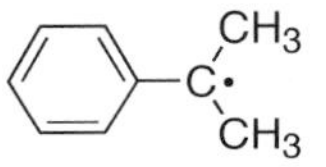

30 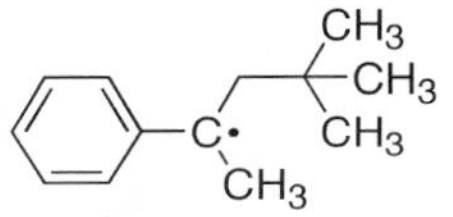 **31***

In extreme cases, suitably bulky substituents at the radical center can render a radical persistent [e.g. di-t-butyl methyl radical (**32**)].[166,167] This radical (**32**) possesses no hydrogens on the α-carbon and therefore cannot decay by the normal disproportionation mechanism.

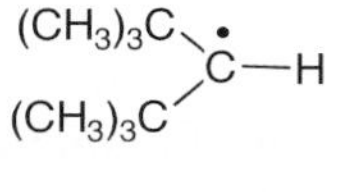

32 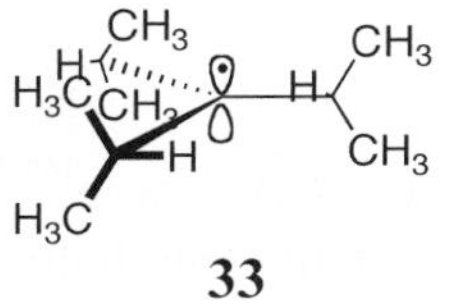 **33**

The triisopropylmethyl radical (**33**) is another example of a persistent radical. In this case, both disproportionation and combination are substantially retarded by steric factors.[168,169] In the preferred conformation shown, the β-hydrogens lie in a plane orthogonal to the orbital bearing the free spin.

The examples considered in this section lead to three conclusions:

(a) Disproportionation and combination can both be dramatically slowed by large β- or γ-substituents.

---

* In the original work[165] the neopentyl substituent is incorrectly shown as a *t*-butyl substituent.

(b) Combination is more sensitive to the presence of bulky β-substituents than disproportionation (*i.e.* k_{td}/k_{tc} is enhanced).
(c) Steric factors can outweigh simple statistical factors (*e.g.* even though **31** has fewer β-hydrogens, it gives more disproportionation than **30**).

Two quite separate influences are important in determining the rate of disproportionation:
(a) Steric hindrance to approach of the attacking radical (important for combination and disproportionation).
(b) Steric hindrance to rotation about the α,β-bond (important for disproportionation).

This latter term is considered in more detail under stereoelectronic factors (Section 1.4.3.4).

1.4.3.3 Polar factors

Minato *et al.*[162] proposed that the transition state for disproportionation has polar character while that for combination is neutral. The finding that polar solvents enhance k_{td}/k_{tc} for ethyl[170] and *t*-butyl radicals (Section 2.5.3.5), the very high k_{td}/k_{tc} seen for alkoxy radicals with α-hydrogens,[171] and the trend in k_{td}/k_{tc} observed for reactions of a series of fluoroalkyl radicals (Scheme 1.13, Table 1.7) have been explained in these terms.[144,162]

$$CH_{3-x}F_x\text{-}C_2H_5 \xleftarrow{k_{tc}} CH_{3-x}F_x\cdot\ +\ C_2H_5\cdot \xrightarrow{k_{td}} CH_{4-x}F_x\ +\ C_2H_4$$

Scheme 1.13

Table 1.7 Values of k_{td}/k_{tc} for the Cross-Reaction between Fluoromethyl and Ethyl Radicals (25 °C) [172-174]

Radical	k_{td}/k_{tc}	Radical	k_{td}/k_{tc}	Radical	k_{td}/k_{tc}
$CH_3\cdot$	0.039	$CHF_2\cdot$	0.068	$C_2F_5\cdot$	0.24
$CH_2F\cdot$	0.038	$CF_3\cdot$	0.11		

1.4.3.4 Stereoelectronic and other factors

The transition state for disproportionation requires overlap of the β C—H bond undergoing scission and the *p*-orbital containing the unpaired electron.[18] This requirement rationalizes the specificity observed in disproportionation of radicals **29** (Section 1.4.2) and provides an explanation for the persistency of the triisopropylmethyl radical (**33**) and related species (Section 1.4.3.2).[166] In the case of **33**, the β-hydrogens are constrained to lie in the nodal plane of the *p*-orbital due to steric buttressing between the methyls of the adjacent isopropyls.

It has been noted by a number of workers that the presence of α-substituents which delocalize the free spin favors combination over disproportionation.[127,148,175] For radicals of structure $(CH_3)_2C(\bullet)$-X, k_{td}/k_{tc} increases as shown in Figure 1.12. A correlation between the degree of exothermicity and the value of k_{td} has also been found but only for the case of resonance stabilized radicals.[144,176,177]

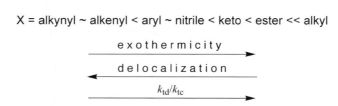

Figure 1.12 Trend in k_{td}/k_{tc} for radicals $(CH_3)_2C(\bullet)$-X.

It has been suggested that benzylic radicals may form a dimeric association complex which may easily collapse to the combination product but be geometrically unfavorable for disproportionation.[178,179] Even if this applies for the aralkyl radicals, it cannot account for the behavior of systems with other *p*-substituents.

Another explanation follows from the above discussion on stereoelectronic factors.[145] If overlap between the semi-occupied orbital and the breaking C-H bond favors disproportionation, then substituents which delocalize the free spin will serve to reduce this interaction and disfavor disproportionation. A proposal along these lines was made originally by Nelson and Bartlett[148] who also noted that diminishment of the spin density at C_α could retard combination. However, it is not necessary that the two effects should cancel one another.

1.4.3.5 *Reaction conditions*

Values of k_{td}/k_{tc} for simple alkyl radicals are sensitive to reaction conditions (solvent, temperature, pressure). However, the effects appear to be generally small (<2-fold).[142,144] Values of k_{td}/k_{tc} for *t*-butyl radicals in solution decrease with increasing temperature (the magnitude of the dependence increases with increasing solvent polarity - Figure 1.13) indicating a difference in activation energy of 3-12 kJ mol^{-1}. Smaller differences (1-2 kJ mol^{-1}) are seen for ethyl radicals.[142] For a given solvent type (alkane or alcohol), a very small dependence on the viscosity of the medium is also observed (Table 1.8). The temperature dependence of k_{td}/k_{tc} has been related to the rate of molecular reorientation and the dependence on viscosity.[180] A very small decrease in k_{td}/k_{tc} with temperature is observed for **28** (Section 5.2.2.1.2).[156,181] This small dependence of k_{td}/k_{tc} on temperature appears in marked contrast with the significant increase in k_{td}/k_{tc} with temperature reported for polymeric species (Section 5.2.2.2.2).

In studies of radical-radical reactions, radicals are typically generated pairwise and the products come from both cage and encounter (non-cage) reactions.

Several studies have indicated that cage *vs* encounter product distributions are the same.[156] However, it has been suggested that influences of pressure and viscosity on k_{td}/k_{tc} are more substantial for radicals which undergo self-reaction within the solvent cage.[149]

Table 1.8 Values of k_{td}/k_{tc} for *t*-Butyl Radicals at 25 °C[180]

Solvent	Temperature (°C)	η cP	k_{td}/k_{tc}
n-C$_8$H$_{18}$	25	0.51	5.4
n-C$_{10}$H$_{22}$	25	0.86	5.7
n-C$_{12}$H$_{26}$	25	1.37	5.9
n-C$_{14}$H$_{30}$	25	2.10	6.4
n-C$_{16}$H$_{34}$	25	3.09	6.9
CH$_3$CN	16.5	7.3	7.5
t-BuOH-pinacol(1:2)	25	-	10.1
3-methyl-3-pentanol	24.5	-	7.5

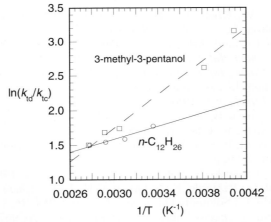

Figure 1.13 Temperature dependence of k_{td}/k_{tc} values for *t*-butyl radicals with dodecane (——) or 3-methyl-3-pentanol (- - -) as solvent.

1.4.4 Summary

The relative importance of combination and disproportionation may be predicted using the following guidelines:[145]

(a) Radical-radical reactions involving carbon-centered radicals give products from both combination and disproportionation.
(b) Simple primary and secondary radicals give predominantly combination. Tertiary radicals give some disproportionation.

(c) The importance of combination is increased by π–substituents at the radical center and decreased by bulky groups at or near the radical center.

1.5 References

1. Bamford, C.H. In *Comprehensive Polymer Science*; Eastmond, G.C.; Ledwith, A.; Russo, S.; Sigwalt, P., Eds.; Pergamon: Oxford, 1989; Vol. 3, p 219.
2. Tedder, J.M. *Angew. Chem., Int. Ed. Engl.* **1982**, *21*, 401.
3. Tedder, J.M. *Tetrahedron* **1982**, *38*, 313.
4. Fischer, H.; Radom, L. *Angew. Chem., Int. Ed. Engl.* **2001**, *40*, 1340.
5. Dolbier, W.R., Jr. *Chem. Rev.* **1996**, *96*, 1557.
6. Kerr, J.A.; Stocker, D.W. In *CRC Handbook of Chemistry and Physics*; Lide, D.R., Ed.; CRC Press: Boca Raton, Florida, 2002; p (9)53.
7. Benson, S.W. *Thermochemical Kinetics*; Wiley: New York, 1976.
8. Fischer, H. In *Free Radicals*; Kochi, J.K., Ed.; Wiley: New York, 1973; Vol. 2, p 435.
9. Leffler, J.E. *An Introduction to Free Radicals*; John Wiley & Sons: New York, 1993.
10. Walling, C. *Free Radicals in Solution*; Wiley: New York, 1957.
11. Hawthorne, D.G.; Johns, S.R.; Solomon, D.H.; Willing, R.I. *Aust. J. Chem.* **1976**, *29*, 1955.
12. Elson, I.H.; Mao, S.W.; Kochi, J.K. *J. Am. Chem. Soc.* **1975**, *97*, 335.
13. Tedder, J.M.; Walton, J.C. *Acc. Chem. Res.* **1976**, *9*, 183.
14. Rüchardt, C. *Angew. Chem., Int. Ed. Engl.* **1970**, *9*, 830.
15. Giese, B. *Angew. Chem., Int. Ed. Engl.* **1989**, *28*, 969.
16. Giese, B. *Angew. Chem., Int. Ed. Engl.* **1983**, *22*, 753.
17. Tedder, J.M.; Walton, J.C. *Tetrahedron* **1980**, *36*, 701.
18. Beckwith, A.L.J. *Tetrahedron* **1981**, *37*, 3073.
19. Ghosez-Giese, A.; Giese, B. *ACS Symp. Ser.* **1998**, *685*, 50.
20. Giese, B.; He, J.; Mehl, W. *Chem. Ber.* **1988**, *121*, 2063.
21. Beranek, I.; Fischer, H. In *Free Radicals in Synthesis and Biology*; Minisci, F., Ed.; Kluwer: Dordrecht, 1989; p 303.
22. Cuthbertson, M.J.; Rizzardo, E.; Solomon, D.H. *Aust. J. Chem.* **1985**, *38*, 315.
23. Jones, M.J.; Moad, G.; Rizzardo, E.; Solomon, D.H. *J. Org. Chem.* **1989**, *54*, 1607.
24. Moad, G.; Rizzardo, E.; Solomon, D.H. *Aust. J. Chem.* **1983**, *36*, 1573.
25. Giese, B.; Lachhein, S. *Angew. Chem., Int. Ed. Engl.* **1981**, *20*, 967.
26. Citterio, A.; Minisci, F.; Vismara, E. *J. Org. Chem.* **1982**, *47*, 81.
27. Rüchardt, C. *Top. Curr. Chem.* **1980**, *88*, 1.
28. Beckwith, A.L.J.; Ingold, K.U. In *Rearrangements in Ground and Excited States*; de Mayo, P., Ed.; Academic Press: New York, 1980; Vol. 1, p 162.
29. Wong, M.W.; Radom, L. *J. Phys. Chem.* **1995**, *99*, 8582.
30. Lorand, J.P. In *Landoldt-Bornstein, New Series, Radical Reaction Rates in Solution*; Fischer, H., Ed.; Springer-Verlag: Berlin, 1984; Vol. II/13a, p 135.
31. Roduner, E.; Crockett, R. In *Landoldt-Bornstein, New Series, Radical Reaction Rates in Solution*; Fischer, H., Ed.; Springer-Verlag: Berlin, 1995; Vol. II/18a, p 79.
32. Howard, J.A.; Scaiano, J.C. In *Landoldt-Bornstein, New Series, Radical Reaction Rates in Solution*; Fischer, H., Ed.; Springer-Verlag: Berlin, 1984; Vol. II/13d, p 5.
33. Lusztyk, J. In *Landoldt-Bornstein, New Series, Radical Reaction Rates in Solution*; Fischer, H., Ed.; Springer-Verlag: Berlin, 1995; Vol. II/18d1, p 1.

34. Hammond, G.S. *J. Am. Chem. Soc.* **1955**, *77*, 334.
35. Gonzalez, C.; Sosa, C.; Schlegel, H.B. *J. Phys. Chem.* **1989**, *93*, 2435.
36. Wong, M.W.; Radom, L. *J. Phys. Chem. A* **1998**, *102*, 2237.
37. Owen, G.E., Jr.; Pearson, J.M.; Szwarc, M. *Trans. Faraday Soc.* **1965**, *61*, 1722.
38. Giese, B.; Meister, J. *Angew. Chem., Int. Ed. Engl.* **1977**, *16*, 178.
39. Giese, B.; Meixner, J. *Chem. Ber.* **1981**, *114*, 2138.
40. Sakurai, H.; Hayashi, S.; Hosomi, A. *Bull. Chem. Soc. Japan* **1971**, *44*, 1945.
41. Pryor, W.A.; Tang, F.Y.; Tang, R.H.; Church, D.F. *J. Am. Chem. Soc.* **1982**, *104*, 2885.
42. Henderson, R.W.; Ward, R.D., Jr. *J. Am. Chem. Soc.* **1974**, *96*, 7556.
43. Pryor, W.A.; Tonellato, U.; Fuller, D.L.; Jumonville, S. *J. Org. Chem.* **1969**, *34*, 2018.
44. Sakurai, H.; Hosomi, A. *J. Am. Chem. Soc.* **1967**, *89*, 458.
45. Howard, J.A.; Chenier, J.H.B. *J. Am. Chem. Soc.* **1973**, *95*, 3054.
46. Michejda, C.J.; Hoss, W.P. *J. Am. Chem. Soc.* **1970**, *92*, 6298.
47. Huyser, E.S. *J. Am. Chem. Soc.* **1960**, *82*, 394.
48. Avila, D.V.; Ingold, K.U.; Lusztyk, J.; Dolbier, W.R., Jr.; Pan, H.-Q.; Muir, M. *J. Am. Chem. Soc* **1994**, *116*, 99.
49. Pryor, W.A.; Lin, T.H.; Stanley, J.P.; Henderson, R.W. *J. Am. Chem. Soc.* **1973**, *95*, 6993.
50. Smart, B.E. In *Molecular Structure and Energetics*; Liebman, J.F.; Greenberg, A., Eds.; VCH: Deerfield Beach, Florida, 1976; Vol. 3, p 141.
51. Arnaud, R.; Subra, R.; Barone, V.; Lelj, F.; Olivella, S.; Solé, A.; Russo, N. *J. Chem. Soc., Perkin Trans. 2* **1986**, 1517.
52. Canadell, E.; Eisenstein, O.; Ohanessian, G.; Poblet, J.M. *J. Phys. Chem.* **1985**, *89*, 4856.
53. Shaik, S.S.; Canadell, E. *J. Am. Chem. Soc.* **1990**, *112*, 1446.
54. Spellmeyer, D.C.; Houk, K.N. *J. Org. Chem.* **1987**, *52*, 959.
55. Beckwith, A.L.J.; Schiesser, C.H. *Tetrahedron* **1985**, *41*, 3925.
56. Beckwith, A.L.J.; Moad, G. *J. Chem. Soc., Perkin Trans. 2* **1980**, 1083.
57. Beckwith, A.L.J. In *Chem. Soc. Spec. Publ. - Essays on Free Radical Chemistry*; Chem. Soc.: London, 1970; Vol. 24, p 239.
58. Wilt, J.W. In *Free Radicals*; Kochi, J.K., Ed.; Wiley: New York, 1973; Vol. 1, p 333.
59. Beckwith, A.L.J.; Easton, C.J.; Serelis, A.K. *J. Chem. Soc., Chem. Commun.* **1980**, 482.
60. Beckwith, A.L.J. *Chem. Soc. Rev.* **1993**, 143.
61. Curran, D.P.; Porter, N.A.; Giese, B. *Stereochemistry of Radical Reactions*; VCH: Weinheim, 1996.
62. Giese, B.; Kopping, B.; Göbel, T.; Dickhaut, J.; Thoma, G.; Kulicke, K.J.; Trach, F. *Organic Reactions* **1996**, *48*, 301.
63. Julia, M. *Pure Appl. Chem.* **1974**, *40*, 553.
64. Julia, M. *Acc. Chem. Res.* **1971**, *4*, 386.
65. Giese, B.; Feix, C. *Isr. J. Chem.* **1985**, *26*, 387.
66. Giese, B. *Acc. Chem. Res.* **1984**, *17*, 438.
67. Spirin, Y., L. *Russ. Chem. Rev. (Engl. Transl.)* **1969**, *38*, 529.
68. Kamachi, M. *Adv. Polym. Sci.* **1981**, *38*, 55.
69. Gromov, V.F.; Khomiskovskii, P.M. *Russ. Chem. Rev. (Engl. Transl.)* **1979**, *48*, 1040.
70. Huyser, E.S. *Adv. Free Radical Chem.* **1965**, *1*, 77.

71. Martin, J.C. In *Free Radicals*; Kochi, J.K., Ed.; Wiley: New York, 1973; Vol. 2, p 493.
72. Shostenko, A.G.; Myshkin, V.E. *Dokl. Phys. Chem. (Engl. Transl.)* **1979**, *246*, 569.
73. Giese, B.; Kretzschmar, G. *Chem. Ber.* **1984**, *117*, 3160.
74. Salikhov, A.; Fischer, H. *Appl. Magn. Reson.* **1993**, *5*, 445.
75. Bednarek, D.; Moad, G.; Rizzardo, E.; Solomon, D.H. *Macromolecules* **1988**, *21*, 1522.
76. Ito, O.; Matsuda, M. *J. Phys. Chem.* **1984**, *88*, 1002.
77. Fong, C.W.; Kamlet, M.J.; Taft, R.W. *J. Org. Chem.* **1983**, *48*, 832.
78. Houk, K.N.; Padden-Row, M.N.; Spellmeyer, D.C.; Rondan, N.G.; Nagase, S. *J. Org. Chem.* **1986**, *51*, 2874.
79. Sosa, C.; Schlegel, H.B. *J. Am. Chem. Soc.* **1987**, *109*, 4193.
80. Arnaud, R.; Barone, V.; Olivella, S.; Russo, N.; Solé, A. *J. Chem. Soc., Chem. Commun.* **1985**, 1331.
81. Delbecq, F.; Ilavsky, D.; Nguyen, T.A.; Lefour, J.M. *J. Am. Chem. Soc.* **1985**, *107*, 1623.
82. Heuts, J.P.A.; Gilbert, R.G.; Radom, L. *Macromolecules* **1995**, *28*, 8771.
83. Coote, M.L.; Davis, T.P.; Radom, L. *Macromolecules* **1999**, *32*, 2935.
84. Heuts, J.P.A.; Gilbert, R.G.; Maxwell, I.A. *Macromolecules* **1997**, *30*, 726.
85. Coote, M.L.; Davis, T.P.; Radom, L. *Macromolecules* **1999**, *32*, 5270.
86. Arnaud, R.; Vetere, V.; Barone, V. *J. Comput. Chem.* **2000**, *21*, 675.
87. Arnaud, R.; Vetere, V.; Barone, V. *Chem. Phys. Lett.* **1998**, *293*, 295.
88. Radom, L.; Wong, M.W.; Pross, A. *ACS Symp. Ser.* **1998**, *685*, 31.
89. Dewar, M.J.S.; Olivella, S. *J. Am. Chem. Soc.* **1978**, *17*, 5290.
90. Arnaud, R.; Douady, J.; Subra, R. *Nouv. J. Chim* **1981**, *5*, 181.
91. Bakken, G.A.; Jurs, P.C. *J. Chem. Inf. Comput. Sci.* **1999**, *39*, 1064.
92. Bakken, G.A.; Jurs, P.C. *J. Chem. Inf. Comput. Sci.* **1999**, *39*, 508.
93. Ito, O.; Matsuda, M. *Prog. Polym. Sci.* **1992**, *17*, 827.
94. Denisov, E.T. *Russ. Chem. Rev.* **2000**, *69*, 153.
95. Flemming, I. *Frontier Orbitals and Organic Chemical Reactions*; Wiley: Chichester, 1976.
96. Shaik, S.S.; Pross, A. *Acc. Chem. Res.* **1983**, *16*, 363.
97. Russell, G.A. In *Free Radicals*; Kochi, J.K., Ed.; Wiley: New York, 1973; Vol. 1, p 275.
98. Poutsma, M.L. In *Free Radicals*; Kochi, J.K., Ed.; Wiley: New York, 1973; Vol. 2, p 113.
99. Hendry, D.G.; Mill, T.; Piszkiewicz, L.; Howard, J.A.; Eigenmann, H.K. *J. Phys. Chem. Ref., Data* **1974**, *3*, 937.
100. Evans, M.G.; Polanyi, M. *Trans. Faraday Soc.* **1938**, *34*, 11.
101. Pross, A.; Yamataka, H.; Nagase, S. *J. Phys. Org. Chem.* **1991**, *4*, 135.
102. Gilliom, R.D. *J. Mol. Struct.* **1986**, *138*, 157.
103. Pryor, W.A.; Davis, W.H., Jr. *J. Am. Chem. Soc.* **1974**, *96*, 7557.
104. Zavitsas, A.A.; Hanna, G.M. *J. Org. Chem.* **1975**, *40*, 3782.
105. Zavitsas, A.A.; Pinto, J.A. *J. Am. Chem. Soc.* **1972**, *94*, 7390.
106. Walling, C.; McGuinness, J.A. *J. Am. Chem. Soc.* **1969**, *91*, 2053.
107. Huang, X.L.; Dannenberg, J.J. *J. Org. Chem.* **1991**, *56*, 5421.
108. Houk, K.N.; Tucker, J.A.; Dorigo, A.E. *Acc. Chem. Res.* **1990**, *23*, 107.
109. Toh, J.S.S.; Huang, D.M.; Lovell, P.A.; Gilbert, R.G. *Polymer* **2001**, *42*, 1915.
110. Filley, J.; McKinnon, J.T.; Wu, D.T.; Ko, G.H. *Macromolecules* **2002**, *35*, 3731.
111. Walling, C.; Padwa, A. *J. Am. Chem. Soc.* **1963**, *85*, 1597.

112. Neale, R.S.; Walsh, M.R.; Marcus, N.L. *J. Org. Chem.* **1965**, *30*, 3683.
113. Busfield, W.K.; Grice, I.D.; Jenkins, I.D. *J. Chem. Soc., Perkin Trans. 2* **1994**, 1079.
114. Busfield, W.K.; Grice, I.D.; Jenkins, I.D.; Monteiro, M.J. *J. Chem. Soc., Perkin Trans. 2* **1994**, 1071.
115. Beckwith, A.L.J.; Easton, C.J. *J. Chem. Soc., Perkin Trans. 2* **1983**, 661.
116. Malatesta, V.; Scaiano, J.C. *J. Org. Chem.* **1982**, *47*, 1455.
117. Beckwith, A.L.J.; Easton, C.J. *J. Am. Chem. Soc.* **1981**, *103*, 615.
118. Malatesta, V.; Ingold, K.U. *J. Am. Chem. Soc.* **1981**, *103*, 609.
119. Griller, D.; Howard, J.A.; Marriott, P.R.; Scaiano, J.C. *J. Am. Chem. Soc.* **1981**, *103*, 619.
120. Roberts, C.; Walton, J.C. *J. Chem. Soc., Perkin Trans. 2* **1985**, 841.
121. Roberts, C.; Walton, J.C. *J. Chem. Soc., Chem. Commun.* **1984**, 1109.
122. Bausch, M.J.; Gostowski, R.; Guadalupe-Fasano, C.; Selmarten, D.; Vaughn, A.; Wang, L.-H. *J. Org. Chem.* **1991**, *56*, 7191.
123. Kanabus-Kaminske, J.M.; Gilbert, B.C.; Griller, D. *J. Am. Chem. Soc.* **1989**, *111*, 3311.
124. Reichardt, C. *Solvent Effects in Organic Chemistry*; Verlag Chemie: Weinheim, 1978.
125. Russell, G.A. *J. Am. Chem. Soc.* **1958**, *80*, 4987.
126. Bunce, N.J.; Ingold, K.U.; Landers, J.P.; Lusztyk, J.; Scaiano, J.C. *J. Am. Chem. Soc.* **1985**, *107*, 5464.
127. Walling. *J. Org. Chem.* **1988**, *53*, 305.
128. Tanko, J.M.; Anderson, F.E., III. *J. Am. Chem. Soc.* **1988**, *110*, 3525.
129. Taylor, C.K.; Skell, P.S. *J. Am. Chem. Soc.* **1983**, *105*, 120.
130. Skell, P.S.; Baxter, H.N.I.; Tanko, J.M.; Chebolu, V. *J. Am. Chem. Soc.* **1986**, *108*, 6300.
131. Ponec, R.; Hajeck, J. *Z. Phys. Chem. (Leipzig)* **1987**, *268*, 1233.
132. Ingold, K.U.; Lusztyk, J.; Raner, K.D. *Acc. Chem. Res.* **1990**, *23*, 219.
133. Walling, C.; Wagner, P.J. *J. Am. Chem. Soc.* **1964**, *86*, 3368.
134. Mendenhall, G.D.; Stewart, L.C.; Scaiano, J.C. *J. Am. Chem. Soc.* **1982**, *104*, 5109.
135. Grant, R.D.; Griffiths, P.G.; Moad, G.; Rizzardo, E.; Solomon, D.H. *Aust. J. Chem.* **1983**, *36*, 2447.
136. Grant, R.D.; Rizzardo, E.; Solomon, D.H. *Makromol. Chem.* **1984**, *185*, 1809.
137. Avila, D.V.; Brown, C.E.; Ingold, K.U.; Lusztyk, J. *J. Am. Chem. Soc* **1993**, *115*, 466.
138. Sheldon, R.A.; Kochi, J.K. *J. Am. Chem. Soc.* **1970**, *92*, 4395.
139. Bertrand, M.P.; Surzur, J.-M. *Tetrahedron Lett.* **1976**, *17*, 3451.
140. Houk, K.N. In *Frontiers in Free Radical Chemistry*; Pryor, W.A., Ed.; Academic Press: New York, 1980; p 43.
141. Abuin, E.; Mujica, C.; Lissi, E. *Rev. Latinoamer. Quim.* **1980**, *11*, 78.
142. Gibian, M.J.; Corley, R.C. *Chem. Rev.* **1973**, *73*, 441.
143. Griller, D. In *Landoldt-Bornstein, New Series, Radical Reaction Rates in Solution*; Fischer, H., Ed.; Springer-Verlag: Berlin, 1984; Vol. II/13a, p 5.
144. Alfassi, Z.B. In *Chemical Kinetics of Small Organic Radicals*; Alfassi, Z.B., Ed.; CRC Press: Boca Raton, Fla., 1988; Vol. 1, p 129.
145. Moad, G.; Solomon, D.H. In *Comprehensive Polymer Science*; Eastmond, G.C.; Ledwith, A.; Russo, S.; Sigwalt, P., Eds.; Pergamon: Oxford, 1989; Vol. 3, p 147.
146. Engel, P.S.; Chen, Y.; Wang, C. *J. Org. Chem.* **1991**, *56*, 3073.
147. Gleixner, G.; Olaj, O.F.; Breitenbach, J.W. *Makromol. Chem.* **1979**, *180*, 2581.

148. Nelsen, S.F.; Bartlett, P.D. *J. Am. Chem. Soc.* **1966**, *88*, 137.
149. Neuman, R.C., Jr.; Amrich, M.J., Jr. *J. Org. Chem.* **1980**, *45*, 4629.
150. Skinner, K.J.; Hochster, H.S.; McBride, J.M. *J. Am. Chem. Soc.* **1974**, *96*, 4301.
151. Langhals, H.; Fischer, H. *Chem. Ber.* **1978**, *111*, 543.
152. Barbe, W.; Rüchardt, C. *Makromol. Chem.* **1983**, *184*, 1235.
153. Zarkadis, A.K.; Neumann, W.P.; Dünnebacke, D.; Penenory, A.; Stapel, R.; Stewen, U. *Chem. Ber.* **1993**, *126*, 1179.
154. Neumann, W.P.; Stapel, R. *Chem. Ber.* **1986**, *119*, 3422.
155. Bartlett, P.D.; McBride, J.M. *Pure Appl. Chem.* **1967**, *15*, 89.
156. Bizilj, S.; Kelly, D.P.; Serelis, A.K.; Solomon, D.H.; White, K.E. *Aust. J. Chem.* **1985**, *38*, 1657.
157. Bradley, J.N.; Rabinovitch, B.S. *J. Chem. Phys.* **1962**, *36*, 3498.
158. Kerr, J.A.; Trotman-Dickenson, A.F. *Prog. React. Kinet.* **1961**, *1*, 107.
159. Benson, S.W. *Acc. Chem. Res.* **1986**, *19*, 335.
160. Dannenberg, J.J.; Baer, B. *J. Am. Chem. Soc.* **1987**, *109*, 292.
161. Imoto, M.; Sakai, S.; Ouchi, T. *J. Chem. Soc. Japan* **1985**, *1*, 97.
162. Minato, T.; Yamabe, S.; Fujimoto, H.; Fukui, K. *Bull. Chem. Soc. Japan* **1978**, *51*, 1.
163. Smith, W.B. *Struct. Chem.* **2001**, *12*, 213.
164. Beckhaus, H.D.; Rüchardt, C. *Chem. Ber.* **1977**, *110*, 878.
165. Fraenkel, G.; Geckle, M.J. *J. Chem. Soc., Chem. Commun.* **1980**, 55.
166. Griller, D.; Ingold, K.U. *Acc. Chem. Res.* **1976**, *9*, 13.
167. Griller, D.; Marriot, P.R. *Int. J. Chem. Kinet.* **1979**, *11*, 1163.
168. Schlüter, K.; Berndt, A. *Tetrahedron Lett.* **1979**, *20*, 929.
169. Griller, D.; Içli, S.; Thankachan, C.; Tidwell, T. *J. Chem. Soc., Chem. Commun.* **1974**, 913.
170. Stefani, A.P. *J. Am. Chem. Soc.* **1968**, *90*, 1694.
171. Druliner, J.D.; Krusic, P.D.; Lehr, G.F.; Tolman, C.A. *J. Org. Chem.* **1985**, *50*, 5838.
172. Pritchard, G.O.; Johnson, K.A.; Nilsson, W.B. *Int. J. Chem. Kinet.* **1985**, *17*, 327.
173. Pritchard, G.O.; Nilsson, W.B.; Kirtman, B. *Int. J. Chem. Kinet.* **1984**, *16*, 1637.
174. Pritchard, G.O.; Kennedy, V.H.; Heldoorn, G.M.; Piasecki, M.L.; Johnson, K.A.; Golan, D.R. *Int. J. Chem. Kinet.* **1987**, *19*, 963.
175. Ingold, K.U. In *Free Radicals*; Kochi, J.K., Ed.; Wiley: New York, 1973; Vol. 1, p 37.
176. Manka, M.J.; Stein, S.E. *J. Phys. Chem.* **1984**, *88*, 5914.
177. Engel, P.S.; Wu, W.-X. *J. Org. Chem.* **1990**, *55*, 2720.
178. Neuman, R.C.; Alhadeff, E.S. *J. Org. Chem.* **1970**, *35*, 3401.
179. Kopecky, K.R.; Yeung, M.-Y. *Can. J. Chem.* **1988**, *66*, 374.
180. Schuh, H.; Fischer, H. *Helv. Chim. Acta* **1978**, *61*, 2463.
181. Trecker, D.J.; Foote, R.S. *J. Org. Chem.* **1968**, *33*, 3527.

3
Initiation

3.1 Introduction

Initiation is defined as the series of reactions that commences with generation of *primary* radicals[*] and culminates in addition to the carbon-carbon double bond of the monomer so as to form *initiating* radicals (Scheme 3.1).[1,2]

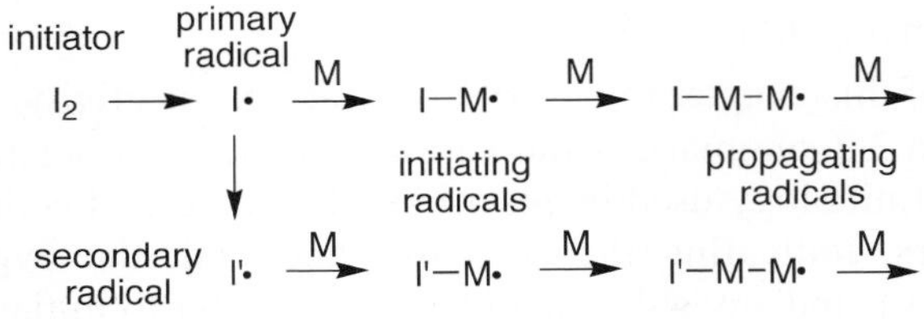

```
           primary
initiator  radical     M            M              M
   I₂  →    I•    →   I—M•    →   I—M-M•      →
                      initiating   propagating
             ↓        radicals     radicals
                       M            M              M
 secondary   I'•  →   I'—M•   →   I'—M-M•      →
  radical
```

Scheme 3.1

Classically, initiation was only considered as the first step in the chain reaction that constitutes radical polymerization. Although the rate and efficiency of initiation were known to be extremely important in determining the kinetics of polymerization, it was generally thought that the detailed mechanism of the process could be safely ignored when interpreting polymer properties. Furthermore, while it was recognized that initiation would lead to formation of structural units different from those which make up the bulk of the chain, the proportion of initiator-derived groups seemed insignificant when compared with total material.[†] This led to the belief that the physical properties and chemistry of polymers could be interpreted purely in terms of the generalized formula - *i.e.* $(CH_2\text{-}CXY)_n$ (see Chapter 1).

This view prevailed until the early 1970s and can still be found in some current-day texts. It is only in recent times that we have begun to understand the complexities of the initiation process and can appreciate the full role of initiation

---

[*] The term primary radical used in this context should be distinguished from that used when describing the substitution pattern of alkyl radicals.

[†] For example, in PS the initiator-derived end groups will account for *ca* 0.2% of units in a sample of molecular weight 100,000 (termination is mainly by combination).

in influencing polymer structure and properties. Four factors may be seen as instrumental in bringing about a revision of the traditional view:

(a) The realization that polymer properties (*e.g.* resistance to weathering, thermal or photochemical degradation) are often not predictable on the basis of the repeat unit structure but are in many cases determined by the presence of "defect groups".[3-6]

(b) The development of techniques whereby details of the initiation and other stages of polymerization can be studied in depth (Section 3.5).

(c) The finding that radical reactions are typically under kinetic rather than thermodynamic control (Section 2.1). Many instances can be cited where the less thermodynamically favored pathway is a significant, or even the major, pathway.

(d) The development of living or controlled radical polymerization (NMP, ATRP, RAFT, see Chapter 9). Lack of specificity in initiation can lead to dead chains and in turn to impure block copolymers or defects in complex architectures (stars, dendrimers, *etc.*).

It is the aim of this chapter to describe the nature, selectivity, and efficiency of initiation. Section 3.2 summarizes the various reactions associated with initiation and defines the terminology used in describing the process. Section 3.3 details the types of initiators, indicating the radicals generated, the byproducts formed (initiator efficiency), and any side reactions (*e.g.* transfer to initiator). Emphasis is placed on those initiators that see widespread usage. Section 3.4 examines the properties and reactions of the radicals generated, paying particular attention to the specificity of their interaction with monomers and other components of a polymerization system. Section 3.5 describes some of the techniques used in the study of initiation.

The intention is to create a greater awareness of the factors that must be borne in mind by the polymer scientist when selecting an initiator for a given polymerization.

3.2 The Initiation Process

The simple initiation process depicted in many standard texts is the exception rather than the rule. The yield of primary radicals produced on thermolysis or photolysis of the initiator is usually not 100%. The conversion of primary radicals to initiating radicals is dependent on many factors and typically is not quantitative. The primary radicals may undergo rearrangement or fragmentation to afford new radical species (secondary radicals) or they may interact with solvent or other species rather than monomer.

The reactions of the radicals (whether primary, secondary, solvent-derived, *etc.*) with monomer may not be entirely regio- or chemoselective. Reactions, such as head addition, abstraction or aromatic substitution, often compete with tail

Initiation

addition. In the sections that follow, the complexities of the initiation process will be illustrated by examining the initiation of polymerization of two commercially important monomers, styrene (S) and methyl methacrylate (MMA), with each of three commonly used initiators, azobisisobutyronitrile (AIBN), dibenzoyl peroxide (BPO), and di-*t*-butyl peroxyoxalate (DBPOX). The primary radicals formed from these three initiators are cyanoisopropyl, benzoyloxy, and *t*-butoxy radicals respectively (Scheme 3.2). BPO and DBPOX may also afford phenyl and methyl radicals respectively as secondary radicals (see 3.2.2).

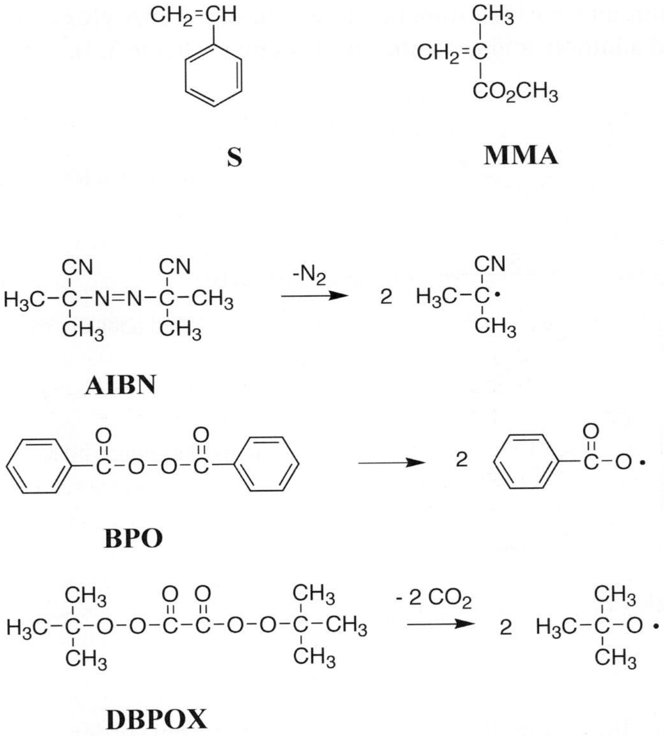

Scheme 3.2

3.2.1 Reaction with Monomer

First consider the interaction of radicals with monomers. Some behave as described in the classic texts and give tail addition as the only detectable pathway (Scheme 3.3). However, tail addition to the double bond is only one of the pathways whereby a radical may react with a monomer. The outcome of the reaction is critically dependent on the structure of both radical and monomer.

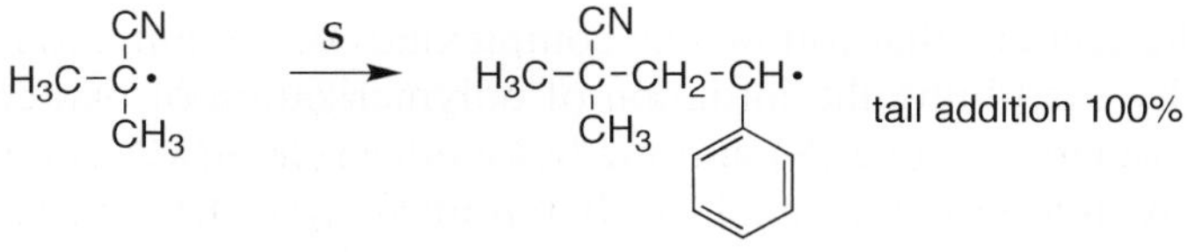

Scheme 3.3

For reactions with S, specificity is found to decrease in the series cyanoisopropyl~methyl~*t*-butoxy>phenyl>benzoyloxy. Cyanoisopropyl (Scheme 3.3),[7] *t*-butoxy and methyl radicals give exclusively tail addition.[8] Phenyl radicals afford tail addition and *ca* 1% aromatic substitution.[8] Benzoyloxy radicals give tail addition, head addition, and aromatic substitution (Scheme 3.4).[8,9]

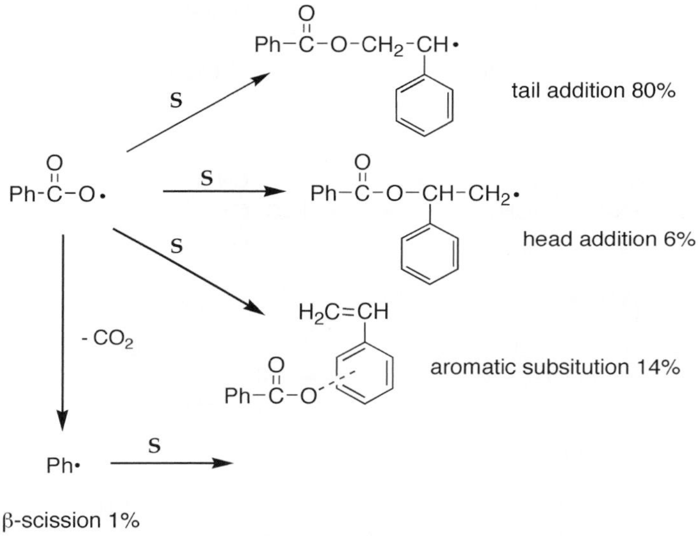

Scheme 3.4

With MMA, these radicals show a quite different order of specificity; regiospecificity decreases in the series cyanoisopropyl~methyl>phenyl >benzoyloxy>*t*-butoxy. Cyanoisopropyl and methyl radicals give exclusively tail addition. Benzoyloxy and phenyl radicals also react almost exclusively with the double bond (though benzoyloxy radicals give a mixture of head and tail addition[10]) and abstraction, while detectable, is a very minor (<1%) pathway.[10,11] On the other hand, only 63% of *t*-butoxy radicals react with MMA by tail addition to give **1** (Scheme 3.5).[12] The remainder abstract hydrogen, either from the α-methyl (predominantly) to give **2** or the ester methyl to give **3**.[12,13] The radicals **1-3** and methyl (formed by β-scission) may then initiate polymerization.

Initiation

These examples clearly show that the initiation pathways depend on the structures of the radical and the monomer. The high degree of specificity shown by a radical (e.g. t-butoxy) in its reactions with one monomer (e.g. S) must not be taken as a sign that a similarly high degree of specificity will be shown in reactions with all monomers (e.g. MMA).

Radicals can be classified according to their tendency to give aromatic substitution, abstraction, double bond addition, or β-scission and further classified in terms of the specificity of these reactions (see 3.4). With this knowledge, it should be possible to choose an initiator according to its suitability for use with a given monomer or monomer system so as to avoid the formation of undesirable end groups or, alternatively, to achieve a desired functionality.

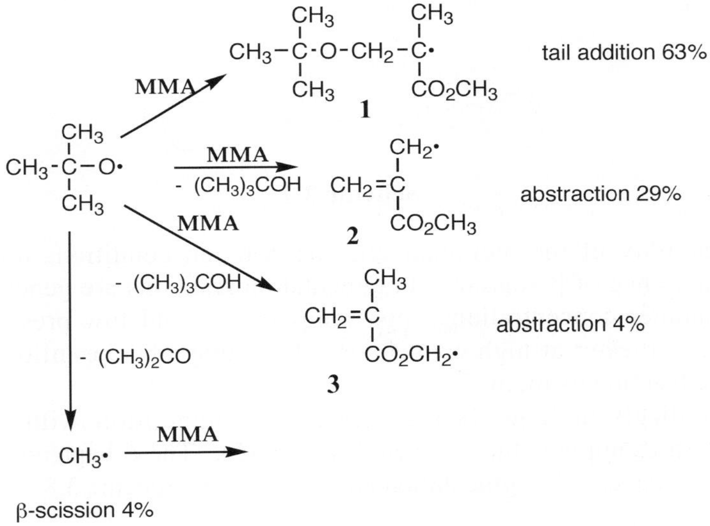

Scheme 3.5

The importance of these considerations can be demonstrated by examining some of the possible consequences for radical-monomer systems. For the case of MMA polymerization initiated by a t-butoxy radical source, chains may be initiated by the radicals **1**, **2** or **3** (Scheme 3.5). A significant proportion of chains will therefore have an olefinic end group rather than an initiator-derived end group. These chain ends may be reactive, either during polymerization, leading to chain branching, or afterwards, possibly leading to an impairment in polymer properties (Section 8.2.2). Polystyrene (PS) formed with BPO as initiator will have a proportion of relatively unstable benzoate end groups formed by benzoyloxy radical reacting by head addition and aromatic substitution (Scheme 3.4).[8,9] There is evidence that PS prepared with BPO as initiator is less thermally stable[14,15] and less resistant to weathering and yellowing[16,17] than that prepared using other initiators (Section 8.2.1).

3.2.2 Fragmentation

Many radicals undergo fragmentation or rearrangement in competition with reaction with monomer. For example, *t*-butoxy radicals undergo β-scission to form methyl radicals and acetone (Scheme 3.6).

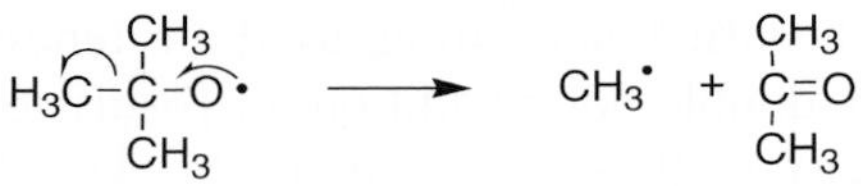

Scheme 3.6

Benzoyloxy radicals decompose to phenyl radicals and carbon dioxide (Scheme 3.7).

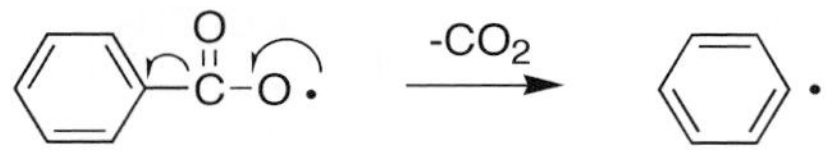

Scheme 3.7

The reactivity of the monomer and the reaction conditions determine the relative importance of β-scission. Fragmentation reactions are generally favored by low monomer concentrations, high temperatures and low pressures. Their significance is greater at high conversion. They may also be influenced by the nature of the reaction medium.

Other radicals undergo rearrangement in competition with bimolecular processes. An example is the 5-hexenyl radical (**5**). The 6-heptenoyloxy radical (**4**) undergoes sequential fragmentation and cyclization (Scheme 3.8).[18]

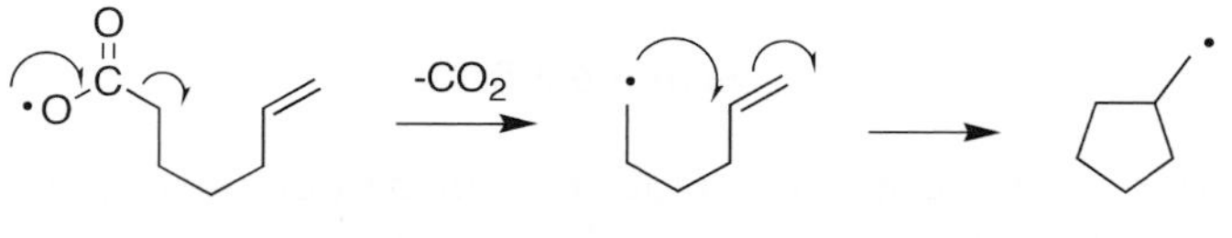

Scheme 3.8

The radicals formed by unimolecular rearrangement or fragmentation of the primary radicals are often termed secondary radicals. Often the absolute rate constants for secondary radical formation are known or can be accurately determined. These reactions may then be used as "radical clocks",[19,20] to calibrate the absolute rate constants for the bimolecular reactions of the primary radicals (*e.g.* addition to monomers - see 3.4). However, care must be taken since the rate constants of some clock reactions (*e.g. t*-butoxy β-scission[21]) are medium dependent (see 3.4.2.1.1).

Initiation

3.2.3 Reaction with Solvents, Additives, or Impurities

A typical polymerization system comprises many components besides the initiators and the monomers. There will be solvents, additives (*e.g.* transfer agents, inhibitors) as well as a variety of adventitious impurities that may also be reactive towards the initiator-derived radicals.

For the case of MMA polymerization with a source of *t*-butoxy radicals (DBPOX) as initiator and toluene as solvent, most initiation may be by way of solvent-derived radicals[21,22] (Scheme 3.9). Thus, a high proportion of chains (>70% for 10% w/v monomers at 60 °C[22]) will be initiated by benzyl rather than *t*-butoxy radicals. Other entities with abstractable hydrogens may also be incorporated as polymer end groups. The significance of these processes increases with the degree of conversion and with the (solvent or impurity):monomer ratio.

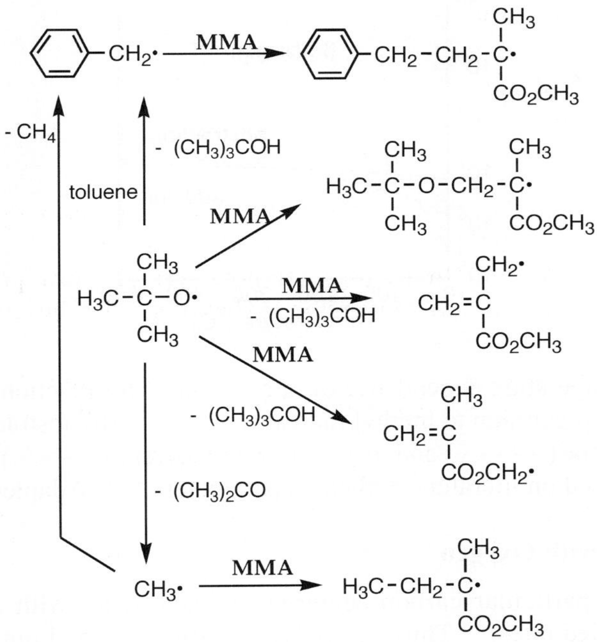

Scheme 3.9

There is potential for this behavior to be utilized in devising methods for the control of the types of initiating radicals formed and hence the polymer end groups.

3.2.4 Effects of Temperature and Reaction Medium on Radical Reactivity

The reaction medium may also modify the reactivity of the primary, or other radicals without directly reacting with them. For example, when *t*-butoxy reacts with MMA (Scheme 3.5), the ratio of addition:abstraction:β-scission varies according to the nature of the solvent[21] and the reaction temperature[23,24] (see 2.3.6 and 3.4.2.1.1).

For *t*-alkoxy radicals, polar and aromatic solvents favor abstraction over addition, and β-scission over either addition or abstraction (3.4.2.1.1). Addition, abstraction and β-scission have quite different Arrhenius parameters. As a further example the temperature dependence of the rate constants for addition of cumyloxy radicals to styrene, abstraction from isopropylbenzene, and β-scission to give methyl radicals is shown in Figure 3.1. Low temperatures favor abstraction over addition and both of these reactions over β-scission.

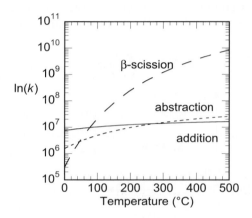

Figure 3.1 Temperature dependence of rate constants for reactions of cumyloxy radicals (a) β-scission to methyl radicals (— — —) (b) abstraction from isopropylbenzene (- - - - -) and (c) addition to styrene (———). Data are an extrapolation based on literature Arrhenius parameters.[25,26] Adapted from Moad.[27]

3.2.5 Reaction with Oxygen

Radicals, in particular carbon-centered radicals, react with oxygen at near diffusion-controlled rates.[28] Thus, for polymerizations carried out either in air or in incompletely degassed media, oxygen is likely to become involved in, and further complicate, the initiation process.

The reaction of oxygen with carbon-centered radicals (*e.g.* cyanoisopropyl, Scheme 3.10) affords an alkylperoxy radical **6**.[29,30] This species may initiate polymerization so forming a relatively unstable peroxidic end group **7**. With respect to most carbon-centered radicals, the alkylperoxy radicals **6** show an enhanced tendency to abstract hydrogen. The alkylperoxy radicals may abstract hydrogen from polymer, monomer, or other components in the system[31] forming a potentially reactive hydroperoxide **8** and a new radical species (R•) which may initiate polymerization. The process is further complicated if **7** or **8** undergo homolysis under the polymerization conditions. The peroxides derived from **7** and **8** may also be active as chain transfer agents.

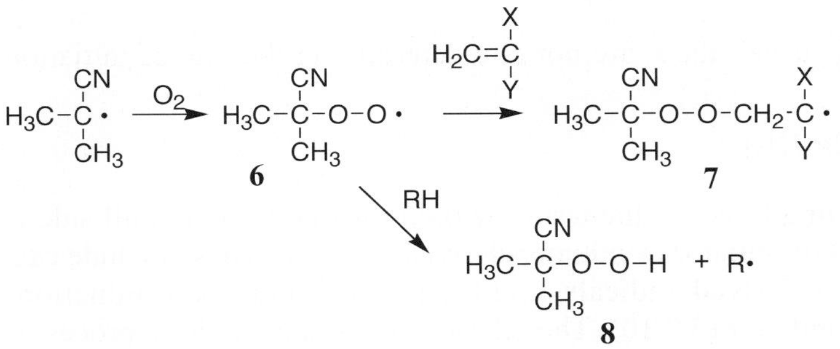

Scheme 3.10

3.2.6 Initiator Efficiency in Thermal Initiation

The proportion of radicals which escape the solvent cage to form initiating radicals is termed the initiator efficiency (f) which is formally defined as follows (eq. 1).*

$$f = \frac{\text{(Rate of initiation of propagating chains)}}{n \text{ (Rate of initiator disappearance)}} \qquad (1)$$

where n is the number of moles of radicals generated per mole of initiator.

The effective rate of initiation (R_i) in the case of thermal decomposition of an initiator (I_2) decomposing by Scheme 3.11 is given by eq. 2

$$I_2 \xrightarrow{k_d} [2\,I\bullet] \longrightarrow I\bullet \qquad \text{primary radical formation}$$

$$I\bullet + M \xrightarrow{k_i} P_1\bullet \qquad \text{initiation}$$

$$I\bullet \longrightarrow I'\bullet \qquad \text{secondary radical formation}$$

$$I'\bullet + M \xrightarrow{k_i'} P_1\bullet \qquad \text{initiation}$$

Scheme 3.11

$$R_i = k_i[I\bullet][M] + k_i'[I'\bullet][M] \qquad (2)$$

---

* In some texts the initiator efficiency (f) is defined simply in terms of the yield of initiator-derived radicals (the fraction of radicals I• that undergo cage escape - Section 3.2.8). This number will always be larger than that obtained by application of eq. 1.

eq. 1 can then be written as follows (eq. 3)

$$f = \frac{(k_i[\text{I}\bullet][\text{M}] + k_i' [\text{I}'\bullet][\text{M}])}{2k_d[\text{I}_2]} \tag{3}$$

If, as is usual, the k_i are not rate determining the rate of initiation is given by eq. 4.

$$R_i = 2k_d f[\text{I}_2] \tag{4}$$

According to eq. 1, the term f should take into account all side reactions that lead to loss of initiator or initiator-derived radicals. These include cage reaction of the initiator-derived radicals (3.2.8), primary radical termination (3.2.9) and transfer to initiator (3.2.10). The relative importance of these processes depends on monomer concentration, medium viscosity and many other factors. Thus f is not a constant and typically decreases with conversion (see 3.3.1.1.3 and 3.3.2.1.3).

3.2.7 Photoinitiation

It is worthwhile to consider some of the special features of photoinitiation. The Jablonski diagram provides a convenient description of the events that follow absorption of light (Figure 3.2). A molecule in its ground state (S_0) absorbs a photon of light to be excited to the singlet state (S_1). As well as being electronically excited, the molecule will be vibrationally and rotationally excited. Certain reactions may take place from the excited singlet state. These will compete with fluorescence, and other deactivation processes that return the molecule to the ground state, and intersystem crossing to the triplet state (T_1). The triplet state is typically of lower energy than the excited singlet state. Chemical reaction then competes with phosphorescence and other deactivation processes.

Azo-compounds and peroxides undergo photodecomposition to radicals when irradiated with light of suitable wavelength. The mechanism appears similar to that of thermal decomposition to the extent that it involves cleavage of the same bonds. The photodecomposition of azo-compounds is discussed in Section 3.3.1.1.2 and peroxides in Sections 3.3.2.1.2 (diacyl peroxides) and 3.3.2.3.2 (peroxyesters). Specific photoinitiators are discussed in Section 3.3.4. It is also worth noting that certain monomers may undergo photochemistry and direct photoinitiation on irradiation of monomer is possible.

Clearly, unless monomer is the intended photoinitiator, it is important to choose an initiator that absorbs in a region of the UV-visible spectrum clear from the absorptions of monomer and other components of the polymerization medium. Ideally, one should choose a monochromatic light source that is specific for the chromophore of the photoinitiator or photosensitizer. It is also important in many experiments that the total amount of light absorbed by the sample is small. Otherwise the rate of initiation will vary with the depth of light penetration into the sample.

Initiation

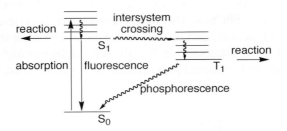

Figure 3.2 Jablonski diagram describing photoexcitation process.

In order to define the rate and efficiency of photoinitiation, consider the simplified reaction Scheme 3.12.

$$I_2 \xrightarrow{h\nu} 2\, I\bullet$$

$$I\bullet + M \xrightarrow{k_i} P_1\bullet$$

Scheme 3.12

The quantum yield (Φ) is the yield of initiating radicals produced per photon of light absorbed (eq. 5)

$$\Phi = \frac{\text{(yield of initiating radicals)}}{n\,\text{(photons absorbed)}} \tag{5}$$

which can also be expressed in terms of the rate of initiation (eq. 6).

$$\Phi = \frac{\text{(rate of initiation of propagating chains)}}{n\,\text{(intensity of incident irradiation absorbed)}} = \frac{R_i}{nI_{abs}} \tag{6}$$

where n is the number of moles of radicals generated per mole of initiator and I_{abs} is the intensity of incident light absorbed.

The Beer-Lambert law (also often called Beer's law) relates I_{abs} to the total incident light intensity (I_0) (eq. 7).

$$\frac{I_{abs}}{I_0} = 1 - 10^{\varepsilon cd} = 1 - e^{\alpha cd} \tag{7}$$

and if αcd is small (<0.1 for <5% error) then this simplifies to eq. 8.

$$\frac{I_{abs}}{I_0} \approx \alpha cd \tag{8}$$

where ε ($=\alpha/2.303$) is the molar extinction coefficient at the given wavelength, c is the concentration of the absorbing substance, and d is the pathlength. It can be seen that the term Φ embraces the same factors as $k_d f$ in thermal initiation. Care must be taken to establish how the molar extinction coefficient (ε or α) was determined since both decadic and natural forms are in common usage.

If the reaction with monomer is not the rate determining step, the rate of radical generation in photoinitiated polymerization is given by eq. 9

$$R_i = 2\Phi\, I_{abs}$$

$$= 2\Phi I_0 (1 - e^{\alpha d [I_2]}) = 2\Phi I_0 (1 - 10^{\varepsilon d [I_2]}) \tag{9}$$

which for small $\alpha d[I_2]$ simplifies to eq. 10.

$$R_i \approx 2\Phi I_0 \alpha d [I_2] \tag{10}$$

3.2.8 Cage Reaction and Initiator-Derived Byproducts

The decomposition of an initiator seldom produces a quantitative yield of initiating radicals. Most thermal and photochemical initiators generate radicals in pairs. The self-reaction of these radicals is often the major pathway for the direct conversion of primary radicals to non-radical products in solution, bulk or suspension polymerization. This cage reaction is substantial even in bulk polymerization at low conversion when the medium is essentially monomer. The importance of the process depends on the rate of diffusion of these species away from one another.

Thus, the size and the reactivity of the initiator-derived radicals and the medium viscosity (or microviscosity) are important factors in determining the initiator efficiency. Thus, the extent of the cage reaction is likely to increase with decreasing reaction temperature and with increasing conversion.[32,33] The cage reaction, as well as lowering the initiation efficiency, can produce a range of byproducts. These materials may be reactive under the polymerization conditions or they may themselves have a deleterious influence on polymer properties. For example, the cage reaction of cyanoisopropyl radicals formed from the decomposition of AIBN produces, amongst other products (Scheme 3.13), MAN, which readily undergoes copolymerization to be incorporated into the final polymer,[7,34] and tetramethylsuccinonitrile (**9**), which is claimed to be toxic and should not be present in polymers used for food contact applications.[35,36]

In other cases, the cage reaction may simply lead to reformation of the initiator. This process is known as cage return and is important during the decomposition of BPO (Section 3.3.2.1.1) and DTBP (Section 3.3.2.4). Cage return lowers the rate of radical generation but does not directly yield byproducts. It is one factor contributing to the solvent and viscosity dependence of k_d and can lead to a reduced k_d at high conversion.

Initiation

A variety of methods may be envisioned to decrease the importance of the cage reaction. One method, given the viscosity dependence of the cage reaction, is to conduct polymerizations in solution rather than in bulk. Another involves carrying out the polymerization in a magnetic field.[37] This is thought to reduce the rate of triplet-singlet intersystem crossing for the geminate pair.[38]

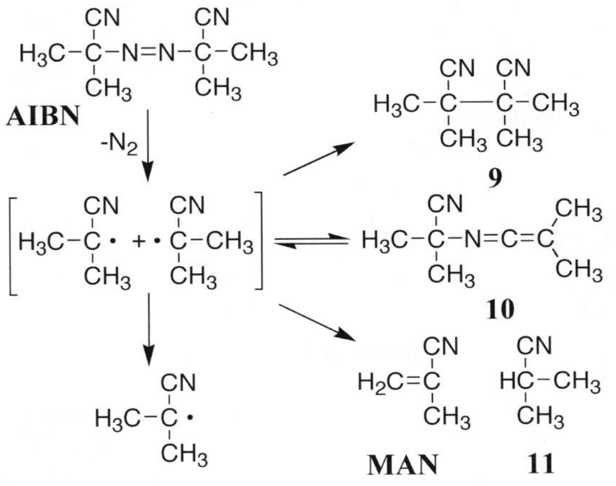

Scheme 3.13

3.2.9 Primary Radical Termination

The primary radicals may also interact with other radicals present in the system after they escape the solvent cage. When this involves a propagating radical, the process is known as primary radical termination. The term also embraces the reactions of other initiator or transfer agent-derived radicals with propagating radicals. Most monomers are efficient scavengers of the initiator-derived radicals and the steady state concentration of propagating radicals is very low (typically $\leq 10^{-7}$M). The concentrations of the primary and other initiator-derived radicals are very much lower (typically $\leq 10^{-9}$M). Thus, with most initiators, primary radical termination has a very low likelihood during the early stages of polymerization.

Primary radical termination may involve combination or disproportionation with the propagating radical. It is often assumed that small radicals give mainly combination even though direct evidence for this is lacking. Both pathways are observed for reaction of cyanoisopropyl radicals with PS• (Scheme 3.14) (Section 7.4.3.2). The end group formed by combination is similar to that formed by head addition to monomer differing only in the orientation of the penultimate monomer unit.

If the rate of addition to monomer is low, primary radical termination may achieve greater importance. For example, in photoinitiation by the benzoin ether **12** both a fast initiating species (**13**, high k_i) and a slow initiating species (**14**, low k'$_i$) are generated (Scheme 3.15). The polymerization kinetics are complicated and the initiator efficiency is lowered by primary radical termination involving the dimethoxybenzyl radical (**14**, see 3.3.4.1.1).[39,40]

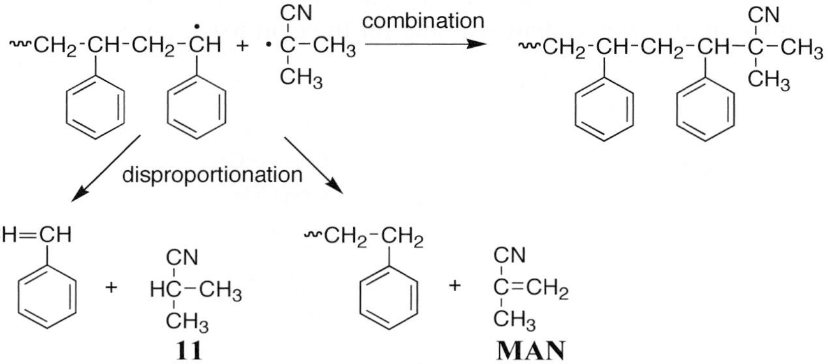

Scheme 3.14

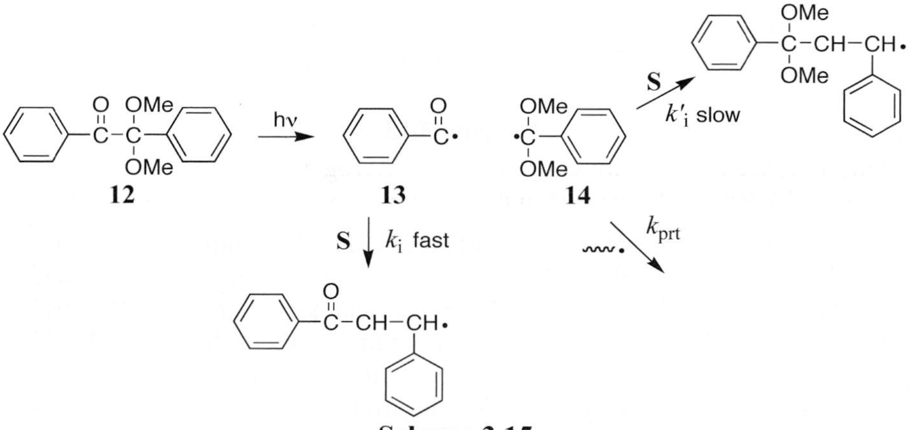

Scheme 3.15

Primary radical termination is also of demonstrable significance when very high rates of initiation or very low monomer concentrations are employed. It should be noted that these conditions pertain in all polymerizations at high conversion and in starved feed processes. Some syntheses of telechelics are based on this process (Section 7.5.1). Reversible primary radical termination by combination with a persistent radical is the desired pathway in many forms of living radical polymerization (Section 9.3).

3.2.10 Transfer to Initiator

Many of the initiators used in radical polymerization are susceptible to induced decomposition by various radical species. When the reaction involves the

Initiation

propagating species, the process is termed transfer to initiator. The importance of this reaction depends on both the initiator and the propagating radical.

Diacyl peroxides are particularly prone to induced decomposition (Scheme 3.16). Transfer to initiator is of greatest importance for polymerizations taken to high conversion or when the ratio of initiator to monomer is high. It has been shown that, during the polymerization of S initiated by BPO, transfer to initiator can be the major pathway for the termination of chains.[7,41]

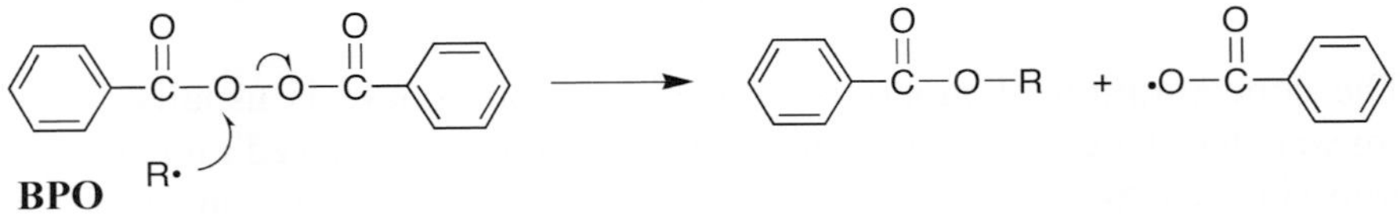

Scheme 3.16

Transfer to initiator introduces a new end group into the polymer, lowers the molecular weight of the polymer, reduces the initiator efficiency, and increases the rate of initiator disappearance. Methods of evaluating transfer constants are discussed in Section 6.2.1.

3.2.11 Initiation in Heterogeneous Polymerization

Many polymerizations are carried out in heterogeneous media, usually water-monomer mixtures, where suspending agents or surfactants ensure proper dispersion of the monomer and control the particle size of the product.

Suspension polymerizations are often regarded as "mini-bulk" polymerizations since ideally all reaction occurs within individual monomer droplets. Initiators with high monomer and low water solubility are generally used in this application. The general chemistry, initiator efficiencies, and importance of side reactions are similar to that seen in homogeneous media.

Emulsion polymerizations most often involve the use of water-soluble initiators (*e.g.* persulfate see 3.3.2.6.1) and polymer chains are initiated in the aqueous phase. A number of mechanisms for particle formation and entry have been described, however, a full discussion of these is beyond the scope of this book. Readers are referred to recent texts on emulsion polymerization by Gilbert[42] and Lovell and El-Aasser[43] for a more comprehensive treatment.

Radicals typically are generated in the aqueous phase and it is now generally believed that formation of an oligomer of average chain length z (z-mer, $P_z\bullet$) occurs in the aqueous phase prior to particle entry.[44] The steps involved in forming a radical in the particle phase from an aqueous phase initiator are summarized in Scheme 3.17. The length of the z-mer depends on the particular monomer and is shorter for more hydrophobic monomers.

64 The Chemistry of Radical Polymerization

$$\text{initiator(aq)} \longrightarrow \text{I}\bullet\text{(aq)} \quad \text{(initiator-derived radical in aqueous phase)}$$

$$\xrightarrow{M} P_1\bullet\text{(aq)} \quad \text{(initiating radical in aqueous phase)}$$

$$\xrightarrow{M} P_z\bullet\text{(aq)} \quad \text{(z-1 monomer additions to give z-mer)}$$

$$\xrightarrow{\rho} P_z\bullet\text{(p)} \quad \text{(transfer to particle phase, entry)}$$

Scheme 3.17

The concentration of monomers in the aqueous phase is usually very low. This means that there is a greater chance that the initiator-derived radicals (I•) will undergo side reactions. Processes such as radical-radical reaction involving the initiator-derived and oligomeric species, primary radical termination, and transfer to initiator can be much more significant than in bulk, solution, or suspension polymerization and initiator efficiencies in emulsion polymerization are often very low. Initiation kinetics in emulsion polymerization are defined in terms of the entry coefficient (ρ) - a pseudo-first order rate coefficient for particle entry.

Microemulsion and miniemulsion polymerization differ from emulsion polymerization in that the particle sizes are smaller (10-30 and 30-100 nm respectively vs 50-300 nm)[42] and there is no monomer droplet phase. All monomer is in solution or in the particle phase. Initiation takes place by the same process as conventional emulsion polymerization.

3.3 The Initiators

Certain polymerizations (*e.g.* S, see 3.3.6.1) can be initiated simply by applying heat; the initiating radicals are derived from reactions involving only the monomer. More commonly, the initiators are azo-compounds or peroxides that are decomposed to radicals through the application of heat, light, or a redox process.

When initiators are decomposed thermally, the rates of initiator disappearance (k_d) show marked temperature dependence. Since most conventional polymerization processes require that k_d should lie in the range 10^{-6}-10^{-5} s^{-1} (half-life *ca* 10 h), individual initiators typically have acceptable k_d only within a relatively narrow temperature range (*ca* 20-30 °C). For this reason initiators are often categorized purely according to their half-life at a given temperature or *vice versa*.[45] For initiators which undergo unimolecular decomposition, the half-life is related to the decomposition rate constant by eq. 11.

$$t_{\frac{1}{2}} = \frac{\ln 2}{k_d} \tag{11}$$

The Arrhenius relationship can be rearranged as follows (eq. 12) to enable calculation of the temperature required to give a desired decomposition rate or half-life.

$$T(°C) = -273.15 - \frac{E_a}{R \ln\left(\frac{k_d}{A}\right)} = -273.15 - \frac{E_a}{R \ln\left(\frac{\ln 2}{A t_{\frac{1}{2}}}\right)} \qquad (12)$$

The temperature at which the half-life is 10h is then given by the following expression (eq. 13).

$$T(°C) = -273.15 - \frac{0.120277 E_a}{-10.8578 + \ln\left(\frac{1}{A}\right)} \qquad (13)$$

The initiator in radical polymerization is often regarded simply as a source of radicals. Little attention is paid to the various pathways available for radical generation or to the side reactions that may accompany initiation. The preceding discussion (see 3.2) demonstrated that in selecting initiators (whether thermal, photochemical, redox, *etc.*) for polymerization, they must be considered in terms of the types of radicals formed, their suitability for use with the particular monomers, solvent, and the other agents present in the polymerization medium, and for the properties they convey to the polymer produced.

Many reviews detailing aspects of the chemistry of initiators and initiation have appeared.[2,45,46] A non-critical summary of thermal decomposition rates is provided in the *Polymer Handbook*.[47,48] The subject also receives coverage in most general texts and reviews dealing with radical polymerization. References to reviews that detail the reactions of specific classes of initiator are given under the appropriate sub-heading below.

Some characteristics of initiators used for thermal initiation are summarized in Table 3.1. These provide some general guidelines for initiator selection. In general, initiators which afford carbon-centered radicals (*e.g.* dialkyldiazenes, aliphatic diacyl peroxides) have lower efficiencies for initiation of polymerization than those that produce oxygen-centered radicals. Exact values of efficiency depend on the particular initiators, monomers, and reaction conditions. Further details of initiator chemistry are summarized in Sections 3.3.1 (azo-compounds) and 3.3.2 (peroxides) as indicated in Table 3.1. In these sections, we detail the factors which influence the rate of decomposition (*i.e.* initiator structure, solvent, complexing agents), the nature of the radicals formed, the susceptibility of the initiator to induced decomposition, and the importance of transfer to initiator and other side reactions of the initiator or initiation system. The reactions of radicals produced from the initiator are given detailed treatment in Section 3.4.

The Chemistry of Radical Polymerization

Table 3.1 Guide to Properties of Polymerization Initiators

Initiator Class	Example	Section
dialkyldiazenes	$H_3C-\underset{CH_3}{\underset{|}{\overset{CN}{\overset{|}{C}}}}-N=N-\underset{CH_3}{\underset{|}{\overset{CN}{\overset{|}{C}}}}-CH_3$ **AIBN**	3.3.1.1
hyponitrites	$H_3C-\underset{CH_3}{\underset{|}{\overset{CH_3}{\overset{|}{C}}}}-O-N=N-O-\underset{CH_3}{\underset{|}{\overset{CH_3}{\overset{|}{C}}}}-CH_3$	3.3.1.2
diacyl peroxides	$CH_3(CH_2)_9CH_2-\overset{O}{\overset{\|}{C}}-O-O-\overset{O}{\overset{\|}{C}}-CH_2(CH_2)_9CH_3$ **LPO**	3.3.2.1
diaroyl peroxides	$Ph\cdot\overset{O}{\overset{\|}{C}}-O-O-\overset{O}{\overset{\|}{C}}\cdot Ph$ **BPO**	3.3.2.1
peroxydicarbonates	$\underset{H_3C}{\overset{H_3C}{\diagdown}}CH-O-\overset{O}{\overset{\|}{C}}-O-O-\overset{O}{\overset{\|}{C}}-O-CH\underset{\diagdown CH_3}{\overset{\diagup CH_3}{}}$	3.3.2.2
peroxyesters	$H_3C-\underset{CH_3}{\underset{|}{\overset{CH_3}{\overset{|}{C}}}}-O-O-\overset{O}{\overset{\|}{C}}-\underset{CH_3}{\underset{|}{\overset{CH_3}{\overset{|}{C}}}}-CH_3$	3.3.2.3
peroxyoxalates	$H_3C-\underset{CH_3}{\underset{|}{\overset{CH_3}{\overset{|}{C}}}}-O-O\overset{O}{\overset{\|}{C}}-\overset{O}{\overset{\|}{C}}O-O-\underset{CH_3}{\underset{|}{\overset{CH_3}{\overset{|}{C}}}}-CH_3$ **DBPOX**	3.3.2.3
dialkyl peroxides	$H_3C-\underset{CH_3}{\underset{|}{\overset{CH_3}{\overset{|}{C}}}}-O-O-\underset{CH_3}{\underset{|}{\overset{CH_3}{\overset{|}{C}}}}-CH_3$ **DTBP**	3.3.2.4
dialkyl ketone peroxides	cyclohexane with $O-O-C(CH_3)_3$ and $O-O-C(CH_3)_3$	3.3.2.5
hydroperoxides	$H_3C-\underset{CH_3}{\underset{|}{\overset{CH_3}{\overset{|}{C}}}}-O-O-H$	3.3.2.5
persulfate	$^-O-\underset{O}{\underset{\|}{\overset{O}{\overset{\|}{S}}}}-O-O-\underset{O}{\underset{\|}{\overset{O}{\overset{\|}{S}}}}-O^-$	3.3.2.6
disulfides	$\underset{C_2H_5}{\overset{C_2H_5}{\diagdown}}N-\overset{S}{\overset{\|}{C}}-S-S-\overset{S}{\overset{\|}{C}}-N\underset{\diagdown C_2H_5}{\overset{\diagup C_2H_5}{}}$	3.3.5

a 1° = primary radical from initiator decomposition, 2° = secondary radical-derived by fragmentation of 1° radical. Species shown in parentheses may be formed under some conditions but are seldom observed in polymerizations of common monomers.

Table 3.1 (continued)

Radicals generated[a]	Efficiency[b]	Transfer[c]
1° alkyl	low	low
1° alkoxy 2° alkyl	high	low
(1° acyloxy) 2° alkyl	low	high
1° aroyloxy 2° aryl	high	high
1° alkoxycarbonyloxy (2° alkoxy)	high	high
1° alkoxy, acyloxy 2° alkyl	med.	med.
1° alkoxy 2° alkyl	high	med.
1° alkoxy 2° alkyl	high	low
1° alkoxy 2° alkyl	med.	low
1° hydroxy, alkoxy 2° alkyl	high	high
1° sulfate radical anion	low	low
1° thiyl	high	high

b Efficiency decreases as the importance of cage reactions increases.   c Susceptibility to radical-induced decomposition.

3.3.1 Azo-Compounds

Two general classes of azo-compound will be considered in this section, the dialkyldiazenes (**15**) (3.3.1.1) and the dialkyl hyponitrites (**16**) (3.3.1.2).

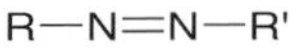

R—N=N—R'   R—O—N=N—O—R'
    15 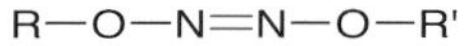   **16**

Polymeric azo-compounds and multifunctional initiators with azo-linkages are discussed elsewhere (see 3.3.3 and 7.6.1) as are azo compounds, which find use as iniferters (see 9.3.4).

3.3.1.1 Dialkyldiazenes

The kinetics and mechanism of the thermal and photochemical decomposition of dialkyldiazenes (**15**) have been comprehensively reviewed by Engel.[49] The use of these compounds as initiators of radical polymerization has been covered by Moad and Solomon[2] and Sheppard.[50] The general chemistry of azo-compounds has also been reviewed by Koga et al.,[51] Koenig,[52] and Smith.[53]

Dialkyldiazenes (**15**, R=alkyl) are sources of alkyl radicals. While there is clear evidence for the transient existence of diazenyl radicals (**17**; Scheme 3.18) during the decomposition of certain unsymmetrical diazenes[49,51] and of cis-diazenes,[54] all isolable products formed in thermolysis or photolysis of dialkyldiazenes (**15**) are attributable to the reactions of alkyl radicals.

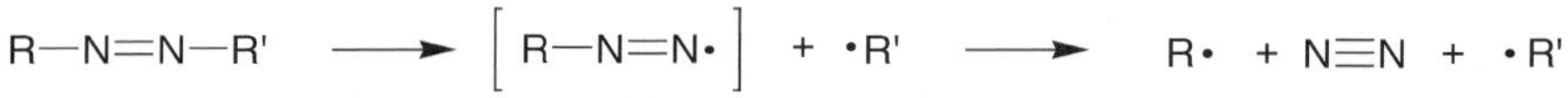

R—N=N—R' ⟶ [R—N=N•] + •R' ⟶ R• + N≡N + •R'
    15   **17**

Scheme 3.18

In the decomposition of symmetrical azo compounds the intermediacy of diazenyl radicals remains a subject of controversy. However, it is clear that diazenyl radicals, if they are intermediates, do not have sufficient lifetime to be trapped or to initiate polymerization. Ayscough et al.[55] photolyzed AIBN in a matrix at -196 °C and observed EPR signals which were attributed to the diazenyl radical, $(CH_3)_2(CN)C–N=N•$ [this assignment has been questioned[51]]. However for AIBN decomposition in solution, at temperatures normally encountered in polymerizations, the finding, that the rate of decomposition is independent of solvent viscosity (i.e. no cage return) is evidence for concerted 2-bond cleavage.[31] Commercially available dialkyldiazenes initiators (**15**) tend to be symmetrical and the R groups are generally tertiary with functionality to stabilize the incipient radical [e.g. cyano AIBN, (**18-29**), ester (AIBMe), amidinium salt (**22, 23**), amide (**24, 25**) or phenyl (**21**)]. Those most commonly encountered are the azonitriles, these include 2,2'-azobis(2-methylpropanenitrile) [better known as azobis-

Initiation

(isobutyronitrile) or AIBN], 2,2'-azobis(2-methylbutanenitrile) (**19**), 1,1'-azobis(1-cyclohexanenitrile) (**20**)). The initiator **18** exists as a mixture of diastereoisomers that have differing k_d (Table 3.2). Azoisooctane **26** and azo-*t*-butane **27** are high temperature initiators.

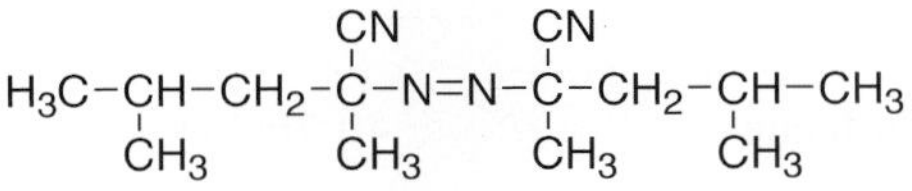

18

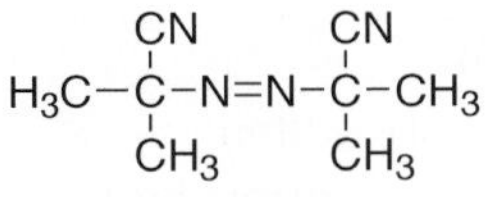

AIBN

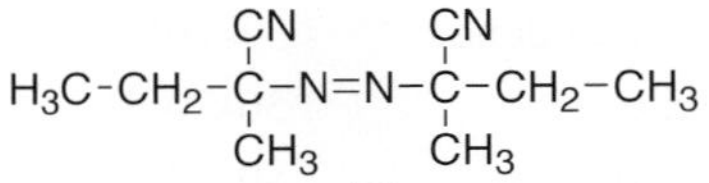

19

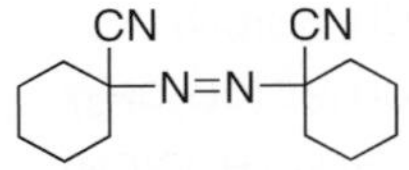

20

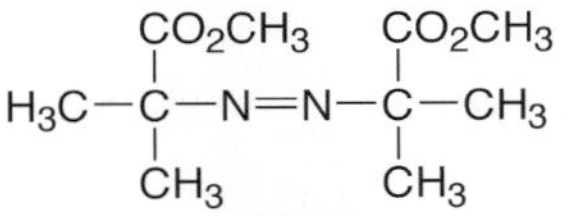

AIBMe

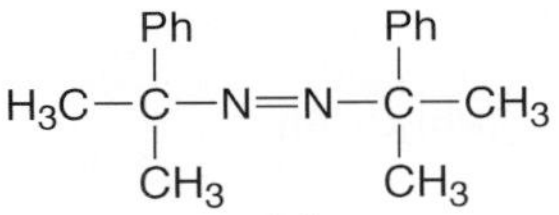

21

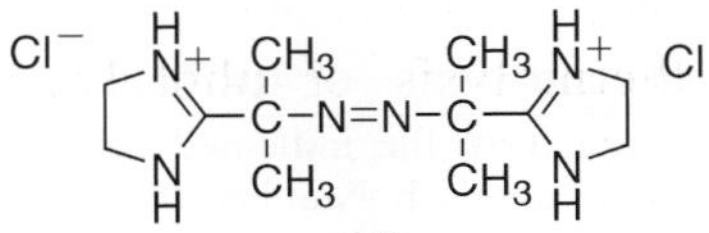

22

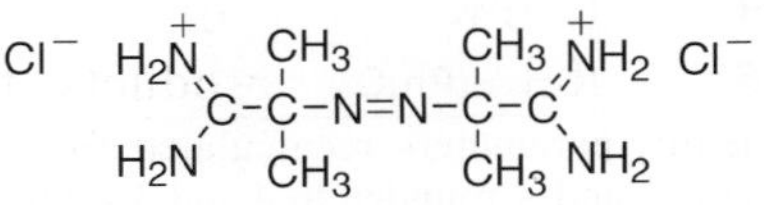

23

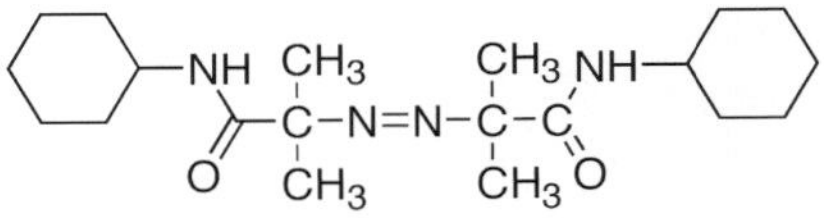

24

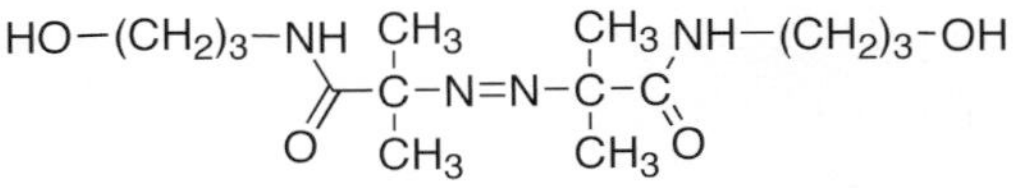

25

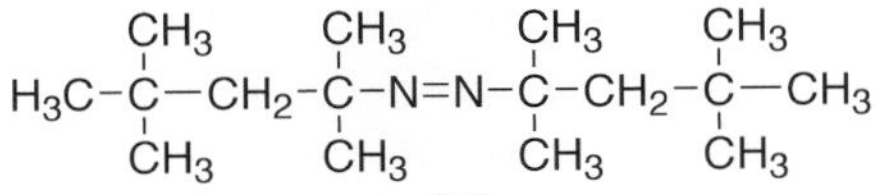

26

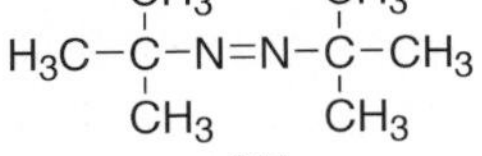

27

Table 3.2 Selected Kinetic Data for Decomposition of Azo-Compounds[a]

Initiator	R (, R')	Solvent	Temp. range[b] °C
diazenes (**15**)			
21	$(CH_3)_2C(Ph)$	toluene	40-70(17)
18	$(CH_3)_2CHCH_2C(CH_3)(CN)$[e]	toluene	60-70(2)
18	$(CH_3)_2CHCH_2C(CH_3)(CN)$[e]	toluene	70-80(2)
AIBN	$(CH_3)_2C(CN)$	benzene/toluene	37-105(13)
AIBMe	$(CH_3)_2C(CO_2CH_3)$	benzene	50-70(4)
19	$(CH_3)(C_2H_5)C(CN)$	ethylbenzene	80-100(3)
20	$(c\text{-}C_6H_{10})C(CN)$	toluene	80-100(3)
26	$(CH_3)_3CHCH_2C(CH_3)_2$	diphenyl ether	130-160(7)
27	$(CH_3)_3C$	diphenyl ether	165-200(6)
30	$(Ph)_3C$, Ph	benzene/toluene	25-75(9)
hyponitrites (**16**)			
34	$(CH_3)_3C$	isooctane	45-75(4)
35	$(CH_3)_2(Ph)C$	cyclohexane	40-70(12)

a Arrhenius parameters recalculated from original data taken from the indicated references. Values of E_a and A rounded to 4 and 3 significant figures respectively.   b Number of data points given in parentheses.   c Calculated from the Arrhenius parameters shown and rounded to 2 significant figures.   d Temperature for ten hour half-life calculated with eq. 13.   e Diastereoisomers.

Table 3.2 (continued)

E_a kJ mol^{-1}	A × 10^{-15} s^{-1}	k_d × 10^6 (60 °C)^c s^{-1}	10 h t$_{1/2}$^{d} °C	Ref.	Initiator
126.7	12.2	170	45	56	**21**
118.9	0.376	86	49	57	**18**^{e}
123.7	1.39	56	52	57	**18**^{e}
131.7	4.31	9.6	65	32,58-61	**AIBN**
124.0	0.248	8.9	66	62,63	**AIBMe**
137.8	20.3	5.0	69	64	**19**
149.1	71.0	0.30	88	65	**20**
137.0	0.10	0.033	109	66	**26**
180.4	91.7	~5×10^{-6}	161	67	**27**
114.6	0.486	522	35	68-70	**30**
119.5	1.17	214	42	71	**34**
113.9	0.99	1370	29	72	**35**

Water-soluble azo compounds include 4,4'-azobis(4-cyanovaleric acid) (**29**) and the amidinium hydrochlorides (**22** and **23**).

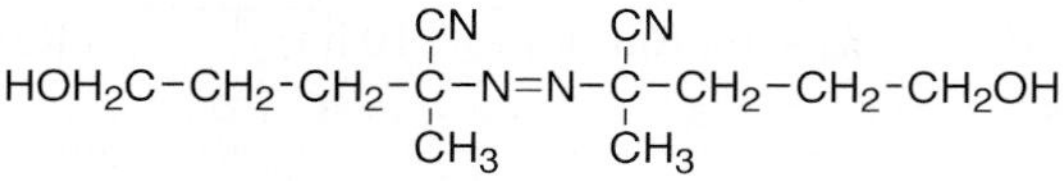
CN
CN
HOH2C-CH2-CH2-C-N=N-C-CH2-CH2-CH2OH
CH3
CH3

28

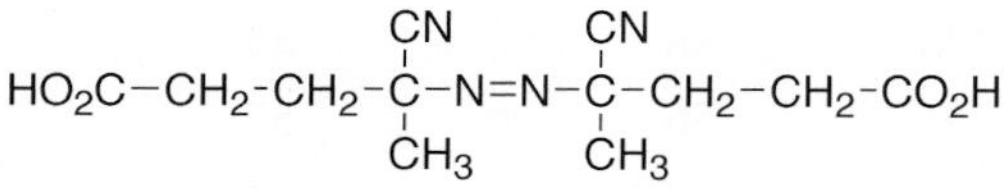
CN
CN
HO2C-CH2-CH2-C-N=N-C-CH2-CH2-CO2H
CH3
CH3

29

Unsymmetrical azo-compounds find application as initiators of polymerization in special circumstances, for example, as initiators of living radical polymerization [*e.g.* triphenylmethylazobenzene (**30**) (see 9.3.4)], as hydroxy radical sources [*e.g.* α-hydroperoxydiazene (**31**) (see 3.3.3.1)], for enhanced solubility in organic solvents [*e.g.* *t*-butylazocyclohexanecarbonitrile (**32**)], or as high temperature initiators [*e.g.* *t*-butylazoformamide (**33**)]. They have also been used as radical precursors in model studies of cross-termination in copolymerization (Section 7.4.3).

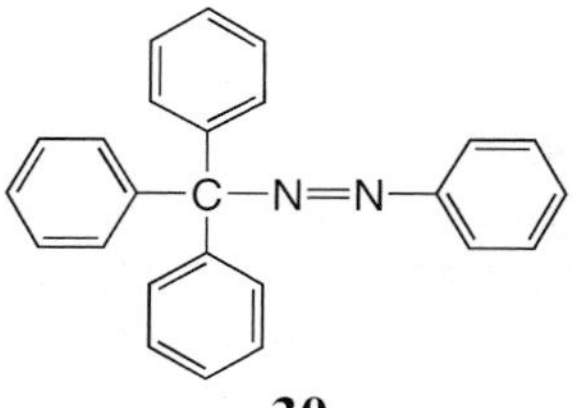
C–N=N

30

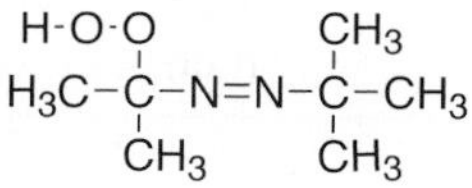
H-O-O
CH3
H3C-C-N=N-C-CH3
CH3
CH3

31

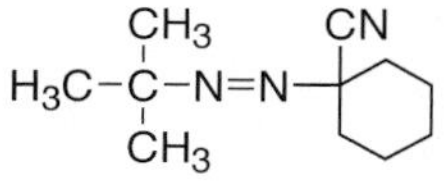
CH3
CN
H3C-C-N=N
CH3

32

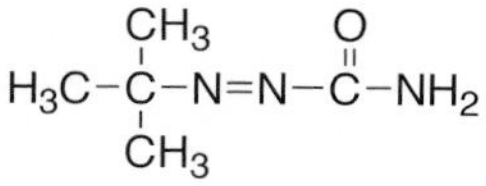
CH3
O
H3C-C-N=N-C-NH2
CH3

33

3.3.1.1.1 Thermal decomposition

While some details of the kinetics of radical production from dialkyldiazenes remain to be unraveled, their decomposition mechanism and behavior as polymerization initiators are largely understood. Kinetic parameters for some common azo-initiators are presented in Table 3.2.

Thermolysis rates (k_d) of dialkyldiazenes (**15**) show a marked dependence on the nature of R (and R'). The values of k_d increase in the series where R (=R') is aryl<primary<secondary<tertiary<allyl. In general, k_d is dramatically accelerated by α-substituents capable of delocalizing the free spin of the incipient radical.[49] For example, Timberlake[73] has found that for the case of dialkyldiazenes,

Initiation

X-C(CH$_3$)$_2$-N=N-C(CH$_3$)$_2$-X that k_d increases in the series where X is CH$_3$<-OCH$_3$ <-SCH$_3$<-CO$_2$R~-CN<-Ph<-CH=CH$_2$ (see also Table 3.2). These results can be rationalized in terms of the relative stability of the radicals generated (R•, R'•).

However, steric factors are also important.[74] Rüchardt and coworkers showed, for a series of acyclic alkyl derivatives, that a good correlation exists between k_d and ground state strain.[75,76] Additional factors are important for bicyclic and other conformationally constrained azo-compounds.[49,51,77] Wolf[78] has described a scheme for calculating k_d based on radical stability (HOMO π-delocalization energies) and ground state strain (steric parameters).

There have been numerous studies on the kinetics of decomposition of AIBN, AIBMe and other dialkyldiazenes.[46] Solvent effects on k_d are small by conventional standards but, nonetheless, significant. Data for AIBMe is presented in Table 3.3. The data come from a variety of sources and can be seen to increase in the series where the solvent is: aliphatic < ester (including MMA) < aromatic (including styrene) < alcohol. There is a factor of two difference between k_d in methanol and k_d in ethyl acetate. The value of k_d for AIBN is also reported to be higher in aromatic than in hydrocarbon solvents and to increase with the dielectric constant of the medium.[31,79,80] The k_d of AIBMe and AIBN show no direct correlation with solvent viscosity (see also 3.3.1.1.3), which is consistent with the reaction being irreversible (*i.e.* no cage return).

Thermolysis rates are enhanced substantially by the presence of certain Lewis acids (*e.g.* boron and aluminum halides), and transition metal salts (*e.g.* Cu^{2+}, Ag$^+$).[46] There is also evidence that complexes formed between azo-compounds and Lewis acids (*e.g.* ethyl aluminum sesquichloride) undergo thermolysis or photolysis to give complexed radicals which have different specificity to uncomplexed radicals.[81-83]

Table 3.3 Solvent Dependence of Rate Constants for AIBMe Decomposition[a]

$k_d \times 10^5$ s^{-1}	Solvent	Temperature °C	Reference
0.58	cyclohexane	60.0	62
0.72	ethyl acetate	60.0	63
0.74	methyl isobutyrate	60.0	63
0.83	1:1 MMA/S	60.0	84
1.18[a]	aliphatic esters	60.0	85
0.88	benzene	60.0	62
0.91	benzene	60.0	63
1.01	acetonitrile	60.0	62
1.13	S	60.0	86
1.20	methanol	60.0	63
1.44	methanol	60.0	62

a Calculated from the expression given: ln(k_d)=33.1-(14800/T); said to be valid for a range of aliphatic ester solvents including MMA.

3.3.1.1.2 *Photochemical decomposition*

The *trans*-dialkyldiazenes have λ_{max} 350-370 nm and ε 2-50 M^{-1} cm^{-1} and are photolabile. They are, therefore, potential photoinitiators.[49,87] The efficiency and rate of radical generation depends markedly on structure.[49] Dialkyldiazenes are often depicted without indicating the stereochemistry about the nitrogen-nitrogen double bond. However, except when constrained in a ring system, the dialkyldiazenes can be presumed to have the *trans*-configuration.

Alicyclic *cis*-dialkyldiazenes are very thermolabile when compared to the corresponding *trans*-isomers, often having only transient existence under typical reaction conditions. It has been proposed[49] that the main light-induced reaction of the dialkyldiazenes is *trans-cis* isomerization. Dissociation to radicals and nitrogen is then a thermal reaction of the *cis*-isomer (Scheme 3.19).

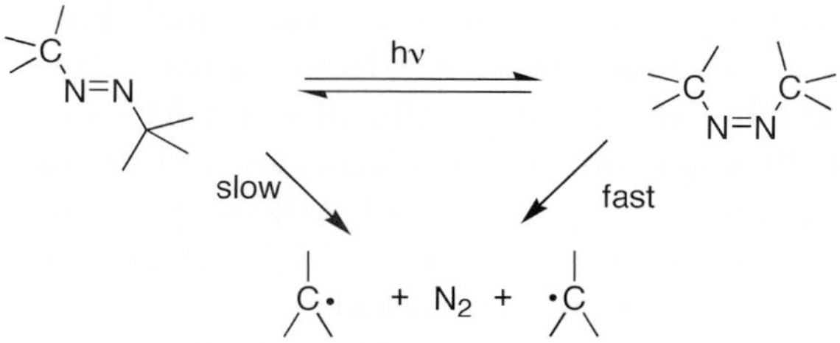

Scheme 3.19

Therefore, the quantum yield for photoisomerization approximates that for nitrogen formation and both are typically *ca* 0.5. Where the *cis* isomer is thermally stable, quantum yields for initiator disappearance are low (ϕ<0.1).[49]

An important ramification of the photolability of azo-compounds is that, when using dialkyldiazenes as thermal initiators, care must be taken to ensure that the polymerization mixture is not exposed to excessive light during its preparation.

3.3.1.1.3 *Initiator efficiency*

The proportion of 'useful' radicals generated from common dialkyldiazenes is never quantitative; typically it is the range 50-70% in media of low viscosity (*i.e.* in low conversion polymerizations).[32,88,89] The main cause of this inefficiency is loss of radicals through self-reaction within the solvent cage.

For dialkyldiazenes where the α-positions are not fully substituted, tautomerization to the corresponding hydrazone may also reduce the initiator efficiency[90] (Scheme 3.20). This rearrangement is catalyzed by light and by acid.

Scheme 3.20

Initiation

There is also evidence for a radical-induced mechanism involving initial hydrogen abstraction (Scheme 3.21).

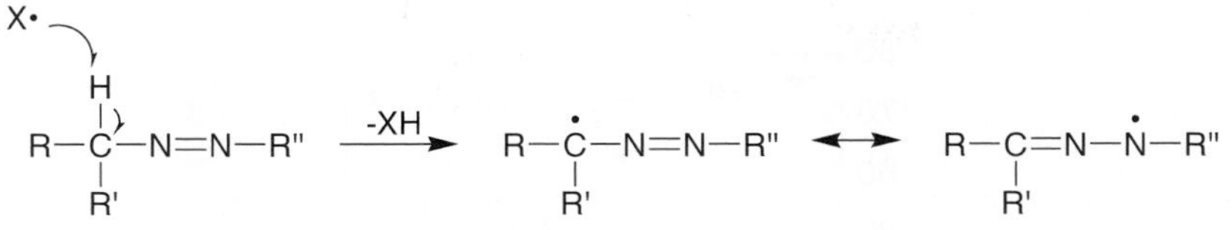

$$X\cdot + R-\underset{R'}{\overset{H}{C}}-N=N-R'' \xrightarrow{-XH} R-\underset{R'}{\overset{\cdot}{C}}-N=N-R'' \longleftrightarrow R-\underset{R'}{C}=N-\overset{\cdot}{N}-R''$$

Scheme 3.21

Conflicting statements have appeared on the sensitivity of f to the nature of the monomer involved. Braun and Czerwinski[91] reported that for low conversion polymerizations, f is essentially the same in MMA, S, and NVP. Fukuda et al.[92] reported that f varies between MMA and S. The solvent dependence of k_d may account for this apparent conflict (Table 3.3).

While the rate of azo-compound decomposition shows only a small dependence on solvent viscosity, the amount of cage reaction (and hence f) varies dramatically with the viscosity of the reaction medium and hence with factors that determine the viscosity (conversion, temperature, solvent, *etc.*).[31]

Most values of f have been measured at zero or low conversions. During polymerization the viscosity of the medium increases and the concentration of monomer decreases dramatically as conversion increases (*i.e.* as the volume fraction of polymer increases). The value of f is anticipated to drop accordingly.[32,33,93-96] For example, with S polymerization in 50% (v/v) toluene at 70 °C initiated by 0.1 M AIBN the 'instantaneous' f was determined to vary from 76% at low conversion to <20% at 90-95% conversion (Figure 3.3).[32] The assumption that the rate of initiation ($k_d f$) is invariant with conversion (common to most pre 1990s and many recent kinetic studies of radical polymerization) cannot be supported.

The viscosity dependence of f may lead to the initiator efficiency being dependent on the molecular weight of the polymer being produced. This, in turn, is a function of the initiator and monomer concentration. For example, initiator efficiencies are expected to be higher during oligomer synthesis than in preparation of high molecular weight polymer. Initiator efficiency has also been shown to depend on the size of the initiator-derived radicals.[33] There is an inverse relationship between the rate of escape from the solvent cage and radical size.

Initiator efficiency increases with reaction temperature (Table 3.4). It is also worth noting that apparent zero-conversion initiator efficiencies depend on the method of measurement. Better scavengers trap more radicals. The data in Table 3.4 suggest that monomers (MMA, S) are not as effective at scavenging radicals as the inhibitors used to measure initiator efficiencies. The finding suggests that in polymerization the initiator-derived radicals have a finite probability of undergoing self-reaction after they escape the solvent cage and numbers obtained by the inhibitor method should be considered as upper limits.

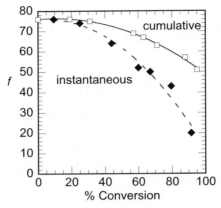

Figure 3.3 Cumulative (□) and instantaneous (◆) initiator efficiency (*f*) of AIBN as initiator in S polymerization (50% v/v toluene, 70 °C) as a function of monomer conversion (lines are a polynomial fit to the datapoints).[1,32]

Table 3.4 Zero-Conversion Initiator Efficiency (*f*) for AIBMe under Various Reaction Conditions

f	Scavenger	Temperature °C	Solvent	Reference
0.81[a]	none	98	S	86
0.72[a]	none	90	S	86
0.77	galvinoxyl	90	chlorobenzene	97
0.76	nitroxide	80	chlorobenzene	98
0.70[b]	triphenylverdazyl	80	MMA	85
0.63[a]	none	80	S	86
0.68-0.60[c]	none	60	MMA/S/ benzene	84
0.56b	triphenylverdazyl	60	MMA	85
0.48	DPPH	60	not specified	99
0.45	none	60	benzene	63
0.40[a]	none	60	S	86

a Estimated by analysis of polymerization kinetics. b Calculated using the expression ln*f* = 0.58-(330/T).[85] c [Polymer end groups]/[total products] with AIBMe-α-^{13}C as initiator. Overall efficiency reduces from 0.68 at <16% conversion to 0.60 at 95% conversion (Figure 3.3).

The byproducts of decomposition of certain dialkyldiazenes can be a concern. Consider the case of AIBN decomposition (Scheme 3.13). The major byproduct is the ketenimine (**10**).[61,100-102] This compound is itself thermally labile and reverts to cyanoisopropyl radicals at a rate constant similar to that for AIBN thermolysis.[59,60,102] This complicates any analysis of the kinetics of initiation.[32,60]

Initiation

Another concern, is the potential reactivity of **10** as a transfer agent under polymerization conditions (see 3.3.1.1.4).[103] Tetramethylsuccinonitrile (**9**) appears to be essentially inert under polymerization conditions.* However, the compound is reported to be toxic and may be a problem in polymers used in food contact applications.[35,36] Methacrylonitrile (MAN) formed by disproportionation readily copolymerizes.[7,34] The copolymerized MAN may affect the thermal stability of polymers. A suggestion[105] that copolymerized MAN may be a "weak link" in PS initiated with AIBN has been disputed.[14]

Some of the complications associated with the use of AIBN may be avoided by use of alternative azo-initiators. Azobis(methyl isobutyrate) (AIBMe) has a decomposition rate only slightly less than AIBN and has been promoted for use in laboratory studies of polymerization[85] because the kinetics and mechanism of its decomposition kinetics are not complicated by ketenimine formation.

The azonitrile **19** also shows similar decomposition kinetics to AIBN (Table 3.2). The initiators **19** and AIBMe also have greater solubility in organic solvents than AIBN.

3.3.1.1.4 Transfer to initiator

Dialkyldiazenes are often preferred over other (peroxide) initiators because of their lower susceptibility to induced decomposition. The importance of transfer to initiator during polymerizations initiated by AIBN has been the subject of some controversy. While the early work of Baysal and Tobolsky,[106] Bevington and Lewis[107] and others suggested that transfer to initiator was insignificant during polymerizations of MMA or S, a number of subsequent studies on polymerization kinetics report a significant transfer constant (C_I ca 0.1).[104,108-112] Studies of S polymerization initiated by ^{13}C-labeled AIBN demonstrate that transfer to initiator has little importance in that system.[7] Thus, other explanations for those irregularities in polymerization kinetics previously attributed to transfer to initiator have to be considered: for example, failure to allow for the variation of initiator efficiency with conversion (see 3.3.1.1.3). There is some evidence that transfer to initiator may be of importance during AIBN-initiated vinyl acetate polymerization.[113]

Even though AIBN has a low transfer constant, the ketenimine formed by combination of cyanoisopropyl radicals (Scheme 3.13) is anticipated to be more susceptible to induced decomposition (Scheme 3.22).[103]

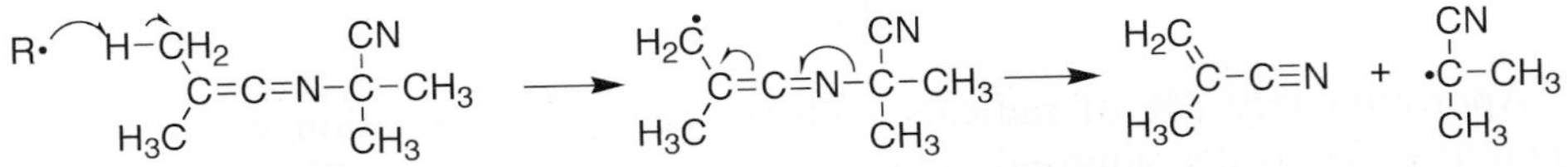

Scheme 3.22

---

* Pryor and Fiske[104] have determined $C_T=3.7\times10^{-5}$ for **9** at 60 °C in S polymerization.

3.3.1.2 *Hyponitrites*

The hyponitrites (**16**), esters of hyponitrous acid (HO–N=N–OH), are low temperature sources of alkoxy or acyloxy radicals. A detailed study of the effect of substituents on k_d for the hyponitrite esters has been reported by Quinga and Mendenhall.[114]

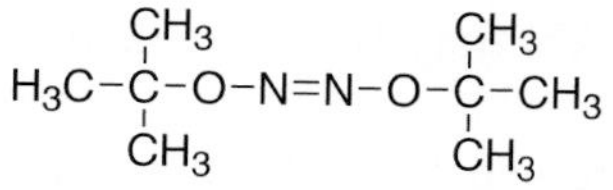

H₃C–C(CH₃)₂–O–N=N–O–C(CH₃)₂–CH₃   Ph–C(CH₃)₂–O–N=N–O–C(CH₃)₂–Ph

34 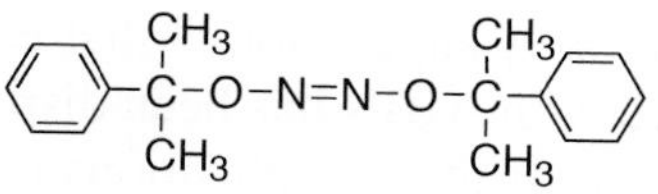   **35**

While di-*t*-butyl (**34**) and dicumyl hyponitrites (**35**) have proved convenient sources of *t*-butoxy and cumyloxy radicals respectively in the laboratory,[71,72,115-117] the utilization of hyponitrites as initiators of polymerization has been limited by difficulties in synthesis and commercial availability. Dialkyl hyponitrites (**16**) show only weak absorption at λ>290 nm and their photochemistry is largely a neglected area. The triplet sensitized decomposition of these materials has been investigated by Mendenhall *et al.*[118]

The hyponitrites generally appear somewhat more efficient with respect to radical generation than the dialkyldiazenes (see 3.3.1.1). However, a proportion of radicals is lost through cage reaction with formation of the corresponding dialkyl peroxides or ketone plus alcohol (Scheme 3.23).[119,120] The disproportionation pathway is open only to hyponitrites with α-hydrogens. Kiefer and Traylor[121] showed that the extent of cage reaction was strongly dependent on the medium viscosity.

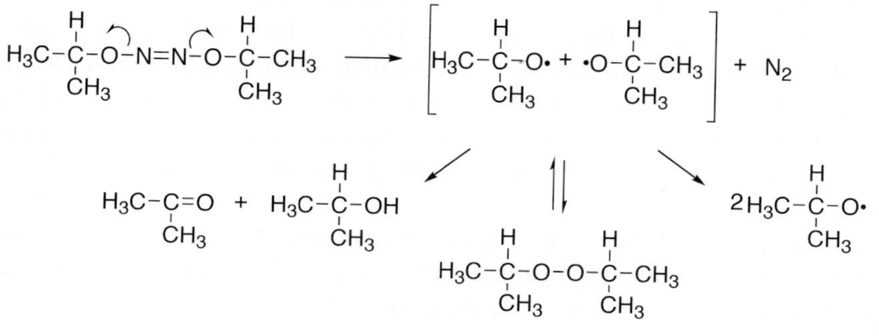

$$H_3C\text{-}CH(CH_3)\text{-}O\text{-}N=N\text{-}O\text{-}CH(CH_3)\text{-}CH_3 \longrightarrow [H_3C\text{-}CH(CH_3)\text{-}O\cdot + \cdot O\text{-}CH(CH_3)\text{-}CH_3] + N_2$$

giving: $H_3C\text{-}C(=O)\text{-}CH_3 + H_3C\text{-}CH(CH_3)\text{-}OH$ ⇌ $H_3C\text{-}CH(CH_3)\text{-}O\text{-}O\text{-}CH(CH_3)\text{-}CH_3$ → $2\,H_3C\text{-}CH(CH_3)\text{-}O\cdot$

Scheme 3.23

Approximately 5% of radicals undergo cage recombination when dicumyl hyponitrite (**35**) is decomposed in bulk MMA or S at 60 °C.[72] Dicumyl peroxide, the product of cage recombination is likely to be stable under the conditions where hyponitrites are usually employed. Nonetheless, its formation is a concern since

Initiation

contamination of a product polymer with peroxide may impair its longer term durability.

Tertiary hyponitrites are not particularly susceptible to induced decomposition. However, the same is not true of primary and secondary hyponitrites.[122] Isopropyl hyponitrite is reported[123] to undergo induced decomposition by a mechanism involving initial abstraction of a α-hydrogen (Scheme 3.24).

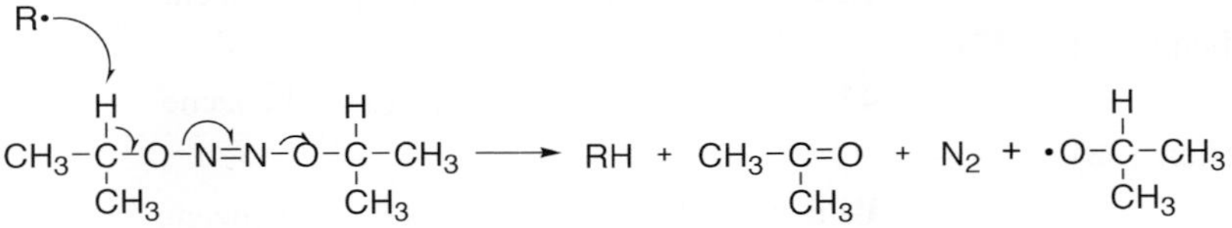

$$CH_3-\underset{CH_3}{\underset{|}{C}}\overset{H}{\underset{|}{-}}O-N=N-\overset{H}{\underset{|}{O}}-\underset{CH_3}{\underset{|}{C}}-CH_3 \longrightarrow RH + CH_3-\underset{CH_3}{\underset{|}{C}}=O + N_2 + \cdot O-\underset{CH_3}{\overset{H}{\underset{|}{C}}}-CH_3$$

with R· abstracting an α-hydrogen.

Scheme 3.24

3.3.2 Peroxides

The general chemistry of the peroxides has been covered in many books and reviews.[2,46,52,124-131] Readers are referred in particular to Swern's Trilogy[127-129] for an excellent background and a comprehensive coverage of the literature through 1970. The chemistry associated with their use as initiators of polymerization was reviewed by Moad and Solomon.[2]

Many types of peroxides (R-O-O-R) are known. Those in common use as initiators include: diacyl peroxides (**36**), peroxydicarbonates (**37**), peroxyesters (**38**), dialkyl peroxides (**39**), hydroperoxides (**40**), and inorganic peroxides [*e.g.* persulfate (**41**)]. Multifunctional and polymeric initiators with peroxide linkages are discussed in Sections 3.3.3 and 6.3.2.1.

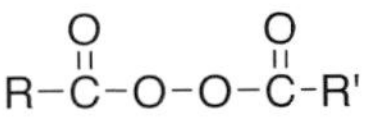

$$\underset{\textbf{36}}{R-\overset{O}{\overset{\|}{C}}-O-O-\overset{O}{\overset{\|}{C}}-R'} \qquad \underset{\textbf{37}}{R-O-\overset{O}{\overset{\|}{C}}-O-O-\overset{O}{\overset{\|}{C}}-O-R'} \qquad \underset{\textbf{38}}{R-\overset{O}{\overset{\|}{C}}-O-O-R'}$$

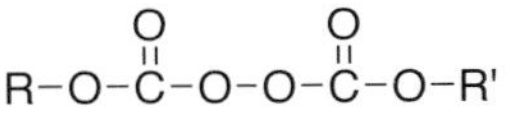

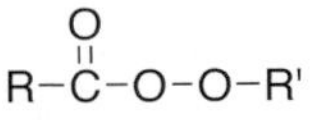

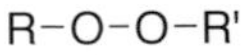

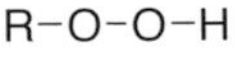

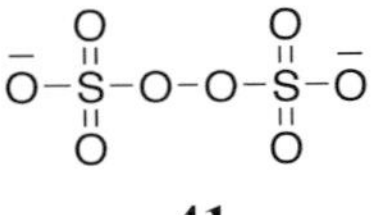

$$\underset{\textbf{39}}{R-O-O-R'} \qquad \underset{\textbf{40}}{R-O-O-H} \qquad \underset{\textbf{41}}{\overset{\overset{O}{\|}}{\underset{\underset{O}{\|}}{O-S-O-O-S-O}}}$$

Peroxides are used most commonly either as thermal initiators or as a component in a redox system. While peroxides are photochemically labile, they seldom find use as photoinitiators other than in laboratory studies because of their poor light absorption characteristics. They generally have low extinction coefficients and absorb in the same region as monomer. Kinetic parameters for decomposition of some important peroxides are given in Table 3.5.

Table 3.5 Selected Kinetic Data for Decomposition of Peroxides[a]

class	initiator	R	R'	solvent	temp. range[b] °C
diacyl peroxides (36)					
	BPO	Ph-	Ph-	benzene[e]	38-80(17)
	LPO	n-$C_{11}H_{23}$-	n-$C_{11}H_{23}$-	benzene	35-70(8)
peroxydicarbonates (37)					
	47	$(CH_3)_2CH$-	$(CH_3)_2CH$-	benzene[e]	35-60(10)
peroxyesters (38)					
	BPB	Ph-	$(CH_3)_3C$-	benzene	110-130(3)
	DBPOX			benzene	35-55(3)
dialkyl peroxides (39)					
	DTBP	$(CH_3)_3C$-	$(CH_3)_3C$-	benzene	100-135(4)
alkyl hydroperoxides (40)					
	59	$(CH_3)_3C$-	-	benzene	155-175(4)
inorganic peroxides					
	41	$K_2S_2O_8$	-	NaOH[f]	50-90(5)

a Kinetic parameters recalculated from original data taken from the references indicated. Values rounded to 3 significant figures. b Number of data points given in parentheses. c Calculated from Arrhenius parameters shown and rounded to two significant figures. d Temperature for ten hour half life – see footnote d to Table 3.2. e In the presence of inhibitor added to prevent induced decomposition. f 0.1 M aqueous NaOH.

Table 3.5 (continued)

E_a kJ mol^{-1}	$A \times 10^{-15}$ s^{-1}	$k_d \times 10^6$ s^{-1} (60 °C)^c	10 h $t_{1/2}^{c,d}$ °C	Ref.	Initiator
139.0	9.34	1.5	78	[132]	**BPO**
125.3	0.393	8.9	66	[133]	**LPO**
126.7	9.75	130	46	[134]	**47**
144.0	1.53	0.04	105	[135]	**BPB**
110.0	0.310	1800	26	[136]	**DBPOX**
152.7	2.16	0.0025	125	[137,138]	**DTBP**
174.2	7.97	-	168	[139]	**59**
148.0	709	4.4	69	[140]	**41**

3.3.2.1 Diacyl or diaroyl peroxides

Diacyl or diaroyl peroxides (**36**, R= alkyl or aryl respectively) are given specific coverage in reviews by Fujimori,[141] Bouillion *et al.*,[142] and Hiatt.[143] They are sources of acyloxy radicals which in turn are sources of aryl or alkyl radicals. Commercially available peroxides of this type include dibenzoyl peroxide (BPO), didodecanoyl or dilauroyl peroxide (LPO), didecanoyl peroxide (**42**) and succinic acid peroxide (**43**).

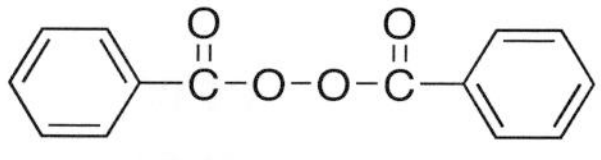

$$\text{Ph-C(O)-O-O-C(O)-Ph} \quad\quad C_{11}H_{23}\text{-C(O)-O-O-C(O)-}C_{11}H_{23} \quad\quad C_9H_{21}\text{-C(O)-O-O-C(O)-}C_9H_{21}$$

BPO 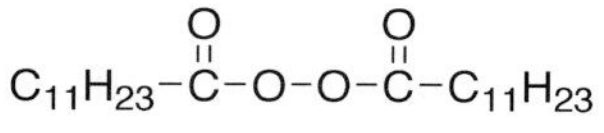   **LPO** 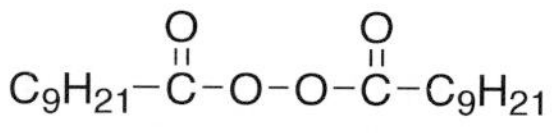   **42**

$$HO_2C\text{-}CH_2CH_2\text{-C(O)-O-O-C(O)-}CH_2CH_2\text{-}CO_2H$$

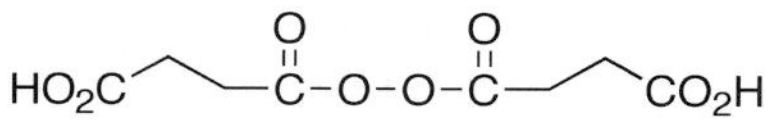

43

3.3.2.1.1 Thermal decomposition

The rates of thermal decomposition of diacyl peroxides (**36**) are dependent on the substituents R. The rates of decomposition increase in the series where R is: aryl~primary alkyl<secondary alkyl<tertiary alkyl. This order has been variously proposed to reflect the stability of the radical (R•) formed on β-scission of the acyloxy radical, the nucleophilicity of R, or the steric bulk of R. For peroxides with non-concerted decomposition mechanisms, it seems unlikely that the stability of R• should by itself be an important factor.

For diaroyl peroxides (**36**, R=aryl), *m*- and *p*-electron withdrawing substituents retard the rate of decomposition while *m*- and *p*-electron donating and all *o*-substituents enhance decomposition rates. The *o*-substituent effect has been attributed to the sensitivity of homolysis to steric factors.

Only a few diacyl peroxides see widespread use as initiators of polymerization. The reactions of the diaroyl peroxides (**36**, R=aryl) will be discussed in terms of the chemistry of BPO (Scheme 3.25). The rate of β-scission of thermally generated benzoyloxy radicals is slow relative to cage escape, consequently, both benzoyloxy and phenyl radicals are important as initiating species. In solution, the only significant cage process is reformation of BPO (*ca* 4% at 80 °C in isooctane);[144,145] only minute amounts of phenyl benzoate or biphenyl are formed within the cage. Therefore, in the presence of a reactive substrate (*e.g.* monomer), the production of radicals can be almost quantitative (see 3.3.2.1.3).

One of the most commonly encountered aliphatic diacyl peroxides (**36**, R=alkyl) is LPO. Lower diacyl peroxides (*e.g.* diacetyl peroxide) cannot be

Initiation

conveniently handled in a pure state due to their susceptibility to induced decomposition. They are shock sensitive and may decompose explosively.

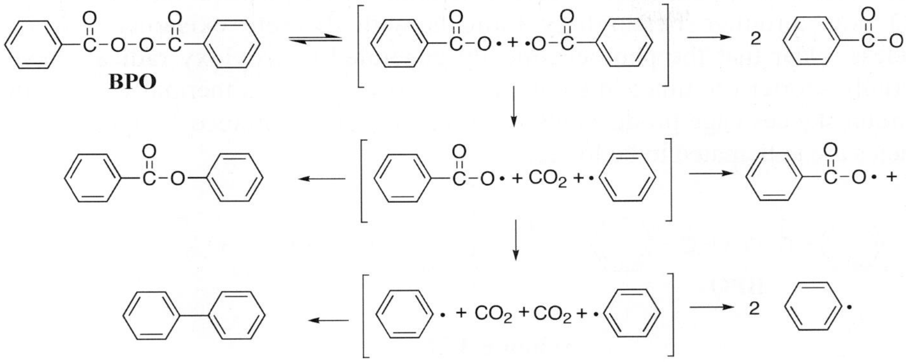

Scheme 3.25

In general, aliphatic diacyl peroxide initiators should be considered as sources of alkyl, rather than of acyloxy radicals. With few exceptions, aliphatic acyloxy radicals have a transient existence at best. For certain diacyl peroxides (**36**) where R is a secondary or tertiary alkyl group there is controversy as to whether loss of carbon dioxide occurs in concert with O-O bond cleavage. Thus, ester end groups observed in polymers prepared with aliphatic diacyl peroxides are unlikely to arise directly from initiation, but rather from transfer to initiator (see 3.3.2.1.4).

The high rate of decarboxylation of aliphatic acyloxy radicals is also the prime reason behind low initiator efficiencies (see 3.3.2.1.3). Decarboxylation occurs within the solvent cage and recombination gives alkane or ester byproducts. Cage return for LPO is 18-35% at 80 °C in *n*-octane as compared to only 4% for BPO under similar conditions.[144]

Observed rates of disappearance for diacyl peroxides show marked dependence on solvent and concentration.[146] In part, this is a reflection of their susceptibility to induced decomposition (see 3.3.2.1.4 and 3.3.2.1.5). However, the rate of disappearance is also a function of the viscosity of the reaction medium. This is evidence for cage return (see 3.3.2.1.3).[145] The observation[144] of slow scrambling of the label in benzoyl-*carbonyl*-^{18}O peroxide between the carbonyl and the peroxidic linkage provides more direct evidence for this process.

3.3.2.1.2 Photochemical decomposition

Diacyl peroxides have continuous weak absorptions in the UV to *ca* 280 nm (ε *ca* 50 M^{-1} cm^{-1} at 234 nm).[147] Although the overall chemistry in thermolysis and photolysis may appear similar, substantially higher yields of phenyl radical products are obtained when BPO is decomposed photochemically. It has been suggested that, during the photodecomposition of BPO, β-scission may occur in concert with O–O bond rupture and give rise to formation of one benzoyloxy radical, one phenyl radical, and a molecule of carbon dioxide (Scheme 3.26).[148] Time resolved EPR experiments[149] have shown that photochemical decomposition of BPO does produce benzoyloxy radicals with discrete existence. It is, nonetheless, clear that the photochemically generated benzoyloxy radicals have substantially shorter life times in solution than those generated thermally.[150,151] In these circumstances cage products also assume greater importance[151] and initiator efficiencies are anticipated to be lower.

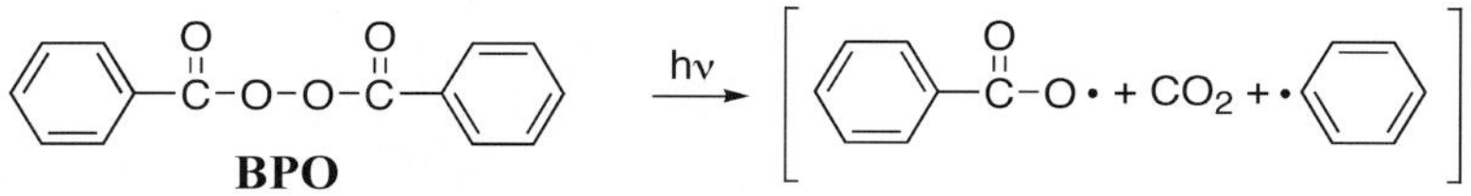

Ph–C(=O)–O–O–C(=O)–Ph →(hv) [Ph–C(=O)–O• + CO_2 + •Ph]

BPO

Scheme 3.26

It has also been suggested that photoexcited benzoyl peroxide is somewhat more susceptible to induced decomposition processes involving electron transfer than the ground state molecule. Rosenthal *et al*.[152] reported on redox reactions with certain salts (including benzoate ion) and neutral molecules (*e.g.* alcohols).

3.3.2.1.3 Initiator efficiency

Ideally all reactions should result from unimolecular homolysis of the relatively weak O–O bond. However, unimolecular rearrangement and various forms of induced and non-radical decomposition complicate the kinetics of radical generation and reduce the initiator efficiency.[46] Peroxide decomposition induced by radicals and redox chemistry is covered in Sections 3.3.2.1.4 and 3.3.2.1.5 respectively.

Cage recombination is also a major factor limiting the efficiency of radical production from aliphatic diacyl peroxides. Initiator efficiency depends on the rate of β-scission of the acyloxy radical formed. If β-scission is slow, the only significant cage reaction involves regeneration of the diacyl peroxide (*e.g.* thermolysis of diaroyl peroxides). Cage return leads to a lowering of the rate of decomposition without reducing the initiator efficiency (see 3.3.2.1.1). However, if β-scission is rapid and decarboxylation occurs within the solvent cage, then combination of the alkyl or aryl radical with another radical to form an ester or alkane will reduce the initiator efficiency (*e.g.* thermolysis or photolysis of aliphatic diacyl peroxides and photolysis of all diacyl peroxides).

The importance of the cage reaction increases according to the viscosity of the reaction medium. This contributes to a decrease in initiator efficiency with conversion.[33,153-155] Stickler and Dumont[156] determined the initiator efficiency during bulk MMA polymerization at high conversions (*ca* 80%) to be in the range 0.1-0.2 depending on the polymerization temperature. The main initiator-derived byproduct was phenyl benzoate.

Diacyl peroxides may also undergo non-radical decomposition *via* the carboxy inversion process to form an acylcarbonate (Scheme 3.27).[46] The reaction is of greatest importance for diaroyl peroxides with electron withdrawing substituents and for aliphatic diacyl peroxides (36) where R is secondary, tertiary or benzyl.[157] The reaction is thought to involve ionic intermediates and is favored in polar solvents[157] and by Lewis acids.[158] Other heterolytic pathways for peroxide decomposition have been described.[159]

$$\text{R-C(=O)-O-O-C(=O)-R} \longrightarrow \text{R-O-C(=O)-O-C(=O)-R}$$

36

Scheme 3.27

3.3.2.1.4 Transfer to initiator and induced decomposition

Transfer to initiator can be a major complication in polymerizations initiated by diacyl peroxides. The importance of the process typically increases with monomer conversion and the consequent increase in the [initiator]:[monomer] ratio.[9,106,160-162] In BPO initiated S polymerization, transfer to initiator may be the major chain termination mechanism. For bulk S polymerization with 0.1 M BPO at 60 °C up to 75% of chains are terminated by transfer to initiator or primary radical termination (<75% conversion).[7] A further consequence of the high incidence of chain transfer is that high conversion PS formed with BPO initiator tends to have a much narrower molecular weight distribution than that prepared with other initiators (*e.g.* AIBN) under similar conditions.

The mechanism of transfer to BPO involves homolytic attack on one of the oxygen atoms of the peroxidic linkage (Scheme 3.16) with formation of an ester end group and expulsion of a benzoyloxy radical. The end group formed (a secondary ester) is distinct from that formed in initiation. Such end groups may contribute to the reduced thermal stability of high conversion PS prepared with benzoyl peroxide (Section 8.2.1).[14,163] In the case of VAc or VC polymerizations the chain end will be a hydrolytically unstable ketal or α-chloroester group respectively (Section 8.2.3).

Other radicals present in the reaction medium may also induce the decomposition of BPO and other diacyl peroxides. These include initiator-derived[146] and stable radicals (*e.g.* galvinoxyl,[132] triphenylmethyl[164,165] and nitroxides[166]).

3.3.2.1.5 Redox reactions

The decomposition of diacyl peroxides (36) is catalyzed by various transition metal salts,[46,167] for example, Cu^+ (Scheme 3.28).[168,169] A side reaction is oxidation of alkyl radicals by the oxidized form of the metal salt (*e.g.* Cu^{2+}).

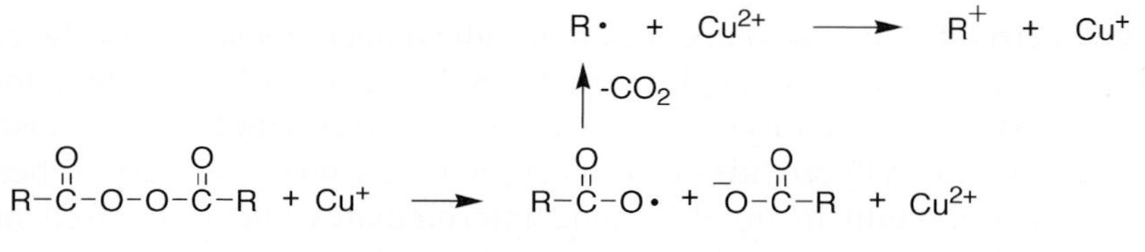

Nitro- and nitroso-compounds,[170,171] amines, and thiols induce the decomposition of diacyl peroxides in what may be written as an overall redox reaction. Certain monomers have been reported to cause induced decomposition of BPO. These include AN,[172] N-vinylcarbazole,[173-177] N-vinylimidazole[178] and NVP.[177]

The mechanism proposed for the production of radicals from the N,N-dimethylaniline/BPO couple[179,180] involves reaction of the aniline with BPO by a S_N2 mechanism to produce an intermediate (**44**). This thermally decomposes to benzoyloxy radicals and an amine radical cation (**46**) both of which might, in principle, initiate polymerization (Scheme 3.29). Pryor and Hendrikson[181] were able to distinguish this mechanism from a process involving single electron transfer through a study of the kinetic isotope effect.

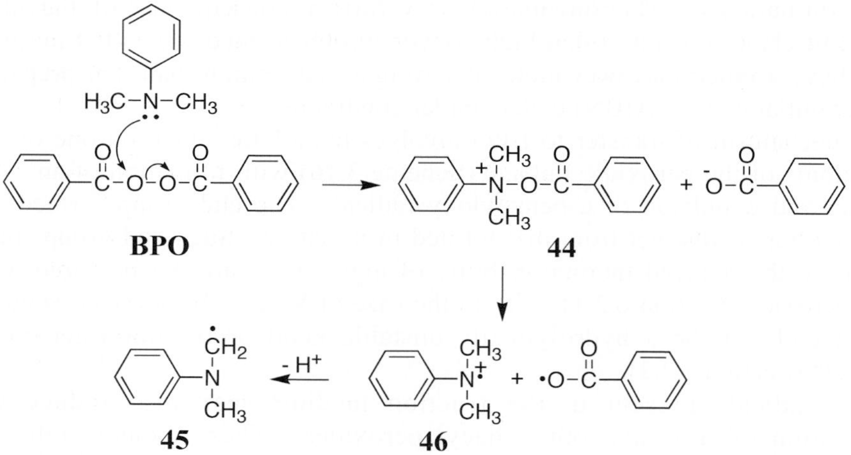

Scheme 3.29

It has been suggested that the amine radical cation (**46**) is not directly involved in initiating chains and that most polymerization is initiated by benzoyloxy radicals.[179] However, Sato et al.[182] employed spin trapping (3.5.2.1) to demonstrate that the anilinomethyl radical (**45**) was formed from the radical cation (**46**) by loss of a proton and proposed that the radical **45** also initiates polymerization. Overall efficiencies for initiation by amine-peroxide redox couples are very low; Imoto and Choe[180] report f ca 25%; Walling[179] reports f=2-5%.

3.3.2.2 Dialkyl peroxydicarbonates

The chemistry of peroxydicarbonates (**37**) and their use as initiators of polymerization has been reviewed by Yamada et al.,[134] Hiatt[143] and Strong.[183]

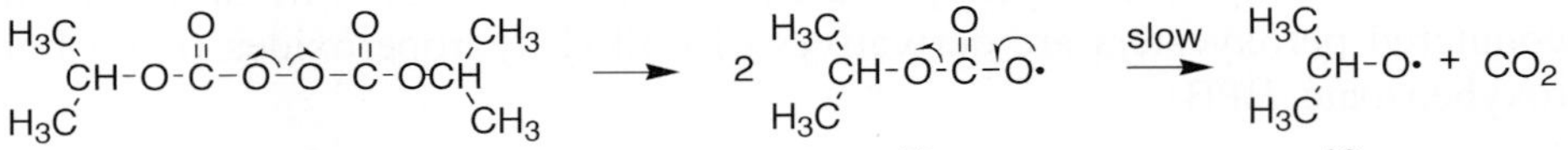

$$\begin{array}{c}H_3C\\ CH-O-C(=O)-O-O-C(=O)-OCH(CH_3)_2\\ H_3C\end{array} \longrightarrow 2 \begin{array}{c}H_3C\\ CH-O-C(=O)-O\cdot\\ H_3C\end{array} \xrightarrow{slow} \begin{array}{c}H_3C\\ CH-O\cdot + CO_2\\ H_3C\end{array}$$

       47   **48**   **49**

Scheme 3.30

Dialkyl peroxydicarbonates have been reported as low temperature sources of alkoxy radicals (Scheme 3.30)[184,185] and these radicals may be formed in relatively inert media. However, it is established, for primary and secondary peroxydicarbonates, that the rate of loss of carbon dioxide is slow compared to the rate of addition to most monomers or reaction with other substrates.[186,187] Thus, in polymerizations carried out with diisopropyl peroxydicarbonate (**47**), chains will be initiated by isopropoxycarbonyloxy (**48**) rather than isopropoxy radicals (**49**) (see 3.4.2.2).[188]

A slow rate of β-scission also means that the main cage recombination process will be cage return to reform the peroxydicarbonate. Dialkyl peroxides are typically not found amongst the products of peroxydicarbonate decomposition. In these circumstances, cage recombination is unlikely to be a factor in reducing initiator efficiency.

Laboratory studies have generally focused on the diisopropyl, dicyclohexyl and di-t-butyl derivatives. These and the s-butyl and 2-ethylhexyl derivatives are commercially available.[189] The rates of decomposition of the peroxydicarbonates show significant dependence on the reaction medium and their concentration. This dependence is, however, less marked than for the diacyl peroxides (**36**) (see 3.3.1.1.4). Induced decomposition may involve a mechanism analogous to that described for diacyl peroxides. However, a more important mechanism for primary and secondary peroxydicarbonates involves abstraction of an α-hydrogen (Scheme 3.31).[190]

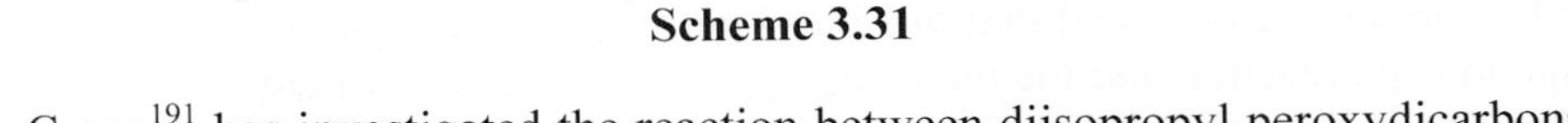

$$R\cdot + H-C(CH_3)_2-O-C(=O)-O-O-C(=O)-O-C(CH_3)_2-H \longrightarrow R-H + (CH_3)_2C=O + CO_2 + \cdot O-C(=O)-O-C(CH_3)_2-H$$

Scheme 3.31

Crano[191] has investigated the reaction between diisopropyl peroxydicarbonate and tertiary amines. These experiments indicate the formation of radicals by loss of a hydrogen from the α-CH_2 of the amine. It seems likely that the mechanism of radical formation is analogous to that observed for diacyl peroxide-amine systems (see 3.3.2.1.5).

3.3.2.3 Peroxyesters

The chemistry of peroxyesters (**38**) also commonly called peresters has been reviewed by Sawaki,[192] Bouillion *et al.*[193] and Singer.[194] The peroxyesters are sources of alkoxy and acyloxy radicals (Scheme 3.32). Most commonly encountered peroxyesters are derivatives of *t*-alkyl hydroperoxides (*e.g.* *t*-butyl peroxybenzoate, BPB).

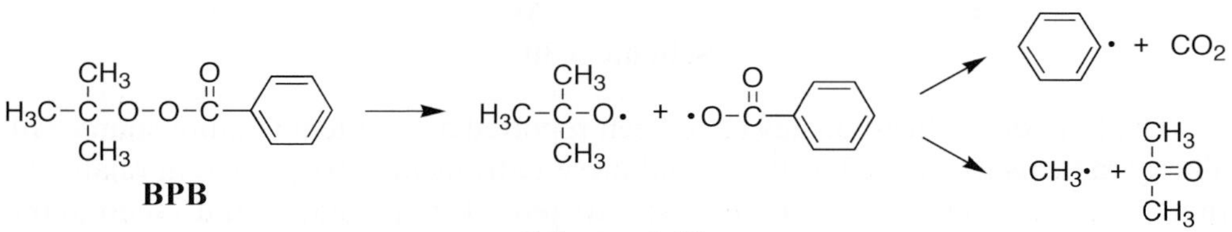

Scheme 3.32

Aryl peroxyesters are generally unsuitable as initiators of polymerization owing to the generation of phenoxy radicals that can inhibit or retard polymerization

3.3.2.3.1 Thermal decomposition

The rates of decomposition of peroxyesters (**38**) are very dependent on the nature of the substituents R and R'. The variation in the decomposition rate with R follows the same trends as have been discussed for the corresponding diacyl peroxides (see 3.3.2.1.1).

Peroxyesters derived from secondary (*e.g.* peroxyisobutyrate esters) and tertiary acids (*e.g.* peroxypivalate esters) are believed to undergo concerted 2-bond cleavage leading to direct production of an alkoxy and an alkyl radical and a molecule of carbon dioxide.[195-198] On the other hand, primary (*e.g.* peroxyacetate and peroxypropionate esters) and aromatic peroxyesters (*e.g.* BPB, Scheme 3.32) are thought to undergo 1-bond scission to generate an acyloxy and an alkoxy radical.[145,196] Evidence for the transient existence of acyloxy radicals includes the observation of substantial cage return.

For *t*-butyl peresters there is also a variation in efficiency in the series where R is primary>>secondary>tertiary. The efficiency of *t*-butyl peroxypentanoate in initiating high pressure ethylene polymerization is >90%, that of *t*-butyl peroxy-2-ethylhexanoate *ca* 60% and that of *t*-butyl peroxypivalate *ca* 40%.[196] Inefficiency is due to cage reaction and the main cage process in the case where R is secondary or tertiary is disproportionation with *t*-butoxy radicals to form *t*-butanol and an olefin.[196]

Initiation

Di-*t*-butyl peroxyoxalate (DBPOX) is a clean, low temperature, source of *t*-butoxy radicals (Scheme 3.33).[136] The decomposition is proposed to take place by concerted 3-bond cleavage to form two alkoxy radicals and two molecules of carbon dioxide.

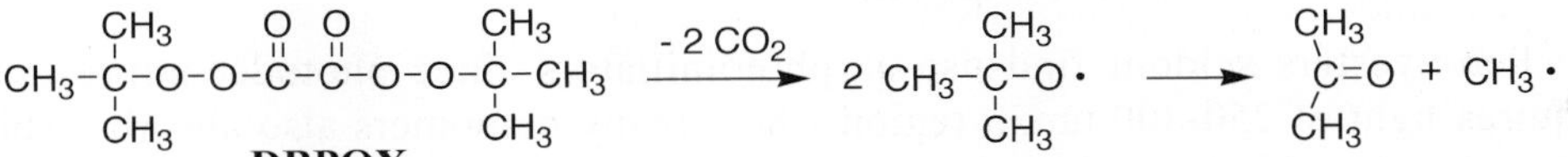

Scheme 3.33

The low conversion initiator efficiency of di-*t*-butyl peroxyoxalate (0.93-0.97)[121] is substantially higher than for other peroxyesters [*t*-butyl peroxypivalate, 0.63; *t*-butyl peroxyacetate, 0.53 (60 °C, isopropylbenzene)[195]]. The dependence of cage recombination on the nature of the reaction medium has been the subject of a number of studies.[121,199,200] The yield of DTBP (the main cage product) depends not only on viscosity but also on the precise nature of the solvent. The effect of solvent is to reduce the yield in the order: aliphatic>aromatic>protic. It has been proposed[199] that this is a consequence of the solvent dependence of β-scission of the *t*-butoxy radical which increases in the same series (Section 3.4.2.1.1).

Transfer to initiator is generally of lesser importance than with the corresponding diacyl peroxides. They are, nonetheless, susceptible to the same range of reactions (see 3.3.2.1.4). Radical-induced decomposition usually occurs specifically to give an alkoxy radical and an ester (Scheme 3.34).

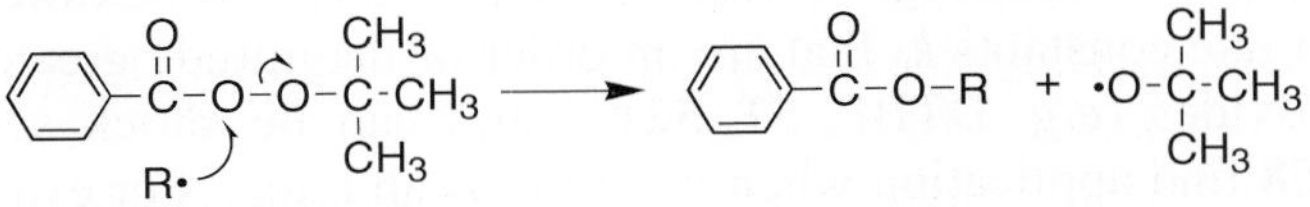

Scheme 3.34

Peroxyesters may undergo non-radical decomposition *via* the Criegee rearrangement (Scheme 3.35). This process is analogous to the carboxy inversion process described for diacyl peroxides (see 3.3.2.1.3) and probably involves ionic intermediates.

Scheme 3.35

The reaction is facilitated when R is electron withdrawing, when R has a high migratory aptitude (ability to stabilize a carbonium ion), and by polar reaction media.

3.3.2.3.2 Photochemical decomposition

Peroxyesters seldom find use as photoinitiators since photodecomposition requires light of 250-300 nm, a region where many monomers also absorb. This situation may be improved by the introduction of a suitable chromophore into the molecule or through the use of sensitizers.[201,202] The peroxyester (**50**) is reported to have λ_{max} 366 nm and ϕ near unity.[201]

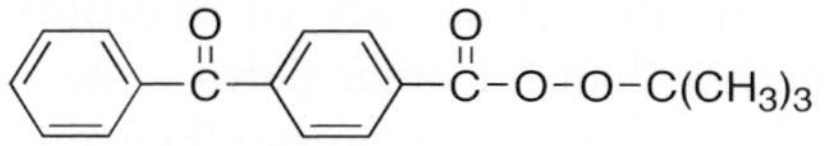

50

3.3.2.4 Dialkyl peroxides

The chemistry of the dialkyl peroxides (**39**) has been reviewed by Matsugo and Saito,[203] Sheldon[204] and Hiatt.[205] Dialkyl peroxides are high temperature sources of alkoxy radicals. Dialkyl peroxides commonly used as initiators have tertiary alkyl substituents. Those available commercially include di-*t*-butyl (DTBP) and dicumyl (**51**) peroxides, sources of *t*-butoxy and cumyloxy radicals respectively, **52** and a variety of dialkyl peroxyketals (*e.g.* **53-55**).[206,207] These latter initiators, including 1,1-di-*t*-butylperoxycyclohexane (**53**), have decomposition rate constants k_d that are an order of magnitude greater than simple di-*t*-alkyl peroxides (*e.g.* DTBP, **51**, **52**)[208] and can be shock sensitive. The peroxides **56-58** find application when volatility is an issue. For example, they are used in graft copolymerization by reactive extrusion (Section 7.6.4).[27]

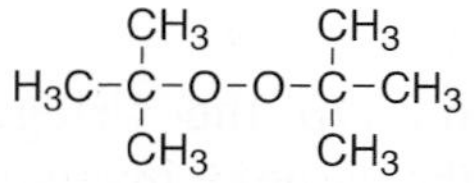

DTBP

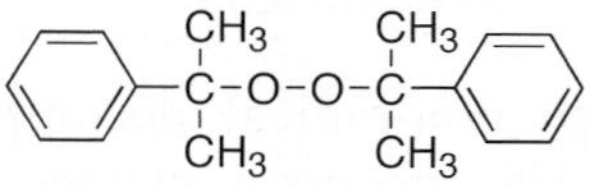

51

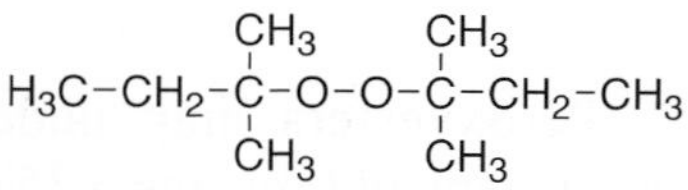

52

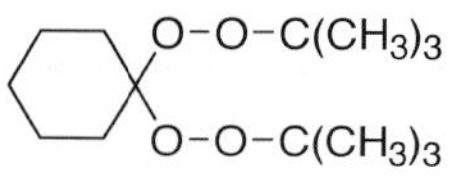

53

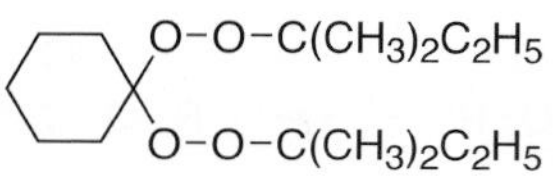

54

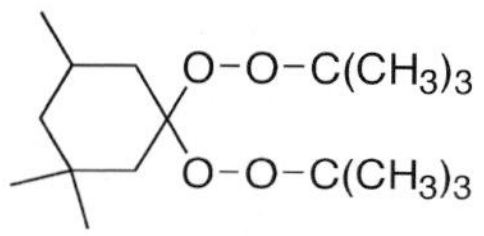

55

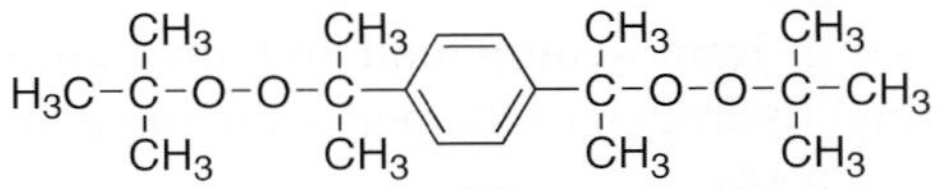

Initiation

$$H_3C-\underset{\underset{CH_3}{|}}{\overset{\overset{CH_3}{|}}{C}}-O-O-\underset{\underset{CH_3}{|}}{\overset{\overset{CH_3}{|}}{C}}-\!\!\!\!\bigcirc\!\!\!\!-\underset{\underset{CH_3}{|}}{\overset{\overset{CH_3}{|}}{C}}-O-O-\underset{\underset{CH_3}{|}}{\overset{\overset{CH_3}{|}}{C}}-CH_3$$

56

$$H_3C-\underset{\underset{CH_3}{|}}{\overset{\overset{CH_3}{|}}{C}}-O-O-\underset{\underset{CH_3}{|}}{\overset{\overset{CH_3}{|}}{C}}-CH_2-CH_2-\underset{\underset{CH_3}{|}}{\overset{\overset{CH_3}{|}}{C}}-O-O-\underset{\underset{CH_3}{|}}{\overset{\overset{CH_3}{|}}{C}}-CH_3 \qquad H_3C-\underset{\underset{CH_3}{|}}{\overset{\overset{CH_3}{|}}{C}}-O-O-\underset{\underset{CH_3}{|}}{\overset{\overset{CH_3}{|}}{C}}-C\equiv C-\underset{\underset{CH_3}{|}}{\overset{\overset{CH_3}{|}}{C}}-O-O-\underset{\underset{CH_3}{|}}{\overset{\overset{CH_3}{|}}{C}}-CH_3$$

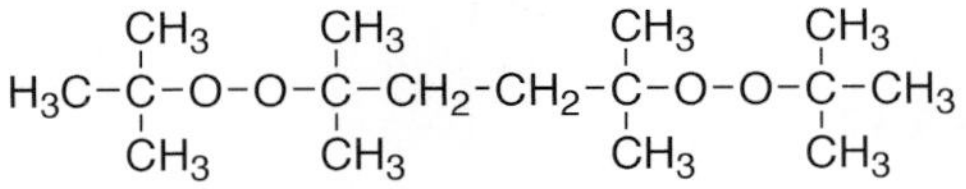

57 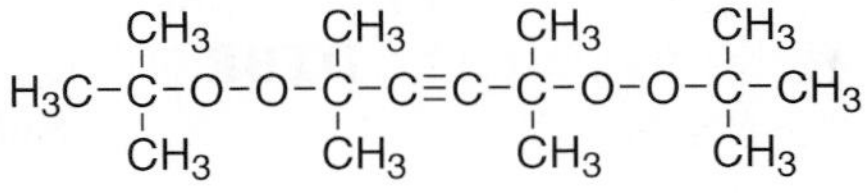   **58**

The decomposition of the peroxyketals (**53**) follows a stepwise, rather than a concerted mechanism. Initial homolysis of one of the O-O bonds gives an alkoxy radical and an α-peroxyalkoxy radical (Scheme 3.36).[206,208-210] This latter species decomposes by β-scission with loss of either a peroxy radical to form a ketone as byproduct or an alkyl radical to form a peroxyester intermediate. The peroxyester formed may also decompose to radicals under the reaction conditions. Thus, four radicals may be derived from the one initiator molecule.

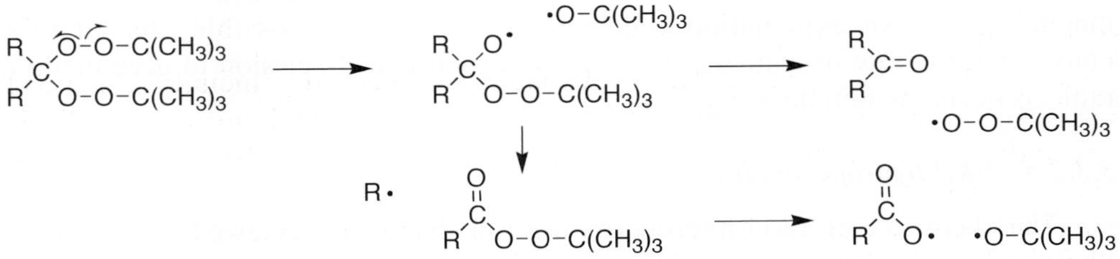

Scheme 3.36

The relative importance of the various pathways depends on the alkyl groups (R). The rate constants for scission of groups (R•) from t-alkoxy radicals ($R^1R^2R^3C-O•$) increase in the order isopropyl<ethyl<t-butylperoxy<methyl.[210] Thus, the pathway affording peroxyester and an alkyl radical is less important when R is methyl than when R is a higher alkyl group. If the pathway to alkylperoxy radicals is dominant, the resultant polymer is likely to have a proportion of peroxy end groups.[206,211]

Solvent dependence of k_d for di-t-alkyl peroxides is small when compared to most other peroxide initiators.[138,212] For di-t-butyl peroxide,[138] k_d is slightly greater (up to two-fold at 125 °C) in protic (t-butanol, acetic acid) or dipolar aprotic solvents than in other media (cyclohexane, triethylamine, tetrahydrofuran).

The chemistry of the di-t-butyl and cumyl peroxides is relatively uncomplicated by induced or ionic decomposition mechanisms. However, induced decomposition of di-t-butyl peroxide has been observed in primary or secondary alcohols[213,214] (Scheme 3.37) and primary or secondary amines.[215] The reaction involves oxidation of an α-hydroxyalkyl or α-aminoalkyl radical, to the corresponding carbonyl- or imino-compound and apparently requires coordination of the hydroxyl or aminyl hydrogen to the peroxidic oxygen.

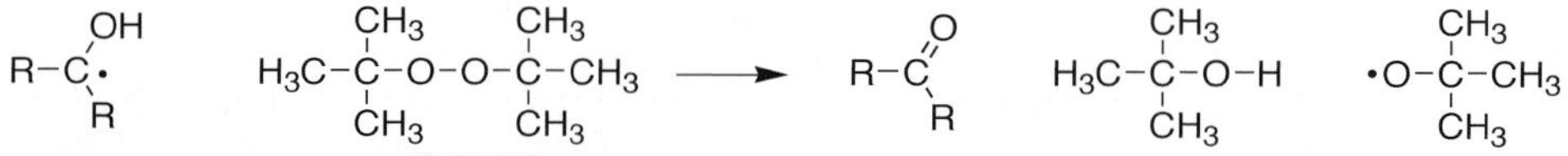

Scheme 3.37

The radical yield from simple di-*t*-alkyl peroxides (*i.e.* dicumyl, di-*t*-butyl) is reported to be almost 100%. The only significant cage reaction is reformation of the peroxide. The efficiencies of dialkyl peroxyketals and primary and secondary peroxides are lower.[207] Lower efficiencies arise when the initially formed radicals undergo β-scission before cage escape or, in the case where primary or secondary alkoxy radicals are formed, by disproportionation within the solvent cage. Primary and secondary peroxides are also susceptible to a variety of induced and non-radical decomposition mechanisms. The initiator efficiency of di-*t*-butyl peroxide in styrene polymerization is reported to remain constant at close to 100% until *ca* 80% when it undergoes a dramatic reduction by more than an order of magnitude.[216] An explanation was not provided. It is possible, that at this conversion the rate of cage escape is slowed such that β-scission to give methyl radicals occurs within the solvent cage.

3.3.2.5 *Alkyl hydroperoxides*

The chemistry of alkyl hydroperoxides (**40**) has been reviewed by Porter,[217] Sheldon[204] and Hiatt.[218] Alkyl hydroperoxides are high temperature sources of alkoxy and hydroxy radicals.[219] They are often encountered as components of redox systems.

The common initiators of this class are *t*-alkyl derivatives, for example, *t*-butyl hydroperoxide (**59**), *t*-amyl hydroperoxide (**60**), cumene hydroperoxide (**61**), and a range of peroxyketals (**62**). Hydroperoxides formed by hydrocarbon autoxidation have also been used as initiators of polymerization.

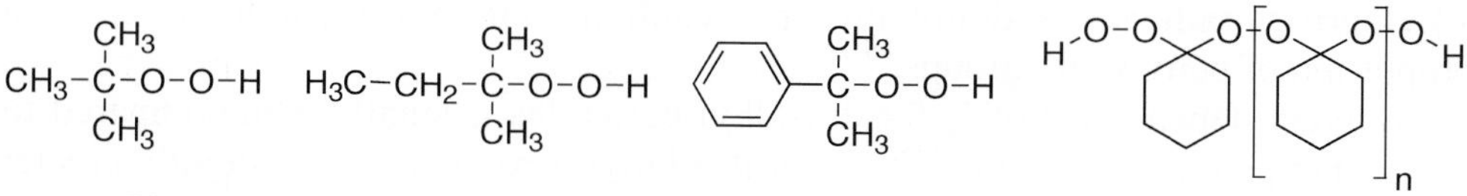

The ROO-H bond of hydroperoxides is weak compared to most other X-H bonds.* Thus, abstraction of the hydroperoxidic hydrogen by radicals is usually an

---

* $D_{ROO-H} \sim 375$ kJ mol^{-1}.[220]

Initiation

exothermic process. The hydroperoxides can therefore be efficient transfer agents and radical-induced decomposition may be a major complication in their use as initiators.[222]

Primary and secondary hydroperoxides are also susceptible to induced decomposition through loss of an α-hydrogen. The radical formed is usually not stable and undergoes β-scission to give a carbonyl compound and hydroxy radical.[223] It is reported that these hydroperoxides may also undergo non-radical decomposition with evolution of hydrogen.[137]

Hydroperoxides react with transition metals in lower oxidation states (Ti^{3+}, Fe^{2+}, Cu^{+}, *etc.*) and a variety of other oxidants to give an alkoxy radical and hydroxide anion (Scheme 3.38).[46,224,225]

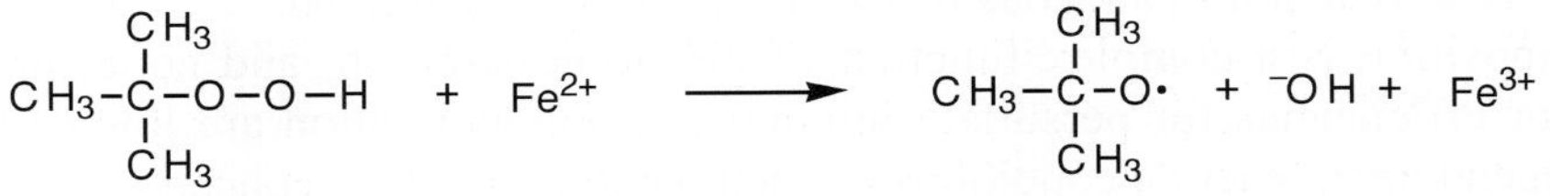

$$CH_3-\underset{\underset{CH_3}{|}}{\overset{\overset{CH_3}{|}}{C}}-O-O-H \ + \ Fe^{2+} \ \longrightarrow \ CH_3-\underset{\underset{CH_3}{|}}{\overset{\overset{CH_3}{|}}{C}}-O\bullet \ + \ {}^-OH \ + \ Fe^{3+}$$

59

Scheme 3.38

With some systems, the hydroperoxide is reduced to hydroperoxy radical by the metal ion in its higher oxidation state (Scheme 3.39). Thus, it is possible to set up a catalytic cycle for hydroperoxide decomposition.

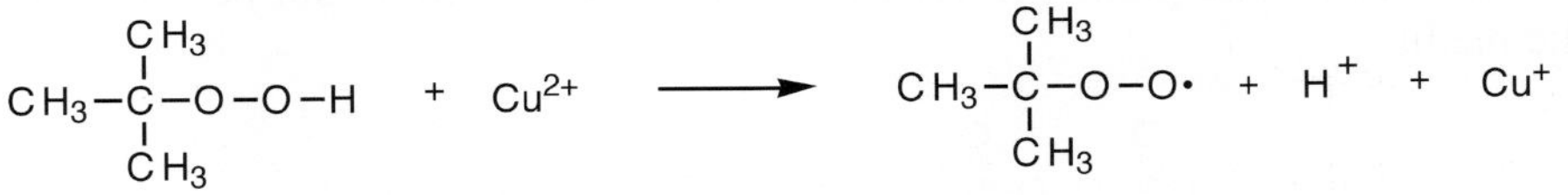

$$CH_3-\underset{\underset{CH_3}{|}}{\overset{\overset{CH_3}{|}}{C}}-O-O-H \ + \ Cu^{2+} \ \longrightarrow \ CH_3-\underset{\underset{CH_3}{|}}{\overset{\overset{CH_3}{|}}{C}}-O-O\bullet \ + \ H^+ \ + \ Cu^+$$

59

Scheme 3.39

With Ti^{4+} and Fe^{3+} this latter pathway is thought not to occur. The formation of ROO•, observed at high hydroperoxide concentrations, is attributed to the occurrence of induced decomposition.[226]

3.3.2.6 *Inorganic peroxides*

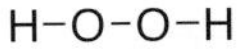

H-O-O-H $^-O-\overset{\overset{O}{\|}}{\underset{\underset{O}{\|}}{S}}-O-O-\overset{\overset{O}{\|}}{\underset{\underset{O}{\|}}{S}}-O^-$ $^-O-\overset{\overset{OH}{|}}{\underset{\underset{O}{\|}}{P}}-O-O-\overset{\overset{OH}{|}}{\underset{\underset{O}{\|}}{P}}-O^-$

63 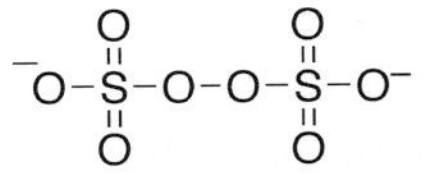   **41** 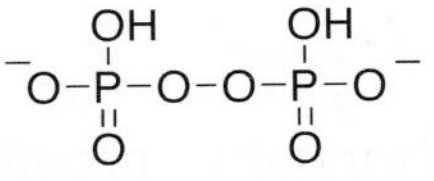   **64**

Inorganic peroxides [hydrogen peroxide (**63**), persulfate (**41**), peroxymonosulfate and peroxydiphosphate (**64**)] generally have limited usefulness as initiators in bulk or solution polymerization due to their poor solubility in

---

$D_{R_3C=H} \sim 385$, $D_{R_2CH=H} \sim 396$, $D_{RCH_2=H} \sim 410$, $D_{RO-H} \sim 435$ kJ mol^{-1}.[221]

organic media. This means that the main use of these initiators is in aqueous[227] or in part-aqueous heterogeneous media (*e.g.* in emulsion polymerization). They are often encountered as one component in a redox initiation system. The history of these systems has been reviewed by Bacon[228] and Sosnovsky and Rawlinson.[229] Their use is also described by Sarac.[230]

The following discussion concentrates on the chemistry of the two most common inorganic peroxides, persulfate and hydrogen peroxide.

3.3.2.6.1 Persulfate

Photolysis or thermolysis of persulfate ion (**41**) (also called peroxydisulfate) results in homolysis of the O-O bond and formation of two sulfate radical anions. The thermal reaction in aqueous media has been widely studied.[231,232] The rate of decomposition is a complex function of pH, ionic strength, and concentration. Initiator efficiencies for persulfate in emulsion polymerization are low (0.1-0.3) and depend upon reaction conditions (*i.e.* temperature, initiator concentration).[233]

A number of mechanisms for thermal decomposition of persulfate in neutral aqueous solution have been proposed.[232] They include unimolecular decomposition (Scheme 3.40) and various bimolecular pathways for the disappearance of persulfate involving a water molecule and concomitant formation of hydroxy radicals (Scheme 3.41). The formation of polymers with negligible hydroxy end groups is evidence that the unimolecular process dominates in neutral solution. Heterolytic pathways for persulfate decomposition can be important in acidic media.

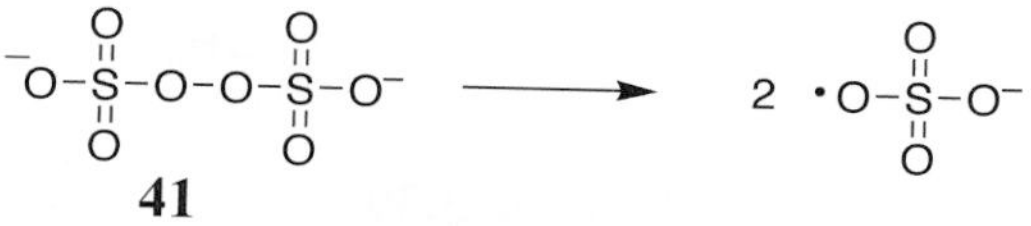

Scheme 3.40

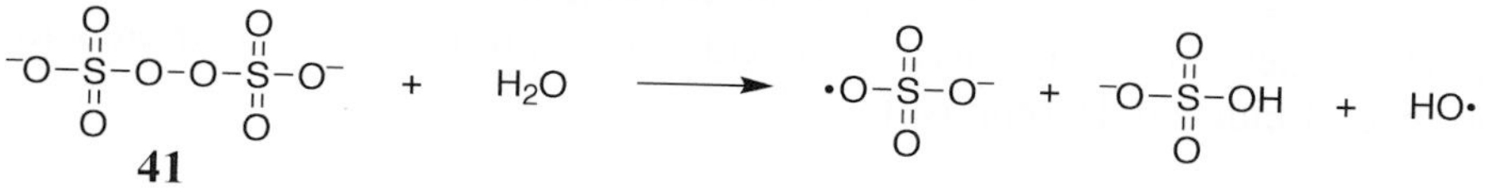

Scheme 3.41

Normally, persulfate (**41**) can only be used to initiate polymerization in aqueous or part aqueous (emulsion) media because it has poor solubility in most organic solvents and monomers. However, it has been reported that polymerizations in organic solvent may be initiated by crown ether complexes of potassium persulfate.[234-237] Quaternary ammonium persulfates can also serve as useful initiators in organic media.[236,238] The rates of decomposition of both the crown ether complexes and the quaternary ammonium salts appear dramatically greater than those of conventional persulfate salts (K^+, Na^+, NH_4^+) in aqueous solution. The crown ether complex can be used to initiate polymerization at ambient temperature.[234]

In part, the accelerated decomposition might be attributed to the occurrence of induced decomposition and primary radical transfer.[239] Persulfate (**41**) is also known to be a strong oxidant and, in this context, has been widely applied in synthetic organic chemistry.[240] It is established that the rate of disappearance of persulfate in aqueous media is accelerated by the presence of organic compounds[231] and induced decomposition is an integral step in the oxidation of organic substrates (including ethers) by persulfate.[241]

Persulfate (**41**) absorbs only weakly in the UV (ε ca 25 M^{-1} cm^{-1} at 250 nm).[242] Nonetheless, direct photolysis of persulfate ion has been used as a means of generating sulfate radical anion in laboratory studies.[242,243]

Persulfate (**41**) reacts with transition metal ions (*e.g.* Ag^+, Fe^{2+}, Ti^{3+}) according to Scheme 3.42. Various other reductants have been described. These include halide ions, thiols (*e.g.* 2-mercaptoethanol, thioglycolic acid, cysteine, thiourea), bisulfite, thiosulfate, amines (triethanolamine, tetramethylethylenediamine, hydrazine hydrate), ascorbic acid, and solvated electrons (*e.g.* in radiolysis). The mechanisms and the initiating species produced have not been fully elucidated for many systems.[244]

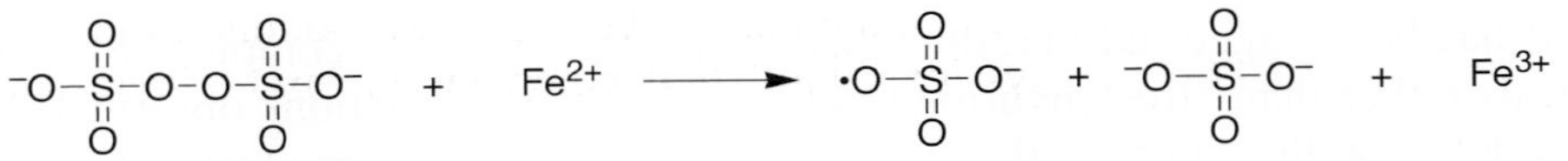

$$^-O-\underset{\underset{O}{\|}}{\overset{\overset{O}{\|}}{S}}-O-O-\underset{\underset{O}{\|}}{\overset{\overset{O}{\|}}{S}}-O^- \;+\; Fe^{2+} \longrightarrow \;\cdot O-\underset{\underset{O}{\|}}{\overset{\overset{O}{\|}}{S}}-O^- \;+\; {}^-O-\underset{\underset{O}{\|}}{\overset{\overset{O}{\|}}{S}}-O^- \;+\; Fe^{3+}$$

41

Scheme 3.42

Various multicomponent systems have also been described. Three component systems in which a second reducing agent (*e.g.* sulfite) acts to recycle the transition metal salt, have the advantage that less metal is used (Scheme 3.43).

$$SO_3^= \;+\; Cu^{2+} \longrightarrow SO_3^{\cdot -} \;+\; Cu^+$$

Scheme 3.43

Redox initiation is commonly employed in aqueous emulsion polymerization. Initiator efficiencies obtained with redox initiation systems in aqueous media are generally low. One of the reasons for this is the susceptibility of the initially formed radicals to undergo further redox chemistry. For example, potential propagating radicals may be oxidized to carbonium ions (Scheme 3.44). The problem is aggravated by the low solubility of the monomers (*e.g.* MMA, S) in the aqueous phase.

3.3.2.6.2 Hydrogen peroxide

Homolytic scission of the O-O bond of hydrogen peroxide may be effected by heat or UV irradiation.[245] The thermal reaction requires relatively high temperatures (>90 °C). Photolytic initiation generally employs 254 nm light. Reactions in organic media require a polar cosolvent (*e.g.* an alcohol).

Hydrogen peroxide also reacts with reducing agents (transition metals, metal complexes, solvated electrons, and some organic reagents) to produce hydroxyl radicals. It reacts with oxidizing agents to give hydroperoxy radicals. The reaction between hydrogen peroxide and transition metal ions in their lower oxidation state is usually represented as the simple process first described by Haber and Weiss (Scheme 3.45).[246] However the mechanism is significantly more complex.

$$H_2O_2 + Fe^{2+} \longrightarrow HO\cdot + {}^-OH + Fe^{3+}$$

Scheme 3.45

It has been suggested that the reactive species are metal complexed hydroxy radicals rather than "free" hydroxyl radicals.[247-250] The reactions observed show dependence on the nature of the metal ion and quite different product distributions can be obtained from reaction of organic substrates with Fe^{2+}-H_2O_2 (Fenton's Reagent) and Ti^{3+}-H_2O_2. However, it is not clear whether these findings reflect the involvement of a different active species or simply the different rates and/or pathways for destruction of the initially formed intermediates.[251] Metal ions in their higher oxidation states (*e.g.* Fe^{3+}) can bring about the destruction of hydrogen peroxide according to Scheme 3.46.

$$H_2O_2 + Fe^{3+} \longrightarrow HOO\cdot + H^+ + Fe^{2+}$$

Scheme 3.46

The Ti^{3+}-H_2O_2 system is preferred over Fenton's reagent because Ti^{4+} is a less powerful oxidizing agent than Fe^{3+} and the above mentioned pathway and other side reactions are therefore of less consequence.[252] Much of the discussion on redox initiation in Section 3.3.2.6.1 is also relevant to hydrogen peroxide.

3.3.3 Multifunctional Initiators

Multifunctional initiators contain two or more radical generating functions within the one molecule. They can be considered in two distinct classes according

Initiation

to whether they undergo concerted (see 3.3.3.1) or non-concerted decomposition (see 3.3.3.2).

3.3.3.1 Concerted decomposition

Multifunctional initiators where the radical generating functions are in appropriate proximity may decompose in a concerted manner or in a way such that the intermediate species can neither be observed nor isolated. Examples of such behavior are peroxyoxalate esters (see 3.3.2.3.1) and α-hydroperoxy diazenes (*e.g.* **31**), derived peroxyesters (**65**)[253,254] and bis- and multi-diazenes such as **66**.[255,256]

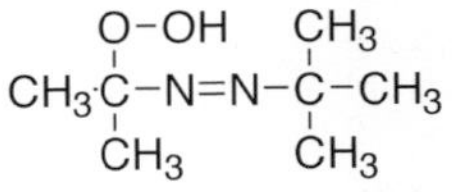

```
    O-OH  CH3              O-O2CR CH3              CH3    CH3 CH3         CH3
CH3-C-N=N-C-CH3        CH3-C-N=N—C-CH3       CH3-C-N=N-C——C-N=N-C-CH3
    CH3   CH3              CH3   CH3              CH3    CH3 CH3         CH3

         31                     65                              66
```

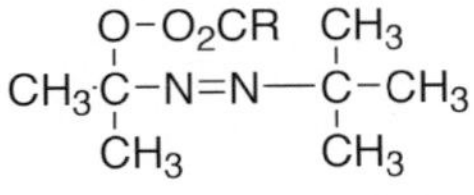

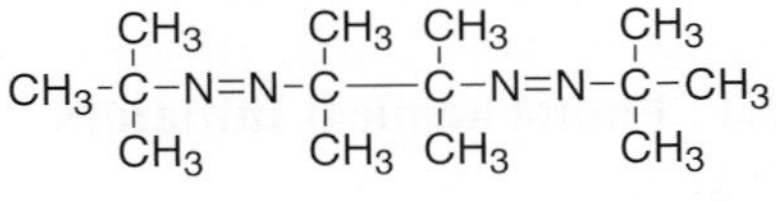

The initiators (**31**) and (**65**) are low temperature sources of alkyl and hydroxy or acyloxy radicals respectively (Scheme 3.47).[253,257,258] The α-hydroperoxy diazenes (*e.g.* **31**) are one of the few convenient sources of hydroxy radicals in organic solution.[253,254]

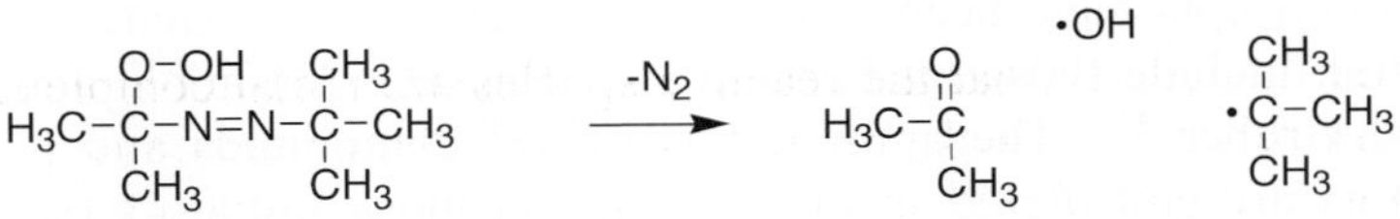

```
     O-OH  CH3                              •OH        CH3
 H3C-C-N=N-C-CH3     -N2      O              •         |
     |     |       ———→   H3C-C             •C-CH3
     CH3   CH3                  |                      |
                               CH3                    CH3
         31
```

Scheme 3.47

It has been reported that the α-hydroperoxy diazenes may undergo induced decomposition either by OH or H transfer.[259]

3.3.3.2 Non-concerted decomposition

Initiators where the radical generating functions are sufficiently remote from each other break-down in a non-concerted fashion. Examples include the azo-peroxide (**68**)[260] and the bis-diazene (**67**).[261] Their chemistry is often understandable in terms of the chemistry of analogous monofunctional initiators.[260] This class also includes the dialkyl peroxyketals (see 3.3.2.4) and hydroperoxyketals (see 3.3.2.5).

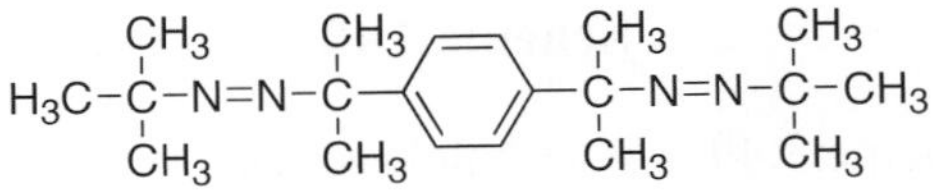

```
              CH3  CH3          CH3   CH3
         H3C-C-N=N-C—⟨   ⟩—C-N=N-C-CH3
              CH3  CH3          CH3   CH3

                          67
```

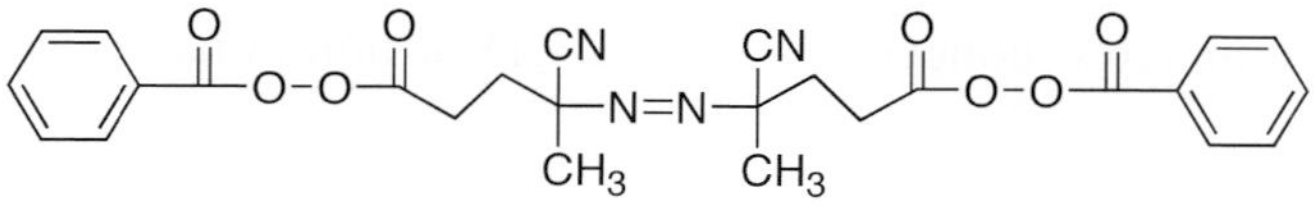

68

The use of initiators such as **68** has been promoted for achieving higher molecular weights or higher conversions in conventional polymerization and for the production of block and graft copolymers. The use and applications of multifunctional initiators in the synthesis of block and graft copolymers is briefly described in Section 7.6.1.

3.3.4 Photochemical Initiators

Photoinitiation is most commonly used in curing or crosslinking processes and in initiating graft copolymerization. Major applications include inks and adhesives and the technologies such as laser direct imaging, holography and stereolithography. Photoinitiation also finds utility in small scale kinetic and mechanistic studies (*e.g.* pulsed laser polymerization, Section 4.5.2). Some approaches to living radical polymerization also make use of photoinitiation (Sections 3.3.4.2, 9.3.2 and 9.3.3).

General concepts have been discussed in Section 3.1.8. General reviews on photoinitiation include those by Pappas,[262-264] Bassi,[265] Mishra[266] and Oster and Yang[267] and Gruber.[268] The applications of azo-compounds and peroxides as photoinitiators are considered in the sections on those initiators (see 3.3.1.1.2, 3.3.2.1.2, & 3.3.2.3.2). References to reviews on specific photoinitiators are given in the appropriate section below.

3.3.4.1 *Aromatic carbonyl compounds*

Many reviews have been written on the photochemistry of aromatic carbonyl compounds[269] and on the use of these compounds as photoinitiators.[270-275] Primary radicals are generated by one of the following processes:

(a) A unimolecular fragmentation involving, most commonly, either α-scission (Scheme 3.48; *e.g.* benzoin ethers, acylphosphine oxides)

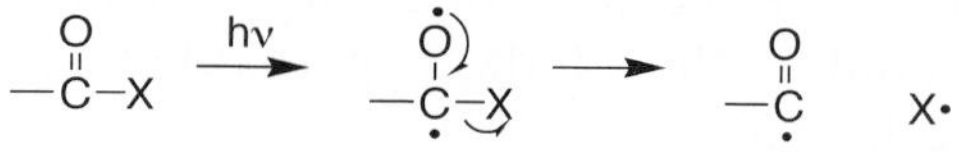

Scheme 3.48

or β-scission (Scheme 3.49; *e.g.* α-haloketones).

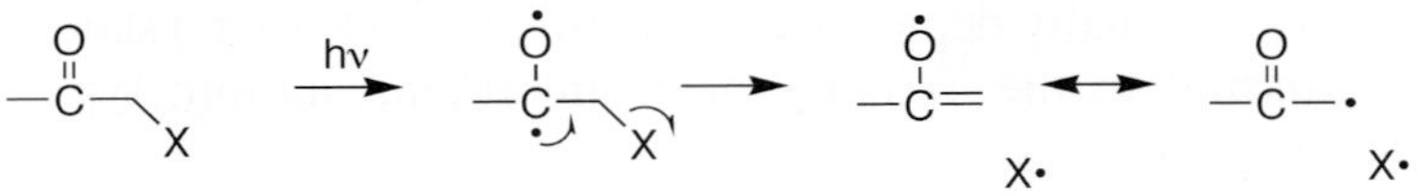

Initiation

Scheme 3.49

Examples of scission of bonds separated from the carbonyl group by a double bond or an aromatic ring are also known. Thus, the benzil monooxime (**69**) undergoes γ-scission (Scheme 3.50) (possibly by consecutive α- then β-scissions).

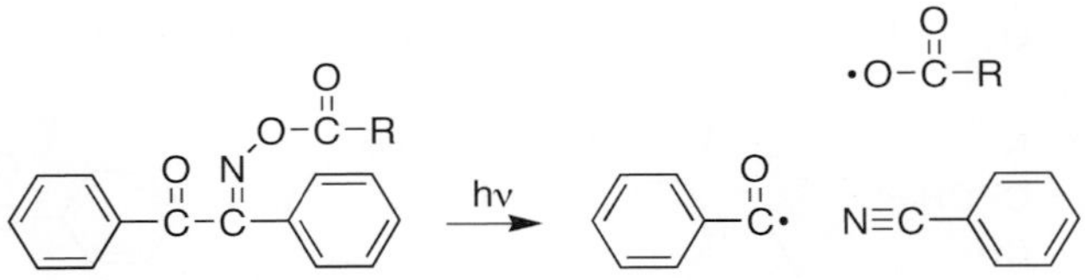

69

Scheme 3.50

(b) A bimolecular process involving direct abstraction of hydrogen from a suitable donor (Scheme 3.51; *e.g.* with hydrocarbons, ethers, alcohols),

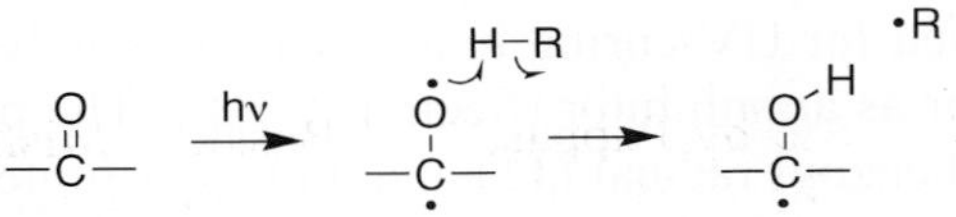

Scheme 3.51

or sequential electron and proton transfer (Scheme 3.52; *e.g.* with amines, thiols). The reaction pathway followed depends on whether H-donors or electron acceptors are present and the relative strengths of the bonds to the α- and β-substituents.

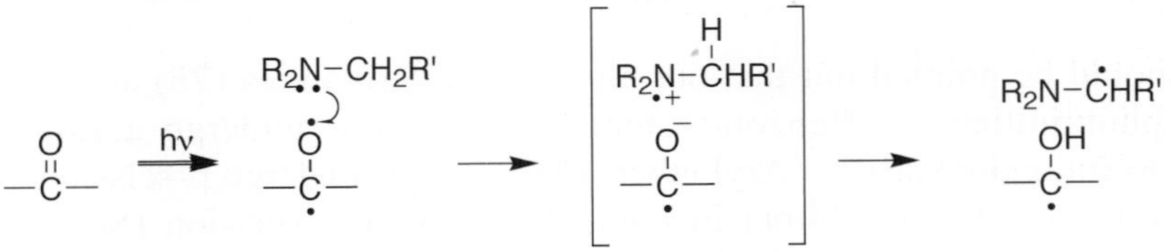

Scheme 3.52

3.3.4.1.1 Benzoin and related compounds

Benzoin and a wide variety of related compounds (*e.g.* **12, 70-74**) have been extensively studied both as initiators of polymerization and in terms of their general photochemistry.[271,273] The acetophenone chromophore absorbs in the near UV (300-400 nm). In the absence of hydrogen atom donors the mechanism of radical generation is usually depicted as excitation to the $S_1(n,\pi^*)$ state followed by intersystem crossing to the $T_1(n,\pi^*)$ state and fragmentation; typically by α-scission (Scheme 3.53).

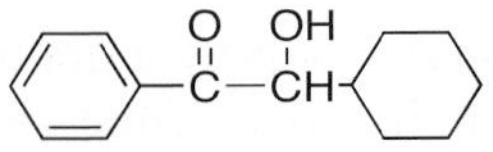

70, 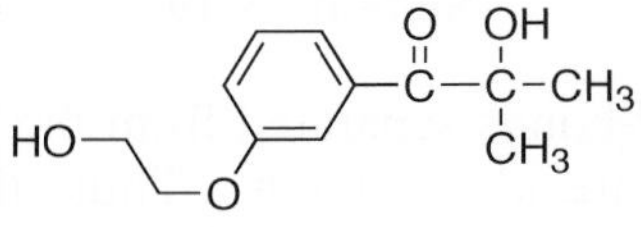 **71**, 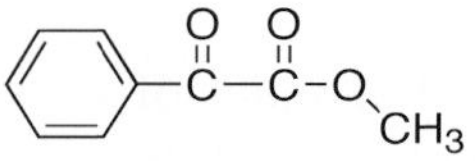 **72**, 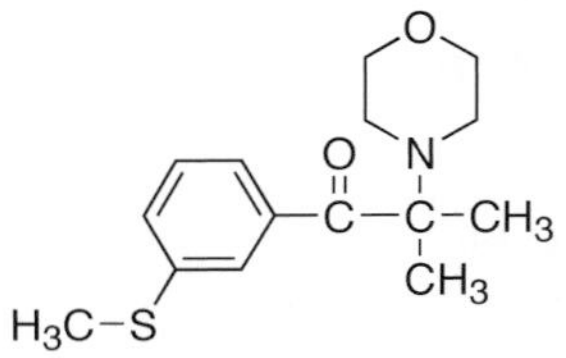 **73**, 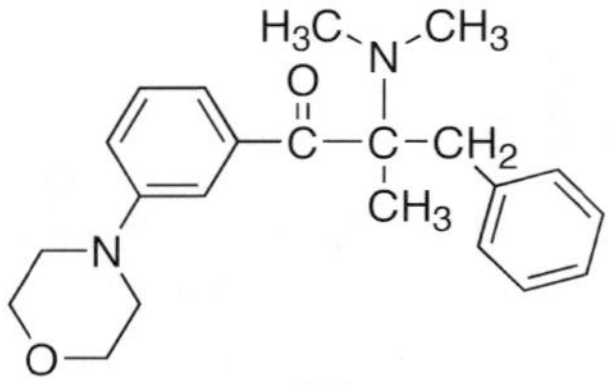 **74**, 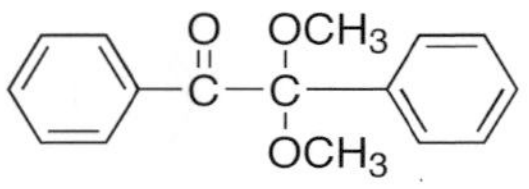 **12**

The benzoin ethers (**75**, R=alkyl; R'=H) and the α-alkyl benzoin derivatives (**75**, R=H, alkyl; R'=alkyl) undergo α-scission with sufficient facility that it is not quenched by oxygen or conventional triplet quenchers.[276] This means that the initiators might be used for UV-curing in air. Unfortunately, it does not mitigate the usual effects of air as an inhibitor (Section 5.3.2). The products of α-scission (Scheme 3.53) are a benzoyl radical (**13**) and an α-substituted benzyl radical (**76**) both of which may, in principle, initiate polymerization.[276,277]

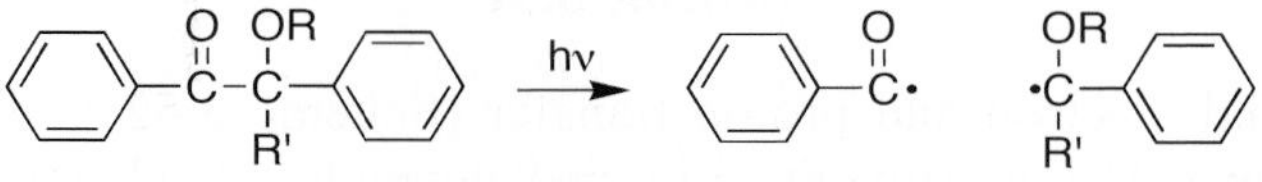

75 → **13** + **76**

Scheme 3.53

It should be pointed out that not all benzoin derivatives (**75**) are suitable for use as photoinitiators. Benzoin esters (**75**, R=acyl) undergo a side reaction leading to furan derivatives. Aryl ethers (**75**, R=aryl) undergo β-scission to give a phenoxy radical (an inhibitor) in competition with α-scission (Scheme 3.54). Benzoin derivatives with α-hydrogens (**75** R'=H) are readily autoxidized and consequently can have poor shelf lives.

There are contradictory reports that phenyl glycolate esters (*e.g.* **72**) undergo photochemistry analogous to the benzoin derivatives. However, a recent study[278] suggests that the α-scission pathway is not significant. Photoinitiation with **72** generally involves hydrogen abstraction from solvent, monomer or other molecules of the initiator to form an initiating species and a relatively unreactive ketyl radical that decays by dimerization.[278]

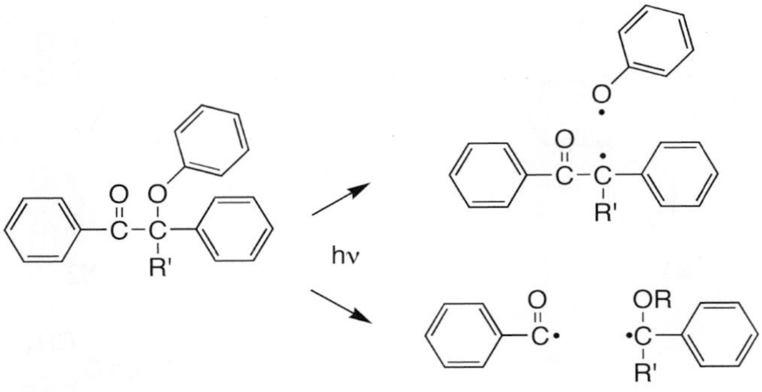

Scheme 3.54

Depending on the nature of the substituent R', the radical **76** (Scheme 3.53) may be slow to add to double bonds and primary radical termination can be a severe complication (see 3.2.9).[39,40,279] The problems associated with formation of a relatively stable radical are mitigated with certain α-alkoxy (**77**) and α-alkanesulfonyl derivatives (**79**).[280] In both cases the substituted benzyl radicals formed by α-scission (**78** and **80** respectively) can themselves undergo a facile fragmentation to form a more reactive radical which is less likely to be involved in primary radical termination (Scheme 3.55, Scheme 3.56).

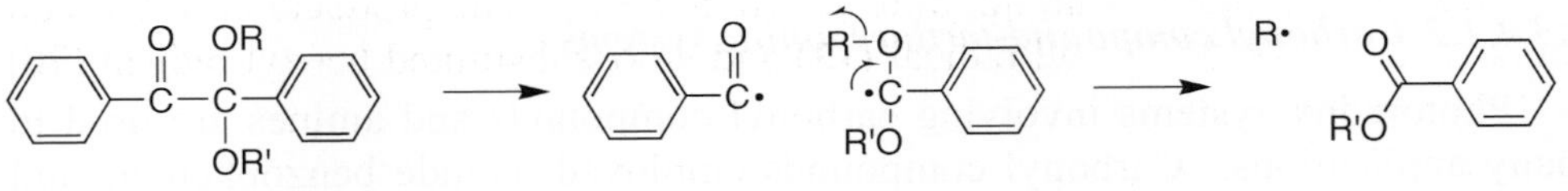

Scheme 3.55

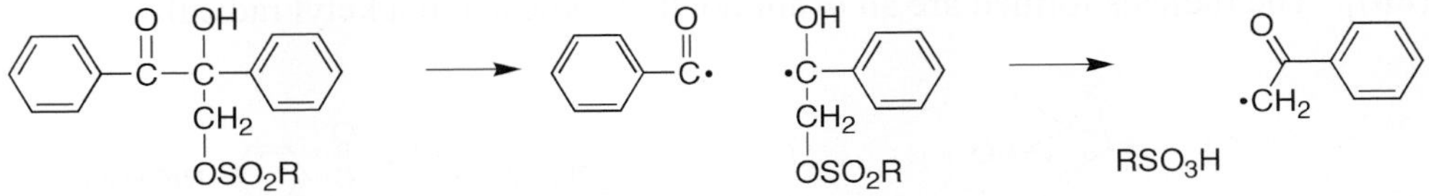

Scheme 3.56

The acyl phosphonates, acyl phosphine oxides and related compounds (*e.g.* **81**, **82**) absorb strongly in the near UV (350-400 nm) and generally decompose by α-scission in a manner analogous to the benzoin derivatives.[281-285] Quantum yields vary from 0.3 to 1.0 depending on structure. The phosphinyl radicals are highly reactive towards unsaturated substrates and appear to have a high specificity for addition *vs* abstraction (see 3.4.3.2).

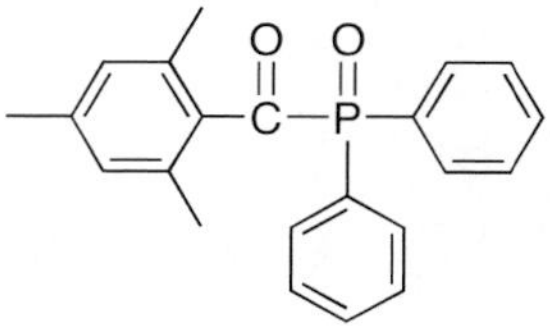

81

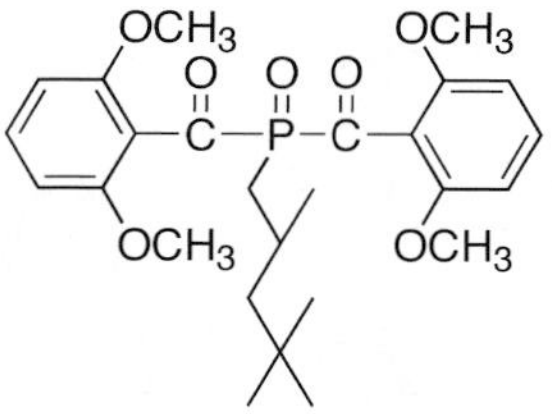

82

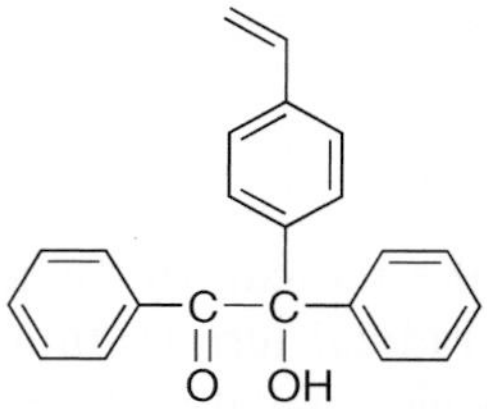

83

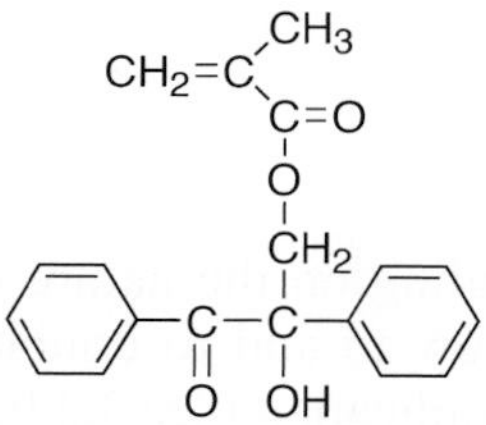

84

Klos et al.[286] described a range of polymerizable benzoin derivatives as photoinitiators (e.g. **83**, **84**). These and other polymeric photoinitiators have advantages as initiators over low molecular weight analogs in circumstances where migratory stability is a problem.[287-289]

3.3.4.1.2 Carbonyl compound-tertiary amine systems

Photoredox systems involving carbonyl compounds and amines are used in many applications. Carbonyl compounds employed include benzophenone and derivatives, α-diketones [e.g. benzil, camphorquinone (**85**),[290,291] 9,10-phenanthrene quinone], and xanthone and coumarin derivatives. The amines are tertiary and must have α-hydrogens [e.g. N,N-dimethylaniline, Michler's ketone (**86**)]. The radicals formed are an α-aminoalkyl radical and a ketyl radical.

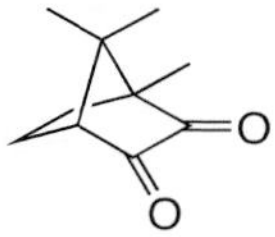

85

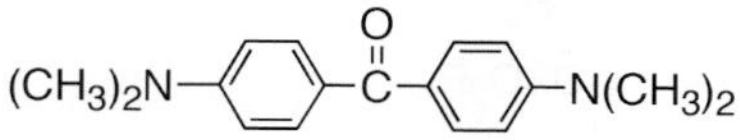

86

The reaction between the photoexcited carbonyl compound and an amine occurs with substantially greater facility than that with most other hydrogen donors. The rate constants for triplet quenching by amines show little dependence on the amine α–C–H bond strength. However, the ability of the amine to release an electron is important.[292] This is in keeping with a mechanism of radical generation which involves initial electron (or charge) transfer from the amine to the photoexcited carbonyl compound. Loss of a proton from the resultant complex (exciplex) results in an α-aminoalkyl radical which initiates polymerization. The

Initiation

concurrently formed ketyl radicals are generally slow to initiate polymerization and consequently primary radical termination is a common complication with these initiator systems.

The electron transfer step is typically fast and efficient. Griller et al.[292] measured absolute rate constants for decay of benzophenone triplet in the presence of aliphatic tertiary amines in benzene as solvent. Values lie in the range $3\text{-}4\times10^9$ $M^{-1}\,s^{-1}$ and quantum yields are close to unity.

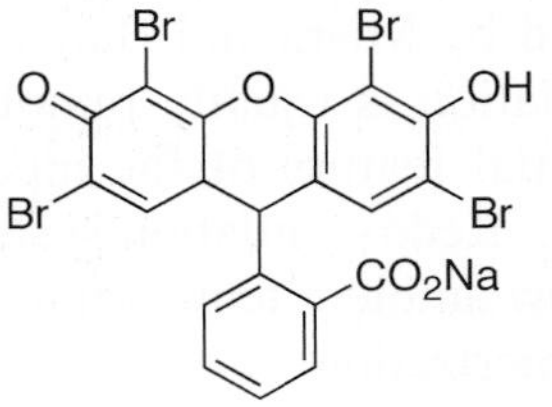

87

Visible light systems comprising a photoreducible dye molecule (e.g. **87**)[293] or an α-diketone (e.g. **85**)[290] and an amine have also been described. The mechanism of radical production is probably similar to that described for the ketone amine systems described above (i.e. electron transfer from the amine to the photoexcited dye molecule and subsequent proton transfer). Ideally, the dye molecule is reduced to a colorless byproduct.

More efficient systems can be constructed by having the two components of the photoredox system in the one molecule.[294]

3.3.4.2 Sulfur compounds

The S-S linkage of disulfides and the C-S linkage of certain sulfides can undergo photoinduced homolysis. The low reactivity of the sulfur-centered radicals in addition or abstraction processes means that primary radical termination can be a complication. The disulfides may also be extremely susceptible to transfer to initiator (C_I for **88** is ca 0.5, Sections 6.2.2.2 and 9.3.2). However, these features are used to advantage when the disulfides are used as initiators in the synthesis of telechelics[295] or in living radical polymerizations.[296] The most common initiators in this context are the dithiuram disulfides (**88**) which are both thermal and photochemical initiators. The corresponding monosulfides [e.g. (**89**)] are thermally stable but can be used as photoinitiators. The chemistry of these initiators is discussed in more detail in Section 9.3.2.

$$\underset{C_2H_5}{\overset{C_2H_5}{N}}-\overset{S}{\overset{\|}{C}}-S-S-\overset{S}{\overset{\|}{C}}-N\underset{C_2H_5}{\overset{C_2H_5}{\diagdown}}$$
88

$$\text{C}_6\text{H}_5-CH_2-S-\overset{S}{\overset{\|}{C}}-N\underset{C_2H_5}{\overset{C_2H_5}{\diagdown}}$$
89

3.3.5 Redox Initiators

The early history of redox initiation has been described by Bacon.[228] The subject has also been reviewed by Misra and Bajpai,[297] Bamford[298] and Sarac.[230] The mechanism of redox initiation is usually bimolecular and involves a single electron transfer as the essential feature of the mechanism that distinguishes it from other initiation processes. Redox initiation systems are in common use when initiation is required at or below ambient temperature and they are frequently used for initiation of emulsion polymerization.

Common components of many redox systems are a peroxide and a transition metal ion or complex. The redox reactions of peroxides are covered in the sections on those compounds. Discussion on specific redox systems can be found in sections on diacyl peroxides (3.3.2.1.5), hydroperoxides (3.3.2.5), persulfate (3.3.2.6.1) and hydrogen peroxide (3.3.2.6.2).

Numerous redox systems have been described which do not involve peroxides including many metal ion free systems such as the photoredox reaction involving carbonyl compounds and tertiary amines (3.3.4.1.2). The following two sections describe redox systems based on the use of metal complexes and simple organic molecules. Various transition metal salts or complexes oxidize or reduce organic substrates by single electron transfer and radicals formed from the organic compound may initiate polymerization.[298] We focus on metal complex-organic halide (3.3.5.1), and ceric ion-organic substrate systems (3.3.5.2).

3.3.5.1 Metal complex-organic halide redox systems

Metal complex-organic halide redox initiation is the basis of ATRP. Further discussion of systems in this context will be found in Section 9.4. The kinetics and mechanism of redox and photoredox systems involving transition metal complexes in conventional radical polymerization have been reviewed by Bamford.[298]

One photoredox system which has seen significant use comprises a transition metal in a low, typically zero, oxidation state (*e.g.* $Mo(CO)_6$, $Re(CO)_6$) and an organic halide. Radical production involves single electron transfer from the metal to the halogen substituent of the alkyl halide which then fragments to form a halide ion and an alkyl radical.[299] Accordingly, the organic fragment of the alkyl halide should be a good electron acceptor, for example, CCl_4, $CHCl_3$, α-haloketones, α-haloesters. The use of polymeric halo compounds allows this chemistry to be used in the preparation of block and graft copolymers (Section 7.6.2).[300,301]

The metal complexes most commonly used in these photoredox systems are manganese and rhenium carbonyls. The proposed mechanism of the photoredox reaction involving $Mn_2(CO)_{10}$ is represented schematically as follows (Scheme 3.57). Quantum yields for photoinitiation are high.[298] Redox couples involving similar metal complexes and an electron deficient monomer (typically a fluoro-olefin) have also been described.[298]

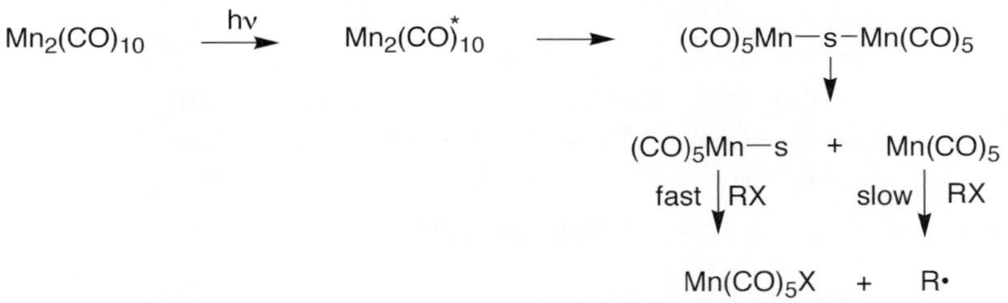

$$Mn_2(CO)_{10} \xrightarrow{h\nu} Mn_2(CO)^*_{10} \longrightarrow (CO)_5Mn-s-Mn(CO)_5$$

$$\downarrow$$

$$(CO)_5Mn-s \ + \ Mn(CO)_5$$

fast $\downarrow$ RX $\quad\quad$ slow $\downarrow$ RX

$$Mn(CO)_5X \ + \ R\cdot$$

s = solvent, monomer or coordinating additive (*e.g.* acetylacetone)

Scheme 3.57

3.3.5.2 *Ceric ion systems*

Ceric ions oxidize various organic substrates and the mechanisms typically involve radical intermediates.[302] When conducted in the presence of a monomer these radicals may initiate polymerization.

The reaction of ceric ion with alcohols,[303] amides and urethanes[304] is thought to involve single electron transfer to the ceric ion and loss of a proton to give the corresponding oxygen- or nitrogen-centered radical (Scheme 3.58). The reaction may involve ligation of cerium. Mechanisms for ceric ion oxidation of alcohols which yield α-hydroxyalkyl radicals as initiating species have also been proposed.

$$XH \ + \ Ce^{4+} \longrightarrow X\cdot \ + \ Ce^{3+} \ + \ H^+$$

Scheme 3.58

Ceric ions react rapidly with 1,2-diols. There is evidence for chelation of cerium and these complexes are likely intermediates in radical generation.[305,306] The overall chemistry may be understood in terms of an intermediate alkoxy radical which undergoes β-scission to give a carbonyl compound and a hydroxyalkyl radical (Scheme 3.59). However, it is also possible that there is concerted electron transfer and bond-cleavage. There is little direct data on the chemical nature of the radical intermediates.

The specificity for reaction with 1,2-diols over mono-ols and 1,3-diols accounts for the finding that oxidation of PVA gives specific cleavage of the 1,2-diol groups present as a consequence of head addition to monomer (see 4.4.3.2). The 1,3-glycol units in PVA also complex ceric ion and, while these complexes decompose only slowly under normal conditions, they undergo a facile photoinduced decomposition to generate initiating species.[307]

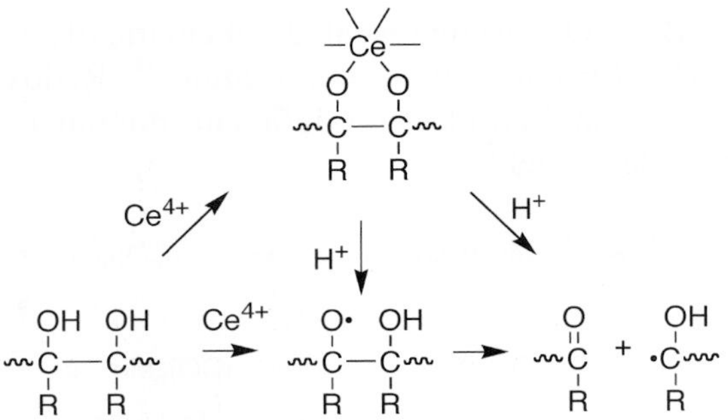

Scheme 3.59

The reaction of ceric ions with polymer-bound functionalities gives polymer-bound radicals. Thus, one of the major applications of ceric ion initiation chemistry has been in grafting onto starch, cellulose,[305,306,308] polyurethanes and other polymers.[304] The advantage of this over conventional initiating systems is that, ideally, no low molecular weight radicals which might give homopolymer contaminant are formed.

The ceric ion also is also known to trap carbon-centered radicals (initiator-derived species, propagating chains) by single electron transfer (Scheme 3.60).

Scheme 3.60

3.3.6 Thermal Initiation

This section describes polymerizations of monomer(s) where the initiating radicals are formed from the monomer(s) by a purely thermal reaction (*i.e.* no other reagents are involved). The adjectives, thermal, self-initiated and spontaneous, are used interchangeably to describe these polymerizations which have been reported for many monomers and monomer combinations. While homopolymerizations of this class typically require above ambient temperatures, copolymerizations involving certain electron-acceptor-electron-donor monomer pairs can occur at or below ambient temperature.

Aspects of thermal initiation have been reviewed by Moad *et al.*,[309] Pryor and Laswell,[310] Kurbatov,[311] and Hall.[312] It is often difficult to establish whether initiation is actually a process involving only the monomer. Trace impurities in the monomers or the reaction vessel may prove to be the actual initiators. Purely thermal homopolymerizations to high molecular weight polymers have only been demonstrated unequivocally for S and its derivatives and MMA. For these and other systems, the identity of the initiating radicals and the mechanisms by which they are formed remain subjects of controversy.

Initiation

3.3.6.1 Styrene homopolymerization

The thermal polymerization of S has a long history.[310] The process was first reported in 1839, though the involvement of radicals was only proved in the 1930s. Carefully purified S undergoes spontaneous polymerization at a rate of *ca* 0.1% per hour at 60 °C and 2% per hour at 100 °C. At 180 °C, 80% conversion of monomer to polymer occurs in approximately 40 minutes. Polymer production is accompanied by the formation of S dimers and trimers which comprise *ca* 2% by weight of total products. The dimer fraction consists largely of *cis*- and *trans*-1,2-diphenylcyclobutanes (**90** and **91**) while the stereoisomeric tetrahydronaphthalenes (**92** and **93**) are the main constituents of the trimer fraction.[313]

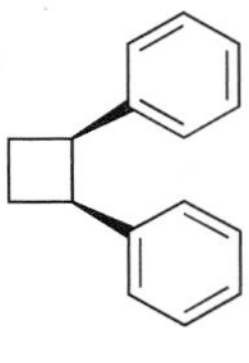

90 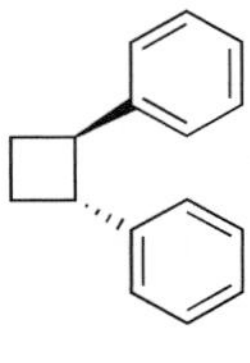   **91** 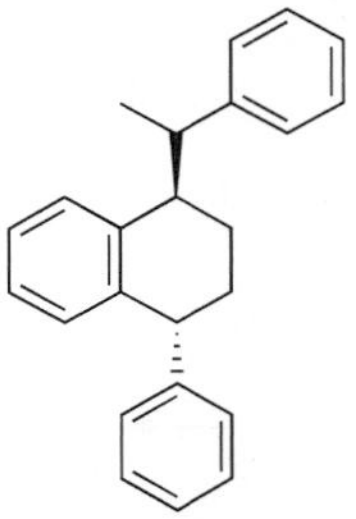   **92** 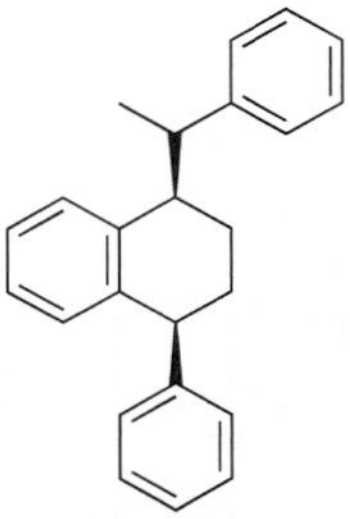   **93**

The two most widely accepted mechanisms for the spontaneous generation of radicals from S are the biradical mechanism (top half of Scheme 3.61) first proposed by Flory[314] and the Mayo[315] or MAH (molecule assisted homolysis) mechanism (lower part of Scheme 3.61).

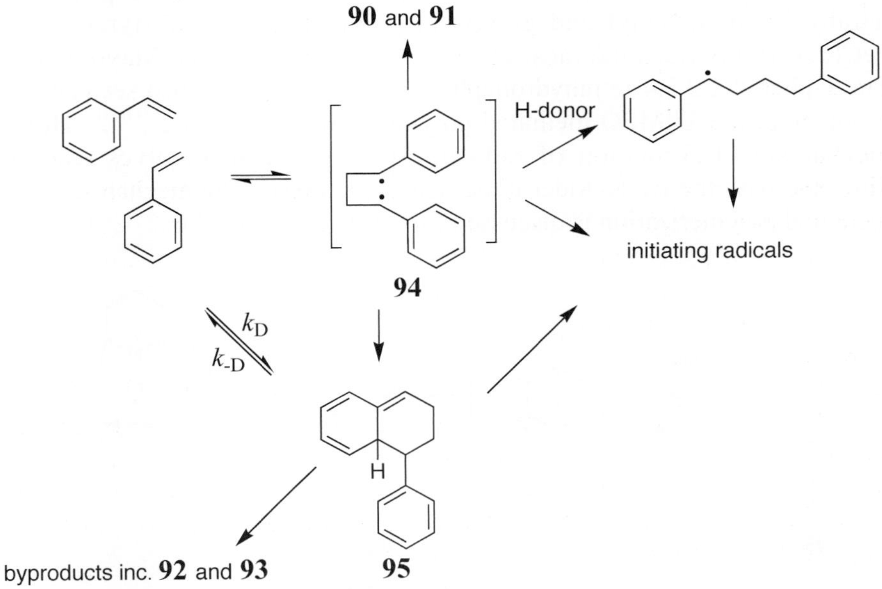

byproducts inc. **92** and **93**   **95**

Scheme 3.61

The Mayo mechanism involves a thermal Diels-Alder reaction between two molecules of S to generate the adduct **95** which donates a hydrogen atom to another molecule of S to give the initiating radicals **96** and **97**. The driving force for the molecule assisted homolysis is provided by formation of an aromatic ring. The Diels-Alder intermediate **95** has never been isolated. However, related compounds have been synthesized and shown to initiate S polymerization.[310]

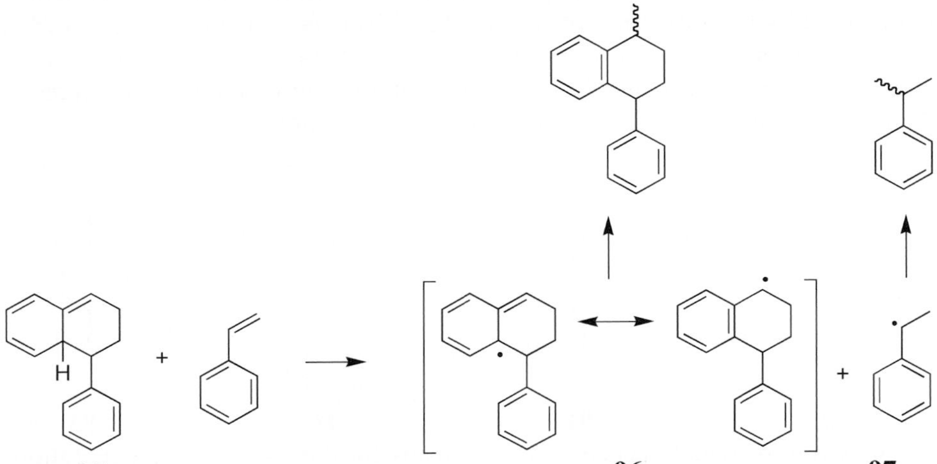

Scheme 3.62

The identification of both phenylethyl and 1-phenyl-1,2,3,4-tetrahydronaphthalenyl end groups in polymerizations of styrene retarded by FeCl$_3$/DMF provides the most compelling evidence for the Mayo mechanism.[316] The 1-phenyl-1,2,3,4-tetrahydronaphthalenyl end group is also seen amongst other products in the TEMPO mediated polymerization of styrene.[317,318] However, the mechanism of formation of radicals **96** in this case involves reaction of the nitroxide with the Diels-Alder dimer (Scheme 3.63). The mechanism of nitroxide mediated polymerization is discussed further in Section 9.3.6.

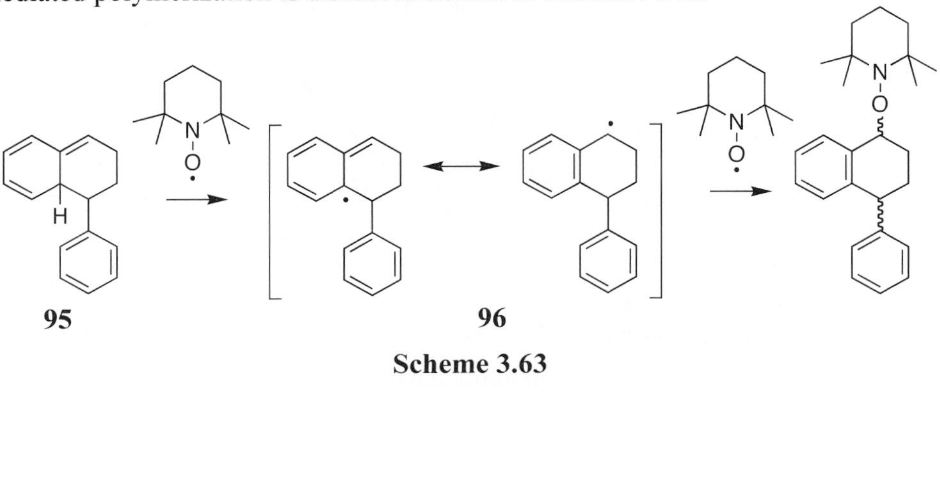

Scheme 3.63

Initiation

The Diels-Alder intermediate (**95**) is also rapidly trapped by aromatization in the presence of acids (Scheme 3.64). Thus, the observation by Buzanowski *et al.*,[319] of dramatically lower rates for S polymerizations carried out in the presence of various acid catalysts, is circumstantial evidence for the Mayo mechanism.

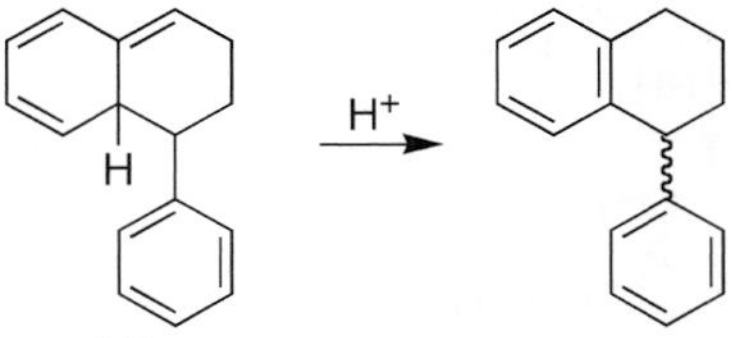

95

Scheme 3.64

Despite the body of evidence in favor of the Mayo mechanism, the formation of diphenylcyclobutanes (**90, 91**) must still be accounted for. It is possible that they arise via the 1,4-diradical **94** and it is also conceivable that this diradical is an intermediate in the formation of the Diels-Alder adduct **95** (Scheme 3.64) and could provide a second (minor) source of initiation. Direct initiation by diradicals is suggested in the thermal polymerization of 2,3,4,5,6-pentafluorostyrene where transfer of a fluorine atom from Diels-Alder dimer to monomer seems highly unlikely (high C-F bond strength) and for derivatives which cannot form a Diels-Alder adduct.

Thermal initiation of styrene has been shown to be third order in monomer. The average rate constants for third order initiation determined by Hui and Hamielec is $k_i = 10^{5.34} e^{(13810/T)}$ $(M^{-2}s^{-1})$.[320] The rate constant for formation of the Mayo dimer determined in trapping experiments with nitroxides (Scheme 3.63) or acid (Scheme 3.64) as $k_D = 10^{4.4} e^{(93500/RT)}$ $(M^{-1}s^{-1})$[321] is substantially higher than is required to account for the rate of initiation. It has been postulated that radical production proceeds mainly through the isomer of **95** in which the phenyl group is axial.[313,322] Both isomers of **95** can give rise to the trimers **92**, possibly by an ene reaction between **95** and S. However, the trimers **92** could also be formed by cage combination of radicals **96** and **97**.

3.3.6.2 Acrylate homopolymerization

Various acrylates, methacrylates and related compounds have been reported to undergo spontaneous polymerization.[310] A complication in studying thermal polymerization of MMA is the difficulty in eliminating impurity initiated polymerization. The monomer is extremely difficult to purify or retain in a "pure" state. These problems have led some to question whether there is any true spontaneous initiation.[323] It is, in any event, clear that the rate of thermal polymerization of MMA is substantially less than that of S at the same temperature (at least 70-fold less at 90 °C).[310,324]

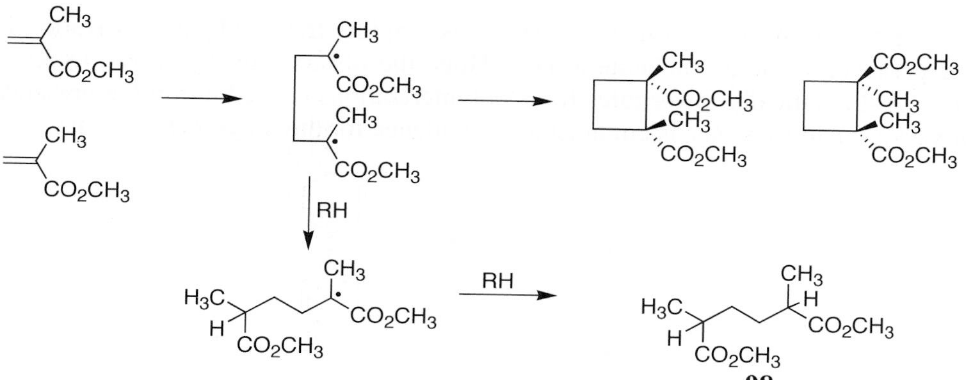

Scheme 3.65

Dimer and trimer byproducts have been isolated from MMA polymerizations and these are suggestive of 1,4-diradical intermediates.[325-328] Lingnau and Meyerhoff[325] found that rates of spontaneous polymerization of MMA were substantially higher in the presence of transfer agents (RH). They were able to isolate the compound (**98**) that might come from trapping of the biradical intermediate (Scheme 3.65).

3.3.6.3 Copolymerization

Monomers that are strong electron donors may undergo spontaneous copolymerization with strong electron acceptor monomers by a radical mechanism. In certain cases homopolymers formed by an ionic mechanism accompany copolymer formation.[312,329]

Examples where radical initiation is believed to be dominant include:

(a) S with MAH,[330,331] AA,[332] AN,[333,334] vinylidene cyanide,[335] or dimethyl 1,1-dicyanoethane-2,2-dicarboxylate.[312]

(b) *p*-Methoxystyrene with trimethyl ethylenetricarboxylate[312] or dimethyl cyanofumarate.[336]

(c) 1,2-Dimethoxyethylene with MAH.[337]

(d) Vinyl sulfides with a range of electrophilic monomers.[338]

Various mechanisms have been proposed to explain the initiation processes. The self-initiated copolymerizations of the monomer pairs S-MMA and S-AN proceed at substantially faster rates than pure S polymerization. For S-AN[334] and S-MAH[331] the mechanism of initiation was proposed to be analogous to that of S homopolymerization (Scheme 3.62) but with acrylonitrile acting as the dienophile in the formation of the Diels-Alder adduct (Scheme 3.66).

Various oligomers formed by Diels-Alder/ene reactions are observed.[333,334] For S-MAH polymerization Sato *et al*.[331] used spin trapping to identify the initiating species. On the other hand, in the case of S-AN copolymerization, the finding that acid catalysts do not affect the rate of polymerization argues against the involvement of this species in the initiation mechanism.[333] Acid catalysts, which effectively trap the Diels-Alder intermediate (**95**) by aromatization (see 3.3.6.1), have been found to lower the rate of thermal S homopolymerization dramatically.[319]

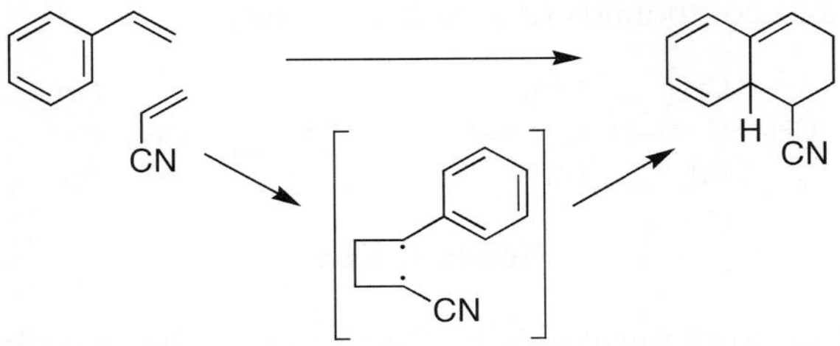

Scheme 3.66

Other postulated mechanisms for spontaneous initiation include electron transfer followed by proton transfer to give two monoradicals,[338] hydrogen atom transfer between a charge-transfer complex and solvent,[330] and formation of a diradical from a charge-transfer complex.[339]

Hall[312,329] has proposed a unifying concept based on tetramethylenes (resonance hybrids of 1,4-diradical and zwitterionic limiting structures - Scheme 3.67) to rationalize all donor-acceptor polymerizations. The predominant character of the tetramethylenes (zwitterionic or diradical) depends on the nature of the substituents.[312,340] However, more evidence is required to prove the more global application of the mechanism.

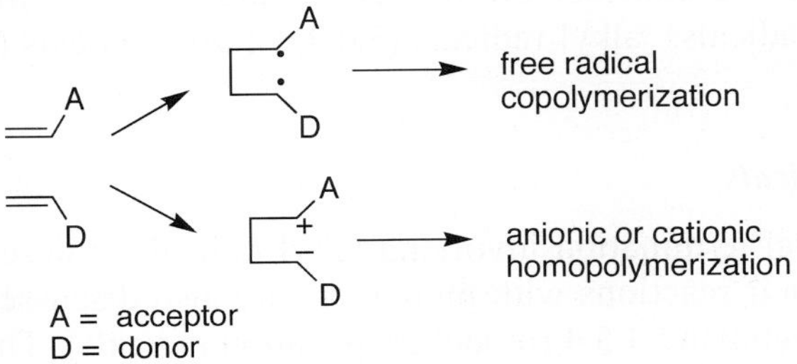

Scheme 3.67

3.4 The Radicals

In this section, the reactions undergone by radicals generated in the initiation or chain transfer processes are detailed. Emphasis is placed on the specificity of radical-monomer reactions and other processes likely to take place in polymerization media under typical polymerization conditions. The various factors important in determining the rate and selectivity of radicals in addition and substitution processes have already been discussed in general terms in Sections 2.3 and 2.4 respectively.

3.4.1 Carbon-Centered Radicals

Carbon-centered radicals are produced as primary radicals in the decomposition of azo-compounds (*e.g.* Scheme 3.68),

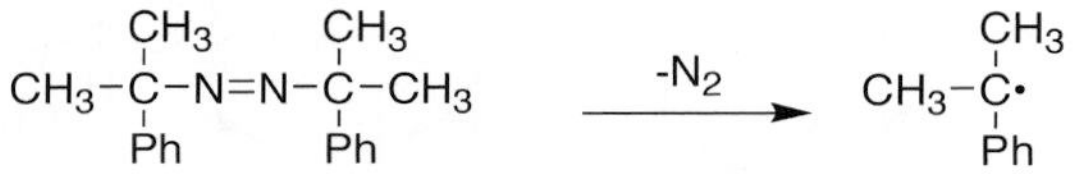

$$CH_3-\underset{\underset{Ph}{|}}{\overset{\overset{CH_3}{|}}{C}}-N=N-\underset{\underset{Ph}{|}}{\overset{\overset{CH_3}{|}}{C}}-CH_3 \quad \xrightarrow{-N_2} \quad CH_3-\underset{\underset{Ph}{|}}{\overset{\overset{CH_3}{|}}{C}}\cdot$$

Scheme 3.68

as secondary radicals from peroxides by β-scission of the initially formed acyloxy or alkoxy radicals (*e.g.* Scheme 3.69),

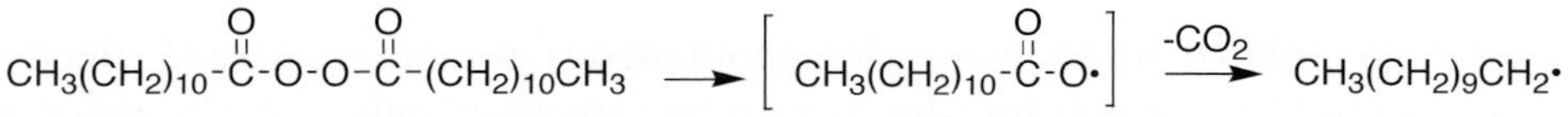

$$CH_3(CH_2)_{10}-\overset{O}{\overset{\|}{C}}-O-O-\overset{O}{\overset{\|}{C}}-(CH_2)_{10}CH_3 \longrightarrow \left[CH_3(CH_2)_{10}-\overset{O}{\overset{\|}{C}}-O\cdot\right] \xrightarrow{-CO_2} CH_3(CH_2)_9CH_2\cdot$$
LPO

Scheme 3.69

and by transfer reactions (*e.g.* Scheme 3.70).

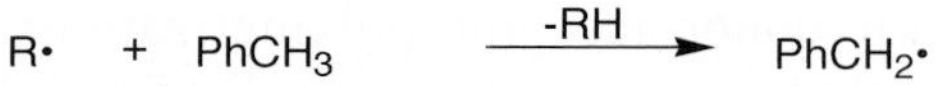

$$R\cdot \;+\; PhCH_3 \quad \xrightarrow{-RH} \quad PhCH_2\cdot$$

Scheme 3.70

In this section we consider the properties and reactions of three classes of carbon-centered radicals: alkyl radicals (3.4.1.1), aryl radicals (3.4.1.2) and acyl radicals (3.4.1.3).

3.4.1.1 Alkyl radicals

Primary radical termination involving alkyl radicals is described in Sections 2.5 and 7.4.3. Their reactions with monomers are also discussed in Sections 2.3 (fundamental aspects) and 4.5.4 (model propagation radicals). Their chemistry has been reviewed by Fischer and Radom,[341] Giese,[342,343] Tedder,[344] Beckwith,[345] Rüchardt,[76] and Tedder and Walton.[346,347]

Alkyl radicals, when considered in relation to heteroatom-centered radicals (*e.g. t*-butoxy, benzoyloxy), show a high degree of chemo- and regiospecificity in their reactions. A discussion of the factors influencing the rate and regiospecificity of addition appears in Section 2.3. Significant amounts of head addition are observed only when addition to the tail-position is sterically inhibited as it is in α,β-disubstituted monomers. For example, with β-alkylacrylates, cyclohexyl radicals give head addition and the proportion can be correlated with the steric size of the β-substituent.[348]

Rate constants for reactions of carbon-centered radicals for the period through 1982 have been compiled by Lorand[349] and Asmus and Bonifacic[350] and for 1982-1992 by Roduner and Crocket.[351] The recent review of Fischer and Radom should also be consulted.[341] Absolute rate constants for reaction with most monomers lie in the range 10^5-10^6 M^{-1} s^{-1}. Rate data for reaction of representative primary, secondary, and tertiary alkyl radicals with various monomers are summarized in Table 3.6.

In the absence of heteroatom containing substituents (*e.g.* halo-, cyano-), at or conjugated with the radical center, carbon-centered radicals have nucleophilic character. Thus, simple alkyl radicals generally show higher reactivity toward electron-deficient monomers (*e.g.* acrylic monomers) than towards electron-rich monomers (*e.g.* VAc, S) – Table 3.6.

Simple alkyl radicals thus seem ideal as initiating species:

(a) They show a high degree of regiospecificity for tail *vs* head addition.
(b) They show a high specificity for addition *vs* abstraction. Rate constants for hydrogen abstraction from monomers and solvents (*e.g.* toluene) are generally much smaller (*ca* 100-fold less) than those for addition to double bonds.
(c) They react rapidly. Side reactions such as primary radical termination are thus minimal.

Thus alkyl radicals do not give unwanted end-group functionality and the kinetics of initiation are comparatively uncomplicated. However, this situation can be perturbed by substitution at or near the radical center.

3.4.1.1.1 α-Cyanoalkyl radicals

Thermal or photochemical decomposition of azonitriles (*e.g.* AIBN) affords α-cyanoalkyl radicals (Scheme 3.71).[29]

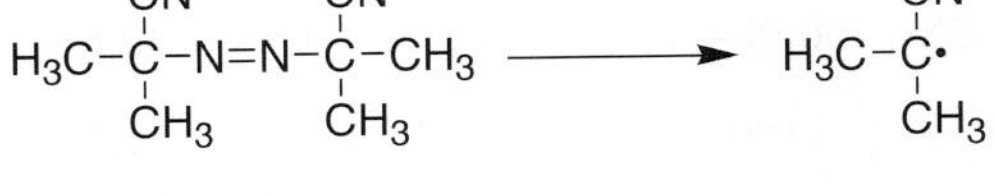

$$H_3C-\underset{CH_3}{\underset{|}{\overset{CN}{\overset{|}{C}}}}-N=N-\underset{CH_3}{\underset{|}{\overset{CN}{\overset{|}{C}}}}-CH_3 \longrightarrow H_3C-\underset{CH_3}{\underset{|}{\overset{CN}{\overset{|}{C}}}}\cdot$$

AIBN

Scheme 3.71

The reactions of cyanoisopropyl radicals with monomers have been widely studied. Methods used include time resolved EPR spectroscopy,[352] radical trapping[353-355] and oligomer[60,356] and polymer end group determination.[60,357-364] Absolute[341] and relative reactivity data obtained using the various methods (Table 3.6) are in broad general agreement.

Table 3.6 Kinetic Data for Reactions of Carbon-Centered Radicals

Radical	Temp °C	$k_S \times 10^{-5}$ M^{-1}s^{-1}	k/k_S AMS	MA	MMA	AA	MAA
$\dot{C}H_2OH$	25	0.23	1.2	31	26	-	-
$\dot{C}H_2Ph$	25	0.011	0.77	0.39	1.9	-	-
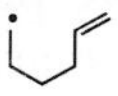 ⌇ (5-hexenyl)	69[b]	3.2[h]	0.6	3.5	-	4.7	4.4
$CH_3\cdot$	65	-	1.16	1.3	1.8	-	-
$CH_3\cdot$	25	2.6	1.2	1.3	1.9	-	-
$\dot{C}H_2C(=O)C(CH_3)_3$	25[b]	19	2.1	0.26	0.68	-	-
$\dot{C}H_2CN$	25	3.8	1.7	0.29	0.63	-	-
$CH_3\dot{C}HPh$	100[f]	-	1.1	1.5	1.9	-	-
$(CH_3)_2\dot{C}H$	60	4.7	-	-	0.3	-	-
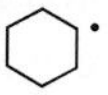 cyclohexyl·	20[i]	-	0.93	6.7	5.0	-	-
$(CH_3)_2\dot{C}OH$	25	7.3	0.27	47	22	-	-
$(CH_3)_3C\cdot$	25	1.3	0.45	8.5	5.1	-	-
$(CH_3)_2\dot{C}CO_2CH_3$	60[f]	-	-	-	0.7	-	-
$(CH_3)_2\dot{C}CO_2C(CH_3)_3$	25	0.055	1.1	0.21	0.67	-	-
$PhCH_2\dot{C}(CO_2Et)_2$	60[a]	-	-	0.0071	-	-	-
$(CH_3)_2\dot{C}CN$	30[d]	-	1.06[c]	-	0.56	-	-
$(CH_3)_2\dot{C}CN$	60[d,f]	0.03	0.95	0.3	0.56	-	-
$(CH_3)_2\dot{C}CN$	100[f]	-	0.87	-	0.56	-	-
$(CH_3)_2\dot{C}CN$	25[f]	0.024	0.96	0.15	0.66	-	-
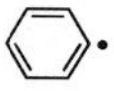 cyclohexadienyl·	25[j]	1100	-	-	1.6	-	-
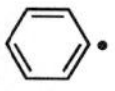 cyclohexadienyl·	60[k]	-	1.31	0.73	1.16	-	-
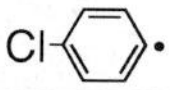 Cl-phenyl·	25[l]	-	-	0.66	1.03	-	-

a In acetic acid. b In acetonitrile. c 40 °C in toluene. d In benzene. Value based on the reported rate constant for addition to MAN[60] and the value of k_{MAN}/k_S shown. e 45 °C. f In toluene. g 30 °C, in ethyl acetate. h Reported values corrected using a more recent rate constant for the 5-hexenyl clock.[365] i In methylene chloride. j In Freon 113. k In carbon tetrachloride. l In aqueous acetone.

Initiation

Table 3.6 (continued)

	k/k_S					Refs.	Radical
AN	MAN	VAc	PAc	VC	PhCH$_3$		
47	29	0.025	0.029	1.2	-	341	$\dot{C}H_2OH$
2.0	6.0	0.013	0.042	-	-	341,366	$\dot{C}H_2Ph$
7.5	-	-	-	-	-	18	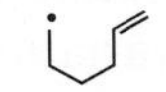 $CH_2=CHCH_2\dot{\,}$
2.2	2.7	0.038	-	-	0.000015	367,368	$CH_3\bullet$
2.4	3.0	0.053	0.046	0.077	-	341	$CH_3\bullet$
0.28	0.49	0.034	0.046	0.037	-	341,369	$\dot{C}H_2C(=O)C(CH_3)_3$
0.29	0.45	0.034	0.031	0.031	-	341	$\dot{C}H_2CN$
5.0	-	-	-	-	-	370,371	$CH_3\dot{C}HPh$
-	-	-	-	-	-	372	$(CH_3)_2\dot{C}H$
24	13	0.12	-	0.016	-	342	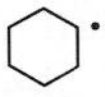 cyclohexyl$\bullet$
205	62	0.010	0.0066	-	-	341	$(CH_3)_2\dot{C}OH$
40	13	0.032	0.013	0.12	-	341,373,374	$(CH_3)_3C\bullet$
-	-	0.03^8	-	-	-	357,375	$(CH_3)_2\dot{C}CO_2CH_3$
0.45	0.81	0.0032	0.011	-	-	341,357,375	$(CH_3)_2\dot{C}CO_2C(CH_3)_3$
0.0088	-	0.0074	-	-	-	376	$PhCH_2\dot{C}(CO_2Et)_2$
-	-	0.02	-	-	-	357,358	$(CH_3)_2\dot{C}CN$
0.44	0.34	0.03	-	0.04^e	-	60,357-363	$(CH_3)_2\dot{C}CN$
-	0.49	0.05	-	-	-	358,362,364	$(CH_3)_2\dot{C}CN$
0.84	0.44	0.017	0.033	0.25	-	341,352	$(CH_3)_2\dot{C}CN$
-	-	-	-	-	0.015	377	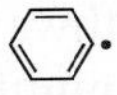 phenyl$\bullet$
1.14	1.30	0.14	0.14	0.18	-	378	 phenyl$\bullet$
0.68	-	-	-	-	-	379	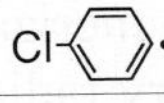 Cl-phenyl$\bullet$

Absolute rate constants for addition reactions of cyanoalkyl radicals are significantly lower than for unsubstituted alkyl radicals falling in the range 10^3-10^4 $M^{-1}s^{-1}$.[341] The relative reactivity data demonstrate that they possess some electrophilic character. The more electron-rich VAc is very much less reactive than the electron-deficient AN or MA. The relative reactivity of styrene and acrylonitrile towards cyanoisopropyl radicals would seem to show a remarkable temperature dependence that must, from the data shown (Table 3.6), be attributed to a variation in the reactivity of acrylonitrile with temperature and/or other conditions.

Cyanoisopropyl radicals generally show a high degree of specificity in reactions with unsaturated substrates. They react with most monomers (*e.g.* S, MMA) exclusively by tail addition (Scheme 3.4). However, Bevington *et al.*[113,362] indicated that cyanoisopropyl radicals give *ca* 10% head addition with VAc at 60 °C and that the proportion of head addition increases with increasing temperature.

α-Cyanoalkyl radicals show relatively little tendency to abstract hydrogen from monomer, solvent, or polymer even in relation to other alkyl radicals.[380] However, these radicals, like other carbon-centered radicals,[28] react with oxygen at diffusion controlled rates (Section 3.2.5). For polymerizations carried out in poorly degassed media, it has been proposed[29,30] that abstraction products, peroxide linkages, and other defect structures may arise through the intermediacy of a alkylperoxy radical (Scheme 3.10).

The α-cyanoalkyl radicals can, in principle, react with substrates either at carbon or at nitrogen (Scheme 3.72). However, reaction at nitrogen to give a ketenimine is usually only observed in cases of reactions with other radicals (Section 5.2.2.1.3) or organometallic reagents.[381] There is a report of a ketenimine structure being formed in a radical substitution reaction (Section 4.4.2). There is as yet no evidence for ketenimine being produced in reactions with monomers or spin traps[7,382] despite several studies aimed specifically at detecting such processes. It is anticipated that reaction through nitrogen would be favored by steric hindrance at the site of attack `and by electron donating substituents on the substrate. It is also likely that addition via the nitrogen will be readily reversible (*i.e.* rapid and irreversible trapping of the initial adduct will be required to observe this pathway).

A number of reports[104,108,383] indicate that primary radical termination can be important during polymerizations initiated by azonitriles. However, for the case of S polymerization initiated by AIBN, NMR end group determination[7] shows that primary radical termination is of little importance except when very high rates of initiation are employed (*e.g.* with high initiator concentrations at high temperatures). Cyanoalkyl radicals give a mixture of combination and disproportionation in their reactions with other radicals (see also Sections 2.5, 7.4.3.2, 7.4.3.3 and 7.4.3.5). This finding is significant for those who use azonitriles as initiators in producing telechelics (Section 7.5.1).

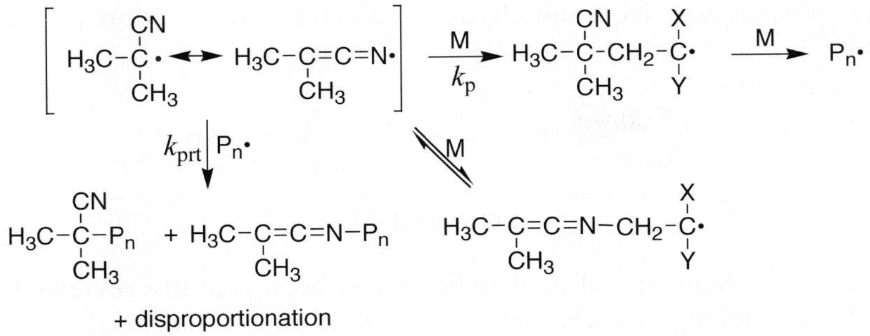

Scheme 3.72

3.4.1.2 Aryl radicals

Aryl radicals are produced in the decomposition of alkylazobenzenes and diazonium salts, and by β-scission of aroyloxy radicals (Scheme 3.73). Aryl radicals have been reported to react by aromatic subsitution (*e.g.* of S^8) or abstract hydrogen (*e.g.* from MMA[10]) in competition with adding to a monomer double bond. However, these processes typically account for ≤1% of the total. The degree of specificity for tail *vs* head addition is also very high. Significant head addition has been observed only where tail addition is retarded by steric factors (*e.g.* methyl crotonate[10] and β-substituted methyl vinyl ketones[379,384]).

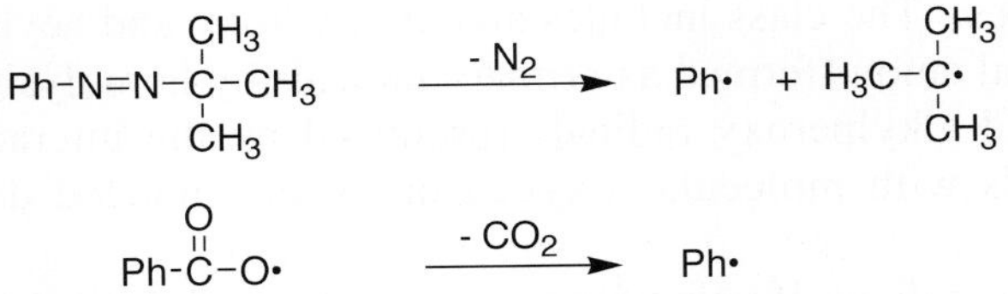

Scheme 3.73

Absolute rate constants for the attack of aryl radicals on a variety of substrates have been reported by Scaiano and Stewart (Ph•)[377] and Citterio *et al.* (*p*-ClPh•).[379,384] The reactions are extremely facile in comparison with additions of other carbon-centered radicals [*e.g.* $k(S) = 1.1 \times 10^8$ M^{-1} s^{-1} at 25 °C].[377] Relative reactivities are available for a wider range of monomers and other substrates (Table 3.6).[377,378,385-387] Phenyl radicals do not show clear cut electrophilic or nucleophilic behavior.

3.4.1.3 Acyl radicals

Phenacyl radicals are produced by photodecomposition of initiators containing the phenone moiety (Scheme 3.74). These initiators include benzoin derivatives and acylphosphine oxides (see 3.3.4.1.1). Acyl radicals can be formed by hydrogen abstraction from aldehydes. Various other sources have been described.[388]

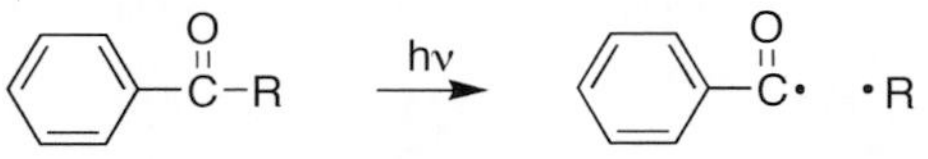

Scheme 3.74

The general chemistry of acyl radicals has been recently reviewed.[388] Acyl radicals have nucleophilic character. Absolute rate constants for substituted phenacyl radical addition to BA have been reported to be in the range $1.3\text{-}5.5\times10^5$ M^{-1} s^{-1} at 25 °C.[285]

Acyl radicals undergo decarbonylation. For aliphatic acyl radicals the rate constant for decarbonylation appears to be correlated with the stability of the alkyl radical formed. Values of the decarbonylation rate constant range from 4 s^{-1} (for $CH_3C(\bullet)O$) to 1.5×10^8 s^{-1} [for $(CH_3)_2C(Ph)C(\bullet)O$] at 298 °C.[388] The loss of carbon monoxide from phenacyl radicals is endothermic and the rate constant is extremely low (ca 10^{-8} s^{-1} at 298 °C).[388] Consequently, the reaction is not observed during polymerization experiments.

3.4.2 Oxygen-Centered Radicals

Oxygen-centered radicals are arguably the most common of initiator-derived species generated during initiation of polymerization and many studies have dealt with these species. The class includes alkoxy, hydroxy and acyloxy radicals and the sulfate radical anion (formed as primary radicals by homolysis of peroxides or hyponitrites) and alkylperoxy radicals (produced by the interaction of carbon-centered radicals with molecular oxygen or by the induced decomposition of hydroperoxides).

There is an excellent, if non critical, compilation of absolute and relative rate data for reactions of oxygen-centered radicals covering the literature through 1982[389] and for 1982-1992.[390] Selected data from these and other sources are summarized in Table 3.7 and Table 3.8. The reactions of oxygen-centered radicals and their use in organic synthesis has been recently reviewed by Hartung et al.[391]

The pathways whereby oxygen-centered radicals interact with monomers show marked dependence on the structure of the radical (Table 3.8). For example, with MMA the proportion of tail addition varies from 66% for t-butoxy to 99% for isopropoxycarbonyloxy radical. The reactions of oxygen-centered radicals are discussed in detail in the following sections.

3.4.2.1 Alkoxy radical

Alkoxy radicals are frequently encountered as initiating species in polymerizations and have been the subject of numerous laboratory studies. Most

Initiation

work has concentrated on the chemistry of *t*-butoxy radical and relatively little attention has been paid to the chemistry of other alkoxy radicals. The chemistry of alkoxy radicals has been the subject of several reviews.[392-395]

Table 3.7 Selected Rate Data for Reactions of Oxygen-Centered Radicals[a]

Radical	Temp °C	$k_S \times 10^{-5}$ $M^{-1}s^{-1}$	\multicolumn{7}{c}{k/k_S}	refs.
			AMS	MA	MMA	AN	MAN	VAc	PhCH$_3$	
(CH$_3$)$_3$CO•	60	~9[b]	1.3	0.06	0.28	0.05	0.12	0.06	0.19	8,12,22,396
(CH$_3$)$_2$(Ph)CO•	60	~30[c]	-	-	0.1	-	-	-	-	72
HO•	60	-	1.2	0.34	0.63	-	-	-	-	397
HO•	25	200000	-	-	1.0	0.27	0.96	-	-	
PhCO$_2$•	24	5100	-	-	-	-	-	-	-	398
PhCO$_2$•	60	-	-	0.05	0.12	<0.05	-	0.36	-	399
PhCO$_2$•	60	-	-	0.02	0.11	0.02	-	0.26	-	10,11,22,400,401

a Overall reactivity. Reaction pathways are shown in Table 3.8.   b Based on rate constant for β-scission as clock reaction[10] and the yield of methyl radical-derived products observed in bulk S polymerization.[8]   c Based on the analysis of Rizzardo *et al.*[72] but assuming a rate constant for β-scission for cumyloxy radical of 1.5×10^6 at 60 °C.

3.4.2.1.1 *t*-Butoxy radicals

The reactions of *t*-butoxy radicals are amongst the most studied of all radical processes.   These radicals are generated by thermal or photochemical decomposition of peroxides or hyponitrites (Scheme 3.75).

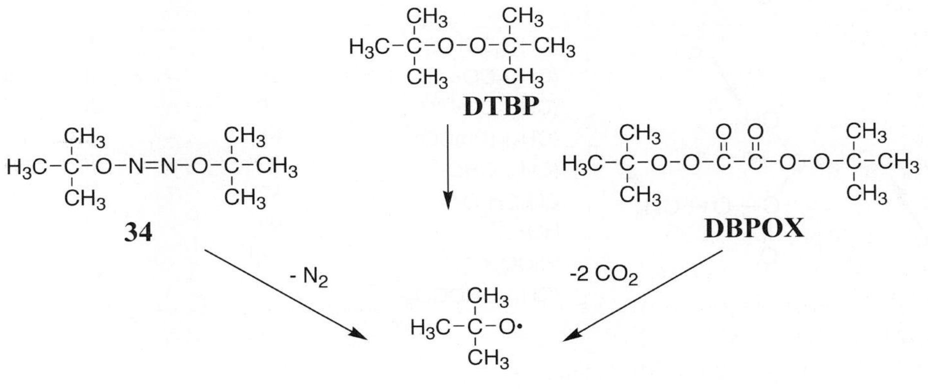

Scheme 3.75

Table 3.8 Specificity Observed in the Reactions of Oxygen-Centered Radicals with Various Monomers at 60 °C

Monomer	Radical	Pathway[a]			
		A	B	C	D
Styrene (CH$_2$=CH-Ph)	(CH$_3$)$_3$CO•[b,c,8]	100	-	-	-
	(CH$_3$)$_2$CH$_2$O•[b,c,402]	100	-	-	-
	CH$_3$CH$_2$O•[b,c,402]	100	-	-	-
	(CH$_3$)$_2$(Ph)CO•[b,c,72]	100	-	-	-
	HO•[b,397]	87	6	-	7
	PhCO$_2$•[c,d,8]	80	6	-	14
	(CH$_3$)$_2$CHOCO$_2$•[b,c,403]	95	-	-	5
α-Methylstyrene (CH$_2$=C(CH$_3$)-Ph)	(CH$_3$)$_3$CO•[b,c,f,396]	85	-	15	-
	HO•[b,397]	83	3	5	9
Methyl acrylate (CH$_2$=CH-C(O)O-CH$_3$)	(CH$_3$)$_3$CO•[b,c,10]	83	2	-	15
	HO•[b,397]	80	17	-	3
	PhCO$_2$•[c,d,10]	84	16	-	-
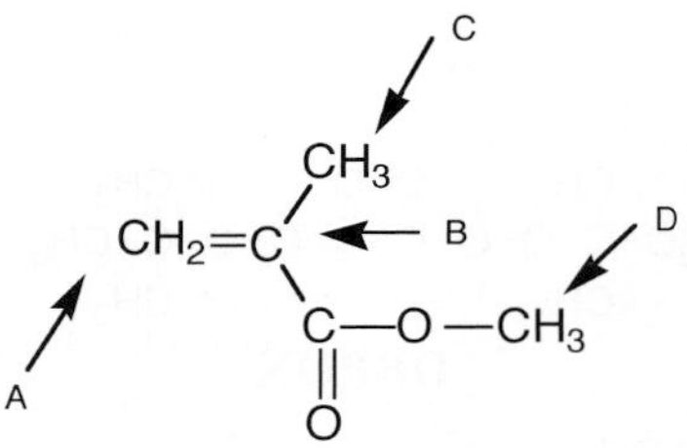 Methyl methacrylate (CH$_2$=C(CH$_3$)-C(O)O-CH$_3$)	(CH$_3$)$_3$CCH$_2$C•	~66[g]	-	~33[g]	[h]
	(CH$_3$)$_2$CO•	66	-	30	4
	(CH$_3$)$_3$CO•[b,c,f,12]	70	-	26	3
	(CH$_3$)$_2$(Ph)CO•[b,c,72]	88	-	12	-
	(CH$_3$)$_2$CHO•[b,c,123]	92	6	8	-
	CH$_3$CH$_2$O•[b,c,402]	87	7	5	2
	HO•[b,397]	93	-	<1[e]	-
	PhCO$_2$•[c,d,10]	>99	-	-	-
	(CH$_3$)$_2$CHOCO$_2$•[b,188]				

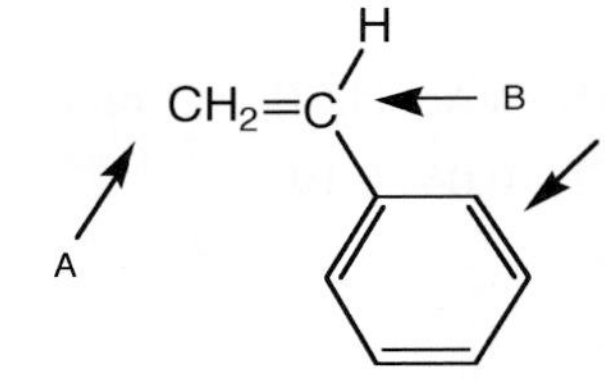

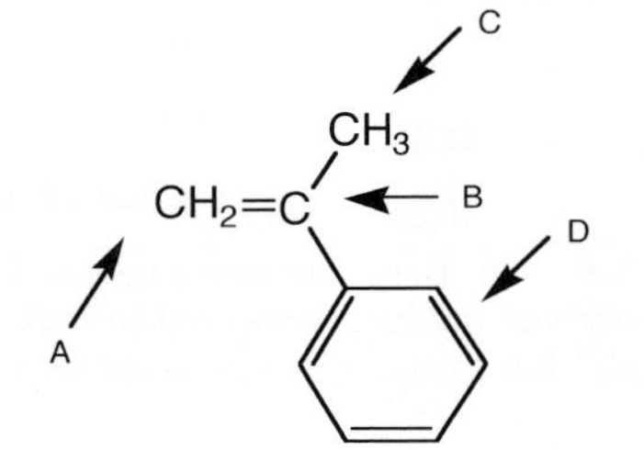

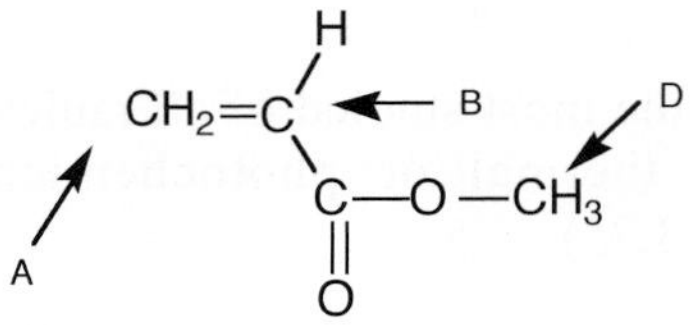

Initiation

Table 3.8 (continued)

Monomer	Radical	Pathway[a]			
		A	B	C	D
acrylonitrile (CH$_2$=CH-CN), pathways A (tail), B (head)	(CH$_3$)$_3$CO•[b,c,404] PhCO$_2$•[c,d,401]	100 98	- 2	- -	- -
methacrylonitrile (CH$_2$=C(CH$_3$)-CN), pathways A (tail), B (head), C (methyl)	(CH$_3$)$_3$CO•[b,c,405]	74	-	26	-
vinyl acetate (CH$_2$=CH-O-C(=O)-CH$_3$), pathways A (tail), B (head), D (acetate methyl)	(CH$_3$)$_3$CO•[b,c,404] PhCO$_2$•[c,d,401]	79 76	15 24	- -	6 -
isopropenyl acetate (CH$_2$=C(CH$_3$)-O-C(=O)-CH$_3$), pathways A (tail), B (head), C (methyl), D (acetate methyl)	(CH$_3$)$_3$CO•[b,c,f,396]	48	-	48	4

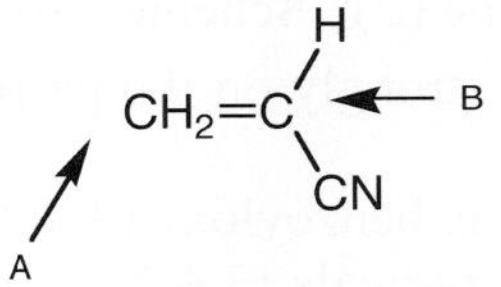

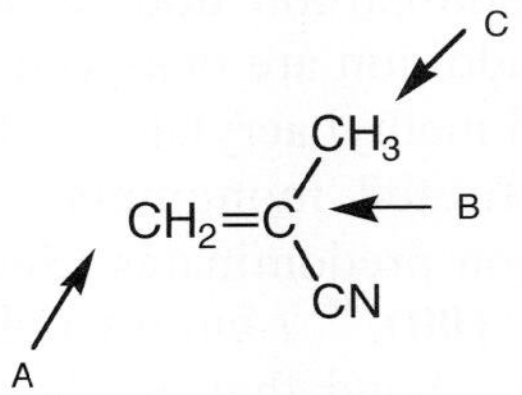

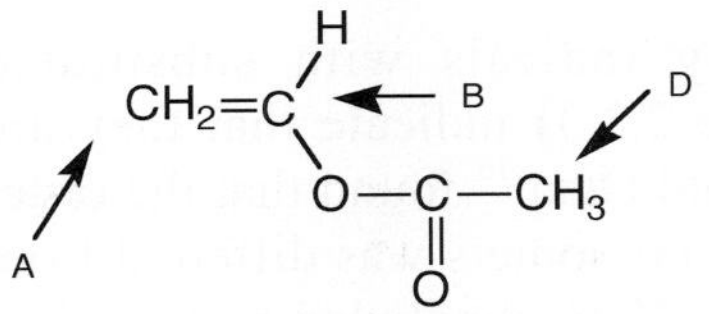

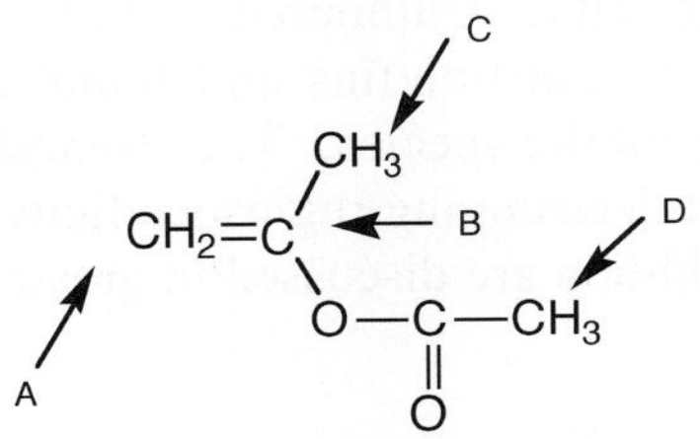

a Relative yields of products formed by pathway indicated. All data rounded to nearest 1%. A dash indicates that the product was not detected.   b In bulk monomer.   c Yields have been normalized to exclude β-scission products.   d In 50% v/v acetone/monomer.   e Total abstraction by benzoyloxy and phenyl radicals.   f Addition:abstraction ratio shows solvent dependence.[21,396]   g Values approximate. Radical gives mainly β-scission and 1,5 H atom transfer   h Product detected

In a polymerization reaction they may:
(a) Initiate a chain by adding to the double bond of a monomer.
(b) Abstract a hydrogen atom from the monomer, solvent, or another component of the reaction mixture to afford a new radical species and *t*-butanol (primary radical transfer).
(c) Undergo β-scission to give methyl radicals and acetone (*e.g.* Scheme 3.6).

The relative importance of these processes depends strongly on the particular monomer(s) and the reaction conditions.

In contrast to most other oxygen-centered radicals [*e.g.* benzoyloxy (3.4.2.2.1), hydroxy (3.4.2.3)], *t*-butoxy radicals and other *t*-alkoxy radicals (3.4.2.1.2) show relatively high regiospecificity in reactions with carbon-carbon double bonds (Table 3.8). Nonetheless, significant amounts of head addition are observed with the halo-olefins,[24,406] simple alkenes,[407] vinyl acetate and methyl acrylate.[404] Head addition is generally not observed with 1,1-disubstituted monomers. The exception is vinylidene fluoride[24,406] where head addition predominates (Section 2.4). With allyl methacrylate (**99**)[408] and allyl acrylate (**100**),[409] *t*-butoxy radicals give substantially more addition to the acrylate double bond than to the allyl double bond (see Figure 3.4).

Studies of the relative reactivity of *t*-butoxy radicals with substituted styrenes,[410] toluenes[411,412] and other substrates (see 2.3.3) indicate that they are slightly electrophilic in character. However, Sato and Otsu [13] found that the order of reactivity of *t*-butoxy radicals towards a series of monomers was different from that of the more electrophilic benzoyloxy radicals. They concluded that product radical stability was important in determining reactivity. Cuthbertson *et al.*[406] examined the reactions of *t*-butoxy radicals toward fluoro-olefins and found a pattern of reactivities more characteristic of a nucleophilic species. The strength of the bond being formed plays an important role in determining regiospecificity. The factors influencing the specificity and rate of addition are discussed in greater detail in Section 2.3.2.

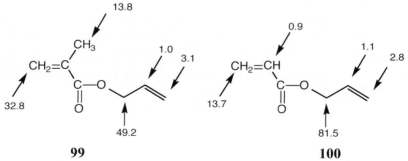

Figure 3.4 Relative reactivity of indicated site towards *t*-butoxy radicals for allyl methacrylate (**99**) and allyl acrylate (**100**)

Initiation

Many investigations[12,13,21,396,413,414] have shown that the reaction of *t*-butoxy radicals with monomers bearing sp^3 hydrogens invariably produces a mixture of initiating radicals arising from hydrogen abstraction and addition (Table 3.8). Simple alkenes (*e.g.* butenes),[407] vinyl ethers[415] and higher acrylates (*e.g.* BMA - Figure 3.5)[12,416] may give predominantly abstraction. The specificity seen in attack on the ester group has been attributed to polar factors.[416] The positions α- and β- to the ester oxygen are strongly deactivated towards attack by *t*-butoxy radicals.

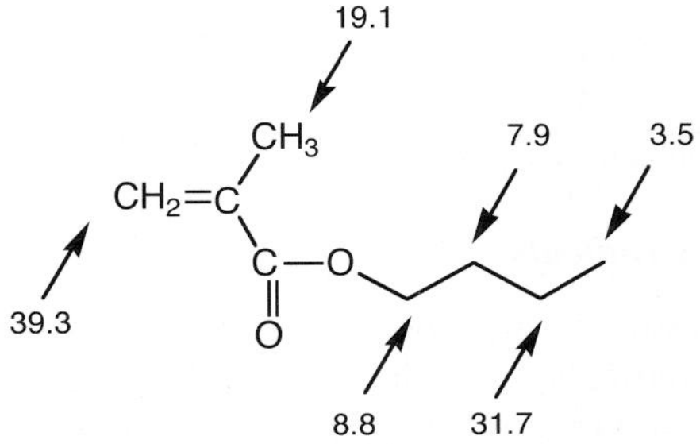

```
              19.1
               ↓
              CH3       7.9    3.5
    CH2=C              ↓      ↓
     ↑     C—O
    39.3   ‖
           O          ↑      ↑
                     8.8    31.7
              BMA
```

Figure 3.5 Relative reactivity of indicated site towards *t*-butoxy radicals.

t-Butoxy radicals also undergo unimolecular fragmentation to produce acetone and methyl radicals (Scheme 3.6). Significant amounts of the β-scission products are obtained in the presence of even the most reactive monomers (*e.g.* S[8]). The reactions of methyl radicals have been discussed above (see 3.4.1.1).

The relative amounts of double bond addition, hydrogen abstraction and β-scission observed are dependent on the reactivity and concentration of the particular monomer(s) employed and the reaction conditions. Higher reaction temperatures are reported to favor abstraction over addition in the reaction of *t*-butoxy radicals with AMS[413] and cyclopentadiene.[417] However, the opposite trend is seen with isobutylene.[23,24]

Pioneering work by Walling[394] established that the specificity shown by *t*-butoxy radical is solvent dependent. Work[21,22,396] on the reactions of *t*-butoxy radicals with a series of α-methylvinyl monomers has shown that polar and aromatic solvents favor abstraction over addition, and β-scission over either addition or abstraction. Recently, Weber and Fischer[418] and Tsentalovich *et al.*[419] reported absolute rate constants for β-scission of *t*-butoxy radicals in various solvents. These studies indicate that β-scission is strongly solvent dependent while abstraction is relatively insensitive to solvent.

Table 3.9. Kinetic Data for Reactions of *t*-Butoxy Radicals in Various Solvents.[418]

solvent	β-Scission			Abstraction from cyclohexane		
	k_β $s^{-1\ b}$	E_a kJ mol^{-1}	$\log(A/s^{-1})$	$k_{abs} \times 10^{-5}$ $M^{-1}s^{-1\ b}$	E_a kJ mol^{-1}	$\log(A/M^{-1}s^{-1})$
Fiigen 113[a]	8050	52.7	13.2	8.3	11.9	8.0
DTBP	12000	50.5	12.9	-	-	-
C$_6$H$_6$	20300	48.7	12.8	9.6	12.1	8.5
C$_6$H$_5$F	21400	47.5	12.7	9.8	14.6	8.2

a 1,1,2-trichloro-1,2,2-trifluoroethane. b Temperature 298 K.

3.4.2.1.2 Other t-alkoxy radicals

Various *t*-alkoxy radicals may be formed by processes analogous to those described for *t*-butoxy radicals. The data available suggest that their propensities for addition *vs* abstraction are similar.[72] However, rate constants for β-scission of *t*-alkoxy radicals show marked dependence on the nature of substituents α to oxygen (Figure 3.6).[210,420,421] Polar, steric and thermodynamic factors are all thought to play a part in favoring this trend.[393]

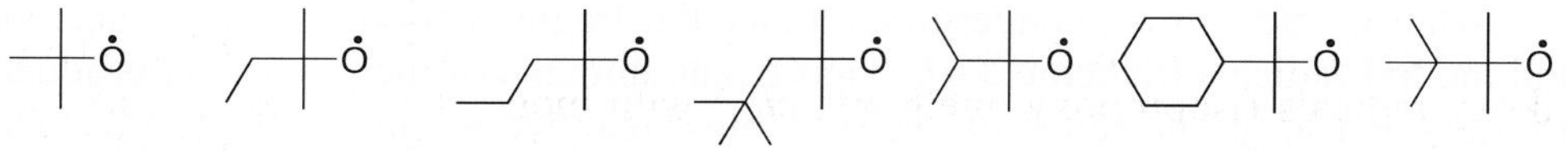

| 1 | 252 | 254 | 2670 | 3300 | 28000 | 86400 |

Figure 3.6 Relative rate constants for β-scission of *t*-alkoxy radicals at 60 °C.[421]

Thus, even if *t*-amyloxy radicals (**101**) show similar specificity for addition *vs* abstraction to *t*-butoxy radicals, abstraction will be of lesser importance.[422,423] The reason is that most *t*-amyloxy radicals do not react directly with monomer. They undergo β-scission and initiation is mainly by ethyl radicals. Ethyl radicals are much more selective and give addition rather than abstraction. This behavior has led to *t*-amyl peroxides and peroxyesters being promoted as superior to the corresponding *t*-butyl derivatives as polymerization initiators.[423]

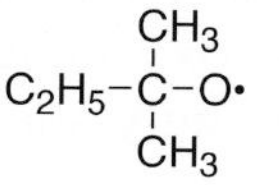

$$C_2H_5-\underset{CH_3}{\overset{CH_3}{\underset{|}{\overset{|}{C}}}}-O\bullet \qquad Ph-\underset{CH_3}{\overset{CH_3}{\underset{|}{\overset{|}{C}}}}-O\bullet$$

101 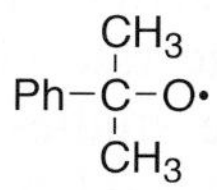 **102**

1,5-H atom transfer is another important unimolecular pathway for *t*-alkoxy radicals that have a suitably disposed hydrogen atom (Scheme 3.76).[421,424,425]

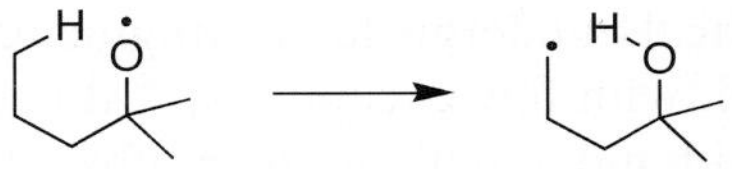

Scheme 3.76

The rate constant of β-scission of cumyloxy radicals (**102**) is also significantly greater than that for *t*-butoxy radicals.[26,420] β-Scission gives exclusively acetophenone and methyl radicals. For the case of S or MMA polymerization initiated by cumyloxy radicals at 60 °C, the proportion of methyl radical initiation is six-fold greater than is seen with *t*-butoxy radicals.[72] The absolute rate constant for β-scission of **102** has been shown to be solvent dependent. The absolute rate constant (2.6×10^5 s^{-1} at 30 °C in CCl$_4$) increases *ca* seven-fold over the series CCl$_4$, C$_6$H$_6$, C$_6$H$_5$Cl, (CH$_3$)$_3$COH, CH$_3$CN, CH$_3$COOH.[426] The rate constant for abstraction from cyclohexane remains at $1.2 \pm 0.1 \times 10^6$ M^{-1} s^{-1} in all solvents. For cumyloxy, and other *t*-alkoxy radicals, β-scission is much more sensitive to temperature than either addition or abstraction (Figure 3.1) such that at high temperatures it is likely to be the major process even in the presence of very reactive substrates.

3.4.2.1.3 Primary and secondary alkoxy radicals

Relatively few studies have dealt with the reactions of primary and secondary alkoxy radicals (isopropoxy, methoxy, *etc.*) with monomers. These radicals are conveniently generated from the corresponding hyponitrites (Scheme 3.77).[123,402]

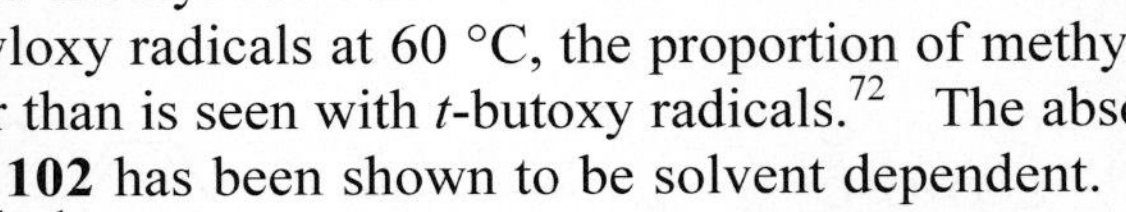

Scheme 3.77

Primary and secondary alkoxy radicals generally show a reduced tendency to abstract hydrogen or to undergo β-scission when compared to the corresponding *t*-alkoxy radical.[123,402] This has been correlated with the lesser nucleophilicity of these radicals.[427]

It has been suggested[123,402] that primary and secondary alkoxy radicals may react with S by donation of a hydrogen atom to the monomer and production of an aldehyde.

3.4.2.2 Acyloxy and alkoxycarbonyloxy radicals

Aroyloxy radicals are formed by thermal or photochemical decomposition of diaroyl peroxides (see 3.3.2.1) and aromatic peroxyesters (3.3.2.3) (Scheme 3.78); alkoxycarbonyloxy radicals are similarly produced from peroxydicarbonates (3.3.2.2).

Aliphatic acyloxy radicals undergo facile fragmentation with loss of carbon dioxide (Scheme 3.69) and, with few exceptions,[428] do not have sufficient lifetime to enable direct reaction with monomers or other substrates. The rate constants for decarboxylation of aliphatic acyloxy radicals are in the range $1\text{-}10\times10^9$ M^{-1} s^{-1} at 20 °C.[429] Ester end groups in polymers produced with aliphatic diacyl peroxides as initiators most likely arise by transfer to initiator (see 3.3.2.1.4). The chemistry of the carbon-centered radicals formed by β-scission of acyloxy radicals is discussed above (see 3.4.1).

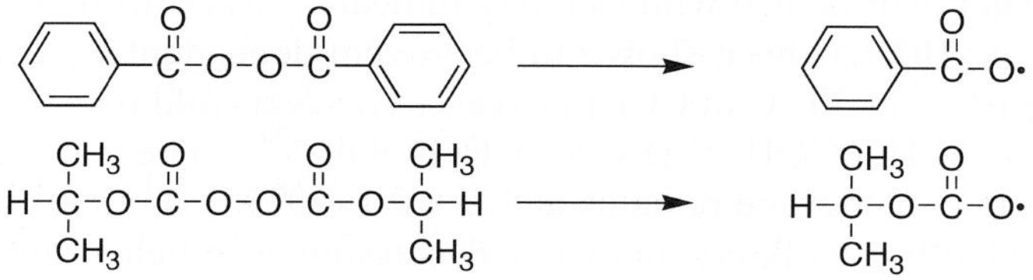

Scheme 3.78

3.4.2.2.1 Benzoyloxy radicals

Benzoyloxy radicals are electrophilic and show higher reactivity towards electron-rich (*e.g.* S, VAc) than electron-deficient (*e.g.* MMA, AN) monomers (Table 3.7).[401,430] Product studies on the reactions of benzoyloxy radicals with simple olefins and monomers[8,10,168,399,401,430-433] show that they have remarkably poor regiospecificity when adding to carbon-carbon double bonds. Their reactions invariably give a mixture of products from head addition and tail addition (Scheme 3.4 and Table 3.8).[8,10,401,433] They also display a marked propensity for aromatic substitution.[8,41,398] On the other hand, compared with alkoxy radicals, they show little tendency to abstract hydrogen.[10]

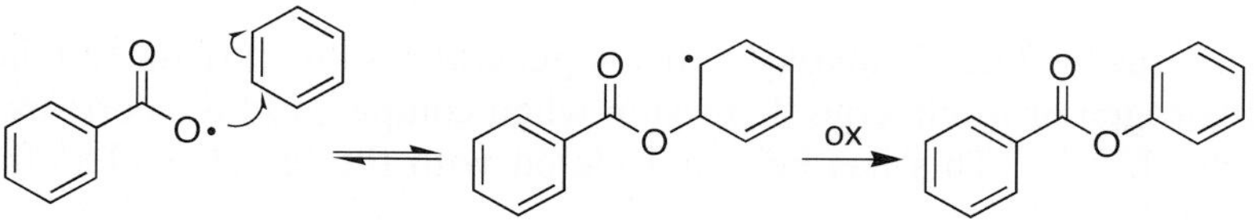

Scheme 3.79

Additions of benzoyloxy radicals to double bonds[434,435] and aromatic rings (Scheme 3.79)[148] are potentially reversible. For double bond addition, the rate constant for the reverse fragmentation step is slow ($k=10^2\text{-}10^3$ s^{-1} at 25 °C) with respect to the rate of propagation during polymerizations. Thus, double bond addition is effectively irreversible. However, for aromatic substrates, the rate of the reverse process is extremely fast. While the aromatic substitution products may be trapped with efficient scavenging agents (*e.g.* a nitroxide[8,41] or a transition metal[169]), they are generally not observed under polymerization conditions.[9] A different situation may pertain when redox initiation is used, as the oxidants employed may be effective radical traps. A small proportion of aromatic benzoate residues can be detected in high conversion PS prepared with benzoyl peroxide. However, it is likely that these arise through attack on PS rather than S.[9,154]

The rate of β-scission of benzoyloxy radicals is such that in most polymerizations initiated by these radicals both phenyl and benzoyloxy end groups will be formed (Scheme 3.4). A reliable value for the rate constant for β-scission would enable the absolute rates of initiation by benzoyloxy radical to be estimated. Various values for the rate constant for β-scission have appeared. Many of the early estimates are low. The activation parameters (in CCl_4 solvent) determined by Chateauneuf et al.[398] are $\log_{10} A = 12.6$ and $E_a = -35.97$ kJ mol^{-1} which corresponds to a rate constant of 9×10^6 s^{-1} at 60 °C.

The rate constant for β-scission is dependent on ring substituents. Rate constants for radicals X-$C_6H_4CO_2$• are reported to increase in the series where X is p-F≤p-CH_3O<p-CH_3~p-Cl<H<m-Cl.[398] There is qualitative evidence that the relative rates for β-scission and addition are insensitive to solvent changes. For benzoyloxy radicals, similar relative reactivities are obtained from direct competition experiments[10] as from studies on individual monomers when β-scission is used as a clock reaction.[399,401]

The rate constants for benzoyloxy and phenyl radicals adding to monomer are high (> 10^7 M^{-1} s^1 for S at 60 °C - Table 3.7). In these circumstances primary radical termination should have little importance under normal polymerization conditions. Some kinetic studies indicating substantial primary radical termination during S polymerization may need to be re-evaluated in this light.[161] Secondary benzoate end groups in PS with BPO initiator may arise by head addition or transfer to initiator (Section 8.2.1).

3.4.2.2.2 Alkoxycarbonyloxy radicals

The chemistry of alkoxycarbonyloxy radicals in many ways parallels that of the aroyloxy radicals (*e.g.* benzoyloxy, see 3.4.2.2.1). Products attributable to the reactions of alkoxy radicals generally are not observed. This indicates that the rate of β-scission is slow relative to the rate of addition to monomers or other substrates.[188,431]

The alkoxycarbonyloxy radicals show little tendency to abstract hydrogen.[188,431] For example, in the reaction of isopropoxycarbonyloxy radicals with MMA, hydrogen abstraction, while observed, is a minor pathway (≤1%). When isopropoxycarbonyloxy radicals abstract hydrogen, isopropanol is the expected byproduct since the intermediate acid undergoes facile decarboxylation. Formation of isopropanol is not evidence for the involvement of isopropoxy radicals (Scheme 3.80).

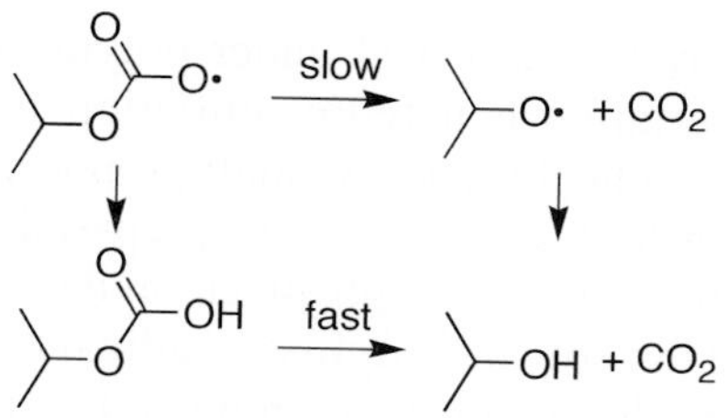

Scheme 3.80

Isopropoxycarbonyloxy radicals undergo facile reaction with aromatic substrates (*e.g.* toluene) by reversible aromatic substitution.[169,436] Isopropoxycarbonyloxy radicals react with S to give ring substitution (*ca* 1%) as well as the expected double bond addition.[403]

3.4.2.3 Hydroxy radicals

Hydroxy radicals are produced by redox reactions involving hydrogen peroxide (see 3.3.2.6.2). They can also be generated in organic solution by thermal decomposition of α-hydroperoxydiazenes (see 3.3.3.1).

The transient radicals produced in reactions of hydroxy radicals with vinyl monomers in aqueous solution have been detected directly by EPR[437-439] or UV spectroscopy.[440,441] These studies indicate that hydroxy radicals react with monomers and other species at or near the diffusion-controlled limit (Table 3.7). However, high reactivity does not mean a complete lack of specificity. Hydroxy radicals are electrophilic and trends in the relative reactivity of the hydroxy radicals toward monomers can be explained on this basis.[397]

Grant et al.[397] examined the reactions of hydroxy radicals with a range of vinyl and α-methylvinyl monomers in organic media. Hydroxy radicals on reaction with AMS give significant yields of products from head addition, abstraction and aromatic substitution (Table 3.8) even though resonance and steric factors combine to favor "normal" tail addition. However, it is notable that the extents of abstraction (with AMS and MMA) are less than obtained with *t*-butoxy radicals and the amounts of head addition (with MMA and S) are no greater than those seen with benzoyloxy radicals under similar conditions. It is clear that there is no direct correlation between reaction rate and low specificity.

Yields of aromatic substitution on S and AMS obtained by Grant et al.[397] should be regarded as minimum yields until the efficiency of trapping of the cyclohexadienyl radicals under their reaction conditions is known. This may help reconcile the finding that, in aqueous media, aromatic substitution is reported to be the main reaction pathway.[441] Grant et al.[397] also found that aromatic substitution on S proceeded by preferential *para* attack. This preference agrees with the calculated relative reactivity of the ring carbons based on frontier electron densities, but is otherwise unprecedented.[442]

3.4.2.4 Sulfate radical anion

The sulfate radical anion is formed by thermal, photochemical or redox decomposition of persulfate salts (**41**, see 3.3.2.6.1). Consequently, it is usually used in aqueous solution. However, crown ether complexes or alkylammonium salts may be used to generate the sulfate radical anion in organic solution (see 3.3.2.6.1).

Two pathways for the reaction of sulfate radical anion with monomers have been described (Scheme 3.81).[252] These are: (A) direct addition to the double bond or (B) electron transfer to generate a radical cation. The radical cation may also be formed by an addition-elimination sequence. It has been postulated that the radical cation can propagate by either cationic or a radical mechanism (both mechanisms may occur simultaneously). However, in aqueous media the cation is likely to hydrate rapidly to give a hydroxyethyl chain end.

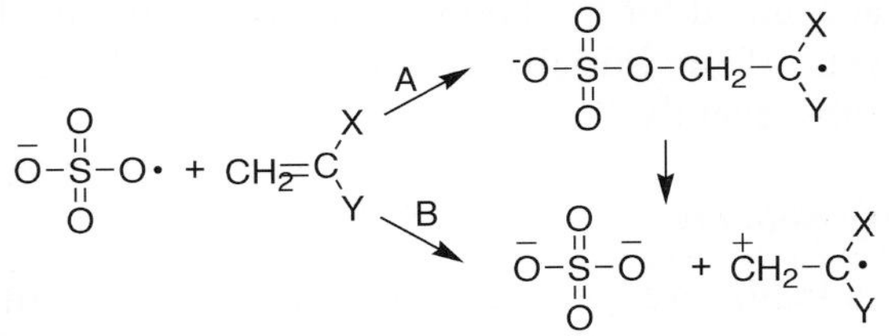

Scheme 3.81

The preferred initiation pathway is dependent on the particular monomer involved and the reaction conditions. Generally radical cation formation (by either mechanism) is facilitated by low pH. The failure to detect an intermediate sulfate adduct led workers to propose that reactions of the sulfate radical anion with electron-rich alkenes and S derivatives proceeded by pathway (B) over a wide range of pH and reaction conditions.[443-445] However, other workers rationalized similar data by allowing the initial formation of a sulfate adduct (pathway A).[446] Detection of an intermediate in the reaction of sulfate radical anion with S[447] or with cyclohexene[242] clearly points to addition being a major pathway in those cases. Moreover, PS formed with persulfate initiation is known to possess a high proportion of sulfate end groups.[448-451] Thus, the bulk of available evidence suggests that in initiation of S polymerization there is initial formation of a sulfate adduct (pathway A) and that, radical cations, if formed, are produced by subsequent elimination (Scheme 3.81).

In the case of electron-deficient monomers (*e.g.* acrylics) it is accepted that reaction occurs by initial addition of the sulfate radical anion to the monomer. Reactions of sulfate radical anion with acrylic acid derivatives have been shown to give rise to the sulfate adduct under neutral or basic conditions but under acidic conditions give the radical cation probably by an addition-elimination process.

Hydroxy radical and sulfate radical anion, though they may sometimes give rise to similar products, show quite different selectivity in their reactions with unsaturated substrates. In particular, the sulfate radical anion has a somewhat lower propensity for hydrogen abstraction than the hydroxyl radical. For example, the sulfate radical anion shows little tendency to abstract hydrogen from methacrylic acid.[252]

Sulfate radical anion may be converted to the hydroxyl radical in aqueous solution. Evidence for this pathway under polymerization conditions is the formation of a proportion of hydroxy end groups in some polymerizations. However, the hydrolysis of sulfate radical anion at neutral pH is slow ($k=10^7$ M^{-1} s^{-1}) compared with the rate of reaction with most monomers ($k=10^8$-10^9 M^{-1} s^{-1}, Table 3.7)[440] under typical reaction conditions. Thus, hydrolysis should only be competitive with addition when the monomer concentration is very low. The formation of hydroxy end groups in polymerizations initiated by sulfate radical anion can also be accounted for by the hydration of an intermediate radical cation or by the hydrolysis of an initially formed sulfate adduct either during the polymerization or subsequently.

3.4.2.5 *Alkylperoxy radicals*

Alkylperoxy radicals are generated by the reactions of carbon-centered radicals with oxygen and in the induced decomposition of hydroperoxides (Scheme 3.82). Their reactions have been reviewed by Howard[452] and rate constants for their self reaction and for their reaction with a variety of substrates including various inhibitors have been tabulated.[453]

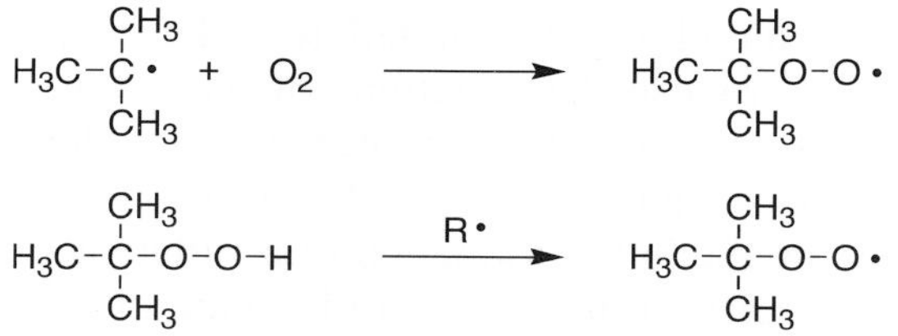

Scheme 3.82

Because of the importance of hydroperoxy radicals in autoxidation processes, their reactions with hydrocarbons are well known. However, reactions with monomers have not been widely studied. Absolute rate constants for addition to common monomers are in the range 0.09-3 M^{-1} s^{-1} at 40 °C. These are substantially lower than k_i for other oxygen-centered radicals (Table 3.7).[454]

Epoxide formation may be a side reaction occurring during initiation by *t*-butylperoxy radicals. The mechanism proposed for this process is as follows (Scheme 3.83).[211]

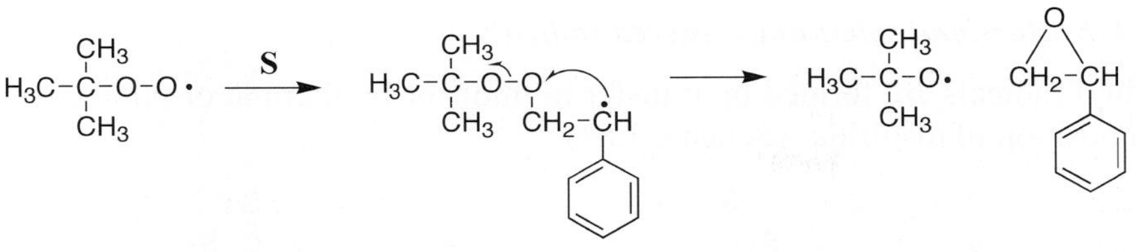

Initiation

Scheme 3.83

3.4.3 Other Heteroatom-Centered Radicals

Various other heteroatom-centered radicals have been generated as initiating species. These include silicon-, sulfur-, selenium- (see 3.4.3.1), nitrogen- and phosphorus-centered species (see 3.4.3.2). Kinetic data for reactions of these radicals with monomers is summarized in Table 3.10.

3.4.3.1 Silicon-centered radicals

Silicon centered radicals can be generated by transfer to silanes and by photolysis of polysilanes. Rate constants for addition to monomer are several orders of magnitude higher than similar carbon centered radicals.[455,456] The radicals have nucleophilic character.

Table 3.10 Selected Rate Data for Reactions of Heteroatom-Centered Radicals

Radical	Temp. °C	k_S M^{-1}s^{-1}	k/k_S[a]						Refs.
			AMS	MA	MMA	AN	MAN	VAc	
(c-C$_6$H$_{13}$)$_2$Si•	c	2.2×10^8	-	-	2.1	-	-	-	294
(C$_2$H$_5$)$_2$Si•	c	1.6×10^8	-	-	4.8	0.63	4.7	-	294
C$_2$H$_5$S•	60	-	-	0.036	-	-	-	-	457
t-C$_4$H$_9$S•	60	-	-	-	1.0[b]	-	0.63[b]	0.13[b]	458
CH$_3$(CH$_2$)$_2$CH$_2$S•	60	-	-	-	0.17	-	-	-	459
PhS•	23	2.0×10^7	-	-	0.16	-	-	-	460
PhS•	60	-	-	-	0.2	-	0.1	0.002	461
p-ClPhS•	23	5.1×10^7	-	-	0.10	0.0090	0.045	0.0009	462
PhC(O)S•	22	-	-	0.03	0.12	0.0091	-	0.0025	463
PhSe•	23	2.2×10^6	0.76	0.0078	0.019	0.0064	0.012	0.0005	464
Ph$_2$P(O)•	c	6.0×10^7	-	0.58	1.33	0.33	0.83	0.027	465
Ph$_2$P(O)•	20	1.1×10^7	1.27	-	1.45	-	-	0.25	466
(CH$_3$O)$_2$P(O)•	c,d	2.2×10^8	-	0.077	0.26	0.26	0.42	0.013	465

a Data rounded to two significant figures.   b k/k_{MMA}.   c Room temperature.   d Similar relative reactivities for VAc and AN have been reported at 60 °C.[467]

3.4.3.2 Sulfur- and selenium-centered radicals

Thiyl radicals are formed by transfer to thiols or by thermal or photochemical decomposition of disulfides (Scheme 3.84).

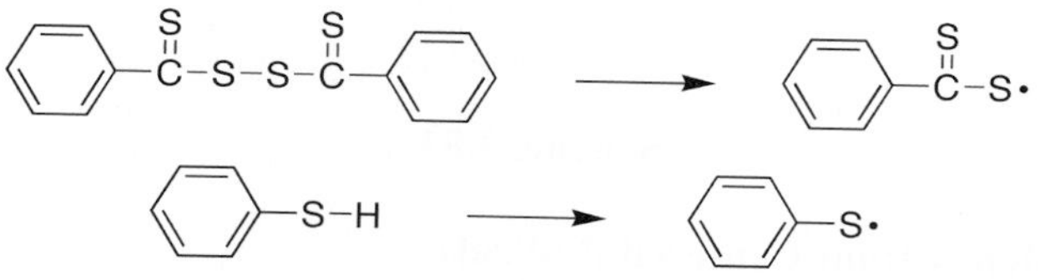

Scheme 3.84

Most studies have concerned the kinetics of arenethiyl radicals with monomers including S and its derivatives[468-472] and MMA.[469,473] The radicals have electrophilic character and add more rapidly to electron-rich systems (Table 3.10). Relative reactivities of the monomers towards the benzoylthiyl radical have also been examined.[463]

It is established that the initial reaction involves predominantly tail addition to monomer.[473] There is no evidence that abstraction competes with addition. It should be noted that the addition of arenethiyl radicals to double bonds is readily reversible.

A study on the kinetics of the reactions of phenylseleno radicals with vinyl monomers has also been reported.[464]

3.4.3.3 Phosphorus-centered radicals

Phosphinyl radicals (*e.g.* **103-107**) are generated by photodecomposition of acyl phosphinates or acyl phosphine oxides (see 3.3.4.1.1)[282,466,474,475] or by hydrogen abstraction from the appropriate phosphine oxide.[467]

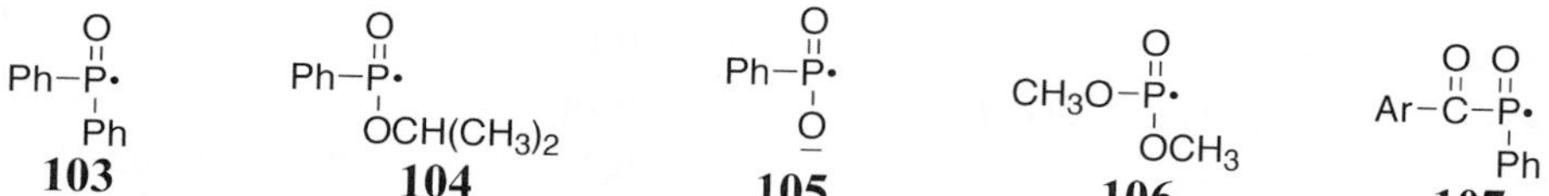

103: Ph-P(=O)(Ph)•
104: Ph-P(=O)(OCH(CH$_3$)$_2$)•
105: Ph-P(=O)(=O)•
106: CH$_3$O-P(=O)(OCH$_3$)•
107: Ar-C(=O)-P(=O)(Ph)•

The reactivities of the various phosphinyl radicals with monomers have been examined (Table 3.10).[283,465,467,475] Absolute rate constants are high, lying in the range 10^6-10^8 M^{-1} s^{-1} and show some solvent dependence. The rate constants are higher in aqueous acetonitrile solvent than in methanol. The high magnitude of the rate constants has been linked to the pyramidal structure of the phosphinyl radicals.[465]

The phosphinyl radicals (**103-107**) all show nucleophilic character (*e.g.* VAc is substantially less reactive than the acrylic monomers). However, the nucleophilicity varies according to the number of oxygen substituents on phosphorous.[465,467]

3.5 Techniques

The low concentration of initiator residues in polymers formed by radical polymerization means that they can usually only be observed directly in exceptional circumstances or in very low molecular weight polymers (Section 3.5.3). Thus, the study of the reactions of initiator-derived radicals with monomers has seen the development of some novel techniques. Three basic approaches have been employed. These involve:

(a) Kinetic studies involving the observation of the disappearance of reactants and/or appearance of products using some time resolved spectroscopic technique (most often EPR spectroscopy or UV-visible spectophotometry – Section 3.5.1).
(b) Isolation of the initiator-monomer reaction by employing a reagent designed to trap the first-formed adduct. This usually involves conducting the polymerization in the presence of an appropriate inhibitor (Section 3.5.2).
(c) Labeling the initiator such that the initiator-derived residues in the polymer can be more readily detected and quantified by chemical or spectroscopic analysis (Section 3.5.4).

3.5.1 Kinetic Studies

Time resolved EPR spectroscopy and UV-visible spectophotometry have proved invaluable in determining the absolute rate constants for radical-monomer reactions. The results of many of these studies are summarized in the Tables included in the previous section (3.4). Absolute rate constants for the reactions of carbon-centered radicals are reported in Table 3.6. These include t-butyl[374] and cyanoisopropyl[352] radicals.

3.5.2 Radical Trapping

Radical traps used for the study of radical monomer reactions should meet a number of criteria:

(a) The trap should ideally show a degree of specificity for reaction with the propagating species as opposed to the initiator-derived radicals.
(b) All products from the reaction with monomer should be trapped with equal efficiency.
(c) The trapped products should be stable under the reaction conditions.

Various reagents have been employed as radical traps. Those most commonly encountered are summarized in Table 3.11. The advantages, limitations and applications of each are considered in the following sections.

3.5.2.1 Spin traps

In spin trapping, radicals are trapped by reaction with a diamagnetic molecule to give a radical product.[476] This feature (*i.e.* that the free spin is retained in the trapped product) distinguishes it from the other trapping methods. The technique involves EPR detection of the relatively stable radicals which result from the trapping of the more transient radicals. No product isolation or separation is required. The use of the technique in studies of polymerization is covered in reviews by Kamachi[477] and Yamada *et al.*[478]

Table 3.11 Radical Trapping Agents for Studying Initiation

Trap	Initiating radicals trapped	Section
spin traps:		
nitroso-compounds	most radicals	3.5.2.1
nitrones	most radicals	3.5.2.1
transition metal ions:		
cupric ions	nucleophilic carbon-centered radicals	3.5.2.2
ferric ions	nucleophilic carbon-centered radicals	3.5.2.2
titanous ions	electrophilic carbon-centered radicals	3.5.2.2
metal hydrides:		
mercuric hydride	electrophilic carbon-centered radicals	3.5.2.3
Group VI hydrides	carbon-centered radicals	3.5.2.3
nitroxides	carbon-centered radicals	3.5.2.4
AMS dimer	most radicals	3.5.2.5

The two most commonly employed spin traps are 2-methyl-2-nitrosopropane (**108**) (more commonly known as nitroso-*t*-butane) and phenyl *t*-butyl nitrone (**109**); both trap radicals to yield nitroxides (Scheme 3.85, Scheme 3.86).

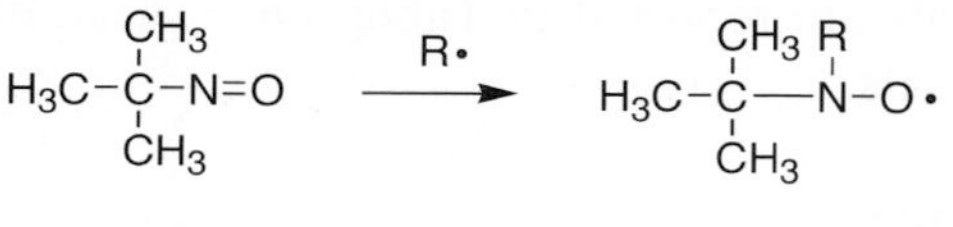

108

Scheme 3.85

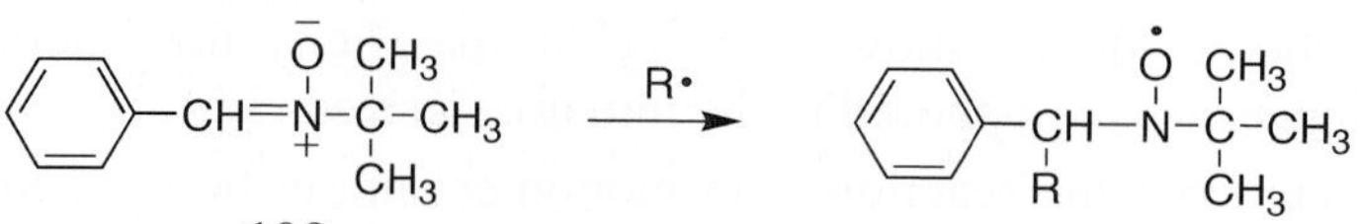

109

Scheme 3.86

Chalfont *et al.*[479] were the first to apply the spin trapping technique in the study of radical polymerization. They studied radicals produced during S polymerization initiated by *t*-butoxy radicals with **108** as the radical trap. Since that time many other systems have been studied using this trap (**108**).[477,478] The use of 2,4,6-tri-*t*-butylnitrosobenzene (**111**) in the study of polymerization, has been advocated by Savedoff and Ranby[480] and by Lane and Tabner.[433] This nitroso-compound is reported to be more thermally and photochemically stable than **108**. However **111** reacts with propagating radicals to give a mixture of anilino radicals (**110**) and nitroxides (**112**) as shown in Scheme 3.87.[481,482] The ratio of **110** to **112** depends on the structure of the propagating radical. Formation of **110** is favored when the radical trapped is more hindered and/or more electron rich.

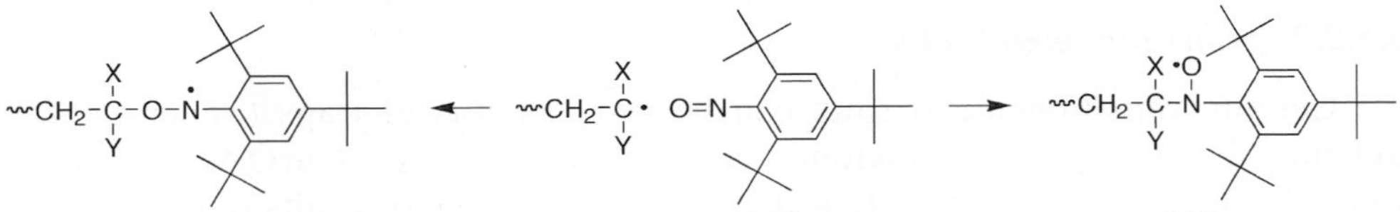

Scheme 3.87

Nitrones are generally more stable than nitroso-compounds and are therefore easier to handle. However, the nitroxides formed by reaction with nitrones [*e.g.* phenyl *t*-butyl nitrone (**109**)][483,484] have the radical center one carbon removed from the trapped radical (Scheme 3.86). The EPR spectra are therefore less sensitive to the nature of that radical and there is greater difficulty in resolving and assigning signals. Nitrones are generally less efficient traps than nitroso-compounds.[476]

There are several limitations on the use of the spin trapping technique when quantitative results are required. These are:

(a) Not all radicals are trapped at equal rates or with equal efficiency.[485]
(b) The product nitroxides may not be stable under the reaction conditions. Nitroxide stability is strongly dependent on the nature of the trapped species. Nitroxides react with radicals at or near diffusion controlled rates and they can also undergo β–scission either to regenerate the trapped radical or to form a new radical.
(c) Side reactions involving the trap and the monomer may give rise to products which complicate the interpretation of the EPR spectra. Various side reactions have been described in the literature:[476] the nitroso-compound (**108**) reacts with α–methylvinyl monomers by an ene reaction (Scheme 3.88);[188] *t*-butyl radicals produced by thermal or photochemical decomposition of (**108**) are trapped as di-*t*-butylnitroxide.

Many of the above-mentioned complications can be avoided or allowed for by carrying out appropriate control experiments. A further difficulty lies with the

The Chemistry of Radical Polymerization

sensitivity of the method. Minor initiation pathways (≤5%) are extremely difficult to determine.

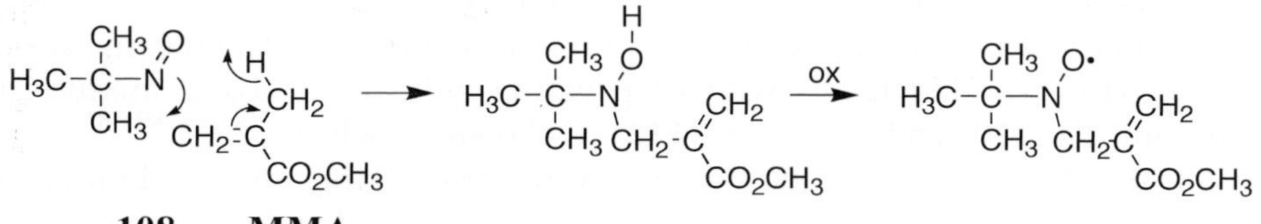

$$\underset{\mathbf{108}}{\overset{\text{CH}_3\ \text{O}}{\underset{\text{CH}_3}{\text{H}_3\text{C}-\overset{|}{\text{C}}-\text{N}}}\underset{\text{CH}_2\text{-}\overset{|}{\text{C}}}{\overset{\text{H}}{\diagdown}\text{CH}_2}}\ \xrightarrow{}\ \overset{\text{CH}_3\ \text{O}}{\underset{\text{CH}_3\ \text{CH}_2\text{-}\overset{|}{\text{C}}}{\text{H}_3\text{C}-\overset{\overset{\text{H}}{|}}{\text{C}}-\text{N}}}\diagdown_{\text{CH}_2}\ \xrightarrow{\text{ox}}\ \overset{\text{CH}_3\ \text{O}\bullet}{\underset{\text{CH}_3\ \text{CH}_2\text{C}}{\text{H}_3\text{C}-\overset{|}{\text{C}}-\text{N}}}\diagdown_{\text{CH}_2}$$
(with CO₂CH₃ groups)

108 **MMA**

Scheme 3.88

3.5.2.2 Transition metal salts

Certain transition metal salts can be used as radical traps (Scheme 3.89, Scheme 3.90).[486] These include various cupric (*e.g.* Cu(OAc)₂, CuCl₂, Cu(SCN)₂),[18,168,393,432,487] ferric (*e.g.* FeCl₃),[316,488] and titanous salts (*e.g.* TiCl₃).[379] These traps react with radicals by ligand- or electron-transfer to give products which can be determined by conventional analytical techniques.

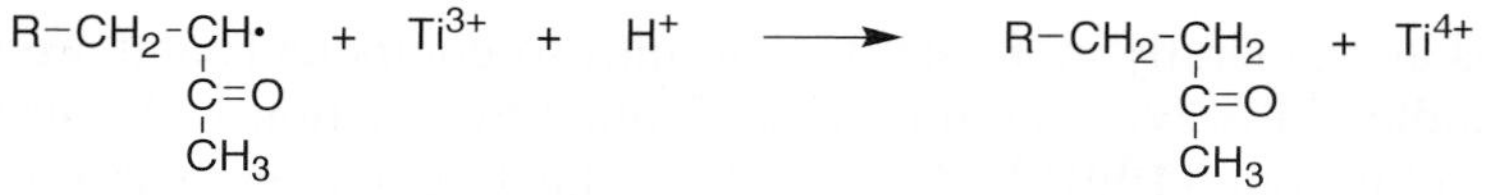

$$\text{R-CH}_2\text{-CH}\bullet\ +\ \text{Ti}^{3+}\ +\ \text{H}^+\ \longrightarrow\ \text{R-CH}_2\text{-CH}_2\ +\ \text{Ti}^{4+}$$
(with C=O, CH₃ substituents on both sides)

Scheme 3.89

The rate of oxidation/reduction of radicals is strongly dependent on radical structure. Transition metal reductants (*e.g.* Ti^{III}) show selectivity for electrophilic radicals (*e.g.* those derived by tail addition to acrylic monomers or alkyl vinyl ketones - Scheme 3.89)[379] while oxidants (Cu^{II}, Fe^{III}) show selectivity for nucleophilic radicals (*e.g.* those derived from addition to S - Scheme 3.90).[18] A consequence of this specificity is that the various products from the reaction of an initiating radical with monomers will not all be trapped with equal efficiency and complex mixtures can arise.

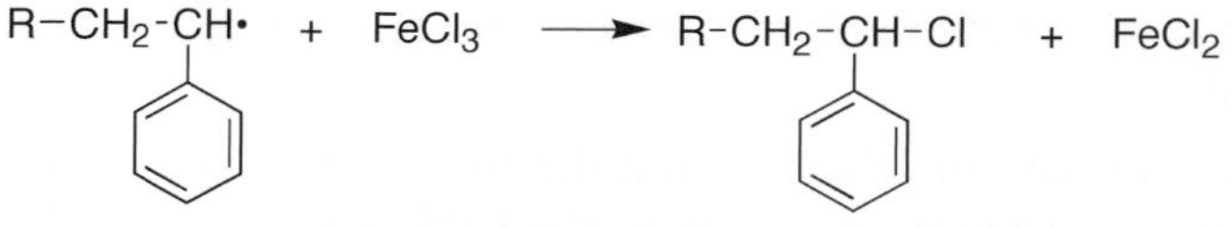

$$\text{R-CH}_2\text{-CH}\bullet\ (\text{Ph})\ +\ \text{FeCl}_3\ \longrightarrow\ \text{R-CH}_2\text{-CH-Cl}\ (\text{Ph})\ +\ \text{FeCl}_2$$

Scheme 3.90

The facile and reversible reaction of propagating species with transition metal halide complexes to form a polymeric halo-compound is one of the key steps in atom transfer radical polymerization (ATRP, see Section 9.4).

3.5.2.3 Metal hydrides

Metal hydride trapping agents have been used extensively in studying the reaction of alkyl radicals with monomers.[489,490]

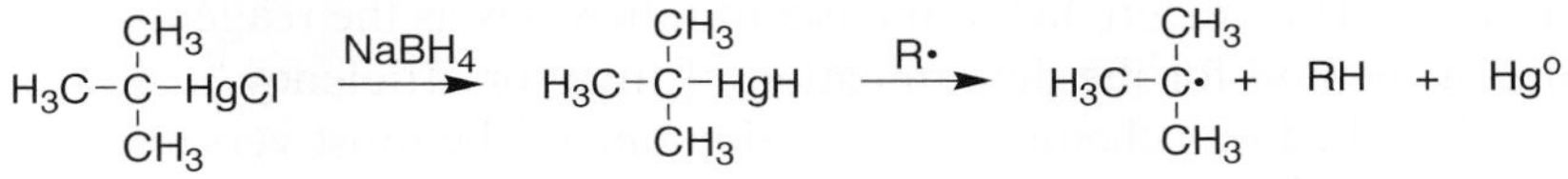

$$H_3C-\underset{\underset{CH_3}{|}}{\overset{\overset{CH_3}{|}}{C}}-HgCl \xrightarrow{NaBH_4} H_3C-\underset{\underset{CH_3}{|}}{\overset{\overset{CH_3}{|}}{C}}-HgH \xrightarrow{R\cdot} H_3C-\underset{\underset{CH_3}{|}}{\overset{\overset{CH_3}{|}}{C}}\cdot + RH + Hg^0$$

Scheme 3.91

Alkyl mercuric hydrides are generated *in situ* by reduction of an alkyl mercuric salt with sodium borohydride (Scheme 3.91). Their use as radical traps was first reported by Hill and Whitesides[491] and developed for the study of radical-olefin reactions by Giese,[489,490] Tirrell[492] and coworkers. Careful choice of reagents and conditions provides excellent yields of adducts of nucleophilic radicals (*e.g.* n-hexyl, cyclohexyl, t-butyl, alkoxyalkyl) to electron-deficient monomers (*e.g.* acrylics).

A consequence of the selectivity for electrophilic radicals is that not all products are trapped with equal efficiency. With electron-rich monomers (*e.g.* S) oligomerization may complicate analysis. Other possible complications in the utilization of this method have been discussed by Russell.[493]

Group IV hydrides (R_3SnH, R_3GeH) have also been used as trapping reagents.[494,495] The reduction of alkyl halides by stannyl or germyl radicals affords alkyl radicals. These react with the group IV hydrides to set up a radical chain (Scheme 3.92).[495] The alkyl radicals may react with a substrate (*e.g.* monomer) in competition with being trapped by the hydride. Absolute rate constants for the reactions of group IV hydrides with radicals are known. Thus the H-atom transfer step may be used as a radical clock to calibrate radical-monomer reactions.[20] This technique has seen widespread use in the study of intramolecular radical reactions.[345] One limitation of the use of the group IV hydrides as radical traps in the study of polymerization is that the stannyl and germyl radicals may themselves add monomer, *albeit* reversibly.

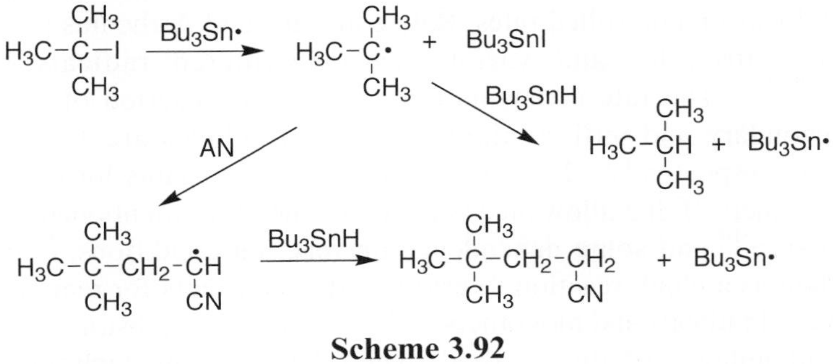

Scheme 3.92

3.5.2.4 Nitroxides

A well-known feature of the chemistry of nitroxides (*e.g.* **113-115**) is that they combine with carbon-centered radicals at near diffusion-controlled rates to give alkoxyamines. This feature led to the use of nitroxides as the reagents of choice in the inhibitor method for the determination of initiator efficiency.[92] Rizzardo and Solomon[496] applied this chemistry to develop one of the most versatile techniques for examining the initiation step of polymerization. The method is reliant on the initiator-derived radicals either not reacting or reacting only slowly with the nitroxide while the propagating radicals are efficiently scavenged to yield stable alkoxyamines (Scheme 3.93). The technique has been successfully used by several groups to study the reactions of heteroatom-centered (ethoxy,[402] isopropoxy,[123,402] *t*-butoxy,[8,10,12,21,22,177,188,396,404,406-410,496,497] cumyloxy,[72] other *t*-alkoxy,[421] benzoyloxy,[8,10,22,41,166,188,401] isopropoxycarbonyloxy,[188] hydroxy,[253,397] thiyl,[458,461,498] phosphinyl[467,474,499]) and more reactive carbon-centered radicals (methyl, undecyl, *t*-butyl, phenyl)[8,10,22,41,177,424,425,500-504] with monomers. The reaction has also been employed to detect radical intermediates in organic reactions and to identify primary radicals produced from photoinitiators.[474]

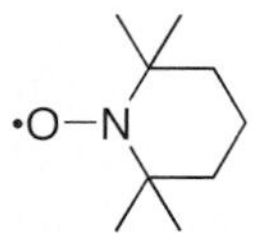

TEMPO 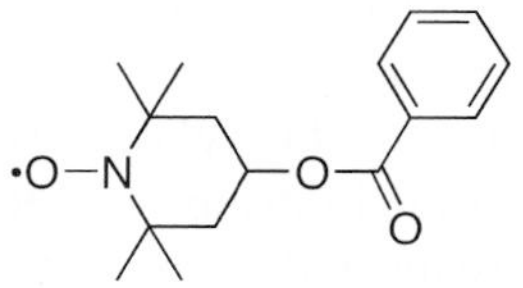   **113** 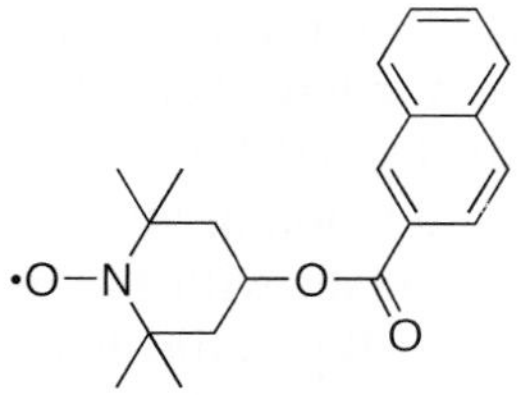   **114** 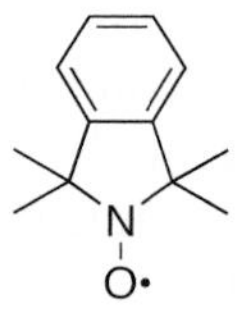   **115**

Busfield and coworkers extended the technique to the study of less reactive carbon-centered radicals (*e.g.* cyanoisopropyl)[353,354] and short propagating radicals[505-507]. The very low concentration of nitroxide required to allow limited propagation was maintained by feeding with a syringe pump.

The reaction between nitroxides and carbon-centered radicals occurs at near (but not at) diffusion controlled rates. Rate constants and Arrhenius parameters for coupling of nitroxides and various carbon-centered radicals have been determined.[508-511] The rate constants (20 °C) for the reaction of TEMPO with primary, secondary and tertiary alkyl and benzyl radicals are 1.2, 1.0, 0.8 and 0.5×10^9 M^{-1} s^{-1} respectively. The corresponding rate constants for reaction of **115** are slightly higher. If due allowance is made for the afore-mentioned sensitivity to radical structure[510] and some dependence on reaction conditions,[511] the reaction can be applied as a clock reaction to estimate rate constants for reactions between carbon-centered radicals and monomers[504,506,507,512] or other substrates.[20]

Major advantages of this method over other trapping techniques are that typical conditions for solution/bulk polymerization can be employed and that a very wide range of initiating systems can be examined. The application of the

Initiation

technique is greatly facilitated by the use of a nitroxide possessing a UV chromophore (*e.g.* **113-115**) which simplifies product analysis by liquid chromatography with UV detection.

Nitroxides have the property of quenching fluorescence. Thus radical trapping with nitroxides containing fluorophores (*e.g.* **114**) can be monitored by observing the appearance of fluorescence.[513-515] The method is highly sensitive and has been applied to quantitatively determine radical yields in PLP experiments (Section 4.5.2).

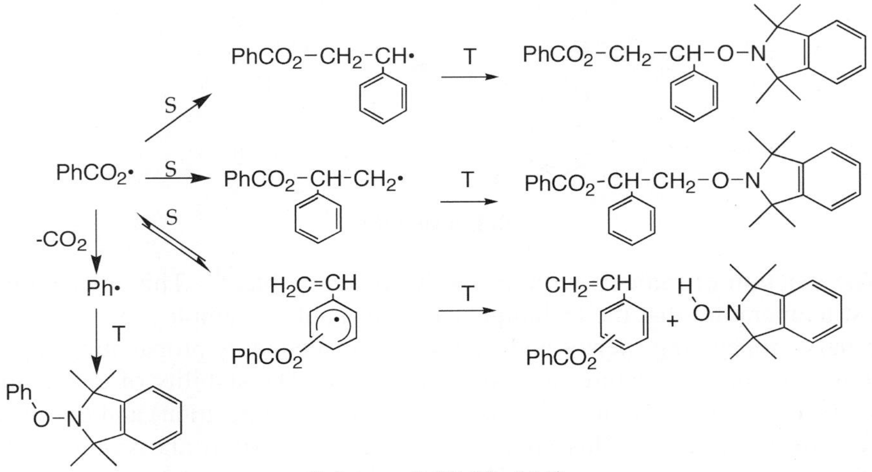

Scheme 3.93 (T=115)

Some limitations of the method arise due to side reactions involving the nitroxide. However, such problems can usually be avoided by the correct choice of nitroxide and reaction conditions. Nitroxides, while stable in the presence of most monomers, may act as oxidants or reductants under suitable reaction conditions.[516] The induced decomposition of certain initiators (*e.g.* diacyl peroxides) can be a problem (Scheme 3.94).[166,177] There is some evidence that nitroxides may disproportionate with alkoxy radicals bearing α-hydrogens.[123] Side reactions with thiols have also been identified.[498]

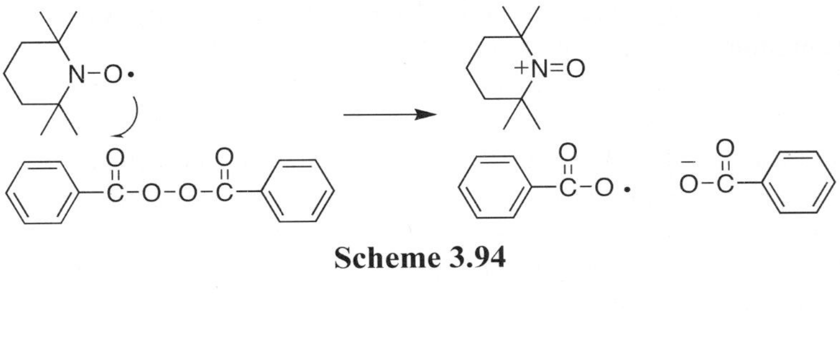

Scheme 3.94

Various light-induced reactions including hydrogen atom abstraction, electron transfer and β-scission occur under the influence of UV light.[517-521] Certain radicals, for example cyclohexadienyl radicals (Scheme 3.95), are trapped by disproportionation rather than coupling.[8] Nitroxides are also reported to react by hydrogen abstraction with molecules that are extremely good hydrogen donors [*e.g.* S dimer (**95**)[318] and the ketenimine (**10**).[103]]

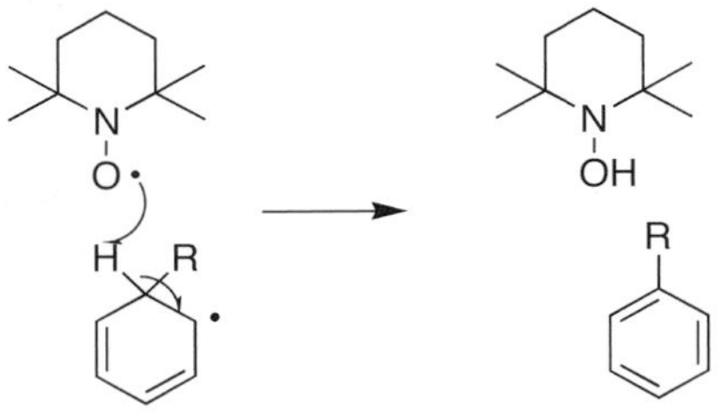

Scheme 3.95

The reaction of radicals with nitroxides is reversible.[309] This means that the highest temperature that the technique can reasonably be employed at is *ca* 80 °C for tertiary propagating species and *ca* 120 °C for secondary propagating species.[22] These maximum temperatures are only guidelines. The stability of alkoxyamines is also dependent on solvent (polar solvents favor decomposition) and the structure of the trapped species. This chemistry has led to certain alkoxyamines being useful as initiators of living polymerization (Section 9.3.6). At elevated temperatures nitroxides are observed to add to monomer albeit slowly.[318,522,523]

3.5.2.5 α-Methystyrene dimer

Watanabe *et al.*[25,524-528] applied AMS dimer (**116**) as a radical trap to examine the reactions of oxygen-centered radicals (*e.g. t*-butoxy, cumyloxy, benzoyloxy). AMS dimer (**116**) is an addition fragmentation chain transfer agent (see 6.2.3.4) and reacts as shown in Scheme 3.96. The reaction products are macromonomers and may potentially react further. The reactivity of oxygen centered radicals towards **116** appears to be similar to that of S.[25] Cumyl radicals are formed as a byproduct of trapping and are said to decay mainly by combination and disproportionation.

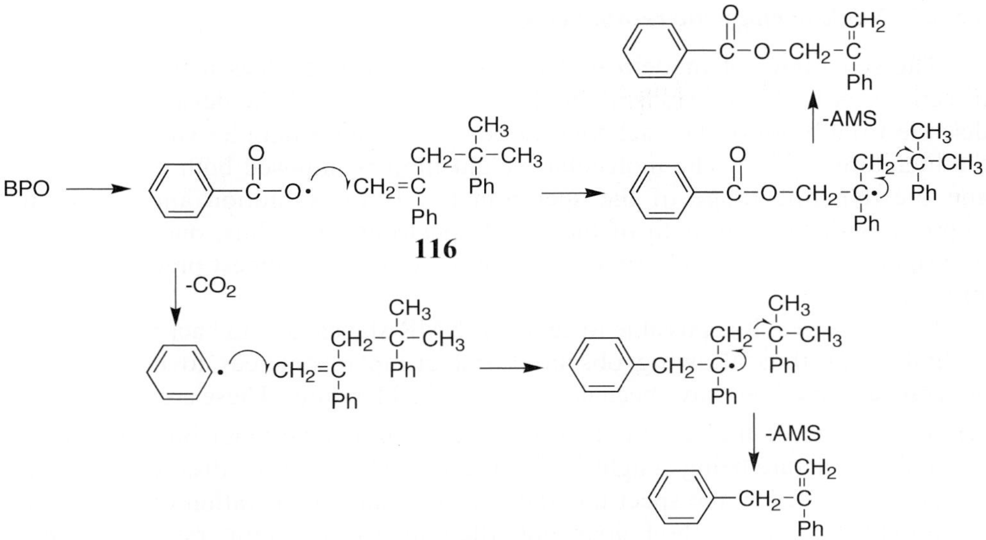

Scheme 3.96

3.5.3 Direct Detection of End Groups

In favorable circumstances initiator-derived end groups may be detected by spectroscopic methods or by chemical or chromatographic analysis. Most of the methods are sensitive only to a given type of end group in a given class of polymer. However, they have the advantage that no special chemistry or isolation steps are required. The main disadvantages associated with these methods are that they require foreknowledge of what the end groups are likely to be and, in general, they can only be applied to low molecular weight polymers.

3.5.3.1 Infra-red and UV-visible spectroscopy

UV[153,529] and IR spectroscopy[94,530,531] have been used for polymer end group determination and to study the kinetics and efficiency of initiation of polymerization. These techniques are not universally applicable. Ideally, it is required (a) that the chromophores are in a clear region of the spectrum and (b) that the positions of the absorptions are sensitive to the chemical environment of the chromophore such that end groups can be distinguished from residual initiator and initiator-derived byproducts.

Garcia-Rubio et al.[153,529] examined S and MMA polymerizations initiated by BPO and have shown that UV can be used to distinguish and quantitatively determine aliphatic and aromatic benzoate groups in MMA and S polymerizations.

Buback et al.[94,531,532] applied FTIR to follow the course of the initiation of S polymerization by AIBN and to determine initiator efficiency. Contributions to the IR signal due to cyanoisopropyl end groups, AIBN, and the ketenimine can be separated using curve resolution techniques.

3.5.3.2 Nuclear magnetic resonance spectroscopy

The sensitivity of modern NMR allows initiator residues to be determined directly in polymers of moderate molecular weight where the desired signals are discrete from those of the backbone carbons.[530] Many examples can be found in the literature.[34,533-538] The molecular weight limit is imposed both by sensitivity and the dynamic range of the spectrometer. Both resolution and sensitivity improve with field strength of the NMR spectrometer. Thus, one strategy for improving the ease of end group detection is to use the highest practicable field strength.[537,538]

In some cases, it is possible to suppress NMR signals due to backbone carbons or hydrogens thus allowing obscured end group resonances to be observed. Several basic methods have been described in the literature. These are:

(a) Subtraction of the spectrum of an exactly similar polymer but without the defect structure being sought.[370,539] The procedure has the disadvantages that noise is added to the spectrum and that it requires preparation of a reference polymer. The method does not alleviate the dynamic range problems discussed above.

(b) Use of a Hahn spin echo experiment to suppress signals from backbone atoms. It has been demonstrated[7,540] that end group signals usually persist longer than backbone signals because of longer T_2 relaxation times. Moad et al.[7] have applied the method to detect obscured cyanoisopropyl end groups in PMMA.

(c) Use of pulse sequences that select for the number of attached hydrogens. For example, for PS prepared with AIBN a 'quaternary only' pulse sequence can be used to better visualize signals due to the quaternary carbons of the AIBN-derived residues.[7]

(d) Analysis of polymers prepared from NMR-inactive monomers. Hatada et al. used ^{1}H NMR to determine end groups in perdeuterated PS[541] and PMMA.[542,543] Similarly, the use of NMR-inactive ^{12}C-enriched monomers has been envisaged as an aid in detecting end groups in ^{13}C NMR experiments.[544]

(e) Use of two (2D) or three dimensional (3D) NMR methods. For example, Bevington and Huckerby[545] applied ^{13}C-^{1}H correlation spectroscopy to advantage to evaluate end groups when ^{13}C signals are discrete yet ^{1}H signals are overlapping. Rinaldi and coworkers[546-548] examined the end group structures of and define the initiation mechanism for polystyrene prepared with an acyl phosphine oxide initiator using 3D NMR.

These five techniques rely on suppressing signals due to the backbone carbons. The end group signals are not enhanced. Therefore, the sensitivity problems associated with detecting end groups in high molecular weight polymers are not entirely solved. However, the methods (b-d) do allow acquisition at higher spectrometer gain settings and assist in overcoming spectrometer dynamic range problems. A drawback of the pulse sequence methods is that quantification may not be a straightforward exercise.

Selective labeling of the initiator with ^{13}C allows substantial enhancement of the signals of the initiator residues relative to signals due to the backbone in ^{13}C NMR spectra. Initiators labeled with or containing NMR active nuclei such as ^{19}F or ^{31}P can also be applied. These methods are described in Section 3.5.4.2.

3.5.3.3 *Electron paramagnetic resonance spectroscopy*

The application of EPR in the detection and quantification of species formed by spin-trapping the products of radical-monomer reactions is described in Section 3.5.2.1. The application of time-resolved EPR spectroscopy to study intermolecular radical-alkene reactions in solution is mentioned in Section 3.5.1.

3.5.3.4 *Mass spectrometry*

Some discussion on the use of mass spectrometry for end group determination can be found in recent texts.[530,549] Traditionally mass spectrometric techniques have required polymers of relatively low molecular weight. Meisters et al.[550] reported that fast atom bombardment mass spectrometry (FAB-MS) can be applied in the analysis of MMA oligomers to at least hexadecamer. For polymers that degrade by unzipping, pyrolysis GCMS has provided extremely useful data on initiation processes. Thus, Farina et al.[459,551] and Ohtani et al.[552,553] described the application of pyrolysis GCMS to determine end groups in PMMA, PS and copolymers.

Two relatively new techniques, matrix assisted laser desorption ionization–time of flight mass spectrometry (MALDI-TOF) and electrospray ionization (ESI), offer new possibilities for analysis of polymers with molecular weights in the tens of thousands. PS molecular weights as high as 1.5 million have been determined by MALDI-TOF. Recent reviews on the application of these techniques to synthetic polymers include those by Hanton[554] and Nielen.[555] The methods have been much used to provide evidence for initiation and termination mechanisms in various forms of living and controlled radical polymerization.[556] Some examples of the application of MALDI-TOF and ESI in end group determination are provided in Table 3.12. The table is not intended to be a comprehensive survey.

MALDI-TOF can be applied to estimate molecular weights of very high molecular weight polymers. However, with the mass resolution of current instruments, molecular weights of less than 5000 are desirable for end groups to be reliably distinguished and determined. There are also issues with sensitivity dependence on molecular weight and composition. Sensitivity depends on volatility and the ease of cationization.[554] For homopolymer samples MALDI-TOF is able to duplicate GPC distributions with reasonable precision when polydispersities are less than about 1.2. For broader molecular weight distributions MALDI-TOF tends to underestimate the molecular weight and the polydispersity. Discrimination according to ease of cationization for low molecular weight polymers may be mitigated by end group derivatization. This, however, requires foreknowledge of the end groups.

Table 3.12 Application of MALDI-TOF or ESI Mass Spectrometry to Polymers Prepared by Radical Polymerization

Polymerization Method	Technique	Polymer
Conventional - AIBN	MALDI-TOF	PMMA,[557] PS[557]
Conventional - with catalytic chain transfer	MALDI-TOF	PMMA, copolymers[556,558,559]
Conventional, AIBN - with transfer to solvent	MALDI-TOF	PNVP[560]
Conventional - photoinitiation	ESI	PMA[561]
RAFT	MALDI-TOF	PNIPAM,[562,563] PS[564] other[565]
RAFT	ESI	PMA,[566]
ATRP	MALDI-TOF	PEA,[567] PMMA,[568]
NMP	MALDI-TOF	PBA,[569] PS[570,571]
NMP	ESI	PAN,[506] PS[507]

3.5.3.5 Chemical methods

Chemical analysis often allows end groups to be determined with high precision though the process is painstaking. A number of techniques have been developed for the chemical derivatization of polymer end groups so they can be more readily measured by spectrophotometric methods. One of the most used is the dye-partition method introduced by Palit.[451,572-574] Variants of this method have been applied to detect hydroxy,[574,575] quaternary ammonium and sulfate end groups.[450,451] A two step dealkylation-derivatization procedure[576] was successfully used for determining t-butoxy end groups in PS. In that case the t-butoxy ends were first cleaved with trifluoroacetic acid to give hydroxy chain ends. This method was not applicable to PMMA. It was found the t-butoxy ends of PMMA could be determined by measuring the release of t-butyl chloride formed on treating the polymer with boron trichloride.[576]

Where the polymer end groups possess reactive functionality, for example hydroxy, amino, thiol or carboxy groups, post-polymerization derivatization may be used to facilitate detection and identification with NMR spectroscopy. Thus, trichloroacetyl isocyanate undergoes a facile reaction with protic end groups when added in slight excess to a solution of the polymer in an NMR tube (Scheme 3.97).[577,578] The imidic hydrogens of the derivatives have a distinctive chemical shift in the region 8-11 ppm depending on the particular functional group. There is also a shift of the hydrogens α-to the chain end.

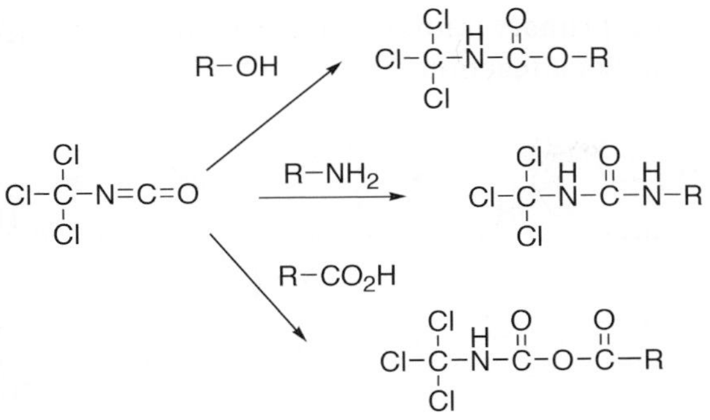

Initiation

Scheme 3.97

3.5.4 Labeling Techniques

Various methods have been described whereby polymers are formed with an initiator that contains chromophores or other functionality to permit ready detection of initiator-derived end groups by chemical or spectroscopic methods.[579,580] A potential disadvantage of this procedure is that the initiator is chemically modified and the specificity shown by the initiator-derived radicals may be different from that of the corresponding unlabeled species.

The best labeling system in this regard is isotopic labeling since it involves the minimum change from the standard initiator. Methods based on radiolabeling and stable isotopes detectable by NMR are described in Sections 3.5.4.1 and 3.5.4.2 respectively.

3.5.4.1 Radiolabeling

Polymer formed using radiolabeled initiators may be isolated and analyzed to determine the concentration of initiator-derived residues and calculate the initiator efficiency. Radiolabeled initiators have also been used extensively to establish the relative reactivity of monomers towards radicals.[88,107,580-582]

Radiolabeling offers greater sensitivity than most other labeling methods. However, the technique has the disadvantage that end groups formed by initiation cannot be directly distinguished from initiator residues produced by other processes (*e.g.* primary radical termination or copolymerization of initiator byproducts) or from residual initiator. In general, the method gives the total initiator residues in the polymer. Analysis of the kinetics of polymerization can help to resolve these problems. A further disadvantage is that polymer isolation and purification is required.

For the case of initiators that produce both primary and secondary radicals (*e.g.* BPO) use of a doubly labeled initiator allows the different types of end groups to be distinguished [*e.g.* **117** and **118** - Scheme 3.98] and the reactivities of monomers towards the primary radicals to be readily established by using the fragmentation step as a clock reaction.[399,583]

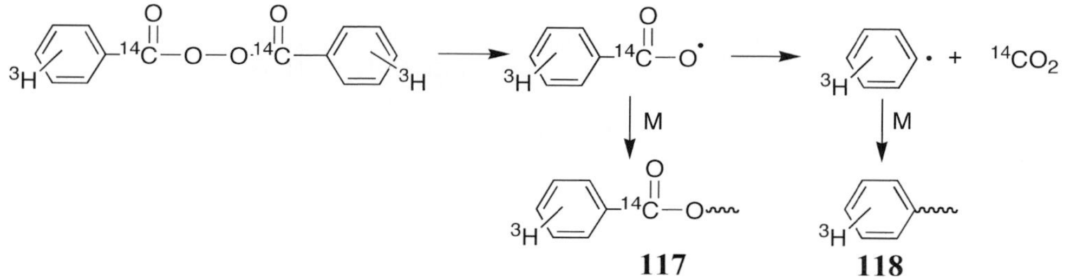

Scheme 3.98

3.5.4.2 Stable isotopes and nuclear magnetic resonance

NMR methods can be applied to give quantitative determination of initiator-derived and other end groups and provide a wealth of information on the polymerization process. They provide a chemical probe of the detailed initiation mechanism and a greater understanding of polymer properties. The main advantage of NMR methods over alternative techniques for initiator residue detection is that NMR signals (in particular ^{13}C NMR) are extremely sensitive to the structural environment of the initiator residue. This means that functionality formed by tail addition, head addition, transfer to initiator or primary radical termination, and various initiator-derived byproducts can be distinguished.

Selective labeling of the initiator allows substantial enhancement of the signals of the initiator residues relative to the signals due to the backbone. Various stable isotopes have been employed in this context (including D, F, ^{15}N and ^{31}P), however, most work has involved the use of ^{13}C-labeling (Table 3.13). The method has been reviewed.[536,584,585] The power of the technique is illustrated by the fact that one experiment allows the determination of:

(a) The total fate of the initiator as a function of conversion (initiator efficiency, nature and amount of byproducts).
(b) The chain ends (reactivity of primary radicals towards monomers, head *vs* tail addition, *etc*.).
(c) The rate of polymerization.
(d) The number average molecular weight = ([end groups]/[monomer used]).

The use of ^{13}C-labeled initiators in assessing the kinetics and efficiency of initiation[2,14,32,60,84] requires that the polymer end groups, residual initiator, and various initiator-derived byproducts should each give rise to discrete signals in the NMR spectrum. So far this method has been demonstrated for homo- and copolymerizations of S and MMA prepared with AIBN-α-^{13}C, AIBMe-α-^{13}C or BPO-*carbonyl*-^{13}C/BPO-*ring*-^{13}C (1:1) as initiator.

Table 3.13 Radical Polymerizations Performed with Initiators Labeled with Stable Isotopes

Initiator	Polymer
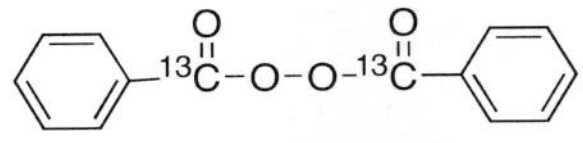 BPO-*carbonyl*-^{13}C (Ph-^{13}C(O)-O-O-^{13}C(O)-Ph)	S,[9] MMA,[11] MMA-*co*-S,[11] other[400,586-594]
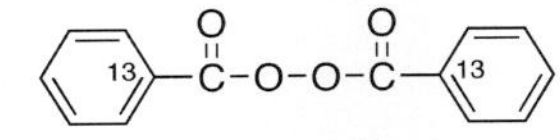 BPO -α-^{13}C (Ph(^{13})-C(O)-O-O-C(O)-Ph(^{13}))	S[14]
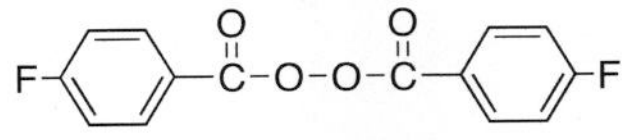 BPO-F (F-Ph-C(O)-O-O-C(O)-Ph-F)	S,[595,596] MMA,[595,596] MMA-*co*-S,[597] other[598]
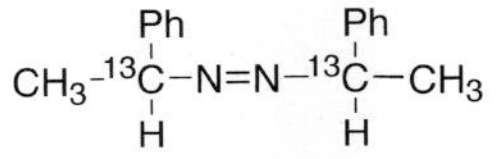 APE-α-^{13}C (CH$_3$-^{13}CH(Ph)-N=N-^{13}CH(Ph)-CH$_3$)	MA,[539] MMA,[539] MPK,[539] VAc,[539] AN-*co*-S,[599] B-*co*-MMA,[600] MAH-*co*-S,[601] MMA-*co*-S,[602] other[371,603]
APE-α-^{13}C/EASC[a]	B-*co*-MMA,[600] MMA-*co*-S[602]
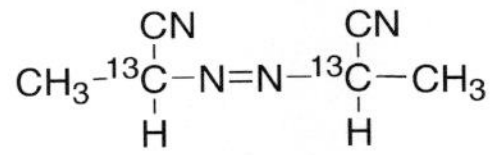 APN-α-^{13}C (CH$_3$-^{13}CH(CN)-N=N-^{13}CH(CN)-CH$_3$)	AN-*co*-S[604]
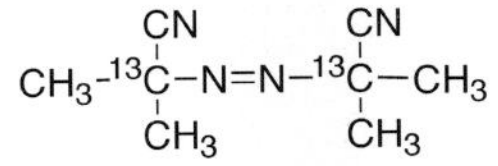 AIBN-α-^{13}C (CH$_3$-^{13}C(CN)(CH$_3$)-N=N-^{13}C(CN)(CH$_3$)-CH$_3$)	MMA,[357] S,[7,32,357] VAc,[357] MMA-*co*-S,[2,357] MMA-*co*-VAc[357]
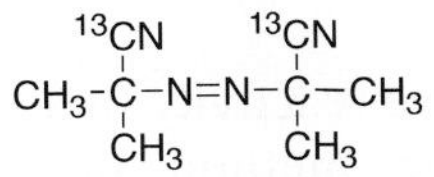 AIBN-*nitrile*-^{13}C (CH$_3$-C(^{13}CN)(CH$_3$)-N=N-C(^{13}CN)(CH$_3$)-CH$_3$)	VAc,[113] VF[113]
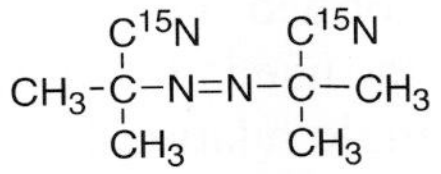 AIBN-^{15}N (CH$_3$-C(C^{15}N)(CH$_3$)-N=N-C(C^{15}N)(CH$_3$)-CH$_3$)	MMA-*co*-S,[382] other[605]
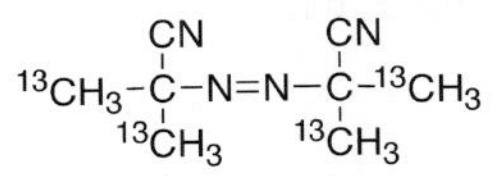 AIBN-ββ-^{13}C (^{13}CH$_3$-C(CN)(^{13}CH$_3$)-N=N-C(CN)(^{13}CH$_3$)-^{13}CH$_3$)	AMS,[358] EA,[360] MMA,[359,606] MAN,[359] S,[359,606] VC,[361] VP,[589] AN-*co*-MMA,[363] B-*co*-MMA,[600] EA-*co*-S[360] MAN-*co*-S,[359] MMA-MPK,[607] MMA-MVK,[607] MMA-*co*-S,[359] MMA-*co*-VC,[361] S-*co*-VC,[361] other[589,600,603,608,609]
AIBN-ββ-^{13}C/EASC[a]	B-*co*-MMA,[600] MMA-*co*-S[81]

Table 3.13 (continued)

Initiator	Polymer
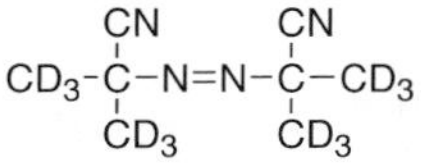 CD$_3$-C(CN)(CD$_3$)-N=N-C(CN)(CD$_3$)-CD$_3$ **AIBN-D**	EA,[360] EA-co-S,[360] MMA-co-S[382]
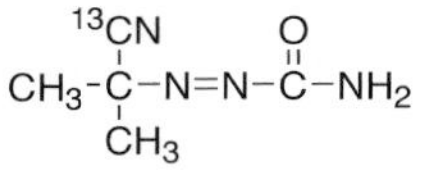 CH$_3$-C(^{13}CN)(CH$_3$)-N=N-C(O)-NH$_2$ **AZOF-nitrile-^{13}C**	VAc[113]
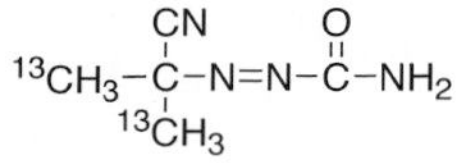 ^{13}CH$_3$-C(CN)(^{13}CH$_3$)-N=N-C(O)-NH$_2$ **AZOF-ββ-^{13}C**	MMA-co-S,[364] MAN-co-S,[362] S-co-VAc[362]
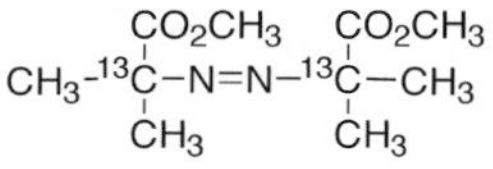 CH$_3$-^{13}C(CO$_2$CH$_3$)(CH$_3$)-N=N-^{13}C(CO$_2$CH$_3$)(CH$_3$)-CH$_3$ **AIBMe-α-^{13}C**	MMA,[357] S,[357] VAc,[357] MMA-co-S,[83,84,357] MMA-co-VAc,[357]
AIBMe-α-^{13}C/EASC^a	MMA-co-S[83]
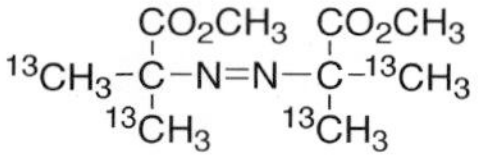 ^{13}CH$_3$-C(CO$_2$CH$_3$)(^{13}CH$_3$)-N=N-C(CO$_2$CH$_3$)(^{13}CH$_3$)-^{13}CH$_3$ **AIBMe-ββ-^{13}C**	AN,[375] MA,[375] MMA,[375] S,[375] B-co-MMA,[600] MMA-co-S,[81] other[600]
AIBMe-ββ-^{13}C/EASC^a	B-co-MMA,[600] MMA-co-S[81]

^{a} EASC = ethyl aluminum sesquichloride

Labeled initiators have been used in evaluating the relative reactivity of a wide range of monomers towards initiating radicals.[359] The method involves determination of the relative concentrations of the end groups formed by addition to two monomers (*e.g.* **119** and **120**) in a binary copolymer formed with use of a labeled initiator. For example, when AIBMe-α-^{13}C is used to initiate copolymerization of MMA and VAc (Scheme 3.99),[357] the simple relationship (eq. 14) gives the relative rate constants for addition to the two monomers. Copolymerizations studied in this way are summarized in Table 3.13.

$$\frac{k_{MMA}}{k_{VAc}} = \frac{[VAc] \cdot [\mathbf{119}]}{[MMA] \cdot [\mathbf{120}]} \qquad (14)$$

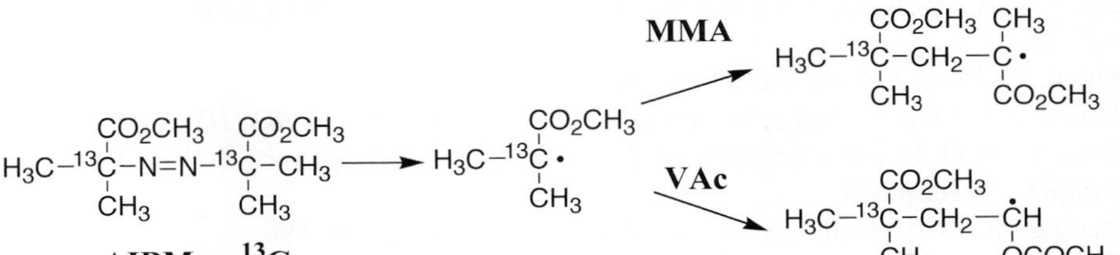

Scheme 3.99

3.6 References

1. Solomon, D.H.; Moad, G. *Makromol. Chem., Macromol. Symp.* **1987**, *10/11*, 109.
2. Moad, G.; Solomon, D.H. In *Comprehensive Polymer Science*; Eastmond, G.C.; Ledwith, A.; Russo, S.; Sigwalt, P., Eds.; Pergamon: Oxford, 1989; Vol. 3, p 97.
3. Solomon, D.H.; Cacioli, P.; Moad, G. *Pure Appl. Chem.* **1985**, *57*, 985.
4. Hwang, E.F.J.; Pearce, E.M. *Polym. Eng. Rev.* **1983**, *2*, 319.
5. Mita, I. In *Aspects of Degradation and Stabilization of Polymers*; Jellineck, H.H.G., Ed.; Elsevier: Amsterdam, 1978; p 247.
6. Solomon, D.H. *J. Macromol. Sci., Chem.* **1982**, *A17*, 337.
7. Moad, G.; Solomon, D.H.; Johns, S.R.; Willing, R.I. *Macromolecules* **1984**, *17*, 1094.
8. Moad, G.; Rizzardo, E.; Solomon, D.H. *Macromolecules* **1982**, *15*, 909.
9. Moad, G.; Solomon, D.H.; Johns, S.R.; Willing, R.I. *Macromolecules* **1982**, *15*, 1188.
10. Moad, G.; Rizzardo, E.; Solomon, D.H. *Aust. J. Chem.* **1983**, *36*, 1573.
11. Moad, G.; Rizzardo, E.; Solomon, D.H. *Polym. Bull.* **1984**, *12*, 471.
12. Griffiths, P.G.; Rizzardo, E.; Solomon, D.H. *J. Macromol. Sci., Chem.* **1982**, *A17*, 45.
13. Sato, T.; Otsu, T. *Makromol. Chem.* **1977**, *178*, 1941.
14. Krstina, J.; Moad, G.; Solomon, D.H. *Eur. Polym. J.* **1989**, *25*, 767.
15. Singh, M.; Nandi, U.S. *J. Polym. Sci., Polym. Lett. Ed.* **1979**, *17*, 121.
16. Schildknecht, C.E. In *Polymerization Processes*; Schildknecht, C.E.; Skeist, I., Eds.; Wiley: New York, 1977; p 88.
17. Boundy, R.H.; Boyer, R.F., Eds. *Styrene, Its Polymers, Copolymers and Derivatives*; Reinhold: New York, 1952.
18. Citterio, A.; Arnoldi, A.; Minisci, F. *J. Org. Chem.* **1979**, *44*, 2674.
19. Griller, D.; Ingold, K.U. *Acc. Chem. Res.* **1980**, *13*, 317.
20. Newcomb, M. *Tetrahedron* **1993**, *49*, 1151.
21. Grant, R.D.; Griffiths, P.G.; Moad, G.; Rizzardo, E.; Solomon, D.H. *Aust. J. Chem.* **1983**, *36*, 2447.
22. Bednarek, D.; Moad, G.; Rizzardo, E.; Solomon, D.H. *Macromolecules* **1988**, *21*, 1522.
23. Walling, C.; Thaler, W. *J. Am. Chem. Soc.* **1961**, *83*, 3877.
24. Elson, I.H.; Mao, S.W.; Kochi, J.K. *J. Am. Chem. Soc.* **1975**, *97*, 335.
25. Watanabe, Y.; Ishigaki, H.; Okada, H.; Suyama, S. *Polym. J.* **1997**, *29*, 693.

26. Baignee, A.; Howard, J.A.; Scaiano, J.C.; Stewart, L.C. *J. Am. Chem. Soc.* **1983**, *105*, 6120.
27. Moad, G. *Prog. Polym. Sci.* **1999**, *24*, 81.
28. Maillard, B.; Ingold, K.U.; Scaiano, J.C. *J. Am. Chem. Soc.* **1983**, *105*, 5095.
29. Hartzler, H.D. In *The Chemistry of the Cyano Group*; Rappoport, Z., Ed.; Wiley: London, 1970; p 671.
30. Bevington, J.C.; Troth, H.G. *Trans. Faraday Soc.* **1962**, *58*, 186.
31. Niki, E.; Kamiya, Y.; Ohta, N. *Bull. Chem. Soc. Japan* **1969**, *42*, 3220.
32. Moad, G.; Rizzardo, E.; Solomon, D.H.; Johns, S.R.; Willing, R.I. *Makromol. Chem., Rapid Commun.* **1984**, *5*, 793.
33. Achilias, D.S.; Kiparissides, C. *Macromolecules* **1992**, *25*, 3739.
34. Starnes, W.H., Jr.; Plitz, I.M.; Schilling, F.C.; Villacorta, G.M.; Park, G.S.; Saremi, A.H. *Macromolecules* **1984**, *17*, 2507.
35. Ishiwata, H.; Inoue, T.; Yoshihira, K. *J. Chromatogr.* **1986**, *370*, 275.
36. Fordham, P.J.; Gramshaw, J.W.; Castle, L. *Food Additives Contaminants* **2001**, *18*, 461.
37. Simionescu, C.I.; Chiriac, A.P.; Chiriac, M.V. *Polymer* **1993**, *34*, 3917.
38. Turro, N.J.; Kraeutler, B. *Acc. Chem. Res.* **1980**, *13*, 369.
39. Fischer, H.; Baer, R.; Hany, R.; Verhoolen, I.; Walbiner, M. *J. Chem. Soc., Perkin Trans. 2* **1990**, 787.
40. Buback, M.; Kowollik, C.; Kurz, C.; Wahl, A. *Macromol. Chem. Phys.* **2000**, *201*, 464.
41. Moad, G.; Rizzardo, E.; Solomon, D.H. *J. Macromol. Sci., Chem.* **1982**, *A17*, 51.
42. Gilbert, R.G. *Emulsion Polymerization: A Mechanistic Approach*; Academic Press: London, 1995.
43. Lovell, P.A.; El-Aasser, M.S., Eds. *Emulsion Polymerization and Emulsion Polymers*; John Wiley & Sons: London, 1997.
44. Morrison, B.R.; Maxwell, I.A.; Gilbert, R.G.; Napper, D.H. *ACS Symp. Ser.* **1992**, *492*, 28.
45. Sheppard, C.S.; Kamath, V.R. *Polym. Eng. Sci.* **1979**, *19*, 597.
46. Barton, J.; Borsig, E. *Complexes in Free Radical Polymerization*; Elsevier: Amsterdam, 1988.
47. Masson, J.C. In *Polymer Handbook*, 3rd ed.; Brandup, J.; Immergut, E.H., Eds.; Wiley: New York, 1989; p II/1.
48. Dixon, K.W. In *Polymer Handbook*, 4th ed.; Brandup, J.; Immergut, E.H.; Grulke, E.A., Eds.; John Wiley and Sons: New York, 1999; p II/1.
49. Engel, P.S. *Chem. Rev.* **1980**, *80*, 99.
50. Sheppard, C.S. In *Encyclopedia of Polymer Science and Engineering*, 2nd ed.; Mark, H.F.; Bikales, N.M.; Overberger, C.G.; Menges, G., Eds.; Wiley: New York, 1985; Vol. 2, p 143.
51. Koga, G.; Koga, N.; Anselme, J.-P. In *The Chemistry of the Hydrazo, Azo and Azoxy Groups*; Patai, S., Ed.; Wiley: London, 1975; Vol. 16, part 2, p 861.
52. Koenig, T. In *Free Radicals*; Kochi, J.K., Ed.; Wiley-Interscience: New York, 1973; Vol. 1, p 113.
53. Smith, P.A.S. In *The Chemistry of Open Chain Organic Nitrogen Compounds*; Benjamin: New York, 1966; Vol. 2, p 269.

54. Neuman, R.C., Jr.; Grow, R.H.; Binegar, G.A.; Gunderson, H.J. *J. Org. Chem.* **1990**, *55*, 2682.
55. Ayscough, P.B.; Brooks, B.R.; Evans, H.E. *J. Phys. Chem.* **1964**, *68*, 3889.
56. Nelsen, S.F.; Bartlett, P.D. *J. Am. Chem. Soc.* **1966**, *88*, 137.
57. Overberger, C.G.; Berenbaum, M.B. *J. Am. Chem. Soc.* **1951**, *73*, 2618.
58. Van-Hook, J.P.; Tobolsky, S. *J. Am. Chem. Soc.* **1958**, *80*, 779.
59. Barbe, W.; Rüchardt, C. *Makromol. Chem.* **1983**, *184*, 1235.
60. Krstina, J.; Moad, G.; Willing, R.I.; Danek, S.K.; Kelly, D.P.; Jones, S.L.; Solomon, D.H. *Eur. Polym. J.* **1993**, *29*, 379.
61. Talat-Erben, M.; Bywater, S. *J. Am. Chem. Soc.* **1955**, *77*, 3712.
62. Otsu, T.; Yamada, B. *J. Macromol. Sci. Chem* **1969**, *3*, 187.
63. Krstina, J.; Moad, G.; Solomon, D.H., unpublished results.
64. Duisman, W.; Rüchardt, C. *Chem. Ber.* **1978**, *111*, 596.
65. Overberger, C.G.; Berenbaum, M.B. *J. Am. Chem. Soc.* **1953**, *75*, 2078.
66. Bandlish, B.K.; Garner, A.W.; Hodges, M.L.; Timberlake, J.W. *J. Am. Chem. Soc.* **1975**, *97*, 5856.
67. Martin, J.C.; Timberlake, J.W. *J. Am. Chem. Soc.* **1970**, *92*, 978.
68. Alder, M.G.; Leffler, J.E. *J. Am. Chem. Soc.* **1954**, *76*, 1425.
69. Cohen, S.G.; Cohen, F.; Wang, C.H. *J. Org. Chem.* **1963**, *28*, 1479.
70. Solomon, S.; Wang, C.H.; Cohen, S.G. *J. Am. Chem. Soc.* **1957**, *79*, 4104.
71. Kiefer, H.; Traylor, T.G. *Tetrahedron Lett.* **1966**, *7*, 6163.
72. Rizzardo, E.; Serelis, A.K.; Solomon, D.H. *Aust. J. Chem.* **1982**, *35*, 2013.
73. Timberlake, J.W. In *Substituent Effects in Radical Chemistry (NATO ASI Ser., Ser. C)*; Viehe, H.G.; Janousek, Z.; Merenyi, R., Eds.; Reidel: Dordecht, 1986; Vol. 189, p 271.
74. Overberger, C.G.; Hale, W.F.; Berenbaum, M.B.; Finestone, A.B. *J. Am. Chem. Soc.* **1954**, *76*, 6185.
75. Duisman, W.; Rüchardt, C. *Tetrahedron Lett.* **1974**, *15*, 4517.
76. Rüchardt, C. *Top. Curr. Chem.* **1980**, *88*, 1.
77. Firestone, R.A. *J. Org. Chem.* **1980**, *45*, 3604.
78. Wolf, R.A. *ACS Symp. Ser.* **1989**, *404*, 416.
79. Henrici-Olivé, G.; Olivé, S. *Makromol. Chem.* **1962**, *58*, 188.
80. Tanaka, H.; Fukuoka, K.; Ota, T. *Makromol. Chem., Rapid. Commun.* **1985**, *6*, 563.
81. Lyons, R.A.; Moad, G.; Senogles, E. *Eur. Polym. J.* **1993**, *29*, 389.
82. Lyons, R.A.; Moad, G.; Senogles, E. In *Pacific Polymer Conference Preprints*; Polymer Division, Royal Australian Chemical Insitute: Brisbane, 1993; Vol. 3, p 249.
83. Krstina, J.; Moad, G.; Solomon, D.H. *Polym. Bull.* **1992**, *27*, 425.
84. Spurling, T.H.; Deady, M.; Krstina, J.; Moad, G. *Makromol. Chem., Macromol. Symp.* **1991**, *51*, 127.
85. Stickler, M. *Makromol. Chem.* **1986**, *187*, 1765.
86. O'Driscoll, K.F.; Huang, J. *Eur. Polym. J.* **1989**, *7/8*, 629.
87. Drewer, R.J. In *The Chemistry of the Hydrazo, Azo and Azoxy Groups*; Patai, S., Ed.; Wiley: London, 1975; Vol. 16, part 2, p 935.
88. Ayrey, G. *Chem. Rev.* **1963**, *63*, 645.
89. Fink, J.K. *J. Polym. Sci., Polym. Chem. Ed.* **1983**, *21*, 1445.

90. Cox, R.A.; Buncel, E. In *The Chemistry of the Hydrazo, Azo and Azoxy Groups*; Patai, S., Ed.; Wiley: London, 1975; Vol. 16, part 2, p 775.
91. Braun, D.; Czerwinski, W.K. *Makromol. Chem.* **1987**, *188*, 2371.
92. Fukuda, T.; Ma, Y.-D.; Inagaki, H. *Macromolecules* **1985**, *18*, 17.
93. Russell, G.T.; Napper, D.H.; Gilbert, R.G. *Macromolecules* **1988**, *21*, 2141.
94. Buback, M.; Huckestein, B.; Kuchta, F.-D.; Russell, G.T.; Schmid, E. *Macromol. Chem. Phys.* **1994**, *195*, 2117.
95. Sack, R.; Schulz, G.V.; Meyerhoff, G. *Macromolecules* **1988**, *21*, 3345.
96. Faldi, A.; Tirrell, M.; Lodge, T.P.; von Meerwall, E. *Macromolecules* **1994**, *27*, 4184.
97. Bizilj, S.; Kelly, D.P.; Serelis, A.K.; Solomon, D.H.; White, K.E. *Aust. J. Chem.* **1985**, *38*, 1657.
98. Trecker, D.J.; Foote, R.S. *J. Org. Chem.* **1968**, *33*, 3527.
99. Kodaira, K.; Ito, K.; Iyoda, S. *Polym. Commun.* **1987**, *28*, 86.
100. Talat-Erben, M.; Bywater, S. *J. Am. Chem. Soc.* **1954**, *77*, 3710.
101. Jaffe, A.B.; Skinner, K.J.; McBride, J.M. *J. Am. Chem. Soc.* **1972**, *94*, 8510.
102. Hammond, G.S.; Trapp, O.D.; Keys, R.T.; Neff, D.L. *J. Am. Chem. Soc.* **1959**, *81*, 4878.
103. Chung, R.P.-T.; Danek, S.K.; Quach, C.; Solomon, D.H. *J. Macromol. Sci., Chem.* **1994**, *A31*, 329.
104. Pryor, W.A.; Fiske, T.R. *Macromolecules* **1969**, *2*, 62.
105. Cascaval, C.N.; Straus, S.; Brown, D.W.; Florin, R.E. *J. Polym. Sci., Polym. Symp.* **1976**, *57*, 81.
106. Baysal, B.; Tobolsky, A.V. *J. Polym. Sci.* **1952**, *8*, 529.
107. Bevington, J.C.; Lewis, T.D. *Polymer* **1960**, *1*, 1.
108. Pryor, W.A.; Coco, J.H. *Macromolecules* **1970**, *3*, 500.
109. May, J.A., Jr.; Smith, W.B. *J. Phys. Chem.* **1968**, *72*, 2993.
110. Ayrey, G.; Haynes, A.C. *Makromol. Chem.* **1974**, *175*, 1463.
111. Athey, R.D., Jr. *J. Polym. Sci., Polym. Chem. Ed.* **1977**, *15*, 1517.
112. Braks, J.G.; Huang, R.Y.M. *J. Appl. Polym. Sci.* **1978**, *22*, 3111.
113. Bevington, J.C.; Breuer, S.W.; Heseltine, E.N.J.; Huckerby, T.N.; Varma, S.C. *J. Polym. Sci., Part A: Polym. Chem.* **1987**, *25*, 1085.
114. Quinga, E.M.Y.; Mendenhall, G.D. *J. Org. Chem.* **1985**, *50*, 2836.
115. Dulog, L.; Klein, P. *Chem. Ber.* **1971**, *104*, 902.
116. Dulog, L.; Klein, P. *Chem. Ber.* **1971**, *104*, 895.
117. Protasiewicz, J.; Mendenhall, G.D. *J. Org. Chem.* **1985**, *50*, 3220.
118. Mendenhall, G.D.; Stewart, L.C.; Scaiano, J.C. *J. Am. Chem. Soc.* **1982**, *104*, 5109.
119. Mendenhall, G.D.; Quinga, E.M.Y. *Int. J. Chem. Kinet.* **1985**, *17*, 1187.
120. Druliner, J.D.; Krusic, P.D.; Lehr, G.F.; Tolman, C.A. *J. Org. Chem.* **1985**, *50*, 5838.
121. Kiefer, H.; Traylor, T.G. *J. Am. Chem. Soc.* **1967**, *89*, 6667.
122. Mendenhall, G.D.; Cary, L.W. *J. Org. Chem.* **1975**, *40*, 1646.
123. Busfield, W.K.; Jenkins, I.D.; Rizzardo, E.; Solomon, D.H.; Thang, S.H. *J. Chem. Soc., Perkin Trans. 1* **1991**, 1351.
124. Ando, W., Ed. *Organic Peroxides*; Wiley: Chichester, 1992.

Initiation 153

125. Sheppard, C.S. In *Encyclopedia of Polymer Science and Engineering*, 2nd ed.; Mark, H.F.; Bikales, N.M.; Overberger, C.G.; Menges, G., Eds.; Wiley: New York, 1987; Vol. 11, p 1.
126. Patai, S., Ed. *The Chemistry of Functional Groups, The Chemistry of Peroxides*; Wiley: Chichester, UK, 1983.
127. Swern, D., Ed. *Organic Peroxides*; Wiley-Interscience: New York, 1970; Vol. 1.
128. Swern, D., Ed. *Organic Peroxides*; Wiley-Interscience: New York, 1971; Vol. 2.
129. Swern, D., Ed. *Organic Peroxides*; Wiley-Interscience: New York, 1971; Vol. 3.
130. Davies, A.G. *Organic Peroxides*; Butterworths: London, 1961.
131. Hawkins, E.G.R. *Organic Peroxides - Their Formation and Reactions*; Van Nostrand: Princeton, 1961.
132. Janzen, E.G.; Evans, C.A.; Nishi, Y. *J. Am. Chem. Soc.* **1972**, *94*, 8236.
133. Bawn, C.E.H.; Halford, R.G. *Trans. Faraday Soc.* **1955**, *51*, 780.
134. Yamada, M.; Kitagawa, K.; Komai, T. *Plast. Ind. News* **1971**, *17*, 131.
135. Blomquist, A.T.; Ferris, A. *J. Am. Chem. Soc.* **1951**, *73*, 3412.
136. Bartlett, P.D.; Benzing, E.P.; Pincock, R.E. *J. Am. Chem. Soc.* **1960**, *82*, 1762.
137. Hiatt, R.; Mill, T.; Irwin, K.C.; Castleman, J.K. *J. Org. Chem.* **1968**, *33*, 1421.
138. Huyser, E.S.; VanScoy, R. *J. Org. Chem.* **1968**, *33*, 3524.
139. Hiatt, R.; Strachan, W.M.J. *J. Org. Chem.* **1963**, *28*, 1893.
140. Kolthoff, I.M.; Miller, I.K. *J. Am. Chem. Soc.* **1951**, *73*, 3055.
141. Fujimori, K. In *Organic Peroxides*; Ando, W., Ed.; Wiley: Chichester, 1992; p 319.
142. Bouillion, G.; Lick, C.; Schank, K. In *The Chemistry of the Peroxides*; Patai, S., Ed.; Wiley: London, 1983; p 279.
143. Hiatt, R. In *Organic Peroxides*; Swern, D., Ed.; Wiley-Interscience: New York, 1971; Vol. 2, p 799.
144. Martin, J.C.; Hargis, J.H. *J. Am. Chem. Soc.* **1969**, *91*, 5399.
145. Pryor, W.A.; Morkved, E.H.; Bickley, H.T. *J. Org. Chem.* **1972**, *37*, 1999.
146. Nozaki, K.; Bartlett, P.D. *J. Am. Chem. Soc.* **1946**, *68*, 1686.
147. Sheldon, R.A.; Kochi, J.K. *J. Am. Chem. Soc.* **1970**, *92*, 4395.
148. Saltiel, J.; Curtis, H.C. *J. Am. Chem. Soc.* **1971**, *93*, 2056.
149. Yamauchi, S.; Hirota, N.; Takahara, S.; Sakuragi, H.; Tokumaru, K. *J. Am. Chem. Soc.* **1985**, *107*, 5021.
150. Grossi, L.; Lusztyk, J.; Ingold, K.U. *J. Org. Chem.* **1985**, *50*, 5882.
151. Nedelec, J.Y.; Lefort, D. *Tetrahedron* **1980**, *36*, 3199.
152. Rosenthal, I.; Mossoba, M.M.; Riesz, P. *J. Magn. Reson.* **1982**, *47*, 200.
153. Garcia-Rubio, L.H.; Mehta, J. *ACS Symp. Ser.* **1986**, 202.
154. Garcia-Rubio, L.H.; Ro, N.; Patel, R.D. *Macromolecules* **1984**, *17*, 1998.
155. Navolokina, R.A.; Zilberman, E.N.; Krasavina, N.B.; Kharitonova, O.A. *Izv. Vyssh. Uchebn. Zaved. Khim. Khim. Tekhnol.* **1986**, *29*, 83.
156. Stickler, M.; Dumont, E. *Makromol. Chem.* **1986**, *187*, 2663.
157. Walling, C.; Waits, H.P.; Milovanovic, J.; Pappiaonnou, C.G. *J. Am. Chem. Soc.* **1970**, *92*, 4927.
158. Sivaram, S.; Singhal, R.K.; Bhardwaj, I.S. *Polym. Bull.* **1980**, *3*, 27.
159. Curci, R.; Edwards, J.O. In *Organic Peroxides*; Swern, D., Ed.; Wiley: New York, 1971; Vol. 1, p 200.
160. Mayo, F.R.; Gregg, R.A.; Matheson, M.S. *J. Am. Chem. Soc.* **1951**, *73*, 1691.
161. Berger, K.C.; Deb, P.C.; Meyerhoff, G. *Macromolecules* **1977**, *10*, 1075.

162. Anisimov, Y.N.; Ivanchev, S.S.; Yurzhenko, A.I. *Polym. Sci. USSR (Engl. Transl.)* **1967**, *9*, 692.
163. Moad, G.; Solomon, D.H.; Willing, R.I. *Macromolecules* **1988**, *21*, 855.
164. Suehiro, T.; Kanoya, A.; Yamauchi, T.; Komori, T.; Igeta, S.-I. *Tetrahedron* **1968**, *24*, 1551.
165. Suehiro, T.; Kanoya, A.; Hara, H.; Nakahama, T.; Komori, T. *Bull. Chem. Soc. Japan* **1967**, *40*, 668.
166. Moad, G.; Rizzardo, E.; Solomon, D.H. *Tetrahedron Lett.* **1981**, *22*, 1165.
167. Sosnovsky, G.; Rawlinson, D.J., Eds. *Organic Peroxides, Metal Ion-Catalyzed Reactions of Symmetric Peroxides*; Wiley-Interscience: New York, 1970; Vol. 1.
168. Kochi, J.K. *J. Am. Chem. Soc.* **1962**, *84*, 1572.
169. Kurz, M.E.; Kovacic, P. *J. Org. Chem.* **1968**, *33*, 1950.
170. Chalfont, G.R.; Hey, D.H.; Liang, K.S.Y.; Perkins, M.J. *Chem. Commun. (London)* **1967**, 367.
171. Perkins, M.J.; Chalfont, G.R.; Hey, D.H.; Liang, K.S.Y. *J. Chem. Soc. (B)* **1971**, 233.
172. Rusakova, A.; Margaritova, M.F. *Vysokomol. Soedin. Ser. B* **1967**, *9*, 515.
173. Jones, R.G.; Catterall, E.; Bilson, R.T.; Booth, R.G. *J. Chem. Soc., Chem. Commun.* **1972**, 22.
174. Bevington, J.C.; Dyball, C.J.; Leech, J. *Makromol. Chem.* **1977**, *178*, 2741.
175. Bevington, J.C.; Dyball, C.J.; Leech, J. *Makromol. Chem.* **1979**, *180*, 657.
176. Sato, T.; Abe, M.; Otsu, T. *Makromol. Chem.* **1977**, *178*, 1259.
177. Bottle, S.; Busfield, W.K.; Jenkins, I.D.; Thang, S.; Rizzardo, E.; Solomon, D.H. *Eur. Polym. J.* **1989**, *25*, 671.
178. Dambatta, B.B.; Ebdon, J.R. *Eur. Polym. J.* **1986**, *22*, 783.
179. Walling, C. *Free Radicals in Solution*; Wiley: New York, 1957.
180. Imoto, M.; Choe, S. *J. Polym. Sci.* **1955**, *15*, 485.
181. Pryor, W.A.; Hendrickson, W.H., Jr. *Tetrahedron Lett.* **1983**, *24*, 1459.
182. Sato, T.; Kita, S.; Otsu, T. *Makromol. Chem.* **1975**, *176*, 561.
183. Strong, W.A. *Ind. Eng. Chem.* **1964**, *56(12)*, 33.
184. McBay, H.C.; Tucker, O. *J. Org. Chem.* **1954**, *19*, 869.
185. Razuvaev, G.A.; Terman, L.M.; Petukhow, G.G. *Dokl. Akad. Nauk. USSR (Engl. Transl.)* **1961**, *136*, 111.
186. Cohen, S.G.; Sparrow, D.B. *J. Am. Chem. Soc.* **1950**, *72*, 611.
187. Van Sickle, D.E. *J. Org. Chem.* **1969**, *34*, 3446.
188. Cuthbertson, M.J.; Moad, G.; Rizzardo, E.; Solomon, D.H. *Polym. Bull.* **1982**, *6*, 647.
189. Pastorino, R.L.; Lewis, R.N. In *Modern Plastics Encyclopedia*; McGraw-Hill: New York, 1988; p 165.
190. Duynstee, E.F.J.; Esser, M.L.; Schellekens, R. *Eur. Polym. J.* **1980**, *16*, 1127.
191. Crano, J. *J. Org. Chem.* **1966**, *31*, 3615.
192. Sawaki, Y. In *Organic Peroxides*; Ando, W., Ed.; Wiley: Chichester, 1992; p 426.
193. Bouillion, G.; Lick, C.; Schank, K. In *The Chemistry of the Peroxides*; Patai, S., Ed.; Wiley: London, 1983; p 287.
194. Singer, L.A. In *Organic Peroxides*; Swern, D., Ed.; Wiley-Interscience: New York, 1970; Vol. 1, p 265.

195. Nakamura, T.; Busfield, W.K.; Jenkins, I.D.; Rizzardo, E.; Thang, S.H.; Suyama, S. *J. Org. Chem.* **2000**, *65*, 16.
196. Buback, M. *Macromol. Symp.* **2002**, *182*, 103.
197. Buback, M.; Klingbeil, S.; Sandmann, J.; Sderra, M.B.; Vogele, H.P.; Wackerbarth, H.; Wittkowski, L. *Z. Phys. Chem.* **1999**, *210*, 199.
198. Buback, M.; Sandmann, J. *Z. Phys. Chem.* **2000**, *214*, 583.
199. Niki, E.; Kamiya, Y. *J. Am. Chem. Soc.* **1974**, *96*, 2129.
200. Hiatt, R.; Traylor, T.G. *J. Am. Chem. Soc.* **1965**, *87*, 3766.
201. Gupta, S.N.; Gupta, I.; Neckers, D.C. *J. Polym. Sci., Polym. Chem. Ed.* **1981**, *19*, 103.
202. Allen, N.S.; Hardy, S.J.; Jacobine, A.; Glaser, D.M.; Catalina, F.; Navaratnam, S.; Parsons, B.J. In *Radiation Curing of Polymers II*; Randell, D.R., Ed.; Royal Society of Chemistry: Cambridge, 1991; p 182.
203. Matsugo, S.; Saito, I. In *Organic Peroxides*; Ando, W., Ed.; Wiley: Chichester, 1992; p 157.
204. Sheldon, R.A. In *The Chemistry of the Peroxides*; Patai, S., Ed.; Wiley: London, 1983; p 161.
205. Hiatt, R. In *Organic Peroxides*; Swern, D., Ed.; Wiley-Interscience: New York, 1971; Vol. 3, p 1.
206. Suyama, S.; Sugihara, Y.; Watanabe, Y.; Nakamura, T. *Polym. J.* **1992**, *24*, 971.
207. Drumright, R.E.; Kastl, P.E.; Priddy, D.B. *Macromolecules* **1993**, *26*, 2246.
208. Matsuyama, K.; Kimura, H. *J. Org. Chem.* **1993**, *58*, 1766.
209. Bischoff, C.; Platz, K.-H. *J. Prakt. Chem* **1973**, *315*, 175.
210. Suyama, S.; Wanatabe, Y.; Sawaki, Y. *Bull. Chem. Soc. Japan* **1990**, *63*, 716.
211. Wanatabe, Y.; Ishigaki, H.; Okada, H.; Suyama, S. *Bull. Chem. Soc. Japan* **1991**, *64*, 1231.
212. Yamamoto, T.; Nakashio, Y.; Onishi; Hirota, M. *Nippon Kagaku Kaishi* **1985**, 2296.
213. Huyser, E.S.; Feng, R.H.C. *J. Org. Chem.* **1969**, *34*, 1727.
214. Huyser, E.S.; Bredeweg, C.J. *J. Am. Chem. Soc.* **1964**, *86*, 2401.
215. Huyser, E.S.; Bredeweg, C.J.; Vanscoy, R.M. *J. Am. Chem. Soc.* **1964**, *86*, 4148.
216. Zetterlund, P.B.; Yamauchi, S.; Yamada, B. *Macromol. Chem. Phys.* **2004**, *205*, 778.
217. Porter, N.A. In *Organic Peroxides*; Ando, W., Ed.; Wiley: Chichester, 1992; p 101.
218. Hiatt, R. In *Organic Peroxides*; Swern, D., Ed.; Wiley-Interscience: New York, 1971; Vol. 2, p 1.
219. Hiatt, R.; Irwin, K.C. *J. Org. Chem.* **1968**, *33*, 1436.
220. Nangia, P.S.; Benson, S.W. *J. Phys. Chem.* **1979**, *83*, 1138.
221. Benson, S.W. *Thermochemical Kinetics*; Wiley: New York, 1976.
222. Hiatt, R.; Mill, T.; Mayo, F.R. *J. Org. Chem.* **1968**, *33*, 1416.
223. Hiatt, R.; Mill, T.; Irwin, K.C.; Castleman, J.K. *J. Org. Chem.* **1968**, *33*, 1428.
224. Hiatt, R.; Irwin, K.C.; Gould, C.W. *J. Org. Chem.* **1968**, *33*, 1430.
225. Sosnovsky, G.; Rawlinson, D.J. In *Organic Peroxides*; Swern, D., Ed.; Wiley-Interscience: New York, 1971; Vol. 2, p 153.
226. Mulcahy, M.F.R.; Steven, J.R.; Ward, J.C. *Aust. J. Chem.* **1965**, *18*, 1177.
227. Hamilton, C.J.; Tighe, B.J. In *Comprehensive Polymer Science*; Eastmond, G.C.; Ledwith, A.; Russo, S.; Sigwalt, P., Eds.; Pergamon: Oxford, 1989; Vol. 3, p 261.

228. Bacon, R.G.R. *Chem. Soc., Quart. Rev.* **1955**, *9*, 287.
229. Sosnovsky, G.; Rawlinson, D.J. In *Organic Peroxides*; Swern, D., Ed.; Wiley-Interscience: New York, 1971; Vol. 2, p 269.
230. Sarac, A.S. *Prog. Polym. Sci.* **1999**, *24*.
231. House, D.A. *Chem. Rev.* **1962**, *62*, 185.
232. Behrman, E.J.; Edwards, J.O. *Rev. Inorg. Chem.* **1980**, *2*, 179.
233. Rudin, A.; Samanta, M.C.; Van Der Hoff, B.M.E. *J. Polym. Sci., Polym. Chem. Ed.* **1979**, *17*, 493.
234. Rasmussen, J.K.; Smith, H.K. *J. Am. Chem. Soc.* **1981**, *103*, 730.
235. Choi, K.Y.; Lee, C.Y. *Ind. Eng. Chem. Res.* **1987**, *26*, 2079.
236. Rasmussen, J.K.; Smith, H.K. *Polym. Prepr. (Am. Chem. Soc., Div. Polym. Chem)* **1982**, *23(1)*, 152.
237. Takeishi, M.; Ohkawa, H.; Hayama, S. *Makromol. Chem., Rapid Commun.* **1981**, *2*, 457.
238. Rasmussen, J.K.; Heilmann, S.M.; Krepski, L.R.; Smith, H.K. *ACS Symp. Ser.* **1987**, *326*, 116.
239. Rasmussen, J.K.; Heilmann, S.M.; Toren, P.E.; Pocius, A.V.; Kotnour, T.A. *J. Am. Chem. Soc.* **1983**, *105*, 6845.
240. Kim, Y.H. In *Organic Peroxides*; Ando, W., Ed.; Wiley: Chichester, 1992; p 387.
241. Curci, R.; Delano, G.; Edwards, J.O.; DiFuria, F.; Gallopo, A.R. *J. Org. Chem.* **1974**, *39*, 3020.
242. Chawla, O.P.; Fessenden, R.W. *J. Phys. Chem.* **1975**, *79*, 2693.
243. Tang, Y.; Thorn, R.P.; Mauldin, R.L.; Wine, P.H. *J. Photochem. Photobiol., A* **1988**, *44*, 243.
244. Ebdon, J.R.; Huckerby, T.N.; Hunter, T.C. *Polymer* **1994**, *35*, 250.
245. Brosse, J.-C.; Derouet, D.; Epaillard, F.; Soutif, J.-C.; Legeay, G.; Dusek, K. *Adv. Polym. Sci.* **1986**, *81*, 167.
246. Haber, F.; Weiss, J.J. *Proc. R. Soc., London* **1934**, *A147*, 332.
247. Shiga, T.; Boukhors, A.; Douzou, P. *J. Phys. Chem.* **1967**, *71*, 4264.
248. Shiga, T.; Boukhors, A.; Douzou, P. *J. Phys. Chem.* **1967**, *71*, 3559.
249. Chiang, Y.S.; Craddock, J.; Mickewich, D.; Turkevich, J. *J. Phys. Chem.* **1966**, *70*, 3509.
250. Dixon, W.T.; Norman, R.O.C. *J. Chem. Soc.* **1963**, 3119.
251. Walling, C. *Acc. Chem. Res.* **1975**, *8*, 125.
252. Norman, R.O.C. In *Chem. Soc. Spec. Publ. - Essays on Free Radical Chemistry*; Chem. Soc.: London, 1970; Vol. 24, p 117.
253. Grant, R.D.; Rizzardo, E.; Solomon, D.H. *J. Chem. Soc., Chem. Commun.* **1984**, 867.
254. Tezuka, T.; Narita, N. *J. Am. Chem. Soc.* **1979**, *101*, 7413.
255. Engel, P.S.; Y., C.; C., W. *J. Am. Chem. Soc.* **1991**, *113*, 4355.
256. Engel, P.S.; Pan, L.; Whitmire, K.H.; Guzman-Jiminez, I.; Willcott, M.R.; Smith, W.B. *J. Org. Chem.* **2000**, *65*, 1016.
257. Nazran, A.S.; Warkentin, J. *J. Am. Chem. Soc.* **1982**, *104*, 6405.
258. Dixon, D.W. *Adv. Oxygenated Processes* **1988**, *1*, 179.
259. Osei-Twum, E.Y.; McCallion, D.; Nazran, A.S.; Pannicucci, R.; Risbood, P.A.; Warkentin, J. *J. Org. Chem.* **1984**, *49*, 336.

Initiation

260. Simionescu, C.; Comanita, E.; Pastravanu, M.; Dumitriu, S. *Prog. Polym. Sci.* **1986**, *12*, 1.
261. Engel, P.S.; Pan, L.; Ying, Y.; Alemany, L.B. *J. Am. Chem. Soc.* **2001**, *123*, 3706.
262. Pappas, S.P. In *Comprehensive Polymer Science*; Eastmond, G.C.; Ledwith, A.; Russo, S.; Sigwalt, P., Eds.; Pergamon: Oxford, 1989; Vol. 4, p 337.
263. Pappas, S.P. In *Encyclopedia of Polymer Science and Engineering*, 2nd ed.; Mark, H.F.; Bikales, N.M.; Overberger, C.G.; Menges, G., Eds.; Wiley: New York, 1987; Vol. 11, p 186.
264. Pappas, S.P. *J. Radiat. Curing* **1987**, *14*, 6.
265. Bassi, G.L. *J. Radiat. Curing* **1987**, *14*, 18.
266. Mishra, M.K. *J. Macromol. Sci., Rev. Macromol. Chem.* **1983**, *C22*, 409.
267. Oster, G.; Yang, N. *Chem. Rev.* **1968**, *68*, 125.
268. Gruber, H.F. *Prog. Polym. Sci.* **1992**, *17*, 953.
269. Wagner, P.J. *Top. Curr. Chem.* **1976**, *66*, 1.
270. Hageman, H.J. *Prog. Org. Coat.* **1985**, *13*, 123.
271. McGinniss, V.D. *Dev. Polym. Photochem.* **1982**, *3*, 1.
272. Berner, G.; Kirchmayr, R.; Rist, G. *J. Oil Colour Chem. Assoc.* **1978**, *61*, 105.
273. Ledwith, A. *J. Oil Colour Chem. Assoc.* **1976**, *59*, 157.
274. Pappas, S.P. *Prog. Org. Coat.* **1973**, *2*, 333.
275. Heine, H.-G.; Rosenkranz, H.-J.; Rudolf, H. *Angew. Chem., Int. Ed. Engl.* **1972**, *11*, 974.
276. Pappas, S.P.; Chattopadhyay, A.K.; Carlblom, L.H. *ACS Symp. Ser.* **1976**, *25*, 12.
277. Hageman, H.J.; Overeem, T. *Makromol. Chem., Rapid Commun.* **1981**, *2*, 719.
278. Hu, S.K.; Wu, X.S.; Neckers, D.C. *Macromolecules* **2000**, *33*, 4030.
279. Lipscomb, N.T.; Tarshiani, Y. *J. Polym. Sci., Part A: Polym. Chem.* **1988**, *26*, 529.
280. Hageman, H.J.; Jansen, L.G.J. *Makromol. Chem.* **1988**, *189*, 2781.
281. Schnabel, W.; Sumiyoshi, T. In *New Trends in the Photochemistry of Polymers*; Allen, N.S.; Rabek, J.F., Eds.; Elsevier Applied Science: London, 1985; p 133.
282. Baxter, J.E.; Davidson, R.S.; Hageman, H.J.; Overeem, T. *Makromol. Chem.* **1988**, *189*, 2769.
283. Majima, T.; Schnabel, W.; Weber, W. *Makromol. Chem.* **1991**, *192*, 2307.
284. Rutsch, W.; Dietliker, K.; Leppard, D.; Kohler, M.; Misev, L.; Kolczak, U.; Rist, G. *Prog. Org. Coat.* **1996**, *27*, 227.
285. Colley, C.S.; Grills, D.C.; Besley, N.A.; Jockusch, S.; Matousek, P.; Parker, A.W.; Towrie, M.; Turro, N.J.; Gill, P.M.W.; George, M.W. *J. Am. Chem. Soc.* **2002**, *124*, 14952.
286. Klos, R.; Gruber, H.; Greber, G. *J. Macromol. Sci., Chem.* **1991**, *A28*, 925.
287. Carlini, C.; Angiolini, L. *Adv. Polym. Sci.* **1995**, *123*, 127.
288. Castelvetro, V.; Molesti, M.; Rolla, P. *Macromol. Chem. Phys.* **2002**, *203*, 1486.
289. Sarker, A.M.; Lungu, A.; Neckers, D.C. *Macromolecules* **1996**, *29*, 8047.
290. Cook, W.D. *Polymer* **1992**, *33*, 600.
291. Andrzejewska, E.; Linden, L.-A.; Rabek, J.F. *Macromol. Chem. Phys.* **1998**, *199*, 441.
292. Griller, D.; Howard, J.A.; Marriott, P.R.; Scaiano, J.C. *J. Am. Chem. Soc.* **1981**, *103*, 619.
293. Alexander, I.J.; Scott, R.J. *Br. Polym. J.* **1983**, *15*, 30.

294. Corrales, T.; Catalina, F.; Peinado, C.; Allen, N.S. *J. Photochem. Photobio. A* **2003**, *159*, 103.
295. Nair, C.P.R.; Clouet, G. *J. Macromol. Sci., Rev. Macromol. Chem. Phys.* **1991**, *C31*, 311.
296. Otsu, T.; Matsumoto, A. *Adv. Polym. Sci.* **1998**, *136*, 75.
297. Misra, N.; Bajpai, U.D.N. *Prog. Polym. Sci.* **1982**, *8*, 61.
298. Bamford, C.H. In *Comprehensive Polymer Science*; Eastmond, G.C.; Ledwith, A.; Russo, S.; Sigwalt, P., Eds.; Pergamon: Oxford, 1989; Vol. 3, p 123.
299. Bamford, C.H. In *Reactivity, Mechanism, and Structure in Polymer Chemistry*; Jenkins, A.D.; Ledwith, A., Eds.; Wiley-Interscience: London, 1974; p 52.
300. Eastmond, G.C.; Grigor, J. *Makromol. Chem., Rapid Commun.* **1986**, *7*, 375.
301. Bamford, C.H.; Eastmond, G.C.; Woo, J.; Richards, D.H. *Polymer* **1982**, *23*, 643.
302. Ho, T.-L. *Synthesis* **1973**, 347.
303. Mino, G.; Kaizerman, S.; Rasmussen, E. *J. Am. Chem. Soc.* **1959**, *81*, 1494.
304. Bamford, C.H.; Middleton, I.P.; Sataka, Y.; Al-Lamee, K.G. In *Advances in Polymer Synthesis*; Culbertson, B.M.; McGrath, J.E., Eds.; Plenum: New York, 1986; p 291.
305. Hebeish, A.; Guthrie, J.T. *The Chemistry and Technology of Cellulosic Copolymers*; Springer: Berlin, 1981.
306. McDowell, D.J.; Gupta, B.S.; Stannett, V.T. *Prog. Polym. Sci.* **1984**, *10*, 1.
307. Hill, D.J.T.; McMillan, A.M.; O'Donnell, J.H.; Pomery, P.J. *Makromol. Chem., Macromol. Symp.* **1990**, *33*, 201.
308. Casinos, I. *Polymer* **1992**, *33*, 1304.
309. Moad, G.; Rizzardo, E.; Solomon, D.H. In *Comprehensive Polymer Science*; Eastmond, G.C.; Ledwith, A.; Russo, S.; Sigwalt, P., Eds.; Pergamon: Oxford, 1989; Vol. 3, p 141.
310. Pryor, W.A.; Lasswell, L.D. *Adv. Free Radical Chem.* **1975**, *5*, 27.
311. Kurbatov, V.A. *Russ. Chem. Rev. (Engl. Transl.)* **1987**, *56*, 505.
312. Hall, H.K., Jr. *Agnew. Chem. Int. Ed. Engl.* **1983**, *22*, 440.
313. Kirchner, K.; Riederle, K. *Angew. Makromol. Chem.* **1983**, *111*, 1.
314. Flory, P.J. *J. Am. Chem. Soc.* **1937**, *59*, 241.
315. Mayo, F.R. *Polym. Prepr. (Am. Chem. Soc., Div. Polym. Chem)* **1961**, *2*, 55.
316. Chong, Y.K.; Rizzardo, E.; Solomon, D.H. *J. Am. Chem. Soc.* **1983**, *105*, 7761.
317. Komber, H.; Gruner, M.; Malz, H. *Macromol. Rapid Commun.* **1998**, *19*, 83.
318. Moad, G.; Rizzardo, E.; Solomon, D.H. *Polym. Bull.* **1982**, *6*, 589.
319. Buzanowski, W.C.; Graham, J.D.; Priddy, D.B.; Shero, E. *Polymer* **1992**, *33*, 3055.
320. Hui, A.; Hamielec, A.E. *J. Appl. Polym. Sci.* **1972**, *16*, 749.
321. Kothe, T.; Fischer, H. *J. Polym. Sci., Part A: Polym. Chem.* **2001**, *39*, 4009.
322. Olaj, O.F.; Kaufmann, H.F.; Breitenbach, J.W. *Makromol. Chem.* **1977**, *178*, 2707.
323. Clouet, G.; Chaumont, P.; Corpart, P. *J. Polym. Sci., Part A: Polym. Chem.* **1993**, *31*, 2815.
324. Walling, C.; Briggs, E.R. *J. Am. Chem. Soc.* **1946**, *68*, 1141.
325. Lingnau, J.; Meyerhoff, G. *Polymer* **1983**, *24*, 1473.
326. Lingnau, J.; Meyerhoff, G. *Macromolecules* **1984**, *17*, 941.
327. Lingnau, J.; Stickler, M.; Meyerhoff, G. *Eur. Polym. J.* **1980**, *16*, 785.
328. Stickler, M.; Meyerhoff, G. *Makromol. Chem.* **1978**, *179*, 2729.
329. Hall, H.K., Jr.; Padias, A.B. *Acc. Chem. Res.* **1990**, *23*, 3.

330. Matsuda, M.; Abe, K. *J. Polym. Sci., Part A-1* **1968**, *6*, 1441.
331. Sato, T.; Abe, M.; Otsu, T. *Makromol. Chem.* **1977**, *178*, 1061.
332. Spychaj, T.; Hamielec, A.E. *J. Appl. Polym. Sci.* **1991**, *42*, 2111.
333. Hasha, D.L.; Priddy, D.B.; Rudolf, P.R.; Stark, E.J.; de Pooter, M.; Van Damme, F. *Macromolecules* **1992**, *25*, 3046.
334. Kirchner, K.; Schlapkohl, H. *Makromol. Chem.* **1976**, *177*, 2031.
335. Stille, J.K.; Chung, D.C. *Macromolecules* **1975**, *8*, 83.
336. Hall, H.K., Jr.; Padias, A.B.; Pandya, A.; Tanaka, H. *Macromolecules* **1987**, *20*, 247.
337. Kokubo, T.; Iwatsuki, S.; Yamashita, Y. *Makromol. Chem.* **1969**, *123*, 256.
338. Sato, T.; Abe, M.; Otsu, T. *J. Macromol. Sci., Chem.* **1981**, *A15*, 367.
339. Gaylord, N.G.; Takahashi, A. *Adv. Chem. Ser.* **1969**, *91*, 94.
340. Jug, K. *J. Am. Chem. Soc.* **1987**, *109*, 3534.
341. Fischer, H.; Radom, L. *Angew. Chem., Int. Ed. Engl.* **2001**, *40*, 1340.
342. Giese, B. *Angew. Chem., Int. Ed. Engl.* **1983**, *22*, 753.
343. Ghosez-Giese, A.; Giese, B. *ACS Symp. Ser.* **1998**, *685*, 50.
344. Tedder, J.M. *Angew. Chem., Int. Ed. Engl.* **1982**, *21*, 401.
345. Beckwith, A.L.J. *Tetrahedron* **1981**, *37*, 3073.
346. Tedder, J.M.; Walton, J.C. *Acc. Chem. Res.* **1976**, *9*, 183.
347. Tedder, J.M.; Walton, J.C. *Tetrahedron* **1980**, *36*, 701.
348. Giese, B.; Lachhein, S. *Angew. Chem., Int. Ed. Engl.* **1981**, *20*, 967.
349. Lorand, J.P. In *Landolt-Bornstein, New Series, Radical Reaction Rates in Solution*; Fischer, H., Ed.; Springer-Verlag: Berlin, 1984; Vol. II/13a, p 135.
350. Asmus, K.-D.; Bonifacic, M. In *Landolt-Börnstein, New Series, Radical Reaction Rates in Solution*; Fischer, H., Ed.; Springer-Verlag: Berlin, 1984; Vol. II/13b.
351. Roduner, E.; Crockett, R. In *Landolt-Bornstein, New Series, Radical Reaction Rates in Solution*; Fischer, H., Ed.; Springer-Verlag: Berlin, 1995; Vol. II/18a, p 79.
352. Heberger, K.; Fischer, H. *Int. J. Chem. Kinet.* **1993**, *25*, 249.
353. Busfield, W.K.; Jenkins, I.D.; Van Le, P. *Polym. Bull.* **1997**, *38*, 149.
354. Busfield, W.K.; Jenkins, I.D.; Van Le, P. *Polym. Bull.* **1996**, *36*, 435.
355. Busfield, W.K.; Jenkins, I.D.; Van Le, P. *J. Polym. Sci., Part A: Polym. Chem.* **1998**, *36*, 2169.
356. Gridnev, A.A.; Ittel, S.D. *Macromolecules* **1996**, *29*, 5864.
357. Krstina, J.; Moad, G.; Solomon, D.H. *Eur. Polym. J.* **1992**, *28*, 275.
358. Behari, K.; Bevington, J.C.; Huckerby, T.N. *Makromol. Chem.* **1987**, *188*, 2441.
359. Bevington, J.C.; Huckerby, T.N.; Hutton, N.W.E. *J. Polym. Sci., Polym. Chem. Ed.* **1982**, *20*, 2655.
360. Bevington, J.C.; Huckerby, T.N.; Hutton, N.W.E. *Eur. Polym. J.* **1984**, *20*, 525.
361. Ayrey, G.; Jumangat, K.; Bevington, J.C.; Huckerby, T.N. *Polym. Commun.* **1983**, *24*, 275.
362. Bevington, J.C.; Huckerby, T.N.; Varma, S.C. *Eur. Polym. J.* **1986**, *22*, 427.
363. Barson, C.A.; Bevington, J.C.; Huckerby, T.N. *Polym. Bull.* **1986**, *16*, 209.
364. Bevington, J.C.; Breuer, S.W.; Huckerby, T.N. *Polym. Commun.* **1984**, *25*, 260.
365. Chatgilialoglu, C.; Ingold, K.U.; Scaiano, J.C. *J. Am. Chem. Soc.* **1981**, *103*, 7739.
366. Heberger, K.; Walbiner, M.; Fischer, H. *Angew. Chem., Int. Ed. Engl.* **1992**, *31*, 635.
367. Szwarc, M. *J. Polym. Sci.* **1955**, *16*, 367.

368. Herk, L.; Stefani, A.; Szwarc, M. *J. Am. Chem. Soc.* **1961**, *83*, 3003.
369. Beranek, I.; Fischer, H. In *Free Radicals in Synthesis and Biology*; Minisci, F., Ed.; Kluwer: Dordrecht, 1989; p 303.
370. Bevington, J.C.; Cywar, D.A.; Huckerby, T.N.; Senogles, E.; Tirrell, D.A. *Eur. Polym. J.* **1990**, *26*, 41.
371. Bevington, J.C.; Cywar, D.A.; Huckerby, T.N.; Senogles, E.; Tirrell, D.A. *Eur. Polym. J.* **1990**, *26*, 871.
372. Nakamura, T.; Suyama, S.; Busfield, W.K.; Jenkins, I.D.; Rizzardo, E.; Thang, S.H. *Polymer* **1999**, *40*, 1395.
373. Russell, G.A.; Jiang, W.; Hu, S.S.; Khanna, R.K. *J. Org. Chem.* **1986**, *51*, 5498.
374. Münger, K.; Fischer, H. *Int. J. Chem. Kinet.* **1985**, *17*, 809.
375. Bevington, J.C.; Lyons, R.A.; Senogles, E. *Eur. Polym. J.* **1992**, *28*, 283.
376. Santi, R.; Bergamini, F.; Citterio, A.; Sebastiano, R.; Nicolini, M. *J. Org. Chem.* **1992**, *57*, 4250.
377. Scaiano, J.C.; Stewart, L.C. *J. Am. Chem. Soc.* **1983**, *105*, 3609.
378. Levin, Y.A.; Abul'khanov, A.G.; Nefedov, A.G.; Skorobogatova, M.S.; Ivanov, B.E. *Dokl. Phys. Chem. (Engl. Transl.)* **1977**, *235*, 728.
379. Citterio, A.; Minisci, F.; Vismara, E. *J. Org. Chem.* **1982**, *47*, 81.
380. Kuwae, Y.; Kamachi, M. *Bull. Chem. Soc. Japan* **1989**, *62*, 2474.
381. Dzhabiyeva, Z.M.; Matkovskii, P.Y.; Pechatnikov, Y.L.; Byrikhina, N.A. *Polym. Sci. USSR (Engl. Transl.)* **1985**, *27*, 2416.
382. Bevington, J.C.; Huckerby, T.N.; Hutton, N.W.E. *Eur. Polym. J.* **1982**, *18*, 963.
383. Mahabadi, H.K.; O'Driscoll, K.F. *Makromol. Chem.* **1977**, *178*, 2629.
384. Citterio, A.; Vismara, E.; Bernardi, R. *J. Chem. Res., Miniprint* **1983**, *4*, 876.
385. Pryor, W.A.; Fiske, T.R. *Trans. Faraday Soc.* **1969**, *65*, 1865.
386. Dickerman, S.C.; Megna, I.S.; Skoultchi, M.M. *J. Am. Chem. Soc.* **1959**, *81*, 2270.
387. Bevington, J.C.; Ito, T. *Trans. Faraday Soc.* **1968**, *64*, 1329.
388. Chatgilialoglu, C.; Crich, D.; Komatsu, M.; Ryu, I. *Chem. Rev.* **1999**, *99*, 1991.
389. Howard, J.A.; Scaiano, J.C. In *Landoldt-Bornstein, New Series, Radical Reaction Rates in Solution*; Fischer, H., Ed.; Springer-Verlag: Berlin, 1984; Vol. II/13d, p 5.
390. Lusztyk, J. In *Landoldt-Bornstein, New Series, Radical Reaction Rates in Solution*; Fischer, H., Ed.; Springer-Verlag: Berlin, 1995; Vol. II/18d1, p 1.
391. Hartung, J.; Gottwald, T.; Spehar, K. *Synthesis* **2002**, 1469.
392. Heicklein, J.P. *Adv. Photochem.* **1988**, *14*, 177.
393. Kochi, J.K. In *Free Radicals*; Kochi, J.K., Ed.; Wiley: New York, 1973; Vol. 2, p 665.
394. Walling, C. *Pure. Appl. Chem.* **1967**, *15*, 69.
395. Ingold, K.U. *Pure Appl. Chem.* **1967**, *15*, 49.
396. Grant, R.D.; Rizzardo, E.; Solomon, D.H. *Makromol. Chem.* **1984**, *185*, 1809.
397. Grant, R.D.; Rizzardo, E.; Solomon, D.H. *J. Chem. Soc., Perkin Trans. 2* **1985**, 379.
398. Chateauneuf, J.; Lusztyk, J.; Ingold, K.U. *J. Am. Chem. Soc.* **1988**, *110*, 2886.
399. Bevington, J.C.; Harris, D.O.; Johnson, M. *Eur. Polym. J.* **1965**, *1*, 235.
400. Bevington, J.C.; Breuer, S.W.; Huckerby, T.N. *Macromolecules* **1989**, *22*, 55.
401. Moad, G.; Rizzardo, E.; Solomon, D.H. *Makromol. Chem., Rapid Commun.* **1982**, *3*, 533.
402. Busfield, W.K.; Jenkins, I.D.; Thang, S.H.; Rizzardo, E.; Solomon, D.H. *Eur. Polym. J.* **1993**, *29*, 397.

403. Cuthbertson, M.C.; Rizzardo, E., personal communication.
404. Griffiths, P.G.; Rizzardo, E.; Solomon, D.H. *Tetrahedron Lett.* **1982**, *23*, 1309.
405. Rizzardo, E., unpublished results.
406. Cuthbertson, M.J.; Rizzardo, E.; Solomon, D.H. *Aust. J. Chem.* **1985**, *38*, 315.
407. Cuthbertson, M.J.; Rizzardo, E.; Solomon, D.H. *Aust. J. Chem.* **1983**, *36*, 1957.
408. Busfield, W.K.; Jenkins, I.D.; Thang, S.H.; Rizzardo, E.; Solomon, D.H. *Aust. J. Chem.* **1985**, *38*, 689.
409. Busfield, W.K.; Jenkins, I.D.; Thang, S.H.; Rizzardo, E.; Solomon, D.H. *J. Chem. Soc., Perkin Trans. 1* **1988**, 485.
410. Jones, M.J.; Moad, G.; Rizzardo, E.; Solomon, D.H. *J. Org. Chem.* **1989**, *54*, 1607.
411. Sakurai, H.; Hosomi, A. *J. Am. Chem. Soc.* **1967**, *89*, 458.
412. Walling, C.; McGuinness, J.A. *J. Am. Chem. Soc.* **1969**, *91*, 2053.
413. Encina, M.V.; Rivera, M.; Lissi, E.A. *J. Polym. Sci., Polym. Chem. Ed.* **1978**, *16*, 1709.
414. Kunitake, T.; Murakami, S. *J. Polym. Sci., Polym. Chem. Ed.* **1974**, *12*, 67.
415. Korth, H.-G.; Sustmann, R. *Tetrahedron Lett.* **1985**, *26*, 2551.
416. Griffiths, P.G.; Rizzardo, E., unpublished data.
417. Wong, P.C.; Griller, D.; Scaiano, J.C. *J. Am. Chem. Soc.* **1982**, *104*, 5106.
418. Weber, M.; Fischer, H. *J. Am. Chem. Soc.* **1999**, *121*, 7381.
419. Tsentalovich, Y.P.; Kulik, L.V.; Gritsan, N.P.; Yurkovskaya, A.V. *J. Phys. Chem. A* **1998**, *102*, 7975.
420. Walling, C.; Padwa, A. *J. Am. Chem. Soc.* **1963**, *85*, 1593.
421. Nakamura, T.; Watanabe, Y.; Suyama, S.; Tezuka, H. *J. Chem. Soc., Perkin Trans. 2* **2002**, 1364.
422. Huyser, E.S.; Jankauskas, K.J. *J. Org. Chem.* **1970**, *35*, 3196.
423. Kamath, V.R.; Sargent, J.D., Jr. *J. Coat. Technol.* **1987**, *59*, 51.
424. Nakamura, T.; Busfield, W.K.; Jenkins, I.D.; Rizzardo, E.; Thang, S.H.; Suyama, S. *Macromolecules* **1997**, *30*, 2843.
425. Nakamura, T.; Busfield, W.K.; Jenkins, I.D.; Rizzardo, E.; Thang, S.H.; Suyama, S. *J. Am. Chem. Soc.* **1997**, *119*, 10987.
426. Avila, D.V.; Brown, C.E.; Ingold, K.U.; Lusztyk, J. *J. Am. Chem. Soc* **1993**, *115*, 466.
427. Bertrand, M.P.; Surzur, J.-M. *Tetrahedron Lett.* **1976**, *17*, 3451.
428. Bertrand, M.P.; Oumar-Mahamat, H.; Surzur, J.M. *Bull. Soc. Chim. Fr.* **1985**, 115.
429. Hilborn, J.W.; Pincock, J.A. *J. Am. Chem. Soc.* **1991**, *113*, 2683.
430. Bevington, J.C. *Angew. Makromol. Chem.* **1991**, *185/186*, 1.
431. Edge, D.J.; Kochi, J.K. *J. Am. Chem. Soc.* **1973**, *95*, 2635.
432. Kochi, J.K. *J. Am. Chem. Soc.* **1962**, *84*, 774.
433. Lane, J.; Tabner, B.J. *J. Chem. Soc., Perkin Trans. 2* **1984**, 1823.
434. Barclay, L.R.C.; Griller, D.; Ingold, K.U. *J. Am. Chem. Soc.* **1982**, *104*, 4399.
435. Beckwith, A.L.J.; Thomas, C.B. *J. Chem. Soc., Perkin Trans. 2* **1973**, 861.
436. Nakata, T.; Tokumaru, K.; Simamura, O. *Tetrahedron Lett.* **1967**, *8*, 3303.
437. Fischer, H. *Z. Naturforsch.* **1964**, *19a*, 866.
438. Fischer, H.; Giacometti, G. *J. Polym. Sci., Polym. Symp.* **1967**, *16*, 2763.
439. Roth, H.K.; Wunsche, P. *Acta Polym.* **1981**, *32*, 491.
440. Maruthamuthu, P. *Makromol. Chem., Rapid. Commun.* **1980**, *1*, 23.
441. McAskill, N.A.; Sangster, D.F. *Aust. J. Chem.* **1984**, *37*, 2137.

442. Sloane, T.M.; Brudzynski, R.J. *J. Am. Chem. Soc.* **1979**, *101*, 1495.
443. Ledwith, A.; Russell, P.J. *J. Polym. Sci., Polym. Lett. Ed.* **1975**, *13*, 109.
444. Citterio, A.; Arnoldi, C.; Giordano, C.; Castaldi, G. *J. Chem. Soc., Perkin Trans. 1* **1983**, 891.
445. Arnoldi, C.; Citterio, A.; Minisci, F. *J. Chem. Soc., Perkin Trans. 2* **1983**, 531.
446. Fristad, W.E.; Peterson, J.R. *Tetrahedron* **1984**, *40*, 1469.
447. McAskill, A.; Sangster, D.F. *Aust. J. Chem.* **1979**, *32*, 2611.
448. Ghosh, N.N.; Mandal, B.M. *Macromolecules* **1986**, *19*, 19.
449. Misra, N.; Mandal, B.M. *Macromolecules* **1984**, *17*, 495.
450. Misra, N.; Mandal, B.M. *J. Polym. Sci., Polym. Lett. Ed.* **1985**, *23*, 63.
451. Banthia, A.K.; Mandal, B.M.; Palit, S.R. *J. Polym. Sci., Polym. Chem. Ed.* **1977**, *15*, 945.
452. Howard, J.A. *Rev. Chem. Intermed.* **1984**, *5*, 1.
453. Neta, P.; Huie, R.E.; Ross, A.B. *J. Chem. Phys. Ref., Data* **1990**, *19*, 413.
454. Howard, J.A. In *Free Radicals*; Kochi, J.K., Ed.; Wiley: New York, 1973; Vol. 2, p 3.
455. Alonso, A.; Peinado, C.; Lozano, A.E.; Catalina, F.; Zimmermann, C.; Schnabel, W. *J. Macromol. Sci., Chem.* **1999**, *A36*, 605.
456. Lozano, A.E.; Alonso, A.; Catalina, F.; Peinado, C. *Macromol. Theory Simul.* **1999**, *8*, 93.
457. Scott, G.P.; Soong, C.C.; Allen, J.L.; Reynolds, J.L. *Polym. Prepr. (Am. Chem. Soc., Div. Polym. Chem)* **1963**, *4(1)*, 67.
458. Busfield, W.K.; Heiland, K.; Jenkins, I.D. *Tetrahedron Lett.* **1994**, *35*, 6541.
459. Farina, M.; Di Silvestro, G.; Sozzani, P. *Makromol. Chem.* **1989**, *190*, 213.
460. Ito, O.; Matsuda, M. *J. Am. Chem. Soc.* **1979**, *101*, 5732.
461. Busfield, W.K.; Jenkins, I.D.; Heiland, K. *Tetrahedron Lett.* **1995**, *36*, 1109.
462. Ito, O.; Matsuda, M. *J. Am. Chem. Soc.* **1979**, *101*, 1815.
463. Sato, T.; Abe, M.; Otsu, T. *Makromol. Chem.* **1977**, *178*, 1951.
464. Ito, O. *J. Am. Chem. Soc* **1983**, *105*, 850.
465. Sumiyoshi, T.; Schnabel, W. *Makromol. Chem.* **1985**, *186*, 1811.
466. Kajiwara, A.; Konishi, Y.; Morishima, Y.; Schnabel, W.; Kuwata, K.; Kamachi, M. *Macromolecules* **1993**, *26*, 1656.
467. Bottle, S.E.; Busfield, W.K.; Grice, I.D.; Heiland, K.; Meutermans, W.; Monteiro, M. *Prog. Pac. Polym. Sci.* **1994**, *3*, 85.
468. Geers, B.N.; Gleicher, G.J.; Church, D.F. *Tetrahedron* **1980**, *36*, 997.
469. Ito, O.; Matsuda, M. *J. Am. Chem. Soc.* **1979**, *101*, 5732.
470. Ito, O.; Matsuda, M. *J. Phys. Chem.* **1984**, *88*, 1002.
471. Ito, O.; Matsuda, M. *J. Org. Chem.* **1983**, *48*, 2748.
472. Ito, O.; Matsuda, M. *J. Org. Chem.* **1983**, *48*, 2410.
473. Bessiere, J.-M.; Boutevin, B.; Sarraf, L. *Polym. Bull.* **1987**, *18*, 253.
474. Baxter, J.E.; Davidson, R.S.; Hageman, H.J.; Overeem, T. *Makromol. Chem., Rapid Commun.* **1987**, *8*, 311.
475. Weber, M.; Turro, N.J. *J. Phys. Chem. A* **2003**, ASAP.
476. Perkins, M.J. *Adv. Phys. Org. Chem* **1981**, *17*, 1.
477. Kamachi, M. *Adv. Polym. Sci.* **1987**, *82*, 207.
478. Yamada, B.; Westmoreland, D.G.; Kobatake, S.; Konosu, O. *Prog. Polym. Sci.* **1999**, *24*, 565.

479. Chalfont, G.R.; Perkins, M.J.; Horsfield, A. *J. Am. Chem. Soc.* **1968**, *90*, 7141.
480. Savedoff, L.G.; Ranby, B. *Polym. Prepr. (Am. Chem. Soc., Div. Polym. Chem)* **1978**, *19(1)*, 629.
481. Yamada, B.; Fujity, M.; Sakamoto, K.; Otsu, T. *Polym. Bull.* **1994**, *33*, 309.
482. Yamada, B.; Yoshikawa, E.; Otsu, T. *Polymer* **1992**, *33*, 3245.
483. Bevington, J.C.; Tabner, B.J.; Fridd, P.F. *Rev. Roum. Chim.* **1980**, *25*, 947.
484. Bevington, J.C.; Fridd, P.F.; Tabner, B.J. *J. Chem. Soc., Perkin Trans. 2* **1982**, 1389.
485. Pichot, C.; Spitz, R.; Guyot, A. *J. Macromol. Sci.- Chem.* **1977**, *A11*, 251.
486. Minisci, F. *Acc. Chem. Res.* **1975**, *8*, 165.
487. Caronna, T.; Citterio, A.; Ghirardini, M.; Minisci, F. *Tetrahedron* **1977**, *33*, 793.
488. Bamford, C.H.; Jenkins, A.D.; Johnston, R. *Proc. R. Soc., London* **1957**, *A239*, 214.
489. Giese, B. *Radicals in Organic Synthesis: Formation of Carbon-Carbon Bonds*; Pergamon Press: Oxford, 1986.
490. Giese, B. *Rev. Chem. Intermed.* **1986**, *7*, 3.
491. Hill, C.L.; Whitesides, G.M. *J. Am. Chem. Soc.* **1974**, *96*, 870.
492. Jones, S.A.; Prementine, G.S.; Tirrell, D.A. *J. Am. Chem. Soc.* **1985**, *107*, 5275.
493. Russell, G.A. *Acc. Chem. Res.* **1989**, *22*, 1.
494. Pike, P.; Hershberger, S.; Hershberger, J. *Tetrahedron Lett.* **1985**, *26*, 6289.
495. Giese, B.; Gonzalez-Gomez, J.A.; Witzel, T. *Angew. Chem., Int. Ed. Engl.* **1984**, *23*, 69.
496. Rizzardo, E.; Solomon, D.H. *Polym. Bull.* **1979**, *1*, 529.
497. Busfield, W.K.; Jenkins, I.D.; Thang, S.H.; Rizzardo, E.; Solomon, D.H. *Tetrahedron Lett.* **1985**, *26*, 5081.
498. Aldabbagh, F.; Busfield, W.K.; Jenkins, I.D. *Aust. J. Chem.* **2001**, *54*, 313.
499. Busfield, W.K.; Grice, I.D.; Jenkins, I.D. *Aust. J. Chem.* **1995**, *48*, 625.
500. Nakamura, T.; Busfield, W.K.; Jenkins, I.D.; Rizzardo, E.; Thang, S.H.; Suyama, S. *J. Org. Chem.* **1997**, *62*, 5578.
501. Nakamura, T.; Watanabe, Y.; Tezuka, H.; Busfield, W.K.; Jenkins, I.D.; Rizzardo, E.; Thang, S.H.; Suyama, S. *Chem. Lett.* **1997**, 1093.
502. Nakamura, T.; Busfield, W.K.; Jenkins, I.D.; Rizzardo, E.; Thang, S.H.; Suyama, S. *Macromolecules* **1996**, *29*, 8975.
503. Nakamura, T.; Busfield, W.K.; Jenkins, I.D.; Rizzardo, E.; Thang, S.H.; Suyama, S. *J. Am. Chem. Soc.* **1996**, *118*, 10824.
504. Beckwith, A.L.J.; Poole, J.S. *J. Am. Chem. Soc* **2002**, *124*, 9489.
505. Busfield, W.K.; Jenkins, I.D.; Monteiro, M.J. *Polymer* **1997**, *38*, 165.
506. Zetterlund, P.B.; Busfield, W.K.; Jenkins, I.D. *Macromolecules* **1999**, *32*, 8041.
507. Zetterlund, P.B.; Busfield, W.K.; Jenkins, I.D. *Macromolecules* **2002**, *35*, 7232.
508. Beckwith, A.L.J.; Bowry, V.W.; Moad, G. *J. Org. Chem.* **1988**, *53*, 1632.
509. Chateauneuf, J.; Lusztyk, J.; Ingold, K.U. *J. Org. Chem.* **1988**, *53*, 1629.
510. Bowry, V.W.; Ingold, K.U. *J. Am. Chem. Soc.* **1992**, *114*, 4992.
511. Beckwith, A.L.J.; Bowry, V.W.; Ingold, K.U. *J. Am. Chem. Soc.* **1992**, *114*, 4983.
512. Moad, G.; Rizzardo, E.; Solomon, D.H.; Beckwith, A.L.J. *Polym. Bull.* **1992**, *29*, 647.
513. Gerlock, J.L.; Zacmanidis, P.J.; Bauer, D.R.; Simpson, D.J.; Blough, N.V.; Salmeen, I.T. *Free Radical Res. Commun.* **1990**, *10*, 119.

514. Moad, G.; Shipp, D.A.; Smith, T.A.; Solomon, D.H. *Macromolecules* **1997**, *30*, 7627.
515. Moad, G.; Shipp, D.A.; Smith, T.A.; Solomon, D.H. *J. Phys. Chem. A* **1999**, *103*, 6580.
516. Golubev, V.A.; Kozlov, Y.N.; Petrov, A.N.; Purmal, A.P. *Prog. React. Kinet.* **1991**, *16*, 35.
517. Anderson, D.R.; Keute, J.; Chapel, H.L.; Koch, T.H. *J. Am. Chem. Soc.* **1979**, *101*, 1904.
518. Keana, J.F.W.; Dinerstein, R.J.; Baitis, F. *J. Org. Chem.* **1971**, *36*, 209.
519. Johnston, L.J.; Tencer, M.; Scaiano, J.C. *J. Org. Chem.* **1986**, *51*, 2806.
520. Coxon, J.M.; Pattsalides, E. *Aust. J. Chem.* **1982**, *35*, 509.
521. Bottle, S.E.; Chand, U.; Micallef, A.S. *Chem. Lett.* **1997**, 857.
522. Aldabbagh, F.; Busfield, W.K.; Jenkins, I.D.; Thang, S.H. *Tetrahedron Lett.* **2000**, *41*, 3673.
523. Connolly, T.J.; Scaiano, J.C. *Tetrahedron Lett.* **1997**, *38*, 1133.
524. Watanabe, Y.; Ishigaki, H.; Okada, H.; Suyama, S. *Polym. J.* **1997**, *29*, 733.
525. Watanabe, Y.; Ishigaki, H.; Okada, H.; Suyama, S. *Polym. J.* **1997**, *29*, 603.
526. Watanabe, Y.; Ishigaki, H.; Okada, H.; Suyama, S. *Polym. J.* **1997**, *29*, 366.
527. Watanabe, Y.; Ishigaki, H.; Okada, H.; Suyama, S. *Polym. J.* **1997**, *29*, 940.
528. Watanabe, Y.; Ishigaki, H.; Okada, H.; Suyama, S. *Polym. J.* **1998**, *30*, 192.
529. Shetty, S.; Garcia-Rubio, L.H. *Polym. Mater. Sci. Eng.* **1991**, *65*, 103.
530. Koenig, J.L. *Spectroscopy of Polymers*; Elsevier: New York, 1999.
531. Buback, M.; Huckestein, B.; Ludwig, B. *Makromol. Chem., Rapid Commun.* **1992**, *13*, 1.
532. Buback, M.; Huckestein, B.; Leinhos, U. *Makromol. Chem., Rapid Commun.* **1987**, *8*, 473.
533. Carduner, K.R.; Carter, R.O.; Zinbo, M.; Gerlock, J.L.; Bauer, D.R. *Macromolecules* **1988**, *21*, 1598.
534. Meijs, G.F.; Morton, T.C.; Rizzardo, E.; Thang, S.H. *Macromolecules* **1991**, *24*, 3689.
535. Meijs, G.F.; Rizzardo, E.; Thang, S.H. *Macromolecules* **1988**, *21*, 3122.
536. Bevington, J.C.; Ebdon, J.R.; Huckerby, T.N. In *NMR Spectroscopy of Polymers*; Ibbett, R.N., Ed.; Blackie: London, 1993; p 51.
537. Hatada, K.; Kitayama, T.; Ute, K.; Terawaki, Y.; Yanagida, T. *Macromolecules* **1997**, *30*, 6754.
538. Hatada, K. *NMR Spectroscopy of Polymers*; Springer-Verlag: Berlin, 2003.
539. Bevington, J.C.; Cywar, D.A.; Huckerby, T.N.; Senogles, E.; Tirrell, D.A. *Eur. Polym. J.* **1988**, *24*, 699.
540. Johns, S.R.; Rizzardo, E.; Solomon, D.H.; Willing, R.I. *Makromol. Chem., Rapid Commun.* **1983**, *4*, 29.
541. Hatada, K.; Kitayama, T.; Masuda, E. *Polym. J.* **1985**, *17*, 985.
542. Kashiwagi, T.; Inaba, A.; Brown, J.E.; Hatada, K.; Kitayama, T.; Masuda, E. *Macromolecules* **1986**, *19*, 2160.
543. Hatada, K.; Kitayama, T.; Masuda, E. *Polym. J.* **1986**, *18*, 395.
544. Bevington, J.C.; Ebdon, J.R.; Huckerby, T.N. *Eur. Polym. J.* **1985**, *21*, 685.
545. Bevington, J.C.; Huckerby, T.N. *Polymer* **1992**, *33*, 1323.
546. Saito, T.; Rinaldi, P.L. *J. Magn. Reson.* **1998**, *130*, 135.

547. Saito, T.; Rinaldi, P.L. *J. Magn. Reson.* **1998**, *132*, 41.
548. Meng, H.H.; Saito, T.; Rinaldi, P.L.; Wyzgoski, F.; Helfer, C.A.; Mattice, W.L.; Harwood, H.J. *Macromolecules* **2001**, *34*, 801.
549. Montaudo, G.; Montaudo, M.S.; Montaudo, G.; Lattimer, R.P., Eds.; CRC Press: Boca Raton, 1999; p 41.
550. Meisters, A.; Moad, G.; Rizzardo, E.; Solomon, D.H. *Polym. Bull.* **1988**, *20*, 499.
551. Farina, M. *Makromol. Chem., Macromol. Symp.* **1987**, *10/11*, 255.
552. Ohtani, H.; Ishiguro, S.; Tanaka, M.; Tsuge, S. *Polym. J.* **1989**, *21*, 41.
553. Ohtani, H.; Suzuki, A.; Tsuge, S. *J. Polym. Sci., Part A: Polym. Chem.* **2000**, *38*, 1880.
554. Hanton, S.D. *Chem. Rev.* **2001**, *101*, 527.
555. Nielen, M.W.F. *Mass Spectrom. Rev.* **1999**, *18*, 309.
556. Maloney, D.R.; Hunt, K.H.; Lloyd, P.M.; Muir, A.V.G.; Richards, S.N.; Derrick, P.J.; Haddleton, D.M. *J. Chem. Soc., Chem. Commun.* **1995**, 561.
557. Zammit, M.D.; Davis, T.P.; Haddleton, D.M.; Suddaby, K.G. *Macromolecules* **1997**, *30*, 1915.
558. Suddaby, K.G.; Hunt, K.H.; Haddleton, D.M. *Macromolecules* **1996**, *29*, 8642.
559. Haddleton, D.M.; Maloney, D.R.; Suddaby, K. *Polymer* **1997**, *38*, 6207.
560. Liu, Z.F.; Rimmer, S. *Macromolecules* **2002**, *35*, 1200.
561. Vana, P.; Davis, T.P.; Barner-Kowollik, C. *Aust. J. Chem.* **2002**, *55*, 315.
562. Schilli, C.; Lanzendoerfer, M.G.; Mueller, A.H.E. *Macromolecules* **2002**, *35*, 6819.
563. Ganachaud, F.; Monteiro, M.J.; Gilbert, R.G.; Dourges, M.A.; Thang, S.H.; Rizzardo, E. *Macromolecules* **2000**, *33*, 6738.
564. Charmot, D.; Corpart, P.; Adam, H.; Zard, S.Z.; Biadatti, T.; Bouhadir, G. *Macromol. Symp.* **2000**, *150*, 23.
565. D'Agosto, F.; Hughes, R.; Charreyre, M.T.; Pichot, C.; Gilbert, R.G. *Macromolecules* **2003**, *36*, 621.
566. Vana, P.; Albertin, L.; Barner, L.; Davis, T.P.; Barner-Kowollik, C. *J. Polym. Sci., Part A: Polym. Chem.* **2002**, *40*, 4032.
567. Norman, J.; Moratti, S.C.; Slark, A.T.; Irvine, D.J.; Jackson, A.T. *Macromolecules* **2002**, *35*, 8954.
568. Borman, C.D.; Jackson, A.T.; Bunn, A.; Cutter, A.L.; Irvine, D.J. *Polymer* **2000**, *41*, 6015.
569. Farcet, C.; Belleney, J.; Charleux, B.; Pirri, R. *Macromolecules* **2002**, *35*, 4912.
570. Dourges, M.A.; Charleux, B.; Vairon, J.P.; Blais, J.C.; Bolbach, G.; Tabet, J.C. *Macromolecules* **1999**, *32*, 2495.
571. Bartsch, A.; Dempwolf, W.; Bothe, M.; Flakus, S.; Schmidt-Naake, G. *Macromol. Rapid. Commun.* **2004**, *24*, 614.
572. Palit, S.R. *Makromol. Chem.* **1959**, *36*, 89.
573. Palit, S.R. *Makromol. Chem.* **1960**, *38*, 96.
574. Rizzardo, E.; Solomon, D.H. *J. Macromol. Sci., Chem.* **1979**, *A13*, 997.
575. Ghosh, N.N.; Sengputa, P.K.; Pramanik, A. *J. Polym. Sci., Part A* **1965**, *3*, 1725.
576. Rizzardo, E.; Solomon, D.H. *J. Macromol. Sci., Chem.* **1979**, *A13*, 1005.
577. Postma, A.; Donovan, R.; Davis, T.P.; Moad, G.; O'Shea, M. *Polymer* **2005**, submitted for publication.
578. Donovan, A.R.; Moad, G. *Polymer* **2005**, 5005.
579. Kern, W. *Chem. Ztg.* **1976**, *100*, 401.

580. Bevington, J.C. *Radical Polymerization*; Academic Press: London, 1961.
581. Bevington, J.C. *Trans. Faraday Soc.* **1955**, *51*, 1392.
582. Bevington, J.C.; Ebdon, J.R. *Developments in Polymerisation* **1979**, *2*, 1.
583. Bevington, J.C. *Makromol. Chem., Macromol. Symp.* **1987**, *10/11*, 89.
584. Moad, G. *Chem. Aust.* **1991**, *58*, 122.
585. Moad, G. In *Annual Reports in NMR Spectroscopy*; Webb, G.A., Ed.; Academic Press: London, 1994; Vol. 29, p 287.
586. Barson, C.A.; Bevington, J.C.; Breuer, S.W. *Eur. Polym. J.* **1989**, *25*, 259.
587. Barson, C.A.; Behari, K.; Bevington, J.C.; Huckerby, T.N. *J. Macromol. Sci., Chem.* **1988**, *A25*, 1137.
588. Barson, C.A.; Bevington, J.C.; Huckerby, T.N. *Polym. Bull.* **1991**, *25*, 83.
589. Bevington, J.C.; Huckerby, T.N.; Varma, S.C. *Eur. Polym. J.* **1987**, *19*, 319.
590. Barson, C.A.; Bevington, J.C.; Huckerby, T.N. *Polym. Bull.* **1989**, *22*, 131.
591. Bevington, J.C.; Huckerby, T.N. *Macromolecules* **1985**, *18*, 176.
592. Barson, C.A.; Bevington, J.C.; Huckerby, T.N. *Polymer* **1991**, *32*, 3415.
593. Barson, C.A.; Bevington, J.C.; Breuer, S.W.; Huckerby, T.N. *Eur. Polym. J.* **1989**, *25*, 527.
594. Barson, C.A.; Bevington, J.C.; Breuer, S.W.; Huckerby, T.N. *Makromol. Chem., Rapid Commun.* **1992**, *13*, 97.
595. Bevington, J.C.; Breuer, S.W.; Huckerby, T.N.; Hunt, B.J.; Jones, R. *Eur. Polym. J.* **1998**, *34*, 539.
596. Bevington, J.C.; Breuer, S.W.; Huckerby, T.N.; Hunt, B.J.; Jones, R. *Eur. Polym. J.* **1997**, *33*, 1225.
597. Bevington, J.C.; Huckerby, T.N.; Vickerstaff, N. *Makromol. Chem., Rapid Commun.* **1983**, *4*, 349.
598. Bevington, J.C.; Huckerby, T.N.; Vickerstaff, N. *Makromol. Chem., Rapid Commun.* **1988**, *9*, 791.
599. Cywar, D.A.; Tirrell, D.A. *Macromolecules* **1986**, *19*, 2908.
600. Fellows, C.M.; Senogles, E. *Eur. Polym. J.* **2001**, *37*, 1091.
601. Fellows, C.M.; Senogles, E. *Eur. Polym. J.* **1998**, *34*, 1249.
602. Fellows, C.M.; Senogles, E. *Eur. Polym. J.* **1999**, *35*, 9.
603. Bevington, J.C.; Bowden, B.F.; Cywar, D.A.; Lyons, R.A.; Senogles, E.; Tirrell, D.A. *Eur. Polym. J.* **1991**, *27*, 1239.
604. Prementine, G.S.; Tirrell, D.A. *Macromolecules* **1987**, *20*, 3034.
605. Kitayama, T.; Kishiro, S.; Masuda, E.; Hatada, K. *Polym. Bull.* **1991**, *25*, 205.
606. Bevington, J.C.; Ebdon, J.R.; Huckerby, T.N.; Hutton, N.W.E. *Polymer* **1982**, *23*, 163.
607. Behari, K.; Bevington, J.C.; Huckerby, T.N. *Polymer* **1988**, *29*, 1867.
608. Bevington, J.C.; Huckerby, T.N. *J. Macromol. Sci., Chem.* **1983**, *A20*, 753.
609. Bevington, J.C.; Huckerby, T.N.; Hunt, B.J. *Br. Polym. J.* **1985**, *17*, 43.

4
Propagation

4.1 Introduction

The propagation step of radical polymerization comprises a sequence of radical additions to carbon-carbon double bonds. The factors that govern the rate and specificity of radical addition have been dealt with in general terms in Section 2.3. In order to produce high molecular weight polymers, a propagating radical must show a high degree of specificity in its reactions with unsaturated systems. It must give addition to the exclusion of side reactions that bring about the cessation of growth of the polymer chain. Despite this limitation, there is considerable scope for structural variation in homopolymers.

The asymmetric substitution pattern of most monomers means that addition gives rise to a chiral center and their polymers will have tacticity (Section 4.2).

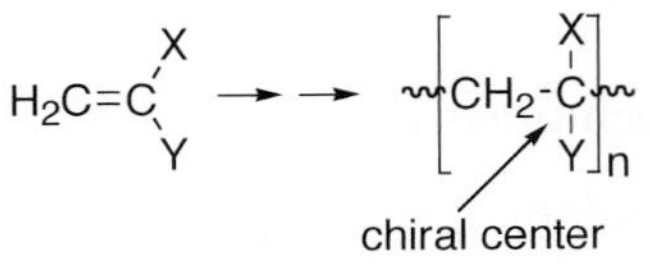

$$H_2C=C\begin{smallmatrix}X\\Y\end{smallmatrix} \longrightarrow \longrightarrow \sim\!\!\left[CH_2-\underset{Y}{\overset{X}{C}}\right]_n$$

chiral center

Addition to double bonds may not be completely regiospecific. The predominant head-to-tail structure may be interrupted by head-to-head and tail-to-tail linkages (Section 4.3).

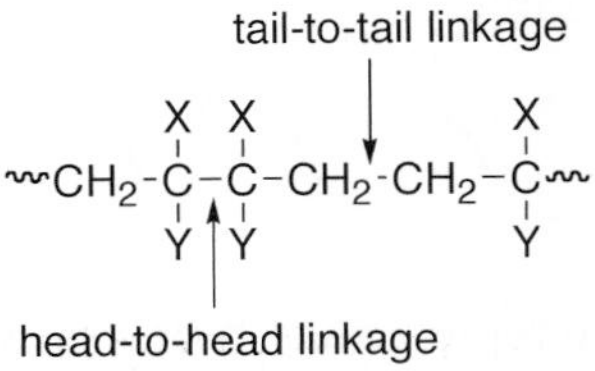

tail-to-tail linkage

$$\sim\!\!CH_2-\underset{Y}{\overset{X}{C}}-\underset{Y}{\overset{X}{C}}-CH_2\text{-}CH_2-\underset{Y}{\overset{X}{C}}\!\!\sim$$

head-to-head linkage

Intramolecular rearrangement of the initially formed radical may occur occasionally (*e.g.* backbiting - Section 4.4.3) or even be the dominant pathway (*e.g.* cyclopolymerization – Section 4.4.1, ring-opening polymerization – Section 4.4.2). These pathways can give rise to branches, rings, or internal unsaturation in the polymer chain.

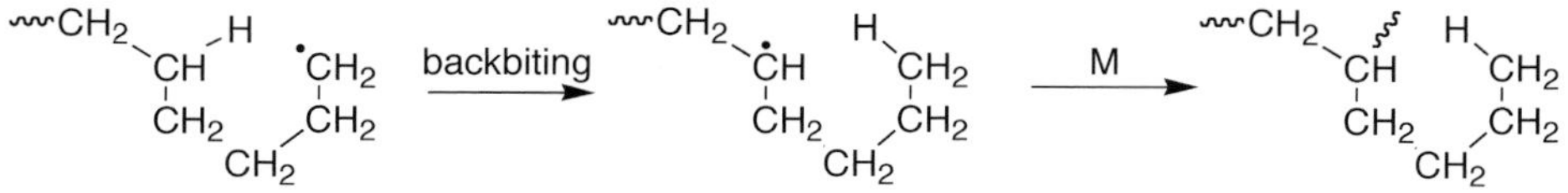

[Scheme at top of page showing backbiting mechanism]

This chapter is primarily concerned with the chemical microstructure of the products of radical homopolymerization. Variations on the general structure $(CH_2\text{-}CXY)_n$ are described and the mechanisms for their formation and the associated rate parameters are examined. With this background established, aspects of the kinetics and thermodynamics of propagation are also considered (Section 4.5).

4.2 Stereosequence Isomerism - Tacticity

The classical representation of a homopolymer chain, in which the end groups are disregarded and only one monomer residue is considered, allows no possibility for structural variation. However, possibilities for stereosequence isomerism arise as soon as the monomer residue is considered in relation to its neighbors and the substituents X and Y are different. The chains have tacticity (Section 4.2.1). Experimental methods for tacticity determination are summarized in 4.2.2 and the tacticity of some common polymers is considered in 4.2.3.

The following discussion is limited to polymers of mono- or 1,1-disubstituted monomers. Other factors become important in describing the types of stereochemical isomerism possible for polymers formed from other monomers (*e.g.* 1,2-disubstituted monomers).[1]

4.2.1 Terminology and Mechanisms

Detailed discussion of polymer tacticity can be found in texts by Randall,[2] Bovey,[1,3] Koenig,[4,5] Tonelli[6] and Hatada.[7] In order to understand stereoisomerism in polymer chains formed from mono- or 1,1-disubstituted monomers, consider four idealized chain structures:

(a) The isotactic chain where the relative configuration of all the substituted carbons in the chain is the same.

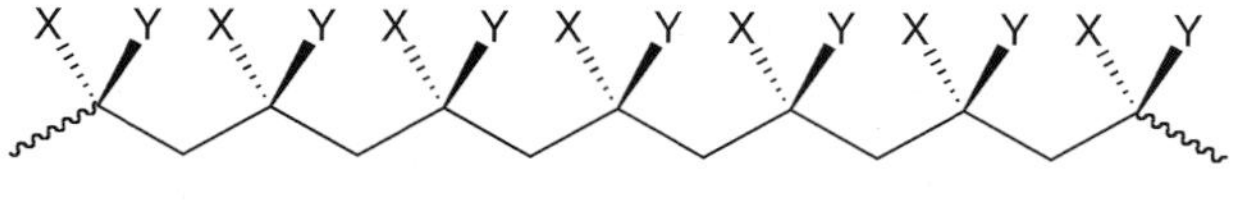

[Structural diagram of isotactic chain in zigzag representation]

For the usual diagrammatic representation of a polymer chain, this corresponds to the situation where similar substituents lie on the same side of a plane perpendicular to the page and containing the polymer backbone.

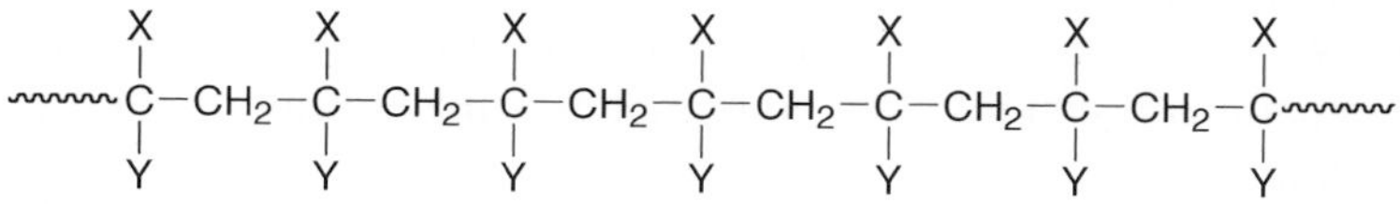

[Structural diagram showing isotactic chain with X substituents above and Y substituents below]

Propagation

(b) The syndiotactic chain where the relative configuration of centers alternates along the chain.

$$\text{wwww}\overset{\overset{X}{|}}{\underset{\underset{Y}{|}}{C}}-CH_2-\overset{\overset{Y}{|}}{\underset{\underset{X}{|}}{C}}-CH_2-\overset{\overset{X}{|}}{\underset{\underset{Y}{|}}{C}}-CH_2-\overset{\overset{Y}{|}}{\underset{\underset{X}{|}}{C}}-CH_2-\overset{\overset{X}{|}}{\underset{\underset{Y}{|}}{C}}-CH_2-\overset{\overset{Y}{|}}{\underset{\underset{X}{|}}{C}}-CH_2-\overset{\overset{X}{|}}{\underset{\underset{Y}{|}}{C}}\text{wwww}$$

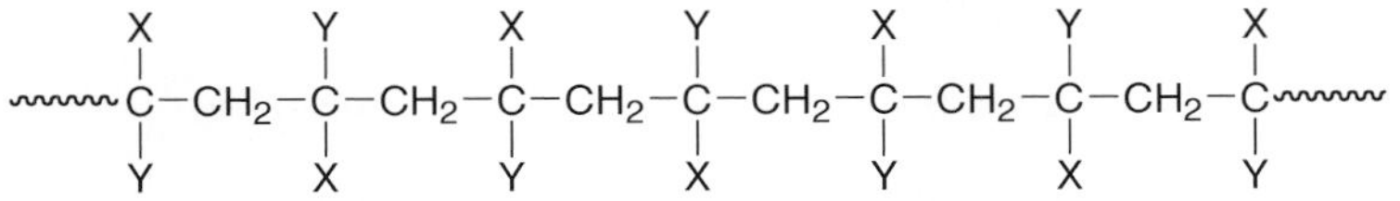

(c) The heterotactic chain where the dyad configuration alternates along the chain.

$$\text{wwww}\overset{\overset{X}{|}}{\underset{\underset{Y}{|}}{C}}-CH_2-\overset{\overset{X}{|}}{\underset{\underset{Y}{|}}{C}}-CH_2-\overset{\overset{Y}{|}}{\underset{\underset{X}{|}}{C}}-CH_2-\overset{\overset{Y}{|}}{\underset{\underset{X}{|}}{C}}-CH_2-\overset{\overset{X}{|}}{\underset{\underset{Y}{|}}{C}}-CH_2-\overset{\overset{X}{|}}{\underset{\underset{Y}{|}}{C}}-CH_2-\overset{\overset{Y}{|}}{\underset{\underset{X}{|}}{C}}\text{wwww}$$

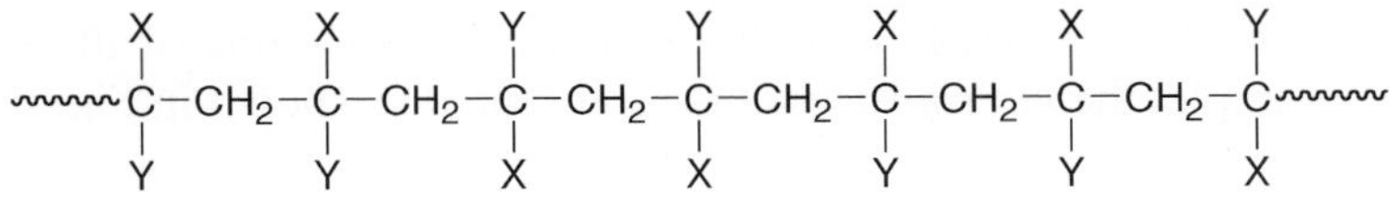

(d) The atactic chain where there is a random arrangement of centers along the chain.*

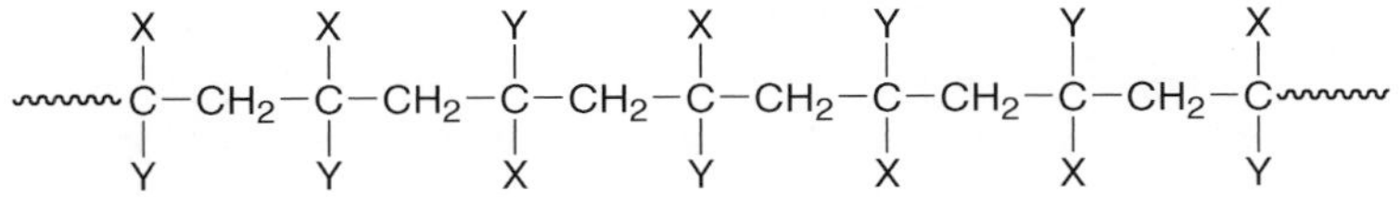

$$\text{wwww}\overset{\overset{X}{|}}{\underset{\underset{Y}{|}}{C}}-CH_2-\overset{\overset{X}{|}}{\underset{\underset{Y}{|}}{C}}-CH_2-\overset{\overset{Y}{|}}{\underset{\underset{X}{|}}{C}}-CH_2-\overset{\overset{X}{|}}{\underset{\underset{Y}{|}}{C}}-CH_2-\overset{\overset{Y}{|}}{\underset{\underset{X}{|}}{C}}-CH_2-\overset{\overset{Y}{|}}{\underset{\underset{X}{|}}{C}}-CH_2-\overset{\overset{X}{|}}{\underset{\underset{Y}{|}}{C}}\text{wwww}$$

For polymers produced by radical polymerization, while one of these structures may predominate, the idealized structures do not occur. It is necessary to define parameters to more precisely characterize the tacticity of polymer chains.

It should be stressed that this treatment of polymer stereochemistry only deals with relative configurations; whether a substituent is "up or down" with respect to that on a neighboring unit. Therefore, the smallest structural unit which contains stereochemical information is the dyad. There are two types of dyad; meso (m), where the two chiral centers have like configuration, and racemic (r), where the centers have opposite configuration (Figure 4.1).

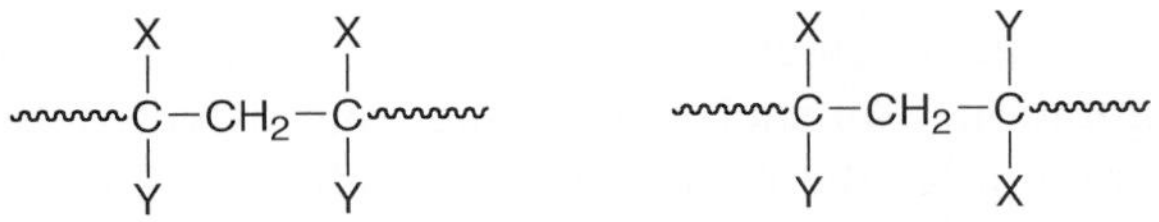

$$\text{wwww}\overset{\overset{X}{|}}{\underset{\underset{Y}{|}}{C}}-CH_2-\overset{\overset{X}{|}}{\underset{\underset{Y}{|}}{C}}\text{wwww} \qquad \text{wwww}\overset{\overset{X}{|}}{\underset{\underset{Y}{|}}{C}}-CH_2-\overset{\overset{Y}{|}}{\underset{\underset{X}{|}}{C}}\text{wwww}$$

          meso (m)   racemic (r)

Figure 4.1 Representation of meso (m) and racemic (r) dyads with polymer chains.

Confusion can arise because of the seemingly contradictory nomenclature established for analogous model compounds with just two asymmetric centers.[8] In such compounds, the diastereoisomers are named as in the following example (Figure 4.2).

---

* In the literature the term atactic is sometimes used to refer to any polymer that is not entirely isotactic or not entirely syndiotactic.

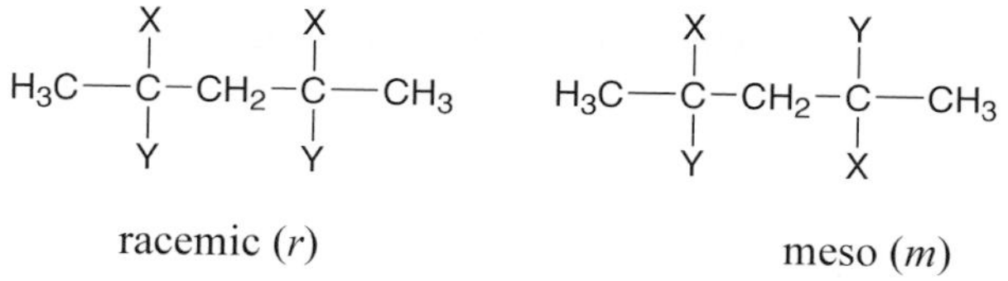

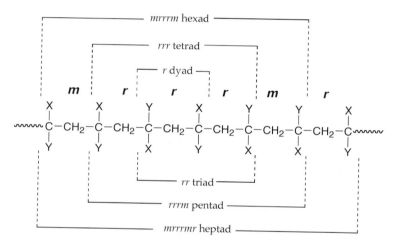

Figure 4.2 Representation of meso (*m*) and racemic (*r*) diastereoisomers of low molecular weight compounds.

It is usual to discuss triads, tetrads, pentads, *etc.* in terms of the component dyads. For example, the *mrrrmr* heptad is represented as shown in Figure 4.3.

Figure 4.3 Representation of *mrrrmr* heptad identifying component *n*-ads.

It is informative to consider how tacticity arises in terms of the mechanism for propagation. The radical center on the propagating species will usually have a planar sp^2 configuration. As such it is achiral and it will only be locked into a specific configuration after the next monomer addition. This situation should be contrasted with that which pertains in anionic or coordination polymerizations where the active center is pyramidal and therefore has chirality. This explains why stereochemical control is more easily achieved in these polymerizations.

The configuration of a center in radical polymerization is established in the transition state for addition of the next monomer unit when it is converted to a tetrahedral sp^3 center. If the stereochemistry of this center is established at random (Scheme 4.1; $k_m = k_r$) then a pure atactic chain is formed and the probability of finding a *meso* dyad, P(*m*), is 0.5.

Polymers formed from monosubstituted monomers (X=H) under the usual reaction conditions (*e.g.* 60 °C, bulk) appear almost atactic with only a slight

Propagation

preference for syndiotacticity and values of $P(m)$ in the range 0.45-0.52 (Table 4.1, Section 4.2.3).

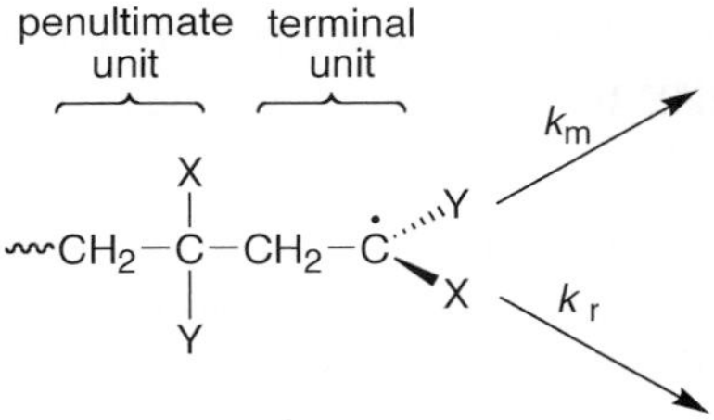

Scheme 4.1

If the reaction center adopts a preferred configuration with respect to the configuration of the penultimate unit in the chain (Scheme 4.1; $k_m \neq k_r$) then Bernoullian statistics apply. The stereochemistry of the chain is characterized by the single parameter, $P(m)$ or $P(r)$ [= $1-P(m)$]. The n-ad concentrations can be calculated simply by multiplying the concentrations of the component dyads. Thus the relative triad concentrations are given by the following expressions (eq. 1-3)

$$mm = P(m)^2 \tag{1}$$

$$mr = rm = 2\,P(m)\,P(r) = 2\,P(m)\,(1 - P(m)) \tag{2}$$

$$rr = P(r)^2 = (1 - P(m))^2 \tag{3}$$

Higher n-ads are calculated similarly. Thus for the *mrrrmr* heptad:

$$mrrrmr = 2\,P(m)\,P(r)\,P(r)\,P(r)\,P(m)\,P(r) = 2\,P(m)^2\,P(r)^4$$

The factor 2 is introduced in the case of asymmetric n-ads which can be formed in two ways (*mrrrmr* = *rmrrrm*).

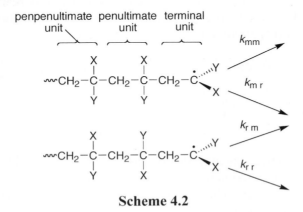

Scheme 4.2

Where the nature of the preceding dyad is important in determining the configuration of the new chiral center (Scheme 4.2), first order Markov statistics apply. Propagation is subject to a penpenultimate unit effect (also called an antepenultimate unit effect). Two parameters are required to specify the stereochemistry, $P(m|r)$ [$=1-P(m|m)$] and $P(r|r)$ [$=1-P(r|m)$], where $P(i|j)$ is the conditional probability that given a j dyad, the next unit in the chain will be an i dyad.* It can be shown that

$$P(m) = P(m|r) / (P(m|r) + P(r|m)) \qquad (4)$$

The relative triad concentrations are then given by the following expressions (eq. 5-7)

$$mm = P(m) \, P(m|m) \qquad (5)$$

$$mr = rm = 2 \, P(m) \, P(r|m) = 2 \, P(m) \, (1 - P(m|m)) \qquad (6)$$

$$rr = P(r) \, P(r|r) \qquad (7)$$

Again the higher n-ads are calculated similarly. Thus for the $mrrrmr$ heptad:

$$mrrrmr = 2 \, P(m) \, P(r|m) \, P(r|r) \, P(r|r) \, P(m|r) \, P(r|m)$$

We can also write expressions to calculate $P(m|r)$ and $P(r|m)$ from the triad concentrations (eq. 8, 9).

$$P(m|r) = mr/(2 \, mm + mr) \qquad (8)$$

$$P(r|m) = rm/(2 \, rr + rm) \qquad (9)$$

The Coleman-Fox two state model describes the situation where there is restricted rotation about the bond to the preceding unit (Scheme 4.3). If this is slow with respect to the rate of addition, then at least two conformations of the propagating radical need to be considered each of which may react independently with monomer. The rate constants associated with the conformational equilibrium and two values of $P(m)$ are required to characterize the process.

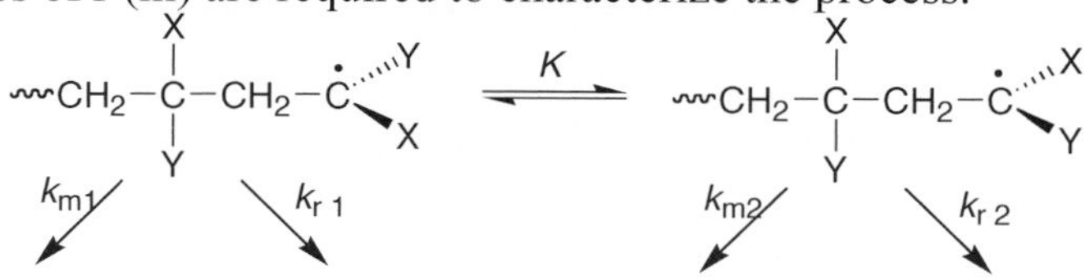

Scheme 4.3

More complex situations may also be envisaged and it should always be borne in mind that the fit of experimental data to a simple model provides support for but does not prove that model. The power of the experiment to discriminate between models has to be considered.

---

* In texts by Bovey[1,3] and Tonelli[6] $P(i|j)$ is written Pj/i.

4.2.2 Experimental Methods for Determining Tacticity

The application of NMR spectroscopy to tacticity determination of synthetic polymers was pioneered by Bovey and Tiers.[9] NMR spectroscopy is the most used method and often the only technique available for directly assessing tacticity of polymer chains.[1,2,7,8,10,11] The chemical shift of a given nucleus in or attached to the chain may be sensitive to the configuration of centers three or more monomer units removed. Other forms of spectroscopy (*e.g.* IR spectroscopy[12,13]) are useful with some polymers and various physical properties (*e.g.* the Kerr effect[14]) may also be correlated with tacticity.

The ambiguity of the NMR peak assignments may cause problems in tacticity determination. The usual method of assigning peaks to configurational sequences involves matching expected and measured peak intensities. There are obvious problems inherent in this approach and these are being redressed by the application of 2D NMR methods which in many cases can provide unambiguous assignments.[15] These methods have been applied to make absolute tacticity assignments for PAA,[16] PMMA,[17-20] PMAN,[21] PVA,[22,23] PVC[24,25] and PVF.[26] In some cases, an *a priori* assignment of chemical shifts using theoretical methods (making use of the rotational isomeric state model and the γ-gauche effect) may also be possible.[6] Such methods have been shown useful for polypropylene, PVC and PVF.

Attention must also be paid to sample preparation methods.[27] The number average molecular weight of the polymer must be sufficiently high that signals due to sequences near the chain ends make no significant contribution to the spectrum. For PMMA with heptad resolution, this requires that $\bar{M}_n$ is in excess of 30,000. Similarly, one must be concerned about structural irregularities introduced through head addition, backbiting and other processes.

4.2.3 Tacticities of Polymers

Many radical polymerizations have been examined from the point of view of establishing the stereosequence distribution. For most systems it is claimed that the tacticity is predictable within experimental error* by Bernoullian statistics [*i.e.* by the single parameter $P(m)$ – see 4.2.1].

Tacticity is most often determined by NMR analysis and usually by looking at the signals associated with the -CXY- group (refer Figure 4.3). The analysis then provides the triad concentrations (*mm, mr* and *rr*) and the value of m or $P(m)$ is given by eq. 10.

$$P(m) = mm + 0.5\ mr \qquad (10)$$

---

* It should be noted that, in some studies, deviations of 5-10% in expected and measured NMR peak intensities have been ascribed to experimental error. Such error is sufficient to hide significant departures from Bernoullian statistics.[28,29]

Most polymers formed by radical polymerization have an excess of syndiotactic over isotactic dyads [*i.e.* $P(m) \leq 0.5$]. $P(m)$ typically lies in the range 0.4-0.5 for vinyl monomers and 0.2-0.5 for 1,1-disubstituted monomers. It is also generally found that $P(m)$ (the fraction of isotactic dyads) decreases with decreasing temperature.[30] Data on tacticity for some common polymers are presented in Table 4.1.

There are exceptions to this general rule. For example, polymerizations of methacrylates with very bulky ester substituents (**1-4**) show a marked preference for isotacticity[31] whereas polymerizations of MMA show a significant preference for syndiotacticity (Table 4.1). Polymerization of the acrylamide derivative **5** which has a bulky substituent on nitrogen also provides a polymer that is highly isotactic.[32,33] AM and simple derivatives (NIPAM, DMAM) give polymers that are slightly syndiotactic (Table 4.1). Tacticity can be influenced by solvent and Lewis acids (Section 8.3).[34]

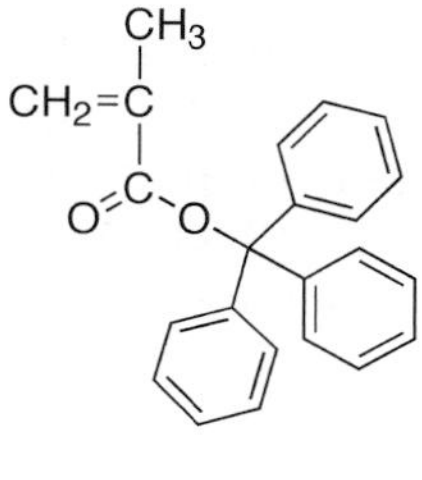

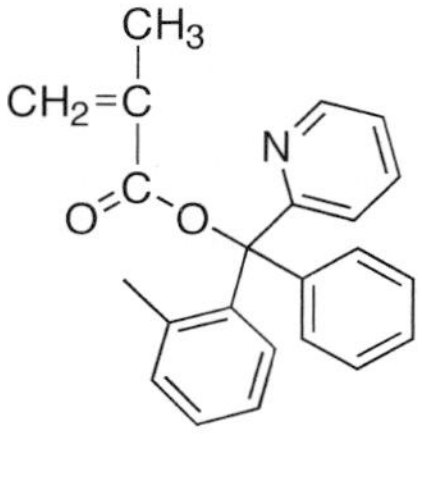

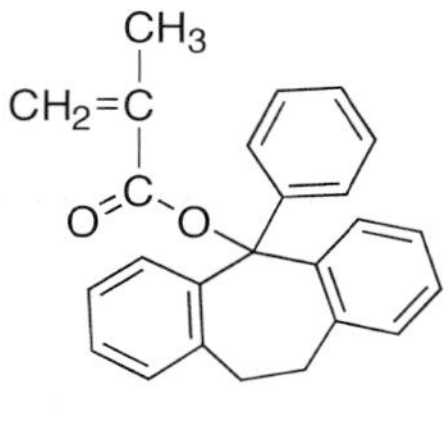

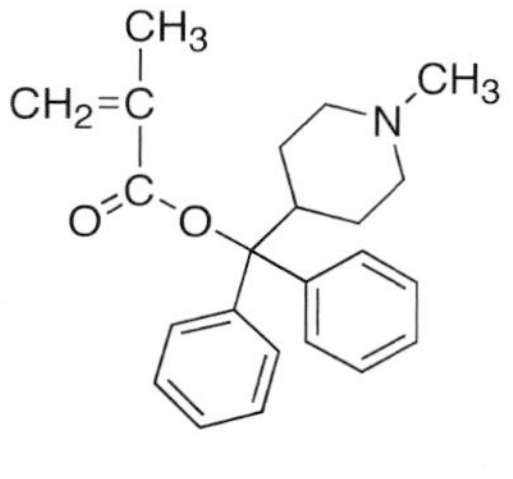

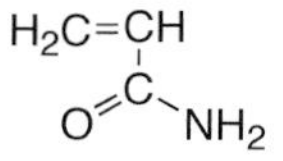

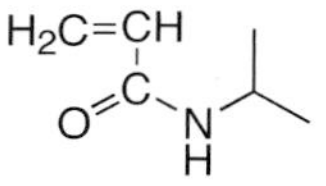

An explanation for the preference for syndiotacticity during MMA polymerization was proposed by Tsuruta *et al.*[35] They considered that the propagating radical should exist in one of two conformations and showed, with models, that attack on the less hindered side of the preferred conformation (where steric interactions between the substituent groups are minimized) would lead to formation of a syndiotactic dyad while similar attack on the less stable conformation would lead to an isotactic dyad.

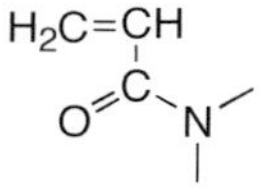

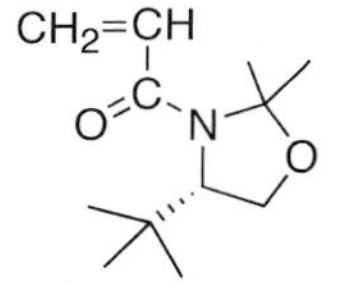

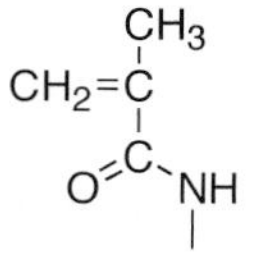

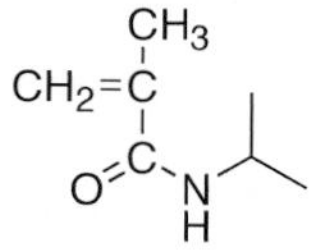

MMA polymerization is one of the most studied systems and was thought to be explicable, within experimental error, in terms of Bernoullian statistics. Moad *et al.*[36] have made precise measurements of the configurational sequence distribution for PMMA prepared from ^{13}C-labeled monomer. It is clear that

Bernoullian statistics do not provide a satisfactory description of the tacticity.[36] This finding is supported by other work.[28,37,38] First order Markov statistics provide an adequate fit of the data. Possible explanations include: (a) penpenultimate unit effects are important; and/or (b) conformational equilibrium is slow (Section 4.2.1). At this stage, the experimental data do not allow these possibilities to be distinguished.

It seems likely that other polymerizations will be found to depart from Bernoullian statistics as the precision of tacticity measurements improves. One study[12] indicated that vinyl chloride polymerizations are also more appropriately described by first order Markov statistics. However, there has been some reassignment of signals since that time.[24,25]

The triad fractions for PVAc[22,39] seem to obey Bernoullian statistics. However, the concentrations of higher order n-ads cannot be explained even by first (or second) order Markov statistics suggesting either that ambiguities still remain in the signal assignments at this level or that there are unresolved complexities in the polymerization mechanism. Tacticities have been shown to be solvent and temperature dependent with the degree of syndiotacticity being significantly enhanced in fluoroalcohol solvents and by lower temperatures.[40,41] Tacticity of vinyl esters is also dependent on the ester group.[42]

Table 4.1 Tacticities of Selected Homopolymers

Monomer	Temp. °C	$P(m)$a	$P(m\|m)$	$P(r\|m)$	Solvent	Conv. %	Ref.
AN	35	0.52	-	-	H$_2$O	-	43
MA	60	0.49	-	-	toluene	<50	44,45
AM	0	ca 0.46b	-	-	methanol	60	34
DMAM	60	ca 0.46b	-	-	methanol	73	34
NIPAM	60	ca 0.45b	-	-	methanol	82	34
S	80	0.46	-	-	benzene	92	46-48
VAc	-	0.46 ± .01c	-	-	d	-	22,49,50
VC	90	(0.454)	0.437	0.465	e	-	12
VC	5	(0.406)	0.391	0.424	e	-	12
VC	-30	(0.377)	0.337	0.391	d	-	12
MAN	60	0.406	-	-	bulk	15	21
6	60	ca 0.14b	-	-	methanol	97	51
6	60	ca 0.28b	-	-	toluene	95	51
7	60	< 0.1c	-	-	methanol	50	51
MMA	60	(0.202)	0.159	0.212	benzene	5	36

a Best fit number for $P(m)$. The polymerization is believed to follow first order Markov statistics. b Bernoullian statistics not established. Values of $P(m)$ estimated from triad distributions given. c See text. d Commercial samples or conditions of preparation unstated. e Suspension polymerization.

Further discussion on the effects of the reaction media and Lewis acids on tacticity appears in Section 7.2. Attempts to control tacticity by template polymerization and by enzyme mediated polymerization are described in Section 7.3. Devising effective means for achieving stereochemical control over propagation in radical polymerization remains an important challenge in the field.

4.3 Regiosequence Isomerism - Head vs Tail Addition

Most monomers have an asymmetric substitution pattern and the two ends of the double bond are distinct. For mono- and 1,1-disubstituted monomers (Section 4.3.1) it is usual to call the less substituted end "the tail" and the more substituted end "the head". Thus the terminology evolved for two modes of addition: head and tail; and for the three types of linkages: head-to-tail, head-to-head and tail-to-tail. For 1,2-di-, tri- and tetrasubstituted monomers definitions of head and tail are necessarily more arbitrary. The term "head" has been used for that end with the most substituents, the largest substituents or the best radical stabilizing substituent (Scheme 4.4).

With 1,3-diene based polymers, greater scope for structural variation is introduced because there are two double bonds to attack and the propagating species is a delocalized radical with several modes of addition possible (see 4.3.2).

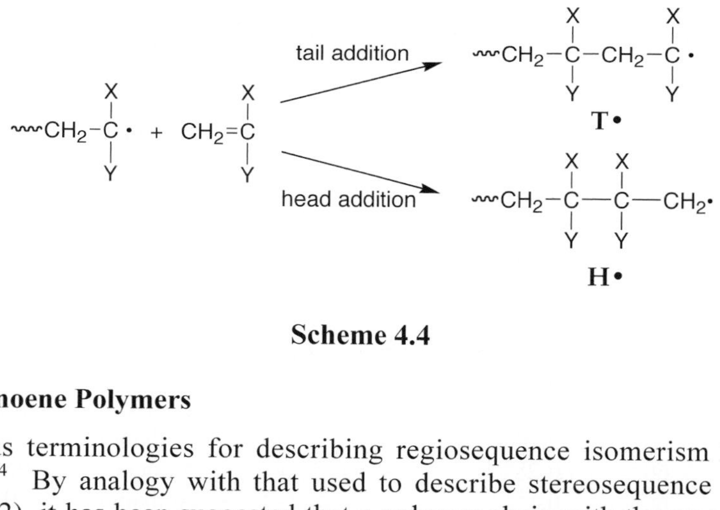

Scheme 4.4

4.3.1 Monoene Polymers

Various terminologies for describing regiosequence isomerism have been proposed.[1,4] By analogy with that used to describe stereosequence isomerism (Section 4.2), it has been suggested that a polymer chain with the monomer units connected by "normal" head-to-tail linkages should be termed isoregic, that with alternating head-to-head and tail-to-tail linkages, syndioregic, and that with a random arrangement of connections, aregic.[1]

For mono- and 1,1-disubstituted monomers, steric, polar, resonance, and bond-strength terms (see Section 2.3) usually combine to favor a preponderance of tail addition; i.e. an almost completely isoregic structure. However, the occurrence of

Propagation

head addition has been unambiguously demonstrated during many polymerizations. During the intramolecular steps of cyclopolymerization, 100% head addition may be obtained (Section 4.4.1).

The tendency for radicals to give tail addition means that a head-to-head linkage will, most likely, be followed by a tail-to-tail linkage (Scheme 4.5). Thus, head-to-head linkages formed by an "abnormal" addition reaction are chemically distinct from those formed in termination by combination of propagating radicals (Scheme 4.6).

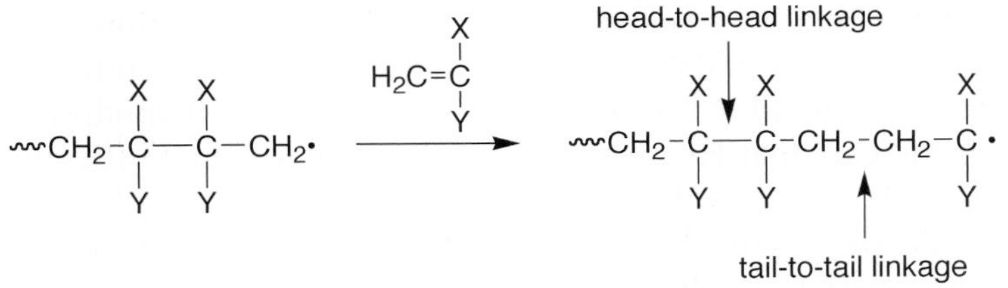

Scheme 4.5

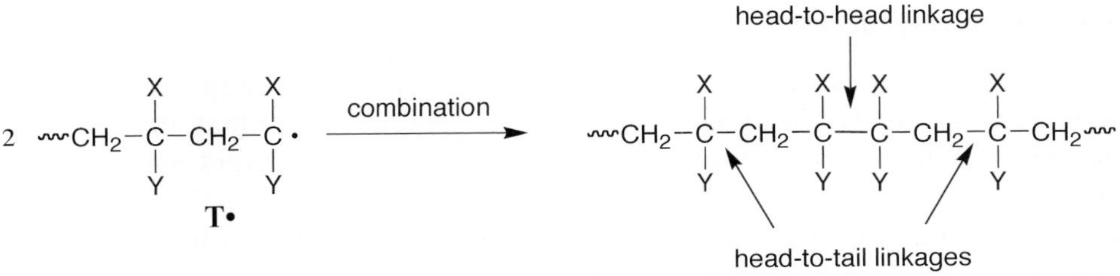

Scheme 4.6

In view of the potential problems associated with discriminating between the various types of head-to-head linkages, it is perhaps curious that, while much effort has been put into finding head-to-head linkages, relatively little attention has been paid to applying spectroscopic methods to detect tail-to-tail linkages where no such difficulty arises.

Even allowing for the above-mentioned complication, the number of head-to-head linkages is unlikely to equate exactly with the number of tail-to-tail linkages. The radicals formed by tail addition (T•) and those formed by head addition (H•) are likely to have different reactivities.

Consideration of data on the reactions for small radicals (Section 2.3) suggest that the primary alkyl radical (H•) is more likely to give head addition than the normal propagating species (T•) for three reasons:

(a) The propensity for head addition, which usually corresponds with attack at the more substituted end of the double bond, should decrease as the steric bulk of the attacking radical increases. Note that H• (a primary alkyl radical in the case of mono- and 1,1-disubstituted monomers) will usually be less sterically bulky than T•.

(b) Most common monomers have some dipolar character. H• and T• will usually be polarized similarly to the head and tail ends of the monomer respectively. This should favor T• adding tail and H• adding head.

(c) The primary alkyl radical (H•) will be more reactive than T• with no α-substituent to stabilize or delocalize the free spin.

However, head addition is usually a very minor pathway and is difficult to determine experimentally. Analysis of the events which follow head addition presents an even more formidable problem. Therefore, there is little experimental data on polymers with which to test the above-mentioned hypothesis. Data for fluoro-olefins indicate that H• gives less head addition than T• (Section 4.3.1.3). No explanation for the observation was proposed.

The primary alkyl radical, H•, is anticipated to be more reactive and may show different specificity to the secondary or tertiary radical, T•. In VAc and VC polymerizations the radical H• appears more prone to undertake intermolecular (Sections 4.3.1.1 and 4.3.1.2) or intramolecular (4.4.3.2) atom transfer reactions.

4.3.1.1 Poly(vinyl acetate)

It is generally agreed that *ca* 1-2% of propagation steps during VAc polymerization involve head addition. There is some evidence that, depending on reaction conditions, a high proportion of the head-to-head linkages may appear at chain ends (Scheme 4.7) and that the number of head-to-head linkages may not equate with tail-to-tail linkages. The extent of head addition in VAc polymerization increases with the polymerization temperature.

The classic method for establishing the proportion of head addition occurring in VAc polymerization involves a two step process.[52] The PVAc is converted to PVA by exhaustive hydrolysis and the number of 1,2-glycol units is determined by periodate cleavage.

The reliability of the chemical method has been assessed by Adelman and Ferguson.[53] They showed that, for low molecular weight PVA, a significant proportion of the 1,2-glycol units appear at chain ends as 2,3-dihydroxybutyl groups (*ca* 20% for $\overline{M}_n$ = 5,000, PVAc prepared in methanol at 75 °C). The inference is that the radical formed by head addition is particularly active in inter- or intramolecular transfer and/or termination reactions. The result suggests that measurements of the decrease in molecular weight caused by periodate cleavage could underestimate the amount of head addition.[52]

Analysis of ^{13}C NMR spectra of PVA provides a direct estimate of the extent of head addition occurring in VAc polymerizations.[39,54,55] Another advantage of the NMR method over chemical methods is that both head-to-head and tail-to-tail linkages can be observed. The polymers examined in these studies[39,54] were of relatively high molecular weight and prepared by emulsion polymerization. They

Propagation

possessed an equal number of head-to-head and tail-to-tail linkages. We have found that NMR can also be used to determine the fraction of head-to-head linkages in PVAc directly.

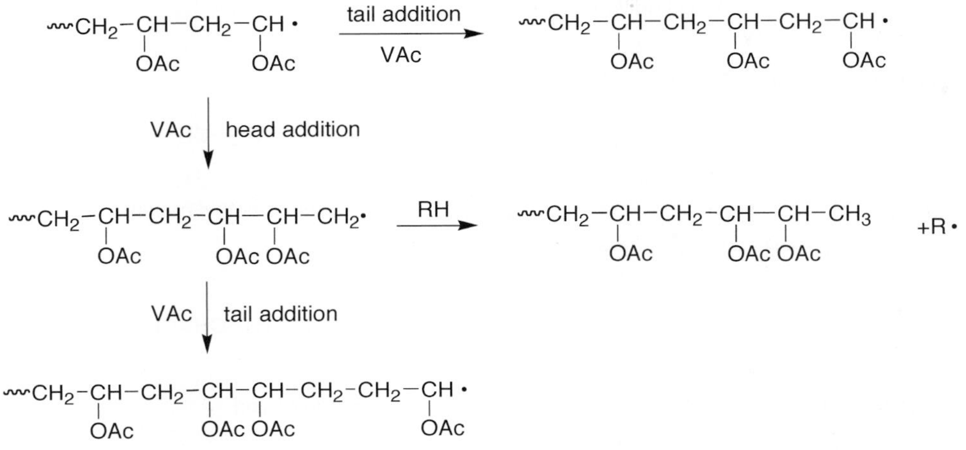

$$\text{\textasciitilde{}CH}_2\text{-CH(OAc)-CH}_2\text{-CH(OAc)}\cdot \xrightarrow{\text{tail addition, VAc}} \text{\textasciitilde{}CH}_2\text{-CH(OAc)-CH}_2\text{-CH(OAc)-CH}_2\text{-CH(OAc)}\cdot$$

$$\downarrow \text{VAc, head addition}$$

$$\text{\textasciitilde{}CH}_2\text{-CH(OAc)-CH}_2\text{-CH(OAc)-CH(OAc)-CH}_2\cdot \xrightarrow{\text{RH}} \text{\textasciitilde{}CH}_2\text{-CH(OAc)-CH}_2\text{-CH(OAc)-CH(OAc)-CH}_3 + \text{R}\cdot$$

$$\downarrow \text{VAc, tail addition}$$

$$\text{\textasciitilde{}CH}_2\text{-CH(OAc)-CH}_2\text{-CH(OAc)-CH(OAc)-CH}_2\text{-CH}_2\text{-CH(OAc)}\cdot$$

Scheme 4.7

The reaction conditions (solvent, temperature) may also influence the amount of head addition and determine whether the radical formed undergoes propagation or chain transfer.

4.3.1.2 Poly(vinyl chloride)

Establishment of the detailed microstructure of PVC has attracted considerable interest. This has been spurred by the desire to rationalize the poor thermal stability of the polymer (Chapter 1). Many reviews have appeared on the chemical microstructure of PVC and the mechanisms of "defect group" formation.[56-60]

Although head addition occurs during PVC polymerization to the extent of ca 1%, it is now thought that PVC contains few, if any, head-to-head linkages (<0.05%).[61,62] Propagation from the radical formed by head addition is not competitive with a unimolecular pathway for its disappearance, namely, 1,2-chlorine atom transfer (see Scheme 4.8).

Rigo et al.[63] were the first to propose that head addition does occur but is immediately followed by a 1,2-chlorine atom shift. The viability of 1,2-chlorine atom shifts is well established in model studies and theoretical calculations.[64] Experimental support for this occurring during VC polymerization has been provided by NMR studies on reduced PVC.[65,66] Starnes et al.[61] proposed that head addition is followed by one or two 1,2-chlorine atom shifts to give chloromethyl or dichloroethyl branch structures respectively (Scheme 4.8). There also is kinetic data to support this hypothesis.

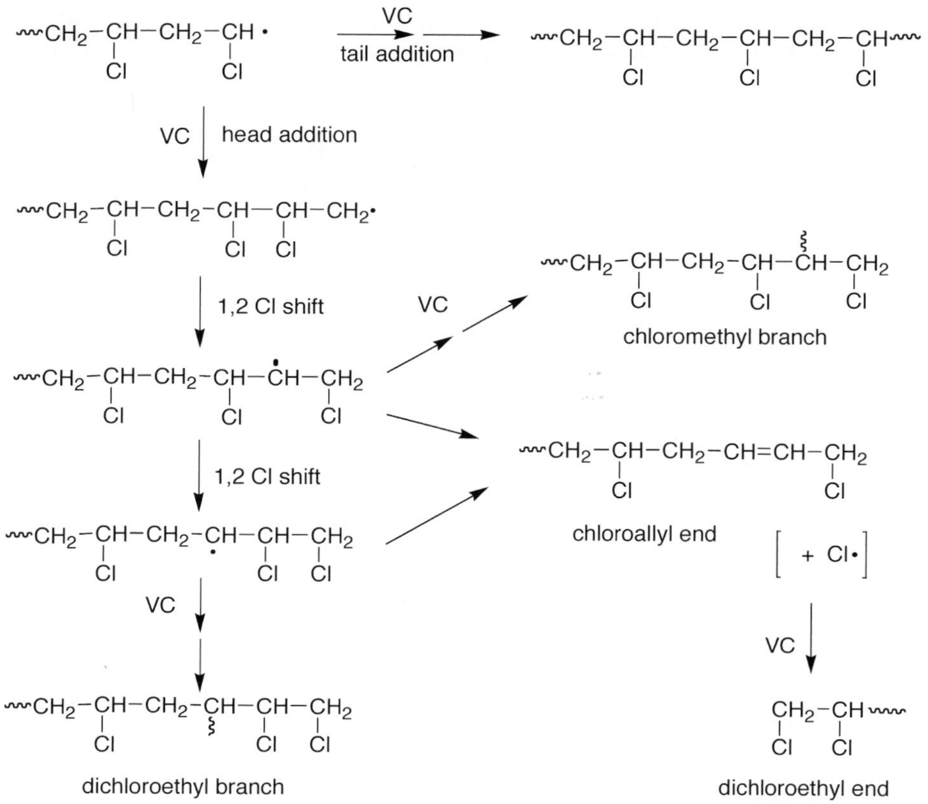

Scheme 4.8

Starnes et al.[67] have also suggested that the head adduct may undergo β-scission to eliminate a chlorine atom which in turn adds VC to initiate a new polymer chain. Kinetic data suggest that the chlorine atom does not have discrete existence. This addition-elimination process is proposed to be the principal mechanism for transfer to monomer during VC polymerization and it accounts for the reaction being much more important than in other polymerizations. The reaction gives rise to terminal chloroallyl and 1,2-dichloroethyl groups as shown in Scheme 4.8.

The presence of 1,2-dichloroethyl end groups and branch structures is likely to confuse attempts to determine head-to-head linkages by chemical methods (*e.g.* iodometric titration[68]).

4.3.1.3 Fluoro-olefin polymers

Propagation reactions involving the fluoro-olefins, vinyl fluoride (VF),[69-72] vinylidene fluoride (VF2)[69,72-74] and trifluoroethylene (VF3),[75] show relatively poor regiospecificity. This poor specificity is also seen in additions of small radicals to the fluoro-olefins (see 2.3). Since the fluorine atom is small, the major factors affecting the regiospecificity of addition are anticipated to be polarity and bond strength.

The fraction of head-to-head linkages in the poly(fluoro-olefins) increases in the series PVF2 < PVF ~ PVF3 (Table 4.2). This can be rationalized in terms of the propensity of electrophilic radicals to add preferentially to the more electron rich end of monomers (*i.e.* that with the lowest number of fluorines). This trend is also seen in the reactions of trifluoromethyl radicals with the fluoro-olefins (see 2.3).

The proportion of head-to-head linkages in fluoro-olefin polymers also depends on the polymerization temperature[69,70,72,73] (Table 4.2).

Table 4.2 Temperature Dependence of Head *vs* Tail Addition for Fluoro-olefin Monomers

temperature	% head addition		
°C	VF3[75]	VF2[73]	VF[70]
100			13.0
80	13.8	5.7	12.5
70			12.5
60			12.5
0	11.8	3.45	-
-80	10.0	3.0	-

^{19}F NMR studies have allowed regiosequence information to be determined at the pentad (VF) or heptad (VF2) level. Early studies[76] found that polymers formed by radical polymerization could be adequately described by Bernoullian statistics. However, Cais and Sloane[74] found that it was more appropriate to use first order Markov statistics to interpret regiospecificity. Their analysis suggests that the -CH$_2$• radical (formed by head addition) is much less likely to add head than the -CFX• radical [by a factor of ~14-18 for VF2 (depending on the polymerization temperature) or ~4 for VF]. No explanation for this selectivity was offered. The findings for fluoro-olefin propagation appear at variance with the considerations discussed above (see 4.3) and observations made for simple models. For example, with VF2, methyl radical is known to give much more head addition than trifluoromethyl radical (see 2.3).

4.3.1.4 Allyl polymers

Matsumoto *et al.*[77-81] have reported that substantial amounts (5-20%) of head addition occur during polymerization of allyl esters and that the proportion increases with polymerization temperature. They report that the proportion of head-to-head linkages in poly(allyl esters) is also dependent on the molecular weight of the polymer chain. For short chains, the fraction is reported to be *ca* 10% irrespective of the nature of the ester group. For longer chains the proportion of head-to-head linkages decreases and the molecular weight dependence of this fraction increases according to the size/polarity of the ester group.

The very high levels of head addition and the substituent effects reported in these studies are inconsistent with expectations based on knowledge of the reactions of small radicals (see 2.3) and are at odds with structures formed in the intermolecular step of cyclopolymerization of diallyl monomers (see 4.4.1.1) where overwhelming tail addition is seen.

4.3.1.5 Acrylic polymers

Before the advent of NMR spectroscopy, a number of reports appeared suggesting the possibility of substantial head addition during polymerization of acrylate ester derivatives. Marvel *et al.*[82,83] reported chemical degradation experiments that suggested that α-haloacrylate polymers contain halogen substituents in a 1,2-relationship. On this basis they proposed that these monomers polymerize in a head-to-head, tail-to-tail fashion. McCurdy and Laidler[84] suggested that irregularities in the heats of polymerization of methyl and higher acrylates and methacrylates could be rationalized if a fraction of units were arranged in head-to-head, tail-to-tail arrangement.

Since that time, many studies by NMR and other techniques on the microstructure of acrylic and methacrylic polymers formed by radical polymerization have proved their predominant head-to-tail structure.

There is, however, some evidence that a small amount of head addition during propagation occurs in the polymerization of acrylic monomers. On the basis of chemical analysis, Sawant and Morawetz[85] suggested that 4.6% of amide groups in PAM may be present as head-to-head linkages. Minigawa[86] has indicated the presence of a small percentage of head-to-head linkages in PAN.

4.3.2 Conjugated Diene Polymers

There is greater scope for structural variation in the diene based polymers than for the monoene polymers already discussed. The polymers contain units from overall 1,2- and *cis*- and *trans*-1,4-addition. Two mechanisms for overall 1,2-addition may be proposed. These are illustrated in Scheme 4.9 and Scheme 4.10:

(a) The delocalized allyl radical produced by addition to the 1- (or 4-) position may react in two ways to give overall 1,2-addition or 1,4-addition (Scheme 4.9).

(b) By analogy with the chemistry seen with monoene monomers the propagating species could, in principle, add to one of the internal (2- or 3-) positions of the diene (Scheme 4.10).

Propagation

Analyses of polymer microstructures do not allow these possibilities to be unambiguously distinguished. However, EPR experiments demonstrate that radicals add exclusively to one of the terminal methylenes.[87]

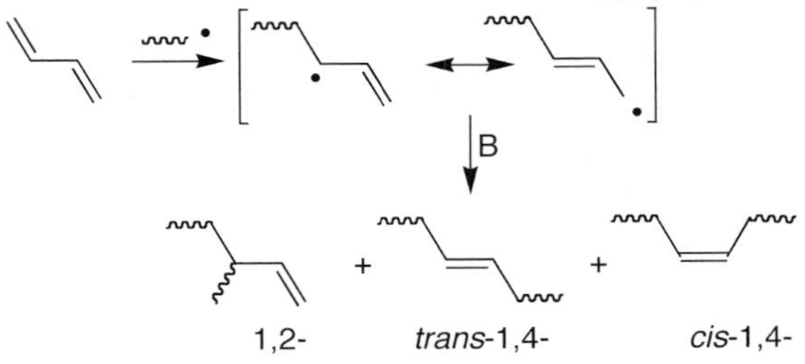

Scheme 4.9

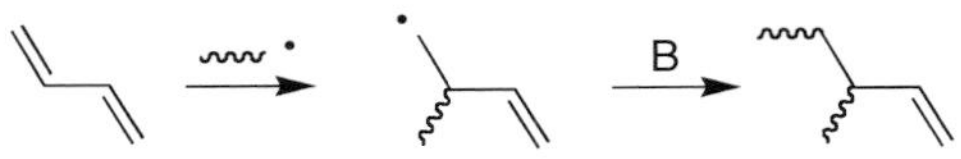

1,2-   trans-1,4-   cis-1,4-

Scheme 4.10

When used in conjunction with unsymmetrical dienes with substituents in the 2-position, the term 'tail addition' has been used to refer to addition to the methylene remote from the substituent. 'Head addition' then refers to addition to the methylene bearing the substituent (*i.e.* head addition ≡ 4,1- or 4,3-addition, tail addition ≡ 1,4- or 1,2-addition) as illustrated below for chloroprene (Scheme 4.11). Note that 1,2- and 4,3-addition give different structures while 1,4- and 4,1-addition give equivalent structures and a chain of two or more monomer units must be considered to distinguish between head and tail addition.

Tacticity is only a consideration for units formed by 1,2-addition. However, units formed by 1,4-addition may have a *cis*- or a *trans*-configuration.

In anionic and coordination polymerizations, reaction conditions can be chosen to yield polymers of specific microstructure. However, in radical polymerization while some sensitivity to reaction conditions has been reported, the product is typically a mixture of microstructures in which 1,4-addition is favored. Substitution at the 2-position (*e.g.* isoprene or chloroprene - Section 4.3.2.2) favors 1,4-addition and is attributed to the influence of steric factors. The reaction temperature does not affect the ratio of 1,2:1,4-addition but does influence the configuration of the double bond formed in 1,4-addition. Lower reaction temperatures favor *trans*-1,4-addition (Sections 4.3.2.1 and 4.3.2.2).

Early work on the microstructure of the diene polymers has been reviewed.[1] While polymerizations of a large number of 2-substituted and 2,3-disubstituted dienes have been reported,[88] little is known about the microstructure of diene polymers other than PB,[89] polyisoprene,[90] and polychloroprene.[91]

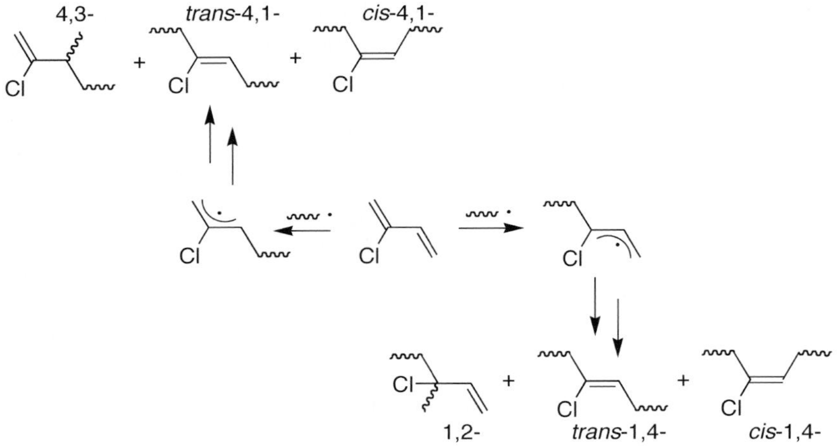

Scheme 4.11

4.3.2.1 Polybutadiene

The mechanism of B polymerization is summarized in Scheme 4.9. 1,2-, and cis- and trans-1,4-butadiene units may be discriminated by IR, Raman, or ^{1}H or ^{13}C NMR spectroscopy.[1,92-94] PB comprises predominantly 1,4-*trans*-units. A typical composition formed by radical polymerization is 57.3:23.7:19.0 for trans-1,4-:cis-1,4-:1,2-. While the ratio of 1,2- to 1,4-units shows only a small temperature dependence, the effect on the *cis-trans* ratio appears substantial. Sato et al.[93] have determined dyad sequences by solution ^{13}C NMR and found that the distribution of isomeric structures and tacticity is adequately described by Bernoullian statistics. Kawahara et al.[94] determined the microstructure (ratio trans-1,4-:cis-1,4-:1,2- and dyad ratios) by performing ^{13}C NMR measurements directly on PB latexes and obtained similar data to that obtained by solution ^{13}C NMR. They[94] also characterized crosslinked PB.

4.3.2.2 Polychloroprene, polyisoprene

The mechanism of chloroprene polymerization is summarized in Scheme 4.11. Coleman et al.[95,96] have applied ^{13}C NMR in a detailed investigation of the microstructure of poly(chloroprene) also known as neoprene. They report a substantial dependence of the microstructure on temperature and perhaps on reaction conditions (Table 4.3). The polymer prepared at −150 °C essentially has a homogeneous 1,4-*trans*-microstructure. The polymerization is less specific at higher temperatures. Note that different polymerization conditions were employed as well as different temperatures and the influence of these has not been considered separately.

Propagation

Table 4.3 Microstructure of Poly(chloroprene) *vs* Temperature

temperature °C	unit				
	4,1-*trans*	1,4-*trans*	1,4-*cis*[a]	1,2-[b]	4,3-
90[c]	75.1	10.3	7.8	2.9	4.1
40[d]	81.6	9.2	5.2	2.5	1.4
0[d]	90.4	5.5	1.8	2.1	1.1
-40[d]	93.2	4.2	0.7	1.4	0.5
-150[e]	98.0	2.0	<0.2	<0.2	<0.2

a 1,4- and 4,1-*cis* not distinguished. b 25-50% of 1,2- are isomerized. c Reaction conditions not stated. d Emulsion polymer. e Polymer prepared by irradiation of crystalline monomer.

Poly(isoprene) can also be prepared by radical polymerization.[97] Although the ratio of 1,4-:1,2-:4,3- units is stated to be *ca* 90:5:5 irrespective of the polymerization temperature (range -20–50 °C), the proportion of *cis*-1,4-addition increases from 0 at –20 °C to 17.6% at 50 °C. EPR studies indicate that radicals add preferentially to the 1-position.[87]

4.4 Structural Isomerism - Rearrangement

During most radical polymerizations, the basic carbon skeleton of the monomer unit is maintained intact. However, in some cases the initially formed radical may undergo intramolecular rearrangement leading to the incorporation of new structural units into the polymer chain. The rearrangement may take the form of ring closure (see 4.4.1), ring-opening (see 4.4.2) or intramolecular atom transfer (see 4.4.3).

The unimolecular rearrangement must compete with normal propagation. As a consequence, for systems where there is <100% rearrangement, the concentration of rearranged units in the polymer chain will be dependent on reaction conditions. The use of low monomer concentrations will favor the unimolecular process and it follows that the rearrangement process will become increasingly favored over normal propagation as polymerization proceeds and monomer is depleted (*i.e.* at high conversion). Higher reaction temperatures generally also favor rearrangement.

4.4.1 Cyclopolymerization

Diene monomers with suitably disposed double bonds may undergo intramolecular ring-closure in competition with propagation (Scheme 4.12). The term cyclopolymerization was coined to cover such systems. Many systems which give cyclopolymerization to the exclusion of "normal" propagation and crosslinking are now known. The subject is reviewed in a series of works by Butler.[98-102]

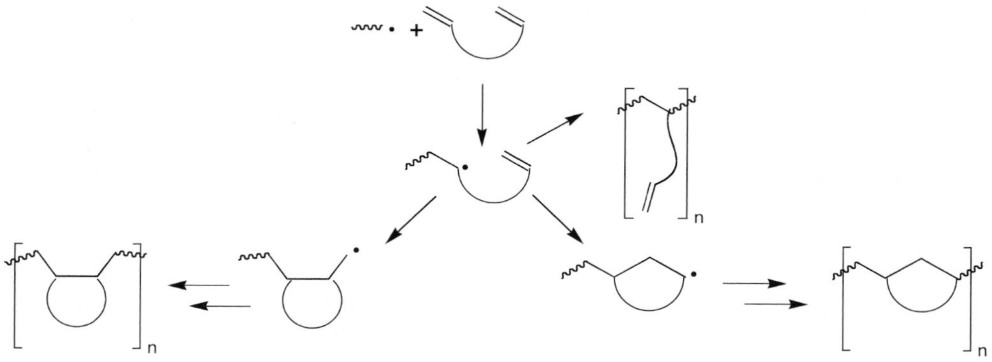

Scheme 4.12

Intramolecular cyclization is subject to the same factors as intermolecular addition (see 2.3). However, stereoelectronic factors achieve greater significance because the relative positions of the radical and double bond are constrained by being part of the one molecule (see 2.3.4) and can lead to head addition being the preferred pathway for the intramolecular step.

Geometric considerations in cyclopolymerization are optimal for 1,6-dienes (see 4.4.1.1). Instances of cyclopolymerization involving formation of larger rings have also been reported (see 4.4.1.4), as have examples where sequential intramolecular additions lead to bicyclic structures within the chain (see 4.4.1.2). Various 1,4- and 1,5-dienes are proposed to undergo cyclopolymerization by a mechanism involving two sequential intramolecular additions (see 4.4.1.3).

4.4.1.1 1,6-Dienes

The polymerization of nonconjugated diene monomers might be expected to afford polymer chains with pendant unsaturation and ultimately, on further reaction of these groups, crosslinked insoluble polymer networks. Thus, the finding by Butler *et al.*,[103-105] that polymerizations of diallylammonium salts, of general structure **8** [*e.g.* diallyldimethylammonium chloride (**9**)] gave linear saturated polymers, was initially considered surprising.

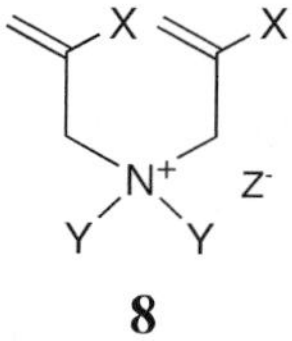

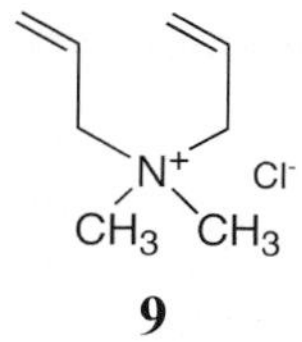

The explanation proposed involved sequential inter- and intramolecular addition steps. The presence of cyclic structures within the polymer chain was soon confirmed by degradation experiments.[106] However, these experiments did

Propagation

not unambiguously define the precise nature of the cyclic units. Their nature was inferred on the basis of the then prevailing theory, that radical additions proceed so as to give the more stable product (a six-membered ring and a secondary radical). As a consequence, the structure of these cyclopolymers was not firmly established until the 1970s when spectroscopic studies showed that five-membered ring formation is the preferred (kinetic) pathway during cyclopolymerization of simple diallyl compounds (**10**).[107-112]

Cyclopolymerizations of other 1,6-dienes afford varying ratios of five- and six-membered ring products depending on the substitution pattern of the starting diene. Substitution of the olefinic methine hydrogen (*e.g.* **11**, R= CH_3) causes a shift from five- to six-membered ring formation. More bulky R substituents can prevent efficient cyclization and cross-linked polymers may result.

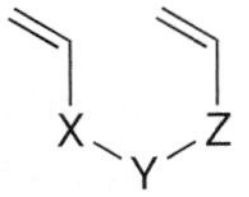

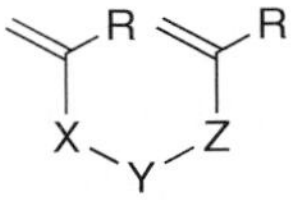

A vast range of symmetrical and unsymmetrical 1,6-diene monomers has now been prepared and polymerized and the generality of the process is well established.[98,109] A summary of symmetrical 1,6-diene structures, known to give cyclopolymerization, is presented in Table 4.4 In many cases, the structure of the repeat units has not been rigorously established. Often the only direct evidence for cyclopolymerization is the solubility of the polymer or the absence of residual unsaturation. In these cases the proposed repeat unit structures are speculative.

The understanding of the mechanism of cyclopolymerization has been one of the initial driving forces responsible for studies on the factors controlling the mode of ring closure of 5-hexenyl radicals and other simple model compounds.[113]

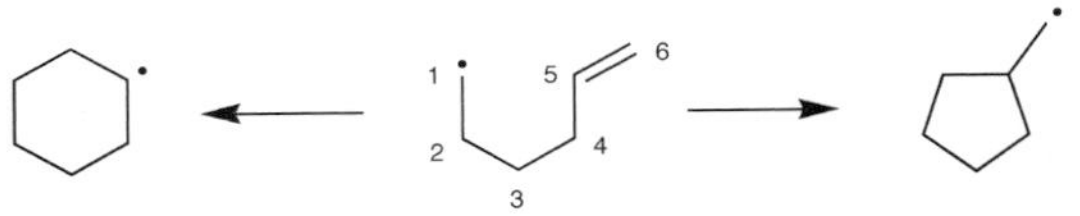

Scheme 4.13

The preferential 1,5-ring closure of unsubstituted 5-hexenyl radicals has been attributed to various factors; these are discussed in greater detail in Section 2.3.4. The mode and rate of cyclization is strongly influenced by substituents. The results may be summarized as follows (Scheme 4.13):

(a) Methyl substitution at C-1 slows the rate of both 1,5 and 1,6-ring closure. Substituents which delocalize the spin into a π-system (CN, CO_2Me) may result in a predominance of six-membered ring products by rendering intramolecular addition readily reversible.

(b) Substitution at C-5 dramatically retards 1,5-ring closure to the extent that 1,6-ring closure may predominate.
(c) Substitution at C-6 retards 1,6-ring closure. If both the 5 and 6 positions are substituted 1,5-ring closure predominates.
(d) Substitution at C-2, C-3, or C-4 facilitates both 1,5- and 1,6- ring closure.
(e) Increased reaction temperatures favor 1,6-ring closure at the expense of 1,5-ring closure.

The presence of heteroatoms and the inclusion of sp^2 centers are also known to affect the rate and mode of cyclization.

Thus, on the basis of model studies, it is possible to reconcile the observation that diallyl monomers that are unsubstituted on the double bond (**10,** X=Z=CH$_2$, Y=CR$_2$, NR, O, *etc.*) give predominantly five-membered rings for the intramolecular step. Dimethallyl monomers and other similarly substituted monomers (**11,** R≠H) generally give predominantly six-membered rings (*e.g.* **11,** X=Z=CH$_2$, R=CH$_3$ or CO$_2$R - Table 4.4). Dimethacrylic anhydride (**11,** Y=O, X=Z=C=O, R=CH$_3$) gives six-membered rings.[114] It is surprising that dimethacrylic imides (**11,** Y=N-alkyl, X=Z=C=O, R= CH$_3$) are reported to give five-membered rings.[114,115]

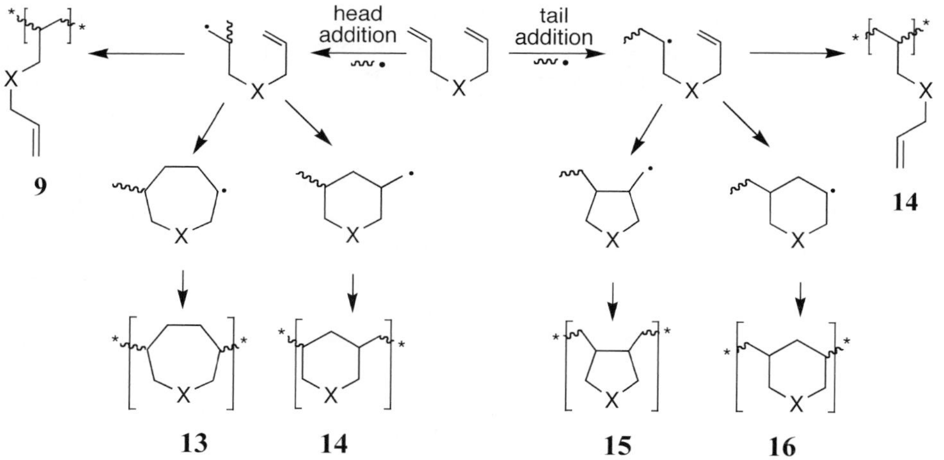

Scheme 4.14

The observation by Matsumoto *et al.* (see 4.3.1.4) that significant amounts of head addition occur in polymerization of simple allyl monomers brings into question the origin of the small amounts of six-membered ring products that are formed in cyclopolymerization of simple diallyl monomers (Scheme 4.14). If the intermolecular addition step were to involve head addition, then the intramolecular step should give predominantly a six-membered ring product (**14**) (by analogy with chemistry seen for 1,7 dienes – see 4.4.1.4). Note that the repeat units **14** and **16**, like **12** and **17** are the same; however, they are oriented differently in the chain.

Propagation

If there is significant intermolecular head addition, the formation of seven-membered units (**13**) might occur.[116]

The stereospecificity of the cyclization step has been examined both for model systems[113,117] and in a few cyclopolymerizations.[111,118,119] In formation of either five- or six-membered rings, there is a preference for the polymeric residues to end up *cis*- to each other. Note that for cyclopolymers with six-membered ring units, the ring stereochemistry is established in the intermolecular addition step (Scheme 4.15). In the case of five-membered ring units, ring stereochemistry is established during the intramolecular step (Scheme 4.16).

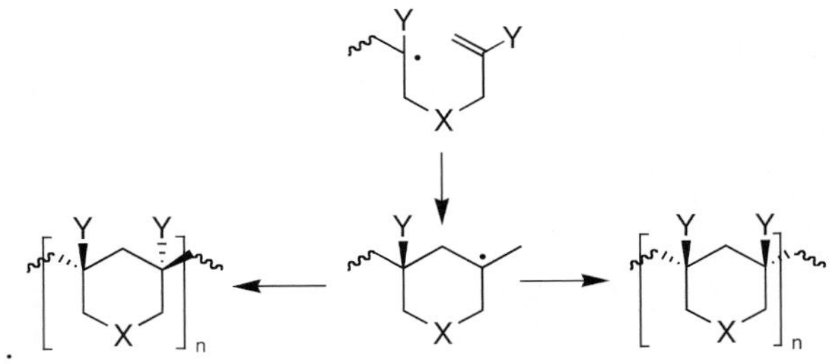

Scheme 4.15

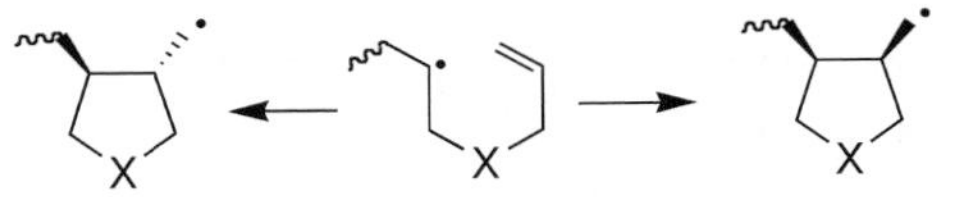

Scheme 4.16

Unsymmetrical 1,6-dienes known to undergo cyclopolymerization include allyl (meth)acrylate (**18** X=H, CH_3; Y=H),[120] (**18** X=CH_3; Y=Ph)[121] and (meth)acrylamide derivatives (**19** X=H, CH_3)[120,122-125] and *o*-allyl (**20** X=H)[126] and *o*-isopropenylstyrene (**20** X=CH_3).[127] With these cyclopolymerizations initial addition is to the double bond with the α−phenyl or carbonyl group and residual double bonds are isopropenyl or allyl groups.[124,125] For these examples, the cyclization step is relatively slow and reaction conditions are extremely important in obtaining soluble (uncrosslinked) polymers.

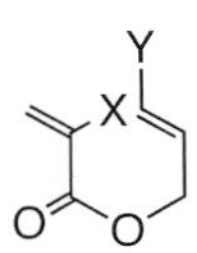

18 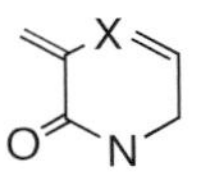   **19** 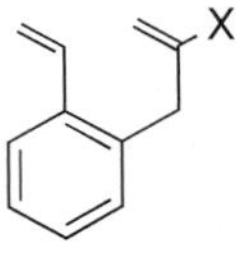   **20**

Table 4.4 Ring Sizes Formed in Cyclopolymerization of Symmetrical 1,6-Diene Monomers

monomer	substituents	ring size[a]	refs.
(a) all carbon skeleton	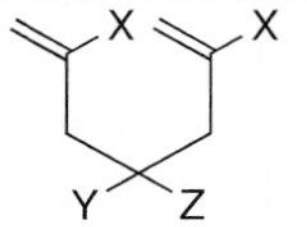 X=Ph; Y=Z=H	6	128,129
	X=CO$_2$H, CO$_2$R, CN; Y=Z=H	6	130,131
	X=CO$_2$Me; Y=Z=CN	6	118
	X=CO$_2$R; Y=Z=CO$_2$R'	6	132
(b) 4-nitrogen	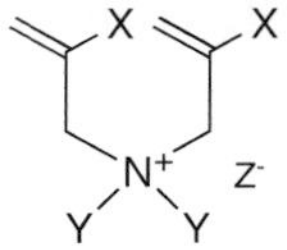 X=H	5	98,99,107,110,133
	X=CH$_3$	6	
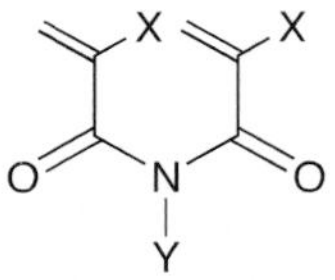	X=H	5	134
	X=CH$_3$	5	114,115
(c) 4-oxygen	X=H	5	80,135-137
	X=CH$_3$	6	
	X=CO$_2$R	6	138-140
(d) 3,5-oxygen	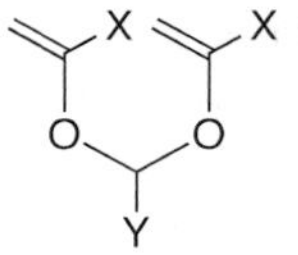 X=Y=H	5+6	141,142
	X=H; Y=C$_3$H$_7$		142
(e) other heteroatom substituents	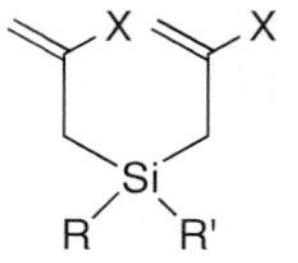 X=H		143-145
	X=CH$_3$		145
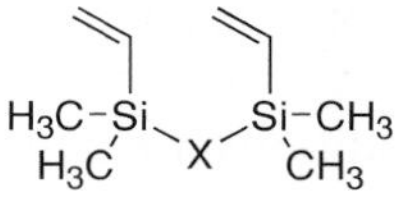	X=O, NH, NCH$_3$	5+6+7?	116
		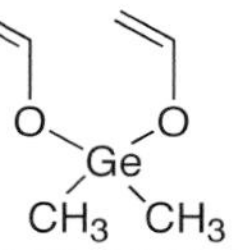 5+6	146

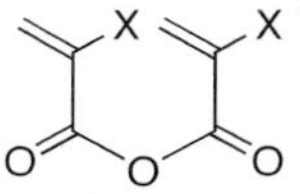

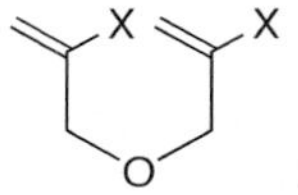

Table 4.4 (continued)

	monomer	substituents	ring size[a]	refs.
(e) other heteroatom substituents		X=H		147
		X=CH$_3$		
		X=H		148,149
		X=CH$_3$		148,150
	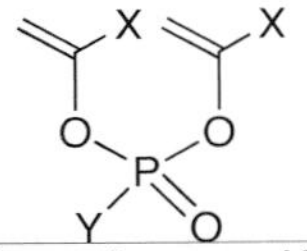	X=H		151

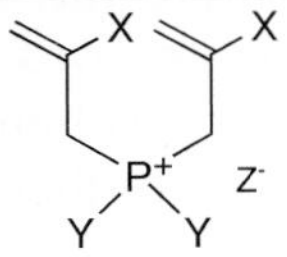

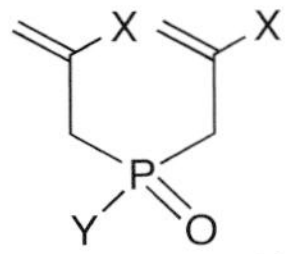

a Predominant ring size. If not specified, it has not been unambiguously determined.

Propagation in cyclopolymerization may be substantially faster than for analogous monoene monomers.[152] The various theories put forward to account for this observation are summarized in Butler's review.[98] A recent theoretical study by Tüzün et al.[133] looks at the effects of substituents on the rate of the cyclization step.

One contributing factor, which seems to have been largely ignored, is that the ring closed radical (in many cases a primary alkyl radical) is likely to be much more reactive towards double bonds than the allyl radical propagating species. This species will also have a different propensity for degradative chain transfer (a particular problem with allylamines and related monomers - see 6.2.6.4) and other processes which complicate polymerizations of the monoenes.

4.4.1.2 Triene monomers

Triallyl monomers [e.g. (**21**) or salts thereof] can potentially undergo two successive intramolecular cyclizations.[153,154] However, in practice these materials give insoluble products.

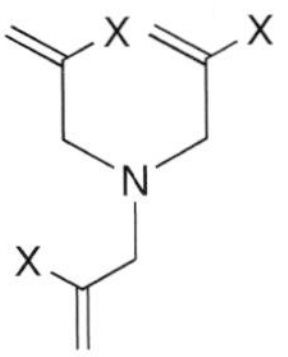

21

A model study has demonstrated the pathways shown in Scheme 4.17. The first cyclization step gave predominantly five-membered rings, the second a mixture of six- and seven-membered rings.[155] Relative rate constants for the individual steps were measured. The first cyclization step was found to be some five-fold faster than for the parent 5-hexenyl system. Although originally put forward as evidence for hyperconjugation in 1,6-dienes, further work showed the rate acceleration to be steric in origin.[113,133]

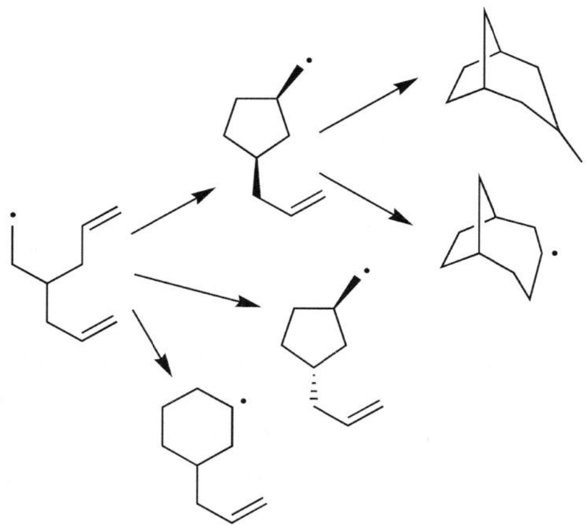

Scheme 4.17

The first cyclization gives a mixture of *cis*- and *trans*-isomers and only the *cis*-isomer goes on to give bicyclic products. The relatively slow rate of the second cyclization step, and the formation of *trans*-product which does not cyclize, provides an explanation for the observation that radical polymerizations of triallyl monomers often give a crosslinked product.

4.4.1.3 1,4- and 1,5-dienes

Geometric considerations would seem to dictate that 1,4- and 1,5-dienes should not undergo cyclopolymerization readily. However, in the case of 1,4-dienes, a 5-hexenyl system is formed after one propagation step. Cyclization via 1,5-backbiting generates a second 5-hexenyl system. Homopolymerization of divinyl ether (**22**) is thought to involve such a bicyclization. The polymer contains a mixture of structures including that formed by the pathway shown in Scheme 4.18.

It has been suggested that certain 1,5-dienes including *o*-divinylbenzene (**23**),[156] vinyl acrylate (**24**, X=H) and vinyl methacrylate (**24**, X=CH$_3$)[120] may also undergo cyclopolymerization with a monomer addition occurring prior to cyclization and formation of a large ring. However, the structures of these cyclopolymers have not been rigorously established.

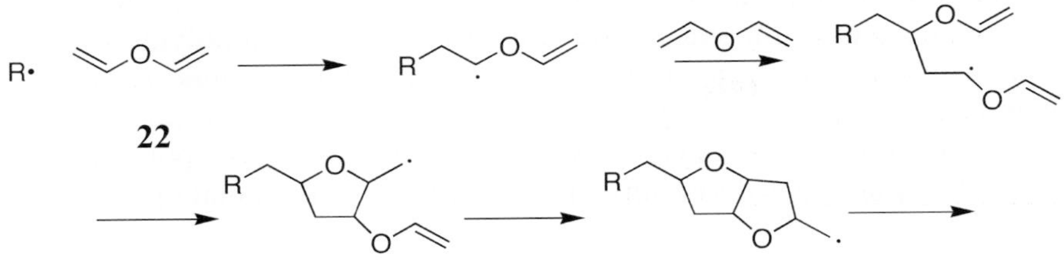

Scheme 4.18

Bicyclo[2,2,1]heptadiene derivatives (**25**) are set up to undergo ring closure to form a three-membered ring and it is proposed that polymers formed from (**25**) contain predominantly nortricyclene units.[157,158]

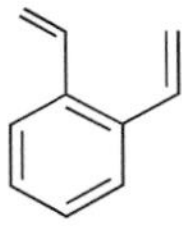

23 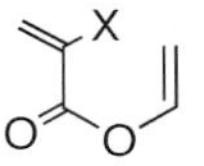   **24** 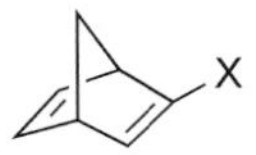   **25**

4.4.1.4 1,7- and higher 1,n-dienes

Several polymerizations of 1,7- and higher diene monomers have been reported. Cyclization to large rings (> six-membered) has been postulated.[159-164] However, in many examples, cyclization is not quantitative and crosslinked polymers are formed. Evidence for ring formation comes from kinetic data and, in particular, from the delay in the gel point from that expected (based on the assumptions that no cyclization occurs and that all pendant double bonds are available for crosslinking reactions). One common monomer that is thought to show such behavior is methylene-bis-acrylamide (ring structure not proven).[159,160]

1,7-dienes give six-membered rings in preference to seven-membered rings; examples include ethylene glycol divinyl ether (**26**) and bis-acryloylhydrazine (**27**).[161] This preference is also seen with model 6-heptenyl radicals.[165] One of the first reported examples of a 'cyclopolymerization' was that of the 1,11-diene, diallyl phthalate (**28**). A significant fraction (30-40%) of repeat units in the low conversion polymer was postulated to have a cyclic structure.[162,163] NMR studies on polymers formed by exhaustive hydrolysis suggest the cyclopolymer contains eleven-membered rings.[164]

Various dimethacrylates have been polymerized in an effort to synthesize a poly(methacrylate) with head-to-head linkages.[114,115] Various 1,6- (*e.g.* dimethacrylamides - see Table 4.4), 1,7- (*e.g.* dimethacrylhydrazines) and 1,8-dienes (*e.g.* dimethacryloylureas) are reported to give head-to-head addition (five-, six- or seven-membered rings respectively) or a mixture of head-to-head and head-to-tail addition. The 1,9-diene, *o*-dimethacryloylbenzene (**29**)[166] and the 1,10-diene 2,4-pentanediol dimethacrylate (**30**) give 100% cyclopolymerization and only head-to-tail addition (nine- and ten-membered rings respectively). Methacrylate derivatives of oligo- and polyhydroxy compounds analogous to **30** have been shown to undergo cyclopolymerization to give ladder polymers. These polymerizations are considered further in the section on template polymerization (see 8.3.5.2).

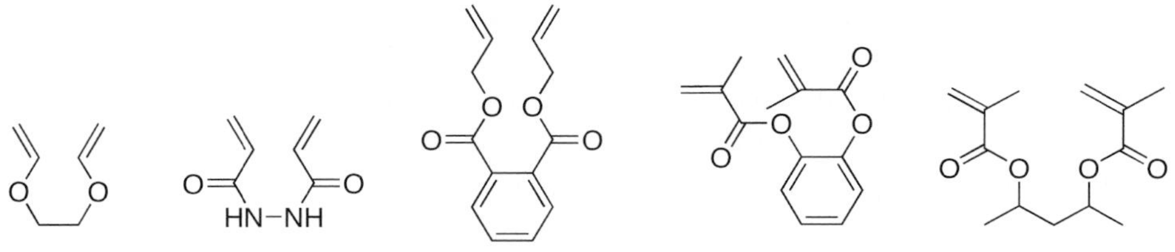

4.4.1.5 Cyclo-copolymerization

In this section we consider systems where the radical formed by propagation can cyclize to yield a new propagating radical. Certain 1,4-dienes undergo cyclo-copolymerization with suitable olefins. For example, divinyl ether and MAH are proposed to undergo alternating copolymerization as illustrated in Scheme 4.19.[167] These cyclo-copolymerizations can be quantitative only for the case of a strictly alternating copolymer. This can be achieved with certain electron donor-electron acceptor pairs, for example divinyl ether-maleic anhydride.

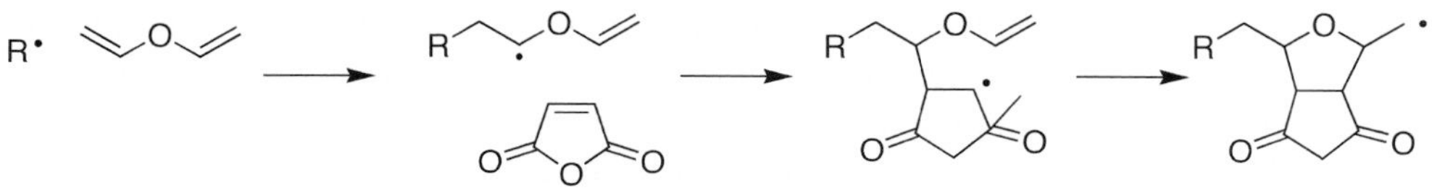

Scheme 4.19

4.4.2 Ring-Opening Polymerization

Much of the interest in ring-opening polymerizations stems from the fact that the polymers formed may have lower densities than the monomers from which they are derived (*i.e.* volume expansion may accompany polymerization).[168-171] This is in marked contrast with conventional polymerizations which typically involve a nett volume contraction. Such polymerizations are therefore of particular interest in adhesive, mold filling, and other applications where volume contraction is undesirable. Their use in dental composite and adhesive compositions has attracted recent attention.[171]

Ring-opening polymerizations and copolymerizations also offer novel routes to polyesters and polyketones (Section 4.4.2.2). These polymers are not otherwise available by radical polymerization. Finally, ring-opening copolymerization can be used to give end functional polymers. For example, copolymerization of ketene acetals with, for example, S, and basic hydrolysis of the ester linkages in the resultant copolymer offers a route to α,ω-difunctional polymers (Section 7.5.4).

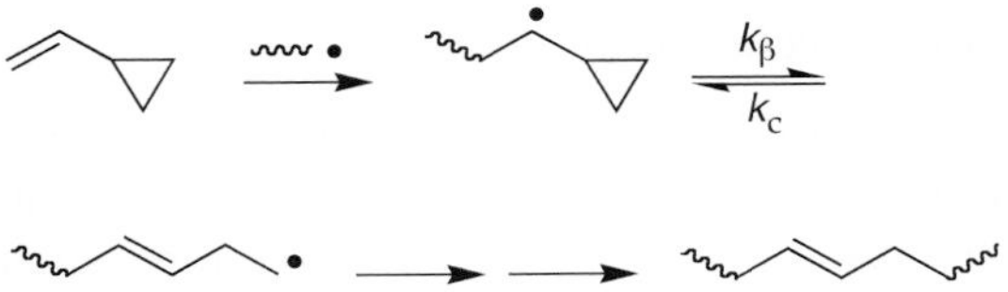

Scheme 4.20

Reviews on radical ring-opening polymerization include those by Sanda and Endo,[172] Klemm and Schultz,[173] Cho,[174] Moszner et al.,[175] Endo and Yokozawa[176] Stansbury[170] and Bailey.[177] A review by Colombani[178] on addition-fragmentation processes is also relevant. Monomers used in ring-opening are typically vinyl (*e.g.* vinylcyclopropane - Scheme 4.20; Section 4.4.2.1) or methylene substituted cyclic compounds (*e.g.* ketene acetals - Section 4.4.2.2) where addition to the double bond is followed by β-scission.

However, there are also examples of addition across a strained carbon-carbon single bond, as occurs with bicyclobutane[179] and derivatives (Scheme 4.21, Scheme 4.22).[180,181] Interestingly, 1-cyano-2,2,4,4-tetramethylbicylobutane (**31**) is reported to provide a polyketenimine (Scheme 4.22).[182] This is the only known examples of a α-cyanoalkyl radical adding monomer *via* nitrogen.

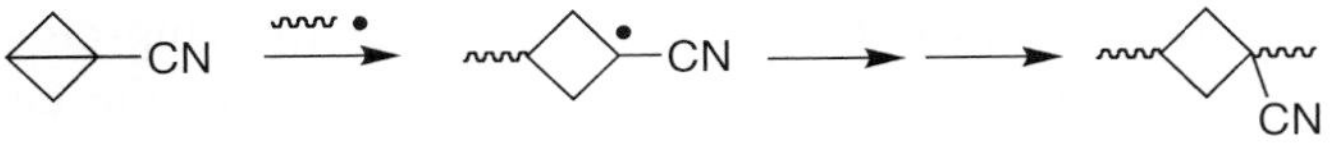

Scheme 4.21

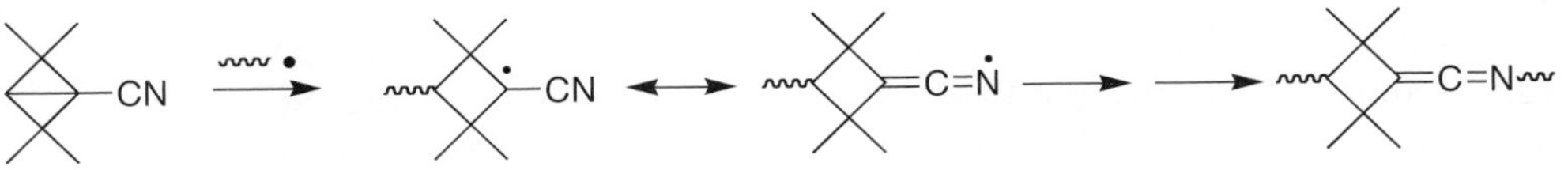

31

Scheme 4.22

For ring-opening to compete effectively with propagation, the former must be extremely facile. For example with $k_p \sim 10^2$-10^3 M^{-1} s^{-1} the rate constant for ring-opening (k_β) must be at least ~10^5-10^6 s^{-1} to give >99% ring-opening in bulk polymerization. The reaction conditions can be chosen so as to favor ring-opening. Ring-opening will be favored by dilute reaction media and, usually, by higher polymerization temperatures.

The ring-opening reaction usually results in the formation of a new unsaturated linkage. When this is a carbon-carbon double bond, the further reaction of this group during polymerization leads to a crosslinked (and insoluble) structure and can be a serious problem when networks are undesirable. In many of the applications mentioned above, crosslinking is desirable.

4.4.2.1 Vinyl substituted cyclic compounds

There must be considerable driving force for ring-opening if it is to compete with propagation. In the case of vinylcyclopropane and derivatives (Scheme 4.20) this is provided by the relief of strain inherent in the three-membered ring. Rates of ring-opening of cyclopropylmethyl radicals are reported to be in the range 10^5-10^8 s^{-1} depending on the substitution pattern.[183-187]

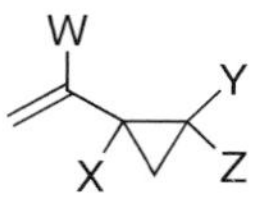

32

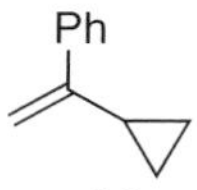

33

Many polymerizations of vinylcyclopropane and substituted derivatives (**32**) have now been reported.[174,175,188-205] All examples give 100% opening of the cyclopropane ring. However, conversions and polymerization rates are often low, even when the double bond is activated towards addition by a phenyl substituent (**33**).[205,206] For this example, the explanation for low polymerization rates probably lies with the reversibility of ring-opening. The reversibility of cyclopropylmethyl radical ring-opening has been established even for the parent system. The α-phenyl substituent reduces the rate of ring-opening by some two to three orders of magnitude[185,207] and the equilibrium lies in favor of the ring-closed radical.[207] Even though the rate constant for ring-opening is slow in the case of **33**, the monomer is unlikely to undergo polymerization without ring-opening. Such a polymerization should have a low ceiling temperature since **33** is structurally analogous to AMS (Section 44.5.1).

In the case of asymmetrically ring-substituted vinylcyclopropane derivatives (**32**, Y and/or Z≠H), two pathways for ring-opening are available (Scheme 4.23).[208] There have been a number of studies on substituent effects on ring-opening of cyclopropylmethyl radicals.[183-187] Steric, polar and stereoelectronic factors are all important in determining the kinetics and preferred mode of ring-opening. Since this is a reversible process, the kinetic and thermodynamic products may be different.[187]

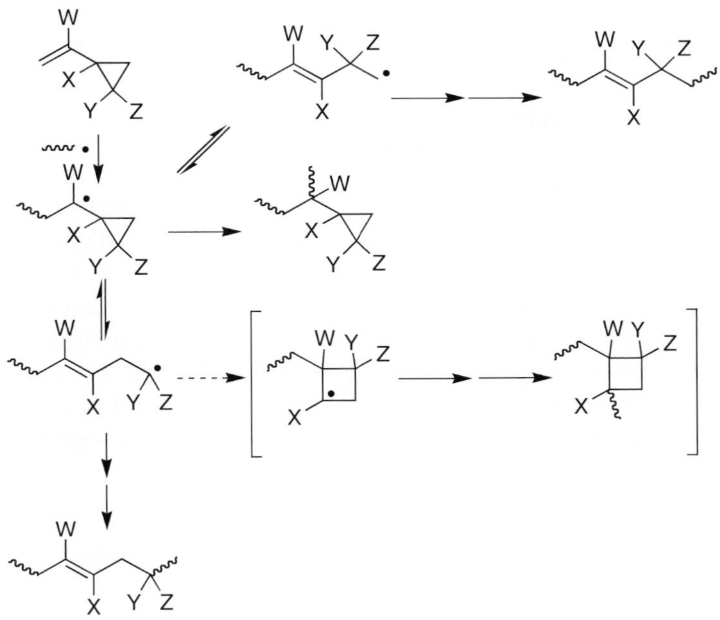

Scheme 4.23

It has also been proposed that the ring-opened radicals may undergo ring-closure to a cyclobutane (Scheme 4.23).[202,208] At this stage the only evidence for this pathway is observation of signals in the NMR spectrum of the polymer that cannot be rationalized in terms of the other structures. There is no precedent for 1,4-ring-closure of a 3-butenyl radical in small molecule chemistry and the result is contrary to expectation based on stereoelectronic requirements for intramolecular addition (Section 2.3.4). However, an alternate explanation has yet to be proposed. The possibility of carbonium ion intermediates should not be discounted.

1,2,2,3,3-Pentafluorovinylcyclopropane (**34**, Scheme 4.24) undergoes facile ring-opening polymerization exclusively as shown (*trans* double bond).[209]

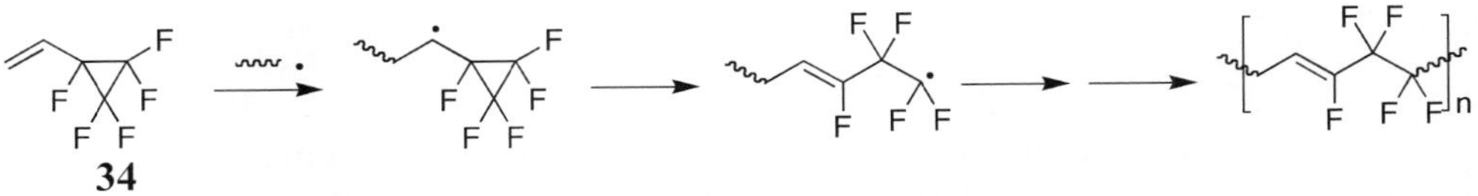

Scheme 4.24

The vinyloxirane (**35**, Scheme 4.25) undergoes ring-opening polymerization to give a polyether structure[210-212] with specific cleavage of the C-C bond. Other oxiranylmethyl radicals (without the phenyl substituent) are reported to give specific cleavage of the C-O bond.[213]

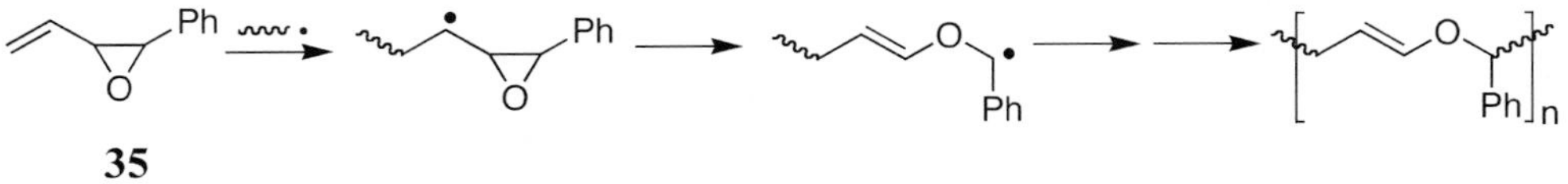

Scheme 4.25

Rate constants for ring-opening of cyclobutylmethyl radicals[214] are less than those for the corresponding cyclopropylmethyl radicals by a factor of *ca* 10^4.[183] This is consistent with the smaller degree of ring strain inherent in the four-membered ring. Model studies have shown that *cis*-β-substituents on the cyclobutane ring lead to a markedly enhanced rate constant for ring-opening and a high specificity for cleavage of the more substituted bond.[214] The substituted vinylcyclobutane (**36**, stereochemistry unspecified) is reported to give >90% ring-opening on polymerization in bulk at 60 °C and a single ring-opened product as shown in Scheme 4.26.[215]

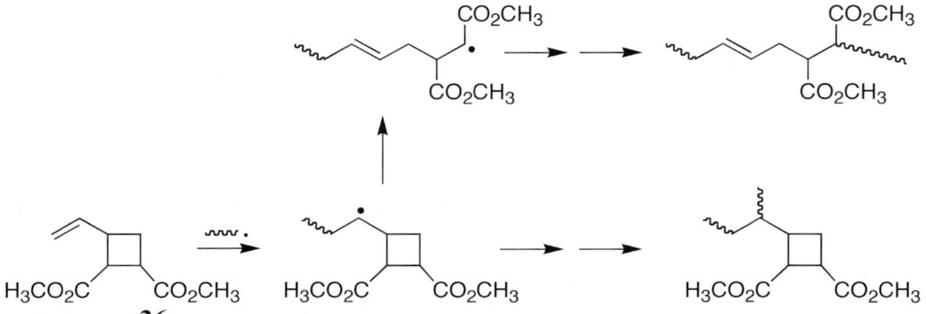

Scheme 4.26

2-Phenyl-1-vinylcyclobutane (**37**) is also reported to give partial ring-opening[174] while the vinylpropiolactones **38** and **39** give 100% ring-opening with loss of carbon dioxide.[174]

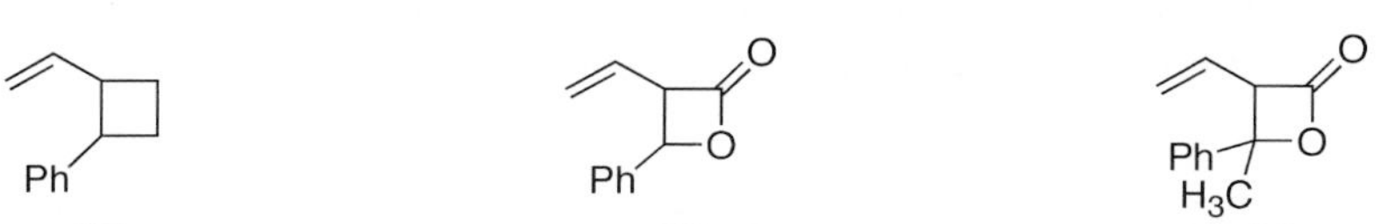

For vinylcyclopentane (cyclopentylmethyl radical) and vinylcyclohexane (cyclohexylmethyl radical) derivatives, ring-opening is generally not a favorable process (Section 4.4.1). However, a number of ring-opening polymerizations involving five- or larger-membered rings have been reported where appropriate substitution is present to provide the driving force for the β-scission step. Examples are the vinylsulfones (**40**, n=0,1,2),[216-218] which undergo ring-opening polymerization by scission of a relatively weak C-S bond and loss of sulfur dioxide, and the spiro derivatives **41**[219], **42**[220] and **43-44**[221] where ring-opening is facilitated by the concomitant aromatization of a cyclohexadiene derivative.

Polymerization of **42** gives between 43% (85 °C, bulk) and 98% (130 °C, bulk) ring-opening depending on reaction temperature.[220] Near quantitative ring-opening has been obtained in the case of polymerizations of **43** and **44** where further driving force for ring-opening is provided by formation of a benzylic radical.[221] These monomers, **43** and **44**, also undergo ring-opening in copolymerization with S.

4.4.2.2 Methylene substituted cyclic compounds

The ring-opening polymerization of ketene acetals (**45**, X=O) provides a novel route to polyesters and many examples have now been reported (Scheme 4.27).[222-227] A disadvantage of these systems is the marked acid sensitivity of the monomers which makes them relatively difficult to handle and complicates characterization. This area is covered by a series of reviews by Bailey et al.[177,228-231]

The main driving force for ring-opening in polymerizations of these compounds is formation of a strong carbon-oxygen double bond. The nitrogen (**45**, X=N-CH$_3$, n=0) and sulfur (**45**, X=S, n=0) analogs undergo ring-opening polymerization (Table 4.5) with selective cleavage of the C-O bond to give polyamides or polythioesters respectively (Scheme 4.27). The specificity is most likely a reflection of the greater bond strength of C=O vs the C=S or C=N double bonds. The corresponding dithianes do not give ring-opening even though this would involve cleavage of a weaker C-S bond.[232,233]

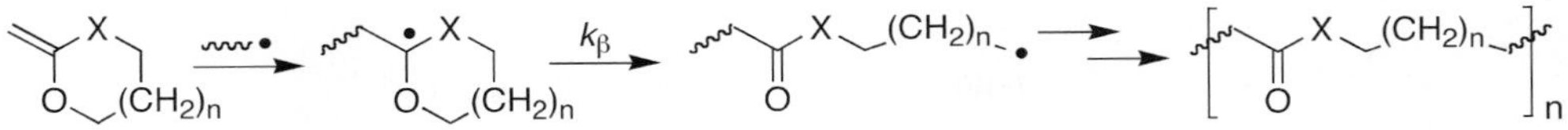

Scheme 4.27

The competition between ring-opening and propagation is dependent on ring size and substitution pattern. For the five-membered ring ketene acetal (**45**, X=O, n=0) ring-opening is not complete except at very high temperatures. However, with the larger-ring system (**45**, X=O, n=2) ring-opening is quantitative. This observation (for the n=2 system) was originally attributed to greater ring strain.

However, it may also reflect the greater ease with which the larger ring systems can accommodate the stereoelectronic requirements for β-scission (Section 2.3.4).[113] Substituents (*e.g.* CH_3, Ph) which lend stabilization to the new radical center, or increase strain in the breaking bond, also favor ring-opening (Table 4.5).

Table 4.5 Extent of Ring-opening During Polymerizations of 2-Methylene-1,3-dioxolane and Related Species

monomer	% ring-opening	conditions	ref.[a]
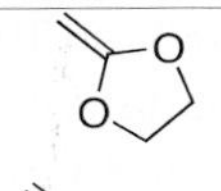 2-methylene-1,3-dioxolane	100 87 50	160 °C, bulk, tBu_2O_2 120 °C 60 °C	
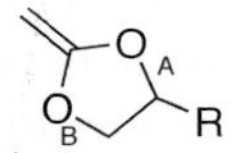 2-methylene-4-R-1,3-dioxolane (bonds A, B)	bond A 61[b] bond B 27	110 °C, bulk, tBu_2O_2	234
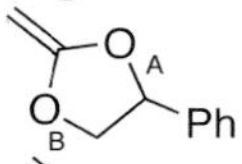 2-methylene-4-Ph-1,3-dioxolane	bond A 100[c]	120 °C, bulk, tBu_2O_2 30 °C hν	224,235 236
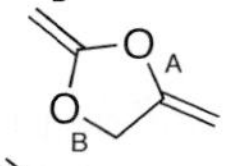 2-methylene-4-vinyl-1,3-dioxolane	bond B 100	120 °C, bulk, tBu_2O_2	237
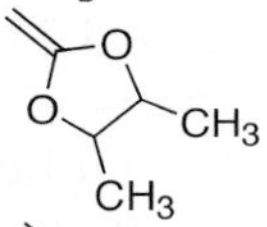 2-methylene-4,4-dimethyl-1,3-dioxolane	100	120 °C, bulk, tBu_2O_2	
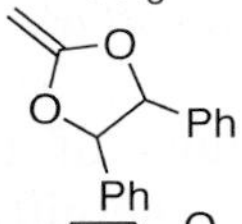 2-methylene-4-methyl-4-phenyl-1,3-dioxolane	100	120 °C, bulk, tBu_2O_2	225
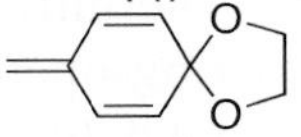 spiro dioxolane-cyclohexadiene	100	65-125 °C, benzene various initiators	226
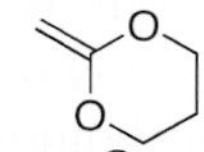 2-methylene-1,3-dioxane	<100	120 °C, bulk, tBu_2O_2	
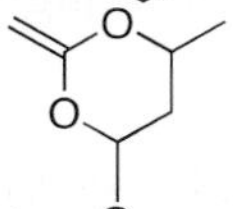 2-methylene-methyl-1,3-dioxane	100	120 °C, bulk, tBu_2O_2	
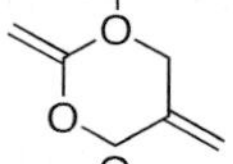 2-methylene-1,3-dioxepane derivative	<100	120 °C, bulk, tBu_2O_2	237
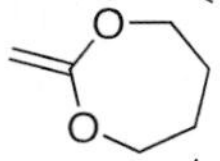 2-methylene-1,3-dioxepane	100	120 °C, bulk, tBu_2O_2	238,239
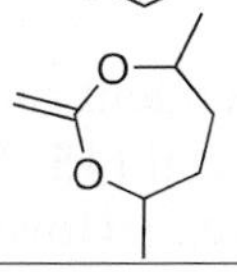 2-methylene-substituted 1,3-dioxepane	100	120 °C, bulk, tBu_2O_2	240

Table 4.5 (continued)

monomer	% ring-opening	conditions	ref.[a]
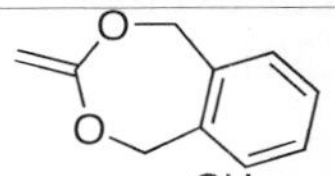 (benzo-fused methylene dioxepane)	100	120 °C, bulk, tBu_2O_2	240
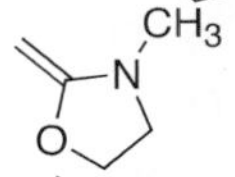 (methyl oxazolidine with vinyl)	100	80 °C, bulk, $(PhCO)_2$	241
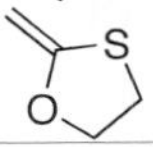 (thio-oxazolidine)	45	120 °C, bulk, tBu_2O_2	242

a Where no reference is given, the examples are taken from Bailey's review.[177] b Data for R=n-decyl. Specificity dependent on R, temperature, and monomer concentration. c Racemization accompanies polymerization of optically active monomer.[224]

Scheme 4.28

The diene shown in Scheme 4.28 is also reported to give 100% ring-opening.[227] However, polymerization had to be carried out in very dilute solution to give a soluble (not crosslinked) product.

Rate constants for ring-opening of dioxolan-2-yl radicals have been measured by Barclay et al.[243] as 10^3-10^4 s^{-1} at 75 °C (Scheme 4.29). There is also evidence that ring-opening is reversible.[243,244] Thus, isomerization of the initially formed product to one more thermodynamically favored is possible if propagation is slow.

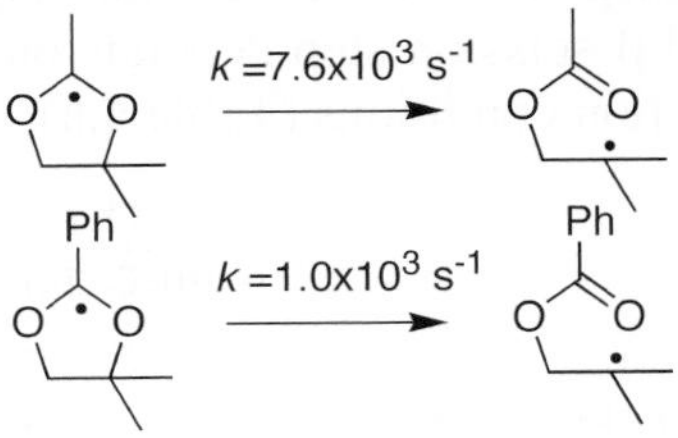

$k = 7.6 \times 10^3$ s^{-1}

$k = 1.0 \times 10^3$ s^{-1}

Scheme 4.29

Bailey et al.[177,245] observed that ring-opening polymerization of the monomers (**39**) and (**40**), which can potentially give rise to the same ring-opened radical, give different polymers. That formed from (**39**) has pendant vinyl groups, while that from (**40**) has in-chain double bonds. They proposed that, in radical polymerization of ketene acetals, ring-opening might be concerted with addition of the next monomer unit and various experiments were suggested to test the hypothesis.[177] One of these was carried out by Acar et al.,[224] who showed that ring-opening polymerization of optically active 4-phenyl-1,3-dioxolane was accompanied by racemization. This is evidence against concerted ring-opening.

It was proposed[177] that radical addition to **46** or **48** should occur exclusively at the respective methylene group to generate radicals **47** (Scheme 4.30).

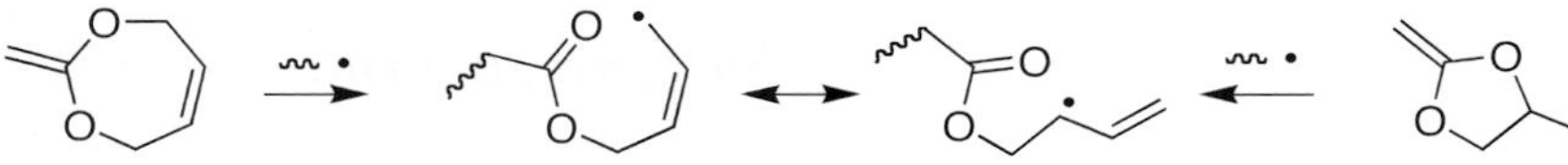

46 → **47** ↔ ↔ **48**

Scheme 4.30

If, however, radicals add preferentially to the vinyl group of **48**, ring-opening polymerization would give the polymer with in-chain double bonds specifically *via* resonance structure **49** (Scheme 4.31). Thus, the two pathways are readily distinguishable. No other ring-opening polymerizations of vinyl dioxolane derivatives appear to have been reported to date.

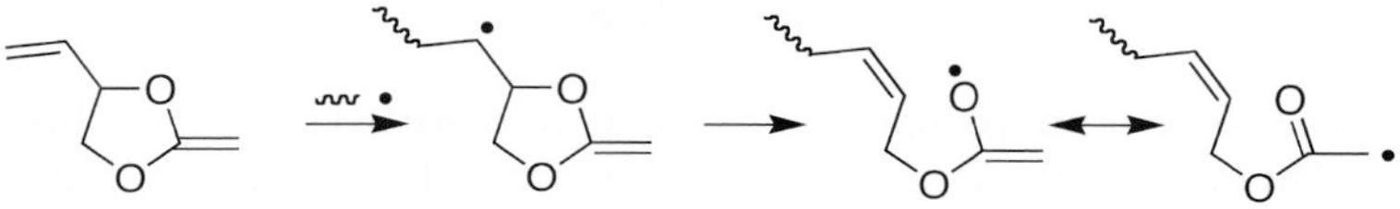

48 → → ↔ **49**

Scheme 4.31

4-Methylenedioxolane derivatives also undergo ring-opening. However, the ring-opened radical may undergo a further β-scission (*e.g.* **50**, Scheme 4.32).[223,246-252] The extent of the second β-scission step depends on the nature of substituents at the 2-position and the reaction conditions (Table 4.6).

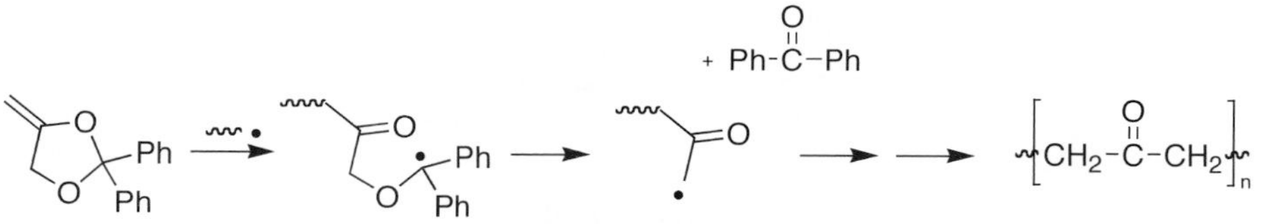

50 → → + Ph-C(=O)-Ph → → $-[CH_2-C(=O)-CH_2]_n-$

Scheme 4.32

Of the 4-methylene-1,3-dioxolanes reported thus far (Table 4.6), only the 2,2-diphenyl derivative (**50**) is reported to give the polyketone quantitatively (Scheme 4.32). This requires temperatures in excess of 120 °C in bulk polymerization.[246,247] The 2-phenyl-2-alkyl derivatives give <100% ring-opening but still give 100% elimination of the ring-opened product at 120 °C.[223] The 2-phenyl derivative is

Propagation

reported to afford ring-opening without elimination of benzaldehyde at temperatures less than 30 °C (photochemical initiation).[249] At higher temperatures terpolymers are formed that comprise units that are non-ring-opened, ring-opened, and ring-opened with β-scission.

Table 4.6 Extent of Ring-Opening During Polymerizations of 4-Methylene-1,3-dioxolane and 2-Methylene-1,4-dioxane Derivatives

monomer	% ring-opening	% elimination	conditions.	ref.[a]
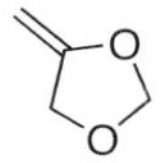 (4-methylene-1,3-dioxolane)	30	100	130 °C, bulk	177
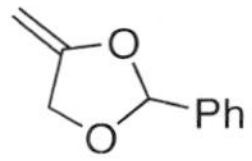 2-Ph derivative	73	36	120 °C, bulk	177,250
	100	0	<30 °C, hv	249
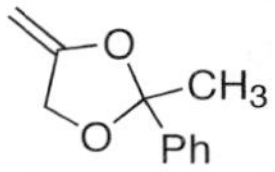 2-CH3, 2-Ph	23	100	120 °C, bulk	b,223
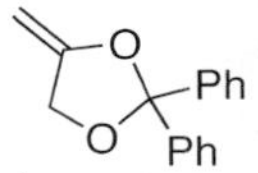 2-Ph, 2-Ph	18	100	60 °C, bulk	248
	100	100	120 °C, bulk	246,247
2-CH3, 2-CH3 (carbonate)	10	0	0-120 °C, bulk	177
			140 °C, bulk	253
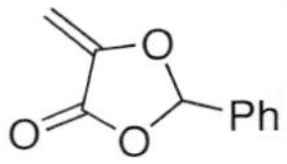 2-Ph (carbonate)	40	0	140 °C, bulk	
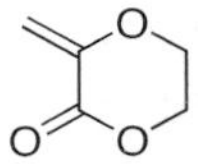 1,4-dioxane derivative	20	0	80 °C, benzene	
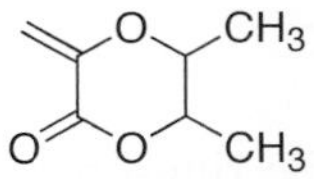 3,3-diCH3 1,4-dioxane	40	0	80 °C, benzene	
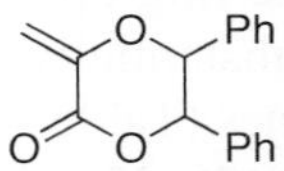 3-Ph 1,4-dioxane	100	0	80 °C, benzene	

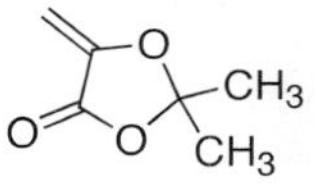

a Where no reference is given, the examples are taken from Bailey's review.[177]   b Other 2-phenyl-2-alkyl derivatives are also reported to give <100% ring-opening and 100% elimination at 120 °C.

The structurally analogous five-membered ring α-alkoxyacrylates (Scheme 4.33) are slow to ring-open and do not undergo β–scission to form an acyl radical propagating species.[177,253-255] This latter observation is probably a reflection of a higher bond strength for the bond α- to the carbonyl group. More ring-opening is observed for six-membered ring systems (Table 4.6).

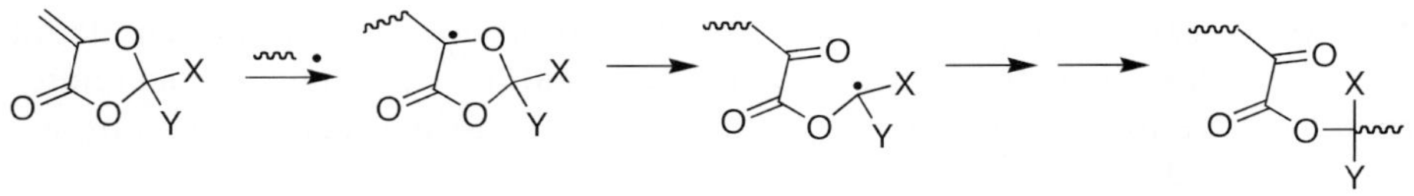

Scheme 4.33

Table 4.7 Extent of Ring-Opening During Polymerizations of 2-Methylenetetrahydrofuran and Related Compounds

monomer	% ring-opening	conditions	ref.[a]
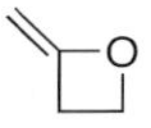 (4-membered oxetane)	40	120 °C, bulk, tBu_2O_2	242
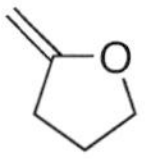 (5-membered, THF)	5	120 °C, bulk, tBu_2O_2	256
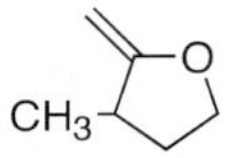 (5-Me substituted)	15-20	120 °C, bulk, tBu_2O_2	
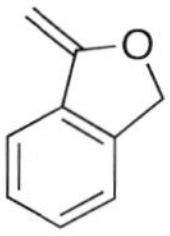 (benzofused)	0	120 °C, bulk, tBu_2O_2	
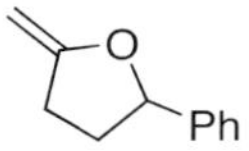 (Ph substituted)	50	120 °C, bulk, tBu_2O_2	256
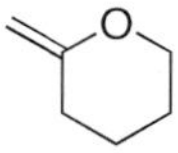 (6-membered)	4-8	120 °C, bulk, tBu_2O_2	

[a] Where no reference is given, the examples are taken from Bailey's review.[177]

Monomers with only a single ring oxygen-atom give less facile ring-opening. For example, the 2-methylenetetrahydrofuran derivatives give substantially less ring-opening than the corresponding 2- or 4-methylene-1,3-dioxolanes (Table 4.7).

Seven- and eight-membered ring cyclic allyl sulfide derivatives (**51, 52, 54-56**) are stable in storage and handling and do not show the acid sensitivity of the cyclic acetal monomers above. They undergo facile ring-opening polymerization even at relatively low temperatures[257-260] with quantitative ring-opening (Scheme 4.34, Scheme 4.35). The monomers also undergo facile ring-opening copolymerization with MMA and S.[261] The corresponding six-membered ring compound (**53**) appears unreactive in homopolymerization.

Ring-opening provides a thiyl radical propagating species. Although the polymers have a double bond on the backbone there is little or no crosslinking (Scheme 4.34, Scheme 4.35). There is, however, evidence of reversible addition

Propagation

and addition-fragmentation involving this double bond.[262] Monomers containing multiple double bonds have been designed to provide ring-opening polymerization with crosslinking.[259]

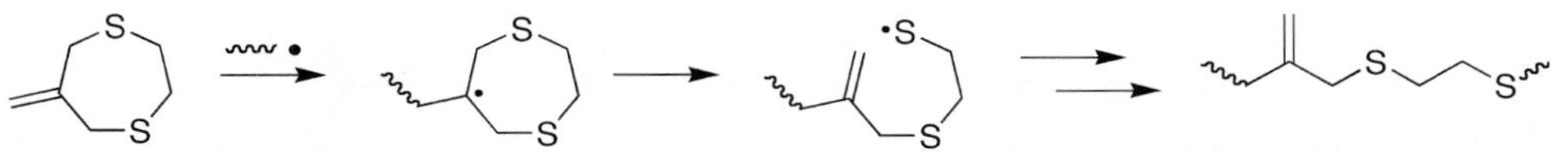

Scheme 4.34[258]

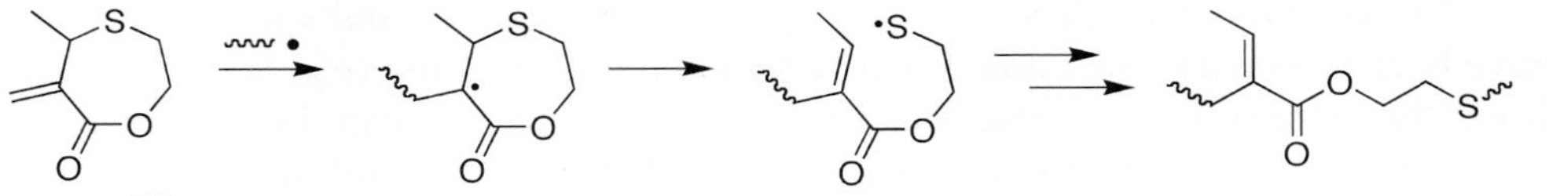

Structure **51**

Scheme 4.35[257]

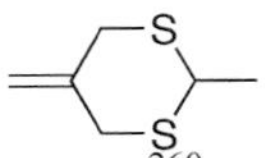

Structure **52**

Structure **53**[260]

Structure **54**
R=CH$_3$[260]
R=CH$_2$O(C=O)CH$_3$[259,260]
R=CH$_2$O(C=O)C(CH$_3$)=CH$_2$[259]

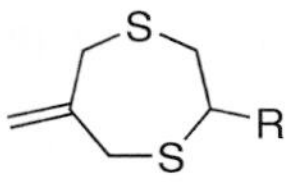

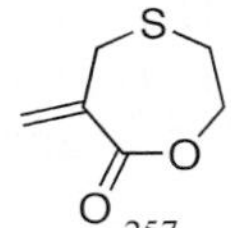

Structure **55**[257]

Structure **56**
R^1, R^2, R^3=H[258]
R^1, R^2=H, R^3=CH$_3$[260,262]
R^1, R^3=H, R^2=O(C=O)CH$_3$[259,260]
R^1, R^3=H, R^2=O(C=O)Ph[259]

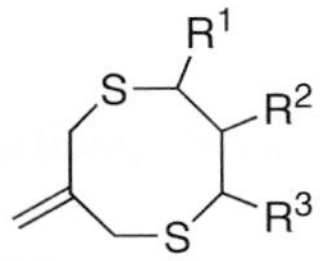

4.4.2.3 Double ring-opening polymerization

While many factors affect the degree of volume change which accompanies polymerization, any volume increase is directly related to the number of rings opened in the propagation step and is inversely related to the size of the rings being broken. Consideration of these factors leads to the conclusion that appreciable volume expansion on polymerization should only be expected when two or more rings are opened[170] and substantial effort has been put into designing systems where two or more rings are opened on polymerization.

It should also be noted that for many of the applications where volume expansion is required (adhesives, composites, *etc.*) a crosslinked product is desirable and some monomers have been designed with this in mind. This does, however, make the products difficult to characterize. Some monomers with potential for double ring-opening are reported in Table 4.8.

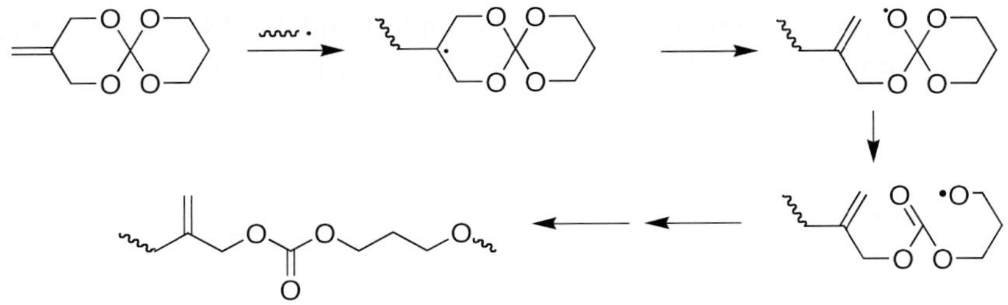

Scheme 4.36

Various methylene derivatives of spiroorthocarbonates and spiroorthoesters have been reported to give double ring-opening polymerization (*e.g.* Scheme 4.36). Like the parent monocyclic systems, these monomers can be sluggish to polymerize and reactivity ratios are such that they do not undergo ready copolymerization with acrylic and styrenic monomers. Copolymerizations with VAc have been reported.[170] These monomers, like other acetals, show marked acid sensitivity.

The vinylcyclopropane derivatives substituted with a five- or six-membered acetal ring give single ring-opening with differing regiospecificity (Scheme 4.37[202,203] and Scheme 4.38[202,203,263]).

Scheme 4.37 (double bond stereochemistry not specified)

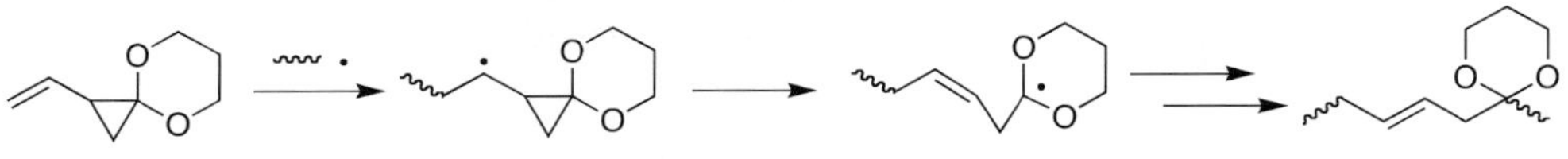

Scheme 4.38 (double bond stereochemistry not specified)

Systems with substituents on the acetal ring[264] or with larger acetal rings may give double ring-opening (*e.g.* Scheme 4.39).[202,203]

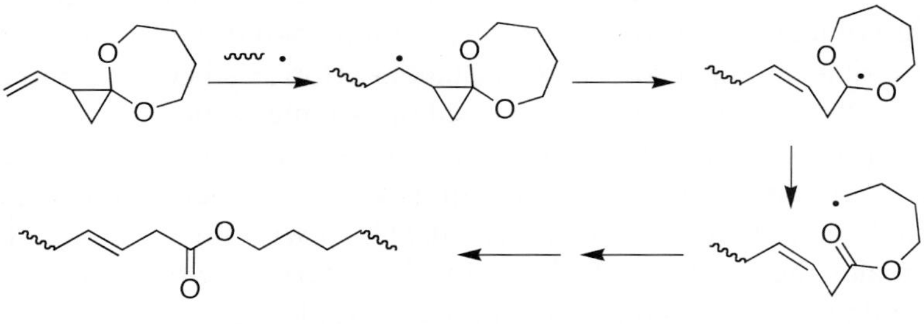

Scheme 4.39 (double bond stereochemistry not specified)

Table 4.8 Extent of Double Ring-Opening During Polymerization of Polycyclic Monomers

Monomer	% ring-opening	conditions	ref.[a]
spiroorthocarbonates			
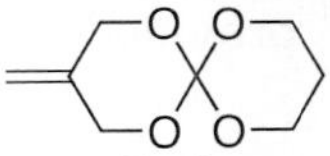	100	130 °C, bulk, 30% conv.[b]	
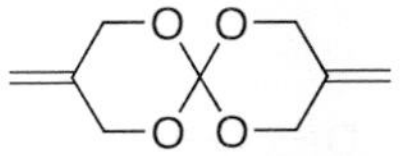	5-100	130 °C, PhCl, <50% conv.[b]	170,206,265
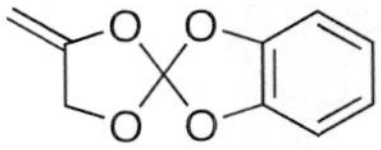	0	165 °C, PhCl	266
spiroorthoesters			
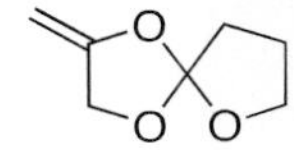	10[c]	120 °C, bulk, tBu_2O_2	267,268
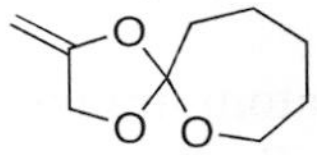	10[d]	120 °C, bulk, tBu_2O_2	269
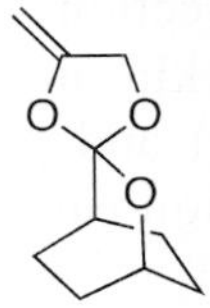	100	120 °C, bulk, tBu_2O_2	
other systems			
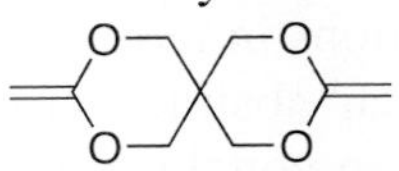	high	100 °C, AIBN	270
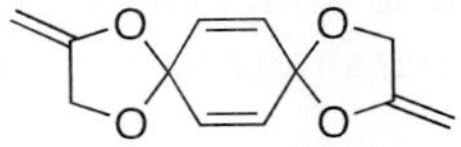	100	130 °C, bulk, tBu_2O_2	271
vinylcyclopropanes			
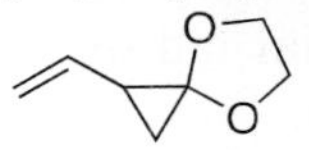	0[e]	60 °C, bulk, AIBN	202,203
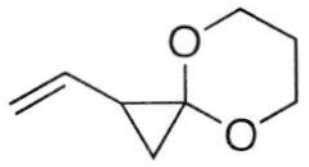	0[e]	60 °C, bulk, AIBN	202,203,263
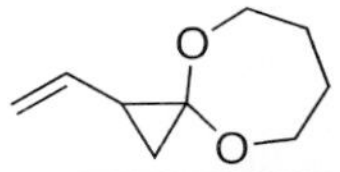	46	60 °C, bulk, AIBN	202,203

a Where no reference is given, the examples are taken from Bailey's review.[177]   b Insoluble and presumably crosslinked polymer formed at higher conversions.   c 50:50 mixture single and double ring-opened products.   d >50% double ring-opened product.   e Single ring-opened product only.

Analogous systems with six-, seven-, or eight-membered spirodithioacetal rings are reported to give single ring-opened products with no olefinic residues. A mechanism involving consecutive cyclopropane ring-opening and cyclization was proposed to rationalize this result.[272]

The spiro monomers **57-59** are reported to give single ring-opening.[273-275] Solution polymerization of **57**[273] and **58 (R=CH$_3$, R'=C$_3$H$_7$)**[274] provided soluble products.

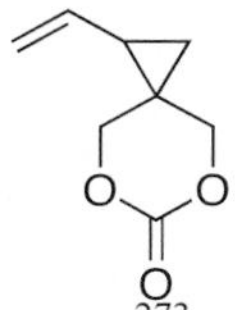

57[273]

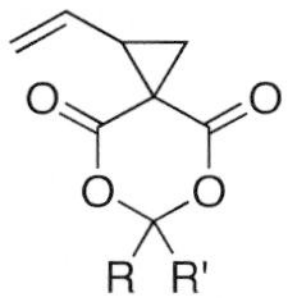

58 (R, R'=alkyl)[274]

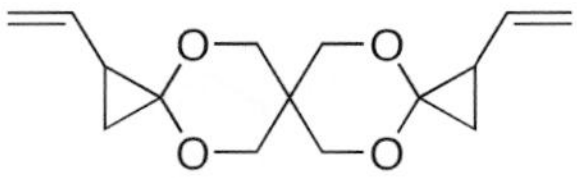

59[275]

4.4.3 Intramolecular Atom Transfer

It has been known for some time that intramolecular atom transfer, or backbiting, complicates polymerizations of E (Scheme 4.40 - Section 4.4.3.1), VAc and VC (see 4.4.3.2). Recent work has shown that backbiting is also prevalent in polymerization of acrylate esters (Section 4.4.3.3) and probably occurs to some extent during polymerizations of most monosubstituted monomers.[276,277]

Viswanadhan and Mattice[278] carried out calculations aimed at rationalizing the relative frequency of backbiting in these and other polymerizations in terms of the ease of adopting the required conformation for intramolecular abstraction (see 2.4.4). More recent theoretical studies generally support these conclusions and provide more quantitative estimates of the Arrhenius parameters for the process.[279,280]

Cases of "addition-abstraction" polymerization have also been reported where propagation occurs by a mechanism involving sequential addition and intramolecular 1,5-hydrogen atom transfer steps (Section 4.4.3.4).

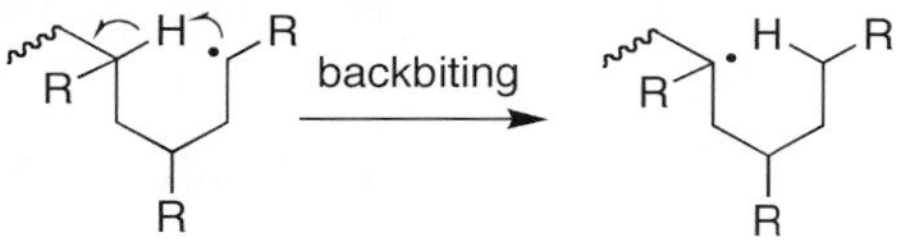

Scheme 4.40

4.4.3.1 Polyethylene and copolymers

The extent of short-chain branching in PE may be quantitatively determined by a variety of techniques including IR,[281,282] pyrolysis-GC,[283] and γ-radiolysis.[284] The most definitive information comes from ^{13}C NMR studies.[285-290] The typical concentration of branch points in PE formed by radical polymerization is 8-25 per 1000 CH_2.[287] These are made up of: ethyl, 1.2-11.3; butyl, 3.9-8.5; pentyl (amyl), 0.6-2.2; hexyl and longer, 0.5-2.8. The range of values for extent and type of short-chain branches arises because the branching process is extremely dependent on the polymerization conditions.[287] High reaction temperatures and low pressures (monomer concentrations) favor the backbiting process.

The backbiting reaction first proposed by Roedel[291] (Scheme 4.40) is generally accepted as the mechanism for short chain branch formation during polymerization of E (for discussion on alternative mechanisms see[292,293]). The preferential formation of butyl [vs propyl, pentyl, or longer branches] branches can be rationalized in terms of the stereoelectronic requirements imposed on the transition state (Section 2.4.4). The preferred coplanar arrangement of atoms is most readily achieved in a six-membered chair-like transition state.[294] 1-Undecyl radicals are a simple model of the PE propagating species and give 1,5- and 1,6-H transfer in the ratio 3:1. Other intramolecular H transfers were not detected.[295] Theoretical studies provide a picture of the transition state and a reasonable estimate of the Arrhenius parameters for backbiting.[279] Both enthalpic and entropic factors favor 1,5-H transfer.

Direct formation of an ethyl branch would require backbiting *via* a highly strained four-membered transition state and, therefore, should have a low probability.[294,296] The relatively large numbers of ethyl branches in PE is accounted for by the occurrence of two successive 1,5-H transfers which leads to either a pair of ethyl branches (Scheme 4.41) or a 2-ethylhexyl branch depending on the site of abstraction.[297] This mechanism for ethyl branch formation requires that the radical formed by backbiting (secondary alkyl) should be substantially more prone to undertake backbiting than the normal propagating species (primary alkyl). This suggests that the former has a reduced rate of propagation (more sterically hindered radical) and/or an increased rate of intramolecular abstraction (Thorpe-Ingold effect).

Backbiting also occurs in ethylene copolymerizations with AN,[298] (meth)acrylate esters[290] and VAc.[280,290,299,300] The structures identified in E-BA copolymerization include **60-63** (X=BA). Structure **60** is formed when the BA terminated chain backbites. Structure **61** is formed when backbiting occurs across a BA unit. Structure **62** and **63** are from backbiting to a BA unit (**63** is from double backbiting). The concentration of comonomer is such that there are few comonomer sequences.

The incidence of the various structures depends strongly on the comonomer. In copolymerization with acrylates structures **62** and **63** dominate. In copolymerization with VAc structure **61** dominates and **62** and **63** are not observed. Structure **60** may be present in VAc copolymers to a very small extent but is not observed in acrylate copolymerizations. Structures **62** and **63** are not observed and cannot be formed in methacrylate copolymerizations.[290] The results were interpreted[290] in terms of the PVAc• propagating radical having a lesser propensity for backbiting. This seems inconsistent with the observation of the products of backbiting during VAc homopolymerization (Section 4.4.3.2).[301] The data might also be rationalized in terms of the influence of polar and enthalpic factors on the facility of the various abstraction reactions (Section 2.4).

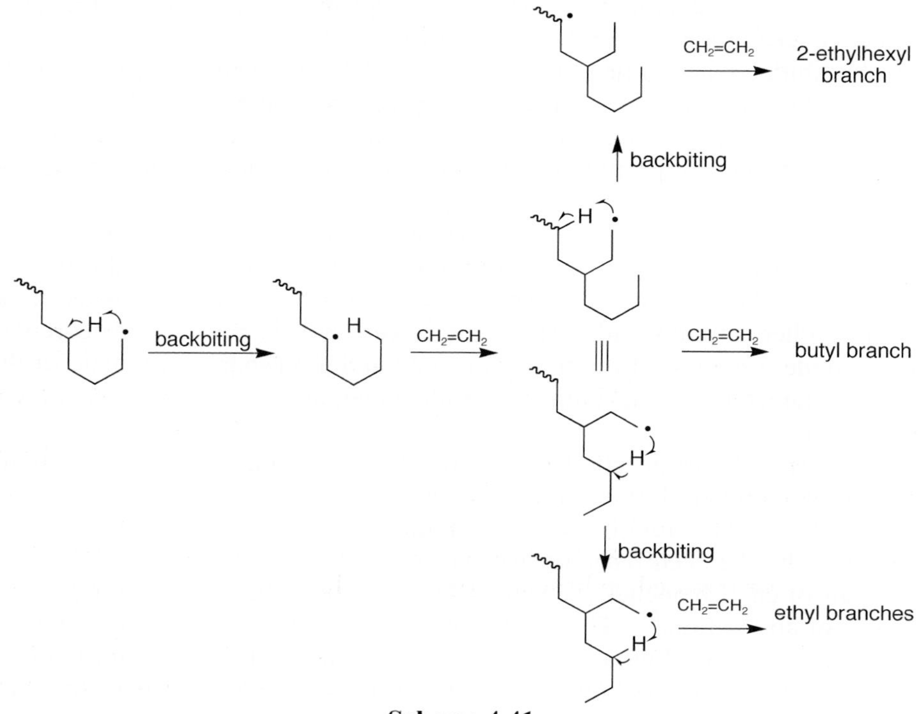

Scheme 4.41

Table 4.9 Structures Formed by Backbiting in Ethylene Copolymerizations[a]

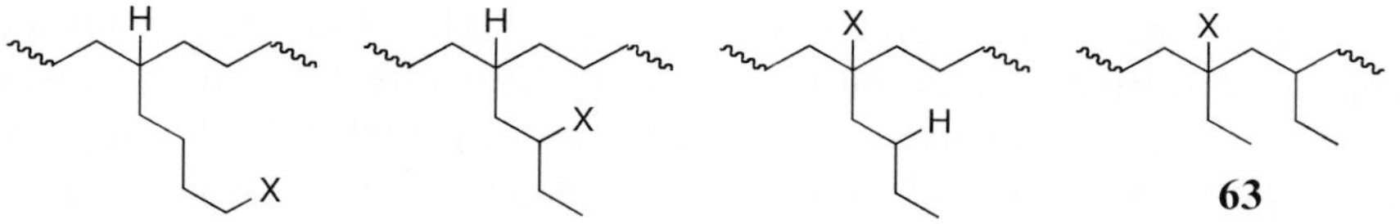

	60	61	62	63
E/BA	0	+	+++	+++
E/AA				
E/VAc	+	+++	0	0
E/BMA	0	+	0	0
E/MAA				

a Legend: +++ prevalent + weak 0 absent.

4.4.3.2 *Vinyl polymers*

There is evidence for backbiting during the polymerizations of VC[67,302] and VAc.[55,301,303-307] The mechanism is believed to be analogous to that discussed for PE above and should lead to the formation of 2,4-dichlorobutyl or 2,4 diacetoxybutyl branches (Scheme 4.42) respectively.

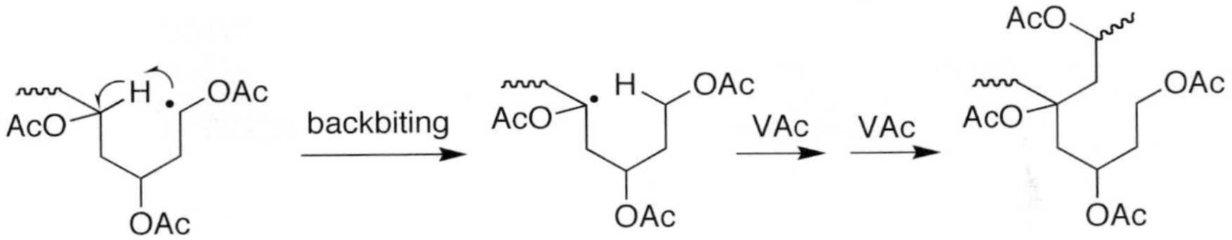

Scheme 4.42

The process is favored by low monomer concentrations as occurs at high conversions and in starved feed polymerizations.[307] Theoretical calculations suggest that the incidence of backbiting should be strongly dependent on the tacticity of the penultimate dyad.[308] Double backbiting in VC or VAc polymerization will lead to 2-chloroethyl or 2-acetoxy ethyl branches respectively (as for E in Scheme 4.41).[302]

There are no proven examples of 1,2-hydrogen atom shifts; this can be understood in terms of the stereoelectronic requirements on the process. The same limitations are not imposed on heavier atoms (*e.g.* chlorine). The postulate[309] that ethyl branches in reduced PVC are all derived from chloroethyl branches formed by sequential 1,5-intramolecular hydrogen atom transfers as described for PE (Section 4.4.3.1) has been questioned.[56,65] It has been shown that many of these ethyl branches are derived from dichloroethyl groups. The latter are formed by sequential 1,2-chlorine atom shifts which follow a head addition (Section 4.3.1.2).

4.4.3.3 *Acrylate esters and other monosubstituted monomers*

Recent work has shown that backbiting is prevalent in polymerizations and copolymerizations of acrylate esters.[276,277,305,306,310-319] It is also observed in styrene polymerization at high temperature[276] and probably occurs to some extent during polymerizations of most monosubstituted monomers. At high temperatures, and at low temperatures in very dilute solution, backbiting may be followed by fragmentation (Scheme 4.43).[276,277,310-312,318] At lower temperatures short chain branch formation dominates.[313-316] The backbiting process complicates the measurement of propagation rate constants for acrylates.[320]

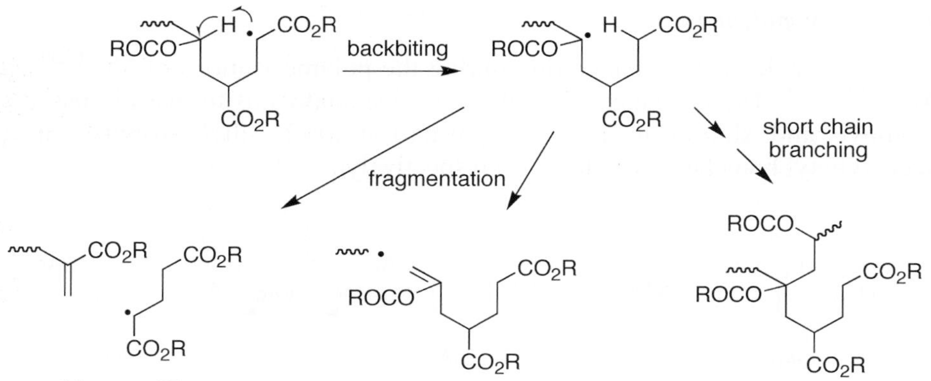

Scheme 4.43

The high temperature polymerization of acrylates with the backbiting-fragmentation process has been used to synthesize macromonomers based on acrylate esters.[276,277,312] Interestingly, fragmentation shows a strong preference for giving the polymeric macromonomer **64** and a small radical **65**.[276,277] An explanation for this specificity has yet to be proposed.

4.4.3.4 Addition-abstraction polymerization

Several examples of addition-abstraction polymerization have been reported. In these polymerizations, the monomers are designed to give quantitative rearrangement of the initially formed adduct *via* 1,5-hydrogen atom transfer (Scheme 4.44). The monomers (**66**) are such that the double bond is electron rich (vinyl ether) and the site for 1,5-H transfer is electron deficient. This arrangement favors intramolecular abstraction over addition. Thus compound **66a** undergoes[321,322] quantitative rearrangement during homopolymerization. For **66b**, where the site of intramolecular attack is less electron deficient, up to 80% of propagation steps involve intramolecular abstraction. As expected, higher reaction temperatures and lower monomer concentrations favor the intramolecular abstraction pathway.

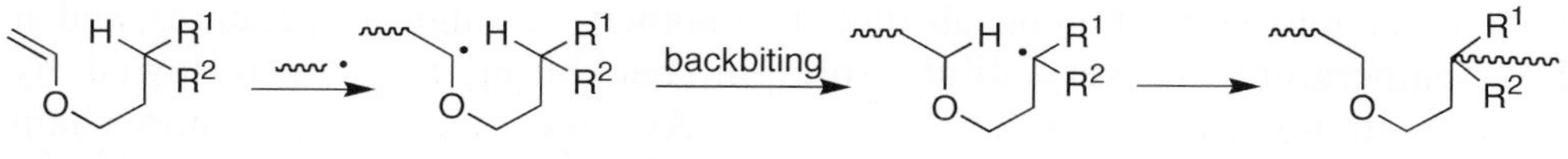

66 a R^1=CN, R^2=CO_2Et
   b R^1= R^2=CO_2Et

Scheme 4.44

4.5 Propagation Kinetics and Thermodynamics

In this section, we consider the kinetics of propagation and the features of the propagating radical ($P_n\bullet$) and the monomer (M) structure that render the monomer polymerizable by radical homopolymerization (Section 4.5.1). The reactivities of monomers towards initiator-derived species (Section 3.3) and in copolymerization (Chapter 6) are considered elsewhere.

$$P_1\bullet + M \rightarrow P_2\bullet \quad\quad k_p(1)[P_n\bullet][M]$$
$$P_2\bullet + M \rightarrow P_3\bullet \quad\quad k_p(2)[P_n\bullet][M]$$
$$P_3\bullet + M \rightarrow P_4\bullet \quad\quad k_p(3)[P_n\bullet][M]$$
$$P_n\bullet + M \rightarrow P_{n+1}\bullet \quad\quad k_p(n)[P_n\bullet][M]$$

Scheme 4.45

In the literature on radical polymerization, the rate constant for propagation, k_p, is often taken to have a single value (*i.e.* $k_p(1) = k_p(2) = k_p(3) = k_p(n)$ - refer Scheme 4.45). However, there is now good evidence that the value of k_p is dependent on chain length, at least for the first few propagation steps (Section 4.5.1), and on the reaction conditions (Section 8.3).

4.5.1 Polymerization Thermodynamics

Polymerization thermodynamics has been reviewed by Allen and Patrick,[323] Ivin,[324] Ivin and Busfield,[325] Sawada[326] and Busfield.[327] In most radical polymerizations, the propagation steps are facile (k_p typically > 10^2 M^{-1} s^{-1} - Section 4.5.2) and highly exothermic. Heats of polymerization (ΔH_p) for addition polymerizations may be measured by analyzing the equilibrium between monomer and polymer or from calorimetric data using standard thermochemical techniques. Data for polymerization of some common monomers are collected in Table 4.10. Entropy of polymerization (ΔS_p) data are more scarce. The scatter in experimental numbers for ΔH_p obtained by different methods appears quite large and direct comparisons are often complicated by effects of the physical state of the monomer and polymers (*i.e* whether for solid, liquid or solution, degree of crystallinity of the polymer).

The addition of radicals and, in particular, propagating radicals, to unsaturated systems is potentially a reversible process (Scheme 4.46). Depropagation is entropically favored and the extent therefore increases with increasing temperature (Figure 4.4). The temperature at which the rate of propagation and depropagation become equal is known as the ceiling temperature (T_c). Above T_c there will be net depolymerization.

$$P_n\bullet + M \underset{}{\overset{K_{eq}}{\rightleftharpoons}} P_{n+1}\bullet$$

Scheme 4.46

With most common monomers, the rate of the reverse reaction (depropagation) is negligible at typical polymerization temperatures. However, monomers with alkyl groups in the α-position have lower ceiling temperatures than monosubstituted monomers (Table 4.10). For MMA at temperatures <100 °C, the value of K_{eq} is <0.01 (Figure 4.4). AMS has a ceiling temperature of <30 °C and is not readily polymerizable by radical methods. This monomer can, however, be copolymerized successfully (Section 7.3.1.4).

The value of T_c and the propagation/depropagation equilibrium constant (K_{eq}) can be measured directly by studying the equilibrium between monomer and polymer or they can be calculated at various temperatures given values of ΔH_p and ΔS_p using eq. 11 and 12 respectively.

$$K_{eq} = \exp\left(\frac{\Delta H_p}{RT} - \frac{\Delta S_p}{R}\right) = \frac{1}{[M]_{eq}} \tag{11}$$

where $[M]_{eq}$ is the equilibrium monomer concentration.

$$T_c = \frac{\Delta H_p}{\Delta S_p + R\ln[M]} \tag{12}$$

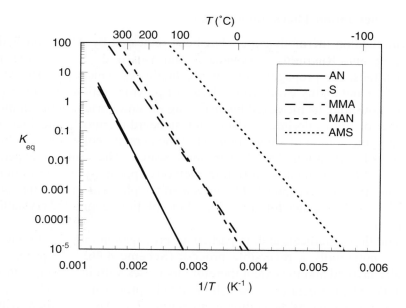

Figure 4.4 Dependence of K_{eq} on temperature for selected monomers. Based on values of ΔH_p and ΔS_p shown in Table 4.10.

Propagation

Table 4.10 Thermodynamic Parameters for Polymerization of Selected Monomers ($CH_2=CRX$)

monomer	X	R	ΔH_p (kJ mol^{-1}) a	b	c	ΔS_p c J mol^{-1} K^{-1}	T_c d °C
AA	CO_2H	H	67	-	-	-	-
MAA	CO_2H	CH_3	43	65	-	-	-
MA	CO_2CH_3	H	78	-	-	-	-
MMA	CO_2CH_3	CH_3	56 (58)	55	56[328,329]	118[328,329]	202
EMA	$CO_2C_2H_5$	CH_3	60 (58)	-	60[330]	124[330]	211
BMA	$CO_2C_4H_9$	CH_3	58 (60)	-	-	-	-
MEA[331]	CO_2CH_3	C_2H_5	32e	-	-	-	22
AN	CN	H	75f	-	-	109[327]	415
MAN	CN	CH_3	57	-	64[332]	142g,[332]	177
S	Ph	H	69 (73)	70	73[333]	104[333]	428
AMS	Ph	CH_3	-	35	45[334]	148[334]	31
VAc	O_2CCH_3	H	88 (90)	-	-	-	-
VC	Cl	H	96	112	-	-	-

a From calorimetry - data are for liquid monomer to amorphous solid polymer or for liquid monomer to polymer in monomer (in parentheses) and are taken from the Polymer Handbook unless otherwise indicated.[327] All data are rounded to the nearest whole number.   b From heat of combustion monomer and polymer - data are for liquid monomer to amorphous solid polymer and are taken from the Polymer Handbook.[327] All data are rounded to the nearest whole number.   c From studies of monomer-polymer equilibria - data are for liquid monomer to amorphous solid polymer. All data are rounded to the nearest whole number.   d Calculated from numbers of ΔH_p (column c except for AN) and ΔS_p shown and [M] = 1.0.   e Based on a measured T_c of 82 °C in bulk monomer and an assumed value for ΔS_p of 105 J mol^{-1} K^{-1}.[331] A more reasonable value of ΔS_p of 120 J mol^{-1} K^{-1} would suggest a ΔH_p of 40 kJ mol^{-1}.   f Partially crystalline polymer.   g In benzonitrile solution.

Note that the value of T_c is dependent on the monomer concentration. In the literature, values of T_c may be quoted for [M] = 1.0 M, for [M] = [M]$_{eq}$ or for bulk monomer. Thus care must be taken to note the monomer concentration when comparing values of T_c. One problem with using the above method to calculate K_{eq} or T_c, is the paucity of data on ΔS_p. A further complication is that literature values of ΔH_p show variation of ±2 kJ mol^{-1} which may in part reflect medium effects.[327] This "error" in ΔH_p corresponds to a significant uncertainty in T_c.

Steric factors appear to be dominant in determining ΔH_p and ΔS_p. The resonance energy lost in converting monomer to polymer is of secondary importance for most common monomers. It is thought to account for ΔH_p for VAc and VC being lower than for acrylic and styrenic monomers.

Evidence for the importance of steric factors comes from a consideration of the effect of α-alkyl substituents. It is found that the presence of an α-methyl substituent raises ΔH_p by at least 20 kJ mol^{-1} (Table 4.10, compare entries for AA and MAA, MA and MMA, AN and MAN, S and AMS). The higher ΔH_p probably reflects the greater difficulty in forming bonds to tertiary centers. This view is supported by the observation that higher alkyl substituents further increase ΔH_p [e.g. ethyl in MEA,[331] Table 4.10). Increasing the chain length of the α-substituent from methyl to ethyl should not greatly increase the thermodynamic stability of the radical, but steric factors will make the new bond both more difficult to form and easier to break.

Limited data suggest that the entropic term may be as important as the enthalpic term in determining polymerizability. The value of ΔS_p is lowered >20 J mol^{-1} K^{-1} by the presence of an α-methyl substituent (Table 4.10, compare entries for AN and MAN, S and AMS). This is likely to be a consequence of the polymers from α-methyl vinyl monomers having a more rigid, more ordered structure than those from the corresponding vinyl monomers.

There have been many studies on the polymerizability of α-substituted acrylic monomers.[331,335-338] It is established that the ceiling temperature for α-alkoxyacrylates decreases with the size of the alkoxy group.[335] However, it is of interest that polymerizations of α-(alkoxymethyl)acrylates (**67**)[335] and α-(acyloxymethyl)acrylates (**68**)[337] and captodative substituted monomers (**69**, **70**)[339] appear to have much higher ceiling temperatures than the corresponding α-alkylacrylates (e.g. methyl ethacrylate, MEA). For example, methyl α-ethoxymethacrylate[335] readily polymerizes at 110 °C whereas MEA[331] has a very low ceiling temperature (Table 4.10). However, values of the thermodynamic parameters for these polymerizations have not yet been reported.

| MEA | 67 | 68 | 69 | 70 |

4.5.2 Measurement of Propagation Rate Constants

Methods for measurement of k_p have been reviewed by Stickler,[340,341] van Herk [342] and more recently by Beuermann and Buback.[343] A largely non critical summary of values of k_p and k_t obtained by various methods appears in the *Polymer Handbook*.[344] Literature values of k_p for a given monomer may span two or more orders of magnitude. The data and methods of measurement have been critically assessed by IUPAC working parties[345-351] and reliable values for most common monomers are now available.[343] The wide variation in values of k_p (and k_t) obtained from various studies does not reflect experimental error but differences in data interpretation and the dependence of kinetic parameters on chain length and polymerization conditions.

Traditionally, measurement of k_p has required determination of the rate of polymerization under steady state (to give k_p/k_t^2) and non-steady state conditions (to give k_p/k_t). The classical techniques in this context are the rotating sector[352-355] and related methods such as spatially intermittent polymerization (SIP).[356]

EPR methods that allow a more direct determination of k_p have been developed. These enable absolute radical concentrations to be determined as a function of conversion. With especially sensitive instrumentation, this can be done by direct measurement.[357-360] An alternative method, applicable at high conversions, involves trapping the propagating species in a frozen matrix[361,362] by rapid cooling of the sample to liquid nitrogen temperatures.

The radical concentration, when coupled with information on the rate of polymerization, allows k_p (and k_t) to be calculated. The EPR methods have been applied to various polymerizations including those of B, DMA, MMA,[361-366] S[367,368] and VAc.[369] Values for k_p are not always in complete agreement with those obtained by other methods (e.g. PLP, SIP) and this may reflect a calibration problem. Problems may also arise because of the heterogeneity of the polymerization reaction mixture,[365] and insufficient sensitivity for the radical concentrations in low conversion polymerizations[362] or very low molecular weights. Some data must be treated with caution. However, the difficulties are now generally recognized and are being resolved.[360]

Pulsed laser photolysis (PLP) has emerged as the most reliable method for extracting absolute rate constants for the propagation step of radical polymerizations.[343] The method can be traced to the work of Aleksandrov et al.[370] PLP in its present form owes its existence to the extensive work of Olaj and coworkers[371] and the efforts of an IUPAC working party.[345-351] The method has now been successfully applied to establish rate constants, k_p(overall), for many polymerizations and copolymerizations.

In PLP the sample is subjected to a series of short (<30 ns) laser pulses at intervals τ. Analysis of the molecular weight distribution gives the length of chain formed between successive pulses (ν) and this yields a value for k_p (eq. 13).

$$\nu = k_p [M] \tau \tag{13}$$

A molecular weight distribution for a PS sample obtained from a PLP experiment with S is shown in Figure 4.5. Olaj et al.[371] found empirically that ν was best estimated from the points of inflection in the molecular weight distribution. Kinetic modeling of PLP has been carried out using Monte Carlo methods[372,373] or by numerical integration.[374,375] These studies confirm that the point of inflection in the molecular weight distribution is usually a good measure of ν. With choice of polymerization conditions the values of ν are relatively insensitive to the termination rate and mechanism and the occurrence of side reactions such as transfer to monomer. Some difficulties are experienced with high k_p monomers (acrylates, VAc) but appear to have been resolved through the use of low reaction temperatures and dilute media.[375] These difficulties may arise through interference from backbiting.[320] Independent determination of the rate of polymerization allows k_p/k_t and hence k_t to be evaluated (Section 5.2).[376]

The Chemistry of Radical Polymerization

There are some reports that values of k_p are conversion dependent and that the value decreases at high conversion due to k_p becoming limited by the rate of diffusion of monomer. While conversion dependence of k_p at extremely high conversions is known, some data that indicate this may need to be reinterpreted, as the conversion dependence of the initiator efficiency was not recognized (Sections 3.3.1.1.3, 3.3.2.1.3 and 5.2.1.4).

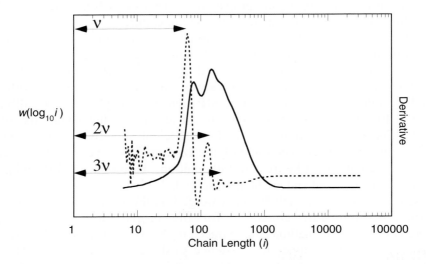

Figure 4.5 Experimental molecular weight distribution obtained by GPC (———) and its first derivative with respect to chain length (---------) for PS prepared by PLP. The vertical scales are in arbitrary units. Polymerization of 4.33 M S at 60 °C with benzoin 0.006 M and laser conditions: λ=350 nm, 80-100 mJ/pulse, τ=0.05 s.[374]

4.5.3 Dependence of Propagation Rate Constant on Monomer Structure

Recent data for k_p are summarized in Table 4.11. Monomers have been grouped into three series according to the α-substituent (hydrogen, methyl, other). Some trends can be seen.

(a) The Arrhenius A factor decreases by almost an order of magnitude in going from monomers with an α-hydrogen (20-80×10^6 M^{-1}s^{-1}) to those with an α-methyl (2-5×10^6 M^{-1}s^{-1}) and decreases further for those with a larger α-substituent, dimethyl itaconate (**71**) and the MA dimer (**72**), (0.2-1×10^6 M^{-1}s^{-1}) (Table 4.11). The same overall trend is seen for analogous reactions of small radicals (Table 4.12, see also Section 2.3) and is predicted by theory.

(b) Within both the α-hydrogen and α-methyl series, the lowest k_p values (for MAN, S, B) are associated with the highest activation energies and the more stable propagating radicals.

Table 4.11 Kinetic Parameters for Propagation in Selected Radical Polymerizations in Bulk Monomer

monomer	k_p (60°C) $M^{-1}s^{-1}$	A $M^{-1}s^{-1} \times 10^6$	E_a kJ mol^{-1}	reference
α-H				
MA	28000	16.6	17.7	377
BA	31000	15.8	17.3	378
DA	39000	17.9	17.0	377
VAc	8300	14.7	20.7	379
S^c	340	42.7	32.5	349
B	200	80.5	35.7	380
α-methyl				
MAA	1200	-	-	381
MAA(MeOH)^d	1000	0.60	17.7	381
MAA(H$_2$O)^e	6700	1.72	15.3	
MMA^c	820	2.67	22.4	348
EMA^c	870	4.06	23.4	347
nBMA^c	970	3.78	22.9	347
iBMA	1000	2.64	21.8	382
EHMA	1200	1.87	20.4	382
DMA^c	1300	2.50	21.0	347
HEMA	3300	8.88	21.9	383
GMA	1600	4.41	21.9	383
MAN	59	2.69	29.7	384
α-other				
71	25	0.20	24.9	385
72	30	1.25	29.5	386

a Values are calculated from the Arrhenius parameters shown and given to two significant figures.   b Values given to three significant figures.   c IUPAC benchmark value.   d 33 vol% MAA in methanol. Values are dependent on solvent and on concentration.   e 15 vol% in water.

(c) Within the series of alkyl acrylates and methacrylates there is a clear tendency for increase in k_p with increase in the length of the alkyl chain. The effect is small and, on the basis of the data shown in Table 4.11, cannot be assigned to a variation in A or E_a. However, there are reasonable theoretical grounds to expect this effect could be assigned to changes in the frequency factor.

(d) The methacrylic monomers with protic substituents (MAA, HEMA) are associated with higher k_p values that are solvent and concentration dependent. The effect is suggestive of monomer-polymer and/or monomer-monomer association through hydrogen-bonding.

(e) The lowering of k_p with the increase in size of the α-substituent (MA>MMA>**71**~**72**) is associated with an increase in A and a decrease in E_a.

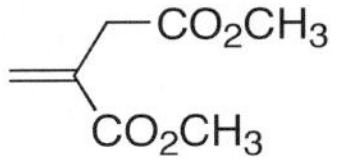

71 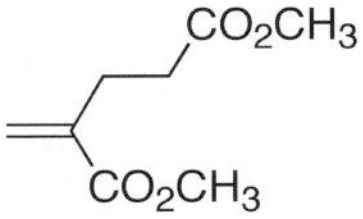   72

4.5.4 Chain Length Dependence of Propagation Rate Constants

It is usually assumed that propagation rate constants in homopolymerization (k_p) are independent of chain length and, for longer chains (length >20), there is experimental evidence to support this assumption.[356,367] However, there is now a body of indirect evidence to suggest that the rate constants for the first few propagation steps $k_p(1)$, $k_p(2)$, etc. can be substantially different from k_p(overall) (refer Scheme 4.45). The effect can be seen as a special case of a penultimate unit effect (Section 7.3.1.2). Evidence comes from a number of sources, for example:

(a) Chain transfer constants (k_p/k_{tr}) often show a marked chain length dependence for very short chain lengths (Section 5.3) indicating that k_p, k_{tr} or both are chain length dependent.[387]

(b) The absolute rate constants for the reaction of small model radicals with monomers are typically at least an order of magnitude greater than the corresponding values of k_p (Table 4.12).[388]

(c) Aspects of the kinetics of emulsion polymerization[389] can be explained by invoking chain length dependence of k_p.

(d) The apparent chain length dependence of k_p(average) in PLP experiments (Section 4.5.2) can be interpreted in this light.[374] However, Olaj et al.[390] have interpreted the same and similar data as suggesting a smaller decrease in k_p over a much longer range of chain lengths. They proposed that chain length dependence was a consequence of a change in the degree of solvation of the polymer chain and thus in the effective monomer concentration in the vicinity of the chain end. The explanation is analogous to that proposed to explain the bootstrap effect in copolymerization. Beuermann[343] has questioned these interpretations pointing out that the interpretation of PLP data can be problematical due to the dependence of the shape of the molecular weight distribution on experimental parameters.

There have been attempts at direct measurements of these important kinetic parameters in AN,[391] MA,[392] MAN,[393,394] MMA[394] and S[395] polymerizations. When the reaction is compared to a reference reaction care must be taken to establish the influence of chain length on the reference reaction.

Propagation

Frequency factors for addition of small radicals to monomers are higher by more than an order of magnitude than those for propagation (Table 4.12). Activation energies are typically lower. However, trends in the data are very similar suggesting that the same factors are important in determining the relative reactivities for both small radicals and propagating species. The same appears to be true with respect to reactivities in copolymerization (Section 7.3.1.2).[388]

$\sim$CH$_2$-ĊH   ḢCH $\sim$CH$_2$-Ċ(CH$_3$)   H$_3$C-Ċ(CH$_3$)
       | | | |
    CO$_2$CH$_3$ CO$_2$C(CH$_3$)$_3$ CO$_2$CH$_3$ CO$_2$CH$_3$

PMA•   **73**   **PMMA•**   **74**

Table 4.12 Rate Constants (25 °C) and Arrhenius Parameters for Propagation of Monomers CH$_2$=CR^{1}R^2 Compared with Rate Constants for Addition of Small Radicals[388]

monomer	k_p^a M^{-1}s^{-1}	log$A$	E_a kJ mol^{-1}	model	k_a^a M^{-1}s^{-1}	logA^b	E_a kJ mol^{-1}
E	77	7.27	34.3	CH$_3$•	12000	8.5	28.2
S	340	7.63	32.5	PhCH$_2$•	4700	8.5	30.8
MA	28000	7.22	17.7	73	1100000	8.5	15.6
AN				CH$_2$CN•	410000	8.5	18.4
MMA	820	6.43	22.4	74	9700	7.5	22.4
MAN	59	6.42	29.7	C(CH$_3$)$_2$CN•	2300	7.5	26.4

a Values at 60 °C calculated from the Arrhenius parameters shown and quoted to two significant figures. b Log A values based on recommendations of Fischer and Radom[388] (refer Section 2.3.7).

4.6 References

1. Bovey, F.A. *Chain Structure and Conformation of Macromolecules*; Wiley: New York, 1982.
2. Randall, J.C. *Polymer Sequence Determination*; Academic Press: New York, 1977.
3. Bovey, F.A. *Polymer Conformation and Configuration*; Academic Press: New York, 1969.
4. Koenig, J.L. *Chemical Microstructure of Polymer Chains*; Wiley: New York, 1980.
5. Koenig, J.L. *Spectroscopy of Polymers*; Elsevier: New York, 1999.
6. Tonelli, A.E. *NMR Spectroscopy and Polymer Microstructure*; VCH: New York, 1989.
7. Hatada, K. *NMR Spectroscopy of Polymers*; Springer-Verlag: Berlin, 2003.
8. Farina, M. *Top. Stereochem.* **1987**, *17*, 1.
9. Bovey, F.A.; Tiers, G.V.D. *J. Polym. Sci.* **1960**, *44*, 173.
10. Hatada, K.; Kitayama, T.; Ute, K. In *Annual Reports in NMR Spectroscopy*; Webb, G.A., Ed.; Academic Press: London, 1993; Vol. 26, p 100.
11. Bovey, F.A.; Mirau, P. *NMR of Polymers*; Academic Press: New York, 1996.
12. King, J.; Bower, D.I.; Maddams, W.F.; Pyszora, H. *Makromol. Chem.* **1983**, *184*, 879.
13. Fox, T.G.; Schnecko, H.W. *Polymer* **1962**, *3*, 575.

14. Khanarian, G.; Cais, R.E.; Kometani, J.M.; Tonelli, A.E. *Macromolecules* **1982**, *15*, 866.
15. Bovey, F.A.; Mirau, P.A. *Acc. Chem. Res.* **1988**, *21*, 37.
16. Beshah, K. *Makromol. Chem.* **1993**, *194*, 3311.
17. Moad, G.; Rizzardo, E.; Solomon, D.H.; Johns, S.R.; Willing, R.I. *Macromolecules* **1986**, *19*, 2494.
18. Berger, P.A.; Kotyk, J.J.; Remsen, E.E. *Macromolecules* **1992**, *25*, 7227.
19. Kotyk, J.J.; Berger, P.A.; Remsen, E.E. *Macromolecules* **1990**, *23*, 5167.
20. Schilling, F.C.; Bovey, F.A.; Bruch, M.D.; Kozlowski, S.A. *Macromolecules* **1985**, *18*, 1418.
21. Dong, L., Hill, D.J.T., O'Donnell, J.H., Whittaker, A.K. *Macromolecules* **1994**, *27*, 1830.
22. Hikichi, K.; Yasuda, M. *Polym. J.* **1987**, *19*, 1003.
23. Gippert, G.P.; Brown, L.R. *Polym. Bull.* **1984**, *11*, 585.
24. Mirau, P.A.; Bovey, F.A. *Macromolecules* **1986**, *19*, 210.
25. Crowther, M.W.; Szeverenyi, N.M.; Levy, G.C. *Macromolecules* **1986**, *19*, 1333.
26. Bruch, M.D.; Bovey, F.A.; Cais, R.E. *Macromolecules* **1984**, *17*, 2547.
27. Hatada, K.; Kitayama, T.; Terawaki, Y.; Chujo, R. *Polym. J.* **1987**, *19*, 1127.
28. Chujo, R.; Hatada, K.; Kitamaru, R.; Kitayama, T.; Sato, H.; Tanaka, Y. *Polym. J.* **1987**, *19*, 413.
29. Chujo, R.; Hatada, K.; Kitamaru, R.; Kitayama, T.; Sato, H.; Tanaka, Y.; Horii, F.; Terawaki, Y. *Polym. J.* **1988**, *20*, 627.
30. Elias, H.G. In *Polymer Handbook*, 3rd ed.; Brandup, J.; Immergut, E.H., Eds.; Wiley: New York, 1989; p II/357.
31. Nakano, T.; Mori, M.; Okamoto, Y. *Macromolecules* **1993**, *26*, 867.
32. Porter, N.A.; Rosenstein, I.J.; Breyer, R.A.; Bruhnke, J.D.; Wu, W.-X.; McPhail, A.T. *J. Am. Chem. Soc.* **1992**, *114*, 7664.
33. Porter, N.A.; Allen, T.; Breyer, R.A. *J. Am. Chem. Soc.* **1992**, *114*, 7676.
34. Habaue, S.; Isobe, Y.; Okamoto, Y. *Tetrahedron* **2002**, *58*, 8205.
35. Tsuruta, T.; Makimoto, T.; Kanai, H. *J. Macromol. Chem.* **1966**, *1*, 31.
36. Moad, G.; Solomon, D.H.; Spurling, T.H.; Johns, S.R.; Willing, R.I. *Aust. J. Chem.* **1986**, *39*, 43.
37. Reinmöller, M.; Fox, T.G. *Polym. Prepr. (Am. Chem. Soc., Div. Polym. Chem)* **1966**, *1*, 999.
38. Ferguson, R.C.; Ovenall, D.W. *Polym. Prepr. (Am. Chem. Soc., Div. Polym. Chem.)* **1985**, *26(1)*, 182.
39. Ovenall, D.W. *Macromolecules* **1984**, *17*, 1458.
40. Nagara, Y.; Yamada, K.; Nakano, T.; Okamoto, Y. *Polymer J.* **2001**, *33*, 534.
41. Nagara, Y.; Nakano, T.; Okamoto, Y.; Gotoh, Y.; Nagura, M. *Polymer* **2001**, *42*, 9679.
42. Yamada, K.; Nakano, T.; Okamoto, Y. *J. Polym. Sci., Part A: Polym. Chem.* **2000**, *38*, 220.
43. Kamide, K.; Yamazaki, H.; Okajima, K.; Hikichi, K. *Polym. J.* **1985**, *17*, 1233.
44. Matsuzaki, K.; Kanai, T.; Kawamura, T.; Matsumoto, S.; Uryu, T. *J. Polym. Sci., Polym. Chem. Ed.* **1973**, *11*, 961.
45. Suzuki, T.; Santee, E.R., Jr; Harwood, H.J.; Vogl, O.; Tanaka, T. *J. Polym. Sci., Polym. Lett. Ed.* **1974**, *12*, 635.
46. Sato, H.; Tanaka, Y.; Hatada, K. *J. Polym. Sci., Polym. Phys. Ed.* **1983**, *21*, 1667.
47. Kawamura, T.; Uryu, T.; Matsuzaki, K. *Makromol. Chem., Rapid Commun.* **1982**, *3*, 661.

48. Kawamura, T.; Toshima, N.; Matsuzaki, K. *Macromol. Rapid Commun.* **1994**, *15*, 479.
49. Bugada, D.C.; Rudin, A. *J. Appl. Polym. Sci.* **1985**, *30*, 4137.
50. Wu, T.K.; Ovenall, D.W. *Macromolecules* **1974**, *7*, 776.
51. Suito, Y.; Isobe, Y.; Habaue, S.; Okamoto, Y. *J. Polym. Sci., Part A: Polym. Chem.* **2002**, *40*, 2496.
52. Flory, P.J.; Leutner, F.S. *J. Polym. Sci.* **1950**, *5*, 267.
53. Adelman, R.L.; Ferguson, R.C. *J. Polym. Sci., Polym. Chem. Ed.* **1975**, *13*, 891.
54. Vercauteren, F.F.; Donners, W.A.B. *Polymer* **1986**, *27*, 993.
55. Amiya, S.; Uetsuki, M. *Macromolecules* **1982**, *15*, 166.
56. Starnes, W.H., Jr.; Wojciechowski, B.J. *Makromol. Chem., Macromol. Symp.* **1993**, *70/71*, 1.
57. Starnes, W.H., Jr. In *Developments in Polymer Degradation.*; Grassie, N., Ed.; Applied Science: London, 1981; Vol. 3, p 135.
58. Caraculacu, A.A. *Pure Appl. Chem.* **1981**, *53*, 385.
59. Hjertberg, T.; Sorvik, E. In *Degradation and Stabilisation of PVC*; Owen, E.D., Ed.; Elsevier Applied Science: Barking, 1984; p 21.
60. Starnes, W.H. *Prog. Polym. Sci.* **2002**, *27*, 2133.
61. Starnes, W.H., Jr.; Schilling, F.C.; Abbas, K.B.; Cais, R.E.; Bovey, F.A. *Macromolecules* **1979**, *12*, 556.
62. Darricades-Llauro, M.F.; Michel, A.; Guyot, A.; Waton, H.; Petiaud, R.; Pham, Q.T. *J. Macromol. Sci., Chem.* **1986**, *A23*, 221.
63. Rigo, A.; Palma, G.; Talamini, G. *Makromol. Chem.* **1972**, *153*, 219.
64. Fossey, J.; Nedelec, J.-Y. *Tetrahedron* **1981**, *37*, 2967.
65. Starnes, W.H., Jr.; Wojciechowski, B.J.; Velazquez, A.; Benedikt, G.M. *Macromolecules* **1992**, *25*, 3638.
66. Park, G.S.; Saleem, M. *Polym. Bull.* **1979**, *1*, 409.
67. Starnes, W.H., Jr.; Schilling, F.C.; Plitz, I.M.; Cais, R.E.; Freed, D.J.; Hartless, R.L.; Bovey, F.A. *Macromolecules* **1983**, *16*, 790.
68. Mitani, K.; Ogata, T.; Awaya, H.; Tomari, Y. *J. Polym. Sci., Polym. Chem. Ed.* **1975**, *13*, 2813.
69. Wilson, C.W., III; Santee, E.R., Jr. *J. Polym. Sci., Part C* **1965**, *8*, 97.
70. Cais, R.E.; Kometani, J.M. *ACS Symp. Ser.* **1984**, *247*, 153.
71. Ovenall, D.W.; Uschold, R.E. *Macromolecules* **1991**, *24*, 3235.
72. Görlitz, V.M.; Minke, R.; Trautvetter, W.; Weisgerber, G. *Angew. Macromol. Chem.* **1973**, *29/30*, 137.
73. Cais, R.E.; Kometani, J.M. *Macromolecules* **1984**, *17*, 1887.
74. Cais, R.E.; Sloane, N.J.A. *Polymer* **1983**, *24*, 179.
75. Cais, R.E.; Kometani, J.M. *Macromolecules* **1984**, *17*, 1932.
76. Ferguson, R.C.; Brame, E.G., Jr. *J. Phys. Chem.* **1979**, *83*, 1397.
77. Matsumoto, A.; Iwanami, K.; Oiwa, M. *J. Polym. Sci., Polym. Lett. Ed.* **1980**, *18*, 211.
78. Matsumoto, A.; Iwanami, K.; Kawaguchi, N.; Oiwa, M. *Technol. Rep. Kansai Univ.* **1983**, *24*, 183.
79. Matsumoto, A.; Kikuta, M.; Oiwa, M. *J. Polym. Sci., Part C: Polym. Lett.* **1986**, *24*, 7.
80. Matsumoto, A.; Terada, T.; Oiwa, M. *J. Polym. Sci., Part A: Polym. Chem.* **1987**, *25*, 775.
81. Matsumoto, A.; Iwanami, K.; Oiwa, M. *J. Polym. Sci., Polym. Lett. Ed.* **1981**, *19*, 497.

82. Marvel, C.S.; Cowan, J.C. *J. Am. Chem. Soc.* **1939**, *61*, 3156.
83. Marvel, C.S.; Dec, J.; Cooke, H.G., Jr.; Cowan, J.C. *J. Am. Chem. Soc.* **1940**, *62*, 3495.
84. McCurdy, K.G.; Laidler, K.J. *Can. J. Chem.* **1964**, *42*, 818.
85. Sawant, S.; Morawetz, H. *J. Polym. Sci., Polym. Lett. Ed.* **1982**, *20*, 385.
86. Minagawa, M. *J. Polym. Sci., Polym. Chem. Ed.* **1980**, *18*, 2307.
87. Kamachi, M.; Kajiwara, A.; Saegusa, K.; Morishima, Y. *Macromolecules* **1993**, *26*, 7369.
88. Henderson, J.N. In *Encyclopedia of Polymer Science and Engineering*, 2nd ed.; Mark, H.F.; Bikales, N.M.; Overberger, C.G.; Menges, G., Eds.; Wiley: New York, 1985; Vol. 2, p 515.
89. Tate, D.P.; Bethea, T.W. In *Encyclopedia of Polymer Science and Engineering*, 2nd ed.; Mark, H.F.; Bikales, N.M.; Overberger, C.G.; Menges, G., Eds.; Wiley: New York, 1985; Vol. 2, p 537.
90. Senyek, M.L. In *Encyclopedia of Polymer Science and Engineering*, 2nd ed.; Mark, H.F.; Bikales, N.M.; Overberger, C.G.; Menges, G., Eds.; Wiley: New York, 1987; Vol. 8, p 487.
91. Stewart, C.A.; Takeshita, T.; Coleman, M.L. In *Encyclopedia of Polymer Science and Engineering*, 2nd ed.; Mark, H.F.; Bikales, N.M.; Overberger, C.G.; Menges, G., Eds.; Wiley: New York, 1986; Vol. 3, p 441.
92. Khachaturov, A.S.; Ivanova, V.P.; Podkorytov, I.S.; Osetrova, L.V. *Vysokomol. Soedin.* **1998**, *40*, 964.
93. Sato, H.; Takebayashi, K.; Tanaka, Y. *Macromolecules* **1987**, *20*, 2418.
94. Kawahara, S.; Bushimata, S.; Sugiyama, T.; Hashimoto, C.; Tanaka, Y. *Rubber Chem. Technol.* **1999**, *72*, 844.
95. Coleman, M.M.; Tabb, D.L.; Brame, E.G. *Rubber Chem. Technol.* **1977**, *50*, 49.
96. Coleman, M.M.; Brame, E.G. *Rubber Chem. Technol.* **1978**, *51*, 668.
97. Sato, H.; Ono, A.; Tanaka, Y. *Polymer* **1977**, *18*, 580.
98. Butler, G.B. In *Comprehensive Polymer Science*; Eastmond, G.C.; Ledwith, A.; Russo, S.; Sigwalt, P., Eds.; Pergamon: Oxford, 1989; Vol. 4, p 423.
99. Butler, G.B. In *Encyclopedia of Polymer Science and Engineering*, 2nd ed.; Mark, H.F.; Bikales, N.M.; Overberger, C.G.; Menges, G., Eds.; Wiley: New York, 1986; Vol. 4, p 543.
100. Butler, G.B. *Acc. Chem. Res.* **1982**, *15*, 370.
101. Butler, G.B. In *Polymeric Amines and Ammonium Salts*; Goethals, E.J., Ed.; Pergamon: New York, 1981; p 125.
102. Butler, G.B. *Cyclopolymerization and Cyclocopolymerization*; Marcel Dekker: New York, 1992.
103. Butler, G.B.; Bunch, R.L. *J. Am. Chem. Soc.* **1949**, *71*, 3120.
104. Butler, G.B.; Ingley, F.L. *J. Am. Chem. Soc.* **1951**, *72*, 894.
105. Butler, G.B.; Angelo, R.J. *J. Am. Chem. Soc.* **1957**, *79*, 3128.
106. Butler, G.B.; Crawshaw, A.; Miller, W.L. *J. Am. Chem. Soc.* **1958**, *80*, 3615.
107. Solomon, D.H. *J. Polym. Sci., Polym. Symp.* **1975**, *49*, 175.
108. Beckwith, A.L.J.; Hawthorne, D.G.; Solomon, D.H. *Aust. J. Chem.* **1976**, *29*, 995.
109. Solomon, D.H. *J. Macromol. Sci., Chem.* **1975**, *A9*, 97.
110. Solomon, D.H.; Hawthorne, D.G. *J. Macromol. Sci., Rev. Macromol. Chem.* **1976**, *C15*, 143.
111. Johns, S.R.; Willing, R.I. *J. Macromol. Sci., Chem.* **1976**, *10*, 875.
112. Lancaster, J.E.; Bacchel, L.; Panzer, H.P. *J. Polym. Sci., Polym. Lett. Ed.* **1976**, *14*, 549.

113. Beckwith, A.L.J. *Tetrahedron* **1981**, *37*, 3073.
114. Xi, F.; Vogl, O. *J. Macromol. Sci., Chem.* **1983**, *A20*, 321.
115. Otsu, T.; Ohya, T. *J. Macromol. Sci., Chem.* **1984**, *A21*, 1.
116. Seyferth, D.; Robison, J. *Macromolecules* **1993**, *26*, 407.
117. Beckwith, A.L.J. *Chem. Soc. Rev.* **1993**, 143.
118. Tsuda, T.; Mathias, L.J. *Macromolecules* **1993**, *26*, 6359.
119. Masterman, T.C.; Dando, N.R.; Weaver, D.G.; Seyferth, D. *J. Polym. Sci., Part A: Polym. Phys.* **1994**, *32*, 2263.
120. Fukuda, W.; Nakao, M.; Okumura, K.; Kakiuchi, H. *J. Polym. Sci., Part A-1* **1972**, *10*, 237.
121. Ichihashi, T.; Kawai, W. *Kobunshi Kagaku* **1971**, *28*, 225.
122. Trossarelli, L.; Guaita, M.; Priola, A. *J. Polym. Sci., Part B* **1967**, *5*, 129.
123. Trossarelli, L.; Guaita, M.; Priola, A. *Makromol. Chem.* **1967**, *100*, 147.
124. Kodaira, T.; Okumura, M.; Urushisaki, M.; Isa, K. *J. Polym. Sci., Part A: Polym. Chem.* **1993**, *31*, 169.
125. Kodaira, T.; Mae, Y. *Polymer* **1992**, *33*, 3500.
126. Yokata, K.; Takada, Y. *Kobunshi Kagaku* **1969**, *26*, 317.
127. Kaye, H. *Macromolecules* **1971**, *4*, 147.
128. Field, N.D. *J. Org. Chem.* **1960**, *25*, 1006.
129. Marvel, C.S.; Gall, E.J. *J. Org. Chem.* **1960**, *25*, 1784.
130. Marvel, C.S.; Vest, R.D. *J. Am. Chem. Soc.* **1959**, *81*, 984.
131. Milford, G.N. *J. Polym. Sci.* **1959**, *41*, 295.
132. Thang, S.H.; Rizzardo, E.; Moad, G. US 5830966, 1996 (*Chem. Abstr.* **1994**, *123*, 229253).
133. Tüzün, N.S.; Aviyente, V.; Houk, K.N. *J. Org. Chem.* **2002**, *67*, 5068.
134. Miyake, T. *Kogyo Kagaku Zasshi* **1961**, *64*, 359.
135. Ohya, T.; Otsu, T. *J. Polym. Sci., Polym. Chem. Ed.* **1983**, *21*, 3503.
136. Matsumoto, A.; Kitamura, T.; Oiwa, M.; Butler, G.B. *J. Polym. Sci., Polym. Chem. Ed.* **1981**, *19*, 2531.
137. Matsumoto, A.; Kitamura, T.; Oiwa, M.; Butler, G.B. *Makromol. Chem., Rapid Commun.* **1981**, *2*, 683.
138. Stansbury, J.W. *Polym. Prepr. (Am. Chem. Soc., Div. Polym. Chem)* **1990**, *31(1)*, 503.
139. Mathias, L.J.; Kusefoglu, S.H.; Ingram, J.E. *Macromolecules* **1988**, *21*, 545.
140. Mathias, L.J.; Colletti, R.F.; Bielecki, A. *J. Am. Chem. Soc.* **1991**, *113*, 1550.
141. Aso, C.; Kunitake, T.; Ando, S. *J. Macromol. Sci., Chem.* **1971**, *A5*, 167.
142. Raymond, M.A.; Dietrich, H.J. *J. Macromol. Sci., Chem.* **1972**, *A6*, 207.
143. Billingham, N.C.; Jenkins, A.D.; Kronfli, E.B.; Walton, D.R.M. *J. Polym. Sci., Polym. Chem. Ed.* **1977**, *15*, 675.
144. Butler, G.B.; Stackman, R.W. *J. Org. Chem.* **1960**, *25*, 1643.
145. Butler, G.B.; Stackman, R.W. *J. Macromol. Sci., Chem.* **1969**, *A3*, 821.
146. Kida, S.; Nozakura, S.-I.; Murahashi, S. *Polym. J.* **1972**, *3*, 234.
147. Butler, G.B.; Skinner, D.L.; Bond, W.C., Jr.; Rogers, C.L. *J. Macromol. Sci., Chem.* **1970**, *A4*, 1437.
148. Berlin, K.D.; Butler, G.B. *J. Am. Chem. Soc* **1960**, *82*, 2712.
149. Benyon, K.I. *J. Polym. Sci., Part A* **1963**, *1*, 3357.
150. Berlin, K.D.; Butler, G.B. *J. Org. Chem.* **1960**, *25*, 2006.
151. Corfield, G.C.; Monks, H.H. *J. Macromol. Sci., Chem.* **1975**, *A9*, 1113.
152. Butler, G.B.; Kimura, S. *J. Macromol. Sci., Chem.* **1971**, *A5*, 181.
153. Matsoyan, S.G.; Pogosyan, G.M.; Elliasyan, M.A. *Vysokomol. Soedin* **1963**, *5*, 777.

154. Hawthorne, D.G.; Solomon, D.H. *J. Macromol. Sci., Chem.* **1975**, *A9*, 149.
155. Beckwith, A.L.J.; Moad, G. *J. Chem. Soc., Perkin Trans. 2* **1975**, 1726.
156. Costa, L.; Chiantore, O.; Guaita, M. *Polymer* **1978**, *19*, 202.
157. Wiley, R.H.; Rivera, W.H.; Crawford, T.H.; Bray, N.F. *J. Polym. Sci.* **1962**, *61*, 538.
158. Graham, P.J.; Buhle, E.L.; Pappas, N. *J. Org. Chem.* **1961**, *26*, 4658.
159. Paulrajan, S.; Gopalan, A.; Subbaratnam, N.R.; Venkatarao, K. *Polymer* **1983**, *24*, 906.
160. Gopalan, A.; Paulrajan, S.; Subbaratnam, N.R.; Rao, K.V. *J. Polym. Sci., Polym. Chem. Ed.* **1985**, *23*, 1861.
161. Nishikubo, T.; Iizawa, T.; Yoshinaga, A.; Nitta, M. *Makromol. Chem.* **1982**, *183*, 789.
162. Simpson, W.; Holt, T.; Zetie, R.J. *J. Polym. Sci.* **1953**, *10*, 489.
163. Haward, R.N. *J. Polym. Sci.* **1953**, *10*, 535.
164. Matsumoto, A.; Iwanami, K.; Oiwa, M. *J. Polym. Sci., Polym. Lett. Ed.* **1980**, *18*, 307.
165. Beckwith, A.L.J.; Moad, G. *J. Chem. Soc., Chem. Commun.* **1974**, 472.
166. Ohya, T.; Otsu, T. *J. Polym. Sci., Polym. Chem. Ed.* **1983**, *21*, 3169.
167. Barton, J.M.; Butler, G.B.; Chapin, E.C. *J. Polym. Sci., Part A* **1965**, *3*, 501.
168. Bailey, W.J.; Sun, R.L. *Polym. Prepr. (Am. Chem. Soc., Div. Polym. Chem)* **1972**, *13*, 281.
169. Brady, R.F., Jr. *J. Macromol. Sci., Rev. Macromol. Chem. Phys.* **1992**, *C32*, 135.
170. Stansbury, J.W. In *Expanding Monomers*; Sadhir, R.K.; Luck, R.M., Eds.; CRC Press: Boca Raton, Florida, 1992; p 153.
171. Moszner, N.; Salz, U. *Prog. Polym. Sci.* **2001**, *26*, 535.
172. Sanda, F.; Endo, T. *J. Polym. Sci., Part A: Polym. Chem.* **2001**, *39*, 265.
173. Klemm, E.; Schulze, T. *Acta Polym.* **1999**, *50*, 1.
174. Cho, I. *Prog. Polym. Sci.* **2000**, *25*, 1043.
175. Moszner, N.; Zeuner, F.; Volkel, T.; Rheinberger, V. *Macromol. Chem. Phys.* **1999**, *200*, 2173.
176. Endo, T.; Yokozawa, T. In *New Methods for Polymer Synthesis*; Mijs, W.J., Ed.; Plenum: New York, 1992; p 155.
177. Bailey, W.J. In *Comprehensive Polymer Science*; Eastmond, G.C.; Ledwith, A.; Russo, S.; Sigwalt, P., Eds.; Pergamon: Oxford, 1989; Vol. 3, p 283.
178. Colombani, D. *Prog. Polym. Sci.* **1999**, *24*, 425.
179. Hall, H.K., Jr.; Ykman, P.J. *J. Polym. Sci., Macromol. Rev.* **1976**, *11*, 1.
180. Bothe, H.; Schluter, A.-D. *Polym. Prepr. (Am. Chem. Soc., Div. Polym. Chem)* **1988**, *29*, 412.
181. Hall, H.K., Jr.; Padias, A.B. *J. Polym. Sci., Part A: Polym. Chem.* **2003**, *41*, 625.
182. Hall, H.K., Jr.; Padias, A.B. *J. Am. Chem. Soc.* **1971**, *4193*, 110.
183. Beckwith, A.L.J.; Moad, G. *J. Chem. Soc., Perkin Trans. 2* **1980**, 1473.
184. Newcomb, M. *Tetrahedron* **1993**, *49*, 1151.
185. Masnovi, J.; Samsel, E.G.; Bullock, R.M. *J. Chem. Soc., Chem. Commun* **1989**, 1044.
186. Ingold, K.U.; Maillard, B.; Walton, J.C. *J. Chem. Soc., Perkin Trans. 2* **1981**, 970.
187. Beckwith, A.L.J.; Bowry, V.W. *J. Org. Chem.* **1989**, *54*, 2681.
188. Takahashi, T.; Yamashita, I.; Miyakawa, T. *Bull. Chem. Soc. Japan* **1964**, *37*, 131.
189. Takahashi, T.; Yamashita, I. *J. Polym. Sci., Part B* **1965**, *3*, 251.
190. Cho, I.; Ahn, K.-D. *J. Polym. Sci., Polym. Chem. Ed.* **1979**, *17*, 3169.

191. Lishanskii, I.S.; Zak, A.G.; Fedorova, E.F.; Khachaturov, A.S. *Vysokomolekul. Soedin* **1965**, *7*, 966.
192. Endo, T.; Watanabe, M.; Suga, K.; Yokozawa, T. *J. Polym. Sci., Part A: Polym. Chem.* **1989**, *27*, 1435.
193. Endo, T.; Suga, K. *J. Polym. Sci., Part A: Polym. Chem.* **1989**, *27*, 1831.
194. Endo, T.; Watanabe, M.; Suga, K.; Yokozawa, T. *Makromol. Chem.* **1989**, *190*, 691.
195. Endo, T.; Watanabe, M.; Suga, K.; Yokozawa, T. *J. Polym. Sci., Part A: Polym. Chem.* **1987**, *25*, 3039.
196. Cho, I.; Ahn, K.-D. *J. Polym. Sci., Polym. Lett. Ed.* **1977**, *15*, 751.
197. Cho, I.; Lee, J.-Y. *Makromol. Chem., Rapid Commun.* **1984**, *5*, 263.
198. Cho, I.; Song, S.S. *J. Polym. Sci., Part A: Polym. Chem.* **1989**, *27*, 3151.
199. Takahashi, T. *J. Polym. Sci., Part A-1* **1968**, *6*, 403.
200. Kennedy, J.P.; Elliot, J.J.; Butler, P.E. *J. Macromol. Sci., Chem.* **1968**, *A2*, 1415.
201. Cho, I.; Song, S.S. *Makromol. Chem., Rapid Commun.* **1989**, *10*, 85.
202. Sanda, F.; Takata, T.; Endo, T. *Macromolecules* **1994**, *27*, 1099.
203. Sanda, F.; Takata, T.; Endo, T. *J. Polym. Sci., Part A: Polym. Chem.* **1993**, *31*, 2659.
204. Sanda, F.; Takata, T.; Endo, T. *Macromolecules* **1993**, *26*, 5748.
205. Sanda, F.; Takata, T.; Endo, T. *Macromolecules* **1992**, *25*, 6719.
206. Sanda, F.; Takata, T.; Endo, T. *Macromolecules* **1993**, *26*, 729.
207. Bowry, V.W.; Lusztyk, J.; Ingold, K.U. *J. Chem. Soc., Chem. Commun* **1990**, 923.
208. Sanda, F.; Takata, T.; Endo, T. *Macromolecules* **1993**, *26*, 1818.
209. Yang, Z.-Y. *J. Am. Chem. Soc.* **2003**, *125*, 870.
210. Cho, I.; Kim, J.-B. *J. Polym. Sci., Polym. Lett. Ed.* **1983**, *21*, 433.
211. Endo, T.; Kanda, N. *Polym. Prepr. Jpn.* **1987**, *36*, 140.
212. Koizumi, T.; Nojima, Y.; Endo, T. *J. Polym. Sci., Part A: Polym. Chem.* **1993**, *31*, 3489.
213. Laurie, D.; Nonhebel, D.C.; Suckling, C.J.; Walton, J.C. *Tetrahderon* **1993**, *49*, 5869.
214. Beckwith, A.L.J.; Moad, G. *J. Chem. Soc., Perkin Trans. 2* **1980**, 1083.
215. Hiraguri, Y.; Endo, T. *J. Polym. Sci., Part C: Polym. Lett.* **1989**, *27*, 333.
216. Cho, I.; Choi, S.Y. *Makromol. Chem., Rapid Commun.* **1991**, *12*, 399.
217. Cho, I.; Kim, S.-K.; Lee, M.-H. *J. Polym. Sci., Polym. Symp.* **1986**, *74*, 219.
218. Cho, I.; Lee, M.-H. *J. Polym. Sci., Part C: Polym. Lett.* **1987**, *25*, 309.
219. Errede, L.A. *J. Polym. Sci.* **1961**, *49*, 253.
220. Bailey, W.J.; Chou, J.L. *Polym. Mater. Sci. Eng.* **1987**, *56*, 30.
221. Bailey, W.J.; Amone, M.J.; Chou, J.L. *Polym. Prepr. (Am. Chem. Soc., Div. Polym. Chem)* **1988**, *29(1)*, 178.
222. Klemm, E.; Schulze, T. *Makromol. Chem.* **1993**, *194*, 2087.
223. Hiraguri, Y.; Endo, T. *J. Polym. Sci., Part A: Polym. Chem.* **1989**, *27*, 4403.
224. Acar, M.H.; Nambu, Y.; Yamamoto, K.; Endo, T. *J. Polym. Sci., Part A: Polym. Chem.* **1989**, *27*, 4441.
225. Bailey, W.J.; Gu, J.M.; Zhou, L.L. *Polym. Prepr. (Am. Chem. Soc., Div. Polym. Chem)* **1990**, *31(1)*, 24.
226. Cho, I.; Song, K.Y. *Makromol. Chem., Rapid Commun.* **1993**, *14*, 377.
227. Cho, I.; Kim, S.-K. *J. Polym. Sci., Part A: Polym. Chem.* **1990**, *28*, 417.
228. Bailey, W.J.; Chen, P.Y.; Chen, S.-C.; Chiao, W.-B.; Endo, T.; Gapud, B.; Kuruganti, Y.; Lin, Y.-N.; Ni, Z.; Pan, C.-Y.; Shaffer, S.E.; Sidney, L.; Wu, S.-R.;

Yamamoto, N.; Yamazaki, N.; Yonezawa, K.; Zhou, L.-L. *Makromol. Chem., Macromol. Symp.* **1986**, *6*, 81.
229. Bailey, W.J. *Polym. J.* **1985**, *17*, 85.
230. Bailey, W.J. *Makromol. Chem., Suppl.* **1985**, *13*, 171.
231. Bailey, W.J.; Chou, J.L.; Feng, P.-Z.; Issari, B.; Kuruganti, V.; Zhou, L.-L. *J. Macromol. Sci., Chem.* **1988**, *A25*, 781.
232. Kobayashi, S.; Kadokawa, J.; Shoda, S.; Uyama, H. *Macromol. Reports* **1991**, *A28 (Suppl. 1)*, 1.
233. Kobayashi, S.; Kadokawa, J.; Matsumura, Y.; Yen, I.F.; Uyama, H. *Macromol. Reports* **1992**, *A29 (Suppl. 3)*, 243.
234. Bailey, W.J.; Wu, S.-R.; Ni, Z. *J. Macromol. Sci., Chem.* **1982**, *A18*, 973.
235. Bailey, W.J.; Wu, S.-R.; Ni, Z. *Makromol. Chem.* **1982**, *183*, 1913.
236. Endo, T.; Yako, N.; Azuma, K.; Nate, K. *Makromol. Chem.* **1985**, *186*, 1543.
237. Yokozawa, T.; Hayashi, R.; Endo, T. *J. Polym. Sci., Part A: Polym. Chem.* **1990**, *28*, 3739.
238. Bailey, W.J.; Ni, Z.; Wu, S.-R. *J. Polym. Sci., Polym. Chem. Ed.* **1982**, *20*, 3021.
239. Endo, T.; Okawara, M.; Bailey, W.J.; Azuma, K.; Nate, K.; Yokona, H. *J. Polym. Sci., Polym. Lett. Ed.* **1983**, *21*, 373.
240. Bailey, W.J.; Ni, Z.; Wu, S.-R. *Macromolecules* **1982**, *15*, 711.
241. Bailey, W.J.; Arfaei, P.Y.; Chen, P.Y.; Chen, S.-C.; Endo, T.; Pan, C.-Y.; Ni, Z.; Shaffer, S.E.; Sidney, L.; Wu, S.-R.; Yamazaki, N. In *Proc. IUPAC 28th Macromol. Symp.*: Amherst, MA, 1982; p 214.
242. Sidney, L.N.; Shaffer, S.E.; Bailey, W.J. *Polym. Prepr. (Am. Chem. Soc., Div. Polym. Chem)* **1981**, *22(2)*, 373.
243. Barclay, L.R.C.; Griller, D.; Ingold, K.U. *J. Am. Chem. Soc.* **1982**, *104*, 4399.
244. Beckwith, A.L.J.; Thomas, C.B. *J. Chem. Soc., Perkin Trans. 2* **1973**, 861.
245. Bailey, W.J.; Zhou, L.L. *Polym. Prepr. (Am. Chem. Soc., Div. Polym. Chem)* **1989**, *30(1)*, 195.
246. Hiraguri, Y.; Endo, T. *J. Polym. Sci., Part A: Polym. Chem.* **1992**, *30*, 689.
247. Hiraguri, Y.; Endo, T. *J. Am. Chem. Soc.* **1987**, *109*, 3779.
248. Hiraguri, Y.; Endo, T. *J. Polym. Sci., Part A: Polym. Chem.* **1989**, *27*, 2135.
249. Cho, I.; Kim, B.-G.; Park, Y.-C.; Kim, C.-B.; Gong, M.-S. *Makromol. Chem., Rapid Commun.* **1991**, *12*, 141.
250. Pan, C.-Y.; Wu, Z.; Bailey, W.J. *J. Polym. Sci., Part C: Polym. Lett.* **1987**, *25*, 243.
251. Sugiyama, J.-I.; Yokozawa, T.; Endo, T. *J. Am. Chem. Soc.* **1993**, *115*, 2041.
252. Morariu, S.; Buruiana, E.C.; Simionescu, B.C. *Polym. Bull.* **1993**, *30*, 7.
253. Miyagawa, T.; Sanda, F.; Endo, T. *J. Polym. Sci., Part A: Polym. Chem.* **2000**, *38*, 1861.
254. Bailey, W.J.; Feng, P.-Z. *Polym. Prepr. (Am. Chem. Soc., Div. Polym. Chem)* **1987**, *28(1)*, 154.
255. Cho, I.; Lee, T.-W. *Macromol. Chem. Rapid Commun.* **1989**, *10*, 453.
256. Tsang, R.; Dickson, J.K., Jr.; Pak, H.; Walton, R.; Fraser-Reid, B. *J. Am. Chem. Soc.* **1987**, *109*, 3484.
257. Evans, R.A.; Moad, G.; Rizzardo, E.; Thang, S.H. *Macromolecules* **1994**, *27*, 7935.
258. Evans, R.A.; Rizzardo, E. *Macromolecules* **1996**, *29*, 6983.
259. Evans, R.A.; Rizzardo, E. *Macromolecules* **2000**, *33*, 6722.
260. Evans, R.A.; Rizzardo, E. *J. Polym. Sci., Part A: Polym. Chem.* **2001**, *39*, 202.
261. Harrisson, S.; Davis, T.P.; Evans, R.A.; Rizzardo, E. *Macromolecules* **2001**, *34*, 3869.

262. Harrisson, S.; Davis, T.P.; Evans, R.A.; Rizzardo, E. *Macromolecules* **2000**, *33*, 9553.
263. Okazaki, T.; Sanda, F.; Endo, T. *J. Polym. Sci., Part A: Polym. Chem.* **1996**, *34*, 2029.
264. Okazaki, T.; Komiya, T.; Sanda, F.; Miyazaki, K.; Endo, T. *J. Polym. Sci., Part A: Polym. Chem.* **1997**, *35*, 2501.
265. Endo, T.; Bailey, W.J. *J. Polym. Sci., Polym. Lett. Ed.* **1975**, *13*, 193.
266. Sugiyama, J.-I.; Yokozawa, T.; Endo, T. *J. Polym. Sci., Part A: Polym. Chem.* **1990**, *28*, 3529.
267. Bailey, W.J.; Zheng, Z.-F. *J. Polym. Sci., Part A: Polym. Chem.* **1991**, *29*, 437.
268. Endo, T.; Bailey, W.J. *J. Polym. Sci., Polym. Lett. Ed.* **1980**, *18*, 25.
269. Tagoshi, H.; Endo, T. *J. Polym. Sci., Part C: Polym. Lett.* **1988**, *26*, 77.
270. Schulze, T.; Klemm, E. *Polym. Bull.* **1993**, *31*, 409.
271. Issari, B.; Bailey, W.J. *Polym. Prepr. (Am. Chem. Soc., Div. Polym. Chem)* **1988**, *29(1)*, 217.
272. Okazaki, T.; Sanda, F.; Endo, T. *Polymer J.* **1998**, *30*, 365.
273. Sanda, F.; Takata, T.; Endo, T. *Macromolecules* **1994**, *27*, 3986.
274. Moszner, N.; Zeuner, F.; Fischer, U.K.; Rheinberger, V. *Polym. Bull.* **1998**, *40*, 447.
275. Okazaki, T.; Sanda, F.; Endo, T. *J. Polym. Sci., Part A: Polym. Chem.* **1997**, *35*, 2487.
276. Chiefari, J.; Jeffery, J.; Moad, G.; Mayadunne, R.T.A.; Rizzardo, E.; Thang, S.H. *Macromolecules* **1999**, *32*, 5559.
277. Chiefari, J.; Jeffery, J.; Mayadunne, R.T.A.; Moad, G.; Rizzardo, E.; Thang, S.H. *ACS Symp. Ser.* **2000**, *768*, 297.
278. Viswanadhan, V.N.; Mattice, W.L. *Makromol. Chem.* **1985**, *186*, 633.
279. Toh, J.S.S.; Huang, D.M.; Lovell, P.A.; Gilbert, R.G. *Polymer* **2001**, *42*, 1915.
280. Filley, J.; McKinnon, J.T.; Wu, D.T.; Ko, G.H. *Macromolecules* **2002**, *35*, 3731.
281. Usami, T.; Takayama, S. *Polym. J.* **1984**, *16*, 731.
282. Blitz, J.P.; McFaddin, D.C. *J. Appl. Polym. Sci.* **1994**, *51*, 13.
283. Ohtani, H.; Tsuge, S.; Usami, T. *Macromolecules* **1984**, *17*, 2557.
284. Bowmer, T.N.; O'Donnell, J.H. *Polymer* **1977**, *18*, 1032.
285. Cutler, D.J.; Hendra, P.J.; Cudby, M.E.A.; Willis, H.A. *Polymer* **1977**, *18*, 1005.
286. Usami, T.; Takayama, S. *Macromolecules* **1984**, *17*, 1756.
287. Axelson, D.E.; Levy, G.C.; Mandelkern, L. *Macromolecules* **1979**, *12*, 41.
288. Bovey, F.A.; Schilling, F.C.; McCrackin, F.L.; Wagner, H.L. *Macromolecules* **1976**, *9*, 76.
289. Bugada, D.C.; Rudin, A. *Eur. Polym. J..* **1987**, *23*, 809.
290. McCord, E.F.; Shaw, W.H.; Hutchinson, R.A. *Macromolecules* **1997**, *30*, 246.
291. Roedel, M.J. *J. Am. Chem. Soc.* **1953**, *75*, 6110.
292. Stoiljkovich, D.; Jovanovich, S. *Makromol. Chem.* **1981**, *182*, 2811.
293. Stoiljkovich, D.; Jovanovich, S. *Br. Polym. J.* **1984**, *16*, 291.
294. Huang, X.L.; Dannenberg, J.J. *J. Org. Chem.* **1991**, *56*, 5421.
295. Nedelec, J.Y.; LeFort, D. *Tetrahedron* **1975**, *31*, 411.
296. Beckwith, A.L.J.; Ingold, K.U. In *Rearrangements in Ground and Excited States*; de Mayo, P., Ed.; Academic Press: New York, 1980; Vol. 1, p 162.
297. Wilbourn, A.H. *J. Polym. Sci.* **1959**, *34*, 569.
298. Randall, J.C.; Buff, C.J.; Keichtermans, M.; Gregory, B.H. *Macromolecules* **1992**, *25*, 2624.
299. Ketels, H.; Beulen, J.; Vandervelden, G. *Macromolecules* **1988**, *21*, 2032.
300. Ketels, H.; Dehaan, J.; Aerdts, A.; Vandervelden, G. *Polymer* **1990**, *31*, 1419.

301. Britton, D.; Heatley, F.; Lovell, P.A. *Macromolecules* **1998**, *31*, 2828.
302. Starnes, W.H.; Zaikov, V.G.; Chung, H.T.; Wojciechowski, B.J.; Tran, H.V.; Saylor, K.; Benedikt, G.M. *Macromolecules* **1998**, *31*, 1508.
303. Morishima, Y.; Nozakura, S. *J. Polym. Sci., Polym. Chem. Ed.* **1976**, *14*, 1277.
304. Melville, H.W.; Sewell, P.R. *Makromol. Chem.* **1959**, *32*, 139.
305. Ahmad, N.M.; Britton, D.; Heatley, F.; Lovell, P.A. *Macromol. Symp.* **1999**, *143*, 231.
306. Britton, D.; Heatley, F.; Lovell, P.A. *Macromolecules* **2001**, *34*, 817.
307. Britton, D.; Heatley, F.; Lovell, P.A. *Macromolecules* **2000**, *33*, 5048.
308. Mattice, W.L.; Viswanadhan, V.N. *Macromolecules* **1986**, *19*, 568.
309. Hjertberg, T.; Sorvik, E. *ACS Symp. Ser.* **1985**, *280*, 259.
310. Yamada, B.; Azukizawa, M.; Yamazoe, H.; Hill, D.J.T.; Pomery, P.J. *Polymer* **2000**, *41*, 5611.
311. Azukizawa, M.; Yamada, B.; Hill, D.J.T.; Pomery, P.J. *Macromol. Chem. Phys.* **2000**, *201*, 774.
312. Hirano, T.; Yamada, B. *Polymer* **2003**, *44*, 347.
313. Plessis, C.; Arzamendi, G.; Alberdi, J.M.; Agnely, M.; Leiza, J.R.; Asua, J.M. *Macromolecules* **2001**, *34*, 6138.
314. Plessis, C.; Arzamendi, G.; Leiza, J.R.; Schoonbrood, H.A.S.; Charmot, D.; Asua, J.M. *Ind. Eng. Chem. Res.* **2001**, *40*, 3883.
315. Plessis, C.; Arzamendi, G.; Leiza, J.R.; Schoonbrood, H.A.S.; Charmot, D.; Asua, J.M. *Macromolecules* **2000**, *33*, 5041.
316. Plessis, C.; Arzamendi, G.; Leiza, J.R.; Schoonbrood, H.A.S.; Charmot, D.; Asua, J.M. *Macromolecules* **2000**, *33*, 4.
317. Ahmad, N.M.; Heatley, F.; Lovell, P.A. *Macromolecules* **1998**, *31*, 2822.
318. Grady, M.C.; Simonsick, W.J.; Hutchinson, R.A. **2002**, *182*, 149.
319. Heatley, F.; Lovell, P.A.; Yamashita, T. *Macromolecules* **2001**, *34*, 7636.
320. Tanaka, K.; Yamada, B.; Willemse, R.; van Herk, A.M. *Polym. J.* **2002**, *34*, 692.
321. Sato, T.; Takahashi, H.; Tanaka, H.; Ota, T. *J. Polym. Sci., Part A: Polym. Chem.* **1988**, *26*, 2839.
322. Sato, T.; Ito, D.; Kuki, M.; Tanaka, H.; Ota, T. *Macromolecules* **1991**, *24*, 2963.
323. Allen, P.E.M.; Patrick, C.R. *Kinetics and Mechanisms of Polymerization Reactions*; Ellis Horwood: Chichester, 1974.
324. Ivin, K.J. In *Reactivity, Mechanism and Structure in Polymer Chemistry*; Jenkins, A.D.; Ledwith, A., Eds.; Wiley: London, 1974; p 514.
325. Ivin, K.J.; Busfield, W.K. In *Encyclopedia of Polymer Science and Engineering*, 2nd ed.; Mark, H.F.; Bikales, N.M.; Overberger, C.G.; Menges, G., Eds.; Wiley: New York, 1987; Vol. 12, p 555.
326. Sawada, H. *J. Macromol. Sci., Rev. Macromol. Chem.* **1969**, *C3*, 313.
327. Busfield, W.K. In *Polymer Handbook*, 3rd ed.; Brandup, J.; Immergut, E.H., Eds.; Wiley: New York, 1989; p II/295.
328. Ivin, K.J. *Trans. Faraday Soc.* **1955**, *51*, 1273.
329. Bywater, S. *Trans. Faraday Soc.* **1955**, *51*, 1267.
330. Cook, R.E.; Ivin, K.J. *Trans. Faraday Soc.* **1957**, *53*, 1273.
331. Penelle, J.; Collot, J.; Rufflard, G. *J. Polym. Sci., Part A: Polym. Chem.* **1993**, *31*, 2407.
332. Bywater, S. *Can. J. Chem.* **1957**, *34*, 552.
333. Bywater, S.; Worsfold, D.J. *J. Polym. Sci.* **1962**, *58*, 571.
334. Ivin, K.J.; leonard, J. *Eur. Polym. J.* **1970**, *6*, 331.
335. Yamada, B.; Satake, M.; Otsu, T. *Makromol. Chem.* **1991**, *192*, 2713.

336. Madruga, E.M. In *Macromolecules 1992*; Kahovec, J., Ed.; VSP: Utrecht, 1992; p 109.
337. Avci, D.; Kusefoglu, S.H.; Thompson, R.D.; Mathias, L.J. *Macromolecules* **1994**, *27*, 1981.
338. Cheng, J.; Yamada, B.; Otsu, T. *J. Polym. Sci., Part A: Polym. Chem.* **1991**, *29*, 1837.
339. Tanaka, H. *Prog. Polym. Sci.* **2003**, *28*, 1171.
340. Stickler, M. In *Comprehensive Polymer Science*; Eastmond, G.C.; Ledwith, A.; Russo, S.; Sigwalt, P., Eds.; Pergamon: London, 1989; Vol. 3, p 59.
341. Stickler, M. In *Comprehensive Polymer Science*; Eastmond, G.C.; Ledwith, A.; Russo, S.; Sigwalt, P., Eds.; Pergamon: Oxford, 1989; Vol. 3, p 85.
342. Van Herk, A.M. *J. Macromol. Sci., Rev. Macromol. Chem. Phys.* **1997**, *37*, 633.
343. Beuermann, S.; Buback, M. *Prog. Polym. Sci.* **2002**, *27*, 191.
344. Kamachi, M.; Yamada, B. In *Polymer Handbook*, 4th ed.; Brandup, J.; Immergut, E.H.; Grulke, E.A., Eds.; John Wiley and Sons: New York, 1999; p II/77.
345. Buback, M.; Gilbert, R.G.; Russell, G.T.; Hill, D.J.T.; Moad, G.; O'Driscoll, K.F.; Shen, J.; Winnik, M.A. *J. Polym. Sci., Part A: Polym. Chem.* **1992**, *30*, 851.
346. Buback, M.; Garcia-Rubio, L.H.; Gilbert, R.G.; Napper, D.H.; Guillot, J.; Hamielec, A.E.; Hill, D.; O'Driscoll, K.F.; Olaj, O.F.; Shen, J.; Solomon, D.H.; Moad, G.; Stickler, M.; Tirrell, M.; Winnik, M.A. *J. Polym. Sci., Part C: Polym. Lett.* **1988**, *26*, 293.
347. Beuermann, S.; Buback, M.; Davis, T.P.; Gilbert, R.G.; Hutchinson, R.A.; Kajiwara, A.; Klumperman, B.; Russell, G.T. *Macromol. Chem. Phys.* **2000**, *201*, 1355.
348. Beuermann, S.; Buback, M.; Davis, T.P.; Gilbert, R.G.; Hutchinson, R.A.; Olaj, O.F.; Russell, G.T.; Schweer, J.; van Herk, A.M. *Macromol. Chem. Phys.* **1997**, *198*, 1545.
349. Buback, M.; Gilbert, R.G.; Hutchinson, R.A.; Klumperman, B.; Kuchta, F.-D.; Manders, B.G.; O'Driscoll, K.F.; Russell, G.T.; Schweer, J. *Macromol. Chem. Phys.* **1995**, *196*, 3267.
350. Gilbert, R.G. *Pure Appl. Chem.* **1996**, *68*, 1491.
351. Gilbert, R.G. *Pure Appl. Chem.* **1992**, *64*, 1563.
352. Burnett, G.M.; Melville, H.W. *Proc. R. Soc., London* **1947**, *A189*, 486.
353. Fukuda, T.; Ma, Y.-D.; Inagaki, H. *Macromolecules* **1985**, *18*, 17.
354. Olaj, O.F.; Kremminger, P.; Schnöll-Bitai, I. *Makromol. Chem., Rapid Commun.* **1988**, *9*, 771.
355. Olaj, O.F.; Schnöll-Bitai, I.; Kremminger, P. *Eur. Polym. J.* **1989**, *25*, 535.
356. O'Driscoll, K.F.; Mahabadi, H.K. *J. Polym. Sci., Polym. Chem. Ed.* **1976**, *14*, 869.
357. Bresler, S.E.; Kazbekov, E.N.; Fomichev, V.N.; Shadrin, V.N. *Makromol. Chem.* **1972**, *157*, 167.
358. Bresler, S.E.; Kazbekov, E.N.; Shadrin, V.N. *Makromol. Chem.* **1974**, *175*, 2875.
359. Kamachi, M. *J. Polym. Sci., Part A: Polym. Chem.* **2001**, *40*, 269.
360. Yamada, B.; Westmoreland, D.G.; Kobatake, S.; Konosu, O. *Prog. Polym. Sci.* **1999**, *24*, 565.
361. Carswell, T.G.; Hill, D.J.T.; Londero, D.I.; O'Donnell, J.H.; Pomery, P.J.; Winzor, C.L. *Polymer* **1992**, *33*, 137.
362. Carswell, T.G.; Hill, D.J.T.; Hunter, D.S.; Pomery, P.J.; O'Donnell, J.H.; Winzor, C.L. *Eur. Polym. J.* **1990**, *26*, 541.
363. Kamachi, M.; Kohno, M.; Kuwae, Y.; Nozakura, S.-I. *Polym. J.* **1982**, *14*, 749.
364. Shen, J.; Tian, Y.; Wang, G.; Yang, M. *Makromol. Chem.* **1991**, *192*, 2669.

365. Zhu, S.; Tian, Y.; Hamielec, A.E.; Eaton, D.R. *Macromolecules* **1990**, *23*, 1144.
366. Tonge, M.P.; Pace, R.J.; Gilbert, R.G. *Macromol. Chem. Phys.* **1994**, *195*, 3159.
367. Yamada, B.; Kageoka, M.; Otsu, T. *Polym. Bull.* **1992**, *28*, 75.
368. Yamada, B.; Kageoka, M.; Otsu, T. *Macromolecules* **1991**, *24*, 5234.
369. Kamachi, M.; Kuwae, Y.; Kohno, M.; Nozakura, S.-I. *Polym. J.* **1985**, *17*, 541.
370. Aleksandrov, H.P.; Genkin, V.N.; Kitai, M.S.; Smirnova, I.M.; Solokov, V.V. *Sov. J. Quantum Electron.* **1977**, *7*, 547.
371. Olaj, O.F.; Bitai, I.; Hinkelmann, F. *Makromol. Chem.* **1987**, *188*, 1689.
372. O'Driscoll, K.F.; Kuindersma, M.E. *Macromol. Theory Simul.* **1994**, *3*, 469.
373. Lu, J.; Zhang, H.; Yang, Y. *Makromol. Chem., Theory Simul.* **1993**, *2*, 747.
374. Deady, M.; Mau, A.W.H.; Moad, G.; Spurling, T.H. *Makromol. Chem.* **1993**, *194*, 1691.
375. Hutchinson, R.A.; Richards, J.R.; Aronson, M.T. *Macromolecules* **1994**, *27*, 4530.
376. Buback, M.; Huckestein, B.; Kuchta, F.-D.; Russell, G.T.; Schmid, E. *Macromol. Chem. Phys.* **1994**, *195*, 2117.
377. Buback, M.; Kurz, C.H.; Schmaltz, C. *Macromol. Chem. Phys.* **1998**, *199*, 1721.
378. Lyons, R.A.; Hutovic, J.; Piton, M.C.; Christie, D.I.; Clay, P.A.; Manders, B.G.; Kable, S.H.; Gilbert, R.G. *Macromolecules* **1996**, *29*, 1918.
379. Hutchinson, R.A.; Paquet, D.A., Jr.; McMinn, J.H. *DECHEMA Monogr.* **1995**, *131*, 467.
380. Deibert, S.; Bandermann, F.; Schweer, J.; Sarnecki, J. *Makromol. Chem., Rapid Commun.* **1992**, *13*, 351.
381. Beuermann, S.; Paquet, D.A., Jr.; McMinn, J.H.; Hutchinson, R.A. *Macromolecules* **1997**, *30*, 194.
382. Hutchinson, R.A.; Beuermann, S.; Paquet, D.A., Jr; McMinn, J.H. *Macromolecules* **1997**, *30*, 3490.
383. Buback, M.; Kurz, C.H. *Macromol. Chem. Phys.* **1998**, *199*, 2301.
384. Shipp, D.A.; Smith, T.A.; Solomon, D.H.; Moad, G. *Macromol. Rapid Commun.* **1995**, *16*, 837.
385. Yee, L.H.; Coote, M.L.; Chaplin, R.P.; Davis, T.P. *J. Polym. Sci., Part A: Polym. Chem.* **2000**, *38*, 2192.
386. Tanaka, K.; Yamada, B.; Fellows, C.M.; Gilbert, R.G.; Davis, T.P.; Yee, L.H.; Smith, G.B.; Rees, M.T.L.; Russell, G.T. *J. Polym. Sci., Part A: Polym. Chem.* **2001**, *39*, 3902.
387. Starks, C.M. *Free Radical Telomerization*; Academic Press: New York, 1974.
388. Fischer, H.; Radom, L. *Angew. Chem., Int. Ed. Engl.* **2001**, *40*, 1340.
389. Morrison, B.R.; Maxwell, I.A.; Gilbert, R.G.; Napper, D.H. *ACS Symp. Ser.* **1992**, *492*, 28.
390. Olaj, O.F.; Vana, P.; Zoder, M. *Macromolecules* **2002**, *35*, 1208.
391. Zetterlund, P.B.; Busfield, W.K.; Jenkins, I.D. *Macromolecules* **1999**, *32*, 8041.
392. Moad, G.; Rizzardo, E.; Solomon, D.H.; Beckwith, A.L.J. *Polym. Bull.* **1992**, *29*, 647.
393. Krstina, J.; Moad, G.; Willing, R.I.; Danek, S.K.; Kelly, D.P.; Jones, S.L.; Solomon, D.H. *Eur. Polym. J.* **1993**, *29*, 379.
394. Gridnev, A.A.; Ittel, S.D. *Macromolecules* **1996**, *29*, 5864.
395. Zetterlund, P.B.; Busfield, W.K.; Jenkins, I.D. *Macromolecules* **2002**, *35*, 7232.

5
Termination

5.1 Introduction

In this chapter we consider reactions that lead to the cessation of growth of one or more polymer chains. Three processes will be distinguished:

(a) The self-reaction of propagating radicals by combination and/or disproportionation (*e.g.* Scheme 5.1) (Section 5.2).

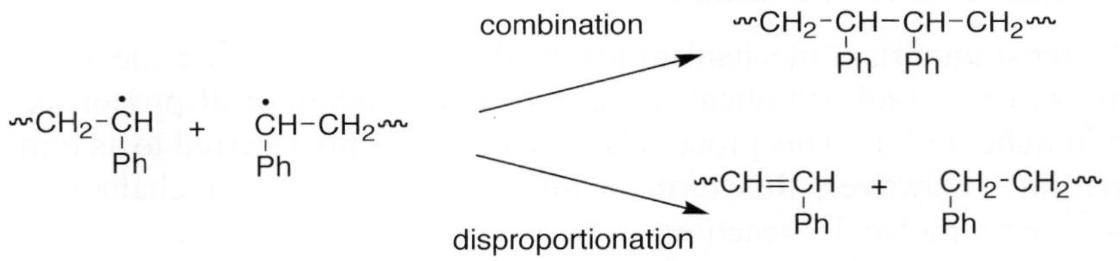

Scheme 5.1

(b) Primary radical termination (Sections 3.2.9, 3.4, 5.2.2.1 and 7.4.3); the reaction of a propagating radical with an initiator-derived (I•, Scheme 5.2) or transfer agent-derived radical. The significance of this process is highly dependent on the structure of the radical (I•).

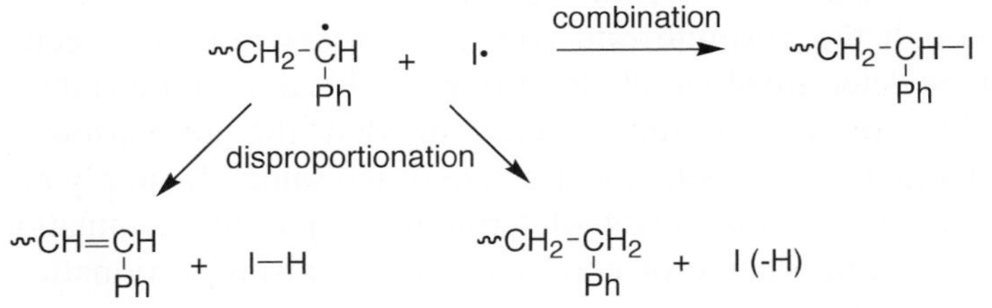

Scheme 5.2

(c) Inhibition (Section 5.3); the reaction of a propagating radical with another species (Z•, Scheme 5.3) to give a dead polymer chain. Z• is usually of low molecular weight. Examples of inhibitors are "stable" radicals (*e.g.* nitroxides, oxygen), non-radical species that react to give "stable" radicals (*e.g.* phenols, quinones, nitroso-compounds) and transition metal salts.

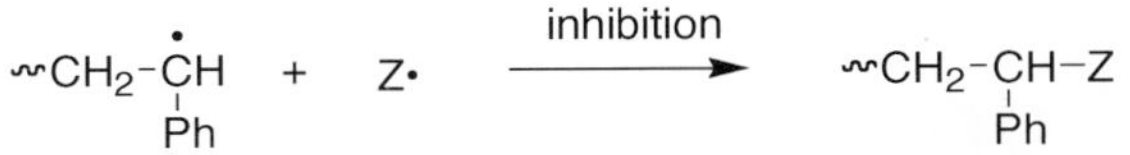

$$\sim CH_2-\overset{\bullet}{\underset{Ph}{CH}} + Z\bullet \xrightarrow{\text{inhibition}} \sim CH_2-\underset{Ph}{CH}-Z$$

Scheme 5.3

Chain transfer, the reaction of a propagating radical with a non-radical substrate to produce a dead polymer chain and a new radical capable of initiating a new polymer chain, is dealt with in Chapter 6. There are also situations intermediate between chain transfer and inhibition where the radical produced is less reactive than the propagating radical but still capable of reinitiating polymerization. In this case, polymerization is slowed and the process is termed retardation or degradative chain transfer. The process is mentioned in Section 5.3 and, when relevant, in Chapter 6.

5.2 Radical-Radical Termination

The most important mechanism for the decay of propagating species in radical polymerization is radical-radical reaction by combination or disproportionation as shown in Scheme 5.1. This process is sometimes simply referred to as bimolecular termination. However, this term is misleading since most chain termination processes are bimolecular reactions.

Before any chemistry can take place the radical centers of the propagating species must come into appropriate proximity and it is now generally accepted that the self-reaction of propagating radicals is a diffusion-controlled process. For this reason there is no single rate constant for termination in radical polymerization. The average rate constant usually quoted is a composite term that depends on the nature of the medium and the chain lengths of the two propagating species. Diffusion mechanisms and other factors that affect the absolute rate constants for termination are discussed in Section 5.2.1.4.

Even though the absolute rate constant for reactions between propagating species may be determined largely by diffusion, this does not mean that there is no specificity in the termination process or that the activation energies for combination and disproportionation are zero or the same. It simply means that this chemistry is not involved in the rate-determining step of the termination process.

The relative importance of combination and disproportionation in relevant model systems and in polymerizations of some common monomers is considered in Sections 5.2.2.1 and 5.2.2.2 respectively. The significance of the termination mechanism on the course of polymerization and on the properties of polymers is discussed briefly in Section 5.2.2 and is further discussed in Section 8.2.

5.2.1 Termination Kinetics

A detailed treatment of termination kinetics is beyond the scope of this book. However, some knowledge is important in understanding the chemistry described in subsequent sections. There are a number of reviews of the kinetics of radical-radical termination of propagating species. Those by North[1] and O'Driscoll[2] provide a useful background. Significant advances in our knowledge of termination kinetics came with the development of pulsed laser methods. Recent reviews include those by Buback *et al.*,[3] Russell[4-7] and de Kock *et al.*.[8,9] Many of the issues surrounding termination have been summarized by one IUPAC working party.[10-12] Values of, and methods of determining, termination rate constants are currently being critically assessed by another working party.[3]

In Section 5.2.1.1 we provide an overview of the classical treatment of polymerization kinetics. Some aspects of termination kinetics are not well understood and no wholly satisfactory unified description is in place, Nonetheless, it remains a fact that many features of the kinetics of radical polymerization can be predicted using a very simple model in which radical-radical termination is characterized by a single rate constant. The termination process determines the molecular weight and molecular weight distribution of the polymer. In section 5.2.1.2, we define the terminology used in describing molecular weights and molecular weight distributions. In Section 5.2.1.3, we provide a simple statistical treatment based on classical kinetics and discuss the dependence of the molecular weight distribution on the termination process. Some of the complexities of termination associated with diffusion control and the dependence on chain length and on conversion are described in Section 5.2.1.4.

Termination in heterogeneous polymerization is discussed in Section 5.2.1.5 and the more controversial subject of termination during living radical polymerization is described in Section 5.2.1.6. Termination in copolymerization is addressed in Section 7.3.

5.2.1.1 Classical kinetics

The overall rate constant for radical-radical termination can be defined in terms of the rate of consumption of propagating radicals. Consider the simplified mechanism for radical polymerization shown in Scheme 5.4.

Ideally, as long as the rate constants for reinitiation (k_{iT}, k_{iM}) are high with respect to that for propagation (k_p), the transfer reactions should not directly affect the rate of polymerization and they need not be considered further in this section. The overall rate constant for radical-radical termination (k_t) can be defined in terms of the rate of consumption of propagating radicals as shown in eq. 1:

$$R_t = -2k_t[P\bullet]^2 \qquad (1)$$

where [P•] is the total concentration of propagating radicals and $k_t = k_{tc} + k_{td}$.

In many works on radical polymerization, the factor 2 is by convention incorporated into the rate constant.[13,14] In this case $R_t = -k_t[P\bullet]^2$. The termination rate constant is then sometimes expressed as $k_t=k_{tc}/2+k_{td}$ to reflect the fact that only one polymer chain is formed when two propagating radicals combine whilst two are formed in disproportionation. In reading the literature and when comparing values of k_t, care must be taken to establish which definitions have been used.[2] In accord with the current IUPAC recommendation,[15] in the following discussion, eq. 1 and $k_t=k_{tc}+k_{td}$ are used.

initiation

$I_2 \rightarrow 2\ I\bullet$ $R_i=2k_d f[I_2]$

$I\bullet + M \rightarrow P_1\bullet$ $k_i \geq k_p$

propagation

$P_n\bullet + M \rightarrow P_{n+1}\bullet$ $R_p=k_p[M][P\bullet]$

termination by disproportionation

$P_n\bullet + P_m\bullet \rightarrow P_n^H + P_m^=$ $R_{td}=2k_{td}[P\bullet]^2$ $R_t = R_{tc} + R_{td}$

termination by combination

$P_n\bullet + P_m\bullet \rightarrow P_{n+m}$ $R_{tc}=2k_{tc}[P\bullet]^2$

termination by chain transfer

$P_n\bullet + I_2 \rightarrow P_n + I\bullet$ $R_{trI}=k_{trI}[I_2][P\bullet]$ $R_{tr}=R_{tr,I}+R_{tr,M}+R_{tr,T}$

$P_n\bullet + M \rightarrow P_n + P_1\bullet$ $R_{trM}=k_{trM}[M][P\bullet]$

$P_n\bullet + T \rightarrow P_n + T\bullet$ $R_{trT}=k_{trT}[T][P\bullet]$

$M\bullet + M \rightarrow P_2\bullet$ $k_{iM} \geq k_p$

$T\bullet + M \rightarrow P_1\bullet$ $k_{iT} \geq k_p$

Scheme 5.4

Application of a steady state approximation (that $R_t = R_i$, eq. 2) and a long chain approximation (negligible monomer consumption in the initiation or reinitiation steps) provides a number of useful relationships.

$$\frac{-d[P\bullet]}{dt} = R_i - R_t = 2k_d f[I_2] - 2k_t[P\bullet]^2 = 0 \quad (2)$$

(a) The total concentration of propagating radicals ($[P\bullet]$) (eq. 3):

$$[P\bullet] = \left(\frac{k_d f}{k_t}\right)^{0.5} [I_2]^{0.5} \quad (3)$$

(b) The mean lifetime of a propagating radical (τ) (eq. 4):

$$\tau = \left(2k_d f[I_2]k_t\right)^{-0.5} \tag{4}$$

(c) The average kinetic chain length ($\bar{v}$) (eq. 5):

$$\bar{v} = \frac{R_p}{R_t} = \frac{R_p}{R_i} = \frac{k_p[M]}{\left(2k_d f[I_2]k_t\right)^{0.5}} \tag{5}$$

(d) The number average degree of polymerization in the absence of chain transfer (eq. 6):

$$\overline{X}_n = \frac{k_p[M]}{\left(1+\frac{k_{td}}{k_t}\right)\left(2k_d f[I_2]k_t\right)^{0.5}} \tag{6}$$

(e) The initiator efficiency (eq. 7):

$$f = \left(1+\frac{k_{td}}{k_t}\right)\frac{R_p}{\overline{X}_n k_d[I_2]} \tag{7}$$

It also enables elimination of the radical concentration in the expression for rate of polymerization (eq. 8):

$$R_p = \frac{-d[M]}{dt} = k_p[P\bullet][M]$$

$$= k_p\left(\frac{k_d f}{k_t}\right)^{0.5}[I_2]^{0.5}[M] \tag{8}$$

In eq. 8, the rate of polymerization is shown as being half order in initiator (I_2). This is only true for initiators that decompose to two radicals both of which begin chains. The form of this term depends on the particular initiator and the initiation mechanism. The equation takes a slightly different form in the case of thermal initiation (S), redox initiation, diradical initiation, etc. Side reactions also cause a departure from ideal behavior.

Eq. 8 can be recast in terms of the fractional conversion of monomer to polymer as in eq. 9:

$$-\frac{d\ln([M]/[M]_o)}{dt} = \left(\frac{k_d f k_p^2}{k_t}\right)^{0.5}[I_2]^{0.5} \tag{9}$$

From this we can see that knowledge of k_df and R_p in a conventional polymerization process readily yields a value of the ratio k_p^2/k_t. In order to obtain a value for k_t we require further information on k_p. Analysis of R_p data obtained under non-steady state conditions (when there is no continuous source of initiator radicals) yields the ratio k_p/k_t. Various non-steady state methods have been developed including the rotating sector method, spatially intermittent polymerization and pulsed laser polymerization (PLP). The classical approach for deriving the individual values of k_p and k_t by combining values for k_p^2/k_t with k_p/k_t obtained in separate experiments can, however, be problematical because the values of k_t are strongly dependent on the polymerization conditions (Section 5.2.1.4). These issues are thought to account for much of the scatter apparent in literature values of k_t.[3,16] PLP and related methods yield absolute values of k_p directly (the methods used for extracting k_p are discussed in Section 4.5.2). These values may be combined with either k_p^2/k_t or k_p/k_t to give k_t.

The SP-PLP[8,17,18] and PS-PLP[17,19] techniques involve following the monomer conversion induced by a single laser pulse or a sequence of laser pulses. These experiments are usually conducted at high pressure because rates of termination are lower and sensitivities are somewhat higher.[17]

EPR methods can be used to determine the radical concentration [P•] either directly[20,21] or *via* trapping methods.[22] Fluorescence experiments have also been designed to give [P•] for a particular conversion.[23-25] Given [P•] and the rate of polymerization, k_p can be evaluated using eq. 8. Given the rate of initiation and [P•], k_t can be calculated using eq. 3.[20,21,26] It is also possible to estimate k_t from the molecular weight distributions given k_p and [P•] using kinetic simulation.[24,25]

For low conversions, values of the rate constants k_t for monosubstituted monomers (S and acrylates) are $\sim 10^8$ M^{-1}s^{-1} and those for methacrylates are $\sim 10^7$ M^{-1}s^{-1} and activation energies are small and in the range 3-8 kJ mol^{-1}.[17] These activation energies relate to the rate-determining diffusion process (Section 5.2.1.4) rather than to radical-radical coupling.

Values of termination constants for sterically hindered monomers may be several orders of magnitude lower than those for S and (methacrylates). Such monomers include various α-substituted methacrylates, itaconates, fumarates, and N-substituted itaconimides and maleimides. Values of k_t for these monomers have been reported to lie in the range 10-10^5 M^{-1}s^{-1} depending on the particular structure.[20]

5.2.1.2 *Molecular weights and molecular weight averages*

The degree of polymerization of a polymer (X_i) is equal to the chain length i (the number of monomer units in the chain). If we neglect end groups,* the number molecular weight (M_n) is given by eq. 10:

---

* By definition, the molar mass of the end groups should be included in the molecular weight of a polymer but the corresponding quantity is not included in the degree of polymerization.

$$M_i = X_i M_0 \tag{10}$$

where M_0 is the molecular weight or molar mass of the monomer or repeat unit.[†]

The number average molecular weight ($\overline{M}_n$) is the average molecular weight of all of the polymer chains that make up a sample and is given by eq. 11:

$$\overline{M}_n = \frac{\sum n_i X_i}{\sum n_i} M_0 \tag{11}$$

where n_i is the concentration of chains of length i (monomer units)

The weight average molecular weight ($\overline{M}_w$) is given by eq. 12:

$$\overline{M}_w = \frac{\sum w_i X_i}{\sum w_i} M_0 = \frac{\sum n_i X_i^2}{\sum n_i X_i} M_0 \tag{12}$$

where w_i is the weight of chains of length i.

The Z average molecular weight ($\overline{M}_z$) is provided by eq. 13:

$$\overline{M}_z = \frac{\sum n_i X_i^3}{\sum n_i X_i^2} M_0 \tag{13}$$

This term gives some information about the asymmetry of the molecular weight distribution and is important in analyzing sedimentation behavior in ultracentrifugation.

It is also useful to define the moments of the chain length distribution. The jth moment is defined in eq. 14:

$$\lambda^j = \sum n_i X_i^j \tag{14}$$

The zeroth moment $\lambda^0 = \sum n_i$ can be recognized as the total concentration of polymer chains and the first moment $\lambda^1 = \sum n_i X_i = \sum w_i$ is the total concentration of repeat or monomer units in those chains. The moments can be related to the molecular weight averages as follows:

$$\overline{M}_n = \frac{\lambda^1}{\lambda^0} M_0, \quad \overline{M}_w = \frac{\lambda^2}{\lambda^1} M_0, \quad \overline{M}_z = \frac{\lambda^3}{\lambda^2} M_0$$

The breadth of the molecular weight distribution is often discussed in terms of the dispersity (D)* and is expressed in terms of the moments as shown in eq. 15:

[†] In this book, in accord with common usage, we use the term molecular weight rather than molar mass when referring to polymers.
* The dispersity is also commonly called the polydispersity index or the polydispersity.

$$D = \frac{\overline{X}_w}{\overline{X}_n} = \frac{\overline{M}_w}{\overline{M}_n} = \frac{\lambda^0 \lambda^1}{(\lambda^1)^2} \tag{15}$$

In calculations the moments can be treated as concentrations. Kinetic simulation of radical polymerization to evaluate dispersities typically involves evaluation of the moments rather than the complete distribution. This method of moments is accurate as long as the kinetics are independent of chain length.

5.2.1.3 *Molecular weight distributions*

The simple statistical treatment of radical polymerization can be traced back to Schultz.[27] Texts by Flory[28] and Bamford *et al.*[29] are useful references.

The probability of a propagation event (ϕ) can be defined as shown in eq. 16:

$$\phi = \frac{R_p}{R_p + R_t + R_{tr}}$$

$$= \frac{k_p[M]}{k_p[M] + 2k_t[P\bullet] + k_{trI}[I_2] + k_{trM}[M] + k_{trT}[T]} \tag{16}$$

A given chain will undergo $i-1$ propagation steps (each with probability ϕ) before terminating (with probability $1-\phi$). Thus, if termination is wholly by chain transfer or disproportionation, the chain length distribution is given by eq. 17 (Figure 5.1):

$$n_i = \phi^{i-1}(1-\phi) \tag{17}$$

This distribution is known as the Schultz-Flory or most probable distribution.[28] The moments of the molecular weight distribution are:

$$\lambda^0 = 1, \; \lambda^1 = (1-\phi)^{-1}, \; \lambda^2 = (1+\phi)(1-\phi)^{-2}$$

and the average degrees of polymerization and dispersity are:

$$\overline{X}_n = \frac{1}{1-\phi}, \; \overline{X}_w = \frac{1+\phi}{1-\phi} \;\; \text{and} \;\; D = \frac{\overline{X}_w}{\overline{X}_n} = 1+\phi$$

and for long chains as $\phi \to 1$, $D \to 2$.

If termination is wholly by combination it can be shown[29] that the number distribution is given by eq. 18 (Figure 5.1):

$$n_i = (i-1)(1-\phi)^2 \phi^{i-2} \tag{18}$$

The moments of the molecular weight distribution are:

$$\lambda^0 = 1, \; \lambda^1 = 2(1-\phi)^{-1}, \; \lambda^2 = (4+2\phi)(1-\phi)^{-2}$$

and the average degrees of polymerization and dispersity are:

$$\overline{X}_n = \frac{2}{1-\phi}, \quad \overline{X}_w = \frac{2+\phi}{1-\phi} \quad \text{and} \quad D = \frac{\overline{X}_w}{\overline{X}_n} = \frac{2+\phi}{2}$$

The molecular weight distribution in this case is significantly narrower. For long chains as $\phi \to 1$ so $D \to 1.5$.

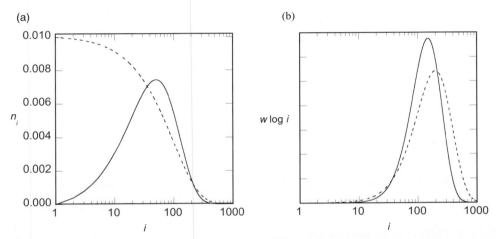

Figure 5.1 (a) Number and (b) GPC distributions for two polymers both with $\overline{X}_n = 100$. The number distribution of chains formed by disproportionation or chain transfer (------, $\sum n_i = 1.0$, $\overline{X}_w / \overline{X}_n = 2.0$) is calculated using eq. 17. The number distribution of chains formed by combination (———, $\sum n_i = 1.0$, $\overline{X}_w / \overline{X}_n = 1.5$) is calculated using eq. 18.

For the more general case, the molecular weight distribution will be described by a weighted average of eqs. 17 and 18 (eq. 19):

$$n_i = \frac{R_{tc}}{R_t + R_{tr}}(i-1)(1-\phi)^2 \phi^{i-2} + \frac{R_{td} + R_{tr}}{R_t + R_{tr}} \phi^{i-1}(1-\phi) \qquad (19)$$

These equations predict that for oligomers with degree of polymerization less than 10, polydispersities significantly less than 1.5 will be obtained - Figure 5.2.

The above treatment only applies to polymerizations where there is negligible conversion of monomer, initiator, and transfer agents. Analytical treatments have been devised to take into account effects of conversion and more complex mechanisms. Discussion of these is beyond the scope of this book.

A common error is to confuse the GPC distribution with the weight distribution. The response of a refractive index detector is proportional to the mass of polymer. The GPC elution volume (V) typically scales according to the logarithm of the degree of polymerization (or the logarithm of the molecular weight). Thus, $V \sim a+b \log i$ (where a and b are constants) and a volume increment (dV) will be proportional to di/i. It follows that the y-axis of the GPC distribution (*e.g.* Figure 5.1b) is proportional to iw_i or $i^2 n_i$.

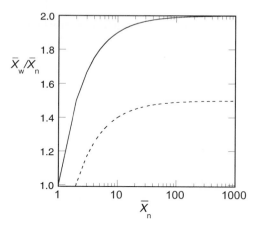

Figure 5.2 Dispersity (*D*) as a function of $\overline{X}_n$ for polymers formed by (a) disproportionation or chain transfer (———) and (b) combination (------).

5.2.1.4 Diffusion controlled termination

Termination by self-reaction of propagating radicals is a diffusion-controlled process even at very low conversion.[3] The evidence for this includes the following:

(a) Analogy with the known chemistry of small radicals. The rate constants for self-reaction of small radicals approach the diffusion-controlled limit and rate constants can be predicted using the Smoluchowski equation.
(b) The value of k_t shows an inverse dependence on medium viscosity as anticipated for a diffusion controlled reaction.
(c) The value of k_t decreases with increasing pressure (positive activation volume). For a reaction involving the combination of two species, the activation volume is expected to be negative.

However, while it is generally accepted that the rate of radical-radical reaction is dependent on how fast the radical centers of the propagating chains ($P_i\bullet$ and $P_j\bullet$) come together, there remains some controversy as to the diffusion mechanism(s) and/or what constitutes the rate-determining step in the diffusion process. The steps in the process as postulated by North and coworkers[30-32] are shown conceptually in Scheme 5.5.

Termination

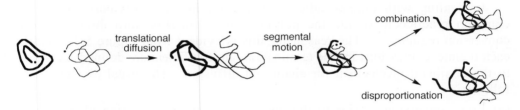

Scheme 5.5

Center of mass or translational diffusion is believed to be the rate-determining step for small radicals[33] and may also be important for larger species. However, other diffusion mechanisms are operative and are required to bring the chain ends together and these will often be the major term in the termination rate coefficient for the case of macromolecular species. These include:

(a) Segmental motion. The internal reorganization of the chain required to bring the reactive ends together.
(b) Reptation. The snaking of the chain through a viscous medium.
(c) Reaction diffusion (also called residual termination). Chain end motion by addition of monomer to the chain end.

The relative importance of these mechanisms, and the value of the overall k_t, depends on the molecular weight and dispersity of the propagating species, the medium and the degree of conversion. The value of k_t is not a constant!

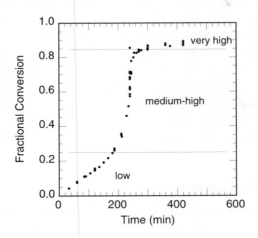

Figure 5.3 Conversion-time profile for bulk MMA polymerization at 50 °C with AIBN initiator illustrating the three conversion regimes. Data are taken from Balke and Hamielec.[34]

In dealing with radical-radical termination in bulk polymerization it is common practice to divide the polymerization timeline into three or more conversion regimes.[2,35] The reason for this is evident from Figure 5.3. Within each regime, expressions for the termination rate coefficient are defined according to the dominant mechanism for chain end diffusion. The usual division is as follows:

(a) Low conversion - prior to the onset of the autoacceleration phenomenon known as the gel or Norrish-Trommsdorff effect[36-38] and characterized by highly mobile propagating species. Center of mass and/or segmental diffusion are the rate-determining mechanisms for chain end movement. Initiator efficiencies are high and approximately constant.

(b) Medium to high conversion – immediately after the onset of the gel effect. The diffusion mechanism is complex. Large chains become effectively immobile (on the timescale of the lifetime of a propagating radical) even though the chain ends may move by segmental diffusion, reptation or reaction diffusion. Monomeric species and short chains may still diffuse rapidly. Short-long termination dominates. Initiator efficiencies may reduce with conversion.

(c) Very high conversion - the polymerization medium is a glassy matrix. Most chains are immobile and reaction diffusion is the rate-determining diffusion mechanism. New chains are rapidly terminated or immobilized. Initiator efficiencies are very low.

The precise conversion ranges are determined by a variety of factors including the particular monomer, the molecular weight of the polymeric species and the solvent (if any). For bulk polymerization of S and MMA (a) is typically <20%, (b) is 20-85% and (c) is >85%. In solution polymerization, or for polymerizations carried out in the presence of chain transfer agents, the duration of the low conversion regime is extended and the very high conversion regime may not occur. Cage escape is also a diffusion controlled process, thus the initiator efficiency (f) and the rate of initiation ($k_d f$) generally decrease with conversion and depend on the conversion regime as indicated above (Sections 3.2.8, 3.3.1.1.3, 3.3.2.1.3. 3.3.2.4).

5.2.1.4.1 Termination at low conversion

Most in depth studies of termination deal only with the low conversion regime. Logic dictates that simple center of mass diffusion and overall chain movement by reptation or many other mechanisms will be chain length dependent. At any instant, the overall rate coefficient for termination can be expressed as a weighted average of individual chain length dependent rate coefficients (eq. 20):[39]

$$k_t = \frac{\sum_{i=1}^{\infty}\sum_{j=1}^{\infty} k_t^{i,j}[P_i\bullet][P_j\bullet]}{[P\bullet]^2} \qquad (20)$$

where $k_t^{i,j}$ is the rate coefficient for reaction between species of chain lengths i and j, and $[P\bullet]$ is the total radical concentration.

Mahabadi and O'Driscoll[39] considered that segmental motion and center of mass diffusion should be the dominant mechanisms at low conversion. They analyzed data for various polymerizations and proposed that $k_t^{i,j}$ should be dependent on chain length such that the overall rate constant obeys the expression:

$$k_t \propto \overline{X}_n^{-\alpha} \qquad (21)$$

where $\overline{X}_n$ is the number average degree of polymerization and $\alpha = 0.5$ for short $\overline{X}_n$ reducing to 0.1 for large $\overline{X}_n$.

Various expressions have been proposed for estimating how the overall rate coefficient k_t and the individual rate coefficients $k_t^{i,j}$ vary with the chain lengths of the reacting species,[2,39-46] simple relationships of the following forms are the most often applied:[32,42,46,47]

(a) The harmonic mean is said to be of the functional form expected if chain end encounter or coil overlap is rate-determining:

$$k_t^{i,j} = k_{to}\left(\frac{2i.j}{i+j}\right)^{-\alpha} \qquad (22)$$

(b) The Smoluchowski mean is of the functional form expected if translational diffusion is rate-determining; it is known to provide a reasonable description of the termination kinetics of small radicals:

$$k_t^{i,j} = 0.5 k_{to}(i^{-\alpha} + j^{-\alpha}) \qquad (23)$$

or:

$$k_t^{i,j} = 2\pi\sigma p_{spin}(D^i + D^j) \qquad (24)$$

where σ is a capture radius, p_{spin} is a spin multiplicity term, and D^i and D^j are chain length dependent diffusion constants. When $\alpha = 1$, the Smoluchowski mean and the harmonic mean approximations are the same

(c) The geometric mean has no physical basis but has been suggested to best approximate the functional form of the segmental diffusion process:

$$k_t^{i,j} = k_{to}(i.j)^{-\alpha/2} \qquad (25)$$

where α and k_{to} are constants.

While many data are suggestive of chain length dependence, the data are not usually suitable for or have not been tested with respect to model discrimination. Values of $k_t^{1,1}$ have been determined for a variety of small "monomeric" radicals to be ca 10^9 M^{-1} s^{-1}.[48] Taking k_{to} as $k_t^{1,1}$ and α as 1.0 in the geometric expression yields values of $k_t^{i,j}$ as shown in Figure 5.4a.[49] Use of the Smoluchowski mean or the harmonic mean approximation predicts a shallower dependence of $k_t^{i,j}$ on the chain length (Figure 5.4b). All expressions yield the same dependence for j=i.

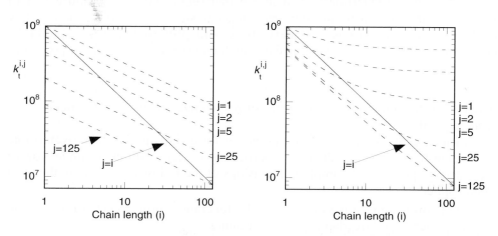

Figure 5.4 Chain length dependence of $k_t^{i,j}$ predicted by (a) the geometric mean (eq, 25) or (b) the harmonic mean approximation (eq. 22) or the Smoluchowski mean (eq. 23) with α=1.0 and k_{to}=10^9; i and j are the lengths of the reacting chains.

However, it has been pointed out that the value of k_{to} in the expressions eqs. 25-23 should not be confused with the small radical $k_t^{1,1}$, rather, the value of k_{to} represents the termination rate constant of a single unit chain if the implied diffusion mechanism was the rate-determining process.

Recent work has allowed values of $k_t^{1,1}$ and α for bulk polymerization in dilute solution to be estimated. This work suggests values of k_{to}=$k_t^{1,1}$ ~ 1×10^8 M^{-1} and α ~ 0.15-0.25 for both MMA and S.[17,50] Some values of $k_t^{1,1}$ and α for S and methacrylates estimated from SP-PLP at high pressure experiments are shown in Table 5.1.

The value of the exponent α obtained in the above-mentioned experiments is in remarkable accord with predictions based on a consideration of excluded kinetic volume effects. Khokhlov[51] proposed, that for a slow, chemically controlled, reaction between the ends of long chains α should be 0.16. The value of α was suggested to increase to 0.28 for chain end-mid chain reaction and to 0.43 for mid-chain-mid chain reaction. The latter provides one possible explanation for the greater exponent for higher acrylates (Table 5.1).[52]

Table 5.1 Parameters Characterizing Chain Length Dependence of Termination Rate Coefficients in Radical Polymerization of Common Monomers[a]

Monomer	T (°C)	P (bar)	k_p (M^{-1}s^{-1})	k_{to} (M^{-1}s^{-1})	α	ref
S	40	1000	1600	7×10^7	0.16	18
MMA	40	2000	1700	4×10^7	0.14	52
DMA	40	1000	1400	3×10^6	0.15	52
MA	40	1000	28600	2×10^8	0.15	52
BA	40	1000	35600	6×10^7	0.14	52
DA	40	1000	39800	8×10^7	0.43	52

a Determined by the SP-PLP technique. Values apply to bulk polymerization at low conversion (up to 15% conversion).

For the situation where the chain length of one or both of the species is "small" (not entangled with itself or other chains) and conversion of monomer to polymer is low, the termination kinetics should be dominated by the rate of diffusion of the shorter chain. While the chain remains short, the time required for the chain reorganization to bring the reacting centers together will be insignificant and center of mass diffusion can be the rate-determining step. As the chain becomes longer, segmental diffusion will become more important. Thus, it is expected that $k_t^{i,j}$ should lie between an upper limit predicted by the Smoluchowski mean (eq. 23) and a lower limit predicted by the geometric mean (eq, 25) with the value being closer to the geometric mean value for higher chain lengths as shown in Figure 5.5.

Smith et al.[50] have recently suggested a composite model based on similar considerations to predict $k_t^{i,j}$ over the entire chain length range. Experimental data for $k_t^{i,i}$ for dodecyl methacrylate polymerization consistent with such a model have been provided by Buback et al.[53]

Since shorter, more mobile, chains diffuse more rapidly (by center of mass diffusion or other mechanisms), they are more likely to be involved in termination. For this reason, most termination involves reaction of a long species with a short species. The lower mobility of long chains ensures that they are unlikely to react with each other. Cardenas and O'Driscoll[54] proposed that propagating species be considered as two populations; those with chain length below the entanglement limit and those above. This basic concept has also been adopted by other authors.[24,55-58] Russell[55] has provided a detailed critique of these concepts. Direct experimental evidence for the importance of the dispersity of the propagating radicals on termination kinetics has been reported by Faldi et al.[56] O'Neil and Torkelson questioned the chain entanglement concept pointing out that for low conversions chain entanglements are unlikely even for chain lengths >100.

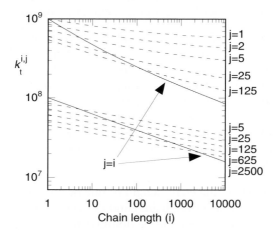

Figure 5.5 Chain length dependence of $k_t^{i,j}$ predicted by the Smoluchowski mean (eq. 23) with $\alpha=0.5$ and $k_{to}=10^9$ (upper series) and the geometric mean (eq. 25) with $\alpha=0.2$ and $k_{to}=10^8$ (lower series); i and j are the lengths of the reacting chains. For low conversions, $k_t^{i,j}$ is expected to lie between the values predicted by eqs. 23 and 25 (see text).

For larger species, even though the chains themselves may be in contact, chain end diffusion by segmental motion, reptation, or reactive diffusion will be required to bring the radical centers together. These terms are likely to be more important than center of mass diffusion. North[1] argued that diffusion of the reactive chain end of longer chains by segmental diffusion should be independent of chain length and has presented some experimental evidence for this hypothesis.

Bamford[45,59-63] has proposed a general treatment for solving polymerization kinetics with chain length dependent k_t and considered in some detail the ramifications with respect to molecular weight distributions and the kinetics of chain transfer, retardation, *etc*.

5.2.1.4.2 Termination at medium to high conversions

Changes in the population of propagating species and the increase in the polymer concentration mean that the rate coefficient for radical-radical termination will decrease with conversion. The moderate conversion regime is characterized by the autoacceleration phenomenon known as the gel or Norrish-Trommsdorf effect.[36-38] Various empirical relationships defining k_t or the rate of diffusion of long chains in terms of either the viscosity[1,64] or the free volume[34,35,44,65-69] have been proposed which enable the onset of the gel effect (Figure 5.3) to be predicted for a number of polymer systems.

Ito,[70] Tulig and Tirrell,[71] and de Gennes[72] have proposed expressions for k_t based on a reptation mechanism. More recently, the manner in which the termination rate coefficient scales with chain length for entangled systems has been considered in some detail in studies by O'Shaughnessy and coworkers.[57,58,73,74] For the situation where both chains are long (entangled), the way in which the termination coefficient (or diffusion rates) should scale with chain length means that a long chain is unlikely to terminate by reaction with another long chain. Short-long termination is dominant. Measurements of the diffusion rate constants of oligomers and polymers provide some support for this theory.

The concept of reaction diffusion (also called residual termination) has been incorporated into a number of treatments.[75,76] Reaction diffusion will occur in all conversion regimes. However at low and intermediate conversions the process is not of great significance as a diffusion mechanism. At high conversion long chains are essentially immobile and reaction diffusion becomes the dominant diffusion mechanism (when i and j are both "large" >100). The termination rate constant is determined by the value of k_p and the monomer concentration. In these circumstances, the rate constant for termination $k_t^{i,j}$ should be independent of the chain lengths i and j and should obey an expression of the form:[75]

$$k_t^{i,j} = k_{t1} k_p [M] \tag{26}$$

where k_{t1} is a constant.

5.2.1.5 *Termination in heterogeneous polymerization*

The kinetics of termination in suspension polymerization is generally considered to be the same as for solution or bulk polymerization under similar conditions and will not be discussed further. A detailed discussion on the kinetics of termination in emulsion polymerization appears in recent texts by Gilbert[77] and Lovell and El-Aasser[78] and readers should consult these for a more comprehensive treatment.

The steps involved in entry of a radical into the particle phase from an aqueous phase initiator have been summarized in Section 3.1.11. Aqueous phase termination prior to particle entry should be described by conventional dilute solution kinetics (Section 5.2.1.4.1). Note that chain lengths of the aqueous soluble species are short (typically <10 units).

Even though the chemical reactions are the same (*i.e.* combination, disproportionation), the effects of compartmentalization are such that, in emulsion polymerization, particle phase termination rates can be substantially different to those observed in corresponding solution or bulk polymerizations. A critical parameter is $\bar{n}$, the average number of propagating species per particle. The value of $\bar{n}$ depends on the particle size and the rates of entry and exit.

Many emulsion polymerizations can be described by so-called zero-one kinetics. These systems are characterized by particle sizes that are sufficiently small that entry of a radical into a particle already containing a propagating radical always causes instantaneous termination. Thus, a particle may contain either zero or one propagating radical. The value of $\bar{n}$ will usually be less than 0.4. In these systems, radical-radical termination is by definition not rate determining. Rates of polymerization are determined by the rates of particle entry and exit rather than by rates of initiation and termination. The main mechanism for exit is thought to be chain transfer to monomer. It follows that radical-radical termination, when it occurs in the particle phase, will usually be between a short species (one that has just entered) and a long species.

Treatments (Smith-Ewart,[79] pseudo-bulk[77]) have been devised which allow for the possibility of greater than one radical per particle and for the effects of chain length dependent termination. Further discussion on these is provided in the references mentioned above.[77,78]

Microemulsion and miniemulsion polymerization processes differ from emulsion polymerization in that the particle sizes are smaller (10-30 and 30-100 nm respectively *vs* 50-300 nm)[77] and there is no discrete monomer droplet phase. All monomer is in solution or in the particle phase. Initiation usually takes place by the same process as conventional emulsion polymerization. As particle sizes reduce, the probability of particle entry is lowered and so is the probability of radical-radical termination. This knowledge has been used to advantage in designing living polymerizations based on reversible chain transfer (*e.g.* RAFT, Section 9.5.2).[80-82]

5.2.1.6 *Termination during living radical polymerization*

It remains a common misconception that radical-radical termination is suppressed in processes such as NMP or ATRP. Another issue, in many people's minds, is whether processes that involve an irreversible termination step, even as a minor side reaction, should be called living. Living radical polymerization appears to be an oxymoron and the heading to this section a contradiction in terms (Section 9.1.1). In any processes that involve propagating radicals, there will be a finite rate of termination commensurate with the concentration of propagating radicals and the reaction conditions. The processes that fall under the heading of living or controlled radical polymerization (*e.g.* NMP, ATRP, RAFT) provide no exceptions.

In conventional radical polymerization, the chain length distribution of propagating species is broad and new short chains are formed continually by initiation. As has been stated above, the population balance means that, termination, most frequently, involves the reaction of a shorter, more mobile, chain with a longer, less mobile, chain. In living radical polymerizations, the chain lengths of most propagating species are similar (*i.e.* $i \sim j$) and increase with conversion. Ideally, in ATRP and NMP no new chains are formed. In practice,

Termination

some new chains may be formed, as, for example, from thermal initiation in S polymerization. In processes such as RAFT new small radicals are continuously formed by initiation as in the conventional process but form a much smaller part of the population as they undergo rapidly equilibration with longer dormant chains.

Diffusion mechanisms depend on chain length as follows:

(a) Very short chains ($X_n < 10$ units). Translational diffusion is the most important diffusion mechanism.
(b) Chains of moderate length ($X_n \sim 10\text{-}100$ units). Segmental motion of the chain ends is the rate-determining diffusion mechanism.
(c) Long chains. Chains immobile, reaction diffusion is rate-determining.

On this basis it might be expected that at low conversions the extent of termination would be higher than in a conventional polymerization since all chains are short. Similarly, for higher conversions the extent of termination should be lower than in a conventional polymerization because most chains are long.[80] It has also been proposed that the molecular weight distribution in living radical polymerization might be analyzed to provide values of $k_t^{1,1}$ as a function of molecular weight. Recently, Vana et al.[83] have analyzed RAFT polymerization in this context. Their data suggests a chain length dependence in general agreement with that suggested by other methods. It can also be noted that the SP-PLP experiment is, in some respects, a good model of a living radical polymerization and also provides values of $k_t^{1,1}$.[17,52,53]

It can also be noted that reversible chain transfer, in RAFT and similar polymerizations, and reversible activation-deactivation, in NMP and ATRP, provide other mechanisms for reaction diffusion.

5.2.2 Disproportionation vs Combination

Even though the rate of radical-radical reaction is determined by diffusion, this does not mean there is no selectivity in the termination step. As with small radicals (Section 2.5), self-reaction may occur by combination or disproportionation. In some cases, there are multiple pathways for combination and disproportionation. Combination involves the coupling of two radicals (Scheme 5.1). The resulting polymer chain has a molecular weight equal to the sum of the molecular weights of the reactant species. If all chains are formed from initiator-derived radicals, then the combination product will have two initiator-derived ends. Disproportionation involves the transfer of a β–hydrogen from one propagating radical to the other. This results in the formation of two polymer molecules. Both chains have one initiator-derived end. One chain has an unsaturated end, the other has a saturated end (Scheme 5.1).

Since the mode of termination clearly plays an important part in determining the polymer end groups and the molecular weight distribution, a knowledge of the disproportionation:combination ratio (k_{td}/k_{tc}) is vital to the understanding of structure-property relationships. Unsaturated linkages at the ends of polymer chains, as may be formed by disproportionation, have long been thought to contribute to polymer instability and it has been demonstrated that both head-to-head linkages and unsaturated ends are weak links during the thermal degradation of PMMA (Section 8.2.2).[84-87] Polymer chains with unsaturated ends may also be reactive during polymerization. Copolymerization of macromonomers formed by disproportionation is a possible mechanism for the formation of long chain branches.[88-90] Such macromonomers may also function as transfer agents (Section 6.2.3.4 and 9.5.2).[90]

Knowledge of k_{td}/k_{tc} is also important in designing polymer syntheses. For example, in the preparation of block copolymers using polymeric or multifunctional initiators (Section 7.6.1), ABA or AB blocks may be formed depending on whether termination involves combination or disproportionation respectively. The relative importance of combination and disproportionation is also important in the analysis of polymerization kinetics and, in particular, in the derivation of rate parameters.

5.2.2.1 Model studies

The determination of k_{td}/k_{tc} by direct analysis of a polymerization or the resultant polymer often requires data on aspects of the polymerization mechanism that are not readily available. For this reason, it is appropriate to consider the self-reactions of low molecular weight radicals which are structurally analogous to the propagating species. These model studies provide valuable insights by demonstrating the types of reaction that are likely to occur during polymerization and the factors influencing k_{td}/k_{tc}. These have been discussed in general terms in Section 2.4.

In these model studies, evaluation of k_{td}/k_{tc} is simplified because reactions that compete with disproportionation or combination are more readily detected and allowed for. However, by their very nature, model studies cannot exactly simulate all aspects of the polymerization process. Consequently, a number of factors must be borne in mind when using model studies to investigate the termination process. These stem from differences inherent in polymerization *vs* simple organic reactions and include:

(a) There may be additional pathways open to the poly- or oligomeric radicals which are not available to the simple model species.[91]

(b) In polymerization particular propagating species have only transient existence since they are scavenged by the addition of monomer or other reactions. Model studies are usually designed such that the self-reaction is the only process. This can lead to a very different and sometimes misleading product distribution. A knowledge of the reaction kinetics is extremely important in analyzing the results.

(c) Reaction conditions (solvent, viscosity, *etc.*) chosen for the model experiment and the polymerization experiment are often very different.

Termination

Model carbon-centered radicals are conveniently generated from azo-compounds. These have the advantage that radicals are generated in pairs and that transfer to initiator is generally not a serious problem. All of the major products from thermal or photochemical decomposition in an inert solvent are the products from radical-radical reaction. One frequently observed complication is polymerization of the unsaturated byproducts of disproportionation. This problem may be circumvented by conducting experiments in the presence of an inhibitor, the concentration of which can be chosen such that all radicals which escape the solvent cage are trapped and reactions of the initiator-derived radicals with other species are eliminated.[89] The value of k_{td}/k_{tc} is determined by analyzing the products of cage reaction. Most data indicate no difference in specificity between the cage and encounter (*i.e.* non-cage) processes.[89]

5.2.2.1.1 *Polystyrene and derivatives*

The self reaction of substituted phenylethyl radicals (**1**) has been widely investigated.[92-96] The findings of these studies are summarized in Table 5.2. Unless R^2 is very bulky (*e.g. t*-butyl, see below), combination is by far the dominant process with the value k_{td}/k_{tc} typically in the range 0.05-0.16. Thus, a small amount of disproportionation is always observed.

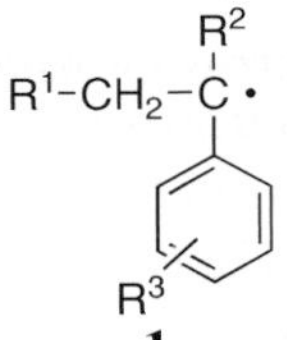

$$R^1-CH_2-\underset{\underset{R^3}{|}}{\overset{\overset{R^2}{|}}{C}}\cdot\text{—Ph}$$

1

The value of k_{td}/k_{tc} shows no significant dependence on chain length for oligostyryl radicals (**4a, b**).[95,96] On the basis of these findings, k_{td}/k_{tc} for PS• should also be small and non-zero.

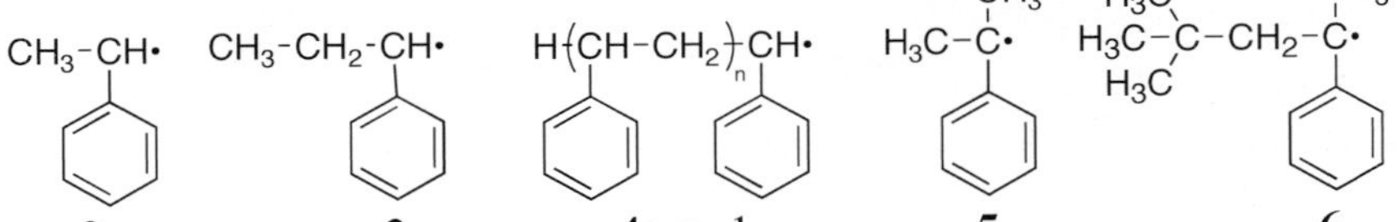

2	3	4a n=1, 4b n=2	5	6
CH₃–CH•–Ph	CH₃–CH₂–CH•–Ph	H–(CH–CH₂)ₙ–CH•–Ph	H₃C–C(CH₃)•–Ph	H₃C–C(CH₃)(CH₃)–CH₂–C(CH₃)•–Ph

For radicals **1**, k_{td}/k_{tc} shows a marked dependence on the bulk of the substituent (R^2). While phenylethyl radicals (**2**) and cumyl radicals (**5**) afford predominantly combination, there are indications of a substantial penultimate unit effect. The radicals **6**, with an α-neopentyl substituent, give predominantly disproportionation. Termination in AMS polymerization might therefore also give substantial disproportionation (However, AMS does not polymerize readily due to a very low ceiling temperature - Section 4.5.1).

Table 5.2 Values of k_{td}/k_{tc} for Polystyryl Radical Model Systems

System	Structure	Temp. (°C)	k_{td}/k_{tc}	System	Structure	Temp. (°C)	k_{td}/k_{tc}
S^{95}	2	20	0.073	S^{96}	4a	141	0.090
S^{95}	2	80	0.081	S^{96}	4a	161	0.078
S^{93}	2	118	0.097	S^{96}	4b	80	0.159
S^{95}	3	20	0.141	S^{96}	4b	90	0.150
S^{95}	3	80	0.146	S^{96}	4b	100	0.134
S^{93}	3	118	0.107	S^{96}	4b	120	0.119
S^{95}	4a	80	0.156	S^{96}	4b	141	0.097
S^{96}	4a	90	0.146	S^{96}	4b	161	0.082
S^{96}	4a	90	0.141	AMS^{97}	5	20-60	0.05
S^{96}	4a	100	0.130	AMS^{98}	5	55	0.1
S^{96}	4a	120	0.109	AMS^{99}	6	55	∞

The value of k_{td}/k_{tc} for oligostyryl radicals (**4**) is reported to decrease with increasing temperature. With 1,3,5-triphenylpentyl radicals (**4b**) k_{td}/k_{tc} halves on increasing the temperature from 80 °C to 160 °C (Table 5.2).[96]

The result indicates that the activation energy for combination is higher than that for disproportionation by ca 10 kJ mol^{-1}. A similar inverse temperature dependence is seen for other small radicals (Section 2.5). However, markedly different behavior is reported for polymeric radicals (Section 5.2.2.2.1).

Benzyl radicals and α- and β- substituted derivatives also undergo unsymmetrical coupling through the aromatic ring (Section 2.5). The formation of the α–*o* and α–*p* coupling products is reversible. Consequently, these materials are often only observed as transient intermediates.

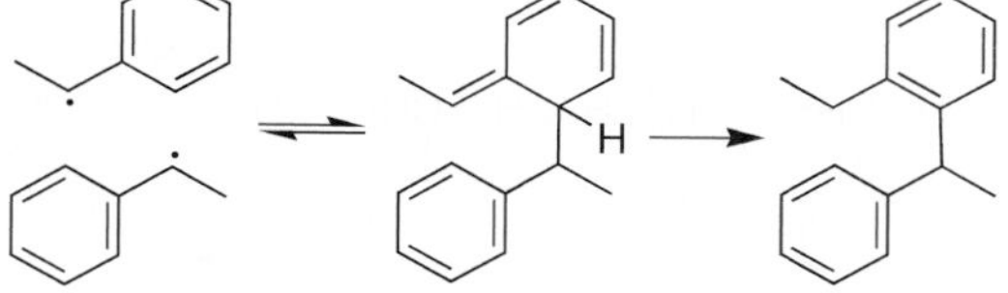

Scheme 5.6

Direct aromatization of the quinonoid intermediates is a photochemically allowed but thermally forbidden rearrangement (Scheme 5.6). When phenylethyl radicals are generated photochemically at 20 °C there is evidence[95] of α–*o* coupling by way of the aromatized product **7**. The products derived from these pathways can be trapped in thermal reactions by radical[98] or acid[100] catalyzed aromatization. With benzyl radicals the ratio of α–o:α–p and [α–o + α–p]:α–α has been shown to increase with increasing temperature.[100] A transient species, presumed to be a quinonoid intermediate, has also been observed when oligomeric radicals **4** are generated thermally.[96]

The formation of the quinonoid species is favored by substitution at the radical center (Section 2.4). Cumyl radicals (**5**)[97,98,101] are reported to give α–α, α–o and α–p coupling products in the ratio 77:8:15. Several studies have examined the reactions of p-substituted phenylethyl radicals. Electron withdrawing substituents favor disproportionation over combination. However, the effect is small.

A report by Businelli et al. suggests a remarkable solvent dependence for the combination:disproportionation ratio.[102] These authors found that 1-phenylpentyl radicals (concentration, temperature unspecified) gave only combination in benzene solvent but combination:disproportionation products in a 1:1 ratio in acetonitrile solvent.

5.2.2.1.2 Poly(alkyl methacrylates)

The self-reactions of 2-carboalkoxy-2-propyl radicals (**8-10**) have been examined.[89,103,104] The results of these studies are reported in Table 5.3. Combination is slightly favored over disproportionation. The value of k_{td}/k_{tc} for **8** was found to be essentially independent of temperature.

```
      CH3              CH3              CH3              CH3   CH3
      |                |                |                |     |
H3C-C•           H3C-C•           H3C-C•           H3C-C-CH2-C•
      |                |                |                |     |
      CO2CH3           CO2C2H5          CO2C4H9          CO2CH3 CO2CH3
      8                9                10                    11
```

Table 5.3 Values of k_{td}/k_{tc} for Methacrylate Ester Model Systems

System	Structure	Temperature (°C)	k_{td}/k_{tc}	ref.
MMA	8	70-90	0.78	89
MMA	8	90	0.62	103
MMA	8	115	0.61	89,103
MMA	8	140	0.60	89,103
MMA	8	165	0.59	103
MMA	11	80	≤1.85	89
EMA	9	80	0.72	89
BMA	10	80	1.17	89
MMA-co-BMA	8, 10	80	1.22	105

Disproportionation increases in the series where the ester is methyl<ethyl<butyl suggesting that this process is favored by increasing the bulk of the ester alkyl group. This trend is also seen for polymeric radicals (Section

5.2.2.2.2). Bizilj et al.[89] reported that disproportionation is more important for oligomeric radicals. While combination products were unequivocally identified, analytical difficulties prevented a precise determination of the disproportionation products. Accordingly, they were only able to state a maximum value of k_{td}/k_{tc}. Their data show that $k_{td}/k_{tc} \leq 1.85$ for the self reaction of **11** and ≤ 1.50 for reaction between **8** and **11**.

An early report[106] indicated that the self reaction of 2-carbomethoxy-2-propyl radicals (**8**), like cyanoisopropyl radicals (**15**) (Section 5.2.2.1.3), affords an unstable coupling product (analogous to a ketenimine). Precedent for a reversible unsymmetrical C-O coupling mode for radicals with a α-carbonyl group has recently been established for the case where normal C-C coupling is sterically very hindered.[107] However, the more recent studies on reactions of 2-carbomethoxy-2-propyl radicals (**8**) and related species provide no evidence for this pathway.[89,103] Bizilj et al.[89] also demonstrated that during disproportionation of oligomeric radicals **12**, the abstraction of a methyl hydrogen (to generate a terminal methylene group - **13**, Scheme 5.7) is preferred $\geq$ 10-fold over abstraction of a methylene hydrogen (to afford an internal double bond **14**). One explanation is that the methyl hydrogens are more sterically accessible than the methylene hydrogens.

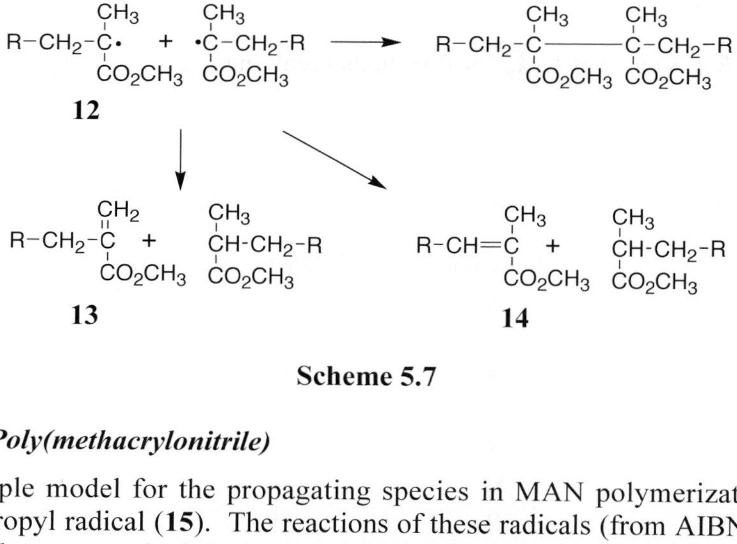

Scheme 5.7

5.2.2.1.3 Poly(methacrylonitrile)

A simple model for the propagating species in MAN polymerization is the cyanoisopropyl radical (**15**). The reactions of these radicals (from AIBN; Scheme 5.8) have been extensively studied. In contrast with the analogous esters **8-10** (Section 5.2.2.1.2), combination is by far the dominant process (Table 5.4).

Serelis and Solomon[108] found that primary radical termination of oligo(MAN) radicals (**16**) with **15** also gives predominantly combination. The ratio k_{td}/k_{tc} was found to have little, if any, dependence on the oligomer chain length (n≤4). As with PMMA•, disproportionation involves preferential abstraction of a methyl

Termination

hydrogen and chains terminated in this way will, therefore, possess a potentially reactive terminal methylene (**17**).

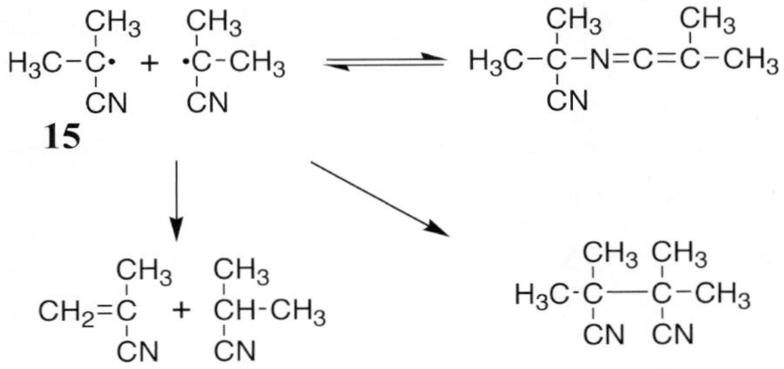

$$\begin{array}{c} CH_3 \quad CH_3 \\ H_3C-\underset{CN}{\overset{|}{C}}\cdot \ + \ \cdot\underset{CN}{\overset{|}{C}}-CH_3 \\ \mathbf{15} \end{array} \rightleftharpoons \begin{array}{c} CH_3 \quad CH_3 \\ H_3C-\underset{CN}{\overset{|}{C}}-N=C=\overset{|}{C}-CH_3 \end{array}$$

$$\begin{array}{c} CH_3 \quad CH_3 \\ CH_2=\underset{CN}{\overset{|}{C}} \ + \ \underset{CN}{\overset{|}{CH}}-CH_3 \end{array} \qquad \begin{array}{c} CH_3 \ CH_3 \\ H_3C-\underset{CN}{\overset{|}{C}}-\underset{CN}{\overset{|}{C}}-CH_3 \end{array}$$

Scheme 5.8

$$H{\left[CH_2-\underset{CN}{\overset{CH_3}{\overset{|}{C}}}\right]}_n CH_2\underset{CN}{\overset{CH_3}{\overset{|}{C}}}\cdot \qquad\qquad H{\left[CH_2-\underset{CN}{\overset{CH_3}{\overset{|}{C}}}\right]}_n CH_2\underset{CN}{\overset{CH_2}{\overset{\|}{C}}}$$

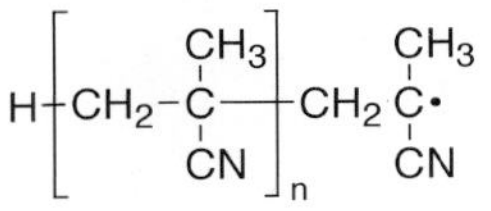

16 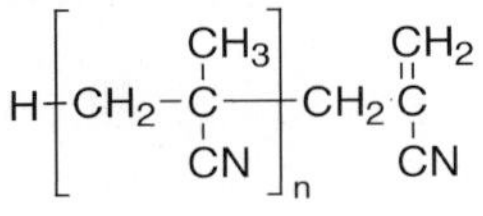   **17**

Table 5.4 Values of k_{td}/k_{tc} for Reactions involving Cyanoisopropyl Radicals

System	Structure	Temperature (°C)	k_{td}/k_{tc}	ref.
MAN	15	80	0.05-0.1	108-110
MAN	16	80	0.1	108
MAN-co-S	4a	90	0.61	111
MAN-co-S	PS•	98	a	112
MAN-co-BMA	PBMA•	25	b	113
MAN-co-E	PE•	80	b	114

a Predominantly combination. b Predominantly disproportionation.

Cyanoisopropyl radicals (**15**) undergo unsymmetrical C-N coupling in preference to C-C coupling.[115] The preferential formation of the ketenimine is a reflection of the importance of polar and steric influences.[116] However, the ketenimine is itself thermally unstable and a source of **15**, thus the predominant isolated product is often from C-C coupling.

Preferential C-N coupling is also observed for oligomeric radicals (Scheme 5.9).[117] A ketenimine (**21**) is the major product from the reaction of the "dimeric" MAN radical **18** with cyanoisopropyl radicals (**15**). Only one of the two possible ketenimines was observed; a result which is attributed to the thermal lability of ketenimine **19**. If this explanation is correct then, although C-N coupling may occur during MAN polymerization, ketenimine structures are unlikely to be found in PMAN by self-reaction of propagating radicals.

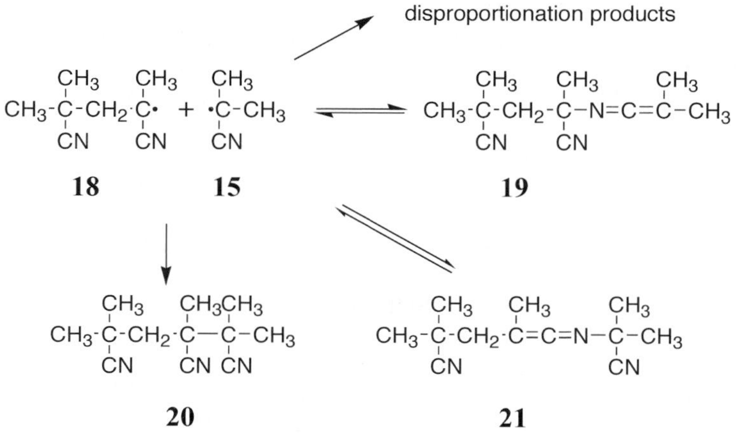

disproportionation products

$$\underset{\textbf{18}}{\underset{\underset{CN}{|}}{CH_3\text{-}\underset{\underset{CH_3}{|}}{C}\text{-}CH_2\text{-}\underset{\underset{CN}{|}}{\overset{\overset{CH_3}{|}}{C}}\bullet}} + \underset{\textbf{15}}{\underset{\underset{CN}{|}}{\bullet\underset{\underset{}{}}{\overset{\overset{CH_3}{|}}{C}}\text{-}CH_3}} \rightleftarrows \underset{\textbf{19}}{\underset{\underset{CN}{|}}{CH_3\text{-}\underset{\underset{}{}}{\overset{\overset{CH_3}{|}}{C}}\text{-}CH_2\text{-}\underset{\underset{CN}{|}}{\overset{\overset{CH_3}{|}}{C}}\text{-}N\text{=}C\text{=}\overset{\overset{CH_3}{|}}{C}\text{-}CH_3}}$$

$$\underset{\textbf{20}}{\underset{\underset{CN}{|}}{CH_3\text{-}\underset{\underset{}{}}{\overset{\overset{CH_3}{|}}{C}}\text{-}CH_2\text{-}\underset{\underset{CN}{|}}{\overset{\overset{CH_3CH_3}{|\ \ |}}{C\text{—}C}}\text{-}CH_3}} \quad\quad \underset{\textbf{21}}{\underset{\underset{CN}{|}}{CH_3\text{-}\underset{\underset{}{}}{\overset{\overset{CH_3}{|}}{C}}\text{-}CH_2\text{-}\underset{}{C}\text{=}C\text{=}N\text{-}\underset{\underset{CN}{|}}{\overset{\overset{CH_3}{|}}{C}}\text{-}CH_3}}$$

Scheme 5.9

5.2.2.1.4 Polyethylene

The self reaction of primary alkyl radicals gives mainly combination.[118] For primary alkyl radicals [$CH_3(CH_2)_nCH_2\bullet$], k_{td}/k_{tc} is reported to lie in the range 0.12-0.14, apparently independent of chain length (n=0-3).[118,119]

5.2.2.2 Polymerization

A substantial number of studies give information on k_{td}/k_{tc} for polymerizations of S (5.2.2.2.1) and MMA (5.2.2.2.2). There has been less work on other systems. One of the main problems in assessing k_{td}/k_{tc} lies with assessing the importance of other termination mechanisms (*i.e.* transfer to initiator, solvent, *etc.*, primary radical termination).

Techniques applied in assessing the relative importance of disproportionation and combination include:

(a) The Gelation technique. This method was developed by Bamford *et al.*[120] In graft copolymerization, termination by combination will give rise to a crosslink while disproportionation (and most other termination reactions) will lead to graft formation. The initiation system based on a polymeric halo-compound [poly(vinyl trichloroacetate)/$Mn_2(CO)_{10}$/hv] was used to initiate polymerization and the time for gelation was used to calculate k_{td}/k_{tc}. In the original work, the results were calibrated with reference to data for S polymerization for which a k_{td}/k_{tc} of 0.0 was assumed. Recent studies suggest that, in S polymerization, disproportionation may account for 10-20% of chains (Section 5.2.2.2.1). Thus the data may require minor adjustment. Systems studied with this technique include AN, MAN, MA, MMA, and S.

(b) Molecular weight measurement. The mode of termination can be calculated by comparing the kinetic chain length (the ratio of the rate of propagation to the rate of initiation or termination) with the measured number average molecular weight.[121-123]

(c) Molecular weight distribution evaluation. This method relies on a precise evaluation of the molecular weight distribution.[124-127] The mode of termination has a significant influence on the shape of the molecular weight distribution with the instantaneous dispersity (D being ~2.0 if termination occurs exclusively by disproportionation of propagating radicals and ~1.5 if termination involves only combination (Section 5.2.1.2).[128] Values of D are conversion dependent so the method should only be applied to very low conversion samples. Truncation of the ends of the distribution as a result of baseline selection difficulties will lead to the dispersity being underestimated.[129] A more precise but related method is to fit the entire molecular weight distribution using kinetic modeling methods.

(d) End group determination. Polymer chains terminated by combination possess two initiator-derived chain ends. Disproportionation affords chains with only one such end. The value of k_{td}/k_{tc} can therefore be determined by evaluating the initiator-derived polymer end groups/molecule by applying eq. 27

$$k_{td}/k_{tc} = (2-x)/2(x-1) \qquad (27)$$

where x is the number of initiator fragments per molecule. The errors inherent in this technique can be large since the polymer end groups typically comprise only a very small fraction of a polymer sample. The initiator-derived ends may be labeled for ease of detection. These techniques are described in Section 3.6. It is necessary to allow for side reactions. If there is transfer to monomer, solvent, *etc.*, the value of k_{td}/k_{tc} will be overestimated. The occurrence of transfer to initiator, primary radical termination, or copolymerization of initiator byproducts will lead to k_{td}/k_{tc} being underestimated.

(e) Mass spectrometry. Matrix-assisted laser desorption ionization time-of-flight mass spectroscopy (MALDI-TOF) has been used to determine k_{td}/k_{tc} in S and MMA polymerization.[130] Chains formed by disproportionation and chains formed by combination form two distinct distributions. Mass spectrometric end group determination is described in Section 3.5.3.4.

Evaluation of molecular weights after ultrasonic scission of high molecular weight polymers (PMMA and PS) in the presence of a radical trap has been claimed to provide evidence of the termination mechanism.[131] However, scission gives radicals as shown in Scheme 5.10.

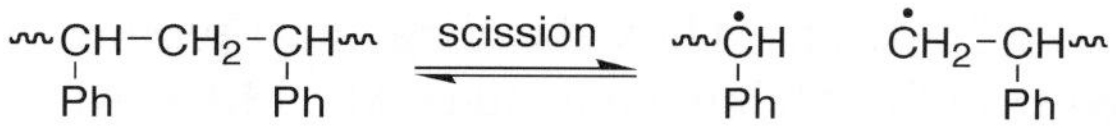

$$\text{\sim CH-CH}_2\text{-CH}\sim \;\; \underset{}{\overset{\text{scission}}{\rightleftarrows}} \;\; \text{\sim }\overset{\bullet}{\text{C}}\text{H} \;\; \overset{\bullet}{\text{C}}\text{H}_2\text{-CH}\sim$$
$$\;\;\;\;|\;\;\;\;\;\;\;\;\;\;\;|\;|\;\;\;\;\;\;\;\;\;|$$
$$\;\;\text{Ph}\;\;\;\;\;\;\;\;\text{Ph}\;\text{Ph}\;\;\;\;\;\;\text{Ph}$$

Scheme 5.10

5.2.2.2.1 Polystyrene

Hensley et al.[132] reported the only direct experimental observation of head-to-head linkages in PS by 2D INADEQUATE NMR on ^{13}C-enriched PS. The method did not enable these groups to be quantified with sufficient precision for evaluation of k_{td}/k_{tc}. Zammit et al.[130] studied chain distribution of low molecular weight PS prepared with AIBN initiator by MALDI-TOF. Separate distributions of chains formed by combination and disproportionation were observed. They estimated k_{td}/k_{tc} at 90 °C to be 0.057.

A wide range of less direct methods has been applied to determine k_{td}/k_{tc} in S polymerization. Most indicate predominant combination.[122,125,133-148] However, distinction between a k_{td}/k_{tc} of 0.0 and one which is non-zero but ≤0.2 is difficult even with the precision achievable with the most modern instrumentation. Therefore, it is not surprising that many have interpreted the experimental finding of predominantly combination as meaning exclusively combination.

Olaj et al.[124] proposed that termination of S polymerization involves substantial disproportionation. They analyzed the molecular weight distribution of PS samples prepared with either BPO or AIBN as initiator at temperatures in the range 20-90 °C and estimated k_{td}/k_{tc} to be ca 0.2. In a more recent study, Olaj et al.[149] determined the molecular weight distribution of PS samples prepared with photoinitiation at 60 and 85 °C and estimated values of k_{td}/k_{tc} of 0.5 and 0.67 respectively. Dawkins and Yeadon[125] discussed the problems associated with estimating k_{td}/k_{tc} on the basis of dispersity measurements and determined that k_{td}/k_{tc} should be "substantially smaller" than suggested by Olaj et al.[149]

Berger and Meyerhoff[150] also reported that termination involves substantial disproportionation. They determined the initiator fragments per molecule in PS prepared with radiolabeled AIBN and conducted a detailed kinetic analysis of the system. They also found a marked temperature dependence for k_{td}/k_{tc}. Values of k_{td}/k_{tc} ranged from 0.168 at 30 °C to 0.663 at 80 °C.

Other determinations of k_{td}/k_{tc} based on end group determination are at variance with these findings. End group analyses by NMR,[146,147] radiotracer techniques,[142-144] or chemical analysis[145] on PS formed with appropriately labeled initiators all indicate predominantly combination. Moad et al.[146,147] used ^{13}C NMR to define and quantify the end groups in samples of PS prepared at 60 °C with either ^{13}C-labeled BPO or AIBN as initiator. This method has the advantage that the end groups from primary radical termination, transfer to initiator, residual initiator and any copolymerized initiator byproducts can be distinguished from the end groups formed by initiation (Section 3.5.3.2). They showed that, under the conditions employed (60 °C, bulk), there are 1.7±0.2 initiator-derived end groups corresponding to a k_{td}/k_{tc} of ca 0.2.   Other NMR end group determinations have yielded similar data.   Barson et al.[151] analyzed PS prepared with ^{13}C-labeled AIBN by ^{13}C NMR.   Bevington et al.[152] analyzed PS prepared with fluorinated BPO by ^{19}F NMR. In each case there were ca 1.6 initiator-derived end groups per molecule (k_{td}/k_{tc} ca 0.3). Yoshikawa et al.[153] formed PS• from narrow dispersity ($\bar{M}_n$=1500, M_w/M_n=1.09) low molecular weight ω-bromopolystyrene by atom transfer to Cu(I) at 110 °C. They used NMR to estimate the fraction of chains formed by disproportionation as 0.07 (k_{td}/k_{tc} ca 0.08) and by GPC peak resolution to be 0.09 (k_{td}/k_{tc} ca 0.1).

The influence of substituents (p-Cl, p-OMe) on k_{td}/k_{tc} was investigated by Ayrey et al.[148] They found disproportionation was favored by the p-OMe substituent and that the extent of disproportionation increased with increasing temperature. This result is contrary to the model studies (Section 5.2.2.1.1) that show k_{td}/k_{tc} has little dependence on substituents and, indeed, suggest the opposite trend.

5.2.2.2.2 Poly(alkyl methacrylates)

Table 5.5 Determinations of k_{td}/k_{tc} for MMA Polymerization

Temperature (°C)	E[133,142]	E[134]	E[155]	G[156]	E[157]	P[158]	E[159]	M[121]	E[85,160]	S[130]	E[151]
-25	-	-	-	-	-	0.14	-	-	-	-	-
0	1.50	-	-	-	-	0.50	-	-	-	-	-
15	-	-	-	-	-	0.76	-	-	-	-	-
25	2.13	-	-	-	2.0	-	-	-	-	-	-
30	-	-	-	-	-	1.18	-	-	-	-	-
40	-	-	0.45	-	-	-	-	-	-	-	-
45	-	-	-	-	-	∞	-	-	-	-	-
60	5.67	1.35	0.75	2.7	2.62	-	2.57	0.44	1.28	-	4.5
80	-	-	1.32	4.0	-	-	-	-	-	-	-
90	-	-	-	-	-	-	-	-	-	4.37	-
100	-	-	-	-	-	-	-	1.5	-	-	-

a Methods used (Section 5.2.2.2): G-gelation technique, M-molecular weight measurement, P-dispersity evaluation, E-end group determination, S-MALDI-TOF mass spectrometry

The nature of the termination reaction in MMA polymerization has been investigated by a number of groups using a wide range of techniques (Table 5.5). There is general agreement that there is substantial disproportionation. However, there is considerable discrepancy in the precise values of k_{td}/k_{tc}. In some cases the difference has been attributed to variations in the way molecular weight data are interpreted or to the failure to allow for other modes of termination under the polymerization conditions (chain transfer, primary radical termination).[154] In other cases the reasons for the discrepancies are less clear.   MALDI-TOF mass spectrometry provides a direct measurement of k_{td}/k_{tc} for low molecular weight MMA and this indicates a value of 4.37 at 90 °C.[130]

Four studies suggest that k_{td}/k_{tc} has a significant temperature dependence (Table 5.5). Although not agreeing on the precise value of k_{td}/k_{tc}, all four studies indicate that the proportion of disproportionation increases with increasing temperature. These results are at variance with model studies that suggest that k_{td}/k_{tc} is independent of temperature. It was also proposed that the preferred termination mechanism is solvent dependent and that disproportionation is favored in more polar media.[161]

Hatada et al.[160,162] showed that the disproportionation-derived unsaturated ends in PMMA can be determined directly by ^{1}H NMR. For PMMA prepared with BPO in toluene at 100 °C they found the number of chain ends per molecule formed from initiation reactions (from BPO and toluene-derived radicals) to be ca 1.25[160] suggesting a k_{td}/k_{tc} of ca 1.5. They also demonstrated the preference for transfer of a methyl vs a methylene hydrogen in disproportionation. This is in line with the studies on model radicals (Section 5.2.2.1.2).

Values of k_{td}/k_{tc} for polymerizations of EMA and BMA and higher methacrylate esters have been determined.[113,120,157,159] The extent of disproportionation increases with the size of the ester alkyl group.

5.2.2.2.3 Poly(methacrylonitrile)

Bamford et al.[120] examined MAN polymerization (25 °C, DMSO) using the gelation technique (Section 5.2.2.2) and have estimated that termination occurs predominantly by disproportionation (k_{td}/k_{tc} = 1.86). This result is at variance with the model studies (Section 5.2.2.1.3).

5.2.2.2.4 Poly(alkyl acrylates)

The termination mechanism in MA polymerization has been variously determined to be predominantly disproportionation[137,157] or predominantly combination.[120,159,163]

Ayrey et al.[163] suggested that transfer reactions may have led to erroneous conclusions being drawn in some of the earlier studies. They concluded that termination is almost exclusively by combination (25 °C, benzene). Bamford et al.[120] came to a similar conclusion using the gelation technique (25 °C, bulk) and determined that the polymerizations of higher acrylate esters also terminate predominantly by combination.

5.2.2.2.5 Poly(acrylonitrile)

There appears to be general agreement that termination in AN polymerization under a variety of conditions (10-90 °C, DMSO, DMF, H_2O) involves mainly combination.[120,123,164,165] It was suggested that this may involve either C-N (ketenimine formation) or C-C coupling.[166]

5.2.2.2.6 Poly(vinyl acetate)

Early reports[137,157,167] suggested that termination during VAc polymerization involved predominantly disproportionation. However, these investigations did not adequately allow for the occurrence of transfer to monomer and/or polymer, which are extremely important during VAc polymerization (Sections 6.2.6.2 and 6.2.7.4 respectively). These problems were addressed by Bamford *et al.*[120] who used the gelation technique (Section 5.2.2.2) to show that the predominant radical-radical termination mechanism is combination (25 °C).

5.2.2.2.7 Poly(vinyl chloride)

Studies on VC polymerization are also complicated by the fact that only a small proportion of termination events may involve radical-radical reactions. Most termination is by transfer to monomer (Sections 4.3.1.2 and 6.2.6.3). Early studies on the termination mechanism which do not allow for this probably overestimate the importance of disproportionation.[168,169]

Park and Smith[170] attempted to allow for chain transfer in their examination of the termination mechanism during VC polymerization at 30 and 40 °C in chlorobenzene. They determined the initiator-derived ends in PVC prepared with radiolabeled AIBN and concluded that $k_{td}/k_{tc} = 3.0$. However, questions have been raised regarding the reliability of these measurements.[171,172] Atkinson *et al.*[172] applied the gelation technique (Section 5.2.2.2) to VC polymerization and proposed that termination involves predominantly combination.

5.2.2.3 Summary

Unequivocal numbers for k_{td}/k_{tc} are not yet available for most polymerizations and there is only qualitative agreement between values obtained in model studies and real polymerizations.

It is tempting to attribute problems in reconciling data from model studies and actual polymerizations to difficulties associated with data interpretation. The polymerization experiments are often complicated by other termination pathways, in particular chain transfer, which must be allowed for when assessing the results. It is notable in this context that the discrepancies are most evident for reactions carried out at higher temperatures (Sections 5.2.2.1.1 and 5.2.2.1.2).

However, some of the differences may be explicable in terms of an effect of molecular size. For many of the model systems at least one of the reaction partners is monomeric (*i.e.* **2, 5, 8-10, 15**). Since combination is known to be more sensitive to steric factors than disproportionation (Section 2.4.3.2), k_{td}/k_{tc} may be anticipated to be higher for the corresponding propagating species. The values of k_{td}/k_{tc} reported for **3** or **4** are significantly greater than those for **2**. Similarly, **6** gives much more disproportionation than **5**. Thus, values of k_{td}/k_{tc} seen for systems involving monomeric model radicals (**2, 5, 8-10**, or **15**) should be considered only as a lower limit for the polymeric system.

Despite these problems in assessing k_{td}/k_{tc}, it is possible to make some generalizations:

(a) Termination of polymerizations involving vinyl monomers (CH_2=CHX) involves predominantly combination.

(b) Termination of polymerizations involving α-methylvinyl monomers (CH_2=C(CH_3)X) always involves a measurable proportion of disproportionation.

(c) During disproportionation of radicals bearing an α-methyl substituent (for example, those derived from MMA), there is a strong preference for transfer of a hydrogen from the α-methyl group rather than the methylene group.

(d) Within a series of vinyl or α-methylvinyl monomers, k_{td}/k_{tc} appears to decrease as the ability of the substituent to stabilize a radical center increases. Thus, k_{td}/k_{tc} for radicals ~C(•)(CH_3)X or ~C(•)HX decreases in the series where X is CO_2R>>CN>Ph.

5.3 Inhibition and Retardation

Inhibitors and retarders are used to stabilize monomers during storage or during processing (*e.g.* synthesis, distillation). They are often used to quench polymerization when a desired conversion has been achieved. They may also be used to regulate or control the kinetics of a polymerization process.

Inhibitors have been defined as species which, when added to a polymerization, react to consume and deactivate the initiator-derived radicals.[173] Retarders have been similarly defined as species which deactivate the propagating radicals.[173] According to this definition, a nitroxide added to a *t*-butoxy radical-initiated polymerization of S should be called a retarder since the *t*-butoxy radicals appear not to react with the nitroxide. However, the initiator-derived and propagating radicals often show similar selectivity in their reactions and the distinction between inhibitors and retarders becomes blurred. In a cyanoisopropyl radical-initiated polymerization of S, an added nitroxide would be called an inhibitor when used in high concentration and a retarder when used at very low concentration. Generally the term inhibitor is used without reference to which radicals are scavenged. With many experimental techniques it is not possible to discriminate between scavenging of initiator-derived and oligomeric propagating radicals. Thus an inhibitor has come to mean any species that is able to rapidly and efficiently scavenge propagating and/or initiator-derived radicals and thus prevent polymer chain formation. The term retarder is commonly used to define species that slows rather than prevents polymerization.

Inhibitors or retarders that give inert products are called 'ideal'.[173] The term 'ideal inhibitor' has also been used to describe a species that stops all polymerization until such time as it is completely consumed (*i.e.* the induction period) and then allows polymerization to proceed at the normal rate. However, in many cases the products formed during inhibition or retardation are not inert. Four main pathways for further reaction following the initial reaction with inhibitor or retarder are distinguished:

(a) Slow reinitiation with reference to propagation following chain transfer (see, for example, Section 5.3.4).
(b) Slow propagation with reference to normal propagation following addition (see, for example, Section 5.3.3).
(c) Further reaction of the initially formed species as an inhibitor or retarder (see, for example, Sections 5.3.4, 5.3.5, 5.3.7).
(d) Reversal of the reaction associated with inhibition or retardation (see, for example, Section 5.3.1 and Chapter 9).

The kinetics and mechanism of retardation and inhibition has been reviewed by Bamford,[173] Tüdos and Földes-Berezsnich,[174] Eastmond,[175] Goldfinger et al.[176] and Bovey and Kolthoff.[177]

Common inhibitors include stable radicals (Section 5.3.1), oxygen (5.3.2), certain monomers (5.3.3), phenols (5.3.4), quinones (5.3.5), phenothiazine (5.3.6), nitro and nitroso-compounds (5.3.7) and certain transition metal salts (5.3.8). Some inhibition constants (k_z/k_p) are provided in Table 5.6. Absolute rate constants (k_z) for the reactions of these species with simple carbon-centered radicals are summarized in Table 5.7.

Table 5.6 Inhibition constants (k_z/k_p, 60 °C, bulk) for Various Inhibitors with Some Common Monomers[a]

Inhibitor	k_z/k_p				
	MMA	MA	AN	S	VAc
$CuCl_2$	1030	-	100[c]	10000	-
$FeCl_3$	5000 k_p[c]	6800 k_p[c]	3.33[c]	536	2300000 k_p
p-benzoquinone	4.5	<0.15 k_p[b]	0.91[b]	520	-
nitrobenzene	0.00464[b]	0.00464[b]	-	0.326	11.2[b]
DPPH	2000	-	-	-	-
oxygen	33000	-	-	14600	-
anthracene	-	0.098[b]	2.67[b]	2[d]	27.8
p-hydroquinone	-	-	-	-	0.7
phenol	-	0.0002[b]	-	-	0.06
styrene	-	-	-	-	40.8[b,174]

a Data taken from Eastmond[13] unless otherwise stated and are rounded to three significant figures. b 50 °C. c in DMF. d 44.4 °C.

Whether a given species functions as an inhibitor, a retarder, a transfer agent or a comonomer in polymerization is dependent on the monomer(s) and the reaction conditions. For example, oxygen acts as an inhibitor in many polymerizations yet it readily copolymerizes with S. Reactivity ratios for VAc-S copolymerization are such that small amounts of S are an effective inhibitor of VAc polymerization (r_S=0.02, r_{VAc}=22.3). The propagating chain with a terminal VAc adds to S preferentially even when VAc is present in large excess over S. The resultant propagating radical with a terminal S adds to VAc only slowly. The reactions of many inhibitors with propagating radicals may become reversible under some reaction conditions. In these circumstances, the reagent may find use as a control agent in living radical polymerization (Chapter 9).

Table 5.7 Absolute Rate Constants (k_z) for the Reaction of Carbon-Centered Radicals with Some Common Inhibitors

Inhibitor	Radical	Temp. (°C)	k_z (M^{-1}s^{-1})	refs.
TEMPO (**23**)	prim. alkyl	60	~1 x 10^9	178-180
oxygen	benzyl	27	2.9 x 10^9	181
p-benzoquinone (**38**)	prim. alkyl	69	2.0 x 10^7	182
CuCl$_2$	prim. alkyl	25	6.5 x 10^5	182

The effectiveness of inhibitors is measured in terms of the rate constant ratio k_z/k_p and the stoichiometric coefficient. The stoichiometric coefficient is the moles of radicals consumed per mole of inhibitor. These parameters may be determined by various methods. A brief description of the classical kinetic treatment for evaluating k_z/k_p follows. Consider the reaction scheme shown which describes ideal inhibition and retardation (Scheme 5.11).

initiation
$$I_2 \rightarrow 2\, I\bullet \qquad R_i = 2\, k_d f[I_2]$$
$$I\bullet + M \rightarrow P_1\bullet \qquad k_i \geq k_p$$

inhibition
$$I\bullet + Z \rightarrow IZ \text{ (dead)} \qquad R_z = k_z[Z][I\bullet]$$

propagation
$$P_n\bullet + M \rightarrow P_{n+1}\bullet \qquad R_p = k_p[M][P\bullet]$$

disproportionation
$$P_n\bullet + P_m\bullet \rightarrow P_n^H + P_m^= \qquad R_{tc} = 2k_{tc}[P\bullet]^2 \qquad R_t = R_{tc} + R_{td}$$

combination
$$P_n\bullet + P_m\bullet \rightarrow P_{n+m} \qquad R_{tc} = 2k_{tc}[P\bullet]^2$$

retardation
$$P_n\bullet + Z \rightarrow P_nZ \text{ (dead polymer)} \qquad R_z = k_z[Z][P\bullet]$$

Scheme 5.11

Termination

With the omission of the reinitiation reaction, this scheme is the same as that for polymerization with chain transfer and an expression (eq 28) for the degree of polymerization similar in form to the Mayo equation can be derived.

$$\frac{1}{\overline{X}_n} = \frac{\left(1+\frac{k_{td}}{k_t}\right)\left(2k_d f[I_2]k_t\right)^{0.5}}{k_p[M]} + \frac{k_z[Z]}{k_p[M]} \qquad (28)$$

If the amount of termination by radical-radical reaction is neglected the degree of polymerization and the kinetic chain length are given by eq. 29:

$$v \approx \overline{X}_n \approx \frac{k_p[M]}{k_z[Z]} \qquad (29)$$

If chains are very short we must include an additional term in the numerator for monomer consumption in the initiation step (eq. 30):

$$\overline{X}_n = \frac{k_p[M]}{k_z[Z]} + 1 \qquad (30)$$

Data on the rate of consumption of the inhibitor as a function of conversion may also be used to obtain k_z/k_p (eq. 31):

$$\frac{k_z}{k_p} = \frac{[M]}{[Z]}\frac{d[Z]}{d[M]} = \frac{d\log[M]}{d\log[Z]} \qquad (31)$$

It is clear that many procedures used to evaluate chain transfer constants can also be used to evaluate the kinetics of inhibition. The following sections will show that the mechanism for inhibition is often more complex than suggested by Scheme 5.11.

5.3.1 'Stable' Radicals

The kinetics and mechanism of inhibition by stable radicals has been reviewed by Rozantsev *et al.*[183] Ideally, for radicals to be useful inhibitors in radical polymerization they should have the following characteristics:

(a) They should not add to, abstract from, or otherwise react with the monomer, solvent, *etc*.
(b) They should not undergo self reaction or unimolecular decomposition.
(c) They must react rapidly with the propagating and/or the initiator-derived radicals to terminate polymer chains.

268 **The Chemistry of Radical Polymerization**

Examples of radicals which are reported to meet these criteria are diphenylpicrylhydrazyl [DPPH, (**22**)], Koelsch radical (**26**), nitroxides [*e.g.* TEMPO (**23**), Fremy's Salt (**24**)], triphenylmethyl (**25**), galvinoxyl (**27**), and verdazyl radicals [*e.g.* triphenylverdazyl (**28**)]. These reagents have seen practical application in a number of contexts. They have been widely utilized in the determination of initiator efficiency (Section 3.3.1.1.3) and in mechanistic investigations (Section 3.5.2).

Stable radicals can show selectivity for particular radicals. For example, nitroxides do not trap oxygen-centered radicals yet react with carbon-centered radicals by coupling at or near diffusion controlled rates.[179,184] This capability was utilized by Rizzardo and Solomon[185] to develop a technique for characterizing radical reactions and has been extensively used in the examination of initiation of radical polymerization (Section 3.5.2.4). In contrast DPPH, while an efficient inhibitor, shows little selectivity and its reaction with radicals is complex.[186]

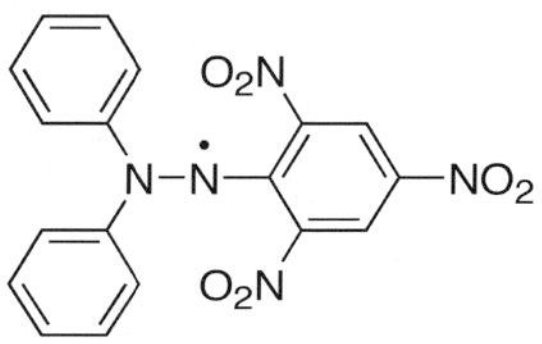

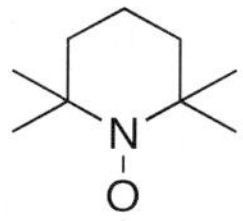

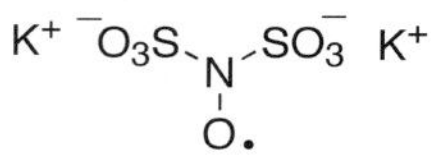

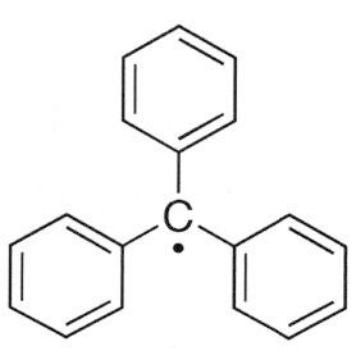

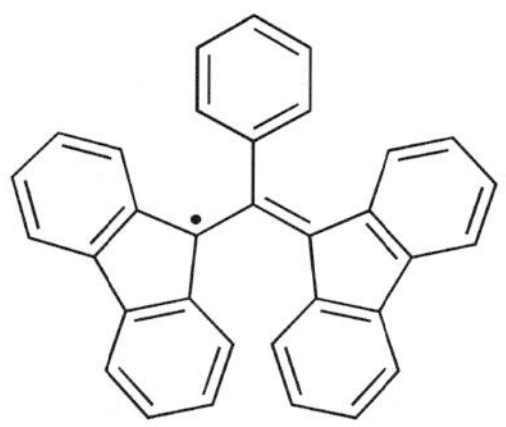

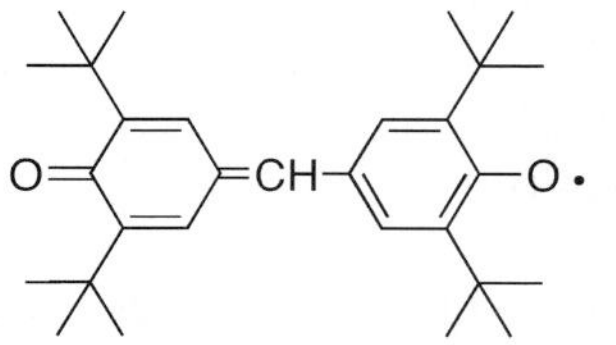

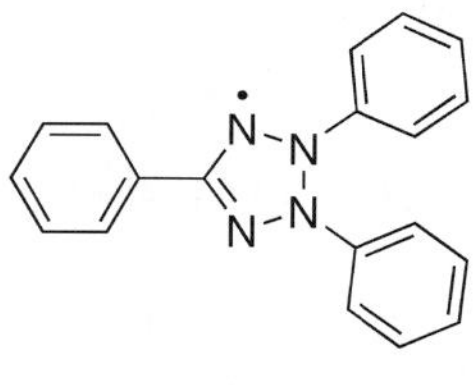

The efficiency of these inhibitors may depend on reaction conditions. For example: the reaction of radicals with stable radicals (*e.g.* nitroxides) may be reversible at elevated temperatures (Section 7.5.3); triphenylmethyl may initiate polymerizations (Section 7.5.2). A further complication is that the products may be capable of undergoing further radical chemistry. In the case of DPPH (**22**) this is attributed to the fact that the product is an aromatic nitro-compound (Section 5.3.7). Certain adducts may undergo induced decomposition to form a stable radical which can then scavenge further.

5.3.2 Oxygen

The role of oxygen in radical and other polymerizations has been reviewed by Bhanu and Kishore.[187] Rate constants for the reaction of carbon-centered radicals with oxygen are extremely fast, generally $\geq 10^9$ M^{-1} s^{-1}.[181,188] The initially formed species are peroxy radicals **29**. These may abstract hydrogen or add monomer (Scheme 5.12).

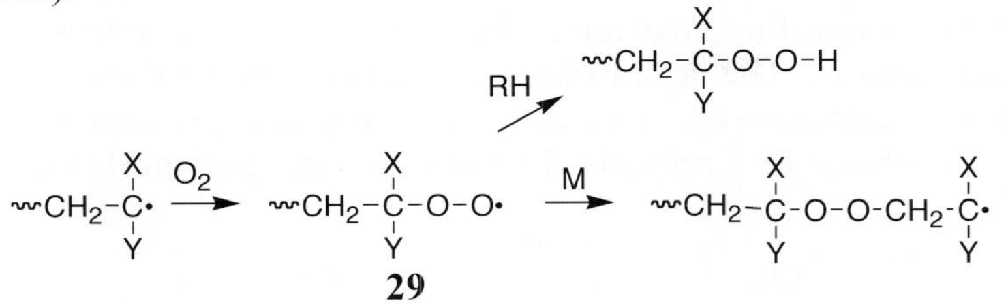

$$\text{\textasciitilde}CH_2-\underset{Y}{\overset{X}{C}}\cdot \xrightarrow{O_2} \text{\textasciitilde}CH_2-\underset{Y}{\overset{X}{C}}-O-O\cdot \xrightarrow{\overset{RH}{}} \text{\textasciitilde}CH_2-\underset{Y}{\overset{X}{C}}-O-O-H$$

$$\text{\textasciitilde}CH_2-\underset{Y}{\overset{X}{C}}-O-O\cdot \xrightarrow{M} \text{\textasciitilde}CH_2-\underset{Y}{\overset{X}{C}}-O-O-CH_2-\underset{Y}{\overset{X}{C}}\cdot$$

29

Scheme 5.12

Thus, while polymerization may proceed in the presence of oxygen, it is an efficient scavenger of both initiating and propagating species in radical polymerization and usually steps must be taken to exclude oxygen or to minimize its effects. Typically, this involves conducting the experiment under vacuum or an inert atmosphere (*e.g.* nitrogen) or in a refluxing solvent. Oxygen may act as an inhibitor or retarder of polymerization, copolymerize (*e.g.* S polymerization), and/or facilitate chain transfer (*e.g.* VAc polymerization) or inhibition with other species (*e.g.* phenols – Section 5.3.4).

The effect observed is dependent on the reactivity of the monomer and other agents present in the polymerization medium towards hydroperoxy radicals **29**. If addition of **29** to monomer is slow, in relation to normal propagation, then retardation or inhibition will be observed. It should also be noted that, polymeric peroxides, one of the products of reaction with oxygen are potentially sources of additional radicals. These may complicate polymerization and can impair the properties of the final polymer (Section 8.2).

5.3.3 Monomers

Certain monomers may act as inhibitors in some circumstances. Reactivity ratios for VAc-S copolymerization (r_S=0.02, r_{VAc}=22.3) and rates of cross propagation are such that small amounts of S are an effective inhibitor of VAc polymerization. The propagating chain with a terminal VAc is very active towards S and adds even when S is present in small amounts. The propagating radical with S adds to VAc only slowly. Other vinyl aromatics also inhibit VAc polymerization.[174]

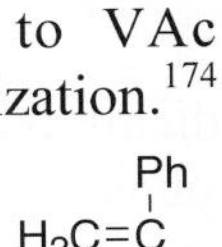

$$\underset{\underset{Ph}{|}}{\overset{\overset{Ph}{|}}{H_2C=C}} \qquad \underset{\underset{CO_2R'}{|}}{\overset{\overset{SR}{|}}{H_2C=C}} \qquad \underset{\underset{CO_2R'}{|}}{\overset{\overset{OR}{|}}{H_2C=C}}$$

30 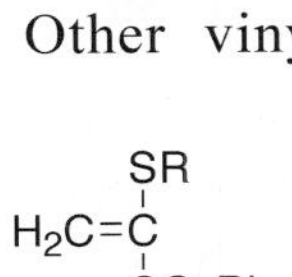   **31** 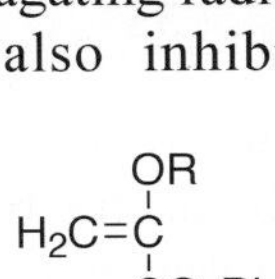   **32**

1,1-diphenylethylene (**30**) acts as a reversible inhibitor in polymerizations of S and MMA (Section 9.3.6).[189] Olefins with captodative substitution such as **31** rapidly scavenge radicals to give new radicals **33** which are unable or slow to reinitiate polymerization (Scheme 5.13).[190,191] Termination is believed to occur exclusively by combination, thus telechelic polymers are available by appropriate choice of the initiator. The head to head coupling product **34** is stable at normal polymerization temperatures. However, at higher temperatures **34** undergoes reversible homolysis and radicals **33** may initiate polymerization (Section 9.3.5).[191,192]

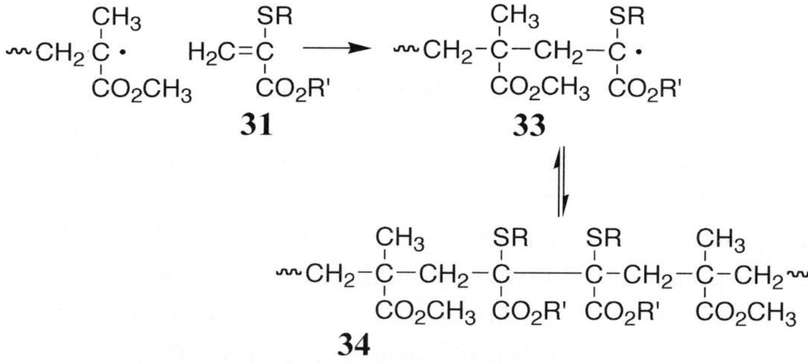

Scheme 5.13

The chemistry is dependent on the particular substituents. Oxygen analogs of **31**, α-alkoxyacylates (**32**), do not inhibit polymerization but readily polymerize and copolymerize with reactivity ratios similar to methacrylate esters.[191-194]

5.3.4 Phenols

Phenolic inhibitors such as hydroquinone (**35**), monomethylhydroquinone (*p*-methoxyphenol) (**36**) and 3,5-di-*t*-butylcatechol (**37**) are added to many commercial monomers to prevent polymerization during transport and storage.

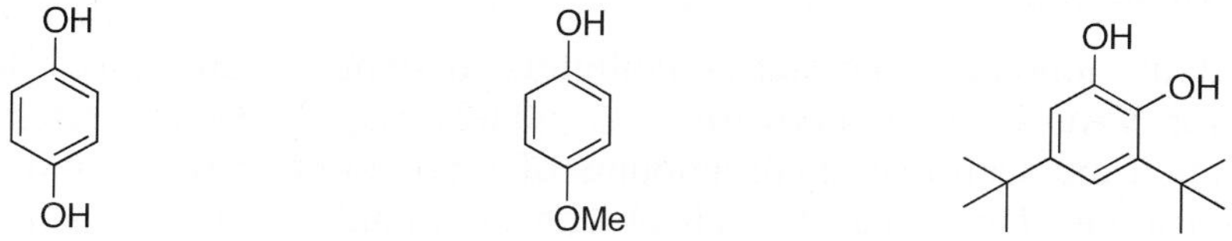

Studies with simple radicals show that carbon-centered radicals react with phenols by abstracting a phenolic hydrogen (Scheme 5.14). The phenoxy radicals may then scavenge a further radical by C-C or C-O coupling or (in the case of hydroquinones) by loss of a hydrogen atom to give a quinone. The quinone may then react further (Section 5.4.4). Thus two or more propagating chains may be terminated for every mole of phenol.[195]

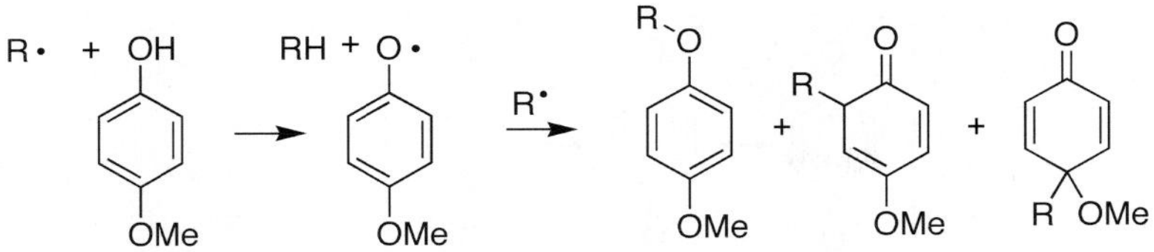

Scheme 5.14

However, by themselves, phenols are poor polymerization inhibitors[196-198] (see also Table 5.6) and are reported to act as accelerants in the ATRP of MMA.[199] They (e.g. hydroquinone) are more effective inhibitors in the presence of oxygen.[196-198,200] The mechanism for inhibition is shown in Scheme 5.15. The reaction of carbon centered radicals (including initiating and propagating radicals) with oxygen is very fast in relation to propagation. Phenols are excellent scavengers of hydroperoxy radicals.

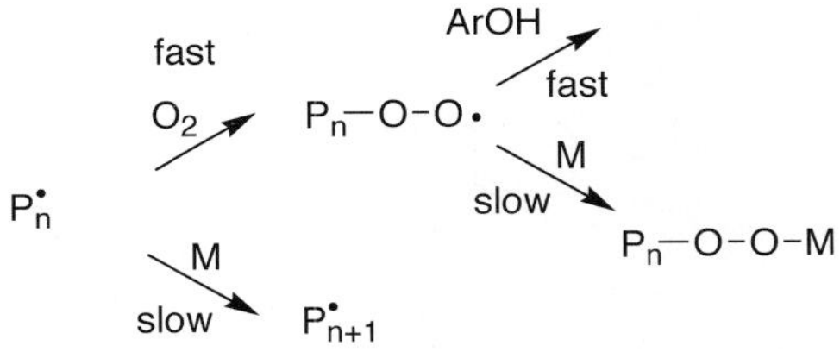

Scheme 5.15

5.3.5 Quinones

Quinones may react with carbon-centered radicals by addition at oxygen or carbon, or by electron transfer (Scheme 5.16).[174,182,195,201,202] The preferred reaction pathway depends both on the attacking radical and the particular quinone (halogenated quinones react preferentially by electron transfer). The radical formed may then scavenge another radical. There is also evidence that certain quinones [e.g. chloranil, benzoquinone (**38**)] may copolymerize under some conditions.[203]

The absolute rate constants for attack of carbon-centered radicals on *p*-benzoquinone (**38**) and other quinones have been determined to be in the range 10^7-10^8 M^{-1} s^{-1}.[182,204] This rate shows a strong dependence on the electrophilicity of the attacking radical and there is some correlation between the efficiency of various quinones as inhibitors of polymerization and the redox potential of the quinone. The complexity of the mechanism means that the stoichiometry of inhibition by these compounds is often not straightforward. Measurements of moles of inhibitor consumed for each chain terminated for common inhibitors of this class give values in the range 0.05-2.0.[176]

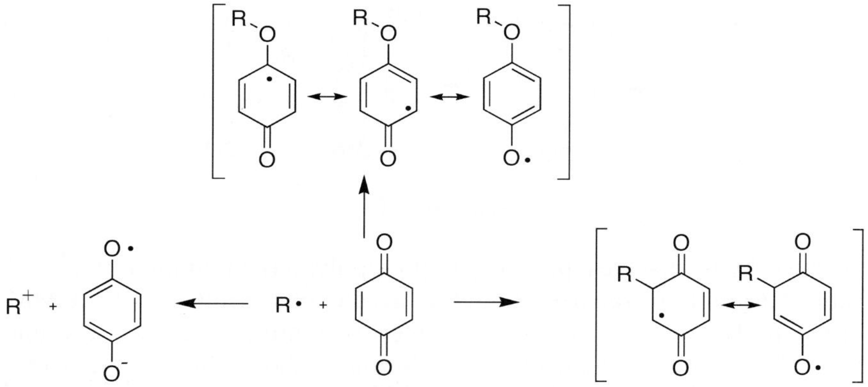

Scheme 5.16

5.3.6 Phenothiazine

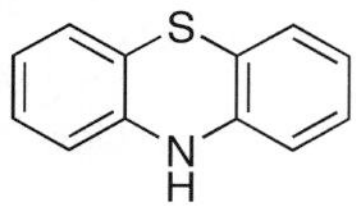

39

In contrast to phenols (Section 5.3.4), phenothiazine (**39**) is reported to be an excellent scavenger of both carbon-centered and oxygen-centered radicals by hydrogen atom transfer and is also used to stabilize monomers in storage.[198]

5.3.7 Nitrones, Nitro- and Nitroso-Compounds

Many nitrones and nitroso-compounds have been exploited as spin traps in elucidating radical reaction mechanisms by EPR spectroscopy (Section 3.5.2.1). The initial adducts are nitroxides which can trap further radicals (Scheme 5.17).

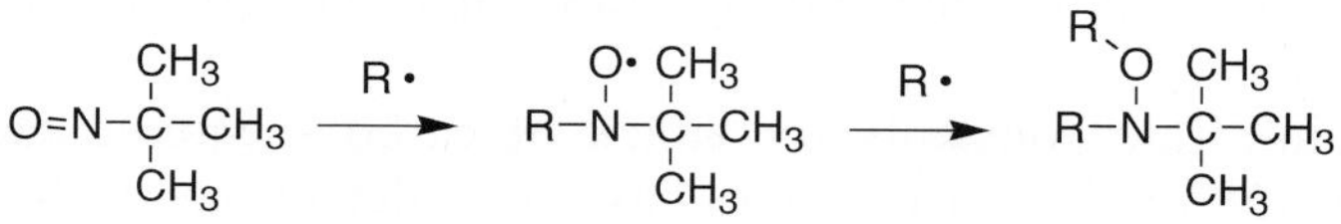

Scheme 5.17

Aromatic nitro-compounds have also seen use as inhibitors in polymerization and as additives in radical reactions. The reactions of these compounds with radicals are very complex and may involve nitroso-compounds and nitroxide intermediates.[205,206] In this case, up to four moles of radicals may be consumed per mole of nitro-compound. The overall mechanism in the case of nitrobenzene has been written as shown in Scheme 5.18. The alkoxyamine **40** can be isolated in good yield from the decomposition of AIBN in the presence of nitrobenzene (Scheme 5.18, R=cyanoisopropyl).[109]

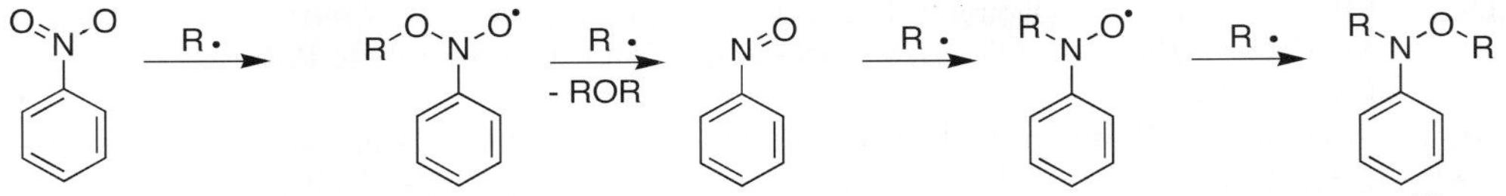

Scheme 5.18

5.3.8 Transition Metal Salts

Transition metal salts trap carbon-centered radicals by electron transfer or by ligand transfer. These reagents often show high specificity for reaction with specific radicals and the rates of trapping may be correlated with the nucleophilicity of the radical (Table 5.6). For example, PS• radicals are much more reactive towards ferric chloride than acrylic propagating species.[207]

Various transition metal salts have been applied in quantitative determination of initiation reactions (Section 3.5.2.2). Under some circumstances, the ligand transfer may be reversible under the polymerization conditions. This chemistry forms the basis of ATRP (Section 9.4).

5.4 References

1. North, A.M. In *Reactivity, Mechanism and Structure in Polymer Chemistry*; Jenkins, A.D.; Ledwith, A., Eds.; Wiley: London, 1974; p 142.
2. O'Driscoll, K.F. In *Comprehensive Polymer Science*; Eastmond, G.C.; Ledwith, A.; Russo, S.; Sigwalt, P., Eds.; Pergamon: Oxford, 1989; Vol. 3, p 161.
3. Buback, M.; Egorov, M.; Gilbert, R.G.; Kaminsky, V.; Olaj, O.F.; Russell, G.T.; Vana, P.; Zifferer, G. *Macromol. Chem. Phys.* **2002**, *201*, 2570.
4. Russell, G.T. *Macromol. Theory Simul.* **1995**, *4*, 519.
5. Russell, G.T. *Macromol. Theory Simul.* **1995**, *4*, 549.
6. Russell, G.T. *Macromol. Theory Simul.* **1995**, *4*, 497.
7. Russell, G.T.; Napper, D.H.; Gilbert, R.G. *Macromolecules* **1988**, *21*, 2133.
8. de Kock, J.B.L.; van Herk, A.M.; German, A.L. *J. Macromol. Sci., Rev. Macromol. Chem. Phys.* **2001**, *C41*, 199.
9. de Kock, J.B.L.; Klumperman, B.; van Herk, A.M.; German, A.L. *Macromolecules* **1997**, *30*, 6743.
10. Buback, M.; Garcia-Rubio, L.H.; Gilbert, R.G.; Napper, D.H.; Guillot, J.; Hamielec, A.E.; Hill, D.; O'Driscoll, K.F.; Olaj, O.F.; Shen, J.; Solomon, D.H.; Moad, G.; Stickler, M.; Tirrell, M.; Winnik, M.A. *J. Polym. Sci., Part C: Polym. Lett.* **1988**, *26*, 293.
11. Buback, M.; Gilbert, R.G.; Russell, G.T.; Hill, D.J.T.; Moad, G.; O'Driscoll, K.F.; Shen, J.; Winnik, M.A. *J. Polym. Sci., Part A: Polym. Chem.* **1992**, *30*, 851.
12. Gilbert, R.G. *Pure Appl. Chem.* **1992**, *64*, 1563.
13. Eastmond, G.C. In *Comprehensive Chemical Kinetics*; Bamford, C.H.; Tipper, C.F.H., Eds.; Elsevier: Amsterdam, 1976; Vol. 14A, p 1.

14. Bamford, C.H. In *Encyclopedia of Polymer Science and Engineering*, 2nd ed.; Mark, H.F.; Bikales, N.M.; Overberger, C.G.; Menges, G., Eds.; Wiley: New York, 1988; Vol. 13, p 708.
15. Mills, I.; Cvitas, T.; Homann, K.; Kallay, N.; Kuchitsu, K. *Quantities, Units and Symbols in Physical Chemistry*, 1988 ed.; Blackwell Scientific Publications: Oxford, 1988.
16. Kamachi, M.; Yamada, B. In *Polymer Handbook*, 4th ed.; Brandup, J.; Immergut, E.H.; Grulke, E.A., Eds.; John Wiley and Sons: New York, 1999; p II/77.
17. Beuermann, S.; Buback, M. *Prog. Polym. Sci.* **2002**, *27*, 191.
18. Buback, M.; Busch, M.; Kowollik, C. *Macromol. Theory Simul.* **2000**, *9*, 442.
19. Beuermann, S.; Buback, M.; Russell, G.T. *Macromol. Chem. Phys.* **1995**, *196*, 2493.
20. Yamada, B.; Westmoreland, D.G.; Kobatake, S.; Konosu, O. *Prog. Polym. Sci.* **1999**, *24*, 565.
21. Kamachi, M. *J. Polym. Sci., Part A: Polym. Chem.* **2001**, *40*, 269.
22. Carswell, T.G.; Hill, D.J.T.; Londero, D.I.; O'Donnell, J.H.; Pomery, P.J.; Winzor, C.L. *Polymer* **1992**, *33*, 137.
23. Moad, G.; Shipp, D.A.; Smith, T.A.; Solomon, D.H. *J. Phys. Chem. A* **1999**, *103*, 6580.
24. Shipp, D.A.; Solomon, D.H.; Smith, T.A.; Moad, G. *Macromolecules* **2003**, *36*, 2032.
25. Moad, G.; Shipp, D.A.; Smith, T.A.; Solomon, D.H. *Macromolecules* **1997**, *30*, 7627.
26. Zetterlund, P.B.; Yamauchi, S.; Yamada, B. *Macromol. Chem. Phys.* **2004**, *205*, 778.
27. Schulz, G.V.; Harborth, G. *Z. Phys. Chem.* **1939**, *B43*, 25.
28. Flory, P.J. *Principles of Polymer Chemistry*; Cornell University Press: Ithaca, New York, 1953.
29. Bamford, C.H.; Barb, W.G.; Jenkins, A.D.; Onyon, P.F. *The Kinetics of Vinyl Polymerization by Radical Mechanisms*; Butterworths: London, 1958.
30. North, A.M.; Reed, G.A. *Trans. Faraday Soc.* **1961**, *57*, 859.
31. North, A.M.; Reed, G.A. *J. Polym. Sci., Part A* **1963**, *1*, 1311.
32. Benson, S.W.; North, A.M. *J. Am. Chem. Soc.* **1962**, *84*, 935.
33. Fischer, H.; Paul, H. *Acc. Chem. Res.* **1987**, *20*, 200.
34. Balke, S.T.; Hamielec, A.E. *J. Appl. Polym. Sci.* **1973**, *17*, 905.
35. Soh, S.K.; Sundberg, D.C. *J. Polym. Sci., Polym. Chem. Ed.* **1982**, *20*, 1345.
36. Norrish, R.G.W.B., E. F. *Proc. Roy. Soc. London* **1939**, *A171*, 147.
37. Trommsdorf, E.; Kohle, H.; Lagally, P. *Makromol. Chem.* **1948**, *1*, 169.
38. Norrish, R.G.W.; Smith, R.R. *Nature* **1942**, *150*, 336.
39. Mahabadi, H.K.; O'Driscoll, K.F. *J. Polym. Sci., Polym. Chem. Ed.* **1977**, *15*, 283.
40. Olaj, O.F.; Zifferer, G. *Macromolecules* **1987**, *20*, 850.
41. Zhu, S.; Hamielec, A.E. *Macromolecules* **1989**, *22*, 3093.
42. Yasukawa, T.; Murakami, K. *Macromolecules* **1981**, *14*, 227.
43. Yasukawa, T.; Murakami, K. *Polymer* **1980**, *21*, 1423.
44. Marten, F.L.; Hamielec, A.E. *J. Appl. Polym. Sci.* **1982**, *27*, 489.
45. Bamford, C.H. *Polymer* **1990**, *31*, 1720.
46. Olaj, O.F.; Zifferer, G.; Gleixner, G. *Makromol. Chem.* **1986**, *187*, 977.
47. Olaj, O.F.; Zifferer, G. *Makromol. Chem., Rapid Commun.* **1982**, *3*, 549.
48. Griller, D. In *Landoldt-Bornstein, New Series, Radical Reaction Rates in Solution*; Fischer, H., Ed.; Springer-Verlag: Berlin, 1984; Vol. II/13a, p 5.

49. Deady, M.; Mau, A.W.H.; Moad, G.; Spurling, T.H. *Makromol. Chem.* **1993**, *194*, 1691.
50. Smith, G.B.; Russell, G.T.; Heuts, J.P.A. *Macromol. Theory Simul.* **2003**, *12*, 299.
51. khokhlov, A.R. *Makromol. Chem., Rapid Commun.* **1981**, *2*, 633.
52. Buback, M.; Egorov, M.; Feldermann, A. *Macromolecules* **2004**, *37*, 1768.
53. Buback, M.; Egorov, M.; Junkers, T.; Panchenko, E. *Macromol. Rapid Commun.* **2004**, *25*, 1004.
54. Cardenas, J.N.; O'Driscoll, K.F. *J. Polym. Sci., Polym. Chem. Ed.* **1976**, *14*, 883.
55. Russell, G.T. *Macromol. Theory Simul.* **1994**, *3*, 439.
56. Faldi, A.; Tirrell, M.; Lodge, T.P. *Macromolecules* **1994**, *27*, 4176.
57. O'Shaughnessy, B.; Yu, J. *Macromolecules* **1994**, *27*, 5079.
58. O'Shaughnessy, B.; Yu, J. *Macromolecules* **1994**, *27*, 5067.
59. Bamford, C.H. *Eur. Polym. J.* **1989**, *25*, 683.
60. Bamford, C.H. *Eur. Polym. J.* **1993**, *29*, 313.
61. Bamford, C.H. *Eur. Polym. J.* **1990**, *26*, 719.
62. Bamford, C.H. *Eur. Polym. J.* **1991**, *27*, 1289.
63. Bamford, C.H. *Eur. Polym. J.* **1990**, *26*, 1245.
64. Verravalli, M.S.; Rosen, S.L. *J. Polym. Sci., Part B: Polym. Phys.* **1990**, *28*, 775.
65. Soh, S.K.; Sundberg, D.C. *J. Polym. Sci., Polym. Chem. Ed.* **1982**, *20*, 1299.
66. Soh, S.K.; Sundberg, D.C. *J. Polym. Sci., Polym. Chem. Ed.* **1982**, *20*, 1315.
67. Soh, S.K.; Sundberg, D.C. *J. Polym. Sci., Polym. Chem. Ed.* **1982**, *20*, 1331.
68. O'Neil, G.A.; Wisnudel, M.B.; Torkelson, J.M. *Macromolecules* **1998**, *31*, 4537.
69. O'Neil, G.A.; Torkelson, J.M. *Macromolecules* **1999**, *32*, 411.
70. Ito, K. *Polym. J.* **1980**, *12*, 499.
71. Tulig, T.J.; Tirrell, M. *Macromolecules* **1981**, *14*, 1501.
72. de Gennes, P.G. *J. Chem. Phys.* **1982**, *76*, 3322.
73. O'Shaughnessy, B.; Yu, J. *Macromolecules* **1998**, *31*, 5240.
74. Kim, J.U.; O'Shaughnessy, B. *Macromolecules* **2004**, *37*, 1630.
75. Buback, M.; Huckestein, B.; Russell, G.T. *Macromol. Chem. Phys.* **1994**, *195*, 539.
76. Chiu, W.Y.; Carrat, G.M.; Soong, D.S. *Macromolecules* **1983**, *16*, 348.
77. Gilbert, R.G. *Emulsion Polymerization: A Mechanistic Approach*; Academic Press: London, 1995.
78. Lovell, P.A.; El-Aasser, M.S., Eds. *Emulsion Polymerization and Emulsion Polymers*; John Wiley & Sons: London, 1997.
79. Smith, W.V.; Ewart, R.H. *J. Chem. Phys.* **1948**, *16*, 592.
80. Krstina, J.; Moad, C.L.; Moad, G.; Rizzardo, E.; Berge, C.T.; Fryd, M. *Macromol. Symp.* **1996**, *111*, 13.
81. Krstina, J.; Moad, G.; Rizzardo, E.; Winzor, C.L.; Berge, C.T.; Fryd, M. *Macromolecules* **1995**, *28*, 5381.
82. Moad, G.; Chiefari, J.; Krstina, J.; Postma, A.; Mayadunne, R.T.A.; Rizzardo, E.; Thang, S.H. *Polym. Int.* **2000**, *49*, 933.
83. Vana, P.; Davis, T.R.; Barner-Kowollik, C. *Macromolecular Rapid Communications* **2002**, *23*, 952.
84. Cacioli, P.; Moad, G.; Rizzardo, E.; Serelis, A.K.; Solomon, D.H. *Polym. Bull.* **1984**, *11*, 325.
85. Kashiwagi, T.; Inaba, A.; Brown, J.E.; Hatada, K.; Kitayama, T.; Masuda, E. *Macromolecules* **1986**, *19*, 2160.
86. Meisters, A.; Moad, G.; Rizzardo, E.; Solomon, D.H. *Polym. Bull.* **1988**, *20*, 499.
87. Manring, L.E. *Macromolecules* **1989**, *22*, 2673.
88. Bamford, C.H.; White, E.F.T. *Trans. Faraday Soc.* **1958**, *54*, 268.

89. Bizilj, S.; Kelly, D.P.; Serelis, A.K.; Solomon, D.H.; White, K.E. *Aust. J. Chem.* **1985**, *38*, 1657.
90. Cacioli, P.; Hawthorne, D.G.; Laslett, R.L.; Rizzardo, E.; Solomon, D.H. *J. Macromol. Sci., Chem.* **1986**, *A23*, 839.
91. Morawetz, H. *J. Polym. Sci., Polym. Symp.* **1978**, *62*, 271.
92. Overberger, C.G.; Finestone, A.B. *J. Am. Chem. Soc.* **1956**, *78*, 1638.
93. Gibian, M.J.; Corley, R.C. *J. Am. Chem. Soc.* **1972**, *94*, 4178.
94. Shelton, J.R.; Liang, C.K. *J. Org. Chem.* **1973**, *38*, 2301.
95. Gleixner, G.; Olaj, O.F.; Breitenbach, J.W. *Makromol. Chem.* **1979**, *180*, 2581.
96. Schreck, V.A.; Serelis, A.K.; Solomon, D.H. *Aust. J. Chem.* **1989**, *42*, 375.
97. Nelsen, S.F.; Bartlett, P.D. *J. Am. Chem. Soc.* **1966**, *88*, 137.
98. Neuman, R.C., Jr.; Amrich, M.J., Jr. *J. Org. Chem.* **1980**, *45*, 4629.
99. Fraenkel, G.; Geckle, M.J. *J. Chem. Soc., Chem. Commun.* **1980**, 55.
100. Langhals, H.; Fischer, H. *Chem. Ber.* **1978**, *111*, 543.
101. Skinner, K.J.; Hochster, H.S.; McBride, J.M. *J. Am. Chem. Soc.* **1974**, *96*, 4301.
102. Businelli, L.; Gnanou, Y.; Maillard, B. *Macromolecular Chemistry and Physics* **2000**, *201*, 2805.
103. Trecker, D.J.; Foote, R.S. *J. Org. Chem.* **1968**, *33*, 3527.
104. Kodaira, K.; Ito, K.; Iyoda, S. *Polym. Commun.* **1987**, *28*, 86.
105. Kelly, D.P.; Serelis, A.K.; Solomon, D.H.; Thompson, P.E. *Aust. J. Chem.* **1987**, *40*, 1631.
106. Mackie, J.S.; Bywater, S. *Can. J. Chem.* **1957**, *35*, 570.
107. Neumann, W.P.; Stapel, R. *Chem. Ber.* **1986**, *119*, 3422.
108. Serelis, A.K.; Solomon, D.H. *Polym. Bull.* **1982**, *7*, 39.
109. Gingras, B.A.; Waters, W.A. *J. Chem. Soc.* **1954**, 1920.
110. Barbe, W.; Rüchardt, C. *Makromol. Chem.* **1983**, *184*, 1235.
111. Serelis, A.K. *Personal Communication*.
112. Konter, W.; Bömer, B.; Köhler, K.H.; Heitz, W. *Makromol. Chem.* **1981**, *182*, 2619.
113. Barton, J.; Capek, I.; Juranicova, V.; Riedel, S. *Makromol. Chem., Rapid Commun.* **1986**, *7*, 521.
114. Guth, W.; Heitz, W. *Makromol. Chem.* **1976**, *177*, 1835.
115. Jaffe, A.B.; Skinner, K.J.; McBride, J.M. *J. Am. Chem. Soc.* **1972**, *94*, 8510.
116. Minato, T.; Yamabe, S.; Fujimoto, H.; Fukui, K. *Bull. Chem. Soc. Japan* **1978**, *51*, 1.
117. Krstina, J.; Moad, G.; Willing, R.I.; Danek, S.K.; Kelly, D.P.; Jones, S.L.; Solomon, D.H. *Eur. Polym. J.* **1993**, *29*, 379.
118. Gibian, M.J.; Corley, R.C. *Chem. Rev.* **1973**, *73*, 441.
119. Heitz, W. In *Telechelic Polymers: Synthesis and Applications*; Goethals, E.J., Ed.; CRC Press: Boca Raton, Florida, 1989; p 61.
120. Bamford, C.H.; Dyson, R.W.; Eastmond, G.C. *Polymer* **1969**, *10*, 885.
121. Stickler, M. *Makromol. Chem.* **1979**, *180*, 2615.
122. Burnett, G.M.; North, A.M. *Makromol. Chem.* **1964**, *73*, 77.
123. Bamford, C.H.; Jenkins, A.D.; Johnston, R. *Trans. Faraday Soc.* **1959**, *55*, 179.
124. Olaj, O.F.; Breitenbach, J.W.; Wolf, B. *Monatsh. Chem.* **1964**, *95*, 1646.
125. Dawkins, J.V.; Yeadon, G. *Polymer* **1979**, *20*, 981.
126. Baker, C.A.; Williams, R.J.P. *J. Chem. Soc.* **1956**, 2352.
127. Henrici-Olive, G.; Olive, S. *Fortschr. Hochpolym. Forsch.* **1961**, *2*, 496.
128. Rudin, A. In *Comprehensive Polymer Science*; Eastmond, G.C.; Ledwith, A.; Russo, S.; Sigwalt, P., Eds.; Pergamon: Oxford, 1989; Vol. 3, p 239.
129. Moad, G.; Moad, C.L. *Macromolecules* **1996**, *29*, 7727.

Termination

130. Zammit, M.D.; Davis, T.P.; Haddleton, D.M.; Suddaby, K.G. *Macromolecules* **1997**, *30*, 1915.
131. Catalgil-Giz, H.; Giz, A.; Oncul-Koc, A. *Polym. Bull.* **1999**, *43*, 215.
132. Hensley, D.R.; Goodrich, S.D.; Harwood, H.J.; Rinaldi, P.L. *Macromolecules* **1994**, *27*, 2351.
133. Bevington, J.C.; Melville, H.W.; Taylor, R.P. *J. Polym. Sci.* **1954**, *14*, 463.
134. Ayrey, G.; Moore, C.G. *J. Polym. Sci.* **1959**, *36*, 41.
135. Hakozaki, J.; Yamada, N. In *J. Chem. Soc., Japan*, 96263r (1968) ed., 1967; Vol. 70, p 1560.
136. O'Driscoll, K.F.; Bevington, J.C. *Eur. Polym. J.* **1985**, *21*, 1039.
137. Bamford, C.H.; Jenkins, A.D. *Nature* **1955**, *176*, 78.
138. Mayo, F.R.; Gregg, R.A.; Matheson, M.S. *J. Am. Chem. Soc.* **1951**, *73*, 1691.
139. Johnson, D.H.; Tobolsky, A.V. *J. Am. Chem. Soc.* **1952**, *74*, 938.
140. Braks, J.G.; Huang, R.Y.M. *J. Appl. Polym. Sci.* **1978**, *22*, 3111.
141. Henrici-Olive, G.; Olive, S. *J. Polym. Sci.* **1960**, *48*, 329.
142. Bevington, J.C.; Melville, H.W.; Taylor, R.P. *J. Polym. Sci.* **1954**, *12*, 449.
143. Kolthoff, I.M.; O'Connor, P.R.; Hansen, J.L. *J. Polym. Sci.* **1955**, *15*, 459.
144. Arnett, L.M.; Peterson, J.H. *J. Am. Chem. Soc.* **1952**, *74*, 2031.
145. Bessiere, J.-M.; Boutevin, B.; Loubet, O. *Polym. Bull.* **1993**, *31*, 673.
146. Moad, G.; Solomon, D.H.; Johns, S.R.; Willing, R.I. *Macromolecules* **1984**, *17*, 1094.
147. Moad, G.; Solomon, D.H.; Johns, S.R.; Willing, R.I. *Macromolecules* **1982**, *15*, 1188.
148. Ayrey, G.; Levitt, F.G.; Mazza, R.J. *Polymer* **1965**, *6*, 157.
149. Olaj, O.F.; Kaufmann, H.F.; Breitenbach, J.W.; Bieringer, H. *J. Polym. Sci., Polym. Lett. Ed.* **1977**, *15*, 229.
150. Berger, K.C.; Meyerhoff, G. *Makromol. Chem.* **1975**, *176*, 1983.
151. Barson, C.A.; Bevington, J.C.; Hunt, B.J. *Polymer* **1998**, *39*, 1345.
152. Bevington, J.C.; Breuer, S.W.; Huckerby, T.N.; Hunt, B.J.; Jones, R. *Eur. Polym. J.* **1998**, *34*, 539.
153. Yoshikawa, C.; Goto, A.; Fukuda, T. *e-Polymers* **2002**, *2002*, 13.
154. Allen, P.W.; Ayrey, G.; Merrett, F.M.; Moore, C.G. *J. Polym. Sci.* **1956**, *22*, 549.
155. Schulz, G.V.; Henrici-Olive, G.; Olive, S. *Makromol. Chem.* **1959**, *31*, 88.
156. Bamford, C.H.; Eastmond, G.C.; Whittle, D. *Polymer* **1969**, *10*, 771.
157. Chaudhuri, A.K.; Palit, S.R. *J. Polym. Sci., Part A-1* **1968**, *6*, 2187.
158. Braks, J.G.; Mayer, G.; Huang, R.Y.M. *J. Appl. Polym. Sci.* **1980**, *25*, 449.
159. Ayrey, G.; Haynes, A.C. *Eur. Polym. J.* **1973**, *9*, 1029.
160. Hatada, K.; Kitayama, T.; Ute, K.; Terawaki, Y.; Yanagida, T. *Macromolecules* **1997**, *30*, 6754.
161. Boudevska, H.; Brutchkov, C.; Platchkova, S. *Makromol. Chem.* **1981**, *182*, 3257.
162. Hatada, K.; Kitayama, T.; Masuda, E. *Polym. J.* **1986**, *18*, 395.
163. Ayrey, G.; Humphrey, M.J.; Poller, R.C. *Polymer* **1977**, *18*, 840.
164. Bailey, B.E.; Jenkins, A.D. *Trans. Faraday Soc.* **1960**, *56*, 903.
165. Bevington, J.C.; Eaves, D.E. *Trans. Faraday Soc.* **1959**, *55*, 1777.
166. Patron, L.; Bastianelli, U. *Appl. Polym. Symp.* **1974**, *25*, 105.
167. Funt, B.L.; Paskia, W. *Can. J. Chem.* **1960**, *38*, 1865.
168. Danusso, F.; Pajaro, G.; Sianesi, D. *Chim. Ind. (Milan)* **1959**, *41*, 1170.
169. Talamini, G.V., G. *Chim. Ind. (Milan)* **1964**, *46*, 16.
170. Park, G.S.; Smith, D.G. *Makromol. Chem.* **1970**, *131*, 1.
171. Starnes, W.H., Jr.; Plitz, I.M.; Schilling, F.C.; Villacorta, G.M.; Park, G.S.; Saremi, A.H. *Macromolecules* **1984**, *17*, 2507.

172. Atkinson, W.H.; Bamford, C.H.; Eastmond, G.C. *Trans. Faraday Soc.* **1970**, *66*, 1446.
173. Bamford, C.H. In *Comprehensive Polymer Science*; Agarwal, S.L.; Russo, S., Eds.; Pergamon: Oxford, 1992; Vol. Suppl. 1, p 1.
174. Tüdos, F.; Földes-Berezsnich, T. *Prog. Polym. Sci.* **1989**, *14*, 717.
175. Eastmond, G.C. In *Comprehensive Chemical Kinetics*; Bamford, C.H.; Tipper, C.F.H., Eds.; Elsevier: Amsterdam, 1976; Vol. 14A, p 153.
176. Goldfinger, G.; Yee, W.; Gilbert, R.D. In *Encyclopedia of Polymer Science and Technology*; Mark, H.F.; Gaylord, N.M.; Bikales, N.M., Eds.; Wiley: New York, 1967; Vol. 7, p 644.
177. Bovey, F.A.; Kolthoff, I.M. *Chem. Rev.* **1948**, *42*, 491.
178. Bowry, V.W.; Ingold, K.U. *J. Am. Chem. Soc.* **1992**, *114*, 4992.
179. Beckwith, A.L.J.; Bowry, V.W.; Moad, G. *J. Org. Chem.* **1988**, *53*, 1632.
180. Beckwith, A.L.J.; Bowry, V.W.; Ingold, K.U. *J. Am. Chem. Soc.* **1992**, *114*, 4983.
181. Maillard, B.; Ingold, K.U.; Scaiano, J.C. *J. Am. Chem. Soc.* **1983**, *105*, 5095.
182. Citterio, A.; Arnoldi, A.; Minisci, F. *J. Org. Chem.* **1979**, *44*, 2674.
183. Rozantsev, E.G.; Gol'dfein, M.D.; Trubnikov, A.V. *Russ. Chem. Rev. (Engl. Transl.)* **1986**, *55*, 1070.
184. Chateauneuf, J.; Lusztyk, J.; Ingold, K.U. *J. Org. Chem.* **1988**, *53*, 1629.
185. Rizzardo, E.; Solomon, D.H. *Polym. Bull.* **1979**, *1*, 529.
186. Hawthorne, D.G.; Solomon, D.H. *J. Macromol. Sci., Chem.* **1972**, *A6*, 661.
187. Bhanu, V.A.; Kishore, K. *Chem. Rev.* **1991**, *91*, 99.
188. Neta, P.; Huie, R.E.; Ross, A.B. *J. Phys. Chem. Ref., Data* **1990**, *19*, 413.
189. Wieland, P.C.; Raether, B.; Nuyken, O. *Macromol. Rapid Commun.* **2001**, *22*, 700.
190. Mignani, S.; Janousek, Z.; Merenyi, R.; Viehe, H.G.; Riga, J.; Verbist, J. *Tetrahedron Lett.* **1984**, *25*, 1571.
191. Tanaka, H. *Prog. Polym. Sci.* **2003**, *28*, 1171.
192. Tanaka, H.; Teraoka, Y.; Sato, T.; Ota, T. *Makromol. Chem.* **1993**, *194*, 2719.
193. Hageman, H.J.; Oosterhoff, P.; Overeem, T.; Polman, R.J.; van der Werf, S. *Makromol. Chem.* **1985**, *186*, 2483.
194. Tanaka, H.; Kameshima, T.; Sasai, K.; Sato, T.; Ota, T. *Makromol. Chem.* **1991**, *192*, 427.
195. Kharasch, M.S.; Kawahara, F.; Nudenberg, W. *J. Org. Chem.* **1954**, *19*, 1977.
196. Levy, L.B. *J. Appl. Polym. Sci.* **1996**, *60*, 2481.
197. Levy, L.B. *J. Polym. Sci., Part A: Polym. Chem.* **1992**, *30*, 569.
198. Schulze, S.; Vogel, H. *Chem. Eng. Technol.* **1998**, *21*, 829.
199. Haddleton, D.M.; Clark, A.J.; Crossman, M.C.; Duncalf, D.J.; Heming, A.M.; Morsley, S.R.; Shooter, A.J. *Chem. Commun.* **1997**, 1173.
200. Chen, S.; Tsai, L. *Makromol. Chem.* **1986**, *187*, 653.
201. Bevington, J.C.; Ghanem, N.A.; Melville, H.W. *Trans. Faraday Soc.* **1955**, *51*, 946.
202. Price, C.C.; Read, D.H. *J. Polym. Res.* **1946**, *1*, 44.
203. Yassin, A.A.; Rizk, N.A. *Eur. Polym. J.* **1977**, *13*, 441.
204. Golubev, V.B.; Mun, G.A.; Zubov, V.P. *Russ. J. Phys. Chem. (Engl. Transl.)* **1986**, *60*, 347.
205. Perkins, M.J.; Chalfont, G.R.; Hey, D.H.; Liang, K.S.Y. *J. Chem. Soc. (B)* **1971**, 233.
206. Chalfont, G.R.; Hey, D.H.; Liang, K.S.Y.; Perkins, M.J. *Chem. Commun. (London)* **1967**, 367.
207. Bamford, C.H.; Jenkins, A.D.; Johnston, R. *Proc. R. Soc., London* **1957**, *A239*, 214.

6
Chain Transfer

6.1 Introduction

Chain transfer is the reaction of a propagating radical with a non-radical substrate (X-Y, Scheme 6.1) to produce a dead polymer chain and a new radical (Y•) capable of initiating a polymer chain. The transfer agent (X-Y) may be a deliberate additive (*e.g.* a thiol) or it may be the initiator, monomer, polymer, solvent or an adventitious impurity.

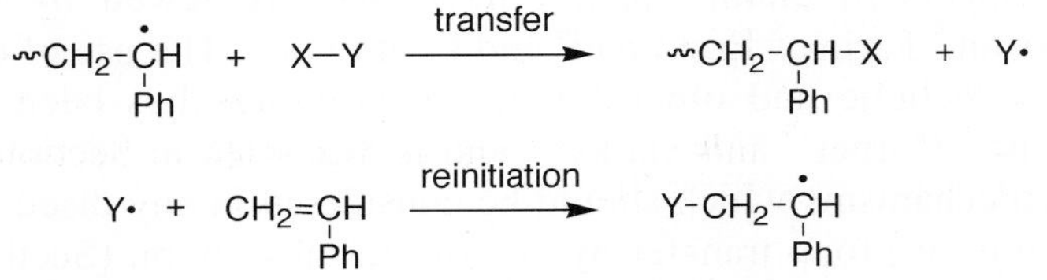

$$\text{\textasciitilde CH}_2\text{-}\overset{\bullet}{\text{CH}}\text{-Ph} + \text{X-Y} \xrightarrow{\text{transfer}} \text{\textasciitilde CH}_2\text{-CH(Ph)-X} + \text{Y}\bullet$$

$$\text{Y}\bullet + \text{CH}_2\text{=CH-Ph} \xrightarrow{\text{reinitiation}} \text{Y-CH}_2\text{-}\overset{\bullet}{\text{CH}}\text{-Ph}$$

Scheme 6.1

Transfer without reinitiation is called inhibition and is discussed in Section 5.3. There are also situations where the reaction produces a dead polymer chain and a radical that is less reactive than the propagating radical but still capable of reinitiating polymerization. The process is then termed retardation or degradative chain transfer.

6.2 Chain Transfer

The general mechanism of chain transfer as first proposed by Flory,[1,2] may be written schematically as shown in Scheme 6.2. The overall process involves a propagating chain ($P_n\bullet$) reacting with a transfer agent (T) to terminate one polymer chain and produce a radical (T•) that initiates a new chain ($P_1\bullet$).

Transfer agents find widespread use in both industrial and laboratory polymer syntheses. They are used to control:
(a) The molecular weight of polymers
(b) The polymerization rate and exotherm (by mitigating the gel or Norrish-Trommsdorff effect)
(c) The polymer end groups.

transfer to transfer agent or solvent

$P_n\bullet + T \rightarrow P_n + T\bullet \qquad R_{trT} = k_{trT} [P\bullet][T]$

reinitiation

$T\bullet + M \rightarrow P_1\bullet \qquad R_{iT} = k_{iT} [T\bullet][M]; k_{iT} \geq k_p$

transfer to initiator

$P_n\bullet + I_2 \rightarrow P_n + I\bullet \qquad R_{trT} = k_{trT} [P\bullet][I_2]$

reinitiation

$I\bullet + M \rightarrow P_1\bullet \qquad R_{iT} = k_{iT} [I\bullet][M]; k_i \geq k_p$

transfer to monomer

$P_n\bullet + M \rightarrow P_n + M\bullet \qquad R_{trT} = k_{trT} [P\bullet][M]$

reinitiation

$M\bullet + M \rightarrow P_1\bullet \qquad R_{iT} = k_{iT} [M\bullet][M]; k_{iM} \geq k_p$

Scheme 6.2

General aspects of chain transfer have been reviewed by Chiefari and Rizzardo,[3] Barson,[4] Farina,[5] Eastmond[6] and Palit et al.[7] The use of chain transfer in producing telechelic and other functional polymers has been reviewed by Boutevin,[8] Heitz,[9] Corner[10] and Starks[11] and is discussed in Section 7.5.2. There are two main mechanisms which should be considered in any discussion of chain transfer: (a) atom or group transfer by homolytic substitution (Section 6.2.2) and (b) addition-fragmentation (Section 6.2.3).

Even in the absence of added transfer agents, all polymerizations may be complicated by transfer to initiator (Sections 3.2.10 and 3.3), solvent (Section 6.2.2.5), monomer (Section 6.2.6) or polymer (Section 6.2.7). The significance of these transfer reactions is dependent upon the particular propagating radicals involved, the reaction medium and the polymerization conditions. Thiol-ene polymerization consists of sequential chain transfer and reinitiation steps and ideally no monomer consumption by propagation (Section 7.5.3).

For efficient chain transfer, the rate constant for reinitiation following transfer (k_{iT}; refer Scheme 6.2) must be greater than or equal to that for propagation (k_p). In these circumstances, the presence of the transfer agent reduces the molecular weight of the polymer without directly influencing the rate of polymerization. If, however, $k_{iT} < k_p$ then polymerization will be retarded and the likelihood that the transfer agent-derived radical (T•) will undergo side reactions such as primary radical termination is increased. Thus, retardation is much more likely in polymerizations of high k_p monomers (*e.g.* MA, VAc) than it is with lower k_p monomers (*e.g.* S, MMA). Retardation is discussed in greater detail in Section 5.3.

Even when $k_{iT} \geq k_p$, the rate of polymerization at higher conversions will often be lower that in the absence of a transfer agent due to a reduced gel or Norrish-

Trommsdorf effect. One cause of this autoacceleration phenomenon is a reduced rate of radical-radical termination brought about by the immobilization of long chains through entanglement at higher conversions (Section 5.2.1.4). In the presence of a transfer agent, the population of short chains is higher and, because the ultimate molecular weight is lower, there are fewer chain entanglements.

The number average degree of polymerization ($\overline{X}_n$) of polymer formed at any given instant during the polymerization can be expressed simply as the rate of monomer usage in propagation divided by the rate of formation of polymer molecules (the overall rate of termination). Thus according to classical kinetics, if termination is only by radical-radical reaction or chain transfer the degree of polymerization is given by eq. 1:

$$\overline{X}_n = \frac{k_p[M]}{\left(1+\frac{k_{td}}{k_t}\right)k_t[P\bullet] + k_{tr,T}[T] + k_{tr,I}[I] + k_{tr,M}[M]} \tag{1}$$

This can be rewritten as eq. 2:

$$\frac{1}{\overline{X}_n} = \frac{\left(1+\frac{k_{td}}{k_t}\right)k_t[P\bullet]}{k_p[M]} + \frac{k_{tr,T}}{k_p}\frac{[T]}{[M]} + \frac{k_{tr,I}}{k_p}\frac{[I]}{[M]} + \frac{k_{tr,M}}{k_p} \tag{2}$$

The ratio k_{tr}/k_p is called the transfer constant (C_{tr}) and C_T, C_I and C_M are the transfer constants for transfer to transfer agent, initiator and monomer respectively. Appropriate substitution gives eq. 3:

$$\frac{1}{\overline{X}_n} = \frac{\left(1+\frac{k_{td}}{k_t}\right)k_t[P\bullet]}{k_p[M]} + C_T\frac{[T]}{[M]} + C_I\frac{[I]}{[M]} + C_M \tag{3}$$

The degree of polymerization in the absence of a chain transfer agent is given by eq. 4:

$$\frac{1}{\overline{X}_{n0}} = \frac{\left(1+\frac{k_{td}}{k_t}\right)k_t[P\bullet]}{k_p[M]} + C_I\frac{[I]}{[M]} + C_M \tag{4}$$

Thus,

$$\frac{1}{\overline{X}_n} = \frac{1}{\overline{X}_{n0}} + C_T\frac{[T]}{[M]} \tag{5}$$

This equation (eq. 5) is commonly known as the Mayo equation.[12] The equation is applicable at low (zero) conversion and is invalidated if the rate constants are chain length dependent.

The magnitude of a transfer constant depends on structural features of both the attacking radical and the transfer agent. A C_{tr} of unity has been called ideal. In these circumstances, the transfer agent:monomer ratio ([T]:[M]) will remain constant throughout the polymerization.[10] This means that $\bar{X}_n$ remains constant with conversion and the dispersity of the molecular weight distribution is thus minimized ($\bar{X}_w/\bar{X}_n$ close to 2.0). If C_{tr} is high (>>1), the transfer agent will be consumed rapidly during the early stages of polymerization and the polymerization will be unregulated at higher conversion. If, on the other hand, C_{tr} is low (<<1), [T]:[M] will increase as the polymerization progresses and there will be a corresponding decrease in $\bar{X}_n$ with conversion. In both circumstances, a broad molecular weight distribution will result from a high conversion batch polymerization. It is often possible to overcome these problems by establishing an incremental protocol for monomer and/or transfer agent addition such that [T]:[M] is maintained at a constant value throughout the polymerization.

The rate constants for chain transfer and propagation may well have a different dependence on temperature (i.e. the two reactions may have different activation parameters) and, as a consequence, transfer constants are temperature dependent. The temperature dependence of C_{tr} has not been determined for most transfer agents. Care must therefore be taken when using literature values of C_{tr} if the reaction conditions are different from those employed for the measurement of C_{tr}. For cases where the transfer constant is close to 1.0, it is sometimes possible to choose a reaction temperature such that the transfer constant is 1.0 and thus obtain ideal behavior.[13]

The value of C_{tr} in homopolymerization can show significant chain length dependence for chain lengths ≤ 5. Some values of transfer constants for homolytic substitution chain transfer agents are shown in Table 6.1.[11] The variation in C_{tr} with chain length can reflect variations in k_p or k_{tr} or (most likely) both. The data provided in Section 4.5.3 show that k_p can be dependent on chain length for at least the first few propagation steps. The magnitude of the effect on C_{tr} for a given monomer varies according to the particular transfer agent. This indicates the sensitivity of k_p and k_{tr} to the penultimate unit is different. Chain transfer constants in addition-fragmentation (Section 6.2.3.4) and catalytic chain transfer have also been shown to be chain length dependent (Section 6.2.5).

Bamford[14] has provided evidence that, in copolymerization, penultimate unit effects can be important in determining the reactivity of propagating radicals toward transfer agents. The magnitude of this effect also depends on the particular monomers and transfer agent involved. The finding that the most pronounced remote unit effects are observed for the most bulky transfer agents (Section 6.2.2.4), has been taken as evidence that the magnitude of the remote unit effect is determined at least in part by steric factors. However, this view has been questioned.[15]

Table 6.1 Chain Length Dependence of Transfer Constants (C_n)

Transfer Agent	Monomer	T (°C)[a]	C_1	C_2	C_3	C_4	$C_{5-\infty}$	Refs.
C_2H_5SH[b]	MA	50	0.94±.07	1.65±.12	1.57±.09	1.52±.06	1.57±.18	16
i-C_3H_7SH[b]	MA	50	0.54±.08	0.67±.07	0.70±.08	0.66±.08	-	17
C_2H_5SH[b]	S	50	7.1±.3	30±10	-	-	17±1	18
CCl_3Br	S	30	0.52±.14	9.4±4.6	37±3	96±12	460±61	19
CCl_4	S	76	0.0006	0.0025	0.0069	0.0115	-	20
CCl_4	VAc	60	-	0.13	0.47	0.67	0.80	21
CCl_4	VC	60	0.00284	0.0184	0.0280	-	-	22
$CHCl_3$	VC	60	0.006	0.0141	0.0292	-	-	23

a Bulk polymerization, medium comprises monomer + transfer agent. b The variation between C_2, C_3, C_4 and $C_{5-\infty}$ is within experimental error.

6.2.1 Measurement of Transfer Constants

Various methods for estimating transfer constants in radical polymerization have been devised. The methods are applicable irrespective of whether the mechanism involves homolytic substitution or addition-fragmentation.

The most used method is based on application of the Mayo equation (eq. 5). For low (zero) conversion polymerizations carried out in the presence of added transfer agent T, it follows from eq. 5 that a plot of $1/\bar{X}_n$ vs $[T]_0/[M]_0$ should yield a straight line with slope C_{tr}.[12] Thus, a typical experimental procedure involves evaluation of the degree of polymerization for low conversion polymerizations carried out in the presence of several concentrations of added transfer agent. The usual way of obtaining $\bar{X}_n$ values is by GPC analysis of the entire molecular weight distribution.

GPC-derived weight average molecular weights are often less prone to error than number average molecular weights. When termination is wholly by disproportionation or chain transfer and chains are long (>10 units), classical kinetics predicts $\bar{X}_n = \bar{X}_w/2$ (Section 5.2.1.3). It follows that C_{tr} can be obtained from the slope of a plot of $2/\bar{X}_w$ vs $[T]_0/[M]_0$.[24,25] The errors introduced even when the dominant process for radical-radical termination is combination (e.g. S polymerization) are small as long as $\bar{X}_n$ is small in relation to $\bar{X}_{n0}$.

It has been shown that equivalent information can be obtained by analysis of log(number chain length distribution) plots (the log CLD method).[24-27] For the case where termination is wholly by disproportionation or chain transfer, it is possible to show that (eq. 6) applies:

$$\frac{d\ln(n_i)}{di} = \frac{d\ln[\phi^{i-1}(1-\phi)]}{di} = \ln(\phi) \qquad (6)$$

For long chains ($\bar{X}_n > 50$ for < 1% error)

$$\ln(\phi) \approx 1 - \frac{1}{\phi} = \frac{1}{(\overline{X}_n - 1)} \tag{7}$$

it is possible to write eq. 8 which is equivalent to the Mayo equation:

$$-\frac{\mathrm{d}\ln(n_i)}{\mathrm{d}i} \approx \frac{\left(1 + \frac{k_{td}}{k_t}\right) k_t [\mathrm{P}\bullet]}{k_p[\mathrm{M}]} + \frac{k_{tr,T}}{k_p} \frac{[\mathrm{T}]}{[\mathrm{M}]} + \frac{k_{tr,I}}{k_p} \frac{[\mathrm{I}]}{[\mathrm{M}]} + \frac{k_{tr,M}}{k_p} \tag{8}$$

$$= \frac{1}{\overline{X}_{n0}} + \frac{k_{tr,T}[\mathrm{T}]}{k_p[\mathrm{M}]} \tag{9}$$

It follows that a plot of the slopes of the log CLD plots vs $[\mathrm{T}]_0/[\mathrm{M}]_0$ should yield a straight line with slope $-C_{tr}$.

In the more general case, where some termination is by combination it can be shown that for sufficiently large chain length (i):

$$\lim_{i \to \infty} \frac{\mathrm{d}\ln(n_i)}{\mathrm{d}i} = \ln(\phi) \tag{10}$$

While it is, in principle, desirable to take the limiting slope of the log CLD plot, in practice the limiting slopes are very susceptible to experimental noise and baseline choice issues. Moad and Moad[24] have shown that very little error is introduced by *systematically* taking the slope over the top 10% or the top 20% of the chain length distribution. The values for the slopes will overestimate $\ln(\phi)$. However, because the discrepancy is systematic, the "Mayo" analysis still provides a good estimate for C_{tr} (~6% error for the example in Figure 6.1).

The log CLD method can sometimes provide better quality data than the conventional Mayo method. It is less sensitive to experimental noise and has application in measuring the transfer constant to polymeric species where the distributions of the transfer agent and the polymer product partially overlap.[24]

Problems arise with any of the abovementioned methods in the measurement of transfer constants for very active transfer agents. Bamford[28] proposed the technique of moderated copolymerization. In these experiments, the monomer of interest is copolymerized with an excess of a moderating monomer that has a much lower (preferably negligible) transfer constant. The method has also been applied to evaluate penultimate unit effects on the transfer constant.[28-30]

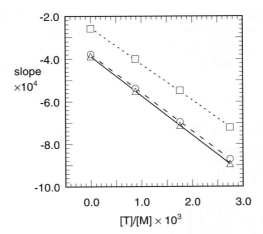

Figure 6.1 "Mayo plots" in which the calculated limiting slopes (triangles, ———, C_{tr} (app) =0.184), "last 10% slopes" (circles, – – –, C_{tr} (app)=0.180) and "top 20% slopes" (squares, -----, C_{tr} (app)=0.169) are graphed as a function of [T]/[M]. Data are for system with $\overline{X}_n$ =5155, $k_{tc}/(k_{tc} + k_{td})$ = 1.0 and C_{tr}=0.184.[24] C_{tr} (app) is the apparent C_{tr} from the slope of the "Mayo plot".

Another classical method for evaluating transfer constants involves evaluation of the usage of transfer agent (or better the incorporation of transfer agent fragments into the polymer) and the monomer conversion:[31]

$$\frac{d[T]}{d[M]} = \frac{k_{trT}[P\bullet][T] + k_{iT}[T\bullet][M]}{k_p[P\bullet][M]} \tag{11}$$

For long chains, consumption of the monomer in the reinitiation step can be neglected and eq. 11 simplifies to eq. 12:

$$\frac{d[T]}{d[M]} = \frac{k_{tr}[T]}{k_p[M]} = C_{tr}\frac{[T]}{[M]} \tag{12}$$

from which eq. 13 follows:

$$\frac{d\ln[T]}{d\ln[M]} = C_{tr} \tag{13}$$

Thus, the slope of a plot of ln[T] vs ln[M] will yield the transfer constant. This method does not rely on molecular weight measurements.

For the situation where short chains cannot be ignored eq. 11 can be transformed to eq. 14:

$$\frac{d[M]}{d[T]} = \frac{[M]}{C_{tr}[T]} + 1 \tag{14}$$

A number of authors have provided integrated forms of the Mayo equation[32-36] which have application when the conversion of monomer to polymer is non-zero. Integration of eq. 12 provides eq. 15:

$$\frac{[T]}{[T]_o} = \left(\frac{[M]}{[M]_o}\right)^{C_{tr}} \tag{15}$$

This enables substitution for [T] in eq. 16 to give eq. 17:[34,36]

$$\frac{1}{\overline{X}_n} = \frac{1}{\overline{X}_{n0}} + \frac{[T]-[T]_o}{[M]-[M]_o} \tag{16}$$

$$\frac{1}{\overline{X}_n} = \frac{1}{\overline{X}_{n0}} + \frac{[T]_o\left(1-\left(\frac{[M]}{[M]_o}\right)^{C_{tr}}\right)}{[M]_o\left(1-\left(\frac{[M]}{[M]_o}\right)\right)} \tag{17}$$

Rearrangement and substitution of 1-x for $[M]/[M]_o$ provides eq. 18:

$$\ln\left(1 - \frac{[M]_o x}{[T]_o}\left(\frac{1}{\overline{X}_n} - \frac{1}{\overline{X}_{n0}}\right)\right) = C_{tr}\ln(1-x) \tag{18}$$

where x is the fractional conversion of monomer into polymer. Thus, a plot of

$$\ln\left(1 - \frac{[M]_o x}{[T]_o}\left(\frac{1}{\overline{X}_n} - \frac{1}{\overline{X}_{n0}}\right)\right) \text{ vs } \ln(1-x)$$

should provide a straight line passing through the origin with slope C_{tr}. Bamford and Basahel[28-30] have reported the derivation of a similar equation for copolymerization. This method is highly dependent on the precision of the conversion measurements since errors in conversions are magnified in C_{tr}.

Cardenas and O'Driscoll[32] and Stickler[33] have shown that, provided that the consumption of transfer agent is negligible with respect to monomer, a plot of

$$\frac{1}{\overline{X}_n} \text{ vs } -\frac{[T]_o}{[M]_o}\frac{\ln(1-x)}{x}$$

should also yield a straight line with slope C_{tr}.[32,33]

Nair et al.[37] have proposed a modified Mayo equation for use when retardation through primary radical termination with transfer agent-derived radicals is significant.

Chain Transfer

Chain transfer is kinetically equivalent to copolymerization. The *Q-e* and 'Patterns of Reactivity' schemes used to predict reactivity ratios in copolymerization (Section 7.3.4) can also be used to predict reactivities (chain transfer constants) in chain transfer and the same limitations apply. Tabulations of the appropriate parameters can be found in the *Polymer Handbook*.[38,39]

6.2.1.1 Addition-fragmentation

Some transfer agents react by addition-fragmentation (Section 6.2.3) or abstraction-fragmentation mechanisms. Both of these processes involve the formation of a short-lived intermediate. The reaction scheme for addition-fragmentation can be summarized schematically as follows (Scheme 6.3).

transfer
   addition
   $P_n\bullet + T \rightarrow [P_nT\bullet]$ $R_{add}=k_{add}[P\bullet][T]$
   fragmentation
   $[P_nT\bullet] \rightarrow P_n + T\bullet$ $R_\beta=k_\beta[PT\bullet]$
reinitiation
   $T\bullet + M \rightarrow P_1\bullet$ $R_{iT}=k_{iT}[T\bullet][M]; k_{iT} \geq k_p$

Scheme 6.3

The reactivity of the transfer agent (T) towards the propagating species and the properties of the adduct ($P_nT\bullet$) are both important in determining the effectiveness of the transfer agent: if the lifetime of the intermediate ($P_nT\bullet$) is significant, it may react by other pathways than β–scission; if it ($P_iT\bullet$) undergoes coupling or disproportionation with another radical species the rate of polymerization will be retarded; if it adds to monomer (T copolymerizes) it will be an inefficient transfer agent.

If both addition and fragmentation are irreversible the kinetics differ little from conventional chain transfer. In the more general case, the rate constant for chain transfer is defined in terms of the rate constant for addition (k_{add}) and a partition coefficient which defines how the adduct is partitioned between products and starting materials (eq. 19).

$$k_{tr} = k_{add} \frac{k_\beta}{k_{-add} + k_\beta} \tag{19}$$

Methods used for evaluating transfer constants are the same as for conventional chain transfer.

6.2.1.2 Reversible chain transfer

In some cases the product of chain transfer (P_n^T) is itself a transfer agent and chain transfer is reversible. Examples include alkyl iodides (Scheme 6.4) and certain addition-fragmentation transfer agents (*e.g.* macromonomers and thiocarbonylthio compounds) (Scheme 6.5).

transfer

$P_n\bullet + T \rightleftharpoons P_n^T + T\bullet \quad R_{tr}=k_{tr}[P\bullet][T]; R_{-tr}=k_{-tr}[T\bullet][P^T]$

reinitiation

$T\bullet + M \rightarrow P_1\bullet \quad R_{iT}=k_{iT}[T\bullet][M]; k_{iT} \gg k_p$

Scheme 6.4

transfer

addition

$P_n\bullet + T \rightleftharpoons [P_nT\bullet] \quad R_{add}=k_{add}[T][P\bullet]; R_{-add}=k_{-add}[PT\bullet]$

fragmentation

$[P_nT\bullet] \rightleftharpoons P_n^T + T\bullet \quad R_\beta=k_\beta[PT\bullet]; R_{-\beta}=k_{-\beta}[T\bullet][P^T]$

reinitiation

$T\bullet + M \rightarrow P_1\bullet \quad R_{iT}=k_{iT}[T\bullet][M]; k_{iT} \gg k_p$

Scheme 6.5

For very active transfer agents, the transfer agent-derived radical (T•) may partition between adding to monomer and reacting with the polymeric transfer agent (P_n^T) even at low conversions. The transfer constant measured according to the Mayo or related methods will appear to be dependent on the transfer agent concentration (and on the monomer conversion).[40-42] A reverse transfer constant can be defined as follows (eq. 20):

$$C_{-tr} = \frac{k_{-tr}}{k_{iT}} \tag{20}$$

and the rate of transfer agent consumption is then given by eq. 21:

$$\frac{d[T]}{d[M]} \approx C_{tr}\frac{[T]}{[M]+C_{tr}[T]+C_{-tr}[P_n^T]}$$

$$= C_{tr}\frac{[T]}{[M]+C_{tr}[T]+C_{-tr}([T_0]-[T])} \tag{21}$$

This equation can be solved numerically to give values of C_{tr} and C_{-tr}.[40,41] For reversible addition-fragmentation chain transfer (RAFT) (Scheme 6.5), the rate constant for the reverse reaction is defined as shown in eq. 22:

$$k_{-tr} = k_{-\beta} \frac{k_{-add}}{k_{-add} + k_{\beta}} \qquad (22)$$

Systems that give reversible chain transfer can display the characteristics of living polymerization. Such systems are discussed in Section 9.5.

6.2.2 Homolytic Substitution Chain Transfer Agents

Chain transfer most commonly involves transfer of an atom or group from the transfer agent to the propagating radical by a homolytic substitution (SH^2) mechanism. The general factors influencing the rate and specificity of these reactions have been dealt with in Section 2.4. Rate constants are determined by a combination of bond strength, steric and polar factors. Transfer agents that react by addition-fragmentation are dealt with in Section 6.2.3. Organometallic species that give catalytic chain transfer are discussed in Section 6.2.5.

The moiety transferred will most often be a hydrogen atom, for example, when the transfer agent is a thiol (*e.g. n*-butanethiol - Scheme 6.6, Section 6.2.2.1), a hydroperoxide (Section 3.3.2.5), the solvent (6.2.2.5), *etc*.

$$P_n^{\bullet} + H-S(CH_2)_3CH_3 \longrightarrow P_n-H + {\bullet}S(CH_2)_3CH_3$$

Scheme 6.6

It is also possible to transfer a heteroatom (*e.g.* a halogen atom from bromotrichloromethane - Scheme 6.7, Section 6.2.2.4),

$$P_n^{\bullet} + Br-CCl_3 \longrightarrow P_n-Br + {\bullet}CCl_3$$

Scheme 6.7

or a group of atoms (*e.g.* from diphenyl disulfide - Scheme 6.8, Section 6.2.2.2).

$$P_n^{\bullet} + PhS-SPh \longrightarrow P_n-SPh + {\bullet}SPh$$

Scheme 6.8

Group transfer processes are of particular importance in the production of telechelic or di-end functional polymers.

The following sections detail the chemistry undergone by specific transfer agents that react by atom or group transfer by a homolytic substitution mechanism. Thiols, disulfides, and sulfides are covered in Sections 6.2.2.1, 6.2.2.2 and 6.2.2.3 respectively, halocarbons in Section 6.2.2.4, and solvents and other agents in Section 6.2.2.5. The transfer constant data provided have not been critically assessed or evaluated but are included to show the order of magnitude of these values and to provide a guide to the relative reactivity of the various reagents.

6.2.2.1 Thiols

Traditionally thiols or mercaptans are perhaps the most commonly used transfer agents in radical polymerization. They undergo facile reaction with propagating (and other) radicals with transfer of a hydrogen atom and form a saturated chain end and a thiyl radical (Scheme 6.6). Some typical transfer constants are presented in Table 6.2. The values of the transfer constants depend markedly on the particular monomer and can depend on reaction conditions.[43,44]

Table 6.2 Transfer Constants (60 °C, bulk) for Thiols (RSH) with Various Monomers[a]

Transfer agent R	C_{tr}				
	MMA	MA	AN	S	VAc
H	-	-	0.30[b]	5[c]	-
n-C_4H_9-	0.67[31]	1.7[d,31]	-	22 [31]	48[e,31]
n-$C_{12}H_{25}$-	0.7[45]	1.5[g,45]	0.73[b]	16 [46]	-
HO-CH_2CH_2- (**3**)	0.62	-	-	-	-
$HOC(=O)CH_2CH_2$-	0.38[f]	-	-	9.4	-
$CH_3OC(=O)CH_2$-	0.30[h,47]	0.64[g,h,47]	-	1.4[h,47]	0.07[h,47]
H_3N^+-CH_2CH_2-	0.11[i,43]		-	11 [43]	-
Ph-	2.7[48]	-	-	0.08	-

a Numbers are taken from the Polymer Handbook[49] unless otherwise stated and have been rounded to two significant figures. b 50 °C. c At 70 °C. d In ethyl acetate solvent. e Substantial retardation observed.[50] f Extrapolated to 60 °C from the data given. The activation energies quoted[51] appear to be calculated incorrectly. g BA. h In benzene solvent. i The corresponding free amine is reported to have a very low transfer constant in MMA polymerization.[43] It may be consumed in a Michael reaction with monomer.

Thiols react more rapidly with nucleophilic radicals than with electrophilic radicals. They have very large C_{tr} with S and VAc, but near ideal transfer constants (C_{tr} ~ 1.0) with acrylic monomers (Table 6.2). Aromatic thiols have higher C_{tr} than aliphatic thiols but also give more retardation. This is a consequence of the poor reinitiation efficiency shown by the phenylthiyl radical. The substitution pattern of the alkanethiol appears to have only a small (<2-fold) effect on the transfer constant. Studies on the reactions of small alkyl radicals with thiols indicate that the rate of the transfer reaction is accelerated in polar solvents and, in particular, water.[52] Similar trends are observed for transfer to **1** in S polymerization with C_{tr} = 1.4 in benzene 3.6 in CH_3CN and 6.1 in 5% aqueous CH_3CN.[44] In copolymerizations, the thiyl radicals react preferentially with electron-rich monomers (Section 3.4.3.2).

Bamford and Basahel[53] have investigated the importance of penultimate unit effects on the reactivity of *n*-butanethiol in a number of copolymerizations (S-MMA, S-MA) using the technique of "moderated copolymerization". Their data indicate that penultimate unit effects are unimportant in these systems. More recently, de la Fuente and Madruga[45] have come to similar conclusions for the reactivity of dodecanethiol in BA-MMA copolymerization. This contrasts with findings for transfer to carbon tetrabromide (Section 6.2.2.4). It has also been found, again in contrast with halocarbons, that C_{tr} for various primary and secondary thiols is essentially independent of chain length for chain lengths ≥ 2 (Table 6.1).

A range of functional thiols [*e.g.* thioglycolic acid (**2**) and mercaptoethanol (**3**)] has been used to produce monofunctional polymers[10,54-56] (Section 7.5.2) and thence as precursors for diblock copolymers.[47]

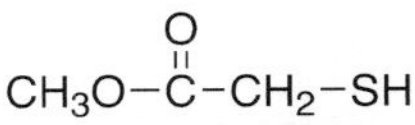

$$CH_3O-\overset{O}{\underset{\|}{C}}-CH_2-SH \qquad HO-\overset{O}{\underset{\|}{C}}-CH_2-SH \qquad HO-CH_2-CH_2-SH$$

$$\textbf{1} \qquad\qquad\qquad \textbf{2} \qquad\qquad\qquad \textbf{3}$$

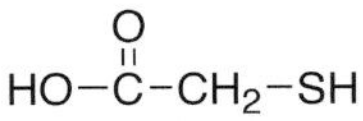

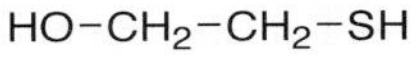

6.2.2.2 Disulfides

A wide range of dialkyl[57] and diaryl disulfides,[58,59] diaroyl disulfides,[60] and xanthogens[61] has been used as transfer agents (Scheme 6.8). Their use ideally leads to the incorporation of functionality at both ends of the polymer chain, thus they find application in the synthesis of telechelics (Section 7.5.2).

The C-S bond of the sulfide end groups can be relatively weak and susceptible to thermal and photo- or radical-induced homolysis. This means that certain disulfides [for example **7-9**] may act as iniferters in living radical polymerization and they can be used as precursors to block copolymers (Sections 7.5.1 and 9.3.2).

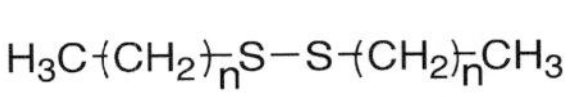

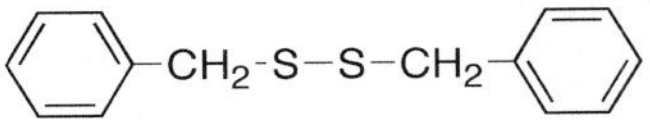

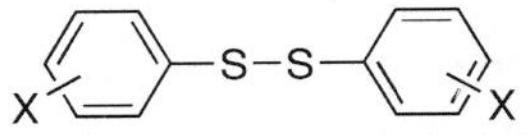

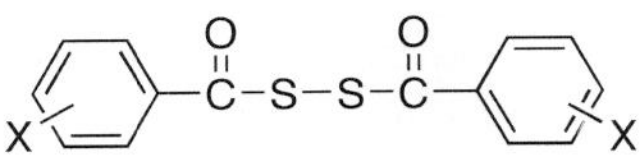

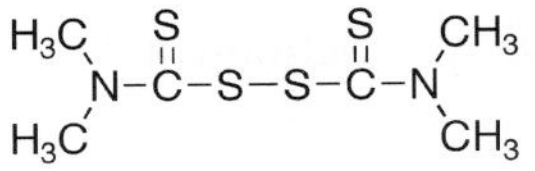

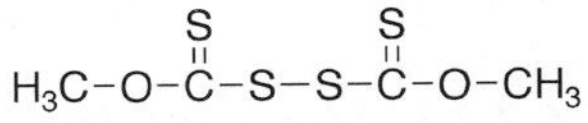

Aliphatic disulfides **4** are not particularly reactive in chain transfer towards MMA and S (Table 6.3). However, they appear to be ideal transfer agents ($C_{tr} \sim 1.0$) for VAc polymerizations.

The reactivity of diphenyl (**6**, X=H) and dibenzoyl (**7**, X=H) disulfide derivatives is higher than aliphatic derivatives. The value of C_{tr} depends markedly on the substituents, X, and on the pattern of substitution. Electron withdrawing substituents (e.g. X = p-CN or p-NO$_2$) may increase C_{tr} by an order of magnitude.[59,60] However, these compounds also give marked retardation.

Compounds with a thiocarbonyl α to the S-S bond such as the dithiuram (e.g. **8**)[62,63] and xanthogen disulfides (e.g. **9**)[64] have transfer constants that are much higher than other disulfides. In part, this may be due to the availability of another mechanism for induced decomposition (Scheme 6.9) involving addition to the C=S double bond and subsequent fragmentation. Thiocarbonyl double bonds are very reactive towards addition and an addition-fragmentation mechanism has been demonstrated for related compounds (Section 6.2.3.5).

Table 6.3 Transfer Constants for Disulfides (R-S-S-R) With Various Monomers[a]

Transfer agent R	C_{tr}			
	MMA	MA	S	VAc
C$_2$H$_5$- (**4**, n=1)	0.00013	-	-	-
n-C$_4$H$_9$- (**4**, n=3)	-	-	0.0024[57]	1.0
PHCH$_2$- (**5**)	0.0063	-	0.01	-
EtOC(C=O)CH$_2$-	0.00065	-	0.015	1.5
Ph- (**6**, X=H)	0.0085[59]	-	0.15	-
PhC(=O)-[b] (**7**, X=H)	0.0010[60]	-	0.0036[60]	-
p-CNC$_6$H$_4$C(=O)-[b] (**7**, X=CN)	0.029[60]	-	0.32[60]	-
(CH$_3$)$_2$NC(=S)-[c] (**8**)	0.53[62]	-	0.57[63]	-
CH$_3$OC(=S)-[d] (**9**)	1.1[64,d]	4.9[64,d,e]	3.1[64,d]	[64,d,f]

a 60 °C, bulk unless indicated otherwise. Numbers are taken from the Polymer Handbook[49] unless otherwise stated, and have been rounded to two significant figures. Where a choice of numbers is available the average value has usually been quoted. b These numbers are reported incorrectly in the Polymer Handbook and many other compilations. c 80 °C. d in benzene. e BA. f inhibition.

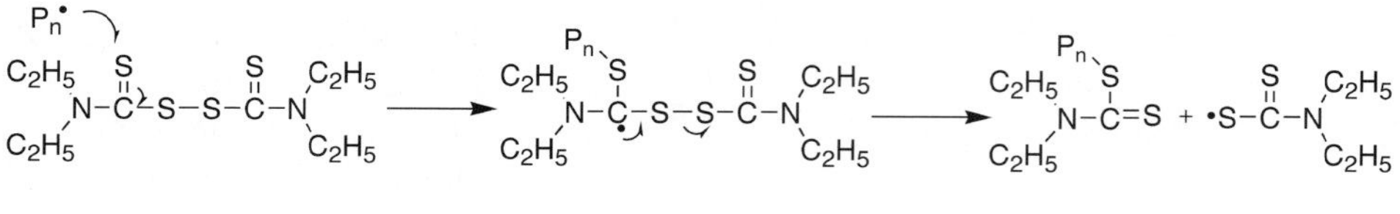

Scheme 6.9

6.2.2.3 Monosulfides

Most monosulfides generally have very low transfer constants. Exceptions to this rule are allyl sulfides (Section 6.2.3.2) and thiocarbonylthio compounds such as the trithiocarbonates and dithioesters (Section 9.5.3) that react by an addition-fragmentation mechanism.

Chain Transfer

t-Butanesulfide (**10**) has a substantially higher transfer constant than other saturated monosulfides (C_{tr} = 0.025 in S polymerization,[57] *n*-butane sulfide has C_{tr} = 0.0022). This result appears counterintuitive if the reaction involves homolytic substitution on sulfur. Pryor and Pickering[57] proposed that this compound may react by hydrogen atom transfer and fragmentation as shown in Scheme 6.10.

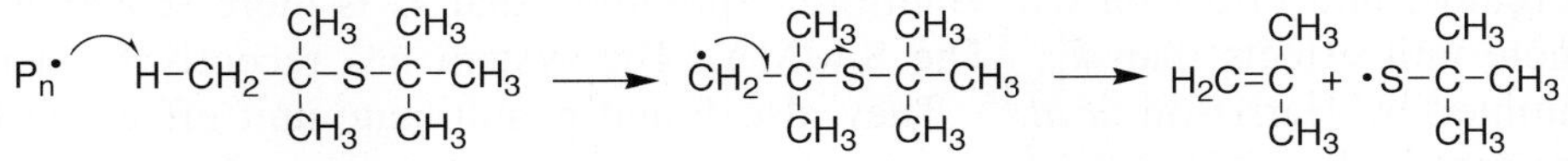

$$P_n^\bullet \quad H-CH_2-\underset{\underset{CH_3}{|}}{\overset{\overset{CH_3}{|}}{C}}-S-\underset{\underset{CH_3}{|}}{\overset{\overset{CH_3}{|}}{C}}-CH_3 \quad \longrightarrow \quad \dot{C}H_2-\underset{\underset{CH_3}{|}}{\overset{\overset{CH_3}{|}}{C}}-S-\underset{\underset{CH_3}{|}}{\overset{\overset{CH_3}{|}}{C}}-CH_3 \quad \longrightarrow \quad H_2C=\underset{\underset{CH_3}{|}}{\overset{\overset{CH_3}{|}}{C}} \;+\; {}^\bullet S-\underset{\underset{CH_3}{|}}{\overset{\overset{CH_3}{|}}{C}}-CH_3$$

10

Scheme 6.10

6.2.2.4 Halocarbons

Halocarbons including carbon tetrachloride, chloroform, bromotrichloromethane[65] (Scheme 6.7) and carbon tetrabromide have been widely used for the production of telomers and transfer to these compounds has been the subject of a large number of investigations.[11] Representative data are shown in Table 6.4. Telomerization involving halocarbons has also been developed as a means of studying the kinetics and mechanism of radical additions.[66]

Table 6.4 Transfer Constants (60 °C, bulk) for Halocarbons with Various Monomers[a]

	$C_{tr} \times 10^4$				
	MMA	BA	AN	S	VAc
CBr$_4$	2700	-	500	2200	7.4×10^6
CCl$_4$	2.4	3.2	0.85	130	9600 [50]
CHCl$_3$	1.8	0.89	5.7	0.5	150 [50]

a Numbers are taken from the Polymer Handbook[49] unless otherwise stated, and have been rounded to two significant figures.

The perhalocarbons, CCl$_4$ and CBr$_4$, react with carbon-centered radicals by halogen-atom transfer to form a perhaloalkyl radical. Halogen atom abstractability decreases in the series iodine>bromine>chlorine. Halohydrocarbons may in principle react by hydrogen-atom, halogen-atom transfer or both. The preferred pathway can often be predicted by considering the relative C-X bond strengths (Section 2.4). For CHCl$_3$, transfer of a hydrogen atom is favored.

The halocarbons react more rapidly with nucleophilic radicals than with electrophilic radicals. Thus, values of C_{tr} with S and VAc are substantially higher than those with acrylic monomers (Table 6.4) where the transfer constant is close to ideal (C_{tr}=1.0). The haloalkyl radicals formed have electrophilic character (Section 2.3.2).

Bamford[14] demonstrated that C_{tr} for transfer to carbon tetrabromide in copolymerization is subject to penultimate unit effects. He found $C_{S.S}=368$, $C_{MA.S}=302$, $C_{MMA.S}=60$ (compare behavior observed with thiols - Section 6.2.2.1). The finding ($C_{MA.S}$~$C_{S.S}$>>$C_{MMA.S}$) suggested that steric factors were more important than either polar or electronic factors in determining the magnitude of the remote unit effect on C_{tr}. Bamford.[14] proposed that k_{tr} is more sensitive to remote unit effects than k_p. The S/MMA/CBr$_4$ system has recently been re-examined by Harrisson *et al.*[15] They also found penultimate unit effects to be important in the S/MMA/CCl$_4$ system. Further evidence for remote unit effects is that C_{tr} in MA and S polymerizations is chain length dependent for chain lengths ≤3 units (Table 6.1). A variation in C_{tr} with chain length for ethylene polymerization has been attributed to polar effects. The electron donating ability of the alkyl chain increases in the series: ethyl<butyl<hexyl.[11]

Ameduri and Boutevin[67] showed that certain transition metal salts and complexes effectively catalyze transfer to the halocarbons. In these cases, initiation/reinitiation involves a redox reaction between the metal and the halocarbon. A transition metal in its oxidized form then reacts with the propagating radical by group transfer to regenerate the metal in its original oxidation state. Transition metal species that are effective in this context, include copper salts and RuCl$_2$(PPh$_3$). Effective transfer constants are substantially higher than when the transfer agent is used alone. Narrow polydispersities were not obtained. Nonetheless, these experiments can be considered to mark the beginnings of ATRP (Section 9.4).

Certain alkyl iodides give reversible chain transfer with S and some fluoro-olefins (Section 9.5.4). In these cases, the polymerization can show some living characteristics.

6.2.2.5 *Solvents and other reagents*

Many solvents and additives have measurable transfer constants (Table 6.5). The accuracy of much of the transfer constant data in the literature is questionable with values for a given system often spanning an order of magnitude. In some cases the discrepancies may be real and reflect differences in experimental conditions. In other cases they are less clear and may be due to difficulties in molecular weight measurements or other problems.

Nonetheless, it is clear that the reactivity of solvents in transfer reactions depends on the nature of the propagating species and some general conclusions can be drawn. The propagating species derived from MMA has relatively little tendency to undertake transfer. That derived from VAc appears extremely reactive towards solvents and other transfer agents (note, however that many reagents give marked retardation with VAc[50]). The factors influencing reactivity in hydrogen atom abstraction reactions are discussed in general terms in Section 2.4.

Table 6.5 Transfer Constants (60 °C, bulk) for Selected Solvents and Additives with Various Monomers[a]

Solvent	$C_{tr} \times 10^4$				
	MMA	MA	AN	S	VAc
benzene	0.04	0.3[b]	2.5	0.02	3.0
toluene	0.20	2.7	5.8	0.12	21
acetone	0.20	0.23	1.1	0.32	12
butan-2-one	0.45	3.2[b]	6.4	5.0	74
ethyl acetate	0.15	-	2.5	5.7	3.0
triethylamine	8.3	400	790	7.1	370

a Numbers have been selected from the Polymer Handbook[49] or references given therein and have been rounded to two significant figures.   b 80 °C.

Mechanisms for chain transfer depend on the particular solvent or reagent. Many solvents have abstractable hydrogens (*e.g.* acetone, butanone, toluene) and may react by loss of those hydrogens (Scheme 6.11).

$$P_n^\bullet + H-CH_2-\overset{O}{\underset{\|}{C}}-CH_3 \longrightarrow P_n-H + {}^\bullet CH_2-\overset{O}{\underset{\|}{C}}-CH_3$$

Scheme 6.11

Benzene may react by addition as shown in Scheme 6.12 (this pathway is also open to other aromatic solvents). The cyclohexadienyl radical is a poor initiating species and may terminate a second chain by hydrogen atom transfer. According to this process, benzene is a retarder rather than a transfer agent.

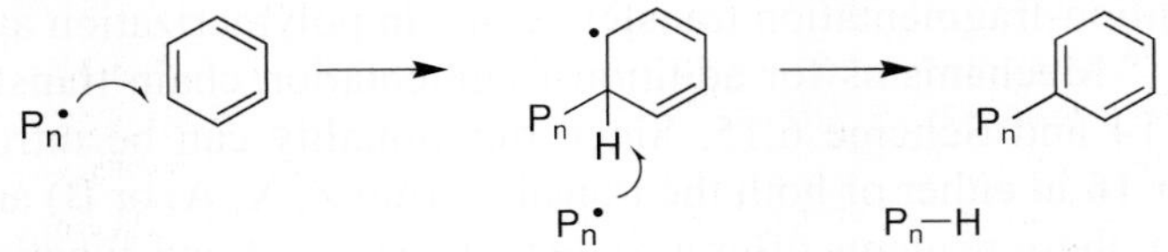

Scheme 6.12

In the case of S, it has been proposed that reinitiation may occur by hydrogen-atom transfer to monomer (Scheme 6.13).[12,68]

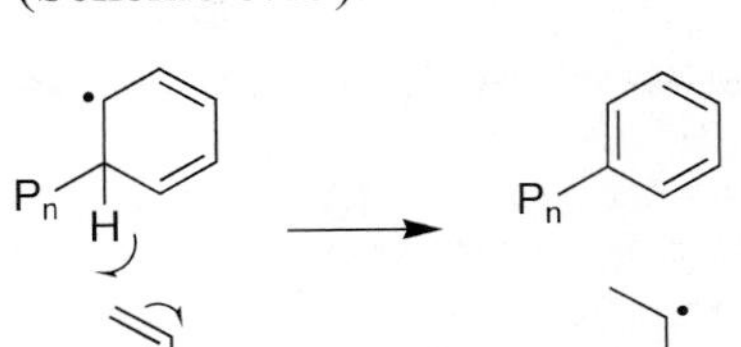

Scheme 6.13

6.2.3 Addition-Fragmentation Chain Transfer Agents

Addition-fragmentation chain transfer has been reviewed by Rizzardo et al,[69] Colombani and Chaumont,[70] Colombani,[71] Yagci and Reetz,[72] and Chiefari and Rizzardo.[3] Certain unsaturated compounds may act as transfer agents by a two-step addition-fragmentation mechanism. All of the compounds discussed in this section have the general structure **11** or **12** where C=X is a reactive double bond (X is most often carbon or sulfur) Z is a group chosen to give the transfer agent an appropriate reactivity with respect to the monomer(s), A is typically CH_2, O or S, B is typically O and R is a radical leaving group. Chain transfer to monomer in VC polymerization (Section 6.2.6.3) and transfer to benzene (6.2.2.5) can also be considered as examples of addition-fragmentation chain transfer.

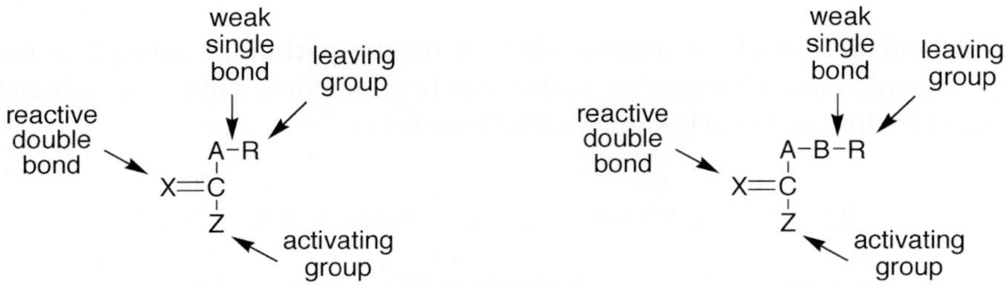

Radical addition-fragmentation processes have been exploited in synthetic organic chemistry since the early 1970's.[73-75] Allyl transfer reactions with allyl stannanes and the Barton-McCombie deoxygenation process with xanthates are two examples of reactions known to involve a S_H2' mechanism. However, the first reports of addition-fragmentation transfer agents in polymerization appeared in the late 1980's.[76-78] Mechanisms for addition-fragmentation chain transfer are shown in Scheme 6.14 and Scheme 6.15. Since functionality can be introduced to the products **14** or **16** in either or both the transfer (from Z, X, A, or B) and reinitiation (from R) steps, these reagents offer a route to a variety of end-functional polymers including telechelics.

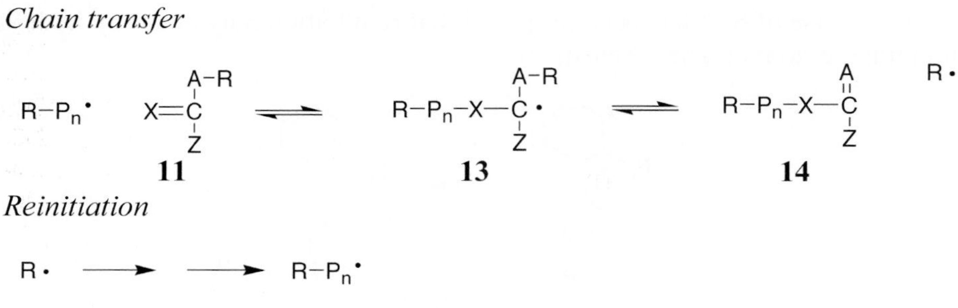

Scheme 6.14

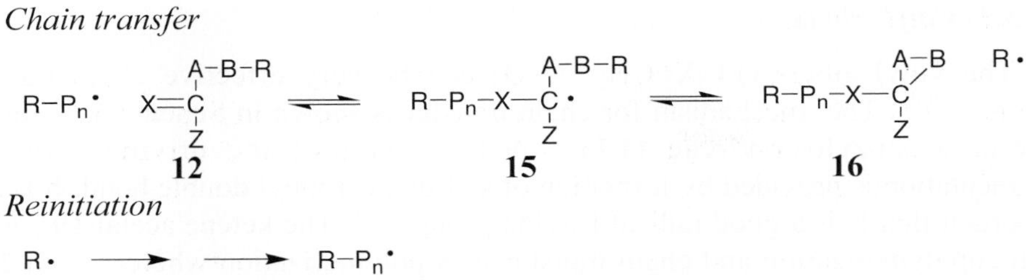

Chain transfer

$$R-P_n^{\bullet} \quad \underset{\underset{Z}{|}}{\overset{\overset{A-B-R}{|}}{X=C}} \quad \rightleftharpoons \quad R-P_n-\underset{\underset{Z}{|}}{\overset{\overset{A-B-R}{|}}{C}}^{\bullet} \quad \rightleftharpoons \quad R-P_n-X-\underset{\underset{Z}{|}}{\overset{\overset{A-B}{\diagup}}{C}} \quad R^{\bullet}$$

$$\qquad\qquad\quad \mathbf{12} \qquad\qquad\qquad\qquad\quad \mathbf{15} \qquad\qquad\qquad\qquad\quad \mathbf{16}$$

Reinitiation

$$R^{\bullet} \longrightarrow \longrightarrow R-P_n^{\bullet}$$

Scheme 6.15

Rates of addition to transfer agents **11**, **12** are determined by the same factors that determine rates of addition to monomers (Section 2.3). Substituents on the remote terminus of a double bond typically have only a minor influence. Thus, in most cases, the double bonds of the transfer agents have a reactivity towards propagating radicals that is comparable with that of the common monomers they resemble. With efficient fragmentation, transfer constants can be close to unity. The radicals formed by addition typically have low reactivity towards further propagation and other intermolecular reactions because of steric crowding about the radical center.

Efficient transfer requires that radicals formed by addition undergo facile β–scission (for **13**) or rearrangement (for **15**) to form a new radical that can reinitiate polymerization. The driving force for fragmentation of the intermediate radical is provided by cleavage of a weak A-R bond and/or formation of a strong C=X bond (for **11**). If fragmentation leads preferentially back to starting materials the transfer constant will be low. If the overall rate of β-scission is slow relative to propagation then retardation may result. The adducts (**13** and **15**) then have the potential to undergo side reactions by addition (*e.g.* copolymerization of the transfer agent) or radical-radical termination. Retardation is an issue particularly for high k_p monomers such as VAc and MA. In designing transfer agents and choosing an R group (see **11**, **12**), a balance must be achieved between the leaving group ability of R and reinitiation efficiency by R•.

When the product of the reaction is itself a potential transfer agent or macromonomer (**11**, X=A=CH$_2$, X=A=S) block, graft or hyperbranched copolymer formation may be an issue particularly at high conversions.[76,79] The design of transfer agents that give reversible addition-fragmentation chain transfer (RAFT) has provided one of the more successful approaches to living radical polymerization (Sections 9.5.2 and 9.5.3). The pathway can be blocked by choice of A (see **11**). For example, when A is oxygen (vinyl ethers, Section 6.2.3.1) or bears an alkyl substituent (*e.g.* A=CH-CH$_3$), the product is unreactive to radical addition.

If R and Z, A or X are connected to form a ring structure the result is a potential ring opening monomer. For many of the transfer agents in this section there are analogous ring-opening monomers described in Section 4.4.2.

6.2.3.1 Vinyl ethers

The vinyl ethers (**11** X=CH$_2$, A=O) can be very effective chain transfer agents.[78,80-82] The mechanism for chain transfer is shown in Scheme 6.16 for the case of α-benzyloxystyrene (**17**). A large part of the driving force for fragmentation is provided by formation of a strong carbonyl double bond. It is also important that R is a good radical leaving group.[81,83] The ketene acetal **19**[83] gives both copolymerization and chain transfer in S polymerization whereas with **20**,[83] and **17**[78] and **21-23**[80] chain transfer is the only reaction detected. Transfer constants for some vinyl ether transfer agents are provided in Table 6.6. Those with a benzyl radical leaving group are designed for use in S or (meth)acrylate ester polymerization and give retardation in VAc polymerization. The polymers formed have a ketone end group (*e.g.* **18**, Scheme 6.16). Additional functionality can be introduced on Z or R (refer **11**) to modify reactivity or to tailor the end groups as in the examples (**24-26**).[82]

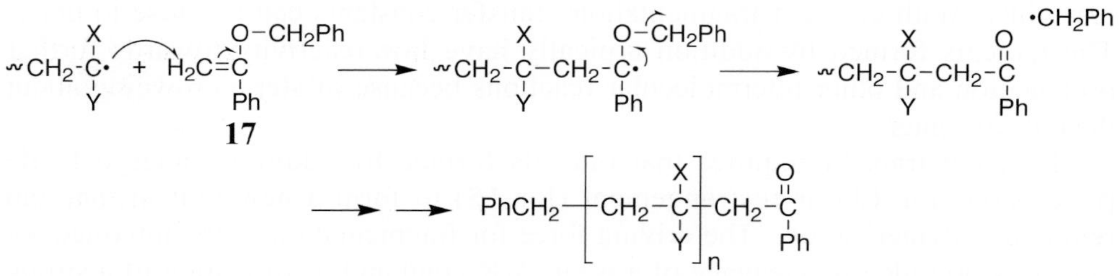

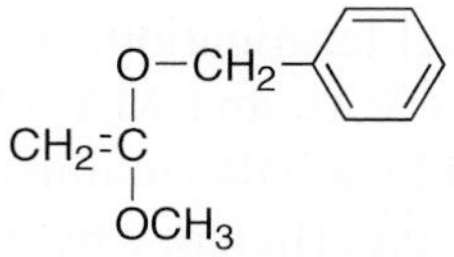

Scheme 6.16

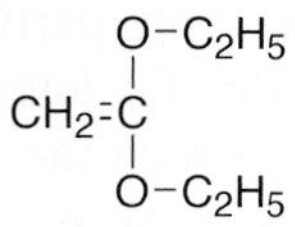

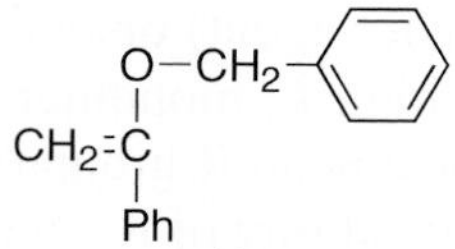

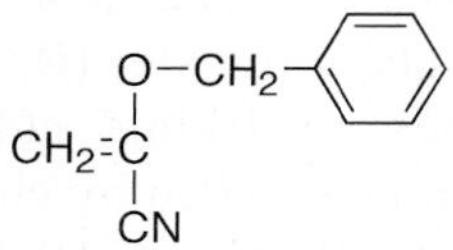

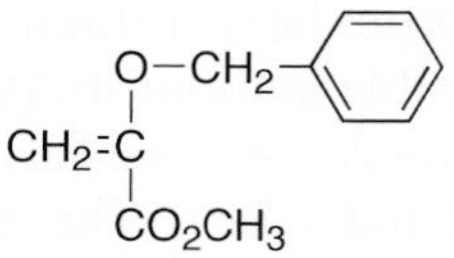

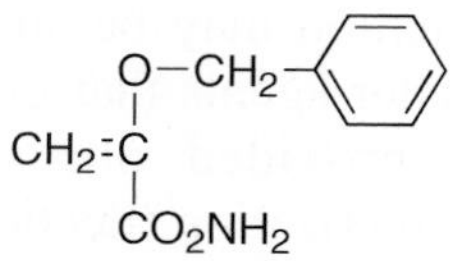

Table 6.6 Transfer Constants for Vinyl Ethers at 60 °C[a]

Transfer agent	C_{tr} for monomer[b]				References
	S	MMA	MA	VAc	
17	0.26	0.76	5.7[c]	9.7[c]	78,80
21	0.036	0.081	0.3[c]	12[c]	80
22	0.046	0.16	0.54[c]	20[c]	80
23	0.2	0.5	1.1[c]	--	80

a Bulk, medium comprises only monomer and transfer agent.   b Transfer constants rounded to two significant figures.   c Significant retardation observed.

The vinyl ether transfer agents, like other vinyl ethers, can show marked acid sensitivity. They are not suited for use with acid monomers. Even traces of acidic impurities in the monomer or the polymerization medium can catalyze decomposition of the transfer agent.

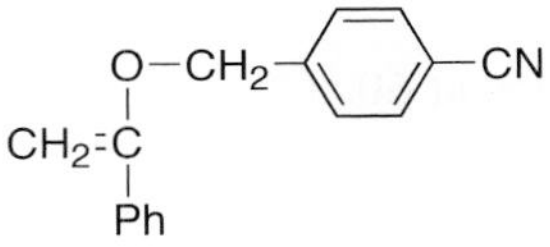

24 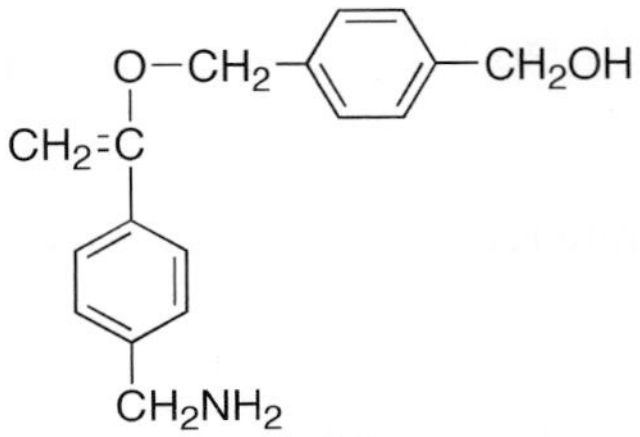 **25** 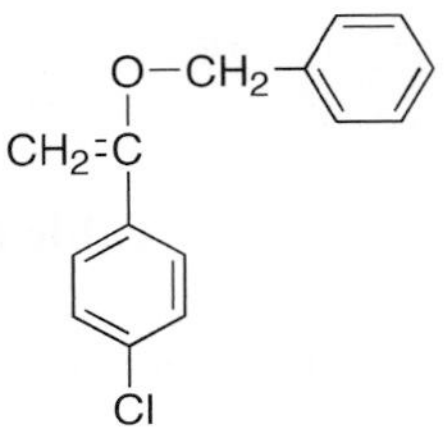 **26**

6.2.3.2 *Allyl sulfides, sulfonates, halides, phosphonates, silanes*

With allyl transfer agents (*e.g.* **11** X=CH$_2$, A=CH$_2$) such as allyl halides,[84-90] sulfides,[77,91,92] sulfones,[84] sulfonates,[84,93] silanes[84] phosphonates[84] and similar compounds,[84] the main driving force is the weak single bond (A-R) of **11**. A similar situation pertains with the corresponding dienyl transfer agents *e.g.* **11** X=CH$_2$, A is CH$_2$=CH$_2$-CH$_2$-.[94,95] The proposed mechanism of chain transfer is shown in Scheme 6.17 for the case of the allyl sulfide **27**. The product will be predominantly a macromonomer (**28**) that may be reactive under the polymerization conditions particularly at high conversion (Section 6.2.3.4).

Some typical transfer constants for allyl sulfides are given in Table 6.7. The values of C_{tr} for these reagents are less dependent on the particular monomer than those for halocarbons (Table 6.2) or thiol transfer agents (Table 6.4). The low transfer constant of **32** demonstrates the importance of the activating group Z (*cf.* **11**).

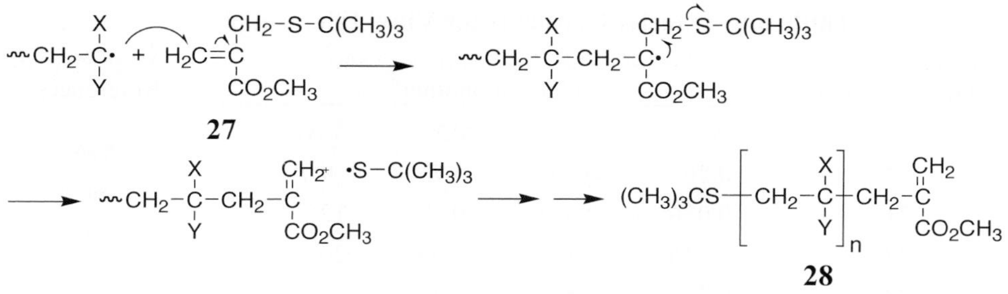

Scheme 6.17

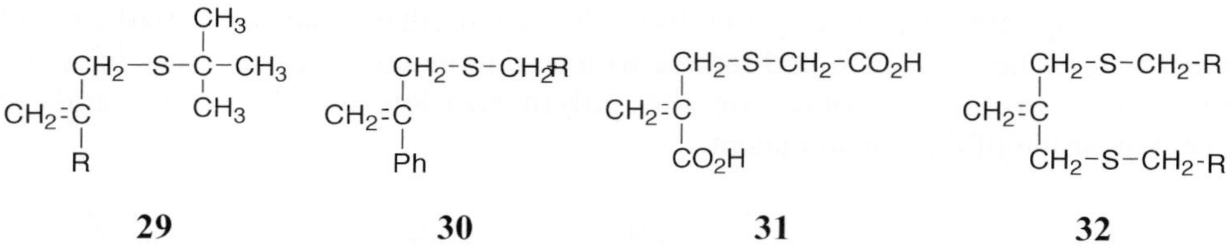

Table 6.7 Transfer Constants for Allyl Sulfides at 60 °C[a]

Transfer agent	C_{tr} for monomer[b]					References
	S	MMA	MA	MAN	VAc	
29, R=Ph	0.80	1.2	4.0[c]	-	~20[c]	77
29, R=CN	1.8	1.4	1.6[c]	-	~60[c]	77
29, R=CO$_2$Et	0.95	0.74	2.2[c]	0.42	~27[c]	77,96,97
30, R=CH$_2$CO$_2$H	0.95	1.1	-	-	-	91
30, R=CH$_2$CH$_2$NH$_2$	0.79	0.91	-	-	-	91
30, R=CH$_2$CH$_2$OH	0.77	1.2	-	-	-	91
31	1.27	0.74	-	-	-	91
32, R=CH$_2$CH$_2$CO$_2$CH$_3$[d]	0.016	-	-	-	-	35
33	0.35	1.11	-	-	-	95
34	1.51c	0.33[c]	-	-	-	94

a Bulk, medium comprises only monomer and transfer agent.   b Transfer constants rounded to two significant figures.   c Significant retardation observed.   d Transfer constants similar for various R.

Allyl sulfonates (**35**, **36**) show analogous behavior. Transfer constants are reported in Table 6.8. Other compounds with weak A-R bonds (*cf.* **11**) that have the capacity to act as transfer agents are listed in Table 6.9. Allyl bromides **43a**, **44**, and **45a** give predominantly chain transfer whereas, the chlorides (*e.g.* **45b**) give copolymerization as well as chain transfer.[84,98] The silane **48** is also able to react as a comonomer.[84] Compounds **11** with R=oxygen are not transfer agents but are comonomers.

33

34

35

36

37

38

39

40

41

42

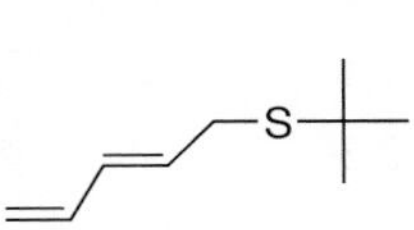

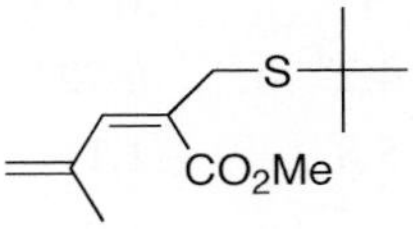

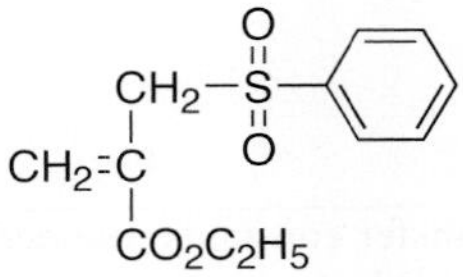

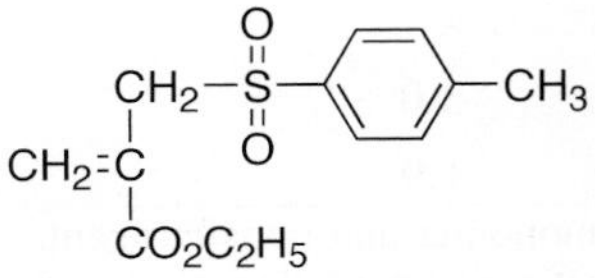

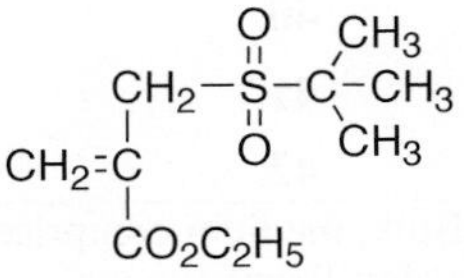

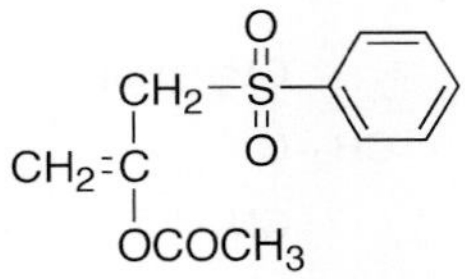

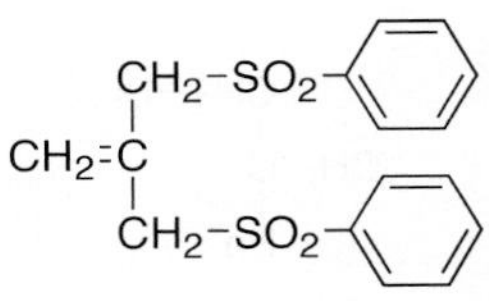

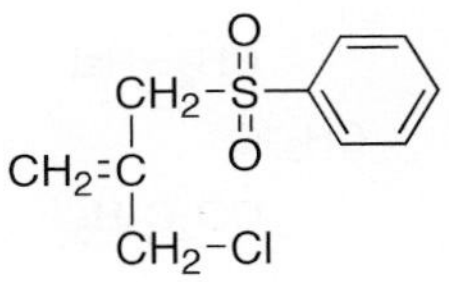

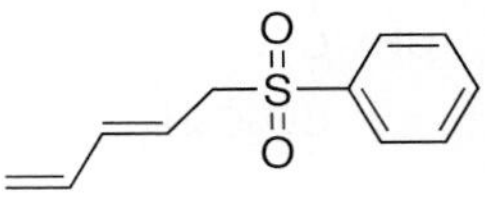

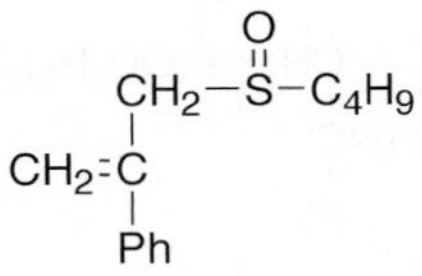

Table 6.8 Transfer Constants for Allyl Sulfonates and Sulfoxides at 60 °C[a]

Transfer agent	C_{tr} for monomer[b]				References
	S	MMA	BA	VAc	
35	4.2[d]	0.72[d]	1.1[d]	-[c]	93
36	5.8	1.1	2.3[d]	-[c]	84
37	-	1.0	-	-	84
38	0.02	0.065	0.20[e]	2.8	84
39	-	-	-	3.9	3
40	-	-	-	0.05	3
41	-	3.0	-	-	99
42	-	1.9	-	-	84

a Bulk, medium comprises only monomer and transfer agent. b Transfer constants rounded to two significant figures. c Significant retardation observed. d 3.46 M monomer in benzene solution[93]. e MA.

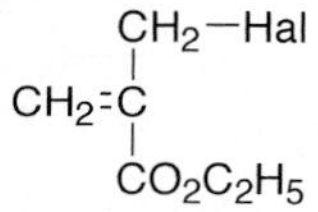

43a Hal=Br
b Hal=Cl

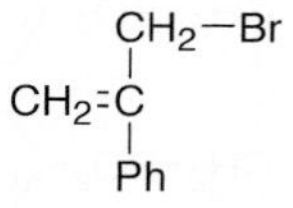

44

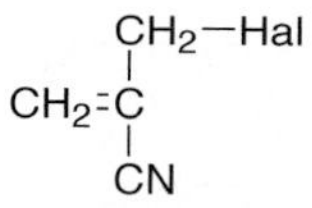

45a Hal=Br
b Hal=Cl

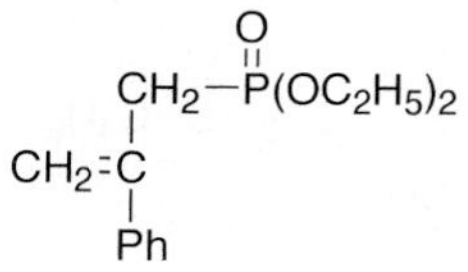

46

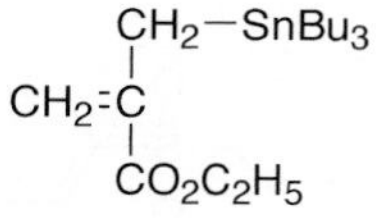

47

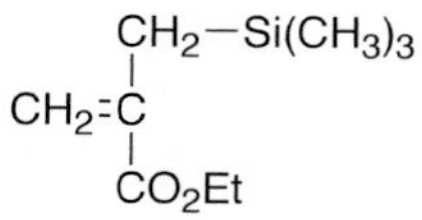

48

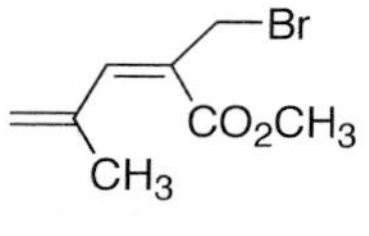

49

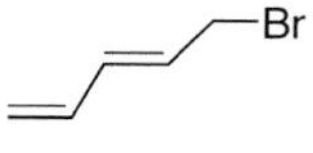

50

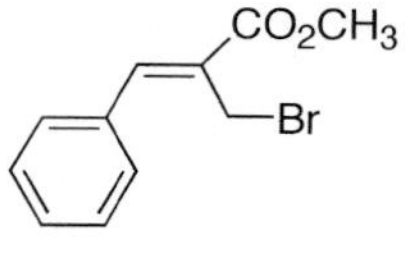

51

Table 6.9 Transfer Constants for Allyl Halides, Phosphonates, Silanes and Stannanes at 60 °C[a]

Transfer agent	C_{tr} for monomer[b]				References
	S	MMA	MA	VAc	
43a, Hal=Br	-	1.5	2.3	-	84
44	2.9	2.3	5.3	-	84
45a, Hal=Br	-	2.2	3.0	-	84
45b, Hal=Cl	-	0.0075[d]	0.046[d]	-	84
46	-	0.4	-	-	84
47	-	3.0	-	-	84
48	-	0.08[d]	-	-	84
49	-	3.4	-	-	99
50	8.1	7.4[c]	-	-	99
51	0.25	-	-	-	100

a Bulk, medium comprises only monomer and transfer agent. b Transfer constants rounded to two significant figures. c Significant retardation observed. d Copolymerization observed.

6.2.3.3 Allyl peroxides

In the case of allyl peroxides (**12** X= CH$_2$, A=CH$_2$, B=O),[101-105] intramolecular homolytic substitution on the O-O bond gives an epoxy end group as shown in Scheme 6.18 (1,3-S$_H$i mechanism). The peroxides **52-59** are thermally stable under the conditions used to determine their chain transfer activity (Table 6.10). The transfer constants are more than two orders of magnitude higher than those for dialkyl peroxides such as di-*t*-butyl peroxide (C_I=0.00023-0.0013) or di-isopropyl peroxide (C_I=0.0003) which are believed to give chain transfer by direct attack on the O-O bond.[49] This is circumstantial evidence in favor of the addition-fragmentation mechanism.

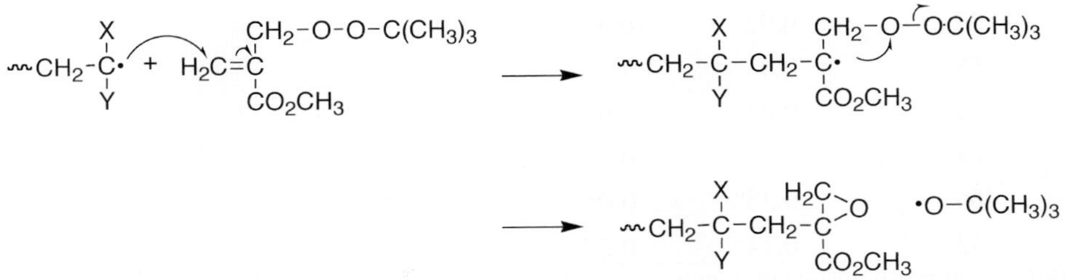

Scheme 6.18

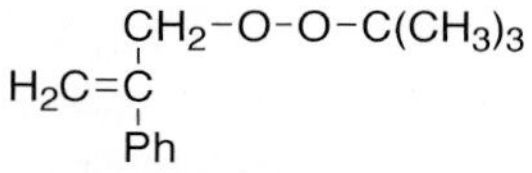

Table 6.10 Transfer Constants for Allyl Peroxide and Related Transfer Agents at 60 °C[a]

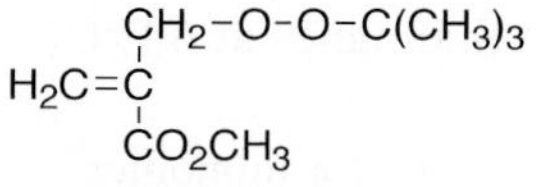

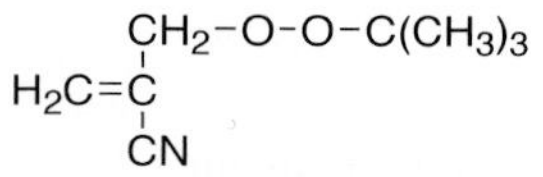

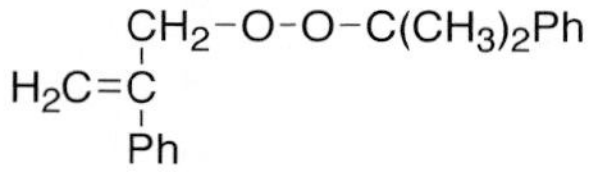

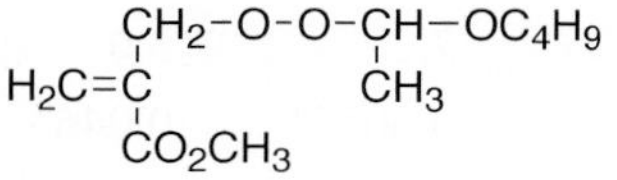

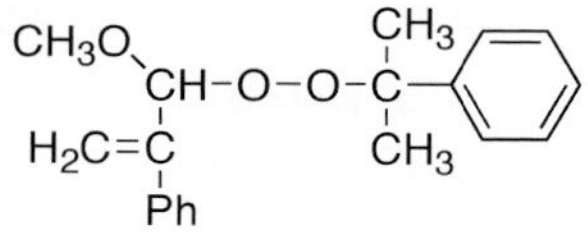

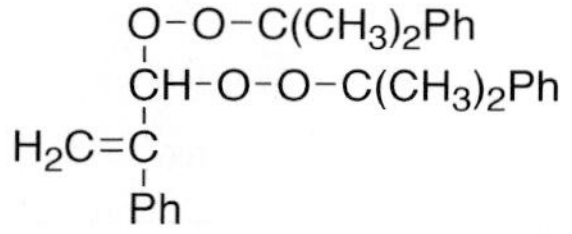

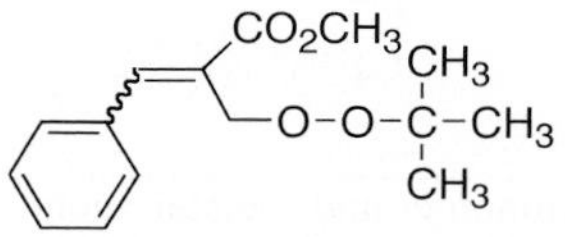

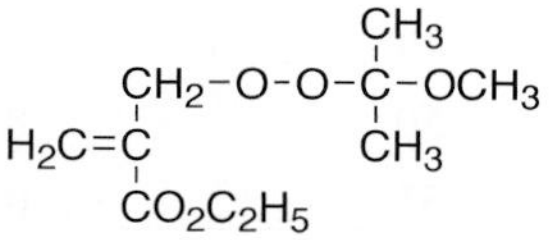

Transfer agent	C_{tr} for monomer[b]				References
	S	MMA	MA	VAc	
52	0.9	0.8	-	-	82,101
53	1.6	0.6	1.0	-	82,101
54	2.0	0.9	0.7	-	82
55	0.8	0.8	-	-	82,101
57	0.92	0.49	1.9	-	106
58	0.9	-	-	-	106
59	0.22	0.012	0.08	3.7	100
61	0.82[c]	0.31[c]	-	-	107
60	0.35[d]	0.05[d]	0.46[e,d]	1.3[d]	44,47
62	0.14	0.57	1.31[e]	-	108

a Bulk, medium comprises only monomer and transfer agent. b Transfer constants rounded to two significant figures. c Compound is also an initiator under the polymerization conditions. Transfer constant obtained using a modified Mayo equation.[107] d In benzene. e BA.

Chain Transfer

Peroxyacetals **58**[106] and peresters such as **61**[107] are also effective transfer agents, however, at typical polymerization temperatures (~60 °C) they are thermally unstable and also act as initiators. Compounds such as **62** which may give addition and 1,5-intramolecular substitution with fragmentation have also been examined for their potential as chain transfer agents (1,5-S_Hi mechanism).[108]

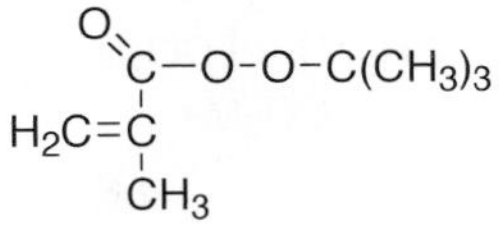

61 H$_2$C=C(CH$_3$)–C(=O)–O–O–C(CH$_3$)$_3$

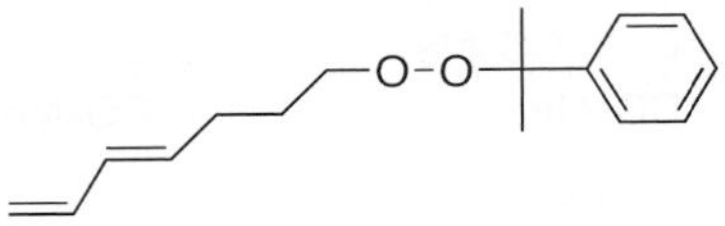

62

6.2.3.4 Macromonomers

The chain transfer agents (**11** X=CH$_2$, A=CH$_2$) are misnamed 'macromonomers' since in this context they do not behave as macromonomers. Copolymerization when it occurs is a side reaction. The mechanism is shown in Scheme 6.19 for MAA 'trimer' (**63**). The final product (**65**) is a also a 'macromonomer' and formation of the adduct (**64**) and chain transfer is reversible (see also Section 6.2.7.2 and Section 9.5.2).[36,76,79,109]

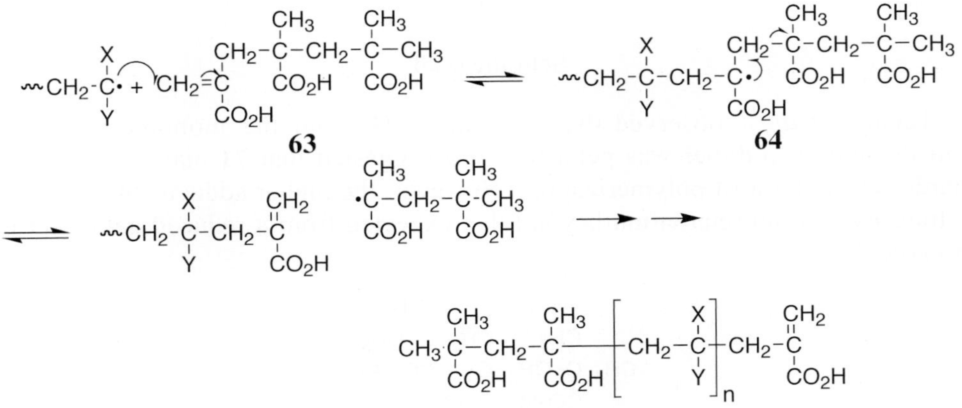

Scheme 6.19

The most used transfer agents in this class are the methacrylate macromonomers (*e.g.* **66-68**) and AMS dimer (**76**). The applications of these compounds are summarized in a review.[110]

The rate constants (k_{add}) for addition of the MMA propagating radical[36] (and other radicals[79]) to **66-68** are believed to be similar. The transfer constant of **66** is thought to be lower than **67** and **68** by more than an order of magnitude because of

306   The Chemistry of Radical Polymerization

an unfavorable partition coefficient. The fragmentation of **70** preferentially gives back **66** and the MMA propagating radical rather than **67** and the monomeric radical **69** (Scheme 6.20). The result has been attributed to steric factors.[36,111]

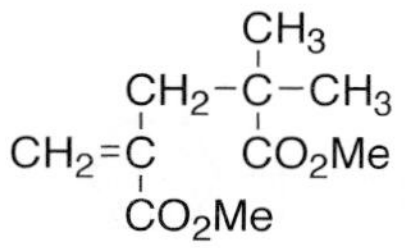

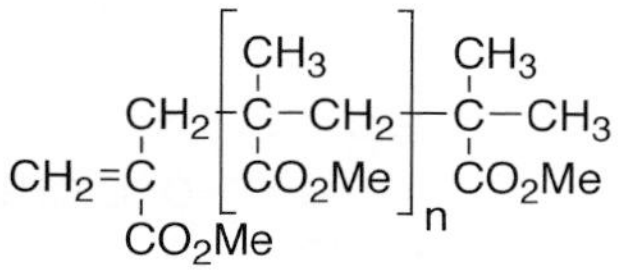

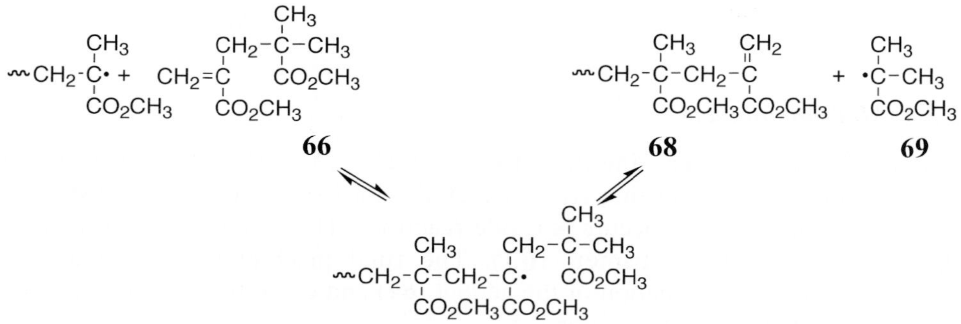

$$
\underset{\mathbf{66}}{\text{CH}_2=\overset{\overset{\displaystyle \text{CH}_2-\overset{\overset{\displaystyle \text{CH}_3}{|}}{\underset{|}{\text{C}}}-\text{CH}_3}{|}}{\underset{|}{\text{C}}}}
\qquad
\underset{\mathbf{67}}{\text{CH}_2=\text{C}(\text{CO}_2\text{Me})-\text{CH}_2-\text{C}(\text{CH}_3)(\text{CO}_2\text{Me})-\text{CH}_2-\text{C}(\text{CH}_3)(\text{CO}_2\text{Me})}
\qquad
\underset{\mathbf{68}}{\text{CH}_2=\text{C}(\text{CO}_2\text{Me})-[\text{CH}_2-\text{C}(\text{CH}_3)(\text{CO}_2\text{Me})]_n-\text{CH}_2-\text{C}(\text{CH}_3)(\text{CO}_2\text{Me})}
$$

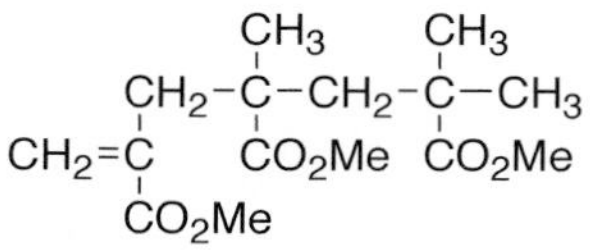

Scheme 6.20

Tanaka *et al.*[109] observed that the adduct **71** from the monomeric MMA radicals adding to dimer was persistent and suggested that **71** may also act as a retarder or inhibitor of polymerization. However, the higher adducts **70** appear to be transient and no retardation beyond that expected from a reduced gel effect is observed.[36]

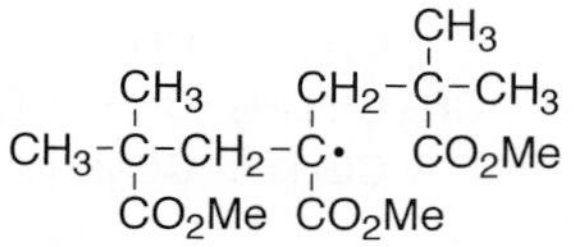

71

Transfer constants of the methacrylate macromonomers in MMA polymerization do not depend on the ester group but are slightly higher for MAA trimer. Compounds **72** and **73** are derived from the MMA trimer (**67**) by selective hydrolysis or hydrolysis and reesterification respectively. They offer a route to telechelic polymers.

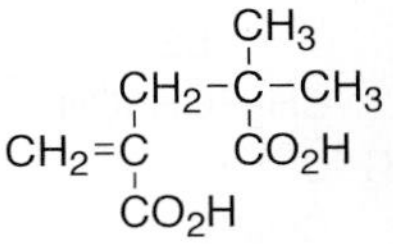

Chain Transfer

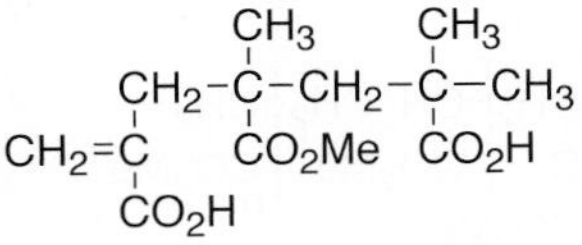

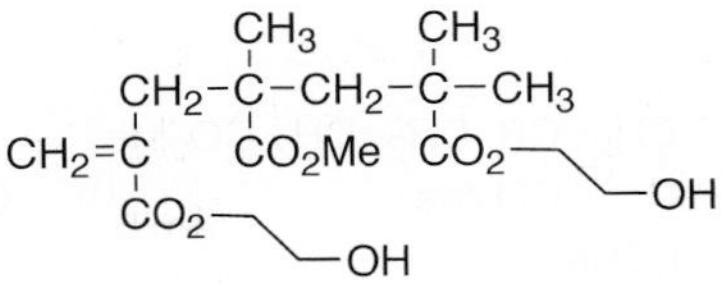

72: $CH_2=C(CO_2H)-CH_2-C(CH_3)_2-CO_2H$

73: $CH_2=C(CO_2H)-CH_2-C(CH_3)(CO_2Me)-CH_2-C(CH_3)_2-CO_2H$

74: $CH_2=C(CO_2-)-CH_2-C(CH_3)(CO_2Me)-CH_2-C(CH_3)(CO_2-)-CH_3$ (with OH groups)

In the case of polymerization of monosubstituted monomers (*e.g.* S, BA) with **66-68**, copolymerization of the macromonomer to form a graft copolymer is a significant side reaction.[76]

Table 6.11 Transfer Constants for Macromonomers[a]

Transfer agent	Temperature (°C)	C_{tr} for monomer[b]			References
		S	MMA	EA	
66	60 °C	-	0.013	0.12[d]	36,112
67	60 °C	0.55[d]	0.19	0.84[d]	36,112
68, n=2	60 °C	-	0.31	-	36
68, av. n=14	60 °C	-	0.21	-	36
72	60 °C	-	0.18	-	111
63	60 °C	-	0.26	-	111
73	60 °C	-	0.18	-	111
74	60 °C	-	0.27	-	111
75	60 °C	-	0.015	-	113
76	110 °C	0.20	0.13	-	114,115
80	60 °C	0.552[c,d]	0.123[c,d]	-	116

a Bulk, medium comprises only monomer and transfer agent. b Transfer constants rounded to two significant figures. c Significant retardation observed. d. Copolymerization observed as side reaction.

For polymerization of MMA in the presence of the macromonomers **77**,[117] **78**[118] and **79**[119] where the leaving group is a primary or secondary radical, the adduct radical partitions between fragmentation and propagation. In the case of **80**, where the leaving group is a more stable secondary radical,[116] fragmentation is the favored pathway but copolymerization is still observed.

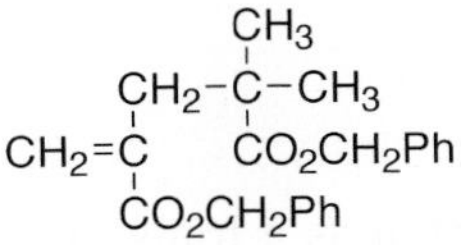

75: $CH_2=C(CO_2CH_2Ph)-CH_2-C(CH_3)_2-CO_2CH_2Ph$

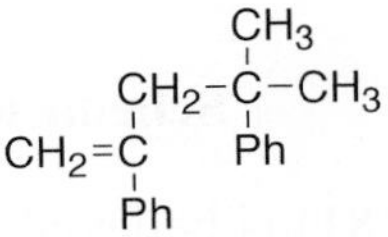

76: $CH_2=C(Ph)-CH_2-C(CH_3)_2-Ph$

77: $CH_2=C(CO_2Me)-CH_2-CH_2-CO_2Me$

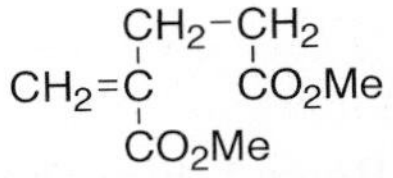

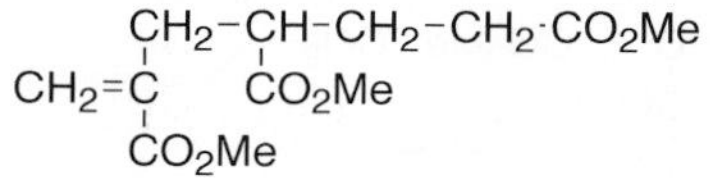

78 — CH₂=C(CO₂Me)–CH₂–CH(CO₂Me)–CH₂–CH₂·CO₂Me (structure)

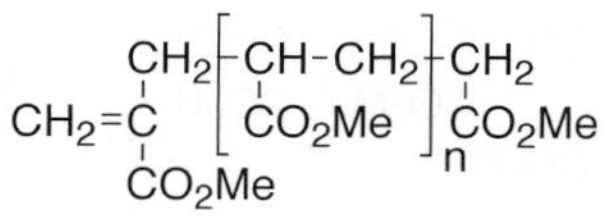

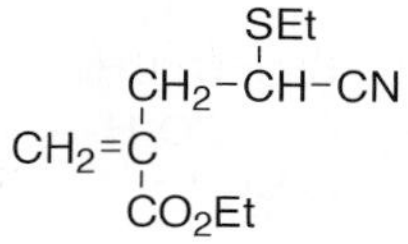

79 — CH₂=C(CO₂Me)–[CH(CO₂Me)–CH₂]ₙ–CH₂ (structure)

80 — CH₂=C(CO₂Et)–CH₂–CH(SEt)–CN (structure)

6.2.3.5 Thionoester and related transfer agents

Other transfer agents which react with propagating species by an addition-fragmentation mechanism include the thione derivatives (**81-83**)[120-122] and RAFT agents (Chapter 9). The thiohydroxamic esters **82** and **83** are sometimes known as Barton esters because of the work of Barton and coworkers who explored their use as radical generators in organic chemistry.[123-125] Transfer constants for some thione derivatives are provided in Table 6.12. The initiating species formed from **82** and **83** are acyloxy radicals which may undergo decarboxylation before initiating a new chain (Scheme 6.21).

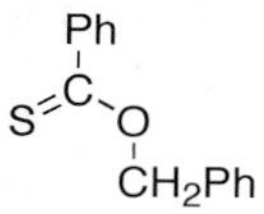

81 — PhC(=S)OCH₂Ph (benzyl thionobenzoate)

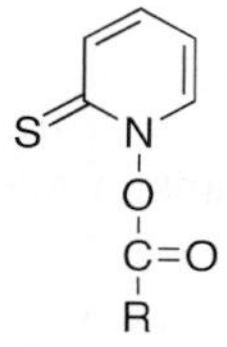

82 — N-acyloxy pyridine-2-thione

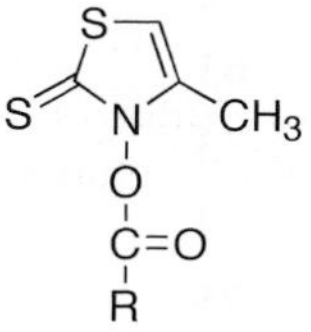

83 — N-acyloxy thiazole-2-thione derivative (4-CH₃)

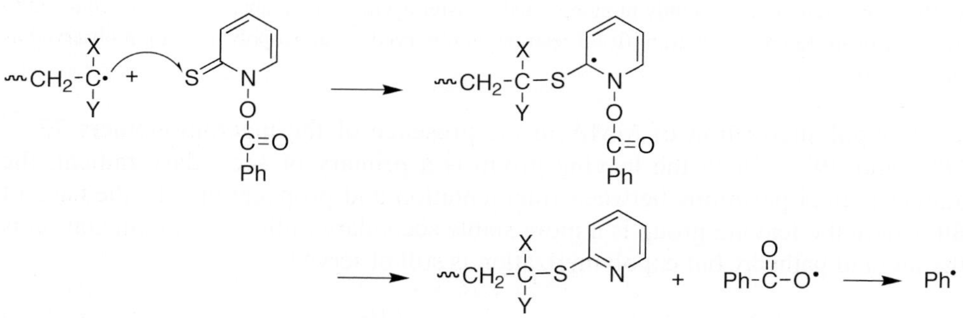

Scheme 6.21

Benzyl thionobenzoate (**81**) is believed to be ineffective as a transfer agent in MMA polymerization because of an unfavorable partition coefficient. PMMA• is a much better radical leaving group than benzyl radical. Analogous benzyl thiocarbonylthio compounds are also ineffective as RAFT agents (Section 9.5.3).

Table 6.12 Transfer Constants for Thionoester and Related Transfer Agents at 60 °C[a]

Transfer agent	C_{tr} for monomer[b]				References
	S	MMA	MA	VAc	
81	1.0	~0	1.2[c]	>20[d]	122
82 R=C$_{15}$H$_{31}$	3.8	4.0	~20[c]	~36[d]	121
82 R=PhCH$_2$	3.9	4.3	-	~80[d]	121
82 R=Ph	-	2.8	-	-	121
83 R=C$_{15}$H$_{31}$	0.3	0.6	3.1	9.7[c]	121
83 R=PhCH$_2$	1.0	1.0	-	18[d]	121

a Bulk, medium comprises only monomer and transfer agent. b Transfer constants rounded to two significant figures. c Significant retardation observed. d Strong retardation observed.

These thiohydroxamic esters have seen use in grafting of PAN onto PE,[126] of PS, PAM and PNIPAM onto cellulose[127,128] and of PS, PMMA, PVP and PAM onto poly(arylene ether sulfone).[129] The process involves derivitization of a parent carboxy functional polymer to form the thiohydoxamic ester **82** (R=polymer) which then behaves as a polymeric transfer agent and/or radical generator.

6.2.4 Abstraction-Fragmentation Chain Transfer

Other multistep mechanisms for chain transfer are possible. An example is abstraction-fragmentation chain transfer shown by silylcyclohexadienes (**84**, Scheme 6.22).[130]

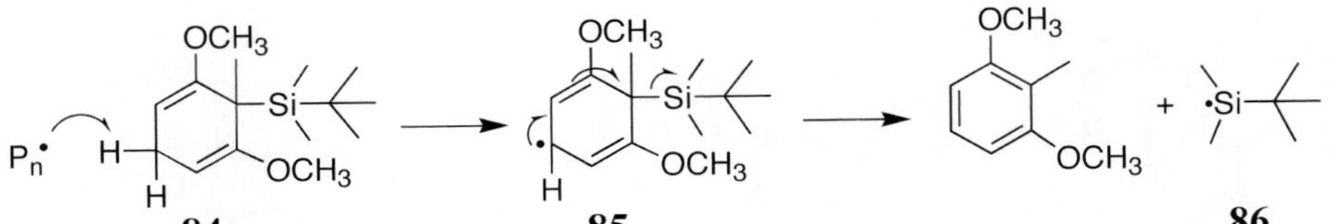

Scheme 6.22

The cyclohexadiene **84** is a good H donor but the cyclohexadienyl radical **85** is slow to react and fragments to provide the silyl radical **86** which initiates polymerization. The reported transfer constant for **84** in styrene polymerization at 80 °C is very low (0.00045).[130]

6.2.5 Catalytic Chain Transfer

Enikolopyan *et al.*[131] found that certain Co^{II} porphyrin complexes (*e.g.* **87**) function as catalytic chain transfer agents. Later work has established that various square planar cobalt complexes (*e.g.* the cobaloximes **88-92**) are effective transfer agents.[132,133] The scope and utility of the process has been reviewed several times,[110,134-138] most recently by Heuts *et al*,[137] Gridnev,[138] and Gridnev and Ittel.[110] The latter two references[110,138] provide a historical perspective of the development of the technique.

The major applications of catalytic chain transfer are in molecular weight control and in synthesis of macromonomers based on methacrylate esters. However, they have also been shown effective in polymerizations and copolymerizations of MAA, MAM, MAN, AMS, S and some other monomers.

A major advantage of catalytic transfer agents over conventional agents is that they have very high transfer constants. The value of C_{tr} in MMA polymerization is in the range 10^3-10^5 (Table 6.13), thus only very small amounts are required to bring about a large reduction in molecular weight. Exact values for C_{tr} are dependent on the reaction conditions (Section 6.2.5.3)[131,132,139,140] and, for chain lengths ≤12, on the molecular weight of the propagating species.[139,140] Ideally, they are not used up during polymerization (Section 6.2.5.1).

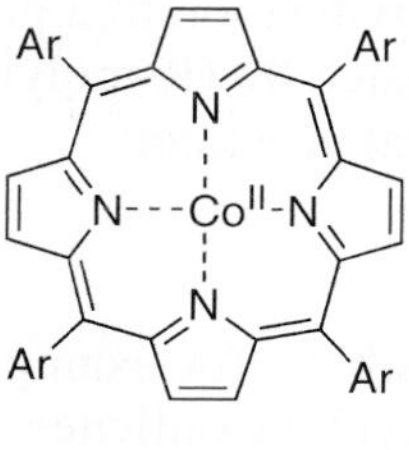

87

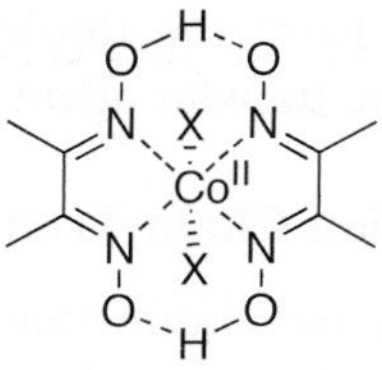

88

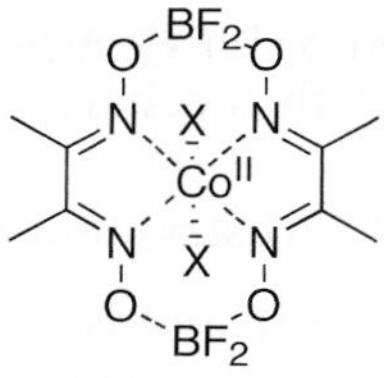

89

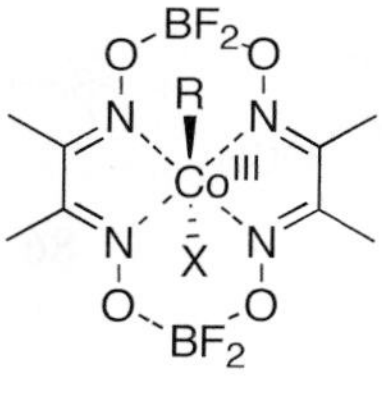

90

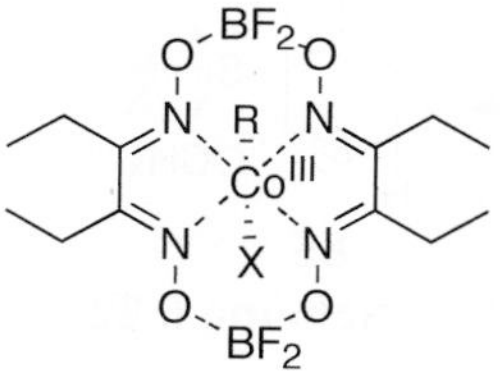

91

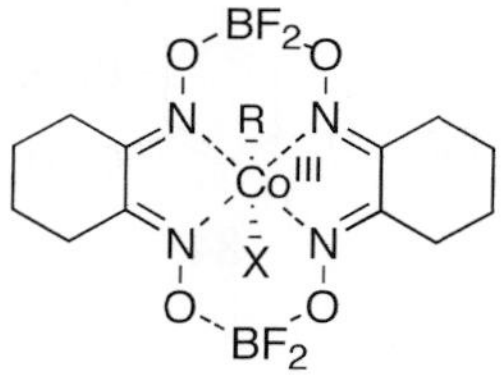

92

X= solvent

6.2.5.1 Mechanism

The mechanism proposed for catalytic chain transfer[132] is shown in Scheme 6.23 for MMA polymerization. The Co^{II} complex (**93**) rapidly and reversibly

Chain Transfer 311

combines with carbon-centered radicals. The product, the alkyl Co^{III} complex (**96**), may eliminate the cobalt hydride (**95**) to form a macromonomer (**94**). Alternatively, the Co^{II} complex (**93**) may undergo disproportionation with the carbon-centered radical to give the same products (**94** and **95**). It is also possible that both mechanisms operate simultaneously. The cobalt hydride (**95**) reinitiates polymerization by donating a hydrogen atom to monomer and in doing so regenerates the cobalt complex (**93**). The majority of chains formed in the presence of these reagents will have one unsaturated end group (**94**).

With S,$^{141-143}$ acrylate esters^{144} and other monosubstituted monomers, the adduct (**98**) has greater intrinsic stability. The overall mechanism proposed for catalytic chain transfer shown in Scheme 6.24 for the case of S polymerization is similar to that for MMA polymerization. However, hydrogen transfer to cobalt gives products (**97**) that have a 1,2-disubstituted double bond and appear inert under the polymerization conditions. The greater stability of **98** is the probable cause of retardation in homopolymerizations involving, in particular, acrylate esters and VAc. Stability is such that certain cobalt complexes have been exploited in living polymerization of acrylate esters (Section 9.3.9.1). Higher temperatures favor chain transfer over coupling and polymerizations of acrylate esters to achieve molecular weight control have been successfully carried out at >110 °C. Molecular weight control with less retardation can also be achieved by carrying out polymerizations in the presence of small amounts of an added α-methyl vinyl monomer (*e.g.* AMS). In this case, the dominant transfer process involves the α-methyl vinyl monomer.^{3,145}

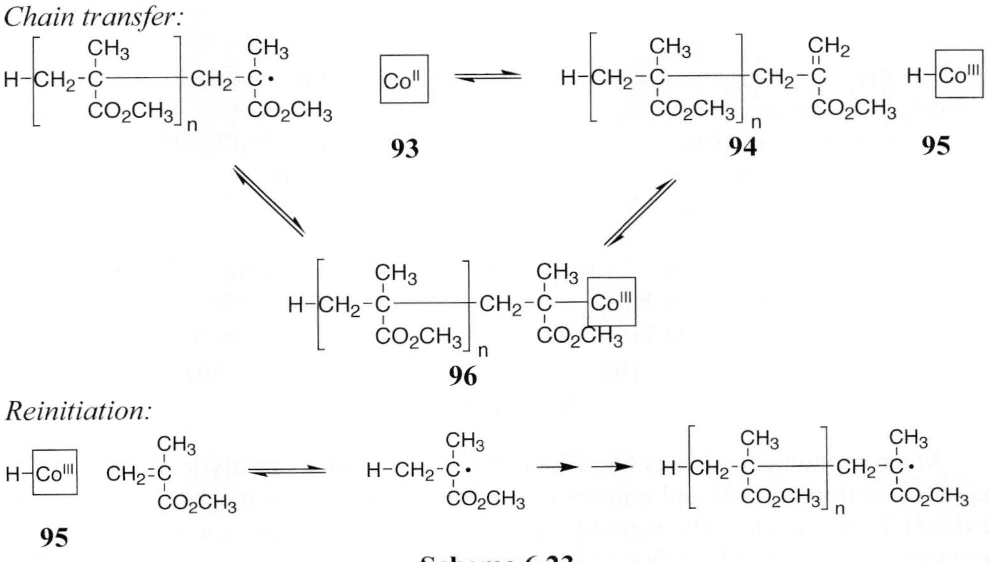

Chain transfer:

Scheme 6.23

Chain transfer:

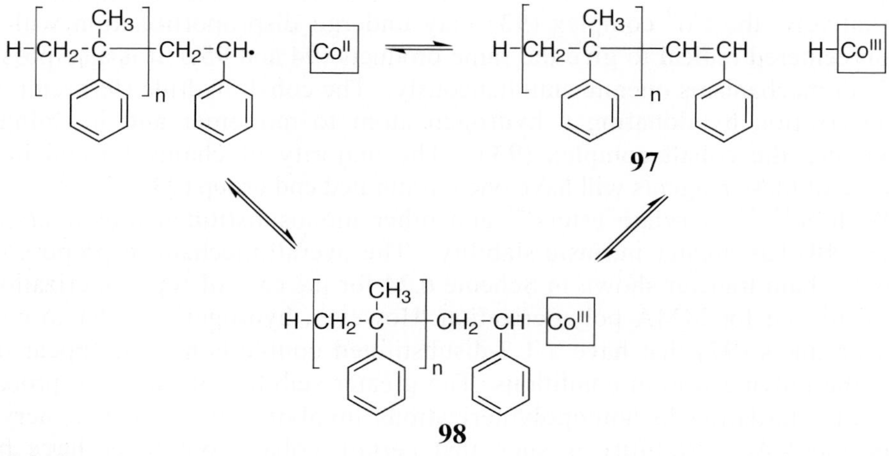

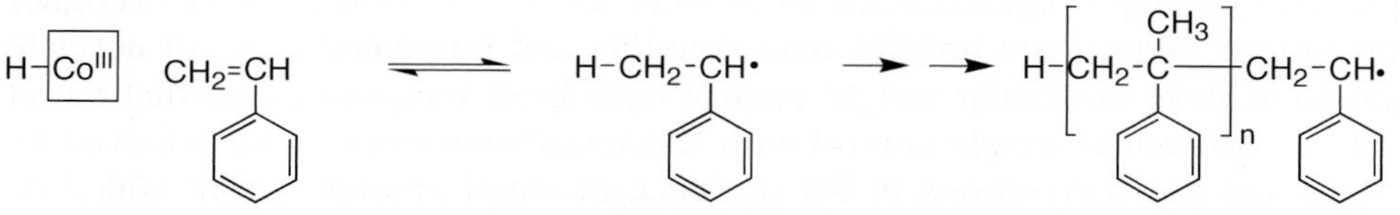

Scheme 6.24

Reinitiation:

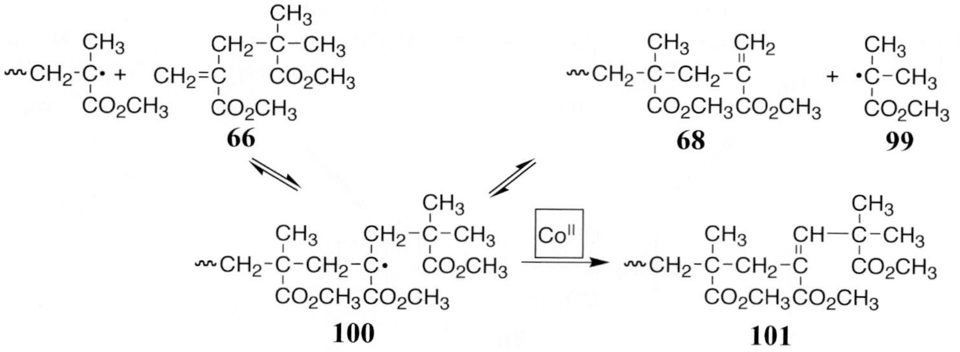

Scheme 6.25

Macromonomers such as **66**, **68** and **94** are themselves catalytic chain transfer agents (Section 6.2.3.4) and transfer to macromonomer is one mechanism for chain extension of the initially formed species. The adduct species in the case of monomeric radical adding dimer (**100**) may also react by chain transfer to give **101** which is inert under polymerization conditions (Scheme 6.25). Polymerizations to give trimer may contain a significant amount of **101** as a byproduct.[111] In the case of higher species scission is fast relative to chain transfer and the corresponding byproducts are not observed. It is also thought that the reaction of **93** with a propagating radical to give cobalt hydride **95** and macromonomer is reversible.[146]

6.2.5.2 Catalysts

Many catalysts have been screened for activity in catalytic chain transfer. A comprehensive survey is provided in Gridnev and Ittel's review.[110] The best known, and to date the most effective, are the cobalt porphyrins (Section 6.2.5.2.1) and cobaloximes (Sections 6.2.5.2.2 and 6.2.5.2.3). There is considerable discrepancy in reported values of transfer constants. This in part reflects the sensitivity of the catalysts to air and reaction conditions (Section 6.2.5.3).

6.2.5.2.1 Cobalt porphyrin and related complexes

Many Co^{II} porphyrins (**87**)[110,131] and phthalocyanine complexes (**102**)[110] have been examined for their ability to function as catalytic chain transfer agents and much mechanistic work has focused on the use of these catalysts. The more widespread application of these complexes has been limited because they often have only sparing solubility and they are highly colored.

While in most complexes the cobalt is coordinated to four nitrogens, there are some exceptions such as **103**.[147]

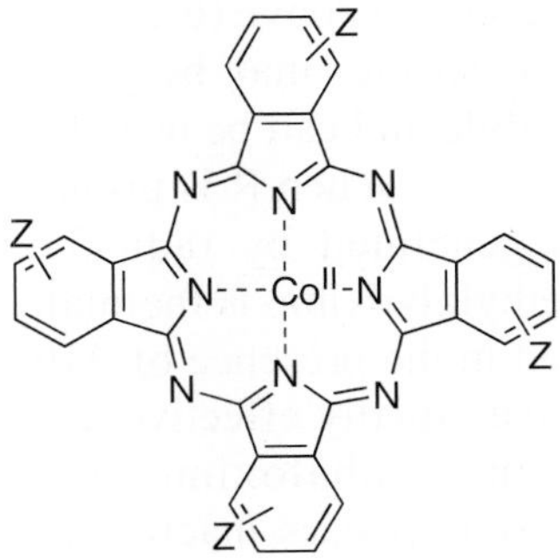

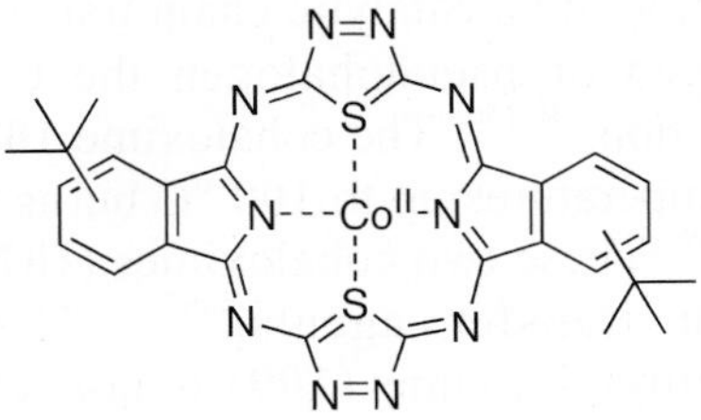

6.2.5.2.2 Cobalt (II) cobaloximes

Much of the recent literature relates to BF_2-bridged Co^{II} cobaloximes based on dimethyl (**89**) or diphenyl glyoxime (**104**).[110] The BF_2-bridged cobaloximes (*e.g.* **89**) show greater stability to hydrolysis than analogous H-bridged species (*e.g.* **88**). The diphenylglyoxime complexes (**104**) show enhanced air and hydrolytic stability

314   The Chemistry of Radical Polymerization

with respect to the corresponding dimethylglyoxime complexes (**89**) but are less active (Table 6.13 on page 316).

The activity in MMA polymerization can be dramatically affected by the apical ligands. Apical aquo or alcohol ligands are labile and rapidly exchange with the polymerization medium. Lewis base ligands (*e.g.* pyridine, triphenyl phosphine) are comparatively stable. In MMA polymerization, it is found that activity increases with the basicity of the ligand. With alkyl Co^{III} complexes, a different order is found possibly because the type of apical ligand also controls the rate of initial generation of the active Co^{II} complex.

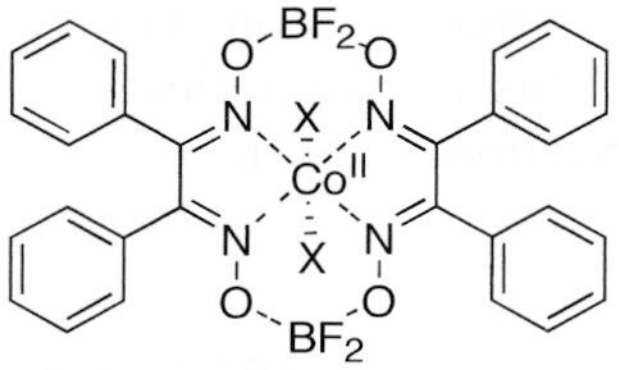

104 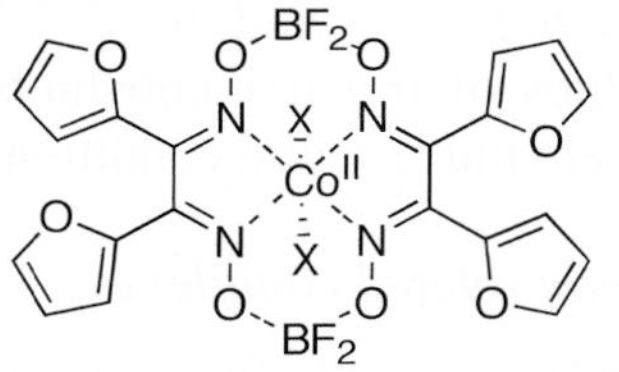   **105** 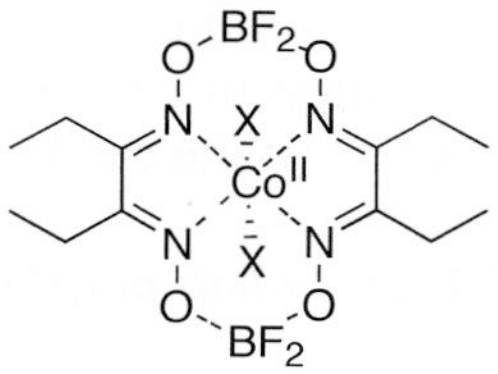   **106**

6.2.5.2.3 Cobalt (III) cobaloximes

Various Co^{III} cobaloximes (**90-92**) have also been used as catalytic chain transfer agents.[133,148,149] To be effective, the complex must be rapidly transformed into the active Co^{II} cobaloximes under polymerization conditions. The mechanism of catalytic chain transfer is then identical to that described above (6.2.5.1).

When R is secondary or tertiary alkyl, the Co^{II} species may be generated by Co^{III}-C bond homolysis. Thus, **107** is thermally labile and can be used both as an initiator and a catalytic chain transfer agent at 60 °C.[148] When R is primary alkyl, halogen or pseudohalogen the Co^{II} species is generated by radical induced reduction.[149,150] The cobaloxime **108** based on diethylglyoxime is thermally stable at temperatures up to 100 °C but is rapidly reduced in the presence of AIBN at 60 °C.[149] These two cobaloximes (**107** and **108**) appear equally effective as catalytic chain transfer agents.[149] The corresponding cobaloxime based on dimethylglyoxime (**109**) is not readily reduced and appears inactive under the same conditions.

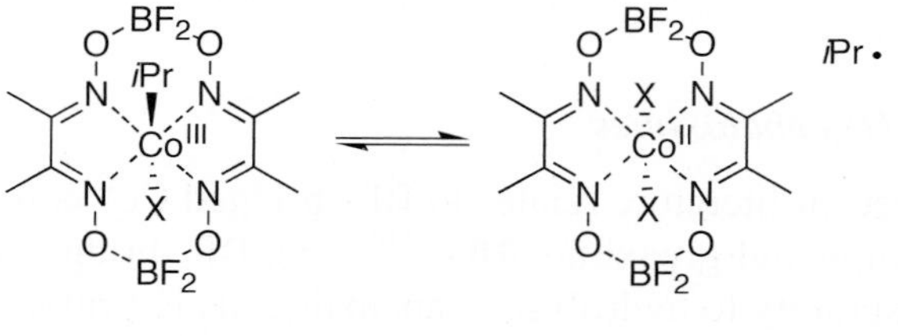

107

Scheme 6.26

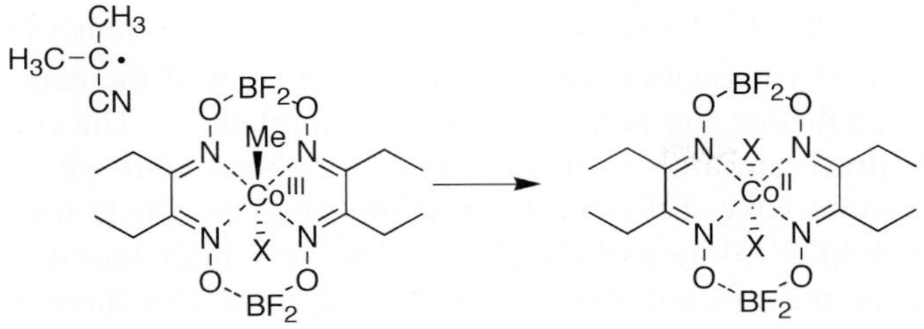

Chain Transfer

[Scheme 6.27 showing cobalt complex 108 with CH3 and CN substituents, and complex 109]

108

Scheme 6.27

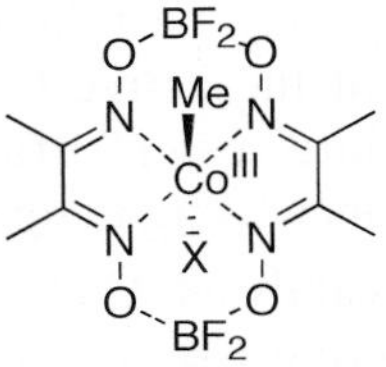

109

6.2.5.2.4 Other catalysts

Other complexes also react with propagating radicals by catalytic chain transfer.[110] These include certain chromium,[151,152] molybdenum[152,153] and iron[154] complexes. To date the complexes described appear substantially less active than the cobaloximes and are more prone to side reactions.

6.2.5.3 Reaction conditions

Catalytic chain transfer has now been applied under a wide range of reaction conditions (solution, bulk, emulsion, suspension) and solvents (methanol, butan-2-one, water). The selection of the particular complex, the initiator, the solvent and the reaction conditions can be critical. For example:

(a) Initiators that generate oxygen centered radicals (*e.g.* BPO) or primary alkyl radicals (*e.g.* LPO) are generally to be avoided. The Co^{II} cobaloximes can react with the initiator-derived radicals to create a species that is inactive or less active under the polymerization conditions. Preferred initiators are those that resemble propagating species and azo compounds that generate tertiary radicals such as AIBN.

(b) The Co^{II} cobaloximes can be extremely air sensitive and rigorous exclusion of air is essential for reproducibility. Co^{III} complexes (**92**) have enhanced air stability with respect to the Co^{II} cobaloximes.[149] Solutions are stable at room temperature even in the presence of air. The active species is generated *in situ* under the polymerization conditions. However, rigorous exclusion of air from the polymerization is still essential.

(c) There are reports of extreme sensitivity to solvent and monomer purity.[155]
(d) In emulsion polymerization, the partition coefficient of the complex between the droplet, aqueous and particle phases is important.[149] The complex should partition preferentially into the particle phase and yet in *ab initio* polymerizations have sufficient water solubility to be able to transfer from the monomer droplet to the particle phase. The very high activity of the cobalt complex, and the concentration typically used, mean that there may be only a few molecules of complex per particle.
(e) In solution polymerization, the apical ligand of cobaloxime complexes may exchange with the medium changing the activity and solubility of the complex.
(f) Intermediate Co^{III} complexes may be relatively stable at low temperatures reducing that concentration of the active Co^{II} complex and the propagating radicals. For S and acrylate esters transfer constants and rates of polymerization increase with increasing temperature.[115]
(g) Co^{III} complexes (alkyl Co^{III} catalysts, Co^{III} intermediates) are light sensitive and will dissociate to the active Co^{II} complex and propagating radicals on irradiation with visible light. For S and acrylate esters higher transfer constants can be achieved by irradiation of the sample.[156]

Catalytic inhibition has been reported for MAM and MMA polymerizations with DMF solvent.[157]

Table 6.13 Transfer Constants for Cobalt Complexes at 60 °C[a]

Transfer agent	C_{tr} for monomer[b]		
	MMA	S[c]	MA[c]
89	32000-37000[155,158,159]	1500-7000 [115,141,142,156,158]	50-1000[156,160]
104	18000[161]	400[161]	-

a Bulk medium comprises only monomer and transfer agent.   b Transfer constants rounded to two significant figures.   c The apparent transfer constant with monosubstituted monomers is strongly dependent on reaction conditions (see text). The lower limit shown is the effective transfer constant in bulk polymerization. The upper limit is the likely actual transfer constant.

6.2.6 Transfer to Monomer

Non-zero transfer constants (C_M) can be found in the literature for most monomers. Values of C_M for some common monomers are given in Table 6.14. For S and the (meth)acrylates the value is small, in the range 10^{-5}-10^{-4}. Transfer to monomer is usually described as a process involving hydrogen atom transfer. While this mechanism is reasonable for those monomers possessing aliphatic hydrogens (*e.g.* MMA, VAc, allyl monomers), it is less acceptable for monomers possessing only vinylic or aromatic hydrogens (*e.g.* VC, S). The details of the mechanisms by which transfer occurs are, in most cases, not proven. Mechanisms

Chain Transfer

for transfer to monomer that involve loss of vinylic hydrogens seem unlikely given the high strength of the bonds involved.

Irrespective of the mechanism by which transfer to monomer occurs, the process will usually produce an unsaturated radical as a byproduct. This species initiates polymerization to afford a macromonomer that may be reactive under typical polymerization conditions.

Table 6.14 Selected Values for Transfer Constants to Monomer[a]

Monomer	Temperature (°C)	$C_M \times 10^4$	Ref.
S	60	0.6	162
MMA	60	0.1	163
MA	60	0.4	164
AN	60	0.3	165
VAc	60	1.8	166
VC	100	50	167,168
allyl acetate	80	1600	169
allyl chloride	80	700	169

a Values rounded to one significant figure and are taken from the references shown. There is considerable scatter in literature values for many monomers.[49]

6.2.6.1 Styrene

The value of C_M has been determined by a number of groups as 6×10^{-5} (Table 6.14).[49] However, the mechanism of transfer has not been firmly established. A mechanism involving direct hydrogen abstraction seems unlikely given the high strength of vinylic and aromatic C-H bonds. The observed value of C_M is only slightly lower than C_{tr} for ethylbenzene ($\sim 7\times 10^{-5}$).[49]

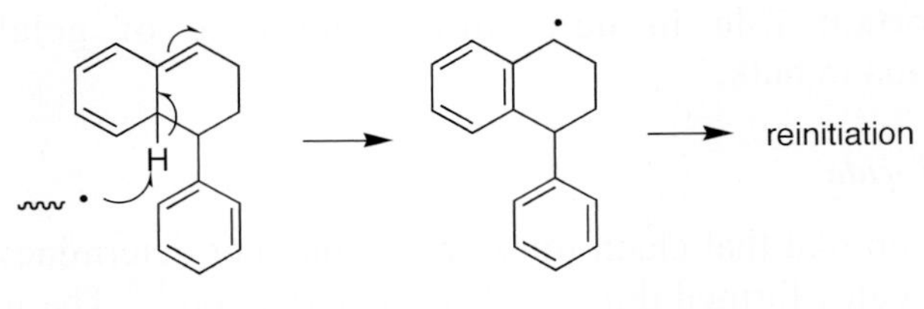

110

Scheme 6.28

It has been proposed that transfer to monomer may not involve the monomer directly but rather the intermediate (**110**) formed by Diels-Alder dimerization (Scheme 6.28).[170] Since **110** is formed during the course of polymerization, its involvement could be confirmed by analysis of the polymerization kinetics.

6.2.6.2 Vinyl acetate

There is a considerable body of evidence (kinetic studies, chemical and NMR analysis) indicating that transfer to VAc monomer involves largely, if not exclusively, the acetate methyl hydrogen to give radical **111** (Scheme 6.29).[171,172] This radical (**111**) initiates polymerization to yield a reactive macromonomer (**112**).

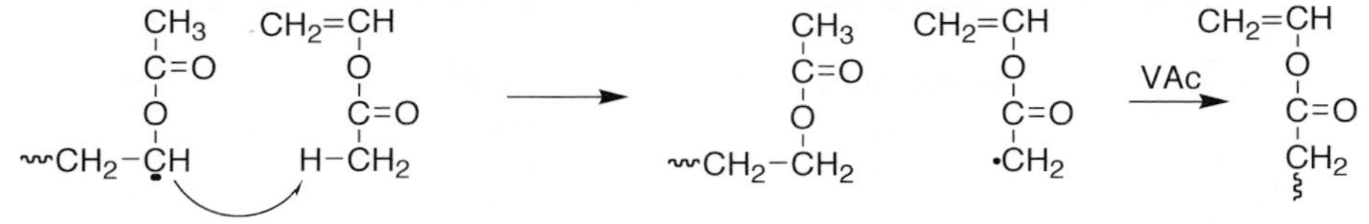

Scheme 6.29

Starnes et al.[173] have provided support for the above mechanism (Scheme 6.29) by determining the unsaturated chain ends (**112**) in low conversion PVAc by ^{13}C NMR. They were able to distinguish (**112**) from chain ends that might have been formed if transfer involved abstraction of a vinylic hydrogen. The number of unsaturated chain ends (**112**) was found to equate with the number of -CH$_2$OAc ends suggesting that most chains are formed by transfer to monomer. Starnes et al.[173] also found an isotope effect k_H/k_D of 2.0 for the abstraction reaction with CH$_2$=CHO$_2$CCD$_3$ as monomer. This result is consistent with the mechanism shown in Scheme 6.28 but is contrary to an earlier finding.[174]

Stein[166] has indicated that the reactivity of the terminal double bond of the macromonomer (**112**) is 80% that of VAc monomer. The kinetics of incorporation of **112** have also been considered by Wolf and Burchard[175] who concluded that **112** played an important role in determining the time of gelation in VAc homopolymerization in bulk.

6.2.6.3 Vinyl chloride

It has been proposed that chain transfer to monomer determines the length of the polymer molecules formed during VC polymerization.[176] The mechanism for transfer, involving an addition-elimination sequence consequent on head addition to monomer (Section 4.3.1.2), was first proposed by Rigo et al.[177] Direct evidence for this pathway has been provided by Starnes et al.[178] and Park and Saleem.[179] This pathway (Scheme 6.30) accounts for C_M for VC being much greater than C_M for other commercially important monomers (Table 6.14) where the analogous pathway is not available. Starnes and Wojciechowski[180] have reported kinetic data which suggest that the chlorine atom does not have a discrete existence but is transferred directly from the β-chloroalkyl radical to VC.

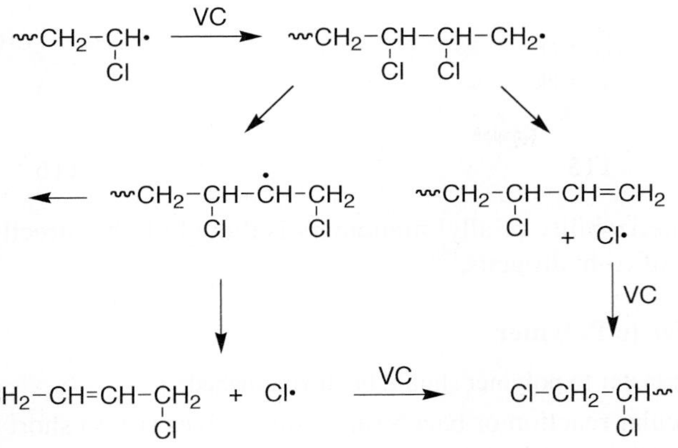

Chain Transfer

$$\text{\textasciitilde CH}_2\text{-CH}\cdot\text{Cl} \xrightarrow{\text{VC}} \text{\textasciitilde CH}_2\text{-CH-CH-CH}_2\cdot$$

Scheme 6.30

6.2.6.4 Allyl monomers

Transfer to monomer is of particular importance during the polymerization of allyl esters (**113**, X=O$_2$CR), ethers (**113**, X=OR), amines (**113**, X=NR$_2$) and related monomers.[169,181,182] The allylic hydrogens of these monomers are activated towards abstraction by both the double bond and the heteroatom substituent (Scheme 6.31). These groups lend stability to the radical formed (**114**) and are responsible for this radical adding monomer only slowly. This, in turn, increases the likelihood of side reactions (*i.e.* degradative chain transfer) and causes the allyl monomers to retard polymerization.

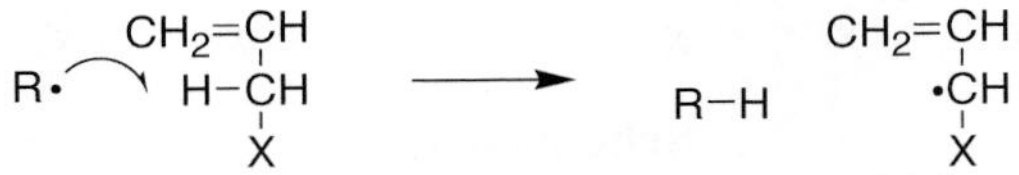

Scheme 6.31

For allyl acetate a significant deuterium isotope effect supports the hydrogen abstraction mechanism (Scheme 6.31).[183] Allyl compounds with weaker CH$_2$-X bonds (**113** X=SR, SO$_2$R, Br, *etc.*) may also give chain transfer by an addition-fragmentation mechanism (Section 6.2.3).

Diallyl monomers find significant use in cyclopolymerization (Section 4.4.1). Transfer to monomer is of greater importance in polymerizations of allyl than it is in diallyl monomers.[184] This might, in part, reflect differences in the nature of the propagating species [*e.g.* a secondary alkyl (**115**) *vs* a primary alkyl radical (**116**)]. Electronic factors may also play a role.[185]

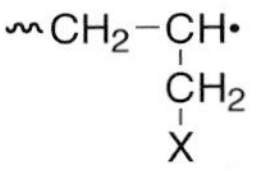

320   The Chemistry of Radical Polymerization

    ~CH$_2$-CH•   ~CH$_2$ CH$_2$•
        CH$_2$
        X   X

     115 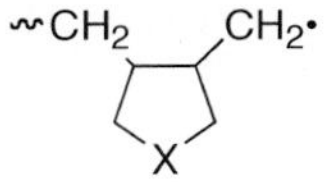   116

The polymerizability of allyl monomers is thought to be directly related to the abstractability of α–hydrogens.[186]

6.2.7 Transfer to Polymer

Two forms of transfer to polymer should be distinguished:
(a) Intramolecular reaction or backbiting, which gives rise to short chain branches (length usually ≤5 carbons).
(b) Intermolecular reaction, which generally results in the formation of long chain branches.

The intramolecular process does not give rise to a new polymer chain and is considered in Section 4.4.3. It will not be considered further in this section.

Available evidence suggests that the main reaction accounting for transfer to vinyl polymers (*e.g.* PMA, PVAc, PVC, PVF) usually involves abstraction of a methine hydrogen (Scheme 6.32) (Sections 6.2.7.3, 6.2.7.4, 6.2.7.5 and 6.2.7.6 respectively). However, definitive evidence for the mechanism is currently only available for a few polymers (*e.g.* PVAc, PVF).

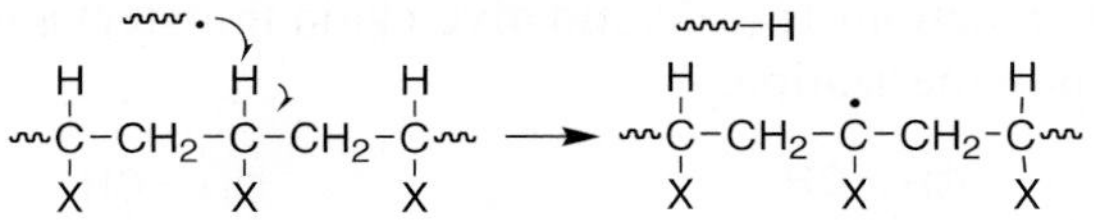

Scheme 6.32

Table 6.15 Transfer Constants to Polymer[a]

Monomer	Temperature (°C)	$C_P \times 10^4$
S	60	1.9-16
MMA	60	0.1-360
MA	60	0.5-1.0
AN	60	3.5
VAc	60	1.4-47
VC	50	5
E	175	110

[a] Numbers are taken from the *Polymer Handbook*[49] and have been rounded to two significant figures.

Transfer constants to polymer (C_P) are not as readily determined as other transfer constants because the process need not lead to an overall lowering of molecular weight. If transfer occurs by hydrogen-atom abstraction from the polymer backbone then, for every polymer chain terminated by transfer, another branched chain is formed. In these circumstances the overall molecular weight remains constant. The extent of chain transfer can then be estimated by measuring the number of long chain branches or by analyzing the molecular weight distribution. As NMR measurement of long chain branching relies on determining the branch points, a major analytical problem is distinguishing the long chain branches from the short chain branches formed by backbiting.

The values of C_P to added polymer are measurable in circumstances where the added material is readily distinguishable from that being formed *in situ*, for example, if it is of significantly different molecular weight or if it is uniquely labeled.[187] Studies with model compounds suggest that oligomers of chain length ≥3 can be used to provide a good estimate of the transfer constant.[188,189]

For some polymers, the value of C_P depends on the polymer molecular weight (*e.g.* Section 6.2.7.2). This may help account for the wide range of values for C_P in the literature (Table 6.15).

6.2.7.1 Polyethylene

The presence of long chain branches in low density polyethylene (LDPE) accounts for the difference in properties (*e.g.* higher melt strength, greater toughness for the same average molecular weight) between LDPE and linear low density polyethylene (LLDPE, made by coordination polymerization).

Long chain branching (>8 carbons) in polyethylene can be detected by ^{13}C NMR analysis.[190-193] However, the length and distribution of the branches are more difficult to determine. Measurements of long chain branching have been made by GPC-light scattering[194-196] or GPC-viscometry.[196-198] The extent of long chain branching is known to be strongly dependent on the reactor design and the reaction conditions employed. These studies indicate that, for a given sample, the branch frequency appears to decrease with increasing molecular weight of PE.[196] An explanation was not given.

6.2.7.2 Poly(alkyl methacrylates)

ω-Unsaturated poly(alkyl methacrylates) (*e.g.* **117**) are produced during radical polymerization of MMA through termination by disproportionation (Sections 5.2.2.1.3 & 5.2.2.2.3). Schulz *et al.*[199] were the first to suggest that reactions of these species (**117**) may complicate MMA homopolymerization. The ω-unsaturated poly(alkyl methacrylates) may act as a chain transfer agent in polymerization by the mechanism shown in Scheme 6.33 (Section 6.2.3.4).

In polymerization of methacrylates, the adducts formed by addition to the macromonomer radicals are relatively unreactive towards adding further monomer and most undergo β–scission. There are two possible pathways for β-scission: one pathway leads back to starting materials; the other gives a new propagating radical and a macromonomer. Transfer is catalytic in macromonomer.

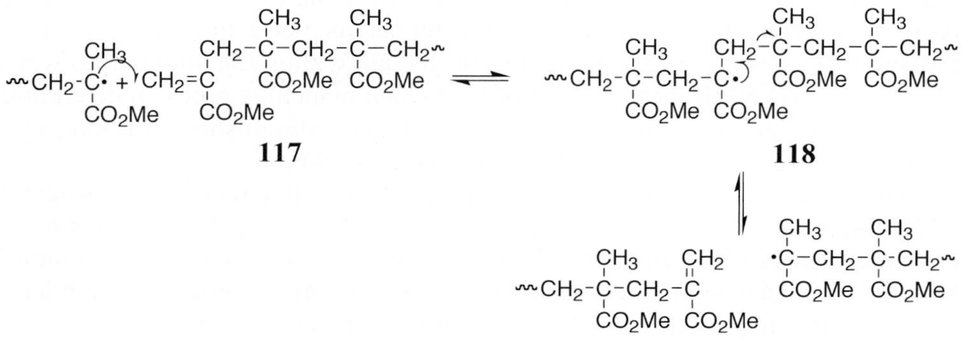

Scheme 6.33

Values of C_P measured in the presence of added PMMA (for example) will depend on how the PMMA was prepared and its molecular weight (*i.e.* on the concentration of unsaturated ends). PMMA formed by radical polymerization in the presence of a good H-donor transfer agent (or by anionic polymerization) would have only saturated chain ends. These PMMA chains should have a different transfer constant to those formed by normal radical polymerization where termination occurs by a mixture of combination and disproportionation. This could account for some of the variation in the values of C_P for this polymer.

6.2.7.3 Poly(alkyl acrylates)

Chain transfer to polymer is reported as a major complication and is thought to be unavoidable in the polymerization of alkyl acrylates.[200-202] The mechanism is believed to involve abstraction of a tertiary backbone hydrogen (Scheme 6.32). It has been proposed that this process and the consequent formation of branches may contribute to the early onset of the gel or Norrish-Trommsdorff effect in the polymerization of these monomers. At high temperatures the radicals formed may undergo fragmentation.

Copolymerization of macromonomers formed by backbiting and fragmentation is a second mechanism for long chain branch formation during acrylate polymerization (Section 4.4.3.3). The extents of long and short chain branching in acrylate polymers in emulsion polymerization as a function of conditions have been quantified.[202]

6.2.7.4 Poly(vinyl acetate)

The degree of branching in PVAc is strongly dependent on the polymerization conditions. Differences in the degree of branching are thought to be one of the main factors responsible for substantial differences in properties between various commercial samples of PVAc or PVA.[203-205]

PVAc is known to contain a significant number of long chain branches. Branches to the acetate methyl may arise by copolymerization of the VAc macromonomer produced as a consequence of transfer to monomer (Section 6.2.6.2). Transfer to polymer may involve either the acetate methyl hydrogens (Scheme 6.34) or the methine (Scheme 6.35) or methylene hydrogens of the polymer backbone.

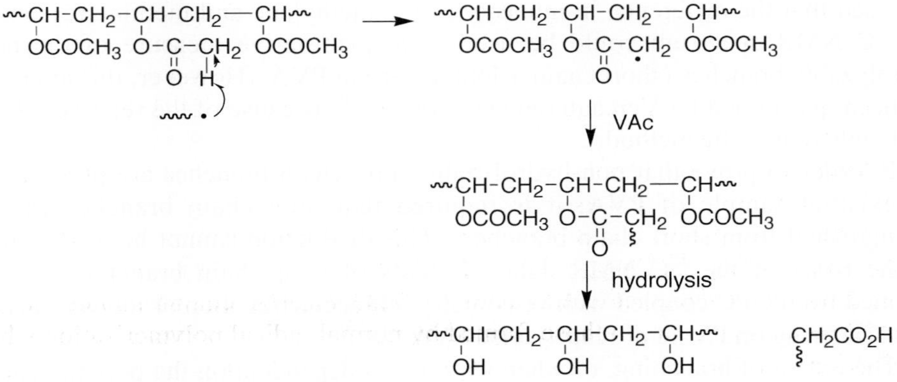

Scheme 6.34 Hydrolyzable branch formation.

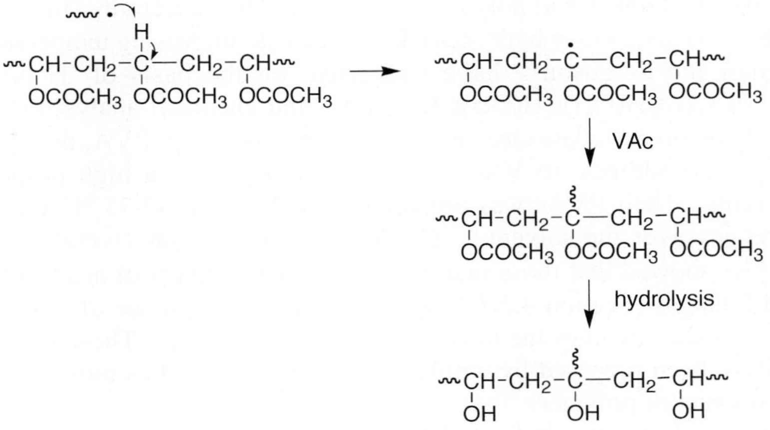

Scheme 6.35 Non-hydrolyzable branch formation.

The presence of hydrolyzable long chain branches in PVAc was established by McDowell and Kenyon[206] in 1940. They observed a reduction in molecular weight obtained on successively hydrolyzing and reacetylating samples of PVAc. Only branches to the acetate methyl will be lost on hydrolysis of the polymer; *i.e.* on conversion of PVAc to PVA.

The proposal that PVAc also has non-hydrolyzable long chain branches stems from the finding that PVA also possesses long chain branches. Nozakura *et al.*[171,207] suggested, on the basis of kinetic measurements coupled with chemical analysis, that chain transfer to PVAc involves preferential abstraction of backbone (methine) hydrogens (*ca* 5:1 *vs* the acetate methyl hydrogens at 60 °C).

^{1}H and ^{13}C NMR studies on PVAc or PVA also provide information on the nature of branches.[203,204,208,209] Dunn and Naravane[203] and Bugada and Rudin[204] proposed that the difference in intensity of the methylene and methine regions of the ^{13}C NMR spectrum could be used as a quantitative measure of the non-hydrolyzable branches (short chain + long chain) in PVA. However, this approach has been questioned by Vercauteren and Donners[204] because of the relatively large errors inherent in the method.

In order to prove that non-hydrolyzable long chain branches are present in a pre-existing sample of PVA, it is required that long chain branches can be distinguished from short chain branches. This distinction cannot be made solely on the basis of the ^{13}C NMR data. Extents of long chain branching can be obtained from GPC coupled with viscometry, ultracentrifugation or low angle laser light scattering on PVAc or reacetylated PVA.[205,210]

The extent of branching, of whatever type, is dependent on the polymerization conditions and, in particular, on the solvent and temperature employed and the degree of conversion. Nozakura *et al.*[171] found that, during bulk polymerization of VAc, the extent of transfer to polymer increased and the selectivity (for abstraction of a backbone *vs* an acetoxy hydrogen) decreases with increasing temperature.

Adelman and Ferguson[208] have suggested, on the basis of ^{1}H NMR data (detection of $CH_3CH(OH)CH(OH)CH_2-$ ends) and chemical analyses (formation of acetaldehyde on periodate cleavage of 1,2-glycol units) on PVA, that the radical formed by head addition to VAc may be responsible for a high proportion of transfer events. Their PVAc was prepared in methanol at 60-75 °C and much of the transfer involves the solvent. ^{13}C NMR[209,211] studies on several commercial PVA samples showed that those materials had equal numbers of head-to-head and tail-to-tail linkages (Section 4.3.1.1) and indicated the presence of $-CH_2OH$ ends (*i.e.* most transfer involves the normal propagating species). These polymers are likely to have been prepared by emulsion polymerization, thus most transfer will involve monomer or polymer.

Hatada *et al.*[212] have indicated that PVAc prepared in aromatic solvents (benzene, chlorobenzene) at 60 °C has fewer branch points than the polymer prepared in ethyl acetate under similar conditions. They attributed this observation to complexation of the propagating radical in the aromatic solvents and the different reactivity of this complexed radical. They have also reported that VAc polymerization is substantially slowed in aromatic solvents and this was also attributed to complexation of the propagating radical[213] (Section 8.3.1.1).

6.2.7.5 Poly(vinyl chloride)

The microstructure of PVC has been the subject of numerous studies (Sections 4.3.1.2 and 6.2.6.3).[214] Starnes *et al.*[168] determined the long chain branch points by NMR studies on PE formed by Bu_3SnH reduction of PVC. They concluded that the probable mechanism for the formation of these branches involved transfer to polymer that occurred by hydrogen abstraction of a backbone methine by the propagating radical (Scheme 6.32).

6.2.7.6 Poly(vinyl fluoride)

Ovenall and Uschold[215] have recently measured the concentration of branch points (tertiary F, Scheme 6.32) in PVF by ^{19}F NMR. These were found to account for between 0.5 to 1.5% of monomer units depending on reaction conditions. Branching was found to be favored by lower reactor pressures or higher reactor temperatures. More branching was observed for polymers produced in batch as opposed to continuous reactors. This effect was attributed to longer residence time of the polymer in the reactor.

6.2.8 Transfer to Initiator

The mechanism and incidence of transfer to various initiators is discussed in Chapter 3. See, in particular, Sections 3.2.10 (introduction), 3.3.2.1.4 (dialkyl diazenes including AIBN), 3.3.2.1.4 (diacyl peroxides including BPO), 3.3.2.3.1 (peroxyesters), 3.3.2.4 (dialkyl peroxides) and 3.3.2.5 (alkyl hydroperoxides).

6.3 References

1. Flory, P.J. *J. Am. Chem. Soc.* **1937**, *59*, 241.
2. Flory, P.J. *Principles of Polymer Chemistry*; Cornell University Press: Ithaca, New York, 1953.
3. Chiefari, J.; Rizzardo, E. In *Handbook of Radical Polymerization*; Davis, T.P.; Matyjaszewski, K., Eds.; John Wiley & Sons: Hoboken, NY, 2002; p 263.
4. Barson, C.A. In *Comprehensive Polymer Science*; Eastmond, G.C.; Ledwith, A.; Russo, S.; Sigwalt, P., Eds.; Pergamon: Oxford, 1989; Vol. 3, p 171.
5. Farina, M. *Makromol. Chem., Macromol. Symp.* **1987**, *10/11*, 255.
6. Eastmond, G.C. In *Comprehensive Chemical Kinetics*; Bamford, C.H.; Tipper, C.F.H., Eds.; Elsevier: Amsterdam, 1976; Vol. 14A, p 153.
7. Palit, S.R.; Chatterjee, S.R.; Mukherjee, A.R. In *Encyclopedia of Polymer Science and Technology*; Mark, H., F.; Gaylord, N.G.; Bikales, N.M., Eds.; Wiley: New York, 1966; Vol. 3, p 575.
8. Boutevin, B. *Adv. Polym. Sci.* **1990**, *94*, 69.

9. Heitz, W. In *Telechelic Polymers: Synthesis and Applications*; Goethals, E.J., Ed.; CRC Press: Boca Raton, Florida, 1989; p 61.
10. Corner, T. *Adv. Polym. Sci.* **1984**, *62*, 95.
11. Starks, C.M. *Free Radical Telomerization*; Academic Press: New York, 1974.
12. Mayo, F.R. *J. Am. Chem. Soc.* **1943**, *65*, 2324.
13. Clouet, G.; Knipper, M. *Makromol. Chem.* **1987**, *188*, 2597.
14. Bamford, C.H. *Polym. Commun.* **1989**, *30*, 36.
15. Harrisson, S.; Kapfenstein-Doak, H.; Davis, T.P. *Macromolecules* **2001**, *34*, 6214.
16. Scott, G.P.; Foster, F.J. *Macromolecules* **1969**, *2*, 428.
17. Scott, G.P.; Elghoul, A.M.R. *J. Polym. Sci., Part A-1* **1970**, *8*, 2255.
18. Scott, G.P.; Wang, J.C. *J. Org. Chem.* **1963**, *28*, 1314.
19. Barson, C.A.; Mather, R.R.; Robb, J.C. *Trans. Faraday Soc.* **1970**, *66*, 2585.
20. Mayo, F.R. *J. Am. Chem. Soc.* **1948**, *70*, 3689.
21. Asahara, T.; Makishima, T. *Kogyo Kagaku Zasshi* **1966**, *69*, 2173.
22. Englin, B.A.; Onishchenko, T.A. *Izv. Akad. Nauk. SSSR, Ser. Khim.* **1969**, 1906.
23. Englin, B.A.; Onishchenko, T.A.; Freidlina, R.K. *Izv. Akad. Nauk SSSR, Ser. Khim.* **1968**, *11*, 2489.
24. Moad, G.; Moad, C.L. *Macromolecules* **1996**, *29*, 7727.
25. Heuts, J.P.A.; Davis, T.P.; Russell, G.T. *Macromolecules* **1999**, *32*, 6019.
26. Whang, B.Y.C.; Ballard, M.J.; Napper, D.H.; Gilbert, R.G. *Aust. J. Chem.* **1991**, *44*, 1133.
27. Clay, P.A.; Gilbert, R.G. *Macromolecules* **1995**, *28*, 552.
28. Bamford, C.H. *J. Chem. Soc., Faraday Trans. 1* **1976**, *72*, 2805.
29. Bamford, C.H.; Basahel, S.N. *J. Chem. Soc., Faraday Trans. 1* **1978**, *74*, 1020.
30. Bamford, C.H.; Basahel, S.N. *Polymer* **1978**, *19*, 943.
31. Walling, C. *J. Am. Chem. Soc.* **1948**, *70*, 2561.
32. Cardenas, J.N.; O'Driscoll, K.F. *J. Polym. Sci., Polym. Chem. Ed.* **1977**, *15*, 2097.
33. Stickler, M. *Makromol. Chem.* **1979**, *180*, 2615.
34. Harwood, H.J.; Medsker, R.E.; Rapo, A. In *MakroAkron 94 Abstracts*; IUPAC, 1994; p 16.
35. Sunder, A.; Muelhaupt, R. *Macromol. Chem. Phys.* **1999**, *200*, 58.
36. Moad, G.; Moad, C.L.; Rizzardo, E.; Thang, S.H. *Macromolecules* **1996**, *29*, 7717.
37. Nair, C.P.R.; Chaumont, P.; Colombani, D. *Macromolecules* **1995**, *28*, 3192.
38. Jenkins, A.D.; Jenkins, J. In *Polymer Handbook*, 4th ed.; Brandup, J.; Immergut, E.H.; Grulke, E.A., Eds.; John Wiley and Sons: New York, 1999; p II/321.
39. Greenley, R.Z. In *Polymer Handbook*, 4th ed.; Brandup, J.; Immergut, E.H.; Grulke, E.A., Eds.; John Wiley and Sons: New York, 1999; p II/309.
40. Moad, G.; Chiefari, J.; Mayadunne, R.T.A.; Moad, C.L.; Postma, A.; Rizzardo, E.; Thang, S.H. In *Macromol. Symp.*, 2002; Vol. 182, p 65.
41. Chiefari, J.; Mayadunne, R.T.A.; Moad, C.L.; Moad, G.; Rizzardo, E.; Postma, A.; Skidmore, M.A.; Thang, S.H. *Macromolecules* **2003**, *36*, 2273.
42. Chong, Y.K.; Krstina, J.; Le, T.P.T.; Moad, G.; Postma, A.; Rizzardo, E.; Thang, S.H. *Macromolecules* **2003**, *36*, 2256.
43. Nair, C.P.R.; Richou, M.C.; Chaumont, P.; Clouet, G. *Eur. Polym. J.* **1990**, *26*, 811.
44. Businelli, L.; Gnanou, Y.; Maillard, B. *Macromol. Chem. Phys.* **2000**, *201*, 2805.
45. de la Fuente, J.L.; Madruga, E.L. *Macromol. Chem. Phys.* **2000**, *201*, 2152.
46. Hutchinson, R.A.; Paquet, D.A.; McMinn, J.H. *Macromolecules* **1995**, *28*, 5655.
47. Businelli, L.; Deleuze, H.; Gnanou, Y.; Maillard, B. *Macromol. Chem. Phys.* **2000**, *201*, 1833.
48. O'Brien, J.L.; Gornick, F. *J. Am. Chem. Soc.* **1955**, *77*, 4757.

49. Ueda, A.; Nagai, S. In *Polymer Handbook*, 4th ed.; Brandup, J.; Immergut, E.H.; Grulke, E.A., Eds.; John Wiley and Sons: New York, 1999; p II/97.
50. Clarke, J.T.; Howard, R.O.; Stockmayer, W.H. *Makromol. Chem.* **1961**, *44*, 427.
51. Roy, K.K.; Pramanick, D.; Palit, S.R. *Makromol. Chem.* **1972**, *153*, 71.
52. Tronche, C.; Martinez, F.N.; Horner, J.H.; Newcomb, M.; Senn, M.; Giese, B. *Tetrahedron Lett.* **1996**, *37*, 5845.
53. Bamford, C.H.; Basahel, S.N. *J. Chem. Soc., Faraday Trans. 1* **1980**, *76*, 112.
54. Boutevin, B.; El Idrissi, A.; Parisi, J.P. *Makromol. Chem.* **1990**, *191*, 445.
55. Boutevin, B.; Lusinchi, J.-M.; Pietrasanta, Y.; Robin, J.-J. *Eur. Polym. J.* **1994**, *30*, 615.
56. Boutevin, B.; Pietrasanta, Y. *Makromol. Chem.* **1985**, *186*, 817.
57. Pryor, W.A.; Pickering, T.L. *J. Am. Chem. Soc.* **1962**, *84*, 2705.
58. Costanza, A.J.; Coleman, R.J.; Pierson, R.M.; Marvel, C.S.; King, C. *J. Polym. Sci.* **1955**, *17*, 319.
59. Otsu, T.; Kinoshita, Y.; Imoto, M. *Makromol. Chem.* **1964**, *73*, 225.
60. Tsuda, K.; Otsu, T. *Bull. Chem. Soc. Japan* **1966**, *39*, 2206.
61. Otsu, T.; Nayatani, K. *Makromol. Chem.* **1958**, *73*, 225.
62. Staudner, E.; Kysela, G.; Beniska, J.; Mikolaj, D. *Eur. Polym. J.* **1978**, *14*, 1067.
63. Beniska, J.; Staudner, E. *J. Polym. Sci., Part C* **1967**, *16*, 1301.
64. Popielarz, R.; Clouet, G. *Makromol. Chem.* **1993**, *194*, 2897.
65. Kimura, T.; Kodaira, T.; Hamashima, M. *Polym. J.* **1983**, *15*, 293.
66. Tedder, J.M. *Angew. Chem., Int. Ed. Engl.* **1982**, *21*, 401.
67. Ameduri, B.; Boutevin, B. *Macromolecules* **1990**, *23*, 2433.
68. Mayo, F.R. *J. Am. Chem. Soc.* **1953**, *75*, 6133.
69. Rizzardo, E.; Chong, Y.K.; Evans, R.A.; Moad, G.; Thang, S.H. *Macromol. Symp.* **1996**, *111*, 1.
70. Colombani, D.; Chaumont, P. *Acta Polym.* **1998**, *49*, 225.
71. Colombani, D. *Prog. Polym. Sci.* **1999**, *24*, 425.
72. Yagci, Y.; Reetz, I. *React. Funct. Polym.* **1999**, *42*, 255.
73. Lewis, S.N.; Miller, J.J.; Winstein, S. *J. Org. Chem.* **1972**, *37*, 1478.
74. Giese, B. *Radicals in Organic Synthesis: Formation of Carbon-Carbon Bonds*; Pergamon Press: Oxford, 1986.
75. Motherwell, W.B.; Crich, D. *Free Radical Chain Reactions in Organic Synthesis*; Academic Press: London, 1992.
76. Cacioli, P.; Hawthorne, D.G.; Laslett, R.L.; Rizzardo, E.; Solomon, D.H. *J. Macromol. Sci., Chem.* **1986**, *A23*, 839.
77. Meijs, G.F.; Rizzardo, E.; Thang, S.H. *Macromolecules* **1988**, *21*, 3122.
78. Meijs, G.F.; Rizzardo, E. *Makromol. Chem., Rapid Commun.* **1988**, *9*, 547.
79. Rizzardo, E.; Harrison, D.; Laslett, R.L.; Meijs, G.F.; Morton, T.C.; Thang, S.H. *Prog. Pacific Polym. Sci.* **1991**, *2*, 77.
80. Meijs, G.F.; Rizzardo, E. *Makromol. Chem.* **1990**, *191*, 1545.
81. Dais, V.A.; Priddy, D.B.; Bell, B.; Sikkema, K.D.; Smith, P. *J. Polym. Sci., Part A: Polym. Chem.* **1993**, *31*, 901.
82. Rizzardo, E.; Meijs, G.F.; Thang, S.H. *Macromol. Symp.* **1995**, *98*, 101.
83. Bailey, W.J.; Endo, T.; Gapud, B.; Lin, Y.-N.; Ni, Z.; Pan, C.-Y.; Shaffer, S.E.; Wu, S.-R.; Yamazaki, N.; Yonezawa, K. *J. Macromol. Sci., Chem.* **1984**, *A21*, 979.
84. Meijs, G.F.; Rizzardo, E.; Thang, S.H. *Polym. Bull.* **1990**, *24*, 501.
85. Yamada, B.; Kobatake, S.; Aoki, S. *Macromol. Chem. Phys.* **1994**, *195*, 581.
86. Yamada, B.; Otsu, T. *Makromol. Chem.* **1991**, *192*, 333.
87. Yamada, B.; Otsu, T. *Makromol. Chem., Rapid Commun.* **1990**, *11*, 513.

88. Yamada, B.; Satake, M.; Otsu, T. *Polym. J.* **1992**, *24*, 563.
89. Yamada, B.; Kato, E.; Kobatake, S.; Otsu, T. *Polym. Bull.* **1991**, *25*, 423.
90. Yamada, B.; Kobatake, S.; Aoki, S. *Polym. Bull.* **1993**, *31*, 263.
91. Meijs, G.F.; Morton, T.C.; Rizzardo, E.; Thang, S.H. *Macromolecules* **1991**, *24*, 3689.
92. Mathias, L.J.; Thompson, R.D.; Lightsey, A.K. *Polym. Bull.* **1992**, *27*, 395.
93. Sato, T.; Seno, M.; Kobayashi, M.; Kohna, T.; Tanaka, H.; Ota, T. *Eur. Polym. J.* **1995**, *31*, 29.
94. Jiang, S.; Viehe, H.G.; Oger, N.; Charmot, D. *Macromol. Chem. Phys.* **1995**, *196*, 2349.
95. Nair, C.P.R.; Chaumont, P.; Charmot, D. *J. Polym. Sci., Part A: Polym. Chem.* **1995**, *33*, 2773.
96. Busfield, W.K.; Zayas-Holdsworth, C.I.; Thang, S.H. *Polymer* **2000**, *41*, 4409.
97. Busfield, W.K.; Zayas-Holdsworth, C.I.; Thang, S.H. *J. Polym. Sci., Part A: Polym. Chem.* **2000**, *39*, 2911.
98. Yamada, B.; Hirano, T.; Kobatake, S. *Polym. Bull.* **2003**, *49*, 305.
99. Zink, M.-O.; Colombani, D.; Chaumont, P. *Eur. Polym. J.* **1997**, *33*, 1433.
100. Chaumont, P.; Colombani, D. *Macromol. Chem. Phys.* **1995**, *196*, 3643.
101. Meijs, G.F.; Rizzardo, E.; Thang, S.H. *Polym. Prepr. (Am. Chem. Soc., Div. Polym. Chem.)* **1992**, *33(1)*, 893.
102. Shanmugananda Murthy, K.; Kishore, K. *J. Polym. Sci., Part A: Polym. Chem.* **1996**, *34*, 1415.
103. Colombani, D.; Chaumont, P. *J. Polym. Sci., Part A: Polym. Chem.* **1994**, *32*, 2687.
104. Colombani, D.; Chaumont, P. *Macromolecules* **1994**, *27*, 5972.
105. Colombani, D.; Chaumont, P. *Polymer* **1995**, *36*, 129.
106. Chaumont, P.; Colombani, D. *Macromolecules* **1995**, *29*, 819.
107. Chaumont, P.; Colombani, D. *Macromol. Chem. Phys.* **1995**, *196*, 947.
108. Colombani, D.; Lamps, J.-P.; Chaumont, P. *Macromol. Chem. Phys.* **1998**, *199*, 2517.
109. Tanaka, H.; Kawa, H.; Sato, T.; Ota, T. *J. Polym. Sci., Part A: Polym. Chem.* **1989**, *27*, 1741.
110. Gridnev, A.A.; Ittel, S.D. *Chem. Rev.* **2001**, *101*, 3611.
111. Hutson, L.; Krstina, J.; Moad, C.L.; Moad, G.; Morrow, G.R.; Postma, A.; Rizzardo, E.; Thang, S.H. *Macromolecules* **2004**, *37*, 4441.
112. Harrison, D.S. MSc Thesis; Swinburne University: Hawthorn, Victoria.
113. Haddleton, D.M.; Topping, C.; Kukulj, D.; Irvine, D. *Polymer* **1998**, *39*, 3119.
114. Yamada, B.; Tagashira, S.; Aoki, S. *J. Polym. Sci., Part A: Polym. Chem.* **1994**, *32*, 2745.
115. Chiefari, J.; Jeffery, J.; Moad, C.L.; Moad, G.; Postma, A.; Rizzardo, E.; Thang, S.H. *Macromolecules* **2005**, in press.
116. Nair, C.P.R.; Chaumont, P. *J. Polym. Sci., Part A: Polym. Chem.* **1999**, *37*, 2511.
117. Kobatake, S.; Yamada, B. *J. Polym. Sci., Part A: Polym. Chem.* **1996**, *34*, 95.
118. Hirano, T.; Yamada, B. *Polymer* **2003**, *44*, 347.
119. Chiefari, J.; Jeffery, J.; Mayadunne, R.T.A.; Moad, G.; Rizzardo, E.; Thang, S.H. *ACS Symp. Ser.* **2000**, *768*, 297.
120. Meijs, G.F.; Morton, T.C.; Le, T.P.T. *Polym. Int.* **1991**, *26*, 239.
121. Meijs, G.F.; Rizzardo, E. *Polym. Bull.* **1991**, *26*, 291.
122. Meijs, G.F.; Rizzardo, E.; Le, T.P.T.; Chong, Y.K. *Macromol. Chem. Phys.* **1992**, *193*, 369.
123. Crich, D.; Quintero, L. *Chem. Rev.* **1989**, *89*, 1413.

124. Barton, D.H.R.; Bridon, D.; Fernandez-Picot, I.; Zard, S.Z. *Tetrahedron* **1987**, *43*, 2733.
125. Barton, D.H.R.; Samadi, M. *Tetrahedron* **1992**, *48*, 7083.
126. Bergbreiter, D.E.; Jing, Z. *J. Polym. Sci., Part A: Polym. Chem.* **1992**, *30*, 2049.
127. Daly, W.H.; Evenson, T.S. *ACS Symp. Ser.* **1998**, *685*, 377.
128. Daly, W.H.; Evenson, T.S.; Iacono, S.T.; Jones, R.W. *Macromol. Symp.* **2001**, *174*, 155.
129. Daly, W.H.; Evenson, T.S. *Polymer* **2000**, *41*, 5063.
130. Studer, A.; Amrein, S.; Scieth, F.; Schulte, T.; Walton, J.C. *J. Amer. Chem. Soc.* **2003**, *125*, 5726.
131. Enikolopyan, N.S.; Smirnov, B.R.; Ponomarev, G.V.; Belgovskii, I.M. *J. Polym. Sci., Polym. Chem. Ed.* **1981**, *19*, 879.
132. Burczyk, A.F.; O'Driscoll, K.F.; Rempel, G.L. *J. Polym. Sci., Polym. Chem. Ed.* **1984**, *22*, 3255.
133. Gridnev, A.A. *Polym. Sci. USSR (Engl. Transl.)* **1989**, *31*, 2369.
134. Karmilova, L.V.; Ponomarev, G.V.; Smirnov, B.R.; Bel'govskii, I.M. *Russ. Chem. Rev. (Engl. Transl.)* **1984**, *53*, 132.
135. Davis, T.P.; Haddleton, D.M.; Richards, S.N. *J. Macromol. Sci., Rev. Macromol. Chem. Phys.* **1994**, *C34*, 243.
136. Parshall, G.W.; Ittel, S.D. *Homogeneous Catalysis*; Wiley: New York, 1992.
137. Heuts, J.P.A.; Roberts, G.E.; Biasutti, J.D. *Aust. J. Chem.* **2002**, *55*, 381.
138. Gridnev, A. *J. Polym. Sci. Pol. Chem.* **2000**, *38*, 1753.
139. Sanayei, R.A.; O'Driscoll, K.F. *J. Macromol. Sci., Chem.* **1989**, *A26*, 1137.
140. Smirnov, B.R.; Marchenko, A.P.; Plotnikov, V.D.; Kuzayev, A.I.; Yenikolopyan, N.S. *Polym. Sci. USSR (Engl. Transl.)* **1981**, *23*, 1169.
141. Roberts, G.E.; Barner-Kowollik, C.; Davis, T.P.; Heuts, J.P.A. *Macromolecules* **2003**, *36*, 1054.
142. Roberts, G.E.; Davis, T.P.; Heuts, J.P.A. *J. Polym. Sci., Part A: Polym. Chem.* **2003**, *41*, 752.
143. Heuts, J.P.A.; Forster, D.J.; Davis, T.P.; Yamada, B.; Yamazoe, H.; Azukizawa, M. *Macromolecules* **1999**, *32*, 2511.
144. Roberts, G.E.; Heuts, J.P.A.; Davis, T.P. *Macromolecules* **2000**, *33*, 7765.
145. Chiefari, J.; Jeffery, J.; Moad, G.; Rizzardo, E.; Thang, S.H. *Polym. Prepr.* **1999**, *40(2)*, 344.
146. Li, Y.; Wayland, B.B. *Macromol. Rapid Commun.* **2003**, *24*, 307.
147. Wang, W.X.; Stenson, P.A.; Marin-Becerra, A.; McMaster, J.; Schroder, M.; Irvine, D.J.; Freeman, D.; Howdle, S.M. *Macromolecules* **2004**, *37*, 6667.
148. Hawthorne, D.G. US 5324879, 1994 (*Chem. Abstr.* **1987**, *107*, 237504).
149. Krstina, J.; Moad, C.L.; Moad, G.; Rizzardo, E.; Berge, C.T.; Fryd, M. *Macromol. Symp.* **1996**, *111*, 13.
150. Gridnev, A.A.; Bel'govskii, I.M.; Enikolopyan, N.S. *Dokl. Akad. Nauk SSSR (Engl. Transl.)* **1986**, *289*, 748.
151. Tang, L.; Norton, J.R.; Edwards, J.C. *Macromolecules* **2003**, *36*, 9716.
152. Tang, L.; Norton, J.R. *Macromolecules* **2004**, *37*, 241.
153. Grognec, E.C.; Claverie, J.; Poli, R. *J. Am. Chem. Soc.* **2001**, *123*, 9513.
154. Gibson, V.C.; O'Reilly, R.K.; Wass, D.F.; White, A.J.P.; Williams, D.J. *Macromolecules* **2003**, *36*, 2591.
155. Pierik, S.C.J.; Vollmerhaus, R.; van Herk, A.M. *Macromol. Chem. Phys.* **2003**, *204*, 1090.

156. Pierik, S.C.J.; Vollmerhaus, R.; van Herk, A.M.; German, A.L. *Macromol. Symp.* **2002**, *182*, 43.
157. Suddaby, K.G.; O'Driscoll, K.F.; Rudin, A. *J. Polym. Sci., Part A: Polym. Chem.* **1992**, *30*, 643.
158. Suddaby, K.G.; Maloney, D.R.; Haddleton, D.M. *Macromolecules* **1997**, *30*, 702.
159. Heuts, J.P.A.; Forster, D.J.; Davis, T.P. *Macromolecules* **1999**, *32*, 3907.
160. Pierik, S.C.J.; van Herk, A.M. *Macromol. Chem. Phys.* **2003**, *204*, 1406.
161. Heuts, J.P.A.; Muratore, L.M.; Davis, T.P. *Macromol. Chem. Phys.* **2000**, *201*, 2780.
162. Mayo, F.R.; Gregg, R.A.; Matheson, M.S. *J. Am. Chem. Soc.* **1951**, *73*, 1691.
163. Baysal, B.; Tobolsky, A.V. *J. Polym. Sci.* **1952**, *8*, 529.
164. Mahadevan, V.; Santhappa, M. *Makromol. Chem.* **1955**, *16*.
165. Das, S.K.; Chatterjee, S.R.; Palit, S.R. *Proc. R. Soc., London* **1955**, *A227*, 252.
166. Stein, D.J. *Makromol. Chem.* **1964**, *76*, 170.
167. Kuchanov, S.I.; Olenin, A.V. *Polym. Sci. USSR (Engl. Transl.)* **1973**, *15*, 2712.
168. Starnes, W.H., Jr.; Schilling, F.C.; Plitz, I.M.; Cais, R.E.; Freed, D.J.; Hartless, R.L.; Bovey, F.A. *Macromolecules* **1983**, *16*, 790.
169. Bartlett, P.D.; Altschul, R. *J. Am. Chem. Soc.* **1945**, *67*, 816.
170. Pryor, W.A.; Coco, J.H. *Macromolecules* **1970**, *3*, 500.
171. Nozakura, S.-I.; Morishima, Y.; Murahashi, S. *J. Polym. Sci., Part A-1* **1972**, *10*, 2853.
172. Melville, H.W.; Sewell, P.R. *Makromol. Chem.* **1959**, *32*, 139.
173. Starnes, W.H., Jr.; Chung, H.; Benedikt, G.M. *Polym. Prepr. (Am. Chem. Soc., Div. Polym. Chem.)* **1993**, *34(1)*, 604.
174. Litt, M.; Chang, K.H.S. *ACS Symp. Ser.* **1981**, *165*, 455.
175. Wolf, C.; Burchard, W. *Makromol. Chem.* **1976**, *177*, 2519.
176. Vidotto, G.; Crosato-Arnaldi, A.; Talamini, G. *Makromol. Chem.* **1968**, *114*, 217.
177. Rigo, A.; Palma, G.; Talamini, G. *Makromol. Chem.* **1972**, *153*, 219.
178. Starnes, W.H., Jr.; Schilling, F.C.; Abbas, K.B.; Cais, R.E.; Bovey, F.A. *Macromolecules* **1979**, *12*, 556.
179. Park, G.S.; Saleem, M. *Polym. Bull.* **1979**, *1*, 409.
180. Starnes, W.H., Jr.; Wojciechowski, B.J. *Makromol. Chem., Macromol. Symp.* **1993**, *70/71*, 1.
181. Zubov, V.P.; Kumar, M.V.; Masterova, M.N.; Kabanov, V.A. *J. Macromol. Sci., Chem.* **1979**, *A13*, 111.
182. Butler, G.B. In *Comprehensive Polymer Science*; Eastmond, G.C.; Ledwith, A.; Russo, S.; Sigwalt, P., Eds.; Pergamon: Oxford, 1989; Vol. 4, p 423.
183. Bartlett, P.D.; Tate, F.A. *J. Am. Chem. Soc.* **1953**, *75*, 91.
184. Butler, G.B. *Cyclopolymerization and Cyclocopolymerization*; Marcel Dekker: New York, 1992.
185. Tüzün, N.S.; Aviyente, V.; Houk, K.N. *J. Org. Chem.* **2002**, *67*, 5068.
186. Vaidya, R.A.; Mathias, L.J. *J. Polym. Sci., Polym. Symp.* **1986**, *74*, 243.
187. Bevington, J.C.; Melville, H.W.; Taylor, R.P. *J. Polym. Sci.* **1954**, *14*, 463.
188. Schulz, G.V.; Stein, D.J. *Makromol. Chem.* **1962**, *52*, 1.
189. Lim, D.; Wichterle, O. *J. Polym. Sci.* **1958**, *29*, 579.
190. Usami, T.; Takayama, S. *Macromolecules* **1984**, *17*, 1756.
191. Axelson, D.E.; Levy, G.C.; Mandelkern, L. *Macromolecules* **1979**, *12*, 41.
192. Bovey, F.A.; Schilling, F.C.; McCrackin, F.L.; Wagner, H.L. *Macromolecules* **1976**, *9*, 76.
193. Bugada, D.C.; Rudin, A. *Eur. Polym. J.* **1987**, *23*, 809.

194. Rudin, A.; Grinshpun, V.; O'Driscoll, K. *J. Liq. Chromatog.* **1984**, *7*, 1809.
195. Grinshpun, V.; Rudin, A.; Russell, K.E.; Scammell, M.V. *J. Polym. Sci., Part B: Polym. Phys.* **1986**, *24*, 1171.
196. Pang, S.; Rudin, A. *Polym. Mat. Sci. Eng.* **1991**, *65*, 95.
197. Bugada, D.C.; Rudin, A. *Eur. Polym. J.* **1987**, *23*, 847.
198. Martin, J. *J. Appl. Polym. Sci.* **1990**, *40*, 1801.
199. Schulz, G.V.; Henrici, G.; Olive, S. *J. Polym. Sci.* **1955**, *17*, 45.
200. Fox, T.G.; Gratch, S. *Ann. New York Acad. Sci.* **1953**, *57*, 367.
201. Lovell, P.A.; Shah, T.H.; Heatley, F. *Polym. Commun.* **1991**, *32*, 98.
202. Ahmad, N.M.; Heatley, F.; Lovell, P.A. *Macromolecules* **1998**, *31*, 2822.
203. Dunn, A.S.; Naravane, S.R. *Br. Polym. J.* **1980**, 75.
204. Bugada, D.C.; Rudin, A. *Polymer* **1984**, *25*, 1759.
205. Bugada, D.C.; Rudin, A. *J. Appl. Polym. Sci.* **1985**, *30*, 4137.
206. McDowell, W.H.; Kenyon, W.O. *J. Am. Chem. Soc.* **1940**, *62*, 415.
207. Nozakura, S.-I.; Morishima, Y.; Murahashi, S. *J. Polym. Sci., Part A-1* **1972**, *10*, 2781.
208. Adelman, R.L.; Ferguson, R.C. *J. Polym. Sci., Polym. Chem. Ed.* **1975**, *13*, 891.
209. Ovenall, D.W. *Macromolecules* **1984**, *17*, 1458.
210. Agarwal, S.H.; Jenkins, R.F.; Porter, R.S. *J. Appl. Polym. Sci.* **1982**, *27*, 113.
211. Vercauteren, F.F.; Donners, W.A.B. *Polymer* **1986**, *27*, 993.
212. Hatada, K.; Terawaki, Y.; Kitayama, T.; Kamachi, M.; Tamaki, M. *Polym. Bull.* **1981**, *4*, 451.
213. Kamachi, M.; Liaw, D.J.; Nozakura, S.-I. *Polym. J.* **1979**, *11*, 921.
214. Starnes, W.H. *Prog. Polym. Sci.* **2002**, *27*, 2133.
215. Ovenall, D.W.; Uschold, R.E. *Macromolecules* **1991**, *24*, 3235.

7
Copolymerization

7.1 Introduction

Copolymerizations are processes that lead to the formation of polymer chains containing two or more discrete types of monomer unit. Several classes of copolymer that differ in sequence distribution and/or architecture will be considered:

(a) Statistical copolymers are formed when a mixture of two or more monomers is polymerized in a single process and where the arrangement of the monomers within the chains is dictated purely by kinetic factors (Section 7.3).

$$\sim CH_2\text{-}CH\text{-}CH_2\text{-}\underset{\underset{CO_2CH_3}{|}}{\overset{\overset{CH_3}{|}}{C}}\text{-}CH_2\text{-}CH\text{-}CH_2\cdot CH\text{-}CH_2\text{-}\underset{\underset{CO_2CH_3}{|}}{\overset{\overset{CH_3}{|}}{C}}\text{-}CH_2\text{-}CH\text{-}CH_2\text{-}\underset{\underset{CO_2CH_3}{|}}{\overset{\overset{CH_3}{|}}{C}}\text{-}CH_2\text{-}CH\text{-}CH_2\cdot CH\text{-}CH_2\text{-}\underset{\underset{CO_2CH_3}{|}}{\overset{\overset{CH_3}{|}}{C}}\sim$$
(Ph substituents on the CH carbons as shown)

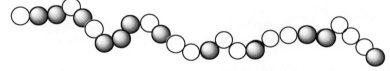

copolymer, *e.g.* poly(methyl methacrylate-*co*-styrene)
random copolymer, *e.g.* poly(methyl methacrylate-*ran*-styrene)
statistical copolymer, *e.g.* poly(methyl methacrylate-*stat*-styrene)

(b) Under some conditions (Section 7.3.1.3) the monomer units alternate in the chain. These copolymers are called alternating copolymers.

$$\sim CH_2\text{-}CH\text{-}CH_2\text{-}\underset{\underset{CO_2CH_3}{|}}{\overset{\overset{CH_3}{|}}{C}}\text{-}CH_2\text{-}CH\text{-}CH_2\cdot \underset{\underset{CO_2CH_3}{|}}{\overset{\overset{CH_3}{|}}{C}}\text{-}CH_2\text{-}CH\text{-}CH_2\text{-}\underset{\underset{CO_2CH_3}{|}}{\overset{\overset{CH_3}{|}}{C}}\text{-}CH_2\text{-}CH\text{-}CH_2\text{-}\underset{\underset{CO_2CH_3}{|}}{\overset{\overset{CH_3}{|}}{C}}\text{-}CH_2\cdot CH\text{-}CH_2\text{-}\underset{\underset{CO_2CH_3}{|}}{\overset{\overset{CH_3}{|}}{C}}\sim$$

alternating copolymer, *e.g.* poly(methyl methacrylate-*alt*-styrene)

(c) In living polymerizations, compositional drift as monomer is converted during the polymerization process and leads to the formation of gradient or tapered copolymers (Section 9.6).

gradient copolymer, e.g. poly(methyl methacrylate-*grad*-styrene)

(d) Block or segmented copolymers are usually prepared by multi-step processes (Section 7.5 and Section 9.7). The blocks may be a homopolymer or may themselves be copolymers.

diblock copolymer, e.g. poly(methyl methacrylate)-*block*-polystyrene

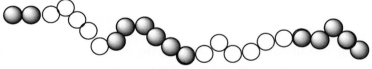

segmented or multiblock copolymer

(e) Graft copolymers and branched (co)polymers are also usually prepared by multi-step processes (Section 7.5). However, they are also formed by copolymerization of macromonomers (Section 7.6.5) and can form as a consequence of intramolecular rearrangement (Section 4.3). The backbone and the pendant chains may be of the same or different composition and may themselves be copolymers.

Copolymerization

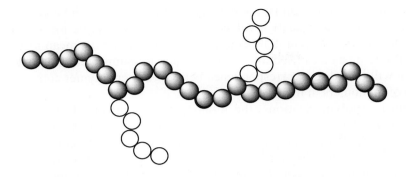

graft or branched copolymer

(f) Special classes of branched copolymers are star polymers, dendrimers, hyperbranched copolymers and microgels (Section 9,8).

In this chapter, we restrict discussion to approaches based on conventional radical polymerization. Living polymerization processes offer greater scope for controlling polymerization kinetics and the composition and architecture of the resultant polymer. These processes are discussed in Chapter 9.

7.2 Copolymer Depiction

IUPAC recommendations suggest that a copolymer structure, in this case poly(methyl methacrylate-*co*-styrene) or copoly(methyl methacrylate/styrene), should be represented as **1**. The most substituted carbon of the configurational repeat unit should appear first. This same rule would apply to the copolymer segments shown in Section 7.1. However, as was mentioned in Chapter 1, in this book, because of the focus on mechanism, we have adopted the more traditional depiction **2** which follows more readily from the polymerization mechanism.

$$\left[\begin{array}{c} CH_3 \\ | \\ C-CH_2 \\ | \\ CO_2Me \end{array} \Big/ \begin{array}{c} \\ HC-CH_2 \\ | \\ Ph \end{array} \right]_n \qquad \left[\begin{array}{c} CH_3 \\ | \\ H_2C-C \\ | \\ CO_2Me \end{array} \Big/ \begin{array}{c} \\ H_2C-CH \\ | \\ Ph \end{array} \right]_n$$

           1   **2**

7.3 Propagation in Statistical Copolymerization

Statistical copolymers are formed when mixtures of two or more monomers are polymerized by a radical process. Many reviews on the kinetics and mechanism of statistical copolymerization have appeared[1-9] and some detail can be found in most text books on polymerization. The term 'random copolymer', often used to describe these materials, is generally not appropriate since the incorporation of monomer units is seldom a purely random process. The arrangement of monomer units in the chains is dictated by the inherent reactivities of the monomers and radicals involved which may, in turn, be influenced by the reaction conditions (solvent, temperature, *etc.*). These factors mean that it is only in special circumstances (Section 7.3.1.1), when monomer reactivities are equal, that there will be a direct correspondence between the copolymer composition and the ratio of monomers in the feed.

In most copolymerizations, the monomers are consumed at different rates dictated by the steric and electronic properties of the reactants. Consequently, both the monomer feed and copolymer composition will drift with conversion. Batch copolymers will generally not be homogeneous in composition at the molecular level. Unfortunately, the detail of the chemical composition of copolymers is not always readily measurable. Many of the traditional techniques only give the average composition (the average ratio of monomers). In living polymerization processes, where ideally all chains grow throughout the polymerization, composition drift is captured within the chain structure. All chains have similar composition and are called gradient or tapered copolymers (Section 9.6).

The detailed microstructure and compositional heterogeneity of copolymers can have a determining influence on copolymer properties. This has been recognized for many years,[10] though the implications are often not fully appreciated. When copolymers with specific properties are required, it is generally not sufficient to control only the average number of functional groups/per polymer molecule.[11-13] It is important to have the functionality distributed in a particular manner along the individual chains (monomer sequence distribution) and amongst the chains (chemical heterogeneity). The microstructure and the degree of heterogeneity can be controlled by designing the monomer feed and/or by selecting the functional monomers according to their inherent reactivity and sometimes by choosing the initiator or transfer agent. The effects of specificity in the initiation and termination steps on the compositional heterogeneity are considered in Section 7.4.5.

Any understanding of the kinetics of copolymerization and the structure of copolymers requires a knowledge of the dependence of the initiation, propagation and termination reactions on the chain composition, the nature of the monomers and radicals, and the polymerization medium. This section is principally concerned with propagation and the effects of monomer reactivity on composition and monomer sequence distribution. The influence of solvent and complexing agents on copolymerization is dealt with in more detail in Section 8.3.1.

Propagation in copolymerization could, in principle, be discussed under the same headings as used for the discussion of propagation in Chapter 4. However, remarkably little information is currently available on the tacticity, extents of head *vs* tail addition, and propensity for rearrangement in copolymerization.

7.3.1 Propagation Mechanisms in Copolymerization

Studies on radical copolymerization and related model systems have demonstrated that many factors can influence the rate and course of propagation in copolymerization. These include:

(a) The structure of the propagating species and the likelihood of significant remote unit effects.
(b) The possibility of complex formation between monomers, between monomer and solvent, *etc.*
(c) The kinetics and thermodynamics of copolymerization and the possibility that depropagation is competitive with propagation.
(d) The nature of the medium and the manner in which it changes during the course of the copolymerization.

The various copolymerization models that appear in the literature (terminal, penultimate, complex dissociation, complex participation, *etc.*) should not be considered as alternative descriptions. They are approximations made through necessity to reduce complexity. They should, at best, be considered as a subset of some overall scheme for copolymerization. Any unified theory, if such is possible, would have to take into account all of the factors mentioned above. The models used to describe copolymerization reaction mechanisms are normally chosen to be the simplest possible model capable of explaining a given set of experimental data. They do not necessarily provide, nor are they meant to be, a complete description of the mechanism. Much of the impetus for model development and drive for understanding of the mechanism of copolymerization comes from the need to predict composition and rates. Developments in models have followed the development and application of analytical techniques that demonstrate the inadequacy of an earlier model.

7.3.1.1 *Terminal model*

The simplest model for describing binary copolymerization of two monomers, M_A and M_B, is the terminal model. The model has been applied to a vast number of systems and, in most cases, appears to give an adequate description of the overall copolymer composition; at least for low conversions. The limitations of the terminal model generally only become obvious when attempting to describe the monomer sequence distribution or the polymerization kinetics. Even though the terminal model does not always provide an accurate description of the copolymerization process, it remains useful for making qualitative predictions, as a starting point for parameter estimation and it is simple to apply.

The terminal model involves a number of approximations:[14]

(a) It is assumed that the copolymer composition is dictated by the relative rates of only four propagation reactions (Scheme 7.1). It is implicit in the model that only the last added monomer unit determines reactivity of the propagating radicals. Note that $P_A\bullet$ and $P_B\bullet$ are propagating species where the terminal (last added) monomer units are M_A and M_B respectively.

$$
\begin{array}{lll}
P_A\bullet + M_A \rightarrow P_A\bullet & \quad & k_{pAA}[P_A\bullet][M_A] \\
P_A\bullet + M_B \rightarrow P_B\bullet & \quad & k_{pAB}[P_A\bullet][M_B] \\
P_B\bullet + M_A \rightarrow P_A\bullet & \quad & k_{pBA}[P_B\bullet][M_A] \\
P_B\bullet + M_B \rightarrow P_B\bullet & \quad & k_{pBB}[P_B\bullet][M_B]
\end{array}
$$

Scheme 7.1

(b) It is assumed that chains are long and therefore the influence of the initiation and termination steps on the rate of monomer consumption can be neglected. The rates of monomer disappearance can then be written as shown in eqs. 1 and 2.

$$R_A = -\frac{d[M_A]}{dt} = k_{pAA}[P_A\bullet][M_A] + k_{pBA}[P_B\bullet][M_A] \tag{1}$$

$$R_B = -\frac{d[M_B]}{dt} = k_{pAB}[P_A\bullet][M_B] + k_{pBB}[P_B\bullet][M_B] \tag{2}$$

The ratio of these equations provides an expression for the instantaneous copolymer composition (eq. 3).

$$\frac{d[M_A]}{d[M_B]} = \frac{k_{pAA}[P_A\bullet][M_A] + k_{pBA}[P_B\bullet][M_A]}{k_{pAB}[P_A\bullet][M_B] + k_{pBB}[P_B\bullet][M_B]} \tag{3}$$

(c) A third assumption is that the concentrations of the two propagating species, $P_A\bullet$ and $P_B\bullet$, achieve a steady state (eq. 4).

$$k_{pAB}[P_A\bullet][M_B] = k_{pBA}[P_B\bullet][M_A] \tag{4}$$

This allows elimination of the radical concentrations from the above equation and the copolymer composition equation (eq. 5),[14-16] also known as the Mayo-Lewis equation, can now be derived.

$$\frac{F_A}{F_B} = \frac{d[M_A]}{d[M_B]} = \frac{[M_A]}{[M_B]}\left(\frac{r_{AB}[M_A] + [M_B]}{[M_A] + r_{BA}[M_B]}\right) = \frac{f_A}{f_B}\left(\frac{r_{AB}f_A + f_B}{f_A + r_{BA}f_B}\right) \tag{5}$$

where F_A (= $1-F_B$) and f_A (= $1-f_B$) are the instantaneous mole fractions of monomer A in the polymer and in the monomer feed respectively, and r_{AB} and r_{BA} are the monomer reactivity ratios which are defined in eqs. 6 and 7. The reactivity ratios, r_{AB} and r_{BA}, are often abbreviated to r_A and r_B. The notation used (r_{AB} and r_{BA}) is preferred since it allows discussion of situations involving more than two monomers (*e.g.* terpolymerization, Section 7.3.2.4).

$$r_{AB} = \frac{k_{pAA}}{k_{pAB}} \tag{6}$$

$$r_{BA} = \frac{k_{pBB}}{k_{pBA}} \tag{7}$$

Other convenient forms of the copolymer composition equation are eq. 8:

$$\frac{F_A}{F_B} = \frac{1 + r_{AB} x}{1 + r_{BA}/x} \tag{8}$$

where $x = f_A/f_B$ and eq. 9:

$$F_A = \frac{r_{AB} f_A^2 + f_A f_B}{r_{AB} f_A^2 + 2 f_A f_B + r_{BA} f_B^2} = \frac{(r_{AB} - 1) f_A^2 + f_A}{(r_{AB} + r_{BA} - 2) f_A^2 + (2 - r_{BA}) f_A + r_{BA}}$$

$$= \frac{1 + r_{AB} x}{2 + r_{AB} x + r_{BA}/x} \tag{9}$$

(d) It is also implicit in this treatment that medium effects are negligible and that there is no participation by monomer-monomer or monomer-solvent complexes.

Table 7.1 Terminal Model Reactivity Ratios for Some Common Monomer Pairs[a]

Monomer B	Monomer A						
	S	MMA	MA	AN	VC	MAH	VAc
S	\	0.51	0.77	0.40	17	0.002	22
MMA	0.49	\	-	2.0	-	5.2	27
MA	0.12	-	\	1.02	-	2.8	9.0
AN	0.05	0.25	0.80	\	3.3	6.0	5.0
VC	0.04	-	-	0.057	\	0.30	1.4
MAH	0.021	0.018	0.011	0.00	0.008	\	0.003
VAc	0.02	0.03	-	0.02	0.73	0.055	\

[a] r_{AB} tabulated vertically, r_{BA} horizontally. Values taken from Laurier et al.[17] or from Greenley's compilation.[18] All values rounded to two significant figures.

Thus, the terminal model allows the copolymer composition for a given monomer feed to be predicted from just two parameters; the reactivity ratios r_{AB} and r_{BA}. Some values of terminal model reactivity ratios for common monomer pairs are given in Table 7.1. Values for other monomers can be found in data compilations.[18] Literature values of reactivity ratios for most monomer pairs can span a considerable range. This can reflect experimental error, uncertain polymerization mechanism and/or inappropriate experimental design. No critical assessment has been made of the data in Table 7.1. Inclusion does not imply that the terminal model adequately describes the system or that the values shown are the best values.

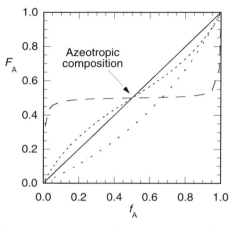

Figure 7.1 Plot of the instantaneous copolymer composition (F_A) vs monomer feed composition (f_A) for the situation where (a) $r_{AB}=r_{BA}=1.0$ (———), (b) $r_{AA}=r_{BA}=0.5$ (········), (c) $r_{AB}=r_{BA}=0.01$ (- - - - -), (d) $r_{AB}=0.5$, $r_{BA}=2.0$ (·····).

It is informative to consider some of the implications of the terminal model and, in particular, how the relative magnitudes of the reactivity ratios affect the copolymer composition (Figure 7.1):

(a) For the special case where $r_{AB} = r_{BA} = 1.0$, the monomers are utilized according to their respective proportions in the monomer feed. The product is a random copolymer. The value of F_A always equals f_A irrespective of the starting f_A. Copolymerizations of structurally similar monomers come closest to achieving this ideal. Examples are, copolymerizations of isotopically labeled monomers or mixtures of (meth)acrylic esters (with non-bulky ester groups) e.g. MMA and BMA).

(b) For many copolymerizations (e.g. S-MMA, S-AN) $r_{AB} < 1$ and $r_{BA} < 1$. In these cases, because cross-propagation is favored over homopropagation, there is a tendency towards alternation. In the extreme, where the values of both r_{AB} and r_{BA} approach zero (e.g. S-MAH), cross propagation occurs to the virtual exclusion of homopropagation and the product is an alternating copolymer.

(c) Where $r_{AB}>1$ and $r_{BA}<1$ (or $r_{AB}<1$ and $r_{BA}>1$), the copolymer will always be richer in one monomer than it is in the other. These copolymerizations have no azeotropic composition. Copolymerizations of VAc, NVP and VC with styrenic and (meth)acrylic monomers are in this class. The special case where the product $r_{AB}r_{BA}$ is unity ($r_{AB}=1/r_{BA}$) has been called ideal because the probabilities of a given monomer adding to the two propagating radicals are identical.[19]

(d) The converse situation, where both r_{AB} and r_{BA} are greater than one, is very rarely encountered. In this case, homopropagation is always favored over cross-propagation and, as a consequence, there will be a degree of blockiness in the copolymer.

In cases where $r_{AB}>1$ and $r_{BA}>1$ or $r_{AB}<1$ and $r_{BA}<1$, there will always be exactly one 'azeotropic composition' or 'critical point' where the copolymer composition will exactly reflect the monomer feed composition (Figure 7.1).

i.e. $\dfrac{d[M_A]}{d[M_B]} = \dfrac{[M_A]}{[M_B]} = x$ or $F_A = f_A$

Substitution into the copolymer composition equation (eq. 8) shows that this condition is satisfied when:

$$x = \dfrac{1-r_{AB}}{1-r_{BA}}$$

The existence of an azeotropic composition has some practical significance. By conducting a polymerization with the monomer feed ratio equal to the azeotropic composition, a high conversion batch copolymer can be prepared that has no compositional heterogeneity caused by drift in copolymer composition with conversion. Thus, the complex incremental addition protocols that are otherwise required to achieve this end, are unnecessary. Composition equations and conditions for azeotropic compositions in ternary and quaternary copolymerizations have also been defined.[20,21]

The overall rate of propagation in copolymerization is given by eq. 10.

$$R_p = -\dfrac{d[M]}{dt} = \bar{k}_p[P\bullet][M]$$
$$= k_{pAA}[P_A\bullet][M_A] + k_{pBA}[P_B\bullet][M_A] + k_{pAB}[P_A\bullet][M_B] + k_{pBB}[P_B\bullet][M_B] \quad (10)$$

where [M] (=[M_A]+[M_B]) is the total monomer concentration and [P•] (=[P_A•]+[P_B•]) is the total concentration of propagating radicals.

An expression (eq. 11) for the overall rate constant for propagation in copolymerization ($\bar{k}_p$) can now be formulated.

$$\bar{k}_p = \dfrac{r_{AB}f_A^2 + f_Af_B^2 + 2f_Af_B}{r_{AB}f_A/k_{pAA} + r_{BA}f_B/k_{pBB}} \quad (11)$$

Note that value of k_p is usually not constant with conversion since it depends on the monomer feed composition.

7.3.1.2 Penultimate model

The general features of the penultimate model in what have become known as the explicit and implicit forms are described in Section 7.3.1.2.1. Evidence for remote unit effects coming from small molecule radical chemistry and experiments other than copolymerization is discussed in Section 7.3.1.2.2. In Sections 7.3.1.2.3 and 7.3.1.2.4 specific copolymerizations are discussed. Finally, in Section 7.3.1.2.5, we consider the origin of the penultimate unit effects. A general recommendation is that when trying to decide on the mechanism of a copolymerization, first consider the explicit penultimate model.[2]

7.3.1.2.1 Model description

The influence of penultimate units on the kinetics of copolymerization and the composition of copolymers was first considered in a formal way by Merz et al.[22] and Ham.[8] They consider eight propagation reactions (Scheme 7.2).

$$P_{AA}\bullet + M_A \rightarrow P_{AA}\bullet \quad k_{pAAA}[P_{AA}\bullet][M_A]$$
$$P_{AA}\bullet + M_B \rightarrow P_{AB}\bullet \quad k_{pAAB}[P_{AA}\bullet][M_B]$$
$$P_{AB}\bullet + M_A \rightarrow P_{BA}\bullet \quad k_{pABA}[P_{AB}\bullet][M_A]$$
$$P_{AB}\bullet + M_B \rightarrow P_{BB}\bullet \quad k_{pABB}[P_{AB}\bullet][M_B]$$
$$P_{BA}\bullet + M_A \rightarrow P_{AA}\bullet \quad k_{pBAA}[B_{BA}\bullet][M_A]$$
$$P_{BA}\bullet + M_B \rightarrow P_{AB}\bullet \quad k_{pBAB}[P_{BA}\bullet][M_B]$$
$$P_{BB}\bullet + M_A \rightarrow P_{BA}\bullet \quad k_{pBBA}[P_{BB}\bullet][M_A]$$
$$P_{BB}\bullet + M_B \rightarrow P_{BB}\bullet \quad k_{pBBB}[P_{BB}\bullet][M_B]$$

Scheme 7.2

From this scheme it can be seen that the copolymer composition is determined by the values of four monomer reactivity ratios.*

$$r_{AAB} = \frac{k_{pAAA}}{k_{pAAB}} \quad r_{BAB} = \frac{k_{pBAA}}{k_{pBAB}} \quad r_{ABA} = \frac{k_{pABB}}{k_{pABA}} \quad r_{BBA} = \frac{k_{pBBB}}{k_{pBBA}}$$

Fukuda et al.[23] were the first to recognize that a further two radical reactivity ratios were required to completely define the polymerization kinetics.

$$s_A = \frac{k_{pAAA}}{k_{pBAA}} \quad s_B = \frac{k_{pBBB}}{k_{pABB}}$$

---

* The reactivity ratios r_{AAB}, r_{BAB}, r_{BBA} and r_{ABA} are sometimes abbreviated to r_{AA}, r_{BA}, r_{BB} and r_{AB} or to r_A, r_A', r_B, r_B' respectively. The notation used (r_{AAB}, r_{BAB}, r_{BBA} and r_{ABA}) is preferred since it allows discussion of situations involving more than two monomers.

Copolymerization

In traditional treatments of copolymerization kinetics, the values of the ratios s_A and s_B are implicitly set equal to unity (Section 7.3.1.2.2). Since they contain no terms from cross propagation, these parameters have no direct influence on either the overall copolymer composition or the monomer sequence distribution; they only influence the rate of polymerization.

The instantaneous copolymer composition is described by the following equation (eq. 12):

$$\frac{F_A}{F_B} = \frac{1 + \dfrac{r_{BAB} x (1 + r_{AAB} x)}{1 + r_{BAB} x}}{1 + \dfrac{r_{ABA}(r_{BBA} + x)}{x(r_{ABA} + x)}} \tag{12}$$

By substituting $\bar{r}_{AB} = r_{BAB} \dfrac{1 + r_{AAB} x}{1 + r_{BAB} x}$ and $\bar{r}_{BA} = r_{ABA} \dfrac{x + r_{BBA}}{x + r_{ABA}}$

eq. 12 may be written in a form similar to the terminal model copolymer composition equation (eq. 8) as eq. 13.

$$\frac{F_A}{F_B} = \frac{1 + \bar{r}_{AB} x}{1 + \bar{r}_{BA}/x} \tag{13}$$

Cases have been reported where the application of the penultimate model provides a significantly better fit to experimental composition or monomer sequence distribution data. In these copolymerizations $r_{AAB} \neq r_{BAB}$ and/or $r_{ABA} \neq r_{BBA}$. These include many copolymerizations of AN,[24-26] B,[27] MAH[28,29] and VC.[30] In these cases, there is no doubt that the penultimate model (or some scheme other than the terminal model) is required. These systems are said to show an explicit penultimate effect. In binary copolymerizations where the explicit penultimate model applies there may be between zero and three azeotropic compositions depending on the values of the reactivity ratios.[31]

It is possible to define average propagation rate constants for copolymerization subject to a penultimate group effect as follows.

$$\bar{k}_{pAA} = k_{pAAA} \frac{1 + r_{AAB} x}{r_{AAB} x + 1/s_B} \quad \text{and} \quad \bar{k}_{pBB} = k_{pBBB} \frac{x + r_{BBA}}{r_{BBA} + x/s_A}$$

Note that the values of $\bar{r}_{AB}$, $\bar{r}_{BA}$, $\bar{k}_{pAA}$, and $\bar{k}_{pBB}$ are dependent on the monomer feed composition and hence on conversion. These parameters may be substituted for r_{AB}, r_{BA}, k_{pAA} and k_{pBB} in eq. 11 to provide an expression for the overall rates of propagation (eq. 14) and of polymerization (eq. 15).

$$\bar{k}_{p} = \frac{\bar{r}_{AB}f_{A}^{2} + f_{A}f_{B}^{2} + 2f_{A}f_{B}}{\bar{r}_{AB}f_{A}/\bar{k}_{pAA} + \bar{r}_{BA}f_{B}/\bar{k}_{pBB}} \tag{14}$$

$$R_{p} = -\frac{d[M]}{dt} = \bar{k}_{p}[P\bullet][M] = \frac{\bar{r}_{AB}f_{A}^{2} + f_{A}f_{B}^{2} + 2f_{A}f_{B}}{\bar{r}_{AB}f_{A}/\bar{R}_{pA} + \bar{r}_{BA}f_{B}/\bar{R}_{pB}} \tag{15}$$

where $\bar{R}_{pA} = R_{pA} \dfrac{1 + r_{AAB}x}{r_{AAB}x + 1/s_{B}}$

For many systems, the copolymer composition appears to be adequately described by the terminal model yet the polymerization kinetics demand application of the penultimate model. These systems where $r_{AAB}=r_{BAB}$ and $r_{ABA}=r_{BBA}$ but $s_A \neq s_B$ are said to show an implicit penultimate effect. The most famous system of this class is MMA-S copolymerization (Section 7.3.1.2.3).

Penpenultimate and higher order remote unit effect models may also affect the outcome of copolymerizations. However, in most cases, experimental data, that are not sufficiently powerful to test the penultimate model, offer little hope of testing higher order models. The importance of remote unit effects on copolymerization will only be fully resolved when more powerful analytical techniques become available.

7.3.1.2.2 Remote substituent effects on radical addition

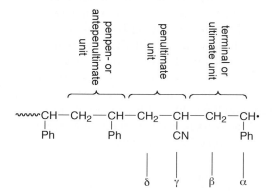

Figure 7.2 Chain end terminology.

In small molecule chemistry it is well established that β- and more remote substituents (Figure 7.2) can have a substantial influence on radical conformation, formation and reactivity. Thus, it should be anticipated that the nature of the penultimate unit of the propagating chain could significantly modify its reactivity towards monomers and other species. However, the magnitude of the effect will be dependent on the exact nature of the remote substituent and the reactants. It is also important to remember that, in copolymerization, we consider the effect of the penultimate unit on a rate constant ratio, not on the rate constant for a particular reaction.

Experimental studies on models of the propagating radicals in S-AN copolymerization[32,33] and a few other systems[34] provide support for an explicit penultimate unit effect. Of particular interest is the data of Tirrell and coworkers. They investigated the relative reactivity of S and AN towards various γ-substituted propyl radicals (Scheme 7.3 and Table 7.2). They found that:

(a) There is only a small effect on radical reactivity when the γ-substituent is a styryl unit (at a PSAN chain end), a phenyl, or an alkyl group.

(b) An electrophilic γ-cyano substituent has a marked effect on radical reactivity.

(c) The relative reactivities of simple model radicals correlate well with the reactivities of propagating species estimated from copolymerization data assuming an explicit penultimate model.

$$R^3\text{-CH-CH}_2\text{-}\overset{\bullet}{\text{CH}} \quad \text{CH}_2=\text{CH} \quad \xrightarrow{k_A}$$
$$\phantom{R^3\text{-CH-}}|\phantom{\text{CH}_2\text{-CH}}| \phantom{\text{CH}_2=} |$$
$$\phantom{R^3\text{-}}R^2 \phantom{\text{-CH}_2\text{-}} R^1 \phantom{\text{CH}_2=} \text{CN}$$

$$R^3\text{-CH-CH}_2\text{-}\overset{\bullet}{\text{CH}} \quad \text{CH}_2=\text{CH} \quad \xrightarrow{k_S}$$
$$\phantom{R^3\text{-CH-}}|\phantom{\text{CH}_2\text{-CH}}| \phantom{\text{CH}_2=} |$$
$$\phantom{R^3\text{-}}R^2 \phantom{\text{-CH}_2\text{-}} R^1 \phantom{\text{CH}_2=} \text{Ph}$$

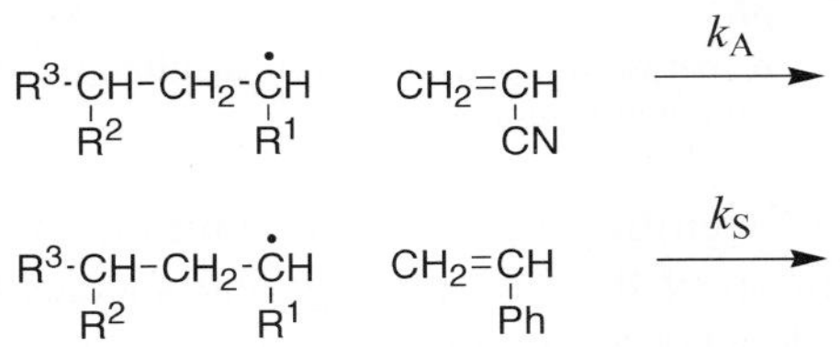

Scheme 7.3

Table 7.2 Relative Rates for Addition of Substituted Propyl Radicals ($R^3R^2\text{CHCH}_2\text{CHR}^1\bullet$) to AN and S at 100 °C[a]

R^1	R^2	R^3	$\dfrac{k_{AN}}{k_S}$
H	H	CH_3	24.5±1.1[32]
H	H	C_3H_7	26.3±2.4[32]
H	H	Ph	22.6±2.0[32]
Ph	H	H	5.0[c]
Ph	H	Ph	4.8±0.3[33]
Ph	PSAN[b]	Ph	4.2[d,25]
H	H	CN	6.8±0.6[32]
Ph	H	CN	1.9±0.1[33]
Ph	PSAN[b]	CN	1.7[d,25]

a refer Scheme 7.3. b Poly(AN-*co*-S) chain. c Value for 1-phenylethyl radical from Table 3.6 Section 3.4.1.1. Value from penultimate model reactivity ratios for AN-S copolymerization at 60 °C.[25]

Other experimental data seem to provide support for an implicit penultimate model. Thus, simple (monomeric) model radicals for the propagating radical chain ends in S-MMA copolymerization show similar (though not the same) chemospecificity to the corresponding propagating radicals in radical addition (Table 7.3). However, rate constants for addition appear more than an order of magnitude higher for the lower molecular weight species. There are many other examples of this type. Additional data on the rate constants for the reactions of small radicals with monomers can be found in Section 3.4.

Table 7.3 Relative Rates for Addition of Substituted Methyl Radicals ($R^3R^2R^1C\bullet$) to MMA and S at ~25 °C

R^1	R^2	R^3	k_S $M^{-1}s^{-1}$	$\dfrac{k_{MMA}}{k_S}$
Ph	H	H	1100 [35]	1.9 [35]
Ph	PSMMA[a]	H	93 [b]	2.2 [c]
CO_2CH_3	CH_3	CH_3	6030 [35]	0.61 [35]
CO_2CH_3	PSMMA[a]	H	180 [b]	0.52 [c]

a Poly(MMA-co-S) chain. b based on k_p for homopropagation. c Value from terminal model reactivity ratios for MMA-S copolymerization at 25 °C.[23]

Further examples of significant penultimate unit effects come from studies of rate constants for addition of the first propagating species to monomer (Scheme 7.4). There is a strong dependence on the particular initiating species. The data in Table 7.4 were provided in Fischer and Radom's review.[35]

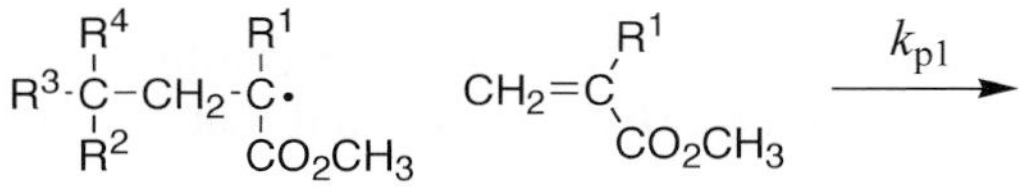

$$R^3-\underset{R^2}{\underset{|}{\overset{R^4}{\overset{|}{C}}}}-CH_2-\underset{CO_2CH_3}{\underset{|}{\overset{R^1}{\overset{|}{C}}}}\bullet \quad CH_2=\underset{CO_2CH_3}{\overset{R^1}{C}} \quad \xrightarrow{k_{p1}}$$

Scheme 7.4

It is known that the penultimate unit influences the conformation of both model radicals and propagating radicals.[35-38] Since addition requires a particular geometric arrangement of the reactants, there are enthalpic barriers to overcome for addition to take place and also potentially significant effects on the entropy of activation. Comparisons of the rate constants and activation parameters for homopropagation with those for addition of simple model radicals to the same monomers also provide evidence for significant penultimate unit effects (Section 4.5.4).

There is also clear evidence that penultimate group effects are important in determining the stereochemistry of addition in many homopolymerizations and copolymerizations. This is made evident from the fact that most homopolymers have tacticity (*i.e.* P(m)≠0.5, Section 4.2). Indeed, for some homopolymerizations there is evidence that the configuration of the penpenultimate unit may also influence the stereochemistry of addition.[39] If penpen- and penultimate units influence the stereospecificity of addition, it is also reasonable to expect that they might affect the rate and chemospecificity of addition.

Table 7.4 Rate Constants (298 K) for Addition of Substituted Propyl Radicals to (Meth)acrylate Esters (Scheme 7.4)[35]

R^1	R^2	R^3	R^4	k_{p1}	$k_{(rel)}$	$k_{(rel)}$
H	PMACH$_2$[a]	CO$_2$CH$_3$	H	19000	3.1	-
H	CH$_3$	CH$_3$	CH$_3$	6120	1.0	-
H	CH$_3$	CH$_3$	OH	3290	0.54	-
H	H	H	Ph	22400	37.	-
H	H	H	OH	18110	3.0	-
CH$_3$	PMMACH$_2$[a]	CO$_2$CH$_3$	H	345	-	0.58
CH$_3$	CH$_3$	CH$_3$	CH$_3$	600	-	1.0
CH$_3$	CH$_3$	CH$_3$	OH	1205	-	2.0
CH$_3$	H	H	Ph	2640	-	4.4
CH$_3$	H	H	OH	3290	-	5.5

a Long chain propagating radical.

Penultimate unit effects are also important in both substitution[40,41] and in addition-fragmentation chain transfer.[42-44] Some examples are provided in Sections 6.2, 6.2.2.4, 6.2.3.4 and 9.5.

Based on the above data, it would seem unusual if reactivity of the propagating species in copolymerization were insensitive to the nature of the last added monomer units. However, while there are ample experimental data to suggest that copolymerizations should be subject to penultimate unit effects that affect the rate and/or copolymer composition, the origin and magnitude of the effect is not always easily predictable.

7.3.1.2.3 MMA-S copolymerization

MMA-S copolymerization has been investigated by many groups.[23,45-49] Fukuda et al.[23] followed established procedure to confirm that the overall composition of MMA-S copolymers was satisfactorily predicted by the terminal model with $r_{AAB}=r_{BAB}=0.52$ and $r_{BBA}=r_{ABA}=0.46$. They applied the rotating sector method to determine absolute values of the overall propagation and termination rate constants. The data showed that the observed dependence of the rate of copolymerization on monomer feed composition, which had previously been attributed to an effect of the kinetics of termination, was in fact due to a composition dependence of the overall propagation rate constant. Fukuda et al.[23] proposed an explanation in terms of an implicit penultimate unit effect. Values of the radical reactivity ratios s_A (=0.52) and s_B (=0.30) were estimated which accounted for the data. Determinations of propagation rate constants using PLP, while suggesting slightly different vales of s_A and s_B (Table 7.5), confirm the basic result.[45,46,50,51]

Table 7.5. Implicit Penultimate Model Reactivity Ratios

M_A	M_2	r_{AB}	r_{BA}	$r_{AB}r_{BA}$	s_A	s_B	$s_A s_B$	Temp. °C	ref.
MMA	S	0.46	0.52	0.24	0.52	0.30	0.16	40	23
MMA	S	(0.46)	(0.52)	0.24	0.65	0.37	0.24^a	40	52
MMA	S	(0.46)	(0.52)	0.24	0.80	0.30	0.24^a	25	52
MMA	S	(0.46)	(0.52)	0.24	0.47	0.18	0.08	25	50
EMA	S	0.35	0.62	0.21	0.21	0.62	0.13	25	53
BMA	S	0.45	0.72	0.32	0.63	0.56	0.35	25	53
LMA	S	0.45	0.57	0.26	0.33	0.59	0.19	25	53
MA	S	0.19	0.73	0.14	0.26	1.10	0.29	25	54
MOS	S	0.82	1.12	0.92	1.0	1.0	1.0	25	55
MMA	MOS	0.29	0.32	0.09	0.60	0.36	0.22	25	55

a Assuming that $r_{AB}r_{BA} = s_A s_B$.

If the terminal model adequately explains the copolymer composition, as is often the case, the terminal model is usually assumed to apply. Even where statistical tests show that the penultimate model does not provide a significantly better fit to experimental data than the terminal model, this should not be construed as evidence that penultimate unit effects are unimportant.[49] It is necessary to test for model discrimination, rather than merely for fit to a given model. In this context, it is important to remember that composition data are of very low power when it comes to model discrimination. For MMA-S copolymerization, even though experimental precision is high, the penultimate model confidence intervals are quite large; $0.4 \leq r_{AAB}/r_{BAB} \leq 2.7$, $0.3 \leq r_{BBA}/r_{ABA} \leq 2.2$.[49] The terminal model ($r_{AAB}=r_{BAB}$, $r_{BBA}=r_{ABA}$) is only one of a number of possible solutions and the experimental composition data do not rule out the possibility of quite substantial penultimate unit effects. The same point was made more recently by Kaim.[47]

Triad information is more powerful, but typically is subject to more experimental error and signal assignments are often ambiguous (Section 7.3.3.2). Triad data for the MMA-S system are consistent with the terminal model and support the view that any penultimate unit effects on specificity are small.[56-58]

Further evidence that penultimate unit effects are small in the MMA-S system comes from comparing the reactivities of small model radicals with the reactivity ratios (Section 7.3.1.2.2 and Table 7.4).

7.3.1.2.4 Other copolymerizations

The kinetics of many copolymerizations have now been examined with absolute (overall) propagation rate constants being determined by the rotating sector, PLP or ESR methods. A similar situation as pertains for the MMA-S system applies in many cases. The terminal model appears to adequately describe copolymer composition but the kinetic data require a penultimate or more complex model. A summary of some recent data to which the implicit terminal model has been applied is provided in Table 7.5.

The values of s_A and s_B are not well defined by kinetic data.[59-61] The wide variation in s_A and s_B for MMA-S copolymerization shown in Table 7.5 reflects the large uncertainties associated with these values, rather than differences in the rate data for the various experiments. Partly in response to this, various simplifications to the implicit penultimate model have been used (*e.g.* $r_{AB}r_{BA}=s_As_B$[52] and $s_A=s_B$). These problems also prevent trends in the values with monomer structure from being established.

It has been pointed out that analysis of terpolymerization data or copolymerization with chain transfer could, in principle, provide a test of the model.[2,3] However, to date experimental uncertainty has prevented this.

7.3.1.2.5 *Origin of penultimate unit effects*

Some theoretical justifications for the prevalence of systems which show an implicit penultimate effect have appeared. These are summarized in the recent reviews by Coote and Davis.[2,3]

$$P_{ij}\bullet \ + \ M_k \ \rightarrow \ P_{jk}\bullet \qquad\qquad k_{pijk}[P_{ij}\bullet][M_k]$$

Scheme 7.5

Fukuda *et al.*[9,62] have argued that, in most copolymerizations, penultimate substituents should mainly influence the enthalpy for addition to monomer. It was proposed that enthalpy change ($-\Delta H_{ijk}$) is given by the eq. 16 (refer Scheme 7.5) which contains a constant term (ΔH_o) and the 'stabilization energies' of the product propagating radical (U_{jk}), the reactant propagating radical (U_{ij}) and the monomer (U_k).

$$-\Delta H_{ijk} = -\Delta H_o + U_{jk} - (U_{ij} + U_k) \tag{16}$$

If the Evans-Polyani rule (Section 2.4.1)[63] applies, the activation energy E_{ijk} will be proportional to the reaction enthalpy ($-\Delta H_{ijk}$) and eq. 17 will hold.

$$E_{ijk} = \beta + \alpha(-\Delta H_{ijk}) = \beta + \alpha[-\Delta H_o + U_{jk} - (U_{ij} + U_k)] \tag{17}$$

where β and α are constants.

If it is assumed that penultimate unit effects on the reaction entropy are insignificant, the terms in eqs. 18 and 19 corresponding to the stabilization energy of the reactant propagating radical will cancel and $r_{iij}=r_{jij}$. There should be no explicit penultimate unit effect on copolymer composition. On the other hand, the radical reactivity ratio s_i (eq. 20) compares two different propagating radicals so there is no cancellation of the penultimate unit effect. On the basis of this argument, the penultimate unit effect is expected to be implicit.

$$r_{iij} = \frac{k_{piii}}{k_{piij}} = \frac{A_{piii}}{A_{piij}} e^{E_{piii} - E_{piij}} \tag{18}$$

$$r_{jij} = \frac{k_{pjii}}{k_{pjij}} = \frac{A_{pjii}}{A_{pjij}} e^{E_{pjii} - E_{pjij}} \tag{19}$$

$$s_i = \frac{k_{pjii}}{k_{piii}} = \frac{A_{pjii}}{A_{piii}} e^{E_{pjii} - E_{piii}} \tag{20}$$

It also follows from this treatment that

$$r_{AB}r_{BA} = s_A s_B \tag{21}$$

However, plots which would demonstrate this relationship show considerable scatter.[44]

The above argument is also at odds with the conventional wisdom that the well-known tendency for monomer alternation in copolymerization can primarily be attributed to polar factors. It was suggested[9] that, in most cases, radical stabilization could provide an alternate explanation. A discussion on the relative importance of steric polar and radical stabilization effects on radical addition appears in Section 2.3.

It has been argued that for a majority of copolymerizations, composition data can be adequately predicted by the terminal model copolymer composition equation (eqs. 5-9). However, in that composition data are not particularly good for model discrimination, any conclusion regarding the widespread applicability of the implicit penultimate model on this basis is premature.

Heuts et al.,[64] while not disputing that penultimate units might influence the activation energies, proposed on the basis of theoretical calculations that penultimate unit effects of the magnitude seen in the S-AN and other systems (i.e. 2-5 fold) can also be explained by variations in the entropy of activation for the process. They also proposed that this effect would mainly influence rate rather than specificity.

7.3.1.3 Models involving monomer complexes

Mechanisms for copolymerization involving complexes between the monomers were first proposed to explain the high degree of alternation observed in some copolymerizations. They have also been put forward, usually as alternatives to the penultimate model, to explain anomalous (not consistent with the terminal model) composition data in certain copolymerizations.[65-74]

While there is clear evidence for complex formation between certain electron donor and electron acceptor monomers, the evidence for participation of such complexes in copolymerization is often less compelling. One of the most studied systems is S-MAH copolymerization.[28,75] However, the models have been applied to many copolymerizations of donor-acceptor pairs. Acceptor monomers have substituents such as carboxy, anhydride, ester, amide, imide or nitrile on the double bond. Donor monomers have substituents such as alkyl, vinyl, aryl, ether, sulfide and silane. A partial list of donor and acceptor monomers is provided in Table 7.6.[65]:

Common features of polymerizations involving such monomer pairs are:

(a) A high degree of monomer alternation in the chain is observed.
(b) The copolymer composition cannot be rationalized on the basis of the terminal model (Section 7.3.1.1).
(c) The rate of copolymerization is usually very much faster than that of either homopolymerization.
(d) Many of the monomers do not readily undergo homopolymerization or copolymerization with monomers of like polarity.
(e) For most systems there is spectroscopic evidence for some form of donor-acceptor interaction.

Table 7.6 List of Donor and Acceptor Monomers

Donors	Acceptors
dienes (*e.g.* B, isoprene)	MAA, itaconic acid
heterocyclic dienes (*e.g.* furan, indole, thiophene)	(meth)acrylate esters (*e.g.* MA and MMA)
vinylbenzene and derivatives (*e.g.* S, AMS)	cinnamate esters
vinyl heteroaromatics (e.g. vinyl pyridine, vinyl carbazole)	methacrylamides
vinyl esters (*e.g.* VAc)	cyanoethylenes (*e.g.* AN and 1,1-dicyanoethylene)
vinyl ethers (*e.g.* ethyl vinyl ether)	maleate, fumarate esters
vinyl sulfides	MAH, citraconic anhydride
vinyl halides	maleimides (*e.g.* N-phenylmaleimide)

However, these observations are not proof of the role of a donor-acceptor complex in the copolymerization mechanism. Even with the availability of sequence information it is often not possible to discriminate between the complex model, the penultimate model (Section 7.3.1.2) and other, higher order, models.[28] A further problem in analyzing the kinetics of these copolymerizations is that many donor-acceptor systems also give spontaneous initiation (Section 3.3.6.3).

Equilibrium constants for complex formation (K) have been measured for many donor-acceptor pairs. Donor-acceptor interaction can lead to formation of highly colored charge-transfer complexes and the appearance of new absorption bands in the UV-visible spectrum may be observed. More often spectroscopic evidence for complex formation takes the form of small chemical shift differences in NMR spectra or shifts in the positions of the UV absorption maxima. In analyzing these systems it is important to take into account that some solvents might also interact with donor or acceptor monomers.

Since intermediates usually cannot be observed directly, the exact nature of the donor-acceptor complex and the mechanisms for their interaction with radicals are speculative. At least three ways may be envisaged whereby complex formation may affect the course of polymerization:

(a) The complex participation model.[75-77] A binary complex is formed that is much more reactive than either of the non-complexed monomers. The monomers are incorporated into the chain in pairs (Scheme 7.6). If reaction with the complexed monomer competes with addition to uncomplexed monomer, the mechanism may be described in terms of six reactivity ratios and one equilibrium constant.

$$P_A\bullet + M_A \rightarrow P_A\bullet \qquad k_{pAA}[P_A\bullet][M_A]$$
$$P_A\bullet + M_B \rightarrow P_B\bullet \qquad k_{pAB}[P_A\bullet][M_B]$$
$$P_B\bullet + M_A \rightarrow P_A\bullet \qquad k_{pBA}[P_B\bullet][M_A]$$
$$P_B\bullet + M_B \rightarrow P_B\bullet \qquad k_{pBB}[P_B\bullet][M_B]$$
$$P_A\bullet + M_BM_A \rightarrow P_A\bullet \qquad k_{pABA}[P_A\bullet][M_AM_B]$$
$$P_A\bullet + M_AM_B \rightarrow P_B\bullet \qquad k_{pAAB}[P_A\bullet][M_AM_B]$$
$$P_B\bullet + M_BM_A \rightarrow P_A\bullet \qquad k_{pBBA}[P_B\bullet][M_AM_B]$$
$$P_B\bullet + M_AM_B \rightarrow P_B\bullet \qquad k_{pBAB}[P_B\bullet][M_AM_B]$$
$$M_A + M_B \rightleftharpoons M_AM_B$$

Scheme 7.6

(b) The complex dissociation model.[78-80] A binary complex is formed that is much more reactive than either of the non-complexed monomers. The complex dissociates after addition and only a single monomer unit is incorporated on reaction with the complex (Scheme 7.7).

(c) Formation of a less reactive complex. This could have the effect of reducing the overall monomer concentration and perhaps altering the ratio of reactive monomers in the feed. However, the fraction of monomer complexed is typically small.

Copolymerization

$$P_A\bullet + M_A \rightarrow P_A\bullet \qquad k_{pAA}[P_A\bullet][M_A]$$
$$P_A\bullet + M_B \rightarrow P_B\bullet \qquad k_{pAB}[P_A\bullet][M_B]$$
$$P_B\bullet + M_A \rightarrow P_A\bullet \qquad k_{pBA}[P_B\bullet][M_A]$$
$$P_B\bullet + M_B \rightarrow P_B\bullet \qquad k_{pBB}[P_B\bullet][M_B]$$
$$P_A\bullet + M_BM_A \rightarrow P_A\bullet + M_B \qquad k_{pABA}[P_A\bullet][M_AM_B]$$
$$P_A\bullet + M_AM_B \rightarrow P_B\bullet + M_A \qquad k_{pAAB}[P_A\bullet][M_AM_B]$$
$$P_B\bullet + M_BM_A \rightarrow P_A\bullet + M_B \qquad k_{pBBA}[P_B\bullet][M_AM_B]$$
$$P_B\bullet + M_AM_B \rightarrow P_B\bullet + M_A \qquad k_{pBAB}[P_B\bullet][M_AM_B]$$
$$M_A + M_B \rightleftharpoons M_AM_B$$

Scheme 7.7

Several studies on the reactivities of small radicals with donor-acceptor monomer pairs have been carried out to provide insight into the mechanism of copolymerizations of donor-acceptor pairs. Tirrell and coworkers[81-83] reported on the reaction of *n*-butyl radicals with mixtures of N-phenylmaleimide and various donor monomers (*e.g.* S, 2-chloroethyl vinyl ether). Jenkins and coworkers[84] have examined the reaction of *t*-butoxy radicals with mixtures of AN and VAc. Both groups have examined the S-AN system (see also Section 7.3.1.2). In each of these donor-acceptor systems only simple (one monomer) adducts are observed. Incorporation of monomers as pairs is not an important pathway (*i.e.* the complex participation model is not applicable). Furthermore, the product mixtures can be predicted on the basis of what is observed in single monomer experiments. The reactivity of the individual monomers (towards initiating radicals) is unaffected by the presence of the other monomer (*i.e.* the complex dissociation model is not applicable). Unless propagating species are shown to behave differently, these results suggest that neither the complex participation nor complex dissociation models apply in these systems.

7.3.1.4 Copolymerization with depropagation

Propagation reactions in radical polymerization and copolymerization are generally highly exothermic and can be assumed to be irreversible. Exceptions to this general rule are those involving monomers with low ceiling temperatures (Section 4.5.1). The thermodynamics of copolymerization has been reviewed by Sawada.[85]

Some of the most important systems known to involve reversible propagation steps are:

(a) Copolymerizations of AMS. Studies on copolymerizations of AMS with AN,[86,87] BA,[88] MMA[87,89-94] and S[86,95] have been reported.
(b) Copolymerizations with sulfur dioxide and carbon monoxide.[85]

Copolymerizations of other monomers may also be subject to similar effects given sufficiently high reaction temperatures (at or near their ceiling temperatures - Section 4.5.1). The depropagation of methacrylate esters becomes measurable at temperatures >100 °C (Section 4.5.1).[96] O'Driscoll and Gasparro[86] have reported on the copolymerization of MMA with S at 250 °C.

The analysis of these systems requires, in addition to reactivity ratios, equilibrium constants for any reversible propagation steps. The reaction scheme is shown in Scheme 7.8. Penultimate unit effects are not considered. In 1960, Lowry[97] developed theory to cover copolymerization involving depropagation of only one monomer. Howell et al.[98] have carried out a more general treatment, allowing for all propagation steps being reversible, and provided expressions for predicting sequence distribution for these systems. Other treatments of copolymerization with depropagation are those of Wittmer[94] and Kruger et al..[99]

$$P_A\bullet + M_A \rightleftharpoons P_{AA}\bullet$$
$$P_A\bullet + M_B \rightleftharpoons P_{AB}\bullet$$
$$P_B\bullet + M_A \rightleftharpoons P_{BA}\bullet$$
$$P_B\bullet + M_B \rightleftharpoons P_{BB}\bullet$$

Scheme 7.8

7.3.2 Chain Statistics

The arrangement of monomer units in copolymer chains is determined by the monomer reactivity ratios which can be influenced by the reaction medium and various additives. The average sequence distribution to the triad level can often be measured by NMR (Section 7.3.3.2) and in special cases by other techniques.[100,101] Longer sequences are usually difficult to determine experimentally, however, by assuming a model (terminal, penultimate, *etc.*) they can be predicted.[7,102] Where sequence distributions can be accurately determined they provide, in principle, a powerful method for determining monomer reactivity ratios.

7.3.2.1 Binary copolymerization according to the terminal model

If chains are long such that the initiation and termination reactions have a negligible effect on the average sequence distribution, then according to the terminal model, P_{AA}, the probability that a chain ending in monomer unit M_A adds another unit M_A, is given by eq. 22:[8]

$$P_{AA} = \frac{k_{pAA}[P_A\bullet][M_A]}{k_{pAA}[P_A\bullet][M_A] + k_{pAB}[P_A\bullet][M_B]} = \frac{r_{AB}x}{r_{AB}x + 1} \quad (22)$$

Similarly, $P_{AB} = \dfrac{1}{r_{AB}x + 1} = 1 - P_{AA}$, $P_{BB} = \dfrac{r_{BA}/x}{r_{BA}/x + 1}$, $P_{BA} = \dfrac{1}{r_{BA}/x + 1} = 1 - P_{BB}$

Copolymerization

The probability of a given sequence is the product of the probabilities of the individual steps that give rise to that sequence. Thus, the fraction of isolated sequences of monomer M_A which are of length n is:

$$N_n^A = P_{AA}^{n-1} P_{AB} = \frac{(r_{AB}x)^{n-1}}{(r_{AB}x+1)^n} \tag{23}$$

while the number average sequence length for monomer units M_A is:

$$\overline{N}^A = \frac{1}{P_{AB}} = r_{AB}x + 1 \tag{24}$$

Expressions for the dyad, triad and higher order n-ad fractions can also be derived in terms of these probabilities. Thus the dyad fractions are given by eqs. 25-27.

$$F_{AA} = P_A P_{AA} = \frac{P_{BA}(1 - P_{AB})}{P_{AB} + P_{BA}} \tag{25}$$

$$F_{AB} = 2 P_A P_{AB} = 2 P_B P_{BA} = \frac{2 P_{BA} P_{AB}}{P_{AB} + P_{BA}} \tag{26}$$

The mirror image sequences, such as the AB and BA dyads, cannot be distinguished.

$$F_{BB} = P_B P_{BB} = \frac{P_{AB}(1 - P_{BA})}{P_{AB} + P_{BA}} = 1 - F_{AA} - F_{AB} \tag{27}$$

The six triad fractions are:

$$F_{AAA} = \frac{P_{BA}(1-P_{AB})^2}{P_{AB} + P_{BA}} \qquad F_{AAB} = \frac{2 P_{BA} P_{AB}(1-P_{AB})}{P_{AB} + P_{BA}} \qquad F_{BAB} = \frac{P_{BA} P_{AB}^2}{P_{AB} + P_{BA}}$$

$$F_{BBB} = \frac{P_{AB}(1-P_{BA})^2}{P_{AB} + P_{BA}} \qquad F_{BBA} = \frac{2 P_{BA} P_{AB}(1-P_{BA})}{P_{AB} + P_{BA}} \qquad F_{ABA} = \frac{P_{BA}^2 P_{AB}}{P_{AB} + P_{BA}}$$

Because $F_{AAA}+F_{AAB}+F_{BAB}+F_{BBB}+F_{BBA}+F_{BBB}=1$ and $2F_{ABA}+F_{BBA}=2F_{BAB}+F_{AAB}$, there are only four independent triad fractions.

7.3.2.2 Binary copolymerization according to the penultimate model

With the penultimate model, the probability that a chain with a terminal M_{BA} dyad will add a M_A unit is given by eq. 28:

$$P_{BAA} = \frac{k_{pBAA}[P_{BA}][M_A]}{k_{pBAA}[P_{BA}][M_A] + k_{pBAB}[P_{BA}][M_B]} = \frac{[M_A]}{[M_A] + [M_B]/r_{BAB}} \quad (28)$$

$$= \frac{r_{BAB}x}{r_{BAB}x + 1} = 1 - P_{BAB}$$

The probability that a chain with a terminal M_{AA} dyad will add a M_A unit is eq. 29:

$$P_{AAA} = \frac{k_{pAAA}[P_{AA}][M_A]}{k_{pAAA}[P_{AA}][M_A] + k_{pAAB}[P_{AA}][M_B]} = \frac{[M_A]}{[M_A] + [M_B]/r_{AAB}} \quad (29)$$

$$= \frac{r_{AAB}x}{r_{AAB}x + 1} = 1 - P_{AAB}$$

Eqs. 30 and 31 are derived similarly:

$$P_{ABB} = \frac{r_{ABA}}{r_{ABA}x + 1} = 1 - P_{ABA} \quad (30)$$

$$P_{BBA} = \frac{x}{r_{BBA}x + x} = 1 - P_{BBB} \quad (31)$$

The probability that a chain with a terminal M_A will add a M_B can be expressed in terms of these probabilities as shown in eq. 32:

$$P_{AB} = \frac{P_{AAB}}{P_{AAB} + P_{BAA}} \quad (32)$$

7.3.2.3 Binary copolymerization according to other models

Expressions for predicting monomer sequence distribution with higher order models[8] and for monomer complex and other models have also been proposed.

There are at least two additional complications that need to be considered when attempting to predict sequence distribution or measure reactivity on the basis of sequence data:

(a) The effects of chain tacticity. Chain ends of differing tacticity may have different reactivity towards monomers.[101] When tacticity is imposed on top of monomer sequence distribution there are then six different dyads and twenty different triads to consider; analytical problems are thus severe. The tacticity of copolymers is usually described in terms of the coisotacticity parameters σ_{AB} and σ_{BA};[103] σ_{AB} is the probability of generating a meso dyad when a chain ending in A adds monomer B. Coisotacticity parameters have to date been reported for only a few copolymers including MMA-S,[104] MMA-MA,[105] and MMA-MAA.[106,107] These data are likely to change due to the complexities associated with data analysis and NMR signal assignment (see also 7.3.3.2).

Copolymers involving only monosubstituted monomers are usually assumed to have random tacticity (*i.e.* $\sigma_{AB} = \sigma_{BA} = 0.5$).

(b) The effects of the reaction medium. Harwood[108,109] observed that copolymers of the same composition have the same monomer sequence distribution irrespective of the solvent used for the copolymerization. He termed this the 'bootstrap effect'. This applies even though estimates of monomer reactivity ratios made on the basis of composition data may be significantly different. Much argument for and against the 'bootstrap effect' has appeared.[1,3,110] Solvent effects on copolymerization and the 'bootstrap effect' are considered in more detail in Section 8.3.1.2.

The full picture of the factors affecting copolymer sequence distribution and their relative importance still needs to be filled in.

7.3.2.4 Terpolymerization

Terpolymerizations or ternary copolymerizations, as the names suggest, are polymerizations involving three monomers. Most industrial copolymerizations involve three or more monomers. The statistics of terpolymerization were worked out by Alfrey and Goldfinger in 1944.[111] If we assume terminal model kinetics, ternary copolymerization involves nine distinct propagation reactions (Scheme 7.9).

$$P_A\bullet + M_A \rightarrow P_A\bullet \qquad k_{pAA}[P_A\bullet][M_A]$$
$$P_A\bullet + M_B \rightarrow P_B\bullet \qquad k_{pAB}[P_A\bullet][M_B]$$
$$P_A\bullet + M_C \rightarrow P_C\bullet \qquad k_{pAC}[P_A\bullet][M_C]$$
$$P_B\bullet + M_A \rightarrow P_A\bullet \qquad k_{pBA}[P_B\bullet][M_A]$$
$$P_B\bullet + M_B \rightarrow P_B\bullet \qquad k_{pBB}[P_B\bullet][M_B]$$
$$P_B\bullet + M_C \rightarrow P_C\bullet \qquad k_{pBC}[P_B\bullet][M_C]$$
$$P_C\bullet + M_A \rightarrow P_A\bullet \qquad k_{pCA}[P_C\bullet][M_A]$$
$$P_C\bullet + M_B \rightarrow P_B\bullet \qquad k_{pCB}[P_C\bullet][M_B]$$
$$P_C\bullet + M_C \rightarrow P_C\bullet \qquad k_{pCC}[P_C\bullet][M_C]$$

Scheme 7.9

Six reactivity ratios are then required to describe the system.

$$r_{AB} = \frac{k_{pAA}}{k_{pAB}} \quad r_{BA} = \frac{k_{pBB}}{k_{pBA}} \quad r_{AC} = \frac{k_{pAA}}{k_{pAC}} \quad r_{BC} = \frac{k_{pBB}}{k_{pBC}} \quad r_{CA} = \frac{k_{pCC}}{k_{pCA}} \quad r_{CB} = \frac{k_{pCC}}{k_{pCB}}$$

Application of a steady state assumption (eqs. 33-35) enables derivation of the composition relationship (eq. 36).

$$k_{pAB}[P_A\bullet][M_B] + k_{pAC}[P_A\bullet][M_C] = k_{pBA}[P_B\bullet][M_A] + k_{pCA}[P_C\bullet][M_A] \quad (33)$$

$$k_{pAB}[P_B\bullet][M_A] + k_{pBC}[P_B\bullet][M_C] = k_{pAB}[P_A\bullet][M_B] + k_{pCB}[P_C\bullet][M_B] \quad (34)$$

$$k_{pCA}[P_C\bullet][M_A] + k_{pCB}[P_C\bullet][M_B] = k_{pAC}[P_A\bullet][M_C] + k_{pBC}[P_B\bullet][M_C] \quad (35$$

$$dM_A : dM_B : dM_C = P_A : P_B : P_C$$

$$= M_A\left[\frac{M_A}{r_{CA}r_{BA}} + \frac{M_B}{r_{BA}r_{CB}} + \frac{M_C}{r_{CA}r_{BC}}\right]\left[M_A + \frac{M_B}{r_{AB}} + \frac{M_C}{r_{AC}}\right]$$

$$: M_B\left[\frac{M_A}{r_{AB}r_{CA}} + \frac{M_B}{r_{AB}r_{CB}} + \frac{M_C}{r_{CB}r_{AC}}\right]\left[M_B + \frac{M_A}{r_{BA}} + \frac{M_C}{r_{BC}}\right]$$

$$: M_C\left[\frac{M_A}{r_{AC}r_{BA}} + \frac{M_B}{r_{BC}r_{AB}} + \frac{M_C}{r_{AC}r_{BC}}\right]\left[M_C + \frac{M_A}{r_{CA}} + \frac{M_B}{r_{CB}}\right] \quad (36)$$

The terpolymer composition can be predicted on the basis of binary copolymerization experiments. If, however, one (or more) monomer is slow to propagate one of the reactivity ratios will approach zero and eq. 36 will become indeterminate. This situation arises in terpolymerizations involving, for example, MAH or AMS. Alfrey and Goldfinger[112] derived eq. 37 for the case where one monomer (C) is slow to propagate (i.e. $k_{pCC} \to 0$ and hence r_{CA} and $r_{CB} \to 0$). Expressions for other cases, for example, where two monomers (B and C) are slow to propagate, were also derived.[112] An equation related to eq. 37 has application in the analysis of binary copolymerizations in the presence of a transfer agent (Section 7.5.6).[113]

$$dM_A : dM_B : dM_C = P_A : P_B : P_C$$

$$= M_A\left[\frac{RM_A}{r_{BA}} + \frac{M_B}{r_{BA}} + \frac{RM_C}{r_{BC}}\right]\left[M_A + \frac{M_B}{r_{AB}} + \frac{M_C}{r_{AC}}\right]$$

$$: M_B\left[\frac{RM_A}{r_{AB}} + \frac{M_B}{r_{AB}} + \frac{M_C}{r_{AC}}\right]\left[M_B + \frac{M_A}{r_{BA}} + \frac{M_C}{r_{BC}}\right]$$

$$: M_C\left[\frac{M_A}{r_{AC}r_{BA}} + \frac{M_B}{r_{BC}r_{AB}} + \frac{M_C}{r_{AC}r_{BC}}\right]\left[RM_A + M_B\right] \quad (37)$$

where $R = \dfrac{k_{pCA}}{k_{pCB}}$.

The value of R can only be evaluated by conducting a terpolymerization.

Copolymerization

The complexity of the terpolymer composition equation (eq. 36) can be reduced to eq. 41 through the use of a modified steady state assumption (eqs. 38-40). However, while these equations apply to component binary copolymerizations it is not clear that they should apply to terpolymerization even though they appear to work well. It can be noted that when applying the Q-e scheme a terpolymer equation of this form is implied.

$$k_{pAB}[P_A\bullet][M_B] = k_{pBA}[P_B\bullet][M_A] \qquad (38)$$

$$k_{pAC}[P_A\bullet][M_C] = k_{pCA}[P_C\bullet][M_A] \qquad (39)$$

$$k_{pBC}[P_B\bullet][M_C] = k_{pCB}[P_C\bullet][M_B] \qquad (40)$$

$$dM_A : dM_B : dM_C = P_A : P_B : P_C$$

$$= M_A \left[M_A + \frac{M_B}{r_{AB}} + \frac{M_C}{r_{AC}} \right]$$

$$: M_B \frac{r_{BA}}{r_{AB}} \left[M_B + \frac{M_A}{r_{BA}} + \frac{M_C}{r_{BC}} \right]$$

$$: M_C \frac{r_{CA}}{r_{AC}} \left[M_C + \frac{M_A}{r_{CA}} + \frac{M_B}{r_{CB}} \right] \qquad (41)$$

Azeotropic compositions are rare for terpolymerization and Ham[114] has shown that it follows from the simplified eqs. 38-40 that ternary azeotropes should not exist. Nonetheless, a few systems for which a ternary azeotrope exists have now been described (this is perhaps a proof of the limitations of the simplified equations) and equations for predicting whether an azeotropic composition will exist for copolymerizations of three or more monomers have been formulated.[20,115] This work also shows that a ternary azeotrope can, in principle, exist even in circumstances where there is no azeotropic composition for any of the three possible binary copolymerizations of the monomers involved.

7.3.3 Estimation of Reactivity Ratios

Methods for evaluation of reactivity ratios comprise a significant proportion of the literature on copolymerization. There are two basic types of information that can be analyzed to yield reactivity ratios. These are (a) copolymer composition/conversion data (Section 7.3.3.1) and (b) the monomer sequence distribution (Section 7.3.3.2). The methods used to analyze these data are summarized in the following sections.

7.3.3.1 Composition data

The traditional method for determining reactivity ratios involves determinations of the overall copolymer composition for a range of monomer feeds at 'zero conversion'. Various methods have been applied to analyze this data. The Fineman-Ross equation (eq. 42) is based on a rearrangement of the copolymer composition equation (eq. 9). A plot of the quantity on the left hand side of eq. 9 vs the coefficient of r_{AB} will yield r_{AB} as the slope and r_{BA} as the intercept.

$$\frac{f_A(1-2F_A)}{F_A(1-f_A)} = r_{AB}\frac{f_A^2(F_A-1)}{F_A(1-f_A)^2} + r_{BA} \qquad (42)$$

Early methods such as the Intersection,[14] and Fineman-Ross[116] methods do not give equal weighting to the experimental points such that there is a non-linear dependence of the error on the composition. Consequently, these methods can give erroneous results.

These problems were addressed by Tidwell and Mortimer[117,118] who advocated numerical analysis by non-linear least squares and Kelen and Tüdos[119,120] who proposed an improved graphical method for data analysis. The Kelen-Tüdos equation is as follows (eq. 43):

$$\eta = \xi\left(r_A + \frac{r_{BA}}{\alpha}\right) - \frac{r_{BA}}{\alpha} \qquad (43)$$

where $\eta = \dfrac{x(y-1)}{y(\alpha + x^2/y)}$, $\xi = \dfrac{x^2/y}{\alpha + x^2/y}$ and α is a constant.

A plot of η vs ξ should yield a straight line with intercepts of $-r_{BA}/\alpha$ and r_{AB} at $\xi=0$ and $\xi=1$ respectively. A value of α corresponding to the highest and lowest values of $(x^2/y)^{0.5}$ used in the experiments results in a symmetrical distribution of experimental data on the plot. Greenley[18,121,122] has re-evaluated much data using the Kelen-Tüdos method and has provided a compilation of these and other results in the Polymer Handbook.[18]

It is also possible to derive reactivity ratios by analyzing the monomer (or polymer) feed composition vs conversion and solving the integrated form of the Mayo Lewis equation.[10,123] The following expression (eq. 44) was derived by Meyer and Lowry:[123]

$$conversion = 1 - \left(\frac{f_A}{f_{Ao}}\right)^\alpha\left(\frac{f_B}{f_{Bo}}\right)^\beta\left(\frac{f_{Ao}-\delta}{f_A-\delta}\right)^\gamma \qquad (44)$$

where $\alpha = \dfrac{r_{BA}}{1-r_{BA}}$ $\quad \beta = \dfrac{r_{AB}}{1-r_{AB}}$ $\quad \delta = \dfrac{1-r_{AB}r_{BA}}{(1-r_{AB})(1-r_{BA})}$ $\quad \gamma = \dfrac{1-r_{BA}}{2-r_{AB}-r_{BA}}$

Numerical approaches for estimating reactivity ratios by solution of the integrated rate equation have been described.[124-126] Potential difficulties associated with the application of these methods based on the integrated form of the Mayo-Lewis equation have been discussed.[124-127] One is that the expressions become undefined under certain conditions, for example, when r_{AB} or r_{BA} is close to unity or when the composition is close to the azeotropic composition. A further complication is that reactivity ratios may vary with conversion due to changes in the reaction medium.

Clearly, great care must be taken in the estimation of reactivity ratios from composition/conversion data. Many papers have been written on the merits of various schemes and comparisons of the various methods for reactivity ratio calculation have appeared.[128-132] Given appropriate design of the experiment, graphical methods for the estimation of reactivity ratios can give reasonable values. They also have the virtue of simplicity and do not require the aid of a computer. However, as a general rule, the use of such methods is not recommended except as an initial guide. It is more appropriate to use some form of non-linear least squares regression analysis to derive the reactivity ratios. The use of "error in variable" methods[6,133-135] which take into account the error structure of the experimental data is highly recommended.

It is also possible to process copolymer composition data to obtain reactivity ratios for higher order models (*e.g.* penultimate model or complex participation, *etc.*). However, composition data have low power in model discrimination (Sections 7.3.1.2 and 7.3.1.3). There has been much published on the subject of the design of experiments for reactivity ratio determination and model discrimination.[49,118,136,137] Attention must be paid to the information that is required; the optimal design for obtaining terminal model reactivity ratios may not be ideal for model discrimination.[49]

One final point should be made. The observation of significant solvent effects on k_p in homopolymerization and on reactivity ratios in copolymerization (Section 8.3.1) calls into question the methods for reactivity ratio measurement which rely on evaluation of the polymer composition for various monomer feed ratios (Section 7.3.2). If solvent effects are significant, it would seem to follow that reactivity ratios in bulk copolymerization should be a function of the feed composition.[138] Moreover, since the reaction medium alters with conversion, the reactivity ratios may also vary with conversion. Thus the two most common sources of data used in reactivity ratio determination (i.e. low conversion composition measurements and composition conversion measurements) are potentially flawed. A corollary of this statement also provides one explanation for any failure of reactivity ratios to predict copolymer composition at high conversion. The effect of solvents on radical copolymerization remains an area in need of further research.

7.3.3.2 Monomer sequence distribution

NMR spectroscopy has made possible the characterization of copolymers in terms of their monomer sequence distribution. The area has been reviewed by Randall,[100] Bovey,[139] Tonelli,[101] Hatada[140] and others. Information on monomer sequence distribution is substantially more powerful than simple composition data with respect to model discrimination.[25,49] Although many authors have used the distribution of triad fractions to confirm the adequacy or otherwise of various models, only a few[25,58,141] have used dyad or triad fractions to calculate reactivity ratios directly.

Terminal model reactivity ratios may be estimated from the initial monomer feed composition and the dyad concentrations in low conversion polymers using the following relationships (eqs. 45, 46).

$$r_{AB} = \frac{f_B}{f_A} \frac{2F_{AA}}{F_{AB}} \tag{45}$$

$$r_{BA} = \frac{f_A}{f_B} \frac{2F_{BB}}{F_{AB}} \tag{46}$$

Note that the dyad concentrations can be easily calculated from the triad concentrations (eqs. 47-49).

$$F_{AA} = F_{AAA} + \frac{F_{AAB}}{2} \tag{47}$$

$$F_{AB} = F_{ABA} + F_{BAB} + \frac{F_{AAB}}{2} + \frac{F_{ABB}}{2} \tag{48}$$

$$F_{BB} = F_{BBB} + \frac{F_{ABB}}{2} \tag{49}$$

Similarly, penultimate model reactivity ratios can be estimated from initial monomer feed composition and triad concentrations using eqs. 50-53.

$$r_{AAB} = \frac{f_B}{f_A} \frac{2F_{AAA}}{F_{AAB}} \tag{50}$$

$$r_{BAB} = \frac{f_B}{f_A} \frac{F_{AAB}}{2F_{BAB}} \tag{51}$$

$$r_{ABA} = \frac{f_A}{f_B} \frac{F_{ABB}}{2F_{ABA}} \tag{52}$$

$$r_{BBA} = \frac{f_B}{f_A} \frac{2F_{BBB}}{F_{ABB}} \tag{53}$$

While sequence distributions are usually subject to more experimental noise than composition data, this is often outweighed by the greater information content. In principle, reactivity ratios can be estimated from a single copolymer sample. The consistency in reactivity ratios estimated with eqs. 45 and 46 for copolymers prepared with different monomer feed compositions and/or obtaining the same result from eqs. 50 and 51 ($r_{AAB}=r_{BAB}$) and eqs. 52 and 53 ($r_{ABA}=r_{BBA}$) are evidence for the applicability of the terminal model.[28,142] Consistent reactivity ratios from application of eqs. 50-53 to copolymers prepared using a range of monomer feed compositions is evidence for the penultimate unit model. A limitation in the use of these equations is the precision of triad distribution data.

Another serious problem in applying these methods is that unambiguous assignments of NMR signals to monomer sequences are, as yet, only available for a few systems. Moreover, assignments are complicated by the fact that the sensitivity of chemical shifts to tacticity may be equal or greater than their sensitivity to monomer sequence.[140,143]

The usual experiment is to prepare a series of copolymers each containing a different ratio of the monomers. A correlation of expected and measured peak intensities may then enable peak assignment.[24,25] However, this method is not foolproof and papers on signal reassignment are not uncommon.[56,104,143] 2D NMR methods,[143] decoupling experiments,[56] special pulse sequences[28] and analyses of isotopically labeled[144,145] or regioregular[56] polymers have greatly facilitated analysis of complex systems. In principle, these methods allow a "mechanism-free" signal assignment.

7.3.4 Prediction of Reactivity Ratios

Various methods for predicting reactivity ratios have been proposed.[146] These schemes are largely empirical although some have offered a theoretical basis for their function. They typically do not allow for the possibility of variation in reactivity ratios with solvent and reaction conditions. They also presuppose a terminal model. Despite their limitations they are extremely useful for providing an initial guess in circumstances where other data is unavailable.

The most popular methods are the *Q-e* (Section 7.3.4.1) and 'Patterns of Reactivity' schemes (Section 7.3.4.2). Both methods may also be used to predict transfer constants (Section 6.2.1). For further discussion on the application of these and other methods to predict rate constants in radical reactions, see Section 2.3.7.

7.3.4.1 *Q-e scheme*

The method for the prediction of reactivity ratios in most widespread usage is the *Q-e* scheme.[17,147] This scheme was devised in 1947 by Alfrey and Price[148] who proposed that the rate constant for reaction of radical (R•) with monomer (M) should be dependent on polarity and resonance terms according to the following expression (eq. 54):

$$k_{RM} = P_R Q_M e^{-e_R e_M} \tag{54}$$

where P_R and Q_M are the 'general reactivity' of the radical and monomer respectively. It has been proposed that these take into account resonance factors. The e values are related to the polarity of the radical or monomer (e_R and e_M are assumed to be the same). The parameters P_R are eliminated in the expressions for the reactivity ratios. The reactivity ratios r_{AB} and r_{BA} depend on Q and e as shown in eqs. 55, 56,

$$r_{AB} = \frac{k_{AA}}{k_{AB}} = \frac{Q_A}{Q_B} e^{-e_A(e_A - e_B)} \tag{55}$$

$$r_{BA} = \frac{k_{BB}}{k_{BA}} = \frac{Q_B}{Q_A} e^{-e_B(e_B - e_A)} \tag{56}$$

S is taken as the reference monomer with $Q=1.0$ and $e = -0.8$. Values for other monomers are derived by regression analysis based on literature or measured reactivity ratios. The Q-e values for some common monomers as presented in the *Polymer Handbook*[149] are given in Table 7.7. The accuracy of Q-e parameters is limited by the quality of the reactivity ratio data and can also suffer from inappropriate statistical treatment employed in their derivation.[17,18] A further problem is that the data analysis makes no allowance for the dependence of reactivity ratios on reaction conditions. Reactivity ratios can be dependent on solvent (Section 7.3.1.2), reaction temperature, pH, *etc*. It follows that values of e and perhaps Q for a given monomer should depend on the medium, the monomer ratio and the particular comonomer. This is especially true for monomers which contain ionizable groups (*e.g.* MAA, AA, vinyl pyridine) or are capable of forming hydrogen bonds (*e.g.* HEMA, HEA).

There have, however, been attempts to correlate Q-e values and hence reactivity ratios to, for example, ^{13}C NMR chemical shifts[150] or the results of MO calculations[151-153] and to provide a better theoretical basis for the parameters. Most recently, Zhan and Dixon[153] applied density functional theory to demonstrate that Q values could be correlated to calculated values of the relative free energy for the radical monomer reaction ($P_A• + M_B \rightarrow P_A•$). The e values were correlated to values of the electronegativities of monomer and radical.

The NMR method of predicting Q-e values appears attractive since spectra can be measured under the particular reaction conditions (solvent, temperature, pH). Thus, it may be possible to predict the dependence of the Q-e values and reactivity ratios on the reaction medium.[150]

Table 7.7 Q-e^{149} and Patterns^{154} Parameters for Some Common Monomers

Monomer	Q	e	$\log r_{1S}$	π	u	v
B	1.70	-0.50	0.1461	-0.100	-0.30	0.41
S	1.0	-0.8	0	0	0	0
MAA	0.98	0.62	-0.2807	0.002	-0.95	0.62
AMS	0.97	-0.81	-0.2219	-0.77	-0.04	-0.03
MAN	0.86	0.69	-0.4815	0.432	-2.08	0.44
BMA	0.82	0.28	-0.2757	0.267	-1.49	0.26
AA	0.83	0.88	-	-	-	-
MMA	0.78	0.40	-0.3372	0.339	-1.18	0.23
AN	0.48	1.23	-1.3980	0.701	-2.6	0.42
MA	0.45	0.64	-0.7447	0.421	-2.34	0.16
BA	0.38	0.85	-0.7447	0.443	-2.22	0.12
VC	0.056	-0.16	-1.26	0.128	-0.90	-1.16
VAc	0.026	-0.88	-1.699	0.315	-0.44	-1.56

7.3.4.2 Patterns of reactivity scheme

Bamford, Jenkins and coworkers[155-157] concluded that many of the limitations of the Q-e scheme stemmed from its empirical nature and proposed a new scheme containing a radical reactivity term, based on experimentally measured values of the rate constant for abstraction of benzylic hydrogen from toluene ($k_{3,T}$), a polar term (the Hammett σ value) and two constants α and β which are specific for a given monomer or substrate (eq. 57):[146]

$$\log k = \log k_{3,T} + \alpha\sigma + \beta \qquad (57)$$

and reactivity ratios are then defined by eqs. 58 and 59:

$$\log r_{AB} = \sigma_A(\alpha_A - \alpha_B) + \beta_A - \beta_B \qquad (58)$$

$$\log r_{BA} = \sigma_B(\alpha_B - \alpha_A) + \beta_B - \beta_A \qquad (59)$$

In the revised Patterns scheme reactivity ratios involving S are used as reference reactions.[154,158] Reactivity ratios are then given by eqs. 60 and 61:

$$\log r_{AB} = \log r_{AS} - u_B\pi_A - v_B \qquad (60)$$

$$\log r_{BA} = \log r_{BS} - u_A\pi_B - v_A \qquad (61)$$

where r_{AS} ($=k_{AA}/k_{aS}$) is the reactivity ratio of the monomer (A) with S ($\log r_{AS}$ is the counterpart of Q in the Q-e scheme), π is a polarity term and is strongly correlated with the Hammet σ parameter (it is the counterpart of e) and u and v are constants. Tabulations of the Patterns parameters can be found in the Polymer Handbook[154] and a subset of this data is reproduced in Table 7.7. The scheme can also be used to predict chain transfer constants.

The Patterns scheme has been tested for its capacity to predict ^{13}C NMR chemical shifts of the CH$_2$= carbon of monomers (CH$_2$=CXY)[159] and in evaluating the reactivities of small radicals towards monomers.[160]

7.4 Termination in Statistical Copolymerization

This section begins with a brief discussion of copolymerization kinetics and various models that have been used to describe termination. These models were derived with the presumption that the terminal model describes propagation in copolymerization. The "chemical control model" (Section 7.4.1) and the various diffusion control models (Section 7.4.2) as originally conceived largely fell from use with the advent of methods that allowed absolute values for the overall propagation rate constant in copolymerization to be reliably determined (*e.g.* PLP). Application of these methods pointed to the failure of terminal model kinetics by demonstrating that the overall propagation rate constant was strongly dependent on the monomer feed composition. Thus, 'anomalies' in copolymerization kinetics previously attributed to variation in the termination rate constant monomer feed composition were in large part associated with variation in the propagation rate constant. The so-called implicit and explicit penultimate models described in Section 7.3.1.2.1 were derived.

More recent work has shown that the observed variation in propagation rate constants with composition is not sufficient to define the polymerization rates.[52,161,162] There remains some dependence of the termination rate constant on the composition of the propagating chain. Thus, the "chemical control" (Section 7.4.1) and the various diffusion control models (Section 7.4.2) have seen new life and have been adapted by substituting the terminal model propagation rate constants (k_{pXY}) with implicit penultimate model propagation rate constants ($\bar{k}_{pXY}$ - Section 7.3.1.2.2).

The chain length dependence of termination rate constants (Section 5.2.1.4) should not be ignored when considering copolymerization kinetics. It has been pointed out that average chain lengths in copolymerization will be a function of the monomer feed composition[161] especially in copolymerizations with disparate propagation rate constants. Factors determining the rate of copolymerization are not fully resolved and copolymerization kinetics remains a topic of discussion and an area in need of further study.

7.4.1 Chemical Control Model

The rate of copolymerization often shows a strong dependence on the monomer feed composition. Many theories have been developed to predict the rate of copolymerization based on the terminal model for chain propagation (Section 7.3.1.1). This usually requires an overall rate constant for termination in copolymerization that is substantially different from that observed in homopolymerization of any of the component monomers.

Copolymerization

In early work, it was assumed that the rate constant for termination was determined by the monomer unit at the reacting chain ends. The kinetics of copolymerization were then dictated by the rate of initiation, the rates of the four propagation reactions (Scheme 7.1) and rates of three termination reactions (Scheme 7.10).[163-165]

$P_A\bullet$	+	$P_A\bullet$	→	products	$k_{tAA}[P_A\bullet][P_A\bullet]$
$P_A\bullet$	+	$P_B\bullet$	→	products	$2k_{tAB}[P_A\bullet][P_B\bullet]$
$P_B\bullet$	+	$P_B\bullet$	→	products	$k_{tBB}[P_B\bullet][P_B\bullet]$

Scheme 7.10

The instantaneous rate of monomer consumption in binary copolymerization is then given by eq. 62:

$$R_p = -\frac{d[M_A + M_B]}{dt}$$

$$= k_{pAA}[P_A\bullet][M_A] + k_{pBA}[P_B\bullet][M_A] + k_{pAB}[P_A\bullet][M_B] + k_{pBB}[P_B\bullet][M_B] \quad (62)$$

Use of the steady state approximation

$$R_t = k_{tAA}[P_A\bullet]^2 + 2k_{tAB}[P_A\bullet][P_B\bullet] + k_{tBB}[P_B\bullet]^2 = R_i = 2k_d f[I_2]$$

allows the concentrations of the active species to be eliminated. Thus eq. 63:

$$-\frac{d[M_A + M_B]}{dt} = \frac{\left(k_{pAA}k_{pBA}[M_A]^2 + 2k_{pAB}k_{pBA}[M_A][M_B] + k_{pBB}k_{pAB}[M_B]^2\right)R_i^{0.5}}{k_{tAA}k_{pBA}^2[M_A]^2 + 2k_{tAB}k_{pAB}k_{pBA}[M_A][M_B] + k_{tBB}k_{pAB}^2[M_B]^2} \quad (63)$$

which can be rewritten as eq. 64:[165]

$$\frac{-d[M_A + M_B]}{dt} = \frac{(r_{AB}[M_A]^2 + 2[M_A][M_B] + r_{BA}[M_B]^2)R_i^{0.5}}{\delta_A^2 r_{AB}^2[M_A]^2 + 2\phi\delta_A\delta_B r_{AB}r_{BA}[M_A][M_B] + \delta_B^2 r_{BA}^2[M_B]^2} \quad (64)$$

where:

$$\phi = \frac{k_{tAB}}{2(k_{tAA}.k_{tBB})^{0.5}} \quad \delta_A = \frac{2k_{tAA}^{0.5}}{k_{pAA}} \quad \delta_B = \frac{2k_{tBB}^{0.5}}{k_{pBB}} \quad r_{AB} = \frac{k_{pAA}}{k_{pAB}} \quad r_{BA} = \frac{k_{pBB}}{k_{pBA}}$$

In evaluating the kinetics of copolymerization according to the chemical control model, it is assumed that the termination rate constants k_{tAA} and k_{tBB} are known from studies on homopolymerization. The only unknown in the above expression is the rate constant for cross termination (k_{tAB}). The rate constant for this reaction in relation to k_{tAA} and k_{tBB} is given by the parameter ϕ.

Values of ϕ required to fit the rate of copolymerization by the chemical control model were typically in the range 5-50 though values <1 are also known. In the case of S-MMA copolymerization, the model requires ϕ to be in the range 5-14 depending on the monomer feed ratio. This "chemical control" model generally fell from favor with the recognition that chain diffusion should be the rate determining step in termination.

However, recent work based on the assumption of the implicit penultimate model suggests a value of ϕ for S-MMA copolymerization to be in the range 2-3.[52,161] This value is in remarkably good agreement with that suggested by experiments with simple model radicals. These experiments also indicate that cross termination is 2-3 times faster than either homotermination reaction (Section 7.4.3.1).

7.4.2 Diffusion Control Models

In the classical diffusion control model it is assumed that propagation occurs according to the terminal model (Scheme 7.1). The rate of the termination step is limited only by the rates of diffusion of the polymer chains. This rate may be dependent on the overall polymer chain composition. However, it does not depend solely on the chain end.[166,167]

$$P_{(AB)}\bullet + P_{(AB)}\bullet \rightarrow \text{products} \qquad k_{t(AB)}[P_{(AB)}\bullet][P_{(AB)}\bullet]$$

North and coworkers[166,168] proposed that chains terminate with a rate constant which is determined by the rate of diffusion. Thus

$$-\frac{d[M_A+M_B]}{dt} = \frac{(r_{AB}[M_A]^2 + 2[M_A][M_B] + r_{BA}[M_B]^2) R_i^{0.5}}{\varepsilon_A r_{AB}[M_A] + \varepsilon_B r_{BA}[M_B]} = \qquad (65)$$

where $\varepsilon_A = k_{t(AB)}^{0.5}/k_{pAA}$, $\varepsilon_B = k_{t(AB)}^{0.5}/k_{pBB}$.

and $k_{t(AB)}$ is the copolymer-composition dependent rate constant for termination. It is not a constant. In eq. 65, the value of $k_{t(AB)}$ is obtained by fitting the experimental data. Various methods have then been proposed to estimate a dependence of $k_{t(AB)}$ on the monomer feed composition and the rate constants for homotermination (eqs. 66-68).[166,169]

$$k_{t(AB)} = F_A k_{tAA} + F_B k_{tBB} \qquad (66)$$

$$k_{t(AB)} = F_A^2 k_{tAA} + F_A F_B k_{tAB} + F_B^2 k_{tBB} \qquad (67)$$

$$k_{t(AB)} = F_A^2 k_{tAA} + 2F_A F_B \phi (k_{tAA} k_{tBB})^{0.5} + F_B^2 k_{tBB} \qquad (68)$$

In eq. 68, ϕ is defined as in the chemical control model but this expression is cast in terms of the monomer feed composition rather than the radical chain end population.

More complex models for diffusion-controlled termination in copolymerization have appeared.[170-173] Russo and Munari[171] still assumed a terminal model for propagation but introduced a penultimate model to describe termination. There are ten termination reactions to consider (Scheme 7.11). The model was based on the hypothesis that the type of penultimate unit defined the segmental motion of the chain ends and their rate of diffusion.

$P_{AA}\bullet$ +	$P_{AA}\bullet$	$\rightarrow$	products	$k_{tAAAA}[P_{AA}\bullet][P_{AA}\bullet]$
$P_{AA}\bullet$ +	$P_{AB}\bullet$	$\rightarrow$	products	$2k_{tAAAB}[P_{AA}\bullet][P_{AB}\bullet]$
$P_{AA}\bullet$ +	$P_{BA}\bullet$	$\rightarrow$	products	$2k_{tAABA}[P_{AA}\bullet][P_{BA}\bullet]$
$P_{AA}\bullet$ +	$P_{AB}\bullet$	$\rightarrow$	products	$2k_{tAABB}[P_{AA}\bullet][P_{AB}\bullet]$
$P_{AB}\bullet$ +	$P_{AB}\bullet$	$\rightarrow$	products	$k_{tABAB}[P_{AB}\bullet][P_{AB}\bullet]$
$P_{AB}\bullet$ +	$P_{BA}\bullet$	$\rightarrow$	products	$2k_{tABBA}[P_{AB}\bullet][P_{BA}\bullet]$
$P_{AB}\bullet$ +	$P_{BB}\bullet$	$\rightarrow$	products	$2k_{tABBB}[P_{AB}\bullet][P_{BB}\bullet]$
$P_{BA}\bullet$ +	$P_{BA}\bullet$	$\rightarrow$	products	$k_{tBABA}[P_{BA}\bullet][P_{BA}\bullet]$
$P_{BA}\bullet$ +	$P_{BB}\bullet$	$\rightarrow$	products	$2k_{tBABB}[P_{BA}\bullet][P_{BB}\bullet]$
$P_{BB}\bullet$ +	$P_{BB}\bullet$	$\rightarrow$	products	$k_{tBBBB}[P_{BB}\bullet][P_{BB}\bullet]$

Scheme 7.11

The rate constants for the cross termination terms are approximated as the geometric mean of the corresponding homotermination terms. Thus:

$$k_{tAAAB}=2(k_{tAAAA}\,k_{tABAB})^{0.5} \quad k_{tAABA}=2(k_{tAAAA}\,k_{tBABA})^{0.5} \quad k_{tAABB}=2(k_{tAAAA}\,k_{tBBBB})^{0.5}$$

$$k_{tABBA}=2(k_{tABAB}\,k_{tBABA})^{0.5} \quad k_{tABBB}=2(k_{tABAB}\,k_{tBBBB})^{0.5} \quad k_{tBABB}=2(k_{tBABA}\,k_{tBBBB})^{0.5}$$

which allows the rate of polymerization to be defined in terms of four termination rate constants (eq. 69).

$$-\frac{d[M_A+M_B]}{dt}$$

$$=\frac{(r_{AB}[M_A]^2+2[M_A][M_B]+r_{BA}[M_B]^2)R_i^{0.5}}{\frac{\delta_{AA}r_{AB}^2[M_A]^2+\delta_{BA}r_{AB}[M_A][M_B]}{r_{AB}[M_A]+[M_B]}+\frac{\delta_{BB}r_{BA}^2[M_B]^2+\delta_{AB}r_{BA}[M_A][M_B]}{r_{BA}[M_B]+[M_A]}} \quad (69)$$

where $\delta_{AA} = \dfrac{k_{tAAAA}^{0.5}}{k_{pAA}}$ $\delta_{AB} = \dfrac{k_{tABAB}^{0.5}}{k_{pAB}}$ $\delta_{BA} = \dfrac{k_{tBABA}^{0.5}}{k_{pBA}}$ $\delta_{BB} = \dfrac{k_{tBBBB}^{0.5}}{k_{pBB}}$

This model provides a better description of the rate of copolymerization for some systems but has been criticized as having too many adjustable parameters.[174]

Fukuda and coworkers[162] have recently derived a model equivalent to the Russo-Munari model but where the implicit penultimate model is used to describe the propagation kinetics.

7.4.3 Combination and Disproportionation during Copolymerization

It is important to realize that, even if the rate of termination is determined by the rates of chain diffusion, the chain end composition and the ratio of combination to disproportionation are not. Knowledge or prediction of the overall rate of termination offers little insight into the detailed chemistry of the termination processes not involved in the rate-determining step.

Even when only the terminal monomer unit is considered, radical-radical termination in binary copolymerization involves at least seven separate reactions (Scheme 7.12). There are two homotermination processes and one cross termination process to consider. In the case of cross termination, there are two pathways for disproportionation. There are then at least three pieces of information to be gained:

(a) The value of k_{td}/k_{tc} for cross termination.
(b) The specificity for hydrogen transfer in disproportionation (*i.e.* from monomer A to monomer B or vice versa).
(c) The relative rates of homo- and cross-termination.

Homotermination of A-ended chains

$P_A\bullet + P_A\bullet \rightarrow P_A\text{-}P_A$
$P_A\bullet + P_A\bullet \rightarrow P_A^H + P_A^=$

Cross termination

$P_A\bullet + P_B\bullet \rightarrow P_A\text{-}P_B$
$P_A\bullet + P_B\bullet \rightarrow P_A^H + P_B^=$
$P_A\bullet + P_B\bullet \rightarrow P_A^= + P_B^H$

Homotermination of B-ended chains

$P_B\bullet + P_B\bullet \rightarrow P_B\text{-}P_B$
$P_B\bullet + P_B\bullet \rightarrow P_B^H + P_B^=$

Scheme 7.12

Perhaps because of this complexity, few studies on determining k_{td}/k_{tc} in cross termination in copolymerization have been reported and most of the available data come from model studies. It is also usually assumed, without specific justification, that penultimate unit effects are unimportant in determining which reactions occur and that values of k_{td}/k_{tc} for the homotermination reactions are similar to those in the corresponding homopolymerizations.

Three types of model study have been performed. The first approach has been to decompose a mixture of two initiators (*i.e.* one to generate radical A, the other to generate radical B). With this method experimental difficulties arise because the two types of radical may not be generated at the same rate and because homotermination products from cage recombination complicate analysis.

A second approach has been to use an unsymmetrical initiator which allows the two radicals of interest to be generated simultaneously in equimolar amounts.[175] In this case, analysis of the cage recombination products provides information on cross termination uncomplicated by homotermination. Analysis of products of the encounter reaction can also give information on the relative importance of cross and homotermination. However, copolymerization of unsaturated products can cause severe analytical problems.

A third technique is to examine the products of primary radical termination in polymerizations carried out with high concentrations of initiator.[176,177] Values of k_{td}/k_{tc} ratios in primary radical termination have been reported for a number of polymerizations carried out with AIBN (model for PMAN•) or AIBMe (model for PMMA•) initiation.

7.4.3.1 *Poly(methyl methacrylate-co-styrene)*

In termination, the rate determining step is the rate at which the chain ends are brought together by diffusion. Since propagation is rapid with respect to termination, the relative radical concentrations are more important than the termination rate constants in determining the products of termination.[178] The relative radical concentrations are in turn determined by the values of the reactivity ratios and the propagation rate constants. These considerations ensure that, during MMA-S copolymerization, the instantaneous concentration of chains ending in S is significantly greater than that of those with a terminal MMA unit.[178] Therefore, homotermination of chains ending in S and cross termination are the most important processes. There is comparatively little homotermination between chains ending in MMA (Table 7.8).

The reaction between the PMMA and PS model radicals (**4** and **5**, generated from the unsymmetrical azo-compound **3**) has been studied as a model for cross-termination in MMA-S copolymerization (Scheme 7.13).[178,179] The value for $k_{td}/k_{tc}(90°C)$ for the cross reaction was 0.56. In disproportionation, transfer of hydrogen from the PS• model **5** to the PMMA• radical **4** was *ca* 5.1 times more prevalent than transfer in the reverse direction (from **4** to **5**). The value of $k_{td}/k_{tc}(90°C)$ is between those of $k_{td}/k_{tc}(90°C)$ for the self-reaction of these radicals under similar conditions (0.13 and 0.78 for **5** and **4** respectively). Analysis of the encounter products indicated a small preference for cross termination over either homotermination process.[178]

Table 7.8 Identity of Chain End Units Involved in Radical-Radical Termination in MMA-S Copolymerization[a]

Reaction	'Chemical Control'[b]			'Diffusion Control'[c]
	$\phi=13$	$\phi=3$	$\phi=1$	
-S• + -S•	0.18	0.47	0.72	0.57
-S• + -MMA•	0.81	0.51	0.26	0.37
-MMA• + -MMA•	0.01	0.02	0.02	0.06

a Calculated by kinetic simulation.[178] b Calculated using the classical chemical control model (7.4.1). c Calculated using the diffusion control model of Russo and Munari[171] (7.4.2).

Both S polymerization initiated by AIBMe[176,180] (*i.e.* PS• + **4**) and MMA polymerization initiated by 1,1'-azobis-1-phenylethane[176] (*i.e.* PMMA• + 1-phenylethyl radical) are reported to give predominantly combination. Ito[176] has concluded that cross termination is not particularly favored over homotermination in S-MMA copolymerization.

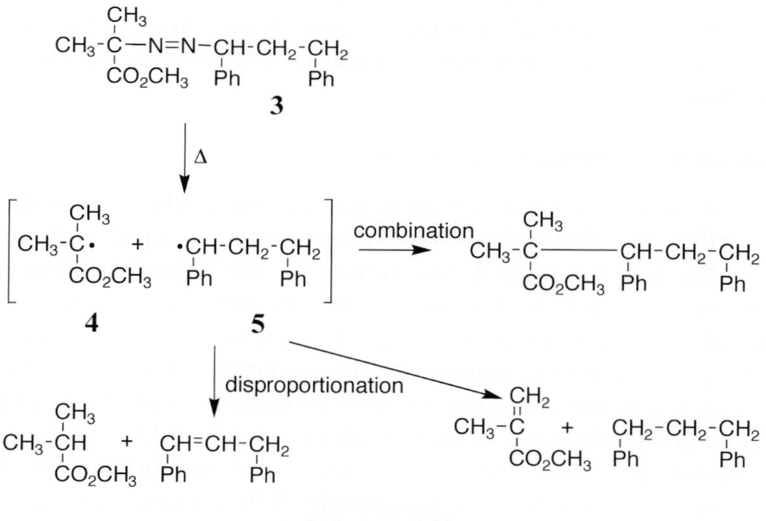

Scheme 7.13

Several experimental studies on S-MMA copolymerization have appeared: all suggest predominant combination.[181-183] Ohtani *et al.*[183] analyzed the end groups of PSMMA (60°C, AIBN, chloroform) by pyrolysis-gas chromatography to find values for the number of end groups per molecule of between 1.56-1.77 (increasing with polymer M_n) which corresponds to an overall k_{td}/k_{tc} of between 0.39 and 0.21. Estimation of k_{td}/k_{tc} for cross termination requires knowledge of the

Copolymerization

extents of homo- and cross termination. Bevington *et al.*[181] examined S-MMA copolymerization (60°C, benzene) using the radiotracer method and found that the cross termination reaction involves predominantly combination (k_{td}/k_{tc} for the homotermination processes were taken to be 0 and 5.67 for chains ending in S and MMA respectively). Chen *et al.*[182] conducted an analysis of polymerization kinetics and came to a similar conclusion. Both groups assumed a "chemical control model" for termination (Section 7.4.1) and the results may need to be reinterpreted.

7.4.3.2 *Poly(methacrylonitrile-co-styrene)*

Analysis of the products from the thermal decomposition of the mixed azo compound **6** showed that in the cross-reaction of radicals **5** and **7** k_{td}/k_{tc}(90°C) is 0.61.[179] This study also found that in disproportionation, hydrogen transfer from **5** to **7** is *ca* 2.2 times more frequent than transfer from **7** to **5**. Both self-reactions involve predominantly combination (Scheme 7.14). The values of k_{td}/k_{tc}(80°C) are 0.16 and 0.05 for radicals **5** (Section 5.2.2.1.1) and **7** (Section 5.2.2.1.3) respectively. It is clear that values of k_{td}/k_{tc} for homotermination cannot be used as a guide to the value for k_{td}/k_{tc} in cross-termination.

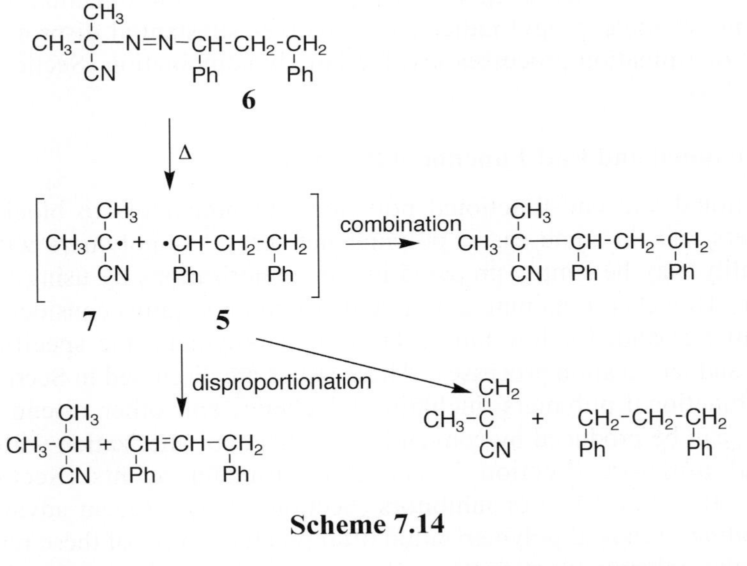

Scheme 7.14

The reaction of oligostyrene radicals with cyanoisopropyl radicals (**7**) has been studied by several groups and reported to give exclusively combination (98°C, toluene),[180,184] or mainly combination (60°C, ethyl acetate;[185] 98°C, toluene[186]). Moad *et al.*[186] examined S oligomerization in toluene at 98°C using high concentrations of AIBN as initiator. While the major products arose from combination, they also isolated and identified small amounts of disproportionation products thus demonstrating that disproportionation does occur.

7.4.3.3 Poly(butyl methacrylate-co-methacrylonitrile)

Barton et al.[187] have reported that primary radical termination between PBMA• and cyanoisopropyl radicals (**7**) involves largely disproportionation.

7.4.3.4 Poly(butyl methacrylate-co-methyl methacrylate)

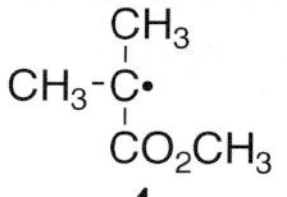

$$\begin{array}{c} \text{CH}_3 \\ | \\ \text{CH}_3\text{-C•} \\ | \\ \text{CO}_2\text{CH}_3 \\ \mathbf{4} \end{array} \qquad \begin{array}{c} \text{CH}_3 \\ | \\ \text{CH}_3\text{-C•} \\ | \\ \text{CO}_2\text{C}_4\text{H}_9 \\ \mathbf{8} \end{array}$$

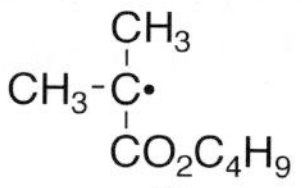

The value of k_{td}/k_{tc}(80°C) in the cross-reaction between radicals **4** and **8** has been examined.[175] This system is a model for cross-termination in MMA-BMA copolymerization. The value of k_{td}/k_{tc} (1.22) is similar to that found for the self-reaction of **8** (1.17) and much larger than that for the self-reaction of **4** (0.78). There is a small preference (*ca* 1.4 fold) for the transfer of hydrogen from the butyl ester (**8**) to the methyl ester (**4**).

7.4.3.5 Poly(ethylene-co-methacrylonitrile)

Guth and Heitz[177] have reported that primary radical termination between PE• radicals and cyanoisopropyl radicals (**7**) involves substantial disproportionation. Both homotermination processes involve largely combination (Sections 5.2.2.1.3 and 5.2.2.1.4).

7.5 Functional and End-Functional Polymers

Functional and end-functional polymers are precursors to block and graft copolymers and, in some cases, polymer networks. Copolymers with in-chain functionality may be simply prepared in copolymerizations by using a functional monomer. However, obtaining a desired distribution requires consideration of the chain statistics and, for low molecular weight polymers, the specificity of the initiation and termination processes. These issues are discussed in Section 7.5.6

End-functional polymers, including telechelic[*] and other di-end functional polymers, can be produced by conventional radical polymerization with the aid of functional initiators (Section 7.5.1), chain transfer agents (Section 7.5.2), monomers (Section 7.5.4) or inhibitors (Section 7.5.5). Recent advances in our understanding of radical polymerization offer greater control of these reactions and hence of the polymer functionality. Reviews on the synthesis of end-functional polymers include those by Colombani,[188] Tezuka,[189] Ebdon,[190] Boutevin,[191] Heitz,[180] Nguyen and Maréchal,[192] Brosse et al.,[193] and French.[194]

---

[*] A telechelic polymer is a di-end-functional polymer where both ends possess the same functionality.

Living polymerization processes lend themselves to the synthesis of end functional polymers; their use in this context is described in Chapter 9. In this section we limit discussion to processes based on conventional radical polymerization.

7.5.1 Functional Initiators

Predominantly di-end-functional polymers may be prepared by conducting polymerizations with high concentrations of a functional initiator. Some of the first commercial products of this class, carboxy and hydroxy-terminated polybutadienes, were produced by this route.[194]

The synthesis of telechelics by what Tobolsky[195] termed dead-end polymerization is described in several reviews.[191,193] In dead-end polymerization very high initiator concentrations and (usually) high reaction temperatures are used. Conversion ceases before complete utilization of the monomer because of depletion of the initiator. Target molecular weights are low (1000-5000) and termination may be mainly by primary radical termination.. The first use of this methodology to prepare telechelic polystyrene was reported by Guth and Heitz.[177]

When a polymer is prepared by radical polymerization, the initiator derived chain-end functionality will depend on the relative significance and specificity of the various chain end forming reactions. Thus, for the formation of telechelic polymers:

(a) The reaction of the initiator-derived radicals with monomer must involve double bond addition (*i.e.* no primary radical transfer).

(b) Secondary radical formation (*e.g.* by β-scission in the case of acyloxy or alkoxy radicals) should either be negligible or not involve loss of the desired functionality.

(c) Chain end formation by chain transfer to monomer, polymer, solvent, *etc.* must be minimal. Chain transfer to initiator may be tolerated if the initiator functionality is transferred.

(d) All radical-radical termination (reaction with primary or propagating radicals) should involve combination.

These conditions severely limit the range of initiators and monomers that can be used and require that attention to reaction conditions is of paramount importance. The relatively low incidence of side reactions associated with the use of azo-compounds (Section 3.3.1) has led to these initiators being favored for this application. Functional azo compounds used in telechelic syntheses include **9**,[196-198] **10**[199,200] and **11**[201,202]. The acylazide end groups formed with initiator **11** may be thermally transformed to isocyanate ends.[201,202]

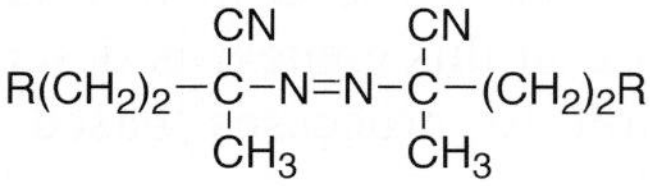

$$R(CH_2)_2-\underset{\underset{CH_3}{|}}{\overset{\overset{CN}{|}}{C}}-N=N-\underset{\underset{CH_3}{|}}{\overset{\overset{CN}{|}}{C}}-(CH_2)_2R$$

9 R=CO$_2$H **10** R=CH$_2$OH **11** R=CON$_3$

Simple azo-compounds (AIBN or AIBMe) have also been used to produce telechelic polymers.[177,184,194] The nitrile and ester functions can be elaborated to reactive carboxy, hydroxy or amino groups and used in polyester or polyurethane formation (*e.g.* Scheme 7.15). Functionalities (number of end groups/molecule) of 1.7 for PE and 2.0 for PS were reported. The latter number seems high given that PS• is known to give some disproportionation both in reaction with cyanoisopropyl radicals ($k_{td}/k_{tc}(90°C)$ = 0.61, Section 7.4.3.2) and in self reaction (Section 5.2.3.1.3). A possible explanation is that the unsaturated by-product from cage-disproportionation (*e.g.* MAN from AIBN, Section 3.3.1.1.3)[186] may copolymerize. This may result in an apparent functionality of ≥2.

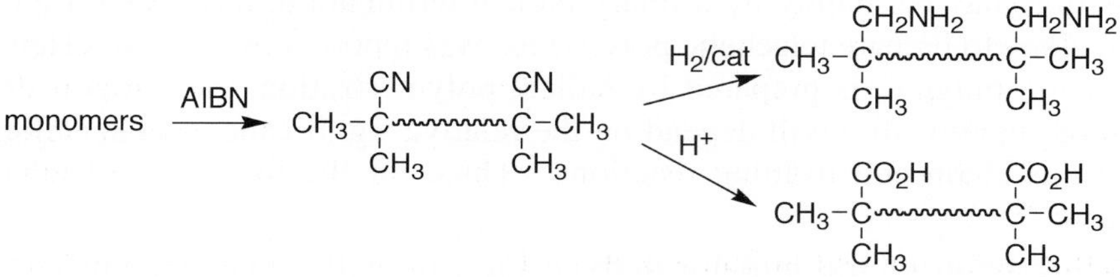

Scheme 7.15

There have been many studies on the applications of peroxide initiators to the synthesis of α,ω-dihydroxy and α,ω-dicarboxy oligomers. Succinic (**12**, n=2) and glutaric acid peroxides (**12**, n=3) have been used to synthesize carboxy end-functional polybutadiene.[194] This use of peroxides is complicated by the tendency of acyloxy radicals and alkoxy radicals to undergo β-scission and by the various pathways that may compete with double bond addition (Section 3.4.2). However, alkoxycarbonyloxy radicals undergo β-scission only slowly (Section 3.4.2.2.2) and peroxydicarbonates have been used to form polymers with carbonate end groups[203] Guth and Heitz[177] reported that ethylene polymerized with peroxydicarbonate initiator has a functionality of only *ca* 1.1. As explanation, they proposed that primary radical termination involving the alkoxycarbonyloxy radical involves disproportionation rather than coupling. The carbonate ends were hydrolyzed to hydroxy ends.[203]

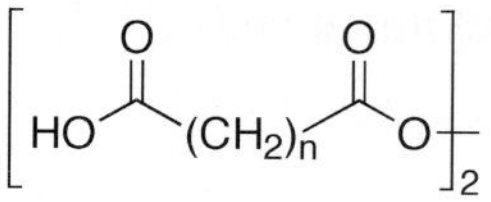

12

The use of ring substituted diacyl peroxides has also been reported.[204] Both the aryl and aroyloxy ends possess the desired functionality. Other initiators used in this context include peroxides (*e.g.* hydrogen peroxide),

Disulfide derivatives and hexasubstituted ethanes[205] may also be used in this context to make end-functional polymers and block copolymers. The use of dithiuram disulfides as thermal initiators was explored by Clouet, Nair and coworkers.[206] Chain ends are formed by primary radical termination and by transfer to the dithiuram disulfide. The chain ends formed are thermally stable under normal polymerization conditions. The use of similar compounds as photo-iniferters, when some living characteristics may be achieved, is described in Section 9.3.2.1.1.

7.5.2 Functional Transfer Agents

Suitably functionalized transfer agents offer a route to end-functional and block and graft polymers.[180,191,207-209] Living polymerization processes involving degenerate or reversible chain transfer (*e.g.* RAFT) are discussed in Section 9.5. For radical polymerization in the presence of a transfer agent, it must be remembered that the initiation and termination steps will always be responsible for a fraction of the chain ends. Therefore, to achieve the highest degree of functionality, an initiator should be chosen which gives the same type of end group as the transfer agent.

Chains with undesired functionality from termination by combination or disproportionation cannot be totally avoided. In attempts to prepare a monofunctional polymer, any termination by combination will give rise to a difunctional impurity. Similarly, when a difunctional polymer is required, termination by disproportionation will yield a monofunctional impurity. The amount of termination by radical-radical reactions can be minimized by using the lowest practical rate of initiation (and of polymerization). Computer modeling has been used as a means of predicting the sources of chain ends during polymerization and examining their dependence on reaction conditions (Section 7.5.6).[210,211] The main limitations on accuracy are the precision of rate constants which characterize the polymerization.

Depending on the choice of transfer agent, mono- or di-end-functional polymers may be produced. Addition-fragmentation transfer agents such as functional allyl sulfides (Scheme 7.16), benzyl ethers and macromonomers have application in this context (Section 6.2.3).[212-216] The synthesis of PEO-block copolymers by making use of PEO functional allyl peroxides (and other transfer agents) has been described by Businelli *et al.*[217] Boutevin *et al.*[218,219] have described the telomerization of unsaturated alcohols with mercaptoethanol or dithiols to produce telechelic diols in high yield.

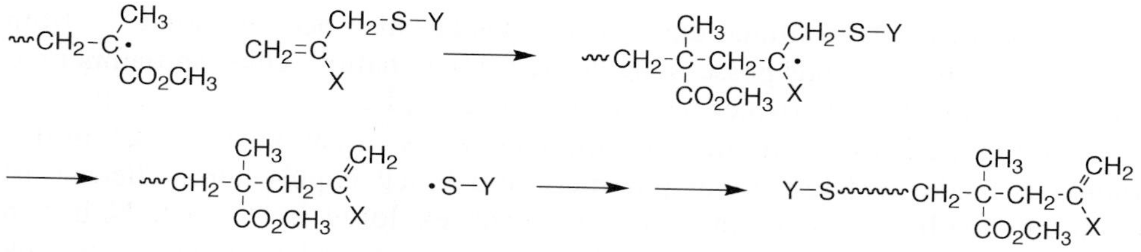

Scheme 7.16

7.5.3 Thiol-ene Polymerization

Thiol-ene polymerization was first reported in 1938.[220] In this process, a polymer chain is built up by a sequence of thiyl radical addition and chain transfer steps (Scheme 7.17). The thiol-ene process is unique amongst radical polymerizations in that, while it is a radical chain process, the rate of molecular weight increase is more typical of a step-growth polymerization. Polymers ideally consist of alternating residues derived from the diene and the dithiol. However, when dienes with high k_p and relatively low k_{tr} monomers (e.g. acrylates) are used, short sequences of units derived from the diene are sometimes formed.

Addition

Chain Transfer

Scheme 7.17

Dithiols and dienes may react spontaneously to afford dithiols or dienes depending on the monomer dithiol ratio.[221] However, the precise mechanism of radical formation is not known. More commonly, photoinitiation or conventional radical initiators are employed. The initiation process requires formation of a radical to abstract from thiol or add to the diene then propagation can occur according to the steps shown in Scheme 7.17 until termination occurs by radical-radical reaction. Termination is usually written as involving the monomer-derived radicals. The process is remarkably tolerant of oxygen and impurities. The kinetics of the thiol-ene photopolymerization have been studied by Bowman and coworkers.[222,223]

The process may be used to form linear polymers. Nuyken and Völkel[224,225] described a method for telechelic production, based on the radical initiated reaction of difunctional transfer agents with dienes (e.g. divinyl benzene (**13**), dimethacrylate esters). However, currently the most common use of thiol-ene polymerization is to form network polymers in a photoinitiated process.[226] Dienes employed include divinyl benzene (**13**), diethylene glycol diacrylate (**15**) and a variety of nonconjugated dienes. The latter include many monomers not commonly used in conventional radical polymerization such as diallyl trimethylolmethane (**16**) and the bis-norbornene derivative (**17**). Diacetylenes (*e.g.* **14**) have also been used. The thiols used include simple aliphatic and aromatic dithiols (*e.g.* octanedithiol). Network polymers typically incorporate a compound with multiple thiol groups, for example, the tetrathiol **18**.

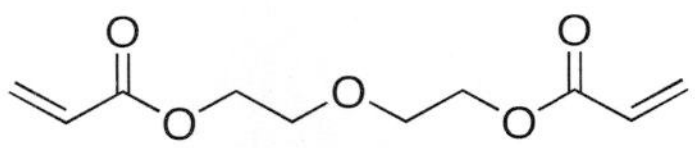

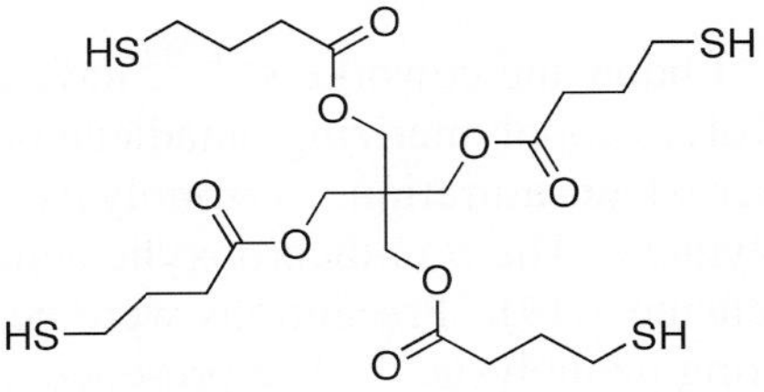

One may envisage polymerizations analogous to the thiol-ene process using other bis- or multi transfer agents (*e.g.* radical-induced hydrosilylation between bis-silanes and dienes). However, none has been described or achieved significance.

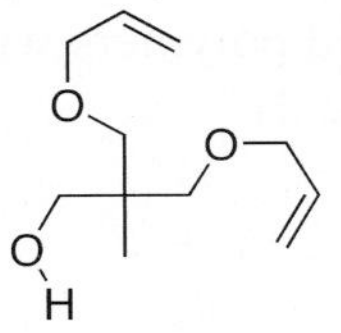

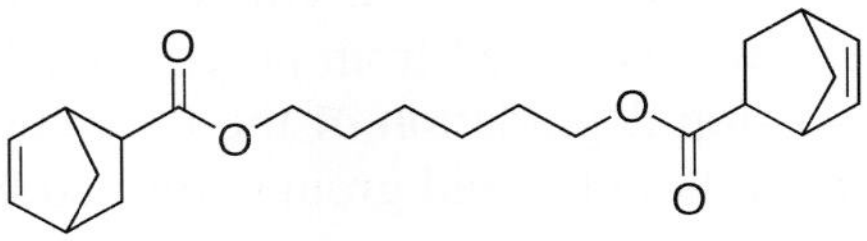

7.5.4 Functional Monomers

Ketene acetals and related monomers undergo ring-opening polymerization to produce polyesters (Section 4.4.2.2). Copolymerization of such monomers with, for example, S (Scheme 7.18), and basic hydrolysis of the ester linkages in the resultant copolymer offers a route to α,ω-difunctional polymers.[227] A limitation on the use of these particular ring-opening monomers is that they are relatively unreactive towards propagating radicals (*e.g.* PS•) thus rates of copolymerization are slow.

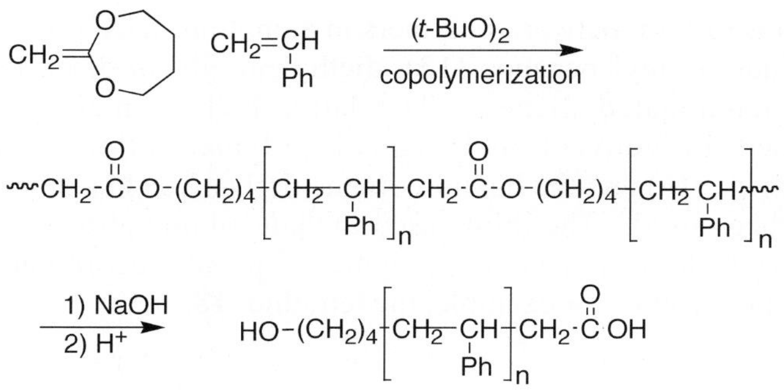

Scheme 7.18

Other ring-opening copolymerizations (of, for example, the cyclic allyl sulfide **19**), also yield polymers with in-chain ester groups and copolymerize more readily (Section 4.4.2.2).

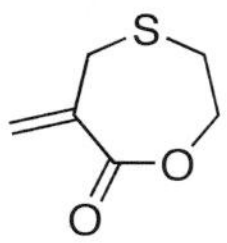

19

Ebdon and coworkers[228-232] have reported telechelic synthesis by a process that involves copolymerizing butadiene or acetylene derivatives to form polymers with internal unsaturation. Ozonolysis of these polymers yields di-end functional polymers. The α,ω-dicarboxylic acid telechelic was prepared from poly(S-*stat*-B) (Scheme 7.19). Precautions were necessary to stop degradation of the PS chains during ozonolysis.[228] The presence of pendant carboxylic acid groups, formed by ozonolysis of 1,2-diene units, was not reported.

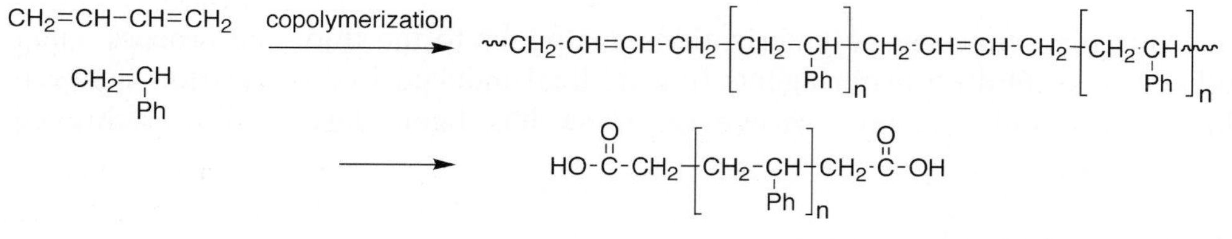

Scheme 7.19

End-functional polymers are also produced by copolymerizations of monosubstituted monomers with α-methylvinyl or other monomers with high transfer constants in the presence of catalytic chain transfer agents (Section 6.2.5).[233-236] Thus, copolymerization of BA with as little as 2% AMS in the presence of cobaloxime provides PBA with AMS at the chain end.[237]

7.5.5 Functional Inhibitors

Inhibitors (Section 5.3), including transition metal complexes and nitroxides, may be used to prepare mono-end-functional polymers. If an appropriate initiator is employed, di-end-functional polymers are also possible.

Only one polymer molecule is produced per mole of inhibitor. The inhibitor must be at least equimolar with the number of chains formed. Concentrations must be chosen (usually very low) to give the desired molecular weight.

7.5.6 Compositional Heterogeneity in Functional Copolymers

The copolymer composition equation only provides the average composition. Not all chains have the same composition. There is a statistical distribution of monomers determined by the reactivity ratios. When chains are short, compositional heterogeneity can mean that not all chains will contain all monomers.

In early work, while compositional heterogeneity was recognized and could be predicted, it was difficult to measure. Now, methods such as GPC combined with NMR and/or MALDI,[238,239] GPC coupled with FTIR[240] and two dimensional HPLC or GPC[241-245] can provide a direct measure of the composition distribution.

Chain compositional heterogeneity is of particular relevance to functional copolymers which find widespread use in the coatings and adhesives industries.[13,240,246] In these applications, the functional copolymer and a crosslinking agent are applied together and are cured to form a network polymer. The functional copolymers are based on functional monomers with reactive groups (e.g. OH). It is desirable that all copolymer molecules have a functionality of at least two. Nonfunctional polymer will not be incorporated and could plasticize the network or be exuded from the polymer. Monofunctional polymers are not involved in crosslink formation and will produce dangling ends.

Various factors are important in determining the composition and molecular weight distribution of multicomponent copolymers (e.g. monomer reactivity ratios, reaction conditions). Stockmayer[247] was one of the first to report on the problem and presented formulae for calculating the instantaneous copolymer composition as a function of chain length. Others[11,12,248-250] have examined the variation in copolymer composition with chain length by computer simulation. One method of ensuring a functionality of at least one is to use a functional initiator or transfer agent.

The influence of selectivity in the initiation, termination or chain transfer steps on the distribution of monomer units within the copolymer chain is usually neglected. Galbraith et al.[11] provided the first detailed analysis of these factors. They applied Monte Carlo simulation to examine the influence of the initiation and termination steps on the compositional heterogeneity and molecular weight distribution of binary and ternary copolymers. Spurling et al.[250] extended this treatment to consider additionally the effects of conversion on compositional heterogeneity.

The ends of polymer chains are often not representative of the overall chain composition. This arises because the initiator and transfer agent-derived radicals can show a high degree of selectivity for reaction with a particular monomer type (Section 3.4). Similarly, there is specificity in chain termination. Transfer agents show a marked preference for particular propagating species (Section 6.2.2 and 6.2.3). The kinetics of copolymerization are such that the probability for termination of a given chain by radical-radical reaction also has a marked dependence on the nature of the last added units (Section 7.4.3).

The effect of the initiation and termination processes on compositional heterogeneity can be seen in data presented in Figure 7.3 and Figure 7.4. The data come from a computer simulation of the synthesis of a hydroxy functional oligomer prepared from S, BA, and HEA with a thiol chain transfer agent. The recipe is similar to those used in some coatings applications.

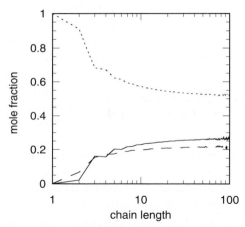

Figure 7.3 Distribution of monomers [HEA(— — —), BA(———), S (-------)] within chains as a function of chain length for a HEA:BA:S copolymer prepared with butanethiol chain transfer agent.[250]

In this copolymerization, most termination is by chain transfer and most chains are initiated by transfer agent-derived radicals. The thiyl radicals generated from the transfer agent react faster with S than they do with acrylate esters (Scheme 7.20).

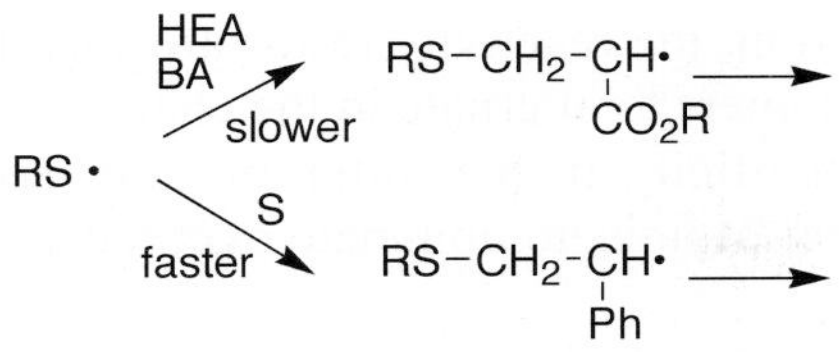

Copolymerization

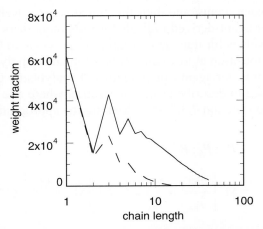

Scheme 7.20

Figure 7.4 Molecular weight distributions for HEA:BA:S copolymer prepared with butanethiol chain transfer agent: (a) all chains (———); (b) chains without HEA (— — —).[250]

The thiol shows a preference to react with propagating radicals with a terminal S unit (Scheme 7.21). This selectivity is due both to chemospecificity in the reaction with thiol and to the relative concentrations of the various propagating species (determined by the reactivity ratios).

A preponderance of chains that both begin and end in S results and this means that short chains are much richer in S than in the acrylic monomers (Figure 7.3). This also has an influence on the fraction of chains that contain the functional monomer (Figure 7.4). The fraction of HEA in very short chains is much less than that in the polymer as a whole and a significant fraction of these short chains contain no functional monomer.

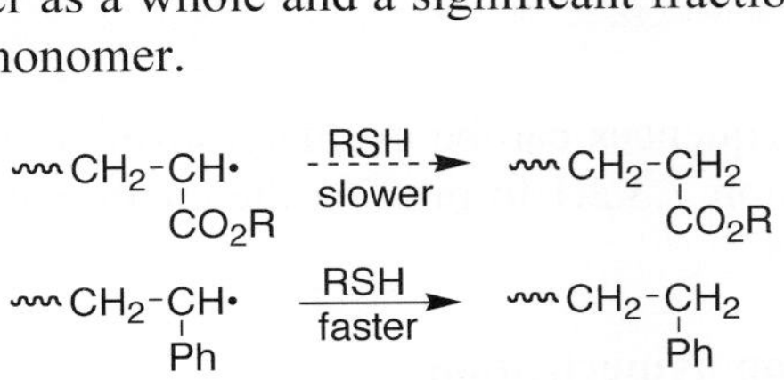

Scheme 7.21

In this copolymerization, the reactivity ratios are such that there is a tendency for S and the acrylic monomers to alternate in the chain. This, in combination with the above-mentioned specificity in the initiation and termination steps, causes chains with an odd number of units to dominate over those with an even number of units.

It is possible to exercise control over this form of compositional heterogeneity (*i.e.* the functionality distribution) by careful selection of the functional monomer and/or the transfer agent taking into account the reactivities of the radical species, monomers, and transfer agents, and their functionality.[11,250] Relative reactivities of initiator and transfer agent-derived radicals towards monomers are summarized in Section 3.4. Some values for transfer constants are provided in Chapter 6.

The overall composition at low conversion of binary copolymers formed in the presence of a chain transfer agent can be predicted analytically using an expression analogous to that used to describe terpolymerization where one monomer does not undergo propagation (Section 7.3.2.4).[236] Making the appropriate substitutions, eq. 37 becomes eq. 70:

$$d\text{M}_A : d\text{M}_B : d\text{T} = P_A : P_B : P_T$$

$$= \text{M}_A \left[\frac{R\text{M}_A}{r_{BA}} + \frac{\text{M}_B}{r_{BA}} + C_B RT \right]\left[\text{M}_A + \frac{\text{M}_B}{r_{AB}} + C_A T \right]$$

$$: \text{M}_B \left[\frac{R\text{M}_A}{r_{AB}} + \frac{\text{M}_B}{r_{AB}} + C_A T \right]\left[\text{M}_B + \frac{\text{M}_A}{r_{BA}} + C_B T \right]$$

$$: \text{T} \left[\frac{\text{M}_A}{C_A r_{BA}} + \frac{\text{M}_B}{r_{BC} r_{AB}} + C_A C_B T \right]\left[R\text{M}_A + \text{M}_B \right] \quad (70)$$

where T is the concentration of transfer agent, C_A and C_B are the transfer constants of the transfer agent in polymerizations of monomer A and B respectively and $R=k_{iA}/k_{iB}$ is the relative rate of initiation by the transfer agent-derived radical. The average molecular weight is given by eq. 71.

$$X_n \approx \frac{1}{P_T} \quad (71)$$

The T containing sequences can be evaluated using expressions analogous to those described in Section 7.3.2.1 to provide the chain end compositions and the chain length distribution.

7.6 Block & Graft Copolymerization

Many block and graft copolymer syntheses involve radical polymerization at some stage of the overall preparation. This section deals with direct syntheses of

Copolymerization

block and graft copolymers by conventional radical processes. Formation of block and graft copolymers by living radical polymerization is discussed in Chapter 9.

In the standard nomenclature [poly(M_A)-*graft*-poly(M_B)] the first named monomer(s) form the backbone while those named second are the grafts or arms. Thus, PMMA-*graft*-PS indicates a backbone of PMMA and grafts of PS.

Graft copolymerizations are categorized according to their method of formation into three main types.[251]

(a) Grafting onto, where reactive functionality on one polymer chain reacts with functionality on a second chain. Condensation of polymer bound functionality with end-functional polymers is a grafting onto process. Processes for the formation of functional polymers are discussed in Section 7.5.

(b) Grafting from, where active sites are created on the polymer chain from which new polymerization is initiated.

(c) Grafting through, where a propagating species reacts with pendant unsaturation on another polymer chain. The copolymerization of macromonomers is a grafting through process (Section 7.6.5).

Four types of 'grafting from' processes are distinguished by the mechanism of radical formation.

(a) Formation of radicals on or at the end of a polymer chain by decomposition of bonded initiator functionality (often an azo or peroxide linkage) (Section 7.6.1).

(b) Formation of radicals by transformation of a polymer bound functionality to radicals typically by some form of redox or multi-step process (Section 7.6.2).

(c) Formation of radicals on non-functional polymer substrates by irradiation with, for example, γ-rays or an electron beam (Section 7.6.3)

(d) Formation of radicals on non-functional polymer substrates by radicals abstracting hydrogen (Section 7.6.4). Transfer to polymer during polymerization also causes branching in a grafting from process (Section 6.2.7).

Specific forms of graft copolymers may go under different names.

(a) Branched polymers where the backbone and the arms are of the same composition

(b) Comb polymers where the arms are of uniform length (*e.g.* PMMA-*comb*-PS)

(c) Hyperbranched polymers when there are branches on branches.

7.6.1 Polymeric and Multifunctional Initiators

Multifunctional initiators contain two or more radical generating functions within the one molecule. The chemistry of these initiators has been the subject of several reviews.[252-255] As long as the radical generating functions are sufficiently remote their decompositions are independent events. If decomposition occurs under sufficiently different reaction conditions, these initiators can be used to form polymers with end groups that contain initiator moieties. The polymeric initiators can be subsequently utilized to yield higher molecular weight polymers, to achieve higher degrees of conversion, and in the production of block and graft copolymers.

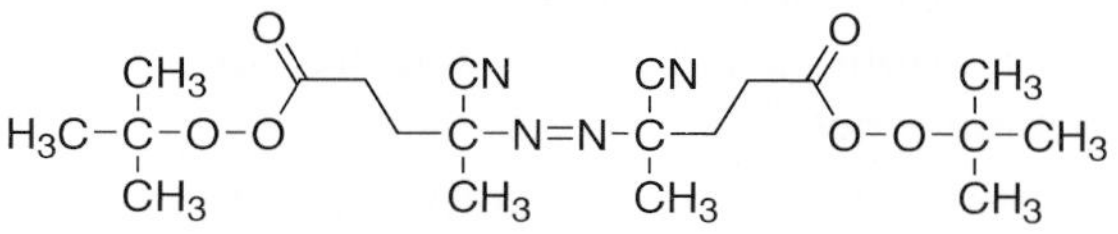

20

The multifunctional initiators may be di- and tri-, azo- or peroxy-compounds of defined structure (e.g. **20**[256]) or they may be polymeric azo- or peroxy-compounds where the radical generating functions may be present as side chains[257] or as part of the polymer backbone.[258-261] Thus, amphiphilic block copolymers were synthesized using the polymeric initiator **21** formed from the reaction between an α,ω-diol and AIBN (Scheme 7.22).[262] Some further examples of multifunctional initiators were mentioned in Section 3.3.3.2. It is also possible to produce less well-defined multifunctional initiators containing peroxide functionality from a polymer substrate by autoxidation or by ozonolysis.[263]

The success of the multifunctional initiators in the preparation of block and graft copolymers depends critically on the kinetics and mechanism of radical production. In particular, the initiator efficiency, the susceptibility to and mechanism of transfer to initiator, and the relative stability of the various radical generating functions. Each of these factors has a substantial influence on the nature and homogeneity of the polymer formed. Features of the kinetics of polymerizations initiated by multifunctional initiators have been modeled by O'Driscoll and Bevington[264] and Choi and Lei.[265]

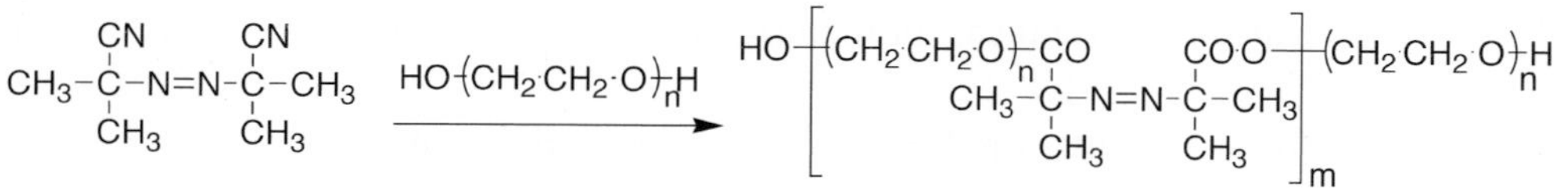

21

Scheme 7.22

A final class of multifunctional initiators is based on the use a (multi)functional polymer and a low molecular weight redox agent. Radicals on the polymer chain are generated from the polymer bound functionality by a redox reaction. Ideally, no free initiating species are formed. The best known of this class are the polyol-redox and related systems. Polymers containing hydroxy or glycol and related functionality are subject to one electron oxidation by species such as ceric ions or periodate (Scheme 7.23).[266,267] Substrates such as cellulose, chitin and poly(vinyl alcohol) provide graft copolymers. The chemistry is briefly discussed in Section 3.3.5.2. Hydroxy end-functional polymers such as poly(ethylene glycol) and poly(4-hydroxybutyrate) yield block copolymers. A further example of this approach, which makes use of a halogen functional polymer, can be found in Section 7.6.2.

Scheme 7.23

7.6.2 Transformation Reactions

Block and graft copolymer syntheses by what have come to be known as 'transformation reactions' involve the preparation of polymeric species by some mechanism which leaves a terminal functionality that allows polymerization to be continued by another mechanism as shown schematically in Scheme 7.24. Examples of transformation of anionic, cationic, Ziegler-Natta, and group transfer polymerization to radical polymerization have been reported. Examples of transformation of radical to ionic polymerization are also known. Additional examples that involving transformation to or from living radical polymerization (NMP, ATRP or RAFT) can be found in Chapter 9. The success of the transformation reactions depends on the efficiency of the transformation process and the avoidance of processes that might lead to concurrent homopolymerization. The general area of block polymer synthesis through 'transformation reactions' has been reviewed by Stewart[268], Schue,[269] Abadie and Ourahmoune.[270] and Eastmond[271] The mechanism of termination also plays an important role in determining the type of block copolymers that may be formed. If standard polymerization conditions are employed, an ABA or AB block may be produced depending on whether termination occurs by combination or disproportionation.

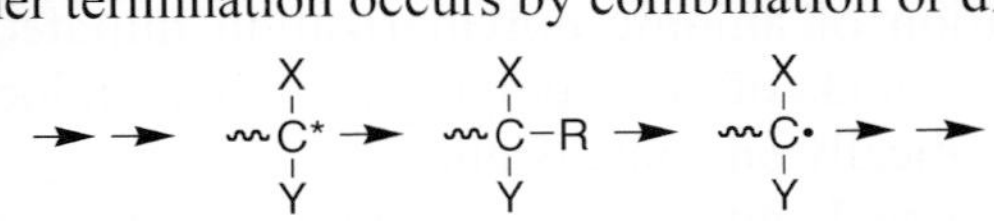

Scheme 7.24 (* = active center; *e.g.* anion, cation, radical)

One of the earliest examples of this methodology involves the reaction of a polymeric anion (formed by living anionic polymerization) with molecular oxygen to form a polymeric hydroperoxide which can be decomposed either thermally or, preferably, in a redox reaction to initiate block polymer formation with a second monomer (Scheme 7.25). However, the usual complications associated with initiation by hydroperoxides apply (Section 3.3.2.5).

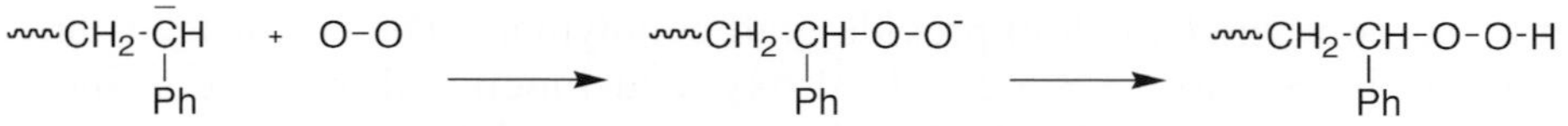

Scheme 7.25

~~CH$_2$-CH(Ph) + O-O → ~~CH$_2$-CH(Ph)-O-O$^-$ → ~~CH$_2$-CH(Ph)-O-O-H

The reactions of polymeric anions with appropriate azo-compounds or peroxides to form polymeric initiators provide other examples of anion-radical transformation (*e.g.* Scheme 7.26).[270,272-274] However, the polymeric azo and peroxy compounds have limited utility in block copolymer synthesis because of the poor efficiency of radical generation from the polymeric initiators (7.5.1).

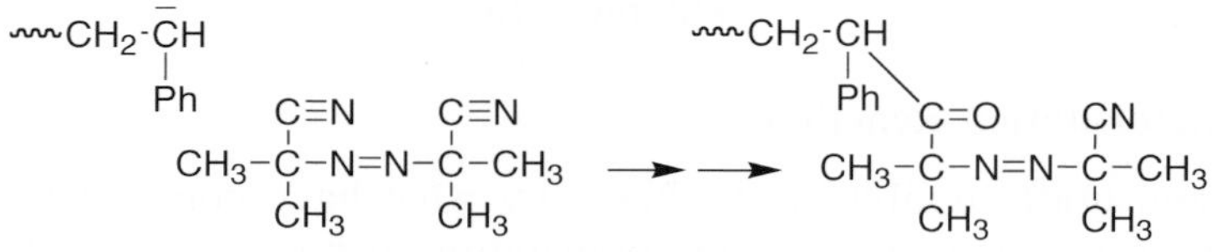

Scheme 7.26

~~CH$_2$-CH(Ph) + CH$_3$-C(C≡N)(CH$_3$)-N=N-C(C≡N)(CH$_3$)-CH$_3$ → → ~~CH$_2$-CH(Ph)-C(=O)-... CH$_3$-C(CN)(CH$_3$)-N=N-C(CN)(CH$_3$)-CH$_3$

Tung *et al.*[275] have reported on the use of a polymeric thiol transfer agent for use in block copolymer production. Various methods have been used for the anion→thiol conversion. Near quantitative yields of thiol are reported to have been obtained by terminating anionic polymerization with ethylene sulfide and derivatives (Scheme 7.27). Transfer constants for the polymeric thiols are reported to be similar to those of analogous low molecular weight compounds.[275]

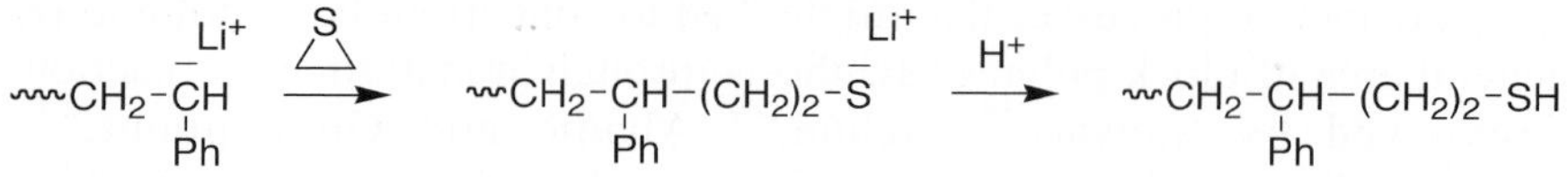

Scheme 7.27

~~CH$_2$-CH(Ph)$^-$ Li$^+$ →[S triangle] ~~CH$_2$-CH(Ph)-(CH$_2$)$_2$-S$^-$ Li$^+$ →[H$^+$] ~~CH$_2$-CH(Ph)-(CH$_2$)$_2$-SH

The preparation of ABA triblock polymers requires use of a telechelic bisthiol prepared by termination of anionic polymerization initiated by a difunctional initiator. The relative yields of homopolymer, di- and triblock obtained in these experiments depend critically on conversion.[275]

Richards *et al.* carried out extensive studies on the use of mercury,[276,277] lead[278,279] and silver compounds to terminate anionic polymerization and form polymeric organometallic species which can be used to initiate polymerization.

Bamford, Eastmond and coworkers[280-285] have employed metal complex-polymeric halide redox systems to initiate block and graft copolymerization. The polymeric halides can be synthesized by a variety of techniques, including radical polymerization,[281] anionic polymerization (Scheme 7.28),[280]

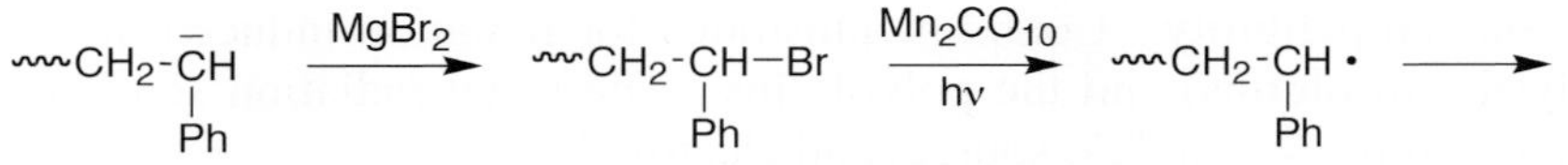

Scheme 7.28

group transfer polymerization (Scheme 7.29),[284]

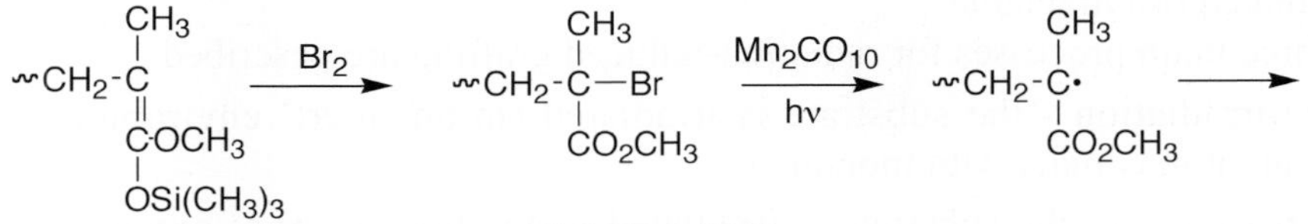

Scheme 7.29

cationic polymerization (Scheme 7.30),[283]

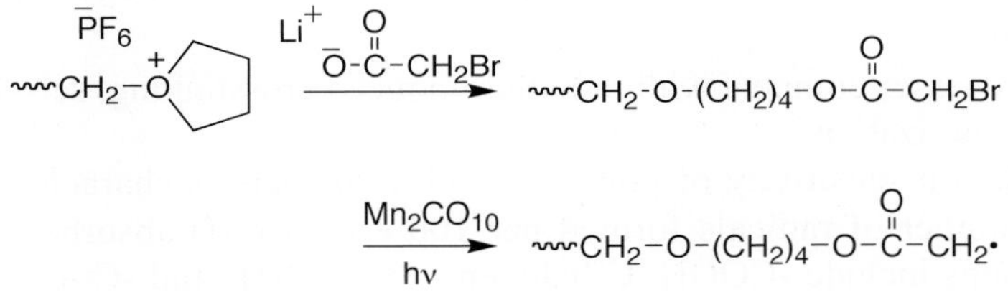

Scheme 7.30

and functionalization of a polymer with carboxylic acid, hydroxy, amino, or ether-urethane groups with a haloisocyanate (Scheme 7.31).[286]

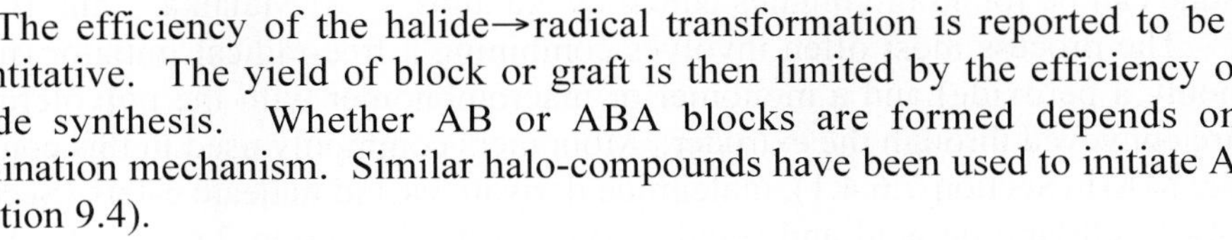

Scheme 7.31

The efficiency of the halide→radical transformation is reported to be near quantitative. The yield of block or graft is then limited by the efficiency of the halide synthesis. Whether AB or ABA blocks are formed depends on the termination mechanism. Similar halo-compounds have been used to initiate ATRP (Section 9.4).

7.6.3 Radiation-Induced Grafting Processes

Radiation-induced grafting and curing processes have been discussed in a number of reviews.[263,287-291] The process is widely used for surface modification. Recent applications are the modification of fuel cell membranes and improving surface biocompatibility. Common substrates for radiation-induced grafting are the poly(fluoro-olefins) and the polyolefins. The usual radiation sources in this context are γ-rays (*e.g.* a ^{60}Co source) and electron beams.

The detailed chemistry of radiation grafting has, in most cases, not been rigorously established. Process characterization is complicated by the fact that often only surface layers are involved and, in other cases, by the substrates being cross-linked or intractable.

Three main processes for radiation-induced grafting are described:

(a) Pre-irradiation - the substrate is irradiated (in an 'inert' environment) then brought in contact with monomer.
(b) Peroxidation - the substrate is irradiated in an atmosphere of oxygen or air to form peroxidic groups, which are then thermally decomposed in the presence of monomer.
(c) Mutual irradiation - the substrate and monomer are brought together then irradiated.

These processes compete with radiation-induced crosslinking, scission and, for case (c), polymerization.

The radiation sensitivity of polymers and monomers is characterized by a *G* value; the number of radicals formed per 100 e.v. (16 aJ) absorbed. Radiation sensitive groups include -COOH, C-halogen, -SO$_2$-, -NH$_2$ and -C=C-. Radiation resistant groups are aromatic rings. It appears that the presence of aromatic moieties also offers some degree of radiation protection to the polymer chain as a whole.

7.6.4 Radical-Induced Grafting Processes

Radical induced grafting may be carried out in solution, in the melt phase,[292-295] or as a solid state process.[296] This section will focus on melt phase grafting to polyolefin substrates but many of the considerations are generic. The direct grafting of monomers onto polymers, in particular polyolefins, in the melt phase by reactive extrusion has been widely studied. Most recently, the subject has been reviewed by Moad[293] and by Russell.[292] More details on reactive extrusion as a technique can be found in volumes edited by Xanthos,[294] Al Malaika[295] and Baker *et al.*[297] The process most often involves combining a free-radical initiator (most commonly a peroxide) and a monomer or macromonomer with the polyolefin as they are conveyed through the extruder. Monomers commonly used in this context include: MAH (Section 7.6.4.1), maleimide derivatives and maleate esters (Section 7.6.4.2), (meth)acrylic acid and (meth)acrylate esters (Section 7.6.4.3), S, AMS and derivatives (Section 7.6.4.4), vinylsilanes (Section 7.6.4.5) and vinyl oxazolines (Section 7.6.4.6).

A major issue is the control of the side reactions that accompany grafting. These reactions include radical-induced degradation of the substrate by cross-linking and/or chain scission and homopolymerization of the graftee monomer.

Copolymerization

Polyethylenes (HDPE, LDPE, LLDPE, high E content - EP) are prone to branching or crosslinking caused by radical-radical combination. This process is characterized by the formation of gels or a partially insoluble product. Polypropylene (PP) and low-density ethylene/α-olefin copolymers may also undergo crosslinking under some conditions. However, the most often-encountered side reaction is degradation caused by the initially formed radical undergoing β-scission. This susceptibility to chain scission is well documented and is used to advantage in the synthesis of controlled rheology PP.

A major challenge is then to devise conditions so as to maximize grafting and minimize or control these side reactions. Some discussion of many of these parameters is provided in the reviews mentioned above. It is significant that many recent publications and patents in the area of reactive extrusion relate, not to the development of new reactions or processes, but to the selection of operating parameters.

The monomer acts to trap radicals that might otherwise undergo chain scission or crosslinking. More degradation is seen with less reactive monomers. Use of a higher monomer concentration may result in less degradation of the polyolefin substrate. However, it is often found that the dependence of grafting yield on monomer concentration passes through a maximum. If the monomer concentration becomes too high, phase separation can occur. This results in reduced grafting yields and an increased likelihood for homopolymerization. In these circumstances, higher graft levels can better be achieved by multipoint/multipass addition of monomer and initiator or by use of a comonomer or other coagent.

It is also necessary to select the initiator according to the particular monomer(s) and the substrate. Factors to consider in this context, aside from initiator half-lives and decomposition rates, are the partition coefficient of the initiator between the monomer and polyolefin phases and the reactivity of the monomer *vs* the polyolefin towards the initiator-derived radicals.

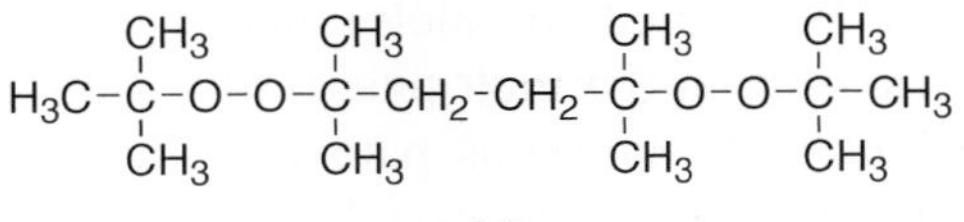

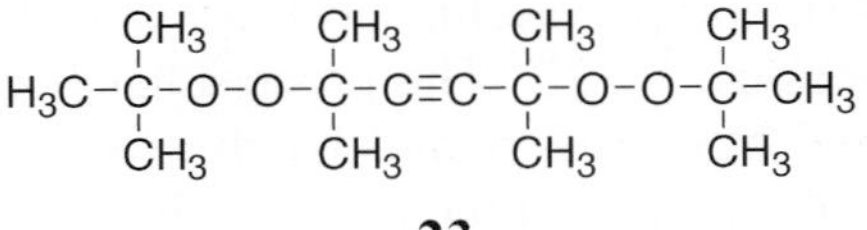

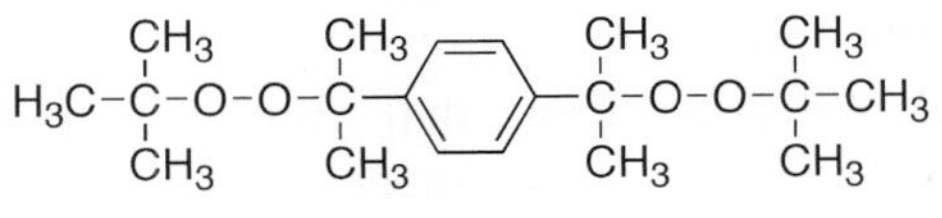

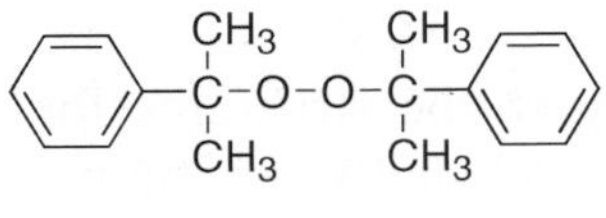

Grafting is most commonly carried out with peroxides that are sources of *t*-alkoxy radicals (*e.g.* **22-25**). At the high temperatures usually used, the extent of β-scission is likely to be significant thus radicals involved in abstraction are likely to be a mixture of *t*-alkoxy and alkyl radicals. Several authors[298,299] have pointed out that R_3CO-H and CH_3-H bond strengths are similar (Section 2.2.2). Even accepting the validity of the Evans-Polyani approach in this context, it must also be noted that C-C bonds are significantly stronger than R_3CO-C bonds (Section 2.4.6). Thus, methyl radical is anticipated to have a greater propensity for addition over abstraction than a *t*-alkoxy radical. The tendency for addition *vs* abstraction is greater for higher alkyl radicals. Abstraction:addition ratios are also temperature dependent (Section 3.2.4). Lower temperatures favor abstraction over addition and, for *t*-alkoxy radicals, both of these reactions are favored over β-scission. The regiospecificity of hydrogen abstraction by *t*-alkoxy and methyl radical is also very different. The methyl radical shows a much greater specificity for methine>methylene>methyl.[293]

While it is important that the initiator-derived radicals react preferentially with the polyolefin substrate, the specificity shown by the initiator-derived radicals may be of only minor importance in determining the ultimate product distribution. The species that abstracts hydrogen is, in many cases, not an initiator-derived radical. This follows from the observation that up to 20 monomer units may be grafted per initiator-derived radical generated.[300,301] Care must be taken in interpreting such data as it is not always clear whether a high number of monomer units grafted per radical generated means a long graft length or a large number of graft sites. Nonetheless, it is clear that in some instances, where graft lengths have been characterized, that most abstraction must occur by way of the propagating species formed by addition of monomer. In these cases, chain transfer is also a major factor in limiting the length of the grafted chain.

An alternative to the direct use of peroxides in monomer grafting is to first functionalize the polymer with initiator or transfer agent functionality.

7.6.4.1 *Maleic anhydride graft polyolefins*

With a history of more than 25 years, the free radical-induced grafting of MAH onto polyolefin substrates is one of the most studied polyolefin modification processes.[293,298,302] The process has been carried out in the melt phase, in various forms of extruders and batch mixers, and there are numerous patents covering various aspects of the process. It has also been carried out successfully in solution and in the solid state. The materials have a range of applications including their use as precursors to graft copolymers, either directly, or during the preparation of blends.[297]

Many of the structures for MAH-modified polyolefins that appear in the literature are wholly speculative, and are based on a proposed mechanism for the grafting reaction rather than an analysis of the reaction or reaction products. In early work, product characterization took the form of determining overall grafting levels by titration or IR spectroscopy. In more recent work, with the availability of

Copolymerization

additional characterization techniques, it has been shown that the structure depends strongly on the particular polyolefin substrate and the synthesis conditions.[303,304]

In early work it was often assumed, without specific proof, that MAH was grafted to polyolefins as single units (Scheme 7.32). This followed from its known sluggishness in homopolymerization and from a consideration of ceiling temperature. Recent NMR studies indicate that MAH is attached to PP[303] and model substrates[292] as single units. However, other studies suggest that a fraction of units may be grouped either as oligo-MAH grafts[305,306] or as adjacent grafts formed by sequential intramolecular abstraction and grafting (Scheme 7.33).[292] Differing reaction conditions used in the various works confuses analysis of the situation.[293]

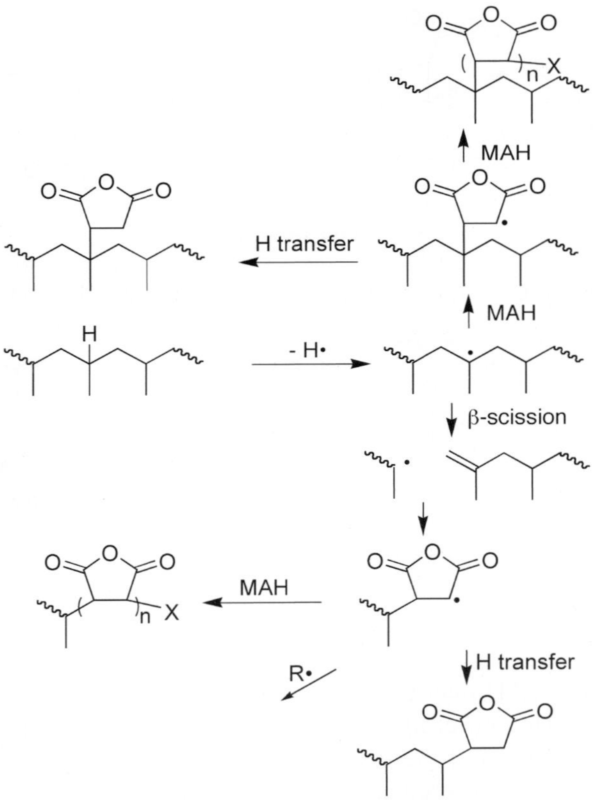

Scheme 7.32

Typical levels of MAH in grafted PP of 0.5-2 wt % correspond to only one or two units per chain. If the MAH units are grouped it follows that many chains may contain no MAH. It has also been suggested that for PP all MAH may appear at the chain ends. This is rationalized in terms of the reaction of mid chain radicals with MAH always being followed by intramolecular chain transfer and chain scission as shown in Scheme 7.34. This pathway would be favored by the slow rate of homopropagation of MAH.

Substantial work has also been carried out on grafting to HDPE,[303,307] LLDPE[303,308-310] and EP copolymers.[303,311] In many early studies, MAH grafting onto PE and ethylene copolymers seemed always to be accompanied by some degree of crosslinking as indicated by a partially insoluble product. However, the recent literature demonstrates that extrusion conditions can be designed to avoid or minimize crosslinking and provide a completely soluble product and still obtain very high grafting yields (~ 80%).[308] The different outcome in these latter studies is attributed to the differences in the effectiveness of mixing, initiator concentrations, and the method of reagent introduction.[312]

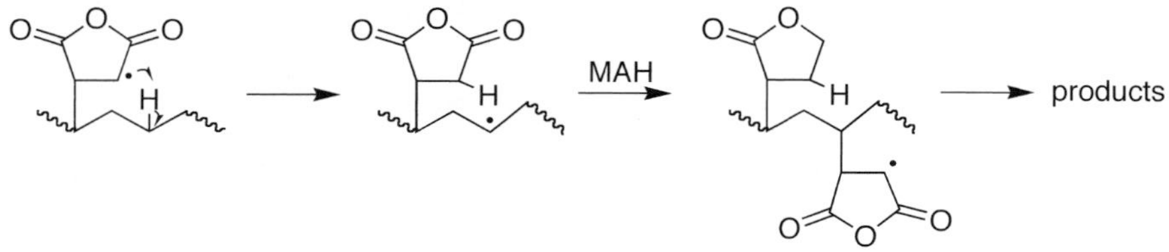

Scheme 7.33

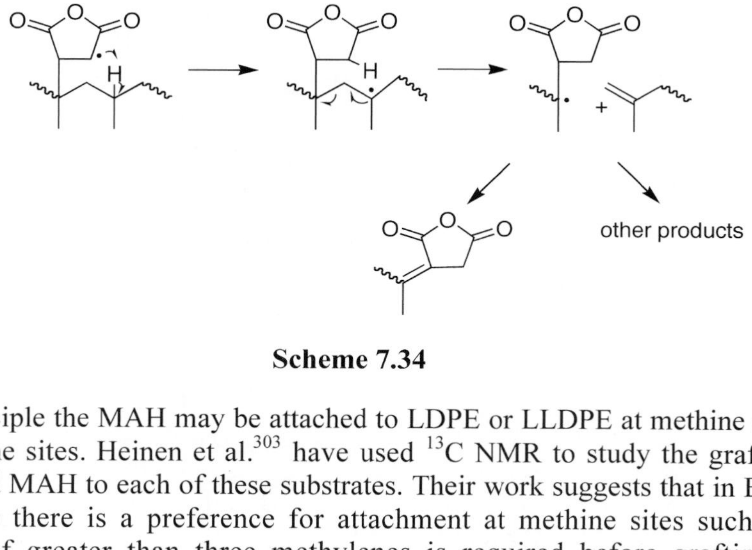

Scheme 7.34

In principle the MAH may be attached to LDPE or LLDPE at methine sites or at methylene sites. Heinen et al.[303] have used ^{13}C NMR to study the grafting of ^{13}C-labelled MAH to each of these substrates. Their work suggests that in EP (and in LLDPE) there is a preference for attachment at methine sites such that a sequence of greater than three methylenes is required before grafting to a methylene site is observed.

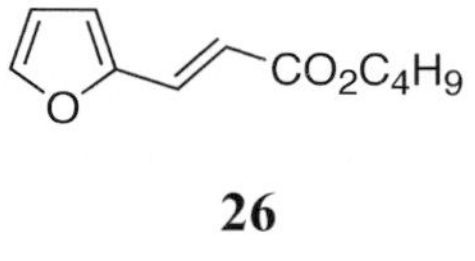

Copolymerization

Various monomers and reagents can be added to improve grafting yields and to decrease the significance of side reactions such as chain scission (PP) or crosslinking (PE). The effect of various comonomers on grafting yield of MAH onto LLDPE (*e.g.* S, MMA, maleate esters)[309] and PP (*e.g.* S, AA, MAA, MMA, NVP)[298,313,314] has been studied. Colai *et al*.[315] have recently reported on the use of furan derivatives (*e.g.* 26) in this context. The use of the coagents can substantially increase grafting yields and reduces the degradation in the case of PP. Grafting yields decrease in the series where the comonomer is S >> AMS > MMA > VAc > (no comonomer) > NVP. Several explanations for the comonomer effect have been proposed. Higher grafting yields have been attributed[298] to formation of a charge transfer complex between the comonomer and MAH (7.3.1.3) and to the greater reactivity of this species. A second explanation is that the comonomer is a more effective trap than MAH for the polyolefin derived radical.[315] However, it is also possible that more efficient grafting may simply be due to attachment of a longer chain length graft rather than a greater number of graft sites. Hu *et al*.[298] have provided NMR data for S-MAH grafts from PP suggesting that the graft is a copolymer chain and not a single S-MAH pair. It would also appear from the copolymer composition that S and MAH do not show the same tendency to alternate in the chain in graft copolymer formation as is seen in conventional free radical copolymerization in solution at lower temperatures (7.3.1.3).[298] It was found the S-MAH ratio in the graft exceeds the initial S-MAH ratio irrespective of that ratio.[298] These observations do not preclude the involvement of a charge transfer complex but do show that the monomers are not incorporated pairwise.

Various solvents, transfer agents and inhibitors have also been used to enhance grafting yields or limit side reactions during polymer modification. If inhibitors can have specificity for the monomer-derived propagating species, it may be possible to prevent homopolymerization while not interfering with abstraction from the polymer backbone by the initiator-derived radicals. Such inhibitors would reduce grafting yields by limiting the length of the grafted chain. Gaylord *et al*. have reported that various 'electron donor additives' are effective in limiting the amount of crosslinking (various PE,[307,310,316] EP[311]) or chain scission (PP[317,318]) that occurs during melt phase maleation. The additives used included various amides (*e.g.* dimethyl acetamide, dimethyl formamide, caprolactam, stearamide),[307,311,316-318] sulfoxides (*e.g.* dimethyl sulfoxide),[307,316] and phosphites (*e.g.* hexamethylphosphoramide, triethyl phosphite).[307,310,316] A mechanism of action for these coagents based on the propensity of MAH to form charge transfer complexes was proposed.[302] Gaylord *et al*.[302] also showed that these agents act as inhibitors of MAH homopolymerization, but not of MMA polymerization, and this may explain why the additives cause a *ca* two-fold reduction in grafting yields with MAH but are not effective in suppressing homopolymerization during grafting of methacrylic monomers. The effectiveness of certain of these coagents has been disputed.[298,319] Wu and Su[319] found that stearamide is only useful for low initiator levels and then does not completely suppress crosslinking during grafting of MAH to EP. It was suggested that, under the process conditions, stearamide acts as a transfer agent. Another report[298] suggests that the effect of these coagents in reducing crosslinking of PP might be duplicated simply by using lower initiator concentrations.

7.6.4.2 Maleate ester and maleimide graft polyolefins

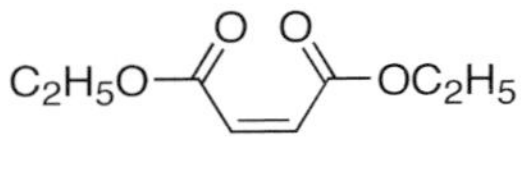

DEM **27** 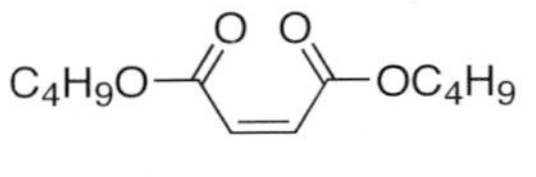 ; **28** 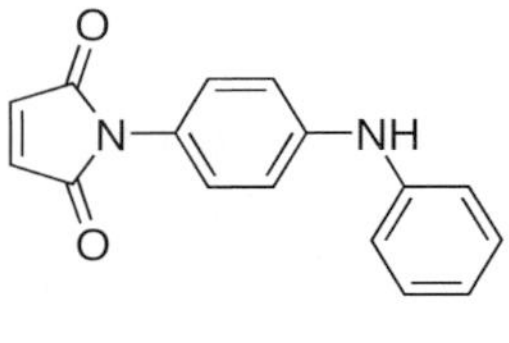 ; **29**

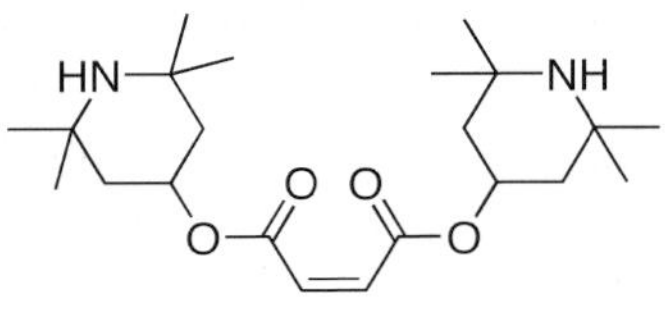

30 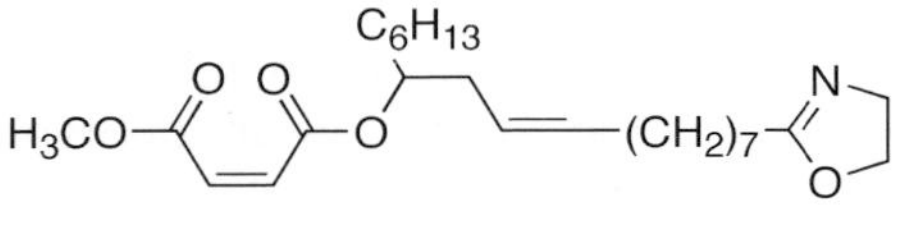 ; **31**

The melt phase grafting of dialkyl maleates, usually the diethyl (**27**) or dibutyl esters (**28**), onto PP,[320,321] LLDPE[320-326] and EP[327-329] has been studied. Their use has been advocated over MAH (Section 7.6.4.1) due to their lower volatility and lower toxicity. All may and have been used as precursors to nylon/polyester grafts. However, the maleate esters are significantly less reactive towards free radical addition than MAH and grafting yields are generally lower. Like MAH, the maleate esters show little tendency to homopolymerize. NMR studies suggest that **27** is grafted onto the PE as isolated units even with relatively high dimethyl maleate:polyolefin ratios (1:1 weight ratio).[323] Even though maleate esters are less reactive than MAH, conditions can be found such that the side reactions associated with peroxide induced grafting (crosslinking, chain scission) appear to be negligible (as indicated by little change in the GPC molecular weight distribution and no insoluble product). This may reflect the greater solubility of the maleate esters in the polyolefin melt.

As with MAH, the extent of grafting varies dramatically with the polyolefin substrate. Some differences have been attributed to variations in the type and amount of stabilizers present in the polyolefins substrate.[326] In the case of isotactic PP, the maximum graft levels attained with **27** were found to correspond to only one unit of DEM per PP molecule[320] This would support a mechanism whereby grafts appear only at the chain ends. Higher graft levels were obtained with atactic PP. The higher reactivity of the atactic PP (and atactic sequences in

Copolymerization

isotactic PP) has been attributed to the greater conformational mobility in atactic sequences and less steric hindrance to grafting.[320,330] It has also been found that grafting yields for EP with a blocky structure are higher than for 'random' copolymers of similar overall composition and molecular weight. This is circumstantial evidence that maleate ester units are preferentially grafted to methylene sites in EP.[328,329]

More elaborate maleate and maleimide derivatives have provided a route to grafting various functionalities onto PP. Examples, include antioxidants (**29** and **30**)[331] and the oxazoline derivative (**31**)[332,333] for which very high grafting yields were reported.

7.6.4.3 (Meth)acrylate graft polyolefins

Various (meth)acrylic monomers have been successfully grafted onto polyolefins. Most studies deal with functional monomers. Grafting yields obtained with PP are usually low (<20%) and are dependent on the particular monomer. Liu et al.[334] carried out a comparative study on the grafting of various functional methacrylates onto PP. The experiments were performed in a batch mixer at 180 °C with 7 wt% monomer and 0.05 wt% **22** as an initiator. Grafting levels (wt%) obtained under these conditions were as follows: HPMA (1), TBAEMA (1), GMA (0.8), HEMA (0.4), DMAEMA (0.3), **32** (0.2). Grafting yields to PE appear generally higher.

AA,[335] and less often MAA or itaconic acid[336] have been successfully grafted onto polyolefins. In the case of AA, grafting is often accompanied by homopolymerization.[335]

Baker and coworkers examined the grafting of methacrylate esters containing secondary or tertiary amino groups such as TBAEMA[299,337,338] or DEAEMA[299,339,340] onto LLDPE. Peroxides undergo induced decomposition in the presence of amino-functional monomers. This problem was overcome by using phenylazotriphenylmethane - a source of phenyl radicals - as initiator.[299] This gave good grafting yields and no discernible side reactions. The mechanism of grafting was explored using squalane and eisocosane as model substrates.[341,342] In these experiments, only single unit grafts were observed and little homopolymerization was detected for temperatures above 130 °C. The findings were rationalized in terms of the occurrence of intra-molecular hydrogen abstraction and a low ceiling temperature for polymerization.[341,342]

Many studies on the melt phase grafting of GMA onto polyolefins (PP,[298,300,301,334,343-347] EP,[343] LDPE,[343,348] LLDPE,[349,350] HDPE[343,349,351,352]) have been reported. The experiments have been conducted in batch mixers and reactive extruders. Grafting efficiencies onto PP obtained in melt phase grafting experiments with GMA alone are typically very low (<20%). However, it is reported that initiator selection is important in determining the grafting yield.[300,301] Use of a short half-life initiator **37** ($t_{1/2}$~6.6 s) gave two to three-fold higher grafting yields than **22**($t_{1/2}$~212 s) under similar processing conditions.[300] The use of comonomers, in particular, styrene, both enhances the grafting efficiency of GMA and reduces PP degradation although some crosslinking occurs when high styrene levels are employed.[298,300,345,347,352]

Grafting of GMA onto LDPE and EP is more efficient.[343] No crosslinking was observed and high grafting yields were attributed in part to the high solubility of GMA in these polyolefins. Grafting efficiencies for GMA are significantly higher than those observed with other methacrylates. Little has been reported on the structure of the GMA graft copolymers. However, Galluci and Going[343] provided circumstantial evidence that GMA is attached to LDPE as oligo(GMA) blocks rather than as single units.

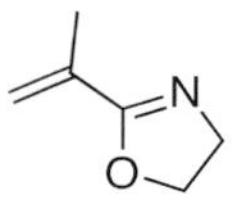

32

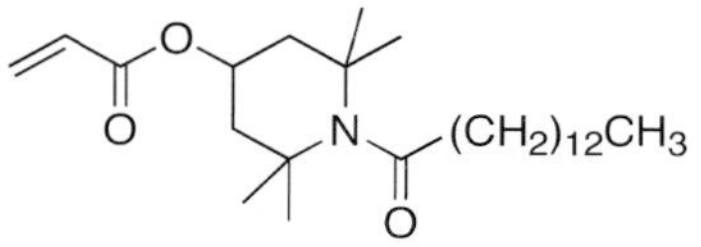

33

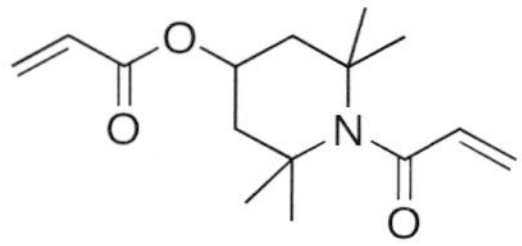

34

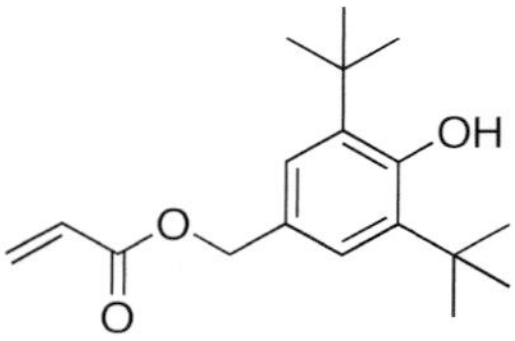

35

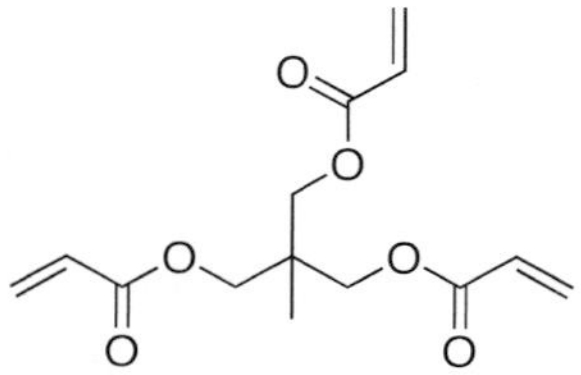

36

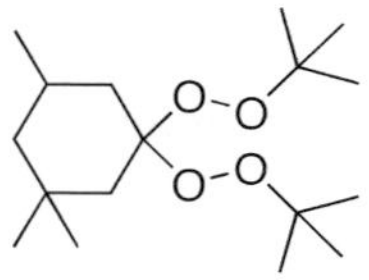

37

Al Malaika et al.[353,354] have reported on the grafting of antioxidant moieties onto PP as mono- (e.g. **33**) or bis-(meth)acrylic derivatives (**34**). Moderate grafting yields (10–40%) and some homopolymerization was observed in the case of the monoacrylate. However, with the bis-acrylate (**34**) close to 100% grafting yield was reported.

In another study, the monoacrylate **35** was grafted onto PP in the presence of tris(acryloylmethyl)propane (**36**) as coagent.[355] Again close to 100% grafting yield was obtained. This was so despite the fact that **35** was anticipated to be an inhibitor of free radical reactions (in fact, phenols are poor inhibitors of (meth)acrylate polymerization - Section 5.3.4). The tris-acrylate **36** and related species have previously been used for producing crosslinked/branched PP.[356,357] The structure of the graft was not established. The remarkable finding was that the final products in the processes involving **36** were not crosslinked and, indeed, were completely soluble in xylene. It was proposed that crosslinking did in fact occur but that the initially formed product underwent *in-situ* degradation by chain scission on further processing to ultimately yield a soluble, gel-free material.[358]

One would not expect this strategy to be useful for grafting onto PE or other polymers less susceptible to shear induced chain scission.

7.6.4.4 Styrenic graft polyolefins

Although there are several reports[345,359] of direct grafting of S onto polyolefins, S and AMS are more often encountered as coagents when grafting MAH and (meth)acrylic and other monomers. Recent reports describe the use of the functional styrene derivative **38** to attach oxazoline groups to ABS and **39**[360,361] to introduce isocyanato groups into PP or PE. Grafting yields with **39** onto PP were improved with use of S as a coagent.[360]

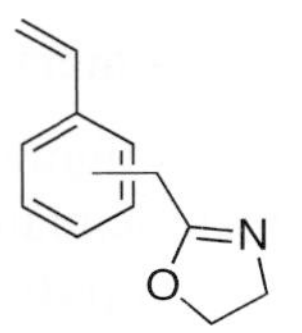

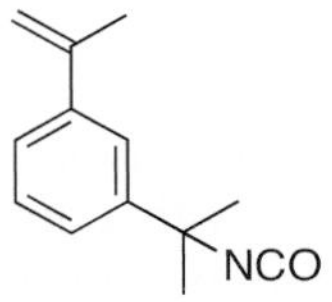

7.6.4.5 Vinylsilane graft polyolefins

The attachment of trialkoxysilane functionality to polyolefins (HDPE, LDPE, PP) though grafting of vinylsilanes (*e.g.* **40, 41**) or silane functional acrylates (*e.g.* **42**) has been widely studied.[362] The principal application of these materials is the preparation of moisture curable crosslinked polyolefins that are widely used in the cable industry.[362] Silane treatment has also been used for surface modification of polyolefins[324] and silane grafted polyolefins might also serve as precursors to graft copolymers.

The vinylsilanes (*e.g.* **40, 41**) do not readily homopolymerize. Forsyth *et al.*[363] explored the mechanism of grafting these monomers using dodecane as a model for PE. Their work suggests that multiple monomer units are attached through a sequence of addition and intramolecular hydrogen atom transfer steps by a mechanism analogous to that shown in Scheme 7.33 on page 394.

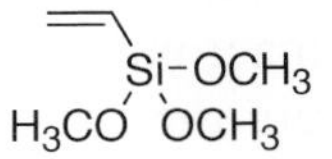

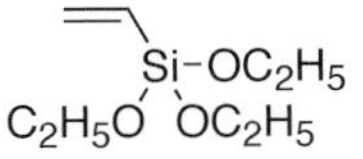

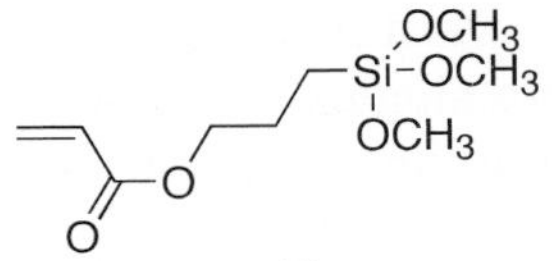

7.6.4.6 Vinyl oxazoline graft polyolefins

The oxazoline moiety has been used in place of anhydride (from MAH) or epoxy groups (from GMA) as a reactive functionality for use in polymer modification by reactive extrusion[364]. Polyolefins containing oxazoline functionality are also used as precursors to graft copolymers or as *in situ* compatibilizers or toughening agents. Several methods have been devised for attaching the oxazoline functionality to polyolefins by free radical-induced grafting. The free radical-induced grafting of 2-isopropenyl-2-oxazoline onto PP was reported by Liu and Baker.[365,366] Vainio et al.[332,333] employed the maleate ester (**31**) to produce an oxazoline functional PP.

7.6.5 Polymerization and Copolymerization of Macromonomers

In the present context, a macromonomer is defined as an oligomer or polymer chain terminated with a double bond or other group such that the material is able to act as a comonomer in a radical copolymerization. The copolymerization of macromonomers with conventional low molecular weight monomers will give a graft copolymer. Since the chain length of the macromonomer determines the chain length of the graft, an important use of these compounds is in the synthesis of graft copolymers with well-defined graft lengths which are also known as polymer brushes.

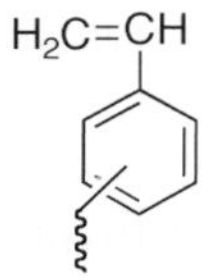

43 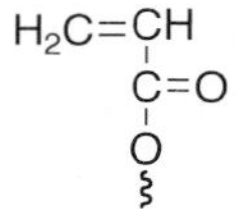   **44** 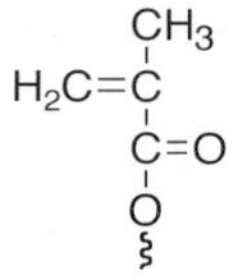   **45** 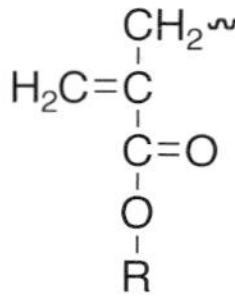   **46**

Various macromonomers have been described in the literature; many are based on polymers of S or (meth)acrylate esters [*e.g.* **43-46**]. The relative merits of macromonomers have been assessed in reviews by Hadjichristidis,[367] Capek and coworkers,[368,369] Ito and coworkers,[370,371] Meijs and Rizzardo,[372] Gnanou and Lutz[373] and Rempp and Franta[374]

Most macromonomers do not readily undergo homopolymerization or do so only sluggishly. The intrinsic reactivity of double bonds of macromonomers is often similar to that of the lower molecular weight monomers they resemble. However, the propagating species generated have low reactivity towards further propagation due to adverse steric factors. Oligomethacrylates (**46**) and similar macromonomers do not undergo homopolymerization or copolymerization with methacrylate esters because of competing addition-fragmentation chain transfer (Scheme 7.35, see also Section 6.2.3.4). On the other hand, with acrylates or S, copolymerization dominates over fragmentation at lower polymerization temperatures (<80 °C).[375] Higher reaction temperatures favor fragmentation to the extent that it is possible to synthesize block copolymers by this form of RAFT polymerization (Section 9.5.2).

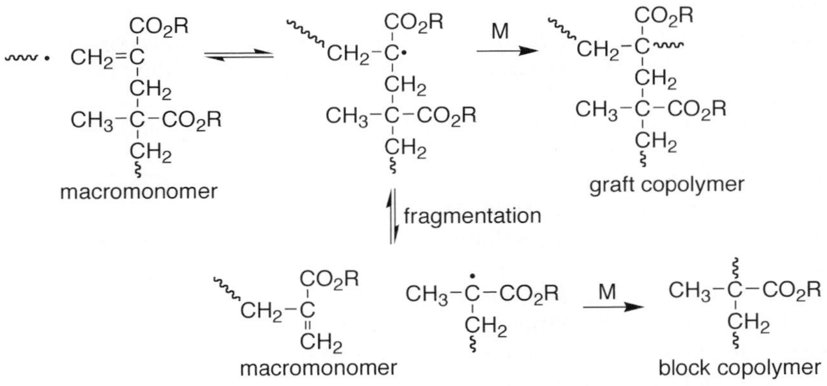

Scheme 7.35

The reactivity of macromonomers in copolymerization is strongly dependent on the particular comonomer-macromonomer pair. Solvent effects and the viscosity of the polymerization medium can also be important. Propagation may become diffusion controlled such that the propagation rate constant and reactivity ratios depend on the molecular weight of the macromonomer and the viscosity or, more accurately, the free volume of the medium.

Primary radical transfer may complicate the initiation process. Due to the low concentration of reactive double bonds, it is even more important than usual to select initiators with a low propensity for hydrogen atom abstraction. The greater viscosity of reaction media containing high concentrations of macromonomer can also cause reduced initiator efficiencies as compared to those for conventional polymerizations. Low rates of diffusion of propagating species may reduce rates of termination. A good solvent for macromonomer and polymer will facilitate interpenetration of the polymer chains by the monomer. The balance between these factors can lead to overall rates of copolymerization that are higher or lower than those of conventional radical copolymerization not involving macromonomers (Section 8.3.1). These factors are also largely responsible for the reactivity ratios of macromonomer showing significant solvent dependence.

7.7 References

1. Madruga, E.L. *Prog. Polym. Sci.* **2002**, *27*, 1879.
2. Coote, M.L.; Davis, T.P. In *Handbook of Radical Polymerization*; Davis, T.P.; Matyjaszewski, K., Eds.; John Wiley & Sons: Hoboken, 2002; p 263.
3. Coote, M.L.; Davis, T.P. *Prog. Polym. Sci.* **1999**, *24*, 1217.
4. Braun, D.; Czerwinski, W.K. In *Comprehensive Polymer Science*; Eastmond, G.C.; Ledwith, A.; Russo, S.; Sigwalt, P., Eds.; Pergamon: Oxford, 1989; Vol. 3, p 207.

5. Tirrell, D.A. In *Comprehensive Polymer Science*; Eastmond, G.C.; Ledwith, A.; Russo, S.; Sigwalt, P., Eds.; Pergamon: Oxford, 1989; Vol. 3, p 195.
6. Hamielec, A.E.; MacGregor, J.F.; Penlidis, A. In *Comprehensive Polymer Science*; Eastmond, G.C.; Ledwith, A.; Russo, S.; Sigwalt, P., Eds.; Pergamon: Oxford, 1989; Vol. 3, p 17.
7. Tirrell, D.A. In *Encyclopedia of Polymer Science and Engineering*, 2nd ed.; Mark, H.F.; Bikales, N.M.; Overberger, C.G.; Menges, G., Eds.; Wiley: New York, 1985; Vol. 4, p 192.
8. Ham, G.E. In *Copolymerization*; Ham, G.E., Ed.; John Wiley and Sons: New York, 1964; p 1.
9. Fukuda, T.; Kubo, K.; Ma, Y.-D. *Prog. Polym. Sci.* **1992**, *17*, 875.
10. Skeist, I. *J. Am. Chem. Soc.* **1946**, *68*, 1781.
11. Galbraith, M.N.; Moad, G.; Solomon, D.H.; Spurling, T.H. *Macromolecules* **1987**, *20*, 675.
12. O'Driscoll, K.F. *J. Coat. Technol.* **1983**, *55*, 57.
13. Hill, L.W.; Wicks, Z.W. *Prog. Org. Coat.* **1982**, *10*, 55.
14. Mayo, F.R.; Lewis, F.M. *J. Am. Chem. Soc.* **1944**, *66*, 1594.
15. Alfrey, T.; Goldfinger, G. *J. Chem. Phys.* **1944**, *12*, 205.
16. Wall, F.T. *J. Am. Chem. Soc.* **1944**, *66*, 2050.
17. Laurier, G.C.; O'Driscoll, K.F.; Reilly, P.M. *J. Polym. Sci., Polym. Symp.* **1985**, *72*, 17.
18. Greenley, R.Z. In *Polymer Handbook*, 4th ed.; Brandup, J.; Immergut, E.H.; Grulke, E.A., Eds.; John Wiley and Sons: New York, 1999; p II/181.
19. Eastmond, G.C. In *Comprehensive Chemical Kinetics*; Bamford, C.H.; Tipper, C.F.H., Eds.; Elsevier: Amsterdam, 1976; Vol. 14A, p 302.
20. Moad, G.; Solomon, D.H.; Spurling, T.H.; Vearing, D.J. *Aust. J. Chem.* **1986**, *39*, 1877.
21. Ham, G.E. *J. Macromol. Sci., Chem.* **1991**, *A28*, 733.
22. Merz, E.; Alfrey, T.; Goldfinger, G. *J. Polym. Sci.* **1946**, *1*, 75.
23. Fukuda, T.; Ma, Y.-D.; Inagaki, H. *Macromolecules* **1985**, *18*, 17.
24. Hill, D.J.T.; O'Donnell, J.H.; O'Sullivan, P.W. *Macromolecules* **1982**, *15*, 960.
25. Hill, D.J.T.; Lang, A.P.; O'Donnell, J.H.; O'Sullivan, P.W. *Eur. Polym. J.* **1989**, *25*, 911.
26. Lin, J.; Petit, A.; Neel, J. *Makromol. Chem.* **1987**, *188*, 1163.
27. Van Der Meer, R.; Alberti, J.M.; German, A.L.; Linssen, H.N. *J. Polym. Sci., Polym. Chem. Ed.* **1979**, *17*, 3349.
28. Hill, D.J.T.; O'Donnell, J.T.H.; O'Sullivan, P.W. *Macromolecules* **1985**, *18*, 9.
29. Brown, A.S.; Fujimora, K.; Craven, I. *Makromol. Chem.* **1988**, *189*, 1893.
30. Guillot, J.; Vialle, J.; Guyot, A. *J. Macromol. Sci., Chem.* **1971**, *A5*, 735.
31. Moad, G.; Solomon, D.H.; Spurling, T.H.; Vearing, D.J. *Aust. J. Chem.* **1985**, *38*, 1287.
32. Jones, S.A.; Prementine, G.S.; Tirrell, D.A. *J. Am. Chem. Soc.* **1985**, *107*, 5275.
33. Cywar, D.A.; Tirrell, D.A. *J. Am. Chem. Soc.* **1989**, *111*, 7544.
34. Giese, B.; Engelbrecht, R. *Polym. Bull.* **1984**, *12*, 55.
35. Fischer, H.; Radom, L. *Angew. Chem., Int. Ed. Engl.* **2001**, *40*, 1340.
36. Tanaka, H.; Sasai, K.; Sato, T.; Ota, T. *Macromolecules* **1988**, *21*, 3534.
37. Tanaka, H.; Sakai, I.; Sasai, K.; Sato, T.; Ota, T. *J. Polym. Sci., Part C: Polym. Lett.* **1988**, *26*, 11.
38. Kajiwara, A.; Nanda, A.K.; Matyjaszewski, K. *Macromolecules* **2004**, *37*, 1378.

39. Moad, G.; Solomon, D.H.; Spurling, T.H.; Johns, S.R.; Willing, R.I. *Aust. J. Chem.* **1986**, *39*, 43.
40. Bamford, C.H. *Polym. Commun.* **1989**, *30*, 36.
41. Harrisson, S.; Kapfenstein-Doak, H.; Davis, T.P. *Macromolecules* **2001**, *34*, 6214.
42. Chong, Y.K.; Krstina, J.; Le, T.P.T.; Moad, G.; Postma, A.; Rizzardo, E.; Thang, S.H. *Macromolecules* **2003**, *36*, 2256.
43. Moad, G.; Moad, C.L.; Rizzardo, E.; Thang, S.H. *Macromolecules* **1996**, *29*, 7717.
44. Fukuda, T.; Yoshikawa, C.; Kwak, Y.; Goto, A.; Tsujii, Y. *ACS Symp. Ser.* **2003**, *854*, 24.
45. Coote, M.L.; Zammit, M.D.; Davis, T.P.; Willet, G.D. *Macromolecules* **1997**, *30*, 8182.
46. Coote, M.L.; Johnston, L.P.M.; Davis, T.P. *Macromolecules* **1997**, *30*, 8191.
47. Kaim, A.; Oracz, P. *Macromol. Theory Simul.* **1997**, *6*, 565.
48. Schweer, J. *Makromol. Chem., Theory Simul.* **1993**, *2*, 485.
49. Moad, G.; Solomon, D.H.; Spurling, T.H.; Stone, R.A. *Macromolecules* **1989**, *22*, 1145.
50. Davis, T.P.; O'Driscoll, K.F.; Piton, M.C.; Winnik, M.A. *J. Polym. Sci., Part C: Polym. Lett.* **1989**, *27*, 181.
51. Olaj, O.F.; Bitai, I.; Hinkelmann, F. *Makromol. Chem.* **1987**, *188*, 1689.
52. Olaj, O.F.; Schnöll-Bitai, I.; Kremminger, P. *Eur. Polym. J.* **1989**, *25*, 535.
53. Davis, T.P.; O'Driscoll, K.F.; Piton, M.C.; Winnik, M.A. *Macromolecules* **1990**, *23*, 2113.
54. Davis, T.P.; O'Driscoll, K.F.; Piton, M.C.; Winnik, M.A. *Polym. Int.* **1991**, *24*, 65.
55. Piton, M.C.; Winnik, M.A.; Davis, T.P.; O'Driscoll, K.F. *J. Polym. Sci., Part A: Polym. Chem.* **1990**, *28*, 2097.
56. Aerdts, A.M.; de Haan, J.W.; German, A.L. *Macromolecules* **1993**, *26*, 1965.
57. Maxwell, I.A.; Aerdts, A.M.; German, A.L. *Macromolecules* **1993**, *26*, 1956.
58. Uebel, J.J.; Dinan, F.J. *J. Polym. Sci., Polym. Chem. Ed.* **1983**, *21*, 917.
59. Beuermann, S.; Buback, M. *Prog. Polym. Sci.* **2002**, *27*, 191.
60. van Herk, A.M.; Droge, T. *Macromol. Theory Simul.* **1997**, *6*, 1263.
61. Heuts, J.P.A.; Coote, M.; Davis, T.P.; Johnston, L.P.M. *ACS Symp. Ser.* **1998**, *685*, 120.
62. Fukuda, T.; Ma, Y.D.; Kubo, K.; Inagaki, H. *Macromolecules* **1991**, *24*, 370.
63. Evans, M.G.; Polanyi, M. *Trans. Faraday Soc.* **1938**, *34*, 11.
64. Heuts, J.P.A.; Clay, P.A.; Christie, D.I.; Piton, M.C.; Hutovic, J.; Kable, S.H.; Gilbert, R.G. *Prog. Pac. Polym. Sci* **1994**, *3*, 203.
65. Cowie, J.M.G. In *Comprehensive Polymer Science*; Eastmond, G.C.; Ledwith, A.; Russo, S.; Sigwalt, P., Eds.; Pergamon: Oxford, 1989; Vol. 4, p 377.
66. Cowie, J.M.G. In *Alternating Copolymers*; Cowie, J.M.G., Ed.; Plenum: New York, 1985; p 19.
67. Cowie, J.M.G. In *Alternating Copolymers*; Cowie, J.M.G., Ed.; Plenum: New York, 1985; p 1.
68. Hill, D.J.T.; O'Donnell, J.H.; O'Sullivan, P.W. *Prog. Polym. Sci.* **1982**, *8*, 215.
69. Furakawa, J. In *Encyclopedia of Polymer Science*, 2nd ed.; Mark, H.F.; Bikales, N.M.; Overberger, C.G.; Menges, G., Eds.; Wiley: New York, 1985; Vol. 4, p 233.
70. Shirota, Y. In *Encyclopedia of Polymer Science and Engineering*, 2nd ed.; Mark, H.F.; Bikales, N.M.; Overberger, C.G.; Menges, G., Eds.; New York: Wiley, 1985; Vol. 3, p 327.
71. Shirota, Y.; Mikawa, H. *J. Macromol. Sci., Rev. Macromol. Chem.* **1977**, *C16*, 129.

72. Ebdon, J.R.; Towns, C.R.; Dodgson, K. *J. Macromol. Sci., Rev. Macromol. Chem. Phys.* **1986**, *26*, 523.
73. Rätzsch, M.; Vogl, O. *Prog. Polym. Sci.* **1991**, *16*, 279.
74. Rzayev, Z.M.O. *Prog. Polym. Sci.* **2000**, *25*, 163.
75. Cais, R.E.; Farmer, R.G.; Hill, D.J.T.; O'Donnell, J.H. *Macromolecules* **1979**, *12*, 835.
76. Seiner, J.A.; Litt, M. *Macromolecules* **1971**, *4*, 308.
77. Pittman, C.U.; Rounsefell, T.D. *Macromolecules* **1975**, *8*, 46.
78. Hill, D.J.T.; O'Donnell, J.H.; O'Sullivan, P.W. *Macromol.* **1983**, *16*, 1295.
79. Tsuchida, E.; Tomono, T. *Makromol. Chem.* **1971**, *141*, 265.
80. Karad, P.; Schneider, C. *J. Polym. Sci., Polym. Chem. Ed.* **1978**, *16*, 1137.
81. Prementine, G.S.; Jones, S.A.; Tirrell, D.A. *Macromolecules* **1989**, *22*, 52.
82. Saito, J.; Tirrell, D.A. *Eur. Polym. J.* **1993**, *29*, 343.
83. Jones, S.A.; Tirrell, D.A. *J. Polym. Sci., Part A: Polym. Chem.* **1987**, *25*, 3177.
84. Bottle, S.E.; Busfield, W.K.; Grice, I.D.; Heiland, K.; Meutermans, W.; Monteiro, M. *Prog. Pac. Polym. Sci.* **1994**, *3*, 85.
85. Sawada, H. *J. Macromol. Sci., Rev. Macromol. Chem.* **1974**, *C11*, 257.
86. O'Driscoll, K.F.; Gasparro, F.P. *J. Macromol. Sci., Chem.* **1967**, *A1*, 643.
87. Wittmer, P. *Makromol. Chem.* **1967**, *103*, 188.
88. McManus, N.T.; Dube, M.A.; Penlidis, A. *Polym. React. Eng.* **1999**, *7*, 131.
89. Palmer, D.E.; McManus, N.T.; Penlidis, A. *J. Polym. Sci., Part A: Polym. Chem.* **2001**, *39*, 1753.
90. Palmer, D.E.; McManus, N.T.; Penlidis, A. *J. Polym. Sci., Part A: Polym. Chem.* **2000**, *38*, 1981.
91. Martinet, F.; Guillot, J. *J. Appl. Polym. Sci.* **1999**, *72*, 1611.
92. Martinet, F.; Guillot, J. *J. Appl. Polym. Sci.* **1997**, *65*, 2297.
93. Izu, M.; O'Driscoll, K.F.; Hill, R.J.; Quinn, M.J.; Harwood, H.J. *Macromolecules* **1972**, *5*, 90.
94. Wittmer, P. *Adv. Chem. Ser.* **1971**, *99*, 140.
95. Fischer, J.P. *Makromol. Chem.* **1972**, *155*, 211.
96. Hutchinson, R.A.; Paquet, D.A.; Beuermann, S.; McMinn, J.H. *Ind. Eng. Chem. Res.* **1998**, *37*, 3567.
97. Lowry, G.G. *J. Polym. Sci.* **1960**, *17*, 463.
98. Howell, J.A.; Izu, M.; O'Driscoll, K.F. *J. Polym. Sci., Part A-1* **1970**, *8*, 699.
99. Kruger, H.; Bauer, J.; Rubner, J. *Makromol. Chem.* **1987**, *188*, 2163.
100. Randall, J.C. *Polymer Sequence Determination*; Academic Press: New York, 1977.
101. Tonelli, A.E. *NMR Spectroscopy and Polymer Microstructure*; VCH: New York, 1989.
102. Koenig, J.L. *Chemical Microstructure of Polymer Chains*; Wiley: New York, 1980.
103. Bovey, F.A. *J. Polym. Sci.* **1962**, *62*, 197.
104. Kale, L.T.; O'Driscoll, K.F.; Dinan, F.J.; Uebel, J.J. *J. Polym. Sci. Part A* **1986**, *24*, 3145.
105. Lopez-Gonzalez, M.M.C.; Fernandez-Garcia, M.; Barreles-Rienda, J.M.; Madruga, E.M.; Arias, C. *Polymer* **1993**, *34*, 3123.
106. Klesper, E.; Johnsen, A.; Gronski, W.; Wehrli, F.W. *Makromol. Chem.* **1975**, *176*, 1071.
107. Johnsen, A.; Klesper, E.; Wirthlin, T. *Makromol. Chem.* **1976**, *177*, 2397.
108. Park, K.Y.; Santee, E.R.; Harwood, H.J. *Eur. Polym. J.* **1989**, *25*, 651.
109. Harwood, H.J. *Makromol. Chem., Macromol. Symp.* **1987**, *10/11*, 331.

110. Davis, T.P.; Heuts, J.P.A.; Barner-Kowollik, C.; Harrisson, S.; Morrison, D.A.; Yee, L.H.; Kapfenstein-Doak, H.M.; Coote, M.L. *Macromol. Symp.* **2002**, *182*, 131.
111. Alfrey, T.; Goldfinger, G. *J. Chem. Phys.* **1944**, *12*, 322.
112. Alfrey, T.; Goldfinger, G. *J. Chem. Phys.* **1946**, *14*, 115.
113. Moad, G.; Chiefari, J.; Mayadunne, R.T.A.; Moad, C.L.; Postma, A.; Rizzardo, E.; Thang, S.H. In *Macromol. Symp.*, 2002; Vol. 182, p 65.
114. Ham, G.E. *J. Macromol. Sci., Chem.* **1967**, *A1*, 93.
115. Tarasov, A.I.; Tskhai, V.A.; Spasski, S.S. *Vysokomol. Soedin.* **1960**, *2*, 1601.
116. Fineman, M.; Ross, S.D. *J. Polym. Sci.* **1950**, *5*, 259.
117. Tidwell, P.W.; Mortimer, G.A. *J. Macromol. Sci., Rev. Macromol. Chem.* **1970**, *C4*, 261.
118. Tidwell, P.W.; Mortimer, G.A. *J. Polym. Sci., Part A* **1965**, *3*, 369.
119. Kelen, T.; Tüdos, F.; Turcsányi, B. *Polym. Bull.* **1981**, *2*, 71.
120. Kelen, T.; Tüdos, F. *J. Macromol. Sci., Chem.* **1975**, *A9*, 1.
121. Greenley, R.Z. *J. Macromol. Sci., Chem* **1980**, *A14*, 445.
122. Greenley, R.Z. In *Polymer Handbook*, 3rd ed.; Brandup, J.; Immergut, E.H., Eds.; Wiley: New York, 1989; p II/153.
123. Meyer, V.E.; Lowry, G.G. *J. Polym. Sci., Part A* **1965**, *3*, 369.
124. Francis, A.P.; Solomon, D.H.; Spurling, T.H. *J. Macromol. Sci., Chem.* **1974**, *A8*, 469.
125. Van den Brink, M.; Van Herk, A.M.; German, A.L. *J. Polym. Sci., Part A: Polym. Chem.* **1999**, *37*, 3793.
126. Giz, A. *Macromol. Theory Simul.* **1998**, *7*, 391.
127. Plaumann, H.P.; Branston, R.E. *J. Polym. Sci., Part A: Polym. Chem.* **1989**, *27*, 2819.
128. Hautus, F.L.M.; Linssen, H.N.; German, A.L. *J. Polym. Sci., Polym. Chem. Ed.* **1984**, *22*, 3661.
129. Hautus, F.L.M.; Linssen, H.N.; German, A.L. *J. Polym. Sci., Polym. Chem. Ed.* **1984**, *22*, 3487.
130. Leicht, R.; Fuhrman, J. *J. Polym. Sci., Polym. Chem. Ed.* **1983**, *21*, 2215.
131. McFarlane, R.C.; Reilly, P.M.; O'Driscoll, K.F. *J. Polym. Sci., Polym. Chem. Ed.* **1980**, *18*, 251.
132. Burke, A.L.; Duever, T.A.; Penlidis, A. *Ind. Eng. Chem. Res.* **1997**, *36*, 1016.
133. O'Driscoll, K.F.; Kale, L.T.; Garcia Rubio, L.H.; Reilly, P.M. *J. Polym. Sci., Polym. Chem. Ed.* **1984**, *22*, 2777.
134. Patino-Leal, H.; Reilly, P.M.; O'Driscoll, K.F. *J. Polym. Sci., Polym. Lett. Ed.* **1980**, *18*, 219.
135. Van Der Meer, R.; Linssen, H.N.; German, A.L. *J. Polym. Sci., Polym. Chem. Ed.* **1978**, *16*, 2915.
136. Burke, A.L.; Duever, T.A.; Penlidis, A. *J. Polym. Sci., Part A: Polym. Chem.* **1993**, *31*, 3065.
137. Kelen, T.; Tüdos, F. *Makromol. Chem.* **1990**, *191*, 1863.
138. Hill, D.J.T.; Lang, A.P.; Munro, P.D.; O'Donnell, J.H. *Eur. Polym. J.* **1992**, *28*, 391.
139. Bovey, F.A. *Chain Structure and Conformation of Macromolecules*; Wiley: New York, 1982.
140. Hatada, K. *NMR Spectroscopy of Polymers*; Springer-Verlag: Berlin, 2003.
141. Rudin, A.; O'Driscoll, K.F.; Rumack, M.S. *Polymer* **1981**, *22*, 740.
142. Brown, P.G.; Fujimori, K. *Makromol. Chem., Rapid Commun.* **1993**, *14*, 677.
143. Moad, G.; Willing, R.I. *Polym. J.* **1991**, *23*, 1401.

144. Moad, G. In *Annual Reports in NMR Spectroscopy*; Webb, G.A., Ed.; Academic Press: London, 1994; Vol. 29, p 287.
145. Moad, G. *Chem. Aust.* **1991**, *58*, 122.
146. Jenkins, A.D. In *Reactivity, Mechanism and Structure in Polymer Chemistry*; Jenkins, A.D.; Ledwith, A., Eds.; Wiley: London, 1974; p 117.
147. Semchikov, Y.D. *Polym. Sci. USSR (Engl. Transl.)* **1990**, *32*, 177.
148. Alfrey, T.; Price, C.C. *J. Polym. Sci.* **1947**, *2*, 101.
149. Greenley, R.Z. In *Polymer Handbook*, 4th ed.; Brandup, J.; Immergut, E.H.; Grulke, E.A., Eds.; John Wiley and Sons: New York, 1999; p II/309.
150. Borchardt, J.K. *J. Macromol. Sci., Chem.* **1985**, *A22*, 1711.
151. Davis, T.P.; Rogers, S.C. *Eur. Polym. J.* **1993**, *29*, 1311.
152. Rogers, S.C.; Mackrodt, W.C.; Davis, T.P. *Polymer* **1994**, *35*, 1258.
153. Zhan, C.G.; Dixon, D.A. *J. Phys. Chem. A* **2002**, *106*, 10311.
154. Jenkins, A.D.; Jenkins, J. In *Polymer Handbook*, 4th ed.; Brandup, J.; Immergut, E.H.; Grulke, E.A., Eds.; John Wiley and Sons: New York, 1999; p II/321.
155. Bamford, C.H.; Jenkins, A.D.; Johnston, R. *Trans. Faraday Soc.* **1959**, *55*, 418.
156. Bamford, C.H.; Jenkins, A.D. *J. Polm. Sci.* **1961**, *59*, 530.
157. Bamford, C.H.; Jenkins, A.D. *Trans. Faraday Soc.* **1963**, *59*, 530.
158. Jenkins, A.D. *Eur. Polym. J.* **1989**, *25*, 721.
159. Jenkins, A.D.; Hatada, K.; Kitayama, T.; Nishiura, T. *J. Polym. Sci., Part A: Polym. Chem.* **2000**, *38*, 4336.
160. Jenkins, A.D. *Polymer* **1999**, *40*, 7045.
161. Olaj, O.F.; Zoder, M.; Vana, P.; Zifferer, G. *Macromolecules* **2004**, *37*, 1544.
162. Fukuda, T.; Goto, A.; Kwak, Y.; Yoshikawa, C.; Ma, Y.D. *Macromol. Symp.* **2002**, *182*, 53.
163. Melville, H.W.; Valentine, L. *Proc. R. Soc., London* **1950**, *A200*, 358.
164. Melville, H.W.; Valentine, L. *Proc. R. Soc., London* **1950**, *A200*, 337.
165. Mayo, F.R.; Walling, C. *Chem. Rev.* **1950**, *46*, 191.
166. Atherton, J.N.; North, A.M. *Trans. Faraday Soc.* **1962**, *58*, 2049.
167. North, A.M. In *Reactivity, Mechanism and Structure in Polymer Chemistry*; Jenkins, A.D.; Ledwith, A., Eds.; Wiley: London, 1974; p 142.
168. North, A.M.; Postlethwaite, D. *Polymer* **1964**, *5*, 237.
169. Chiang, S.S.M.; Rudin, A. *J. Macromol. Sci., Chem.* **1975**, *A9*, 237.
170. O'Driscoll, K.F.; Wertz, W.; Husar, A. *J. Polym. Sci., Part A-1* **1967**, *5*, 2159.
171. Russo, S.; Munari, S. *J. Macromol. Sci., Chem.* **1968**, *A2*.
172. Bonta, G.; Gallo, B.M.; Russo, S. *J. Chem. Soc., Faraday Trans. 1* **1975**, *69*, 1727.
173. Bonta, G.; Gallo, B.M.; Russo, S. *J. Chem. Soc., Faraday Trans. 1* **1973**, *69*, 329.
174. O'Driscoll, K.F.; Huang, J. *Eur. Polym. J.* **1989**, *7/8*, 629.
175. Kelly, D.P.; Serelis, A.K.; Solomon, D.H.; Thompson, P.E. *Aust. J. Chem.* **1987**, *40*, 1631.
176. Ito, K. *Polymer* **1985**, *26*, 1253.
177. Guth, W.; Heitz, W. *Makromol. Chem.* **1976**, *177*, 1835.
178. Moad, G.; Serelis, A.K.; Solomon, D.H.; Spurling, T.H. *Polym. Commun.* **1984**, *25*, 240.
179. Serelis, A.K. *Personal Communication*.
180. Heitz, W. In *Telechelic Polymers: Synthesis and Applications*; Goethals, E.J., Ed.; CRC Press: Boca Raton, Florida, 1989; p 61.
181. Bevington, J.C.; Melville, H.W.; Taylor, R.P. *J. Polym. Sci.* **1954**, *14*, 463.
182. Chen, C.Y.; Wu, Z.Z.; Kuo, J.F. *Polym. Eng. Sci.* **1987**, *27*, 553.

183. Ohtani, H.; Suzuki, A.; Tsuge, S. *J. Polym. Sci., Part A: Polym. Chem.* **2000**, *38*, 1880.
184. Konter, W.; Bömer, B.; Köhler, K.H.; Heitz, W. *Makromol. Chem.* **1981**, *182*, 2619.
185. Kodaira, T.; Ito, K.; Iyoda, S. *Polym. Commun.* **1988**, *29*, 83.
186. Moad, G.; Solomon, D.H.; Johns, S.R.; Willing, R.I. *Macromolecules* **1984**, *17*, 1094.
187. Barton, J.; Capek, I.; Juranicova, V.; Riedel, S. *Makromol. Chem., Rapid Commun.* **1986**, *7*, 521.
188. Colombani, D.; Chaumont, P. *Acta Polym.* **1998**, *49*, 225.
189. Tezuka, Y. *Prog. Polym. Sci.* **1992**, *17*, 471.
190. Ebdon, J.R. In *New methods of Polymer Synthesis*; Ebdon, J.R., Ed.; Blackie: Glasgow, 1991; p 162.
191. Boutevin, B. *Adv. Polym. Sci.* **1990**, *94*, 69.
192. Nguyen, H.A.; Marechal, E. *J. Macromol. Sci., Rev. Macromol. Chem. Phys.* **1988**, *C28*, 187.
193. Brosse, J.C.; Derouet, D.; Epaillard, F.; Soutif, J.-C.; Legeay, G.; Dusek, K. *Adv. Polym. Sci.* **1987**, *81*, 167.
194. French, D.A. *Rubber. Chem. Technol.* **1969**, *42*, 71.
195. Tobolsky, A.V. *J. Am. Chem. Soc.* **1958**, *80*, 5927.
196. David, G.; Robin, J.J.; Boutevin, B. *J. Polym. Sci., Part A: Polym. Chem.* **2001**, *39*, 2740.
197. David, G.; Boutevin, B.; Robin, J.J.; Loubat, C.; Zydowicz, N. *Polym. Int.* **2002**, *51*, 800.
198. Bamford, C.H.; Jenkins, A.D.; Wayne, R.P. *Trans. Faraday Soc.* **1960**, *56*, 932.
199. Reed, S.F. *J. Polym. Sci., Polym. Chem. Ed.* **1973**, *11*, 55.
200. Bamford, C.H.; Jenkins, A.D.; Johnston, R. *Trans. Faraday Soc.* **1959**, *55*, 179.
201. Idage, B.B.; Vernekar, S.P.; Ghatge, N.D. *J. Polym. Sci., Polym. Chem. Ed.* **1983**, *21*, 385.
202. Ghatge, N.D.; Vernekar, S.P.; Wadgaonkar, P.P. *Makromol. Chem., Rapid Commun.* **1983**, *4*, 307.
203. Friedlander, H.N. *J. Polym. Sci.* **1962**, *58*, 455.
204. Bresler, L.S.; Barantsevich, E.N.; Polyansky, V.I. *Makromol. Chem.* **1982**, *183*, 2479.
205. Edelmann, D.; Ritter, H. *Makromol. Chem.* **1993**, *194*, 2375.
206. Nair, C.P.R.; Clouet, G. *J. Macromol. Sci., Rev. Macromol. Chem. Phys.* **1991**, *C31*, 311.
207. Corner, T. *Adv. Polym. Sci.* **1984**, *62*, 95.
208. Starks, C.M. *Free Radical Telomerization*; Academic Press: New York, 1974.
209. Boutevin, B.; Lusinchi, J.-M.; Pietrasanta, Y.; Robin, J.-J. *Eur. Polym. J.* **1994**, *30*, 615.
210. Pryor, W.A.; Coco, J.H. *Macromolecules* **1970**, *3*, 500.
211. Deady, M.; Mau, A.W.H.; Moad, G.; Spurling, T.H. *Makromol. Chem.* **1993**, *194*, 1691.
212. Meijs, G.F.; Morton, T.C.; Rizzardo, E.; Thang, S.H. *Macromolecules* **1991**, *24*, 3689.
213. Colombani, D. *Prog. Polym. Sci.* **1997**, *22*, 1649.
214. Chiefari, J.; Rizzardo, E. In *Handbook of Radical Polymerization*; Davis, T.P.; Matyjaszewski, K., Eds.; John Wiley & Sons: Hoboken, NY, 2002; p 263.

215. Rizzardo, E.; Chong, Y.K.; Evans, R.A.; Moad, G.; Thang, S.H. *Macromol. Symp.* **1996**, *111*, 1.
216. Hutson, L.; Krstina, J.; Moad, C.L.; Moad, G.; Morrow, G.R.; Postma, A.; Rizzardo, E.; Thang, S.H. *Macromolecules* **2004**, *37*, 4441.
217. Businelli, L.; Deleuze, H.; Gnanou, Y.; Maillard, B. *Macromol. Chem. Phys.* **2000**, *201*, 1833.
218. Boutevin, B.; El Idrissi, A.; Parisi, J.P. *Makromol. Chem.* **1990**, *191*, 445.
219. Boutevin, B.; Pietrasanta, Y. *Makromol. Chem.* **1985**, *186*, 817.
220. Kharasch, M.S.; Read, J.; Mayo, F.R. *Chem. Ind. (London)* **1938**, *57*, 752.
221. Klemm, E.; Sensfuss, S. *J. Macromol. Sci., Chem.* **1991**, *A28*, 875.
222. Cramer, N.B.; Bowman, C.N. *J. Polym. Sci., Part A: Polym. Chem.* **2001**, *39*, 3311.
223. Cramer, N.B.; Reddy, S.K.; O'Brien, A.K.; Bowman, C.N. *Macromolecules* **2003**, *36*, 7964.
224. Nuyken, O.; Völkel, T. *Makromol. Chem.* **1990**, *191*, 2465.
225. Nuyken, O.; Völkel, T. *Makromol. Chem., Rapid Commun.* **1990**, *11*, 365.
226. Morgan, C.R.; Magnotta, F.; Ketley, A.D. *J. Polym. Sci., Part A: Polym. Chem.* **1977**, *15*, 627.
227. Bailey, W.J.; Endo, T.; Gapud, B.; Lin, Y.-N.; Ni, Z.; Pan, C.-Y.; Shaffer, S.E.; Wu, S.-R.; Yamazaki, N.; Yonezawa, K. *J. Macromol. Sci., Chem.* **1984**, *A21*, 979.
228. Rimmer, S.; Ebdon, J.R. *J. Polym. Sci., Part A: Polym. Chem.* **1996**, *34*, 3573.
229. Ebdon, J.R.; Flint, N.J. *Eur. Polym. J.* **1996**, *32*, 289.
230. Ebdon, J.R.; Flint, N.J. *J. Polym. Sci., Part A: Polym. Chem.* **1995**, *33*, 593.
231. Ebdon, J.R.; Flint, N.J.; Hodge, P. *Eur. Polym. J.* **1989**, *25*, 759.
232. Ebdon, J.R. *Macromol. Symp.* **1994**, *84*, 45.
233. Gridnev, A.A.; Simonsick, W.J.; Ittel, S.D. *J. Polym. Sci. Pol. Chem.* **2000**, *38*, 1911.
234. Chiefari, J.; Jeffery, J.; Mayadunne, R.T.A.; Moad, G.; Rizzardo, E.; Thang, S.H. *ACS Symp. Ser.* **2000**, *768*, 297.
235. Kukulj, D.; Heuts, J.P.A.; Davis, T.P. *Macromolecules* **1998**, *31*, 6034.
236. Moad, G.; Chiefari, J.; Moad, C.L.; Postma, A.; Mayadunne, R.T.A.; Rizzardo, E.; Thang, S.H. *Macromol. Symp.* **2002**, *182*, 65.
237. Chiefari, J.; Jeffery, J.; Moad, C.L.; Moad, G.; Postma, A.; Rizzardo, E.; Thang, S.H. *Macromolecules* **2005**, in press.
238. Montaudo, M.S. *Macromolecules* **2001**, *34*, 2792.
239. Montaudo, M.S. *Polymer* **2002**, *43*, 1587.
240. Provder, T.; Whited, M.; Huddleston, D.; Kuo, C.Y. *Prog. Org. Coat.* **1997**, *32*, 155.
241. Dawkins, J.V. *Adv. Chem. Ser.* **1995**, *247*, 197.
242. Pasch, H. *Adv. Polym. Sci.* **1997**, *128*, 1.
243. Berek, D. *Prog. Polym. Sci.* **2000**, *25*, 873.
244. Macko, T.; Hunkeler, D. *Adv. Polym. Sci.* **2003**, *163*, 61.
245. Pasch, H.; Mequanint, K.; Adrian, J. *e-Polymers* **2002**, *005*.
246. Spinelli, H.J. *Am. Chem. Soc., Org. Coat. Plas. Chem., Reprints.* **1982**, 529.
247. Stockmayer, W.H. *J. Chem. Phys.* **1945**, *13*, 199.
248. Fueno, T.; Fukawa, J. *J. Polym. Sci., Part A* **1964**, *2*, 3681.
249. Mirabella, F.M., Jr. *Polymer* **1977**, *18*, 705.
250. Spurling, T.H.; Deady, M.; Krstina, J.; Moad, G. *Makromol. Chem., Macromol. Symp.* **1991**, *51*, 127.
251. Remmp, P.F.; Lutz, P.J. In *Comprehensive Polymer Science*; Eastmond, G.C.; Ledwith, A.; Russo, S.; Sigwalt, P., Eds.; Pergamon: Oxford, 1989; Vol. 4, p 403.

252. Simionescu, C.; Comanita, E.; Pastravanu, M.; Dumitriu, S. *Prog. Polym. Sci.* **1986**, *12*, 1.
253. Nuyken, O.; Weidner, R. *Adv. Polym. Sci.* **1986**, *73*, 145.
254. Kuchanov, S.I. In *Comprehensive Polymer Science*; Agarwal, S.L.; Russo, S., Eds.; Pergamon: Oxford, 1992; Vol. Suppl. 1, p 23.
255. Ameduri, B.; Boutevin, B.; Gramain, P. *Adv. Polym. Sci.* **1997**, *127*, 87.
256. Piirma, I.; Chou, L.P.H. *J. Appl. Polym. Sci.* **1979**, *24*, 2051.
257. Nukyen, O.; Weidner, R. *Makromol. Chem.* **1988**, *189*, 1331.
258. Qiu, X.-Y.; Ruland, W.; Heitz, W. *Angew. Makromol. Chem.* **1984**, *125*, 69.
259. Yagci, Y.; Tunca, U.; Biçak, N. *J. Polym. Sci., Part C: Polym. Lett.* **1986**, *24*, 49.
260. Yagci, Y.; Onen, A.; Schnabel, W. *Macromolecules* **1991**, *24*, 4620.
261. Hazar, B.; Baysal, B.M. *Polymer* **1986**, *27*, 961.
262. Walz, R.; Bomer, B.; Heiz, W. *Makromol. Chem.* **1977**, *178*, 2527.
263. Hoffmann, A.S.; Backsai, R. In *Copolymerization*; Ham, G.E., Ed.; John Wiley and Sons: New York, 1964; p 335.
264. O'Driscoll, K.F.; Bevington, J.C. *Eur. Polym. J.* **1985**, *21*, 1039.
265. Choi, K.Y.; Lee, G.D. *AIChE J.* **1987**, *33*, 2067.
266. Jenkins, D.W.; Hudson, S.M. *Chem. Rev.* **2001**, *101*, 3245.
267. McDowell, D.J.; Gupta, B.S.; Stannett, V.T. *Prog. Polym. Sci.* **1984**, *10*, 1.
268. Stewart, M.J. In *New Methods of Polymer Synthesis*; Ebdon, J.R., Ed.; Blackie: Glasgow, 1991; p 107.
269. Schue, F. In *Comprehensive Polymer Science*; Eastmond, G.C.; Ledwith, A.; Russo, S.; Sigwalt, P., Eds.; Pergamon: Oxford, 1989; Vol. 4, p 359.
270. Abadie, M.J.M.; Ourahmoune, D. *Br. Polym. J.* **1987**, *19*, 247.
271. Eastmond, G.C. *Pure & Appl. Chem.* **1981**, *53*, 657.
272. Riess, G.; Reeb, R. *ACS Symp. Ser.* **1981**, *166*, 477.
273. Abadie, M.J.M.; Ourahmoune, D.; Mendjel, H. *Eur. Polym. J.* **1990**, *26*, 515.
274. Ren, Q.; Zhang, H.; Zhang, X.; Huang, B. *J. Polym. Sci., Part A: Polym. Chem.* **1993**, *31*, 847.
275. Tung, L.H.; Lo, G.Y.S.; Griggs, J.A. *J. Polym. Sci., Polym. Chem. Ed.* **1985**, *23*, 1551.
276. Lindsell, W.E.; Service, D.M.; Soutar, I.; Richards, D.H. *Br. Polym. J.* **1987**, *19*, 255.
277. Cunliffe, A.V.; Hayes, G.F.; Richards, D.H. *J. Polym. Sci., Polym. Lett. Ed.* **1976**, *14*, 483.
278. Abadie, M.J.M.; Schue, F.; Souel, T.; Richards, D.H. *Polymer* **1981**, *22*, 1076.
279. Abadie, M.; Burgess, F.J.; Cunliffe, A.V.; Richards, D.H. *J. Polym. Sci., Polym. Lett. Ed.* **1976**, *14*, 477.
280. Bamford, C.H.; Eastmond, G.C.; Woo, J.; Richards, D.H. *Polymer* **1982**, *23*, 643.
281. Bamford, C.H.; Dyson, R.W.; Eastmond, G.C. *Polymer* **1969**, *10*, 885.
282. Eastmond, G.C.; Parr, K.J.; Woo, J. *Polymer* **1988**, *29*, 950.
283. Eastmond, G.C.; Woo, J. *Polymer* **1990**, *31*, 358.
284. Eastmond, G.C.; Grigor, J. *Makromol. Chem., Rapid Commun.* **1986**, *7*, 375.
285. Alimoglu, A.K.; Bamford, C.H.; Ledwith, A.; Mullik, S.U. *Polym. Sci. USSR (Engl. Transl.)* **1980**, *21*, 2651.
286. Bamford, C.H.; Middleton, I.P.; Al-Lamee, K.G.; Paprotny, J. *Br. Polym. J.* **1987**, *19*, 269.
287. Dargaville, T.R.; George, G.A.; Hill, D.J.T.; Whittaker, A.K. *Prog. Polym. Sci.* **2003**, *28*, 1355.
288. Kabanov, V.Y.; Kudryavtsev, V.N. *High Energy Chem.* **2003**, *37*, 1.

289. Kato, K.; Uchida, E.; Kang, E.T.; Uyama, Y.; Ikada, Y. *Prog. Polym. Sci.* **2003**, *28*, 209.
290. Bhattacharya, A. *Prog. Polym. Sci.* **2000**, *25*, 371.
291. Kabanov, V.Y.; Aliev, R.E.; Kudryavtsev, V.N. *Radiat. Phys. Chem.* **1991**, *37*, 175.
292. Russell, K.E. *Prog. Polym. Sci.* **2002**, *27*, 1007.
293. Moad, G. *Prog. Polym. Sci.* **1999**, *24*, 81.
294. Xanthos, M. *Reactive Extrusion*; Hanser: Munich, 1992.
295. Al-Malaika, S., Ed. *Reactive Modifiers for Polymers*; Chapman & Hall: London, 1996.
296. Ratzsch, M.; Arnold, M.; Borsig, E.; Bucka, H.; Reichelt, N. *Prog. Polym. Sci.* **2002**, *27*, 1195.
297. Baker, W.E.; Scott, C.E.; Hu, G.-H. *Reactive Polymer Blending*; Hanser: Munich, 2001.
298. Hu, G.H.; Flat, J.-J.; Lambla, M. In *Reactive Modifiers for Polymers*; Al-Malaika, S., Ed.; Chapman & Hall: London, 1996; p 1.
299. Xie, H.-Q.; Baker, W.E. In *New Advances in Polyolefins*; Chung, T.C., Ed.; Plenum: New York, N. Y., 1993; p 101.
300. Wong, B.; Baker, W.E. *Annu. Tech. Conf. - Soc. Plast. Eng.* **1996**, *54(1)*, 283.
301. Huang, H.; Liu, N.C. *J. Appl. Polym. Sci.* **1998**, *67*, 1957.
302. Gaylord, N.G. In *Reactive Extrusion*; Xanthos, M., Ed.; Hanser: Munich, 1992; p 55.
303. Heinen, W.; Rosenmöller, C.H.; Wenzel, C.B.; de Groot, H.J.M.; Lugtenburg, J. *Macromolecules* **1996**, *29*, 1151.
304. Zhang, M.Z.; Dhuamel, J.; van Duin, M.; Meessen, P. *Macromolecules* **2004**, *37*, 1877.
305. De Roover, B.; Sclavons, M.; Carlier, V.; Devaux, J.; Legras, R.; Momatz, A. *J. Polym. Sci., Part A: Polym. Chem.* **1995**, *33*, 829.
306. De Roover, B.; Devaux, J.; Legras, R. *J. Polym. Sci., Part A: Polym. Chem.* **1996**, *34*, 1195.
307. Gaylord, N.G.; Mehta, R. *J. Appl. Polym. Sci.* **1989**, *38*, 359.
308. Bray, T.; Damiris, S.; Grace, A.; Moad, G.; O'Shea, M.; Rizzardo, E.; van Diepen, G. *Macromol. Symp.* **1997**, *129*, 109.
309. Samay, G.; Nagy, T.; White, J.L. *J. Appl. Polym. Sci.* **1995**, *56*, 1423.
310. Gaylord, G.N.; Mehta, R.; Mohan, D.R.; Kumar, V. *J. Appl. Polym. Sci.* **1992**, *44*, 1941.
311. Gaylord, N.G.; Mehta, N.; Mehta, R. *J. Appl. Polym. Sci.* **1987**, *33*, 2549.
312. Kowalski, R.C. In *Reactive Extrusion*; Xanthos, M., Ed.; Hanser: Munich, 1992; p 7.
313. Hu, G.H.; Flat, J.J.; Lambla, M. *Makromol. Chem., Macromol. Symp.* **1993**, *75*, 137.
314. Hu, G.-H.; Flat, J.-J.; Lambla, M. *Annu. Tech. Conf. - Soc. Plast. Eng.* **1994**, *52(3)*, 2775.
315. Coiai, S.; Passaglia, E.; Aglietto, M.; Ciardelli, F. *Macromolecules* **2004**, *37*, 8414.
316. Gaylord, N.G.; Mehta, R. *J. Polym. Sci., Part A: Polym. Chem.* **1988**, *26*, 1189.
317. Gaylord, N.G.; Deshpande, A.B. *Polym. Mater. Sci. Eng.* **1992**, *67*, 109.
318. Gaylord, G.N.; Mishra, M.K. *J. Polym. Sci., Polym. Lett. Ed.* **1983**, *21*, 23.
319. Wu, C.H.; Su, A.C. *Polymer* **1992**, *33*, 1987.
320. Ruggeri, G.; Aglietto, M.; Petragnani, A.; Ciardelli, F. *Eur. Polym. J.* **1983**, *19*, 863.
321. Benedetti, E.; Aldo, D.A.; Aglietto, M.; Ruggeri, G.; Vergamini, P.; Ciadelli, F. *Polym. Eng. Sci.* **1986**, *26*, 9.

322. Aglietto, M.; Bertani, R.; Ruggeri, G.; Segre, A.L. *Macromolecules* **1990**, *23*, 1928.
323. Aglietto, M.; Bertani, R.; Ruggeri, G. *Makromol. Chem.* **1992**, *193*, 179.
324. Konar, J.; Sen, A.K.; Bhowmick, A.K. *J. Appl. Polym. Sci.* **1993**, *48*, 1579.
325. Rosales, C.; Marques, L.; Gonzalez, J.; Perera, R.; Rojas, B.; Vivas, M. *Polym. Eng. Sci.* **1996**, *36*, 2247.
326. Rosales, C.; Perera, R.; Ichazo, M.; Gonzalez, J.; Rojas, H.; Sanchez, A.; Barrios, A.D. *J. Appl. Polym. Sci.* **1998**, *70*, 161.
327. Greco, R.; Musto, P.; Scarinzi, G. *Polym. Mater. Sci. Eng.* **1987**, *57*, 770.
328. Greco, R.; Riva, F.; Musto, P.V.; Maglio, G. *J. Appl. Polym. Sci.* **1989**, *37*, 789.
329. Greco, R.; Maglio, G.; Musto, P.V.; Scarinzi, G. *J. Appl. Polym. Sci.* **1989**, *37*, 777.
330. Garcia-Martinez, J.M.; Laguna, O.; Collar, E.P. *J. Appl. Polym. Sci.* **1997**, *65*, 1333.
331. Al-Malaika, S. *Polym.-Plast. Technol. Eng.* **1990**, *29*, 73.
332. Vainio, T.; Hu, G.-H.; Lambla, M.; Seppala, J.V. *J. Appl. Polym. Sci.* **1997**, *63*, 883.
333. Vainio, T.; Hu, G.-H.; Lambla, M.; Seppala, J.V. *J. Appl. Polym. Sci.* **1996**, *61*, 843.
334. Liu, N.C.; Xie, H.Q.; Baker, W.E. *Polymer* **1993**, *34*, 4680.
335. Oromehie, A.R.; Hashemi, S.A.; Meldrum, I.G.; Waters, D.N. *Polym. Int.* **1997**, *42*, 117.
336. Pesetskii, S.S.; Jurkowski, B.; Krivoguz, Y.M.; Urbanowicz, R. *J. Appl. Polym. Sci.* **1997**, *65*, 1493.
337. Song, Z.; Baker, W.E. *Polymer* **1992**, *33*, 3266.
338. Song, Z.; Baker, W.E. *J. Appl. Polym. SciJ. Polym. Sci., Part A: Polym. Chem.* **1992**, *44*, 2167.
339. Simmons, A.; Baker, W.E. *Polym. Eng. Sci.* **1989**, *29*, 1117.
340. Oliphant, K.E.; Baker, W.E. *Annu. Tech. Conf. - Soc. Plast. Eng.* **1994**, *52(2)*, 1524.
341. Wong Shing, J.B.; Baker, W.E.; Russell, K.E.; Whitney, R.A. *J. Polym. Sci., Part A: Polym. Chem.* **1994**, *32*, 1691.
342. Wong Shing, J.B.; Baker, W.E.; Russell, K.E. *J. Polym. Sci., Part A: Polym. Chem.* **1995**, *33*, 633.
343. Galluci, R.R.; Going, R.C. *J. Appl. Polym. Sci.* **1982**, *27*, 425.
344. Chen, L.-F.; Wong, B.; Baker, W.E. *Polym. Eng. Sci.* **1996**, *36*, 1594.
345. Sun, Y.-J.; Hu, G.-H.; Lambla, M. *Angew. Makromol. Chem* **1995**, *229*, 1.
346. Sun, Y.-J.; Hu, G.-H.; Lambla, M. *J. Appl. Polym. Sci.* **1995**, *57*, 1043.
347. Cartier, H.; Hu, G.H. *J. Polym. Sci., Part A: Polym. Chem.* **1998**, *36*, 1053.
348. Zhang, X.; Yin, Z.; Li, L.; Yin, J. *J. Appl. Polym. Sci.* **1996**, *61*, 2253.
349. Liu, T.M.; Evans, R.; Baker, W.E. *Annu. Tech. Conf. - Soc. Plast. Eng.* **1995**, *53(2)*, 1564.
350. Pesneau, I.; Champagne, M.F.; Huneault, M.A. *J. Appl. Polym. Sci.* **2004**, *91*, 3180.
351. Torres, N.; Robin, J.J.; Boutevin, B. *J. Appl. Polym. Sci.* **2001**, *81*, 581.
352. Cartier, H.; Hu, G.H. *J. Polym. Sci., Part A: Polym. Chem.* **1998**, *36*, 2763.
353. Al-Malaika, S.; Ibrahim, A.Q.; Rao, J.; Scott, G. *J. Appl. Polym. Sci.* **1992**, *44*, 1287.
354. Al-Malaika, S.; Scott, G.; Wirjosentono, B. *Polym. Degrad. Stab.* **1993**, *40*, 233.
355. Al-Malaika, S.; Suharty, N. *Polym. Degrad. Stab.* **1995**, *49*, 77.
356. Wang, X.; Tzoganakis, C.; Rempel, G.L. *J. Appl. Polym. Sci.* **1996**, *61*, 1395.
357. Kim, B.K. *Korea Polym. J.* **1996**, *4*, 215.
358. Al-Malaika, S. In *Reactive Modifiers for Polymers*; Al-Malaika, S., Ed.; Chapman & Hall: London, 1996; p 266.
359. Kim, B.S.; Kim, S.C. *J. Appl. Polym. Sci.* **1998**, *69*, 1307.
360. Hu, G.H.; Li, H.X.; Feng, L.F.; Pessan, L.A. *J. Appl. Polym. Sci.* **2003**, *88*, 1799.
361. Braun, D.; Schmitt, M.W. *Polym. Bull.* **1998**, *40*, 189.

362. Munteanu, D. In *Reactive Modifiers for Polymers*; Al-Malaika, S., Ed.; Chapman & Hall: London, 1996; p 196.
363. Forsyth, J.C.; Baker, W.E.; Russell, K.E.; Whitney, R.A. *J. Polym. Sci., Part A: Polym. Chem.* **1997**, *35*, 3517.
364. Rösch, J.; Holger, W.; Müller, P.; Schäfer, R.; Wörner, C.; Friedrich, C.; Kressler, J.; Mülhaupt, R. *Macromol. Symp.* **1996**, *102*, 241.
365. Liu, N.C.; Baker, W.E. In *Reactive Modifiers for Polymers*; Al-Malaika, S., Ed.; Chapman & Hall: London, 1996; p 163.
366. Liu, N.C.; Baker, W.E. *Polymer* **1994**, *35*, 988.
367. Hadjichristidis, N.; Pitsikalis, M.; Iatrou, H.; Pispas, S. *Macromol. Rapid Commun.* **2003**, *24*, 979.
368. Capek, I. *Adv Polym. Sci.* **1999**, *145*, 1.
369. Capek, I.; Akashi, M. *J. Macromol. Sci., Rev. Macromol. Chem. Phys.* **1993**, *C33*, 369.
370. Ito, K.; Kawaguchi, S. *Adv. Polym. Sci.* **1999**, *142*, 129.
371. Ito, K. *Prog. Polym. Sci.* **1998**, *23*, 581.
372. Meijs, G.F.; Rizzardo, E. *J. Macromol. Sci., Rev. Macromol. Chem. Phys.* **1990**, *C30*, 305.
373. Gnanou, Y.; Lutz, P. *Makromol. Chem.* **1989**, *190*, 577.
374. Rempp, P.F.; Franta, E. *Adv. Polym. Sci.* **1984**, *58*, 1.
375. Cacioli, P.; Hawthorne, D.G.; Laslett, R.L.; Rizzardo, E.; Solomon, D.H. *J. Macromol. Sci., Chem.* **1986**, *A23*, 839.

8
Controlling Polymerization

8.1 Introduction

Radical polymerization is often the preferred mechanism for forming polymers and most commercial polymer materials involve radical chemistry at some stage of their production cycle. From both economic and practical viewpoints, the advantages of radical over other forms of polymerization are many (Chapter 1). However, one of the often-cited "problems" with radical polymerization is a perceived lack of control over the process: the inability to precisely control molecular weight and distribution, limited capacity to make complex architectures and the range of undefined defect structures and other forms of "structure irregularity" that may be present in polymers prepared by this mechanism. Much research has been directed at providing answers for problems of this nature. In this, and in the subsequent chapter, we detail the current status of the efforts to redress these issues. In this chapter, we focus on how to achieve control by appropriate selection of the reaction conditions in "conventional" radical polymerization.

Minor (by amount) functionality is introduced into polymers as a consequence of the initiation, termination and chain transfer processes (Chapters 3, 5 and 6 respectively). These groups may either be at the chain ends (as a result of initiation, disproportionation, or chain transfer,) or they may be part of the backbone (as a consequence of termination by combination or the copolymerization of byproducts or impurities). In Section 8.2 we consider three polymers (PS, PMMA and PVC) and discuss the types of defect structure that may be present, their origin and influence on polymer properties, and the prospects for controlling these properties through appropriate selection of polymerization conditions.

Structural irregularity is also introduced in the propagation step either through a lack of regio- or stereochemical specificity in radical addition to monomer or by rearrangement of the propagating species (Section 4.4). In Section 8.3 the influence of the reaction media and added reagents on the stereochemistry and rate of radical polymerization is explored. With this knowledge we consider the prospects for controlling polymer structure and properties by appropriate choice of reaction conditions (solvent, temperature, pressure) or through the use of complexing agents and templates to direct the course of polymerization.[1]

8.2 Controlling Structural Irregularities

The functional groups introduced into polymer chains as a consequence of the initiation or termination processes can be of vital importance in determining certain polymer properties. Some such functionality is generally unavoidable. However, the types of functionality can be controlled through selection of initiator, solvent and reaction conditions and should not be ignored.

Such functionality can also be of great practical importance since functional initiators, transfer agents, *etc.* are applied to prepare end-functional polymers (see Section 7.5) or block or graft copolymers (Section 7.6). In these cases the need to maximize the fraction of chains that contain the reactive or other desired functionality is obvious. However, there are also well-documented cases where "weak links" formed by initiation, termination, or abnormal propagation processes impair the thermal or photochemical stability of polymers.

Thus, it is important to know, understand and control the kinetics and mechanism of the entire polymerization process so that desirable aspects of the polymer structure can be maximized while those reactions that lead to an impairment of properties or a less than ideal functionality can be avoided or minimized. A corollary is that it is important to know how a particular polymer was prepared before using it in a critical application.

8.2.1 "Defect Structures" in Polystyrene

There is a substantial literature on the thermal and photochemical degradation of PS and it is well established that polymer properties are sensitive to the manner in which a particular sample of PS is prepared. For example, it has been reported that PS prepared by anionic polymerization shows enhanced stability with respect to that prepared by a radical mechanism.[2-10] This has often been attributed to the presence of "weak links" in the latter polymers. However, the precise nature of the "weak links" remains the subject of some controversy. The situation is further confused by all PS prepared by radical mechanisms often being considered as a class without reference to the particular polymerization conditions employed in their preparation. In many cases the polymers are "commercial samples" with details of the method of preparation incomplete or unstated.

In some cases the "weak links" in radical PS may be peroxidic linkages.[11,12] Such groups may become incorporated in polymers formed by radical polymerization through copolymerization of adventitious oxygen (Section 5.3.2). Peroxidic linkage may be avoided by paying careful attention to monomer purification and rigorous exclusion of oxygen from the polymerization. Head-to-head linkages, such as those formed by termination by combination, have been proposed as a source of thermal instability.[8] However, there is also evidence that thermal behavior depends on the particular radical initiator or reaction conditions (solvent, temperature, conversion) employed in polymer preparation. It also appears that in some cases the thermal degradation of radical PS can be interpreted in terms of initiation by random chain scission uncomplicated by processes initiated at weak links.[12]

BPO is commonly used as an initiator for S polymerizations and copolymerizations and it has been reported that its use can lead to yellowing and impaired stability in PS.[13,14] The initiation and termination pathways observed for S polymerization when BPO is used as initiator have been discussed in Sections 3.2 and 3.4.2.2.1. These give rise to benzoyloxy and phenyl end-groups as follows (Scheme 8.1).

Initiation:

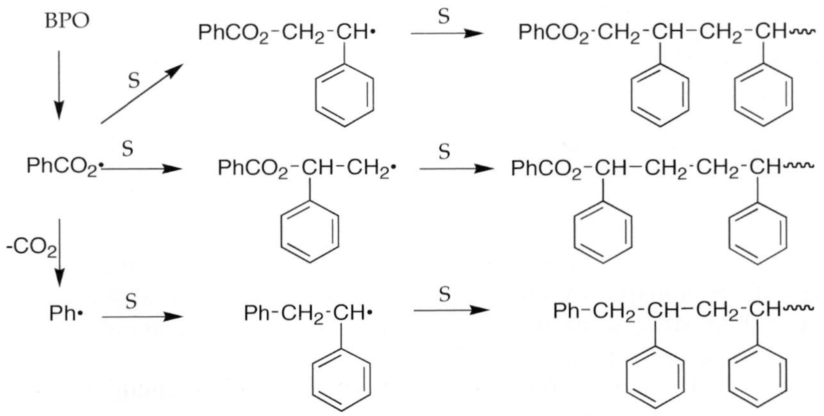

Transfer to initiator/primary radical termination:

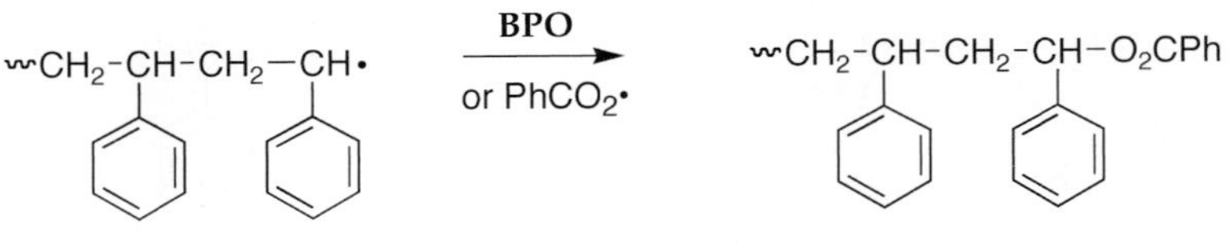

Scheme 8.1

NMR studies[15,16] on polymers prepared with ^{13}C-labeled BPO have shown that the primary benzoyloxy and phenyl end groups formed by tail addition to monomer are thermally stable under conditions where the polymer degrades. They persist to > 50% weight loss at 300°C under nitrogen. Thus, these groups are unlikely to be directly responsible for the poor thermal stability of PS prepared with BPO as initiator. On the other hand, the secondary benzoate end groups, formed by head addition or transfer to initiator, appear extremely labile under these conditions. Their half life at 300°C is <5 min.

Studies with model compounds show that secondary benzoate esters eliminate benzoic acid to form unsaturated chain ends as shown in Scheme 8.2.[15]

Unsaturation has long been thought to be a "weak link" in PS.[4,17] It has been found that for BPO initiated S polymerization at high conversion most chain termination may be by way of transfer to initiator or primary radical termination.[18] Therefore, if these groups are responsible for initiating the chain degradation process, it provides a plausible explanation for high conversion PS formed with BPO initiator being less thermally stable than either a similar low conversion polymer or a polymer prepared with a different initiator.

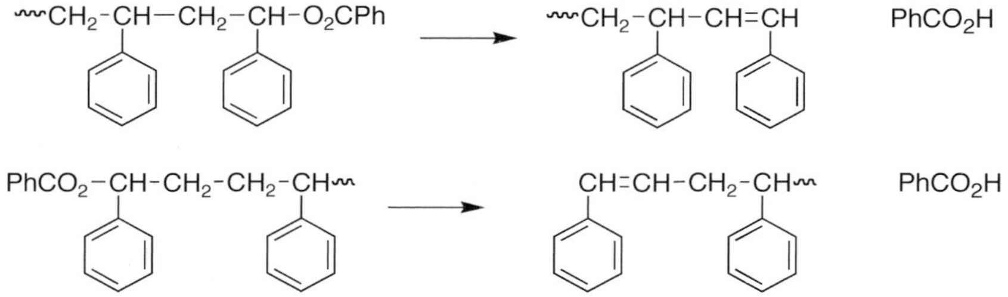

Scheme 8.2

These examples show how initiator selection can be critical in determining the properties of PS prepared by radical polymerization. If thermal stability were of importance, then, since some initiator-derived ends cannot be avoided, a preferred initiator would be one which gives rise to end groups that do not readily eliminate or dissociate. End groups formed with AIBN initiator appear stable with respect to the polymer backbone,[19] Many other systems remain to be studied.

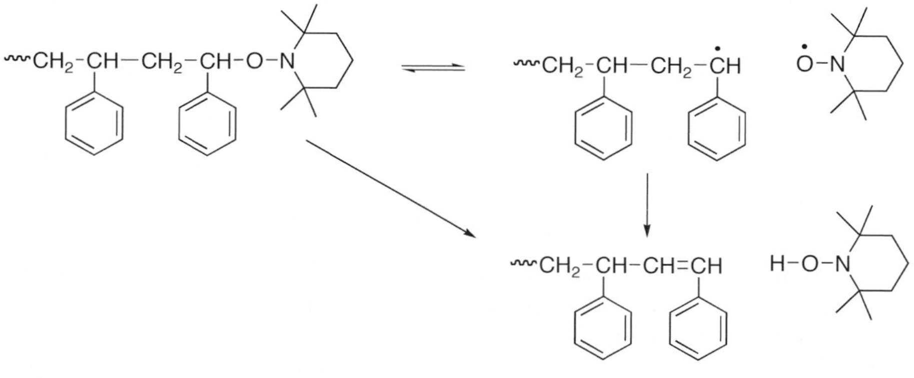

Scheme 8.3

The majority of polymers formed by living radical polymerization (NMP, ATRP, RAFT) will possess labile functionality at chain ends. Recent studies have examined the thermal stability of polystyrene produced by NMP with TEMPO (Scheme 8.3),[20,21] ATRP and RAFT (Scheme 8.4).[22] In each case, the end groups are observed to degrade at relatively low temperatures (~200 °C) by cross disproportionation or thermal elimination to leave an unsaturated chain end. Thermal elimination has been proposed as a simple and convenient method of removing reactive chain ends when this is desirable. For each method of polymerization, various methods of replacing the chain functionality with hydrogen or a more desirable functionality have been devised (Chapter 9).

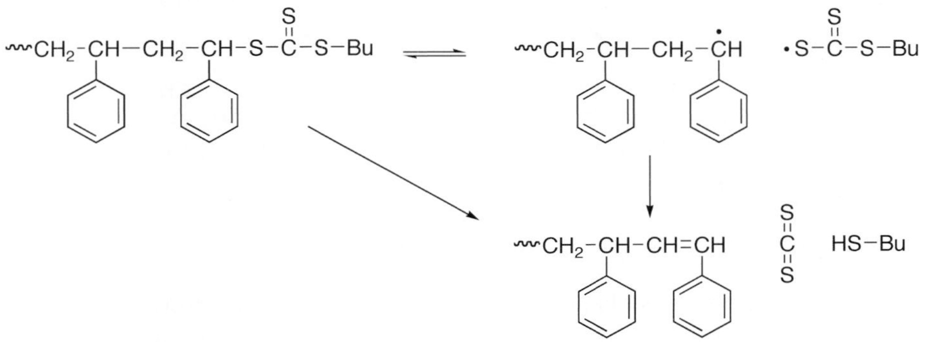

Scheme 8.4

8.2.2 "Defect Structures" in Poly(methyl methacrylate)

There have been many studies on the thermal and thermo-oxidative degradation of PMMA.[23,24] It is well established that the polymer formed by radical polymerization can be substantially less stable than predicted by consideration of the idealized structure and that the kinetics of polymer degradation are dependent on the conditions used for its preparation. There is still some controversy surrounding the details of thermal degradation mechanisms and, in particular, the initiation of degradation.[23]

The thermal degradation of 'ideal' PMMA chains, such as might be formed by anionic polymerization, is thought to be initiated by a random scission process involving cleavage of backbone or side chain bonds.[25-27] The polymer formed by radical polymerization contains weak links. PMMA degrades by unzipping or depropagation (*i.e.* the reverse of radical polymerization). Any structures that are less stable than the backbone or side chain bonds and which give rise to propagating radicals constitute weak links.

Unstable structures are known to arise by chain termination. Mechanisms for radical-radical termination in MMA polymerization have been discussed in Sections 5.2.2.1.2 and 5.2.2.2.2 and these are summarized in Scheme 8.5. It is established that both disproportionation and combination occur to substantial extents. The head-to-head linkages **1** and the unsaturated chain ends **2** both constitute weak links in PMMA.[26,28-33] The presence of these groups account for

PMMA formed by radical polymerization being significantly less stable than that formed by anionic polymerization.

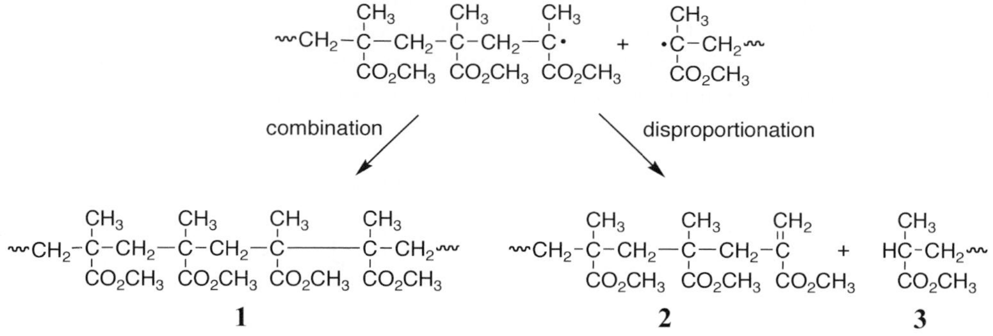

Scheme 8.5

Head-to-head linkages (**1**) are thermally unstable at temperatures above 180°C and may undergo spontaneous scission to form propagating radicals (Scheme 8.6).[29,31-33]

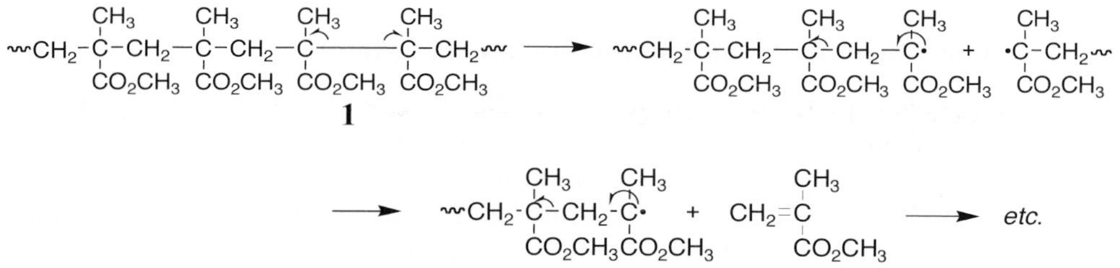

Scheme 8.6

The bond β- to the double bond of the unsaturated disproportionation product **2** is also weaker than other backbone bonds.[10,30,32,33] However, it is now believed that the instability of unsaturated linkages is due to a radical-induced decomposition mechanism (Scheme 8.7).[30] This mechanism for initiating degradation is analogous to the addition-fragmentation chain transfer observed in polymerizations carried out in the presence of **2** at lower temperatures (see 6.2.3.4, 7.6.5 and 9.5.2).

To avoid these stability problems, it is necessary to minimize the proportion of chains that terminate by radical-radical reaction. One way of achieving this is to conduct the polymerization in the presence of an appropriate chain transfer agent. For example, if polymerization is performed in the presence of a H-donor chain transfer agent, conditions can be chosen such that most chains terminate by hydrogen-atom transfer. Bagby et al.[34] examined the thermal stability of PMMA formed with dodecanethiol. These polymer chains will then possess, more thermally stable, saturated end groups (**3**, see Scheme 8.5).[34] If terminated by a proton source, anionic PMMA also has saturated chain ends (**3**).

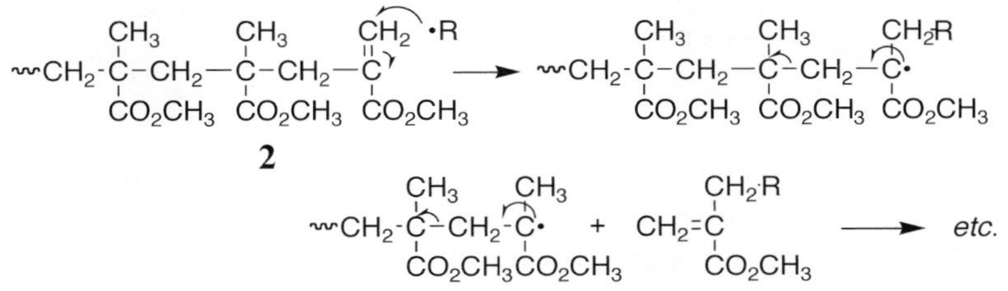

$$\begin{array}{c} \text{CH}_3 \quad \text{CH}_3 \quad \text{CH}_2 \cdot \text{R} \\ \text{\textasciitilde CH}_2\text{-C-CH}_2\text{-C-CH}_2\text{-C} \\ \text{CO}_2\text{CH}_3 \; \text{CO}_2\text{CH}_3 \; \text{CO}_2\text{CH}_3 \end{array} \longrightarrow \begin{array}{c} \text{CH}_3 \quad \text{CH}_3 \quad \text{CH}_2\text{R} \\ \text{\textasciitilde CH}_2\text{-C-CH}_2\text{-C-CH}_2\text{-C}\cdot \\ \text{CO}_2\text{CH}_3 \; \text{CO}_2\text{CH}_3 \; \text{CO}_2\text{CH}_3 \end{array}$$

2

$$\begin{array}{c} \text{CH}_3 \quad \text{CH}_3 \\ \text{\textasciitilde CH}_2\text{-C-CH}_2\text{-C}\cdot \\ \text{CO}_2\text{CH}_3 \text{CO}_2\text{CH}_3 \end{array} + \begin{array}{c} \text{CH}_2\text{R} \\ \text{CH}_2\text{=C} \\ \text{CO}_2\text{CH}_3 \end{array} \longrightarrow \text{etc.}$$

Scheme 8.7

It has also been suggested that, for polymers formed in the presence of air, peroxidic linkages may be weak links.[23] However, in this context, it is of interest that PMMA appears more thermally stable under air than it is under nitrogen (higher initial decomposition temperature).[24,32,35,36] Various explanations have been suggested. Peterson *et al.*[24,36] have attributed this to the propagating radicals PMMA• formed as a consequence of weak link scission being trapped by oxygen to form as hydroperoxy radicals (Scheme 8.8). Other radical traps (nitric oxide) also stabilize the polymer.[24,36]

$$\begin{array}{c} \text{CH}_3 \quad \text{CH}_3 \\ \text{\textasciitilde CH}_2\text{-C-CH}_2\text{-C}\cdot \\ \text{CO}_2\text{CH}_3\text{CO}_2\text{CH}_3 \end{array} + \;\cdot\text{O-O}\cdot \; \rightleftharpoons \; \begin{array}{c} \text{CH}_3 \quad \text{CH}_3 \\ \text{\textasciitilde CH}_2\text{-C-CH}_2\text{-C-O-O}\cdot \\ \text{CO}_2\text{CH}_3\text{CO}_2\text{CH}_3 \end{array}$$

Scheme 8.8

There are other sources of unsaturated chain ends in PMMA formed by radical polymerization:
(a) End groups similar to those formed by disproportionation (**2**) are formed in chain transfer to certain addition-fragmentation transfer agents (*e.g.* allyl sulfides, see 6.2.3.2) or cobalt chain transfer agents (see 6.2.5).
(b) Unsaturated chain ends can arise by primary radical transfer or transfer to MMA. This involves abstraction of the α-methyl or the ester methyl hydrogens. If the monomer-derived radicals so-formed initiate polymerization, the polymer will contain end groups **4** and **5**. The *t*-butoxy and other *t*-alkoxy radicals show a propensity for abstraction (see 3.4.2.1).[37-39]

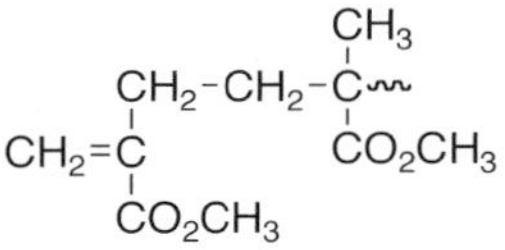

```
            CH₃                              CH₃
 CH₂-CH₂-C∼                        CH₂=C              CH₃
CH₂=C     CO₂CH₃                        CO₂CH₂-CH₂-C∼
 CO₂CH₃                                              CO₂CH₃
    4                                      5
```

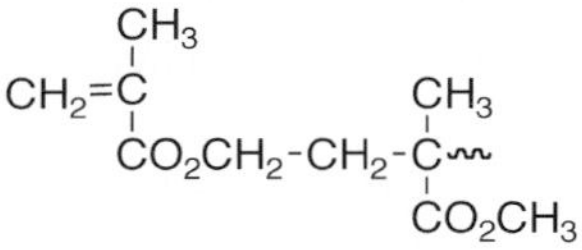

Note, however, that chain ends **4** and **5** may give different chemistry to those formed in termination by disproportionation (**2**, see Scheme 8.5) or the processes under (a) above. Chain scission β to the double bond will not lead to a MMA propagating species. It is not established whether the presence of these ends will give impaired thermal stability.

However, the presence of unsaturated chain ends can have other consequences for polymer properties:

(a) Propagating radicals initiated by abstraction products will not contain an initiator residue at one chain end.[39] Experiments which depend on determination of initiator-derived chain ends may be in error and some literature data may need to be reinterpreted in this light.[40] Syntheses of telechelic or end-functional polymers based on the use of functional initiators will also be detrimentally affected (see 7.5.1).

(b) The unsaturated end groups (**2**, **4** and **5**) may be reactive under polymerization conditions (*i.e.* the polymer chains can be considered as macromonomers) and may copolymerize leading to graft formation (see 7.6.5).[41] The end groups (**2**) may also give chain transfer by an addition-fragmentation mechanism (see 6.2.3.4 and 9.5.2).

It is of interest that thermogravimetric analysis has been used as a means of determining end group purity of PMMA macromonomers formed by catalytic chain transfer.

As in the case of PS (Section 8.2.1) polymers formed by living radical polymerization (NMP, ATRP, RAFT) have thermally unstable labile chain ends. Although PMMA can be prepared by NMP, it is made difficult by the incidence of cross disproportionation.[42] Thermal elimination, possibly by a homolysis-cross disproportionation mechanism, provides a route to narrow polydispersity macromonomers.[43] Chemistries for end group replacement have been devised in the case of polymers formed by NMP (Section 9.3.6), ATRP (Section 9.4) and RAFT (Section 9.5.3).

8.2.3 "Defect Structures" in Poly(vinyl chloride)

Mechanisms of thermal degradation of PVC, the structure of PVC and the stabilization of PVC have been the subject of many reviews. Those by Starnes,[44] Endo[45] and Ivan[46] are some of the more recent. Defect structures in PVC arise during the propagation and chain transfer steps. As with PMMA, PVC formed by anionic polymerization is much more stable than that formed by radical polymerization. The relative stabilities of structures that may be formed by anomalous reactions have been established by comparing the stabilities of low molecular weight model compounds.[44]

PVC formed with diacyl peroxide or peroxydicarbonate initiators will contain a proportion of potentially labile α-haloester chain ends (**6**, Scheme 8.9). However, it is believed that most chain ends in PVC are formed by transfer to monomer as is discussed in Sections 4.3.1.2 and 6.2.6.3.[47]

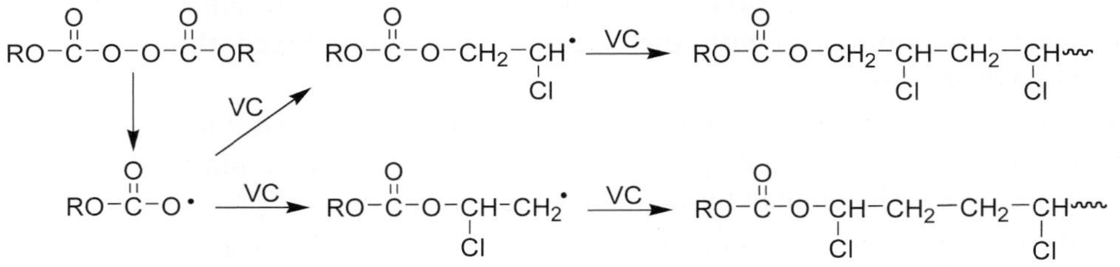

Scheme 8.9

8.3 Controlling Propagation

Given the important role that steric and polar factors play in determining the rate and regiospecificity of radical additions (see 2.3), it might be anticipated that reagents which coordinate with the propagating radical and/or the monomer and thereby modify the effective size, polarity, or inherent stability of that species, could alter the outcome of propagation.

The aspects of polymer structure to be controlled have already been discussed in Chapter 4. For the case of a homopolymer, these are:

(a) Stereosequence isomerism (Section 4.2); the tacticity of the polymer chain. Most polymers formed by radical polymerization have an excess of syndiotactic over isotactic dyads. $P(m)$ typically lies in the range 0.4-0.5 for vinyl monomers and 0.2-0.5 for 1,1-disubstituted monomers. The physical properties of polymers depend on chain stereochemistry. If tacticity control can be achieved, a further challenge is to control the chirality of the chain.

(b) Regiosequence isomerism (Section 4.3); the extent of head *vs* tail addition

(c) Structural isomerism (Section 4.4); rearrangement during propagation. A particular challenge is to control the incidence of short chain branching in PE and in polyacrylates

For the case of copolymers, it is also possible to control the arrangement of monomer units in the chain.

The reagents used for controlling polymer structure may be low molecular weight (*e.g.* the solvent - Sections 8.3.1-8.3.3, Lewis acids - Section 8.3.4) or polymeric (*e.g.* template polymers - Section 8.3.5, enzymes - Section 8.3.6). Control over polymer structure may also be achieved in a topological polymerization where the monomer is crystalline or organized such that the spatial arrangement on the monomer is appropriately constrained (Section 8.3.7).

For greatest effect propagation involving the complexed or constrained species should dominate over normal propagation. For this to occur one of the following should apply:

(a) Either the monomer or propagating species is completely complexed. This requires that concentration of the reagent whether mono- or polymeric to be at least stoichiometric with the species to be complexed throughout the polymerization.

(b) The reactivity of the complexed species is many-fold greater than that of any remaining uncomplexed species and that the equilibrium and rate constants associated with complex formation are high.

Bearing these requirements in mind, the more desirable way of controlling propagation would appear to be to complex the propagating radical (P•). Whereas the initial monomer concentrations are typically in the range ~2-10 M, the typical "steady state" concentration of P• is usually very low ($\sim 10^{-6}$-10^{-7} M) (Scheme 8.10). Therefore, only a small concentration of a catalytic reagent would be required to complex all radicals. However, for this strategy to be successful, the reagent should interact specifically with P• and not associate strongly with either the monomer or the polymer. In any competitive equilibrium, the difference in concentrations (up to 10^8-fold) would clearly favor interaction with monomer or polymer over P•.

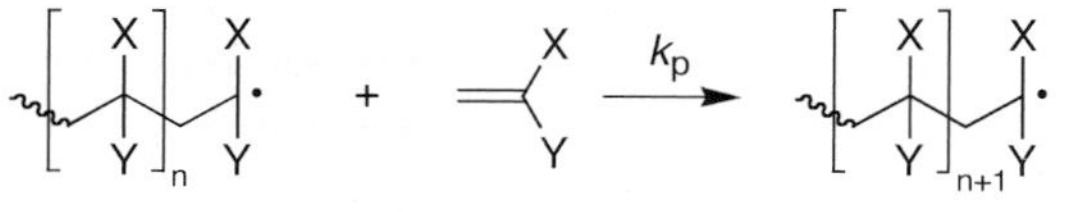

$$[P\bullet] \sim 10^{-6}\text{-}10^{-7}\text{ M} \qquad [M]_o \sim 2\text{-}10 \text{ M}$$

Scheme 8.10

In seeking a suitable complexing agent for the propagating species, one approach is to consider the various species (X•) that are known to reversibly add carbon-centered radicals (Scheme 8.11). Many such reagents have been described in the organic literature. Such species find use as mediators in living radical polymerization. Notable examples are nitroxides (in NMP, Section 9.3.6), dithioesters (in RAFT, Section 9.5.3) and various organometallic complexes (in ATRP, Section 9.4). These species (Z•) react with carbon-centered radicals (R•) at near diffusion controlled rates yet the Z–R bonds of the adduct are relatively weak. The bond strength depends on the nature of R and the functionality on Z. Under the appropriate reaction conditions, the X-R bond may undergo reversible homolysis allowing monomer insertion by a radical mechanism. Could the proximity of X influence the course of propagation?

Most of the studies on polymerization have been concerned with studying the utility of reagents conferring living characteristics on the polymerization (*e.g.* achieving narrow polydispersities, making block copolymers, *etc.*) and at controlling the rate of polymerization. Only a few have explicitly looked for effects on polymer structure. Several studies have explored NMP utilizing chiral and bulky nitroxides.[48,49] The use of chiral metal complexes in ATRP has also been explored.[50,51] No significant influence on polymer structure (on chirality or tacticity) was observed. All evidence suggests that propagating radicals in these processes (NMP, RAFT, ATRP) behave as free (uncomplexed) propagating radicals. To date there is little evidence that complexation of radicals by the reagents discussed above occurs or, if there is, that the complexation influences the regio- or stereospecificity of radical addition.

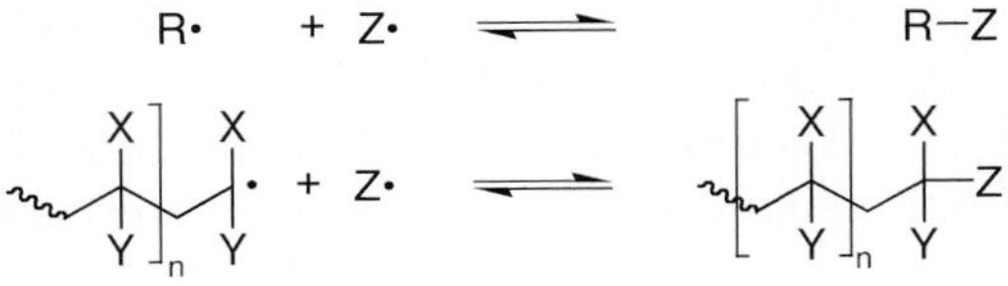

Scheme 8.11

An early report[52] that the stereoregularity of MMA propagation is influenced by Co-porphyrin has not been confirmed by subsequent studies. Giese *et al.*[53] reported that cyclohexyl radicals generated from alkylcobaloximes and cyclohexyl radicals generated from other sources show different specificity in atom transfer reactions. However, they[53] and Clarke and Jones[54] have also provided evidence that the radicals generated from square planar cobalt complexes behave as "normal" radicals in simple radical additions. The utility of cobalt complexes as complexing agents in controlling propagation is limited by side reactions that give chain transfer (these may be used to advantage in macromonomer preparation - Section 6.2.5). The importance of these reactions can be controlled by limiting the application to monosubstituted monomers and by changing the ligands on cobalt (Section 9.3.9.1). Radical polymerization of bulky methacrylamide derivatives (*e.g.* **7**),[55,56] maleimides[57] or methacrylate esters (*e.g.* **8**)[58] provides stereospecific polymerization (Section 4.2.3). More recent work[59] has shown that the polymerization of **8** in the presence of a chiral cobalt(II) salophen complex (**9**) leads to isotactic chains with a one-handed helical structure. It was proposed that **9** selectively retards chain growth of one helix leading to an excess of the one-handed helices. Chiral initiators and transfer agents have also been used to induce chirality.[60]

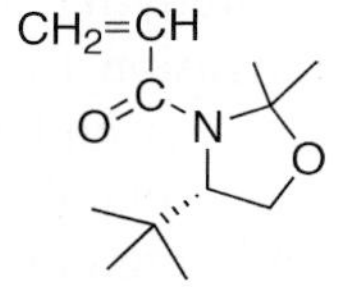

7 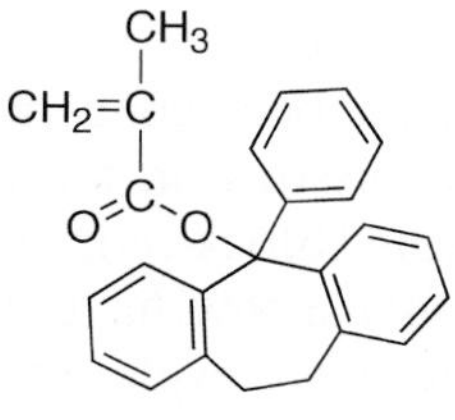 **8** 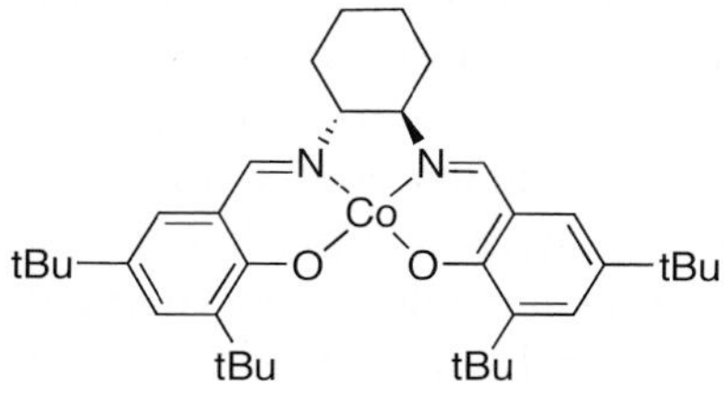 **9**

Puzin et al.[61] reported that the tacticity of PMMA prepared in bulk is influenced (slight increase in syndiotacticity) by very small amounts of titanocene dichloride (10^{-3} M). Selective complexation of the propagating radical was postulated.

Cyclopolymerization of the bis-methacrylates (**10**, **11**)[62,63] or bis-styrene derivatives (**12**)[64] has been used to produce heterotactic polymers and optically active atactic polymers. Cyclopolymerization of racemic **13** by ATRP with a catalyst based on a chiral ligand (Scheme 8.12) gave preferential conversion of the (S,S)-enantiomer.[65,66]

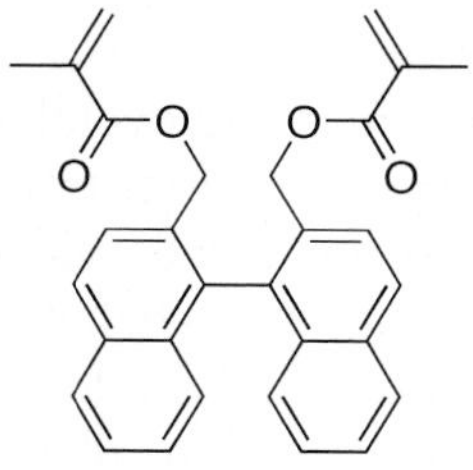

10 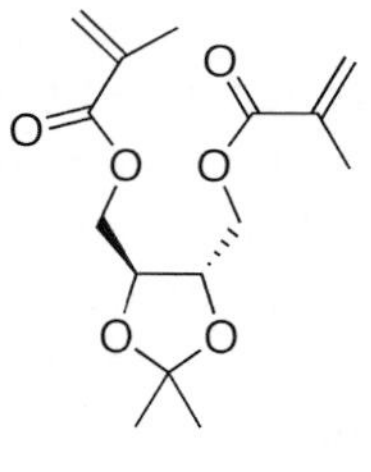 **11** 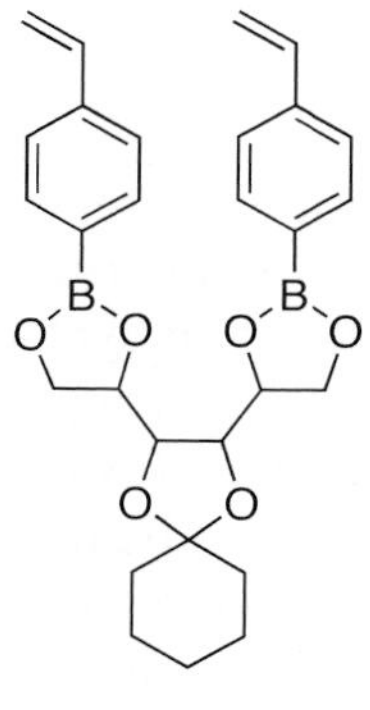 **12**

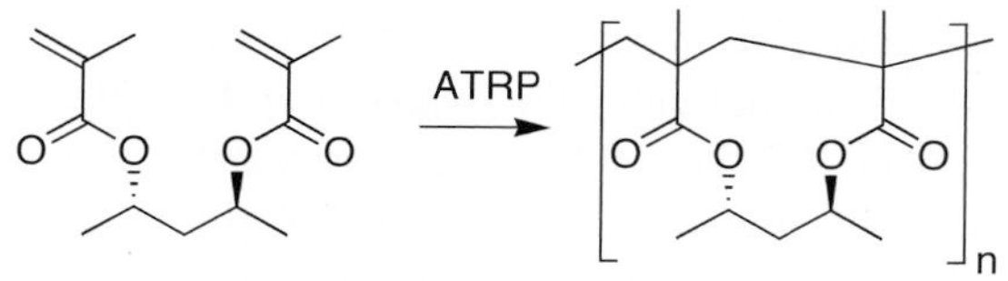

13

Scheme 8.12

Control of the polymerization process by changing the reaction medium, through the use of Lewis acids or with templates, has been studied by various groups since the 1950s. Most studies have focused on control of polymerization kinetics or control of reactivity ratios and hence composition in copolymerization. A lesser number of studies have focused on controlling the stereochemistry of the polymer chains. A survey of these studies is provided in the sections that follow. This section is entitled controlling propagation, however, not surprisingly, many of the reagents/reaction conditions mentioned also have an influence on termination kinetics and with conventional methods these effects are not always easy to distinguish. Stereocontrol in radical polymerization was recently reviewed by Habaue and Okamoto[67] and Matsumoto.[68]

8.3.1 Organic Solvents and Water

Solvent effects on radical polymerization have been reviewed by Coote and Davis,[69] Coote et al.,[70] Barton and Borsig,[71] Gromov,[72] and Kamachi[73] A summary of kinetic data is also included in Beuermann and Buback's review.[74] Most literature on solvent effects on the propagation step of radical polymerization deals with influences of the medium on rate of polymerization.

Solvent effects for polymerizations in supercritical CO_2 and in ionic liquids are considered separately in Sections 8.3.2 and 8.3.3 respectively. In this section, we concentrate on effects of organic solvents and water on the rate and stereospecificity of the propagation step of radical polymerization. We exclude from consideration effects where the solvent is itself a reactant providing byproducts by acting as a comonomer or chain transfer agent (chain transfer to solvent is considered in Section 6.2.2.5). We also exclude differences between bulk and solution polymerization that can be ascribed to a simple concentration effect. In solution polymerization, the rate of propagation should be slowed with reference to that seen in bulk monomer simply because of dilution of monomer. Other reactions of the propagating radicals that do not depend on the monomer concentration can proceed at the same or a similar rate notwithstanding any influence of chain length. These include, radical-radical termination, chain transfer to species other than monomer and intra-molecular rearrangement by cyclization, ring opening or backbiting.

An attractive feature of using the solvent as an agent to control propagation in solution polymerization is that solvents when used are usually present in very large excess in relation to any radical species. Of course, economic, solubility, toxicity, waste disposal, and other considerations limit the range of solvents that can be employed in an industrial polymerization process.

Solvent effects on the reactions of small radicals have been discussed in general terms in Chapter 2 (see 2.3.6.2 & 2.4.5). Small, yet easily discernible, solvent effects have been reported for many reactions involving neutral radicals. These effects on the rates of radical reactions often appear insignificant when compared with the much larger effects observed for similar reactions involving ionic species which may range to orders of magnitude.[75]

Where monomers or radicals are charged, readily ionizable or capable of forming hydrogen bonds, mechanisms whereby the solvent could affect radical reactivity by disruption or involvement of hydrogen bonding may seem obvious. For other systems mechanisms are often still a matter of controversy even in the case of small radicals (Section 2.3.6.2). There are at least three mechanisms whereby the solvent might modify the outcome of a radical process:

(a) Formation of a monomer or radical complex with different reactivity and/or specificity than the uncomplexed species.
(b) Solvation of a transition state or intermediate that may have polar character.
(c) Preferential solvation of one or more reactants leading to local concentrations being different from those in the medium as a whole.

Furthermore, at least three forms of radical-solvent interaction should be considered:

(a) Reversible addition to the solvent molecule. For example, formation of a cyclohexadienyl radical in the case of aromatic solvents.
(b) Formation of a charge transfer complex.
(c) Orbital interaction with a C-H σ-bond or a π-system but without development of charge separation or bond formation.[76]

8.3.1.1 *Homopolymerization*

The values of the rate parameters for many homopolymerizations have been shown to be solvent dependent.[71-74] Large solvent effects are reported for monomers which are ionizable (*e.g.* MAA, AA), give precipitation polymerization (AN), or contain hydroxy or amide groups (*e.g.* HEA, HEMA, AM, NIPAM) which can form hydrogen bonds. Some of the biggest solvent effects are reported for water *vs* other solvents. Substantial dependence of the propagation rate constants on monomer concentration has also been reported with water as solvent. For example, in MAA polymerization at 25 °C the propagation rate constant increases from 600 to 3900 $M^{-1} s^{-1}$ on lowering the monomer concentration from 9.34 to 1.71 M.[74] No pronounced concentration dependence is seen with non-polar solvents.

Very large solvent effects are also observed for systems where the monomers can aggregate either with themselves or another species. For example, the apparent k_p for polymerizable surfactants, such as certain vinyl pyridinium salts and alkyl salts of dimethylaminoalkyl methacrylates, in aqueous solution above the critical micelle concentration (cmc) are dramatically higher than they are below the cmc in water or in non-aqueous media.[77] This does not mean that the value for the k_p is higher. The heterogeneity of the medium needs to be considered. In the micellar system, the effective concentration of double bonds in the vicinity of the propagating species can be up to 100-fold greater than the concentration of monomer in the medium considered as a whole. The number of surfactant molecules per micelle can also influence the molecular weight. However, the microstructure (tacticity) of the polymer chains is claimed to be the same as that obtained in bulk polymerization (see also Section 8.3.7).

For less polar monomers, the most extensively studied homopolymerizations are vinyl esters (*e.g.* VAc), acrylate and methacrylate esters and S. Most of these studies have focused wholly on the polymerization kinetics and only a few have examined the microstructures of the polymers formed. Most of the early rate data in this area should be treated with caution because of the difficulties associated in separating effects of solvent on k_p, k_t and initiation rate and efficiency.

One of the most dramatic examples of a solvent effect on propagation taken from the early literature is for vinyl acetate polymerization.[78,79] Kamachi *et al.*[78] reported a *ca.* 80-fold reduction in k_p (30°C) on shifting from ethyl acetate to benzonitrile solvent (Table 8.1). Effects on polymer structure were also reported. Hatada *et al.*[80] conducted a ^{1}H NMR study on the structure of the PVAc formed in various solvents. They found that PVAc ($\bar{M}_n$~20000) produced in ethyl acetate solvent has ~0.7 branches/chain while that formed in aromatic solvents is essentially unbranched.

Table 8.1 Solvent Effect on Homopropagation Rate Constants for VAc at 30°C[78]

Solvent	$k_p \times 10^{-2}$ (M^{-1} s^{-1})	Solvent	$k_p \times 10^{-2}$ (M^{-1} s^{-1})
benzonitrile	8	fluorobenzene	97
phenyl acetate	37	benzene-d_6	113
anisole	48	benzene	117
chlorobenzene	61	ethyl acetate	637
ethyl benzoate	37		

Solvent effects on k_p in polymerizations of MMA[81-87] and S[81,84-86,88] have been widely studied but are generally small by comparison and there appears to be no clear correlation with solvent dielectric constant or other solvent properties. When solvent effects are observed, does the solvent modify the reactivity of the propagating radicals, the reactivity of the monomer, the homogeneity of the reaction medium or all of these? Experimental data from, for example, PLP experiments (Section 4.5.2) can be used to calculate the propagation rate constants as a function of the reaction medium. Equally one can assume that k_p remains constant and calculate the effective monomer concentration in the proximity of the chain end. The experimental data typically do not allow easy discrimination between whether either or both are varying nor should one necessarily expect a universal rule to apply. Explanations for apparent conversion and chain length dependence of k_p can also be formulated in terms of effects on local monomer concentrations. Termination rate constants may also be affected by solvent quality.[70,74]

The heterogeneity of the reaction medium is also important in determining the molecular weight and k_p in solution polymerization of macromonomers.[89] The magnitude of the effect varies according to the solvent quality. PS macromonomer chains in good solvents (*e.g.* toluene) have an extended conformation whereas in poor solvents (*e.g.* methylcyclohexane) chains are tightly coiled.[89] As a consequence, the radical center may see an environment that is medium dependent (see also Sections 7.6.5 and 8.3.7).

The tacticity of polymers formed by radical polymerization can also be influenced by solvent and by temperature.[90] Fluoro-alcohol solvents have been shown to have a significant influence on the tacticity of PVAc and other vinyl esters.[91] Different effects are seen for VAc (more syndiotactic, fraction of *rr* dyads enhanced), vinyl propionate and other vinyl alkanoates (more heterotactic, fraction of *mr* dyads enhanced) and vinyl benzoate (more isotactic, fraction of mm dyads enhanced).[92] The effect is greater for lower polymerization temperatures and for more bulky fluoro-alcohols. The effect of fluoro-alcohol solvents on polymerization of methacrylate esters has also been investigated[93,94] and data for -40 °C are shown in Table 8.2. Polymerization in fluoro-alcohol solvents enhances syndiotacticity of PMMA and PEMA.[94] For PtBMA, syndiotacticity is reduced.[94] Again, the effect is greatest at the lowest reaction temperature. These solvent effects were attributed to steric factors associated with hydrogen bonding to the ester C=O. The solvent is said to enhance the bulkiness of the ester group of both the propagating radical and the monomer.[93,94]

Table 8.2 Effect of Solvent on Tacticity of Poly(alkyl methacrylate) at -40 °C[94]

Solvent[a]	MMA	EMA	tBMA
	mm:mr:rr	*mm:mr:rr*	*mm:mr:rr*
toluene	1.0:23.0:76.0	4.7:18.2:77.1	2.7:22.8:74.4
methanol	1.6:23.4:74.9	11.8:16.2:72.0	-
HFIP	1.3:19.6:79.1	4.4:15.3:80.3	-
PFTB	0.5:16.6:82.9	0.9:14.7:84.4	1.4:33.2:65.4

a HFIP = hexafluoro-isopropanol, PFTB = perfluoro-*t*-butanol

Tacticity of MAA is influenced by solvent,[90,95] the presence of amines (Table 8.3)[90] and complexation. PMAA appears more isotactic when formed in a non-hydrogen-bonding solvent.[90,95] Polymerization of MAA in CHCl$_3$ in the presence of **14** or **15** also yields a more isotactic polymer.[90] Polymerization of zinc complexes of MAA also yields more isotactic polymers.[96]

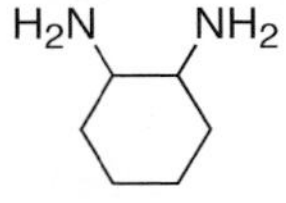

                    14 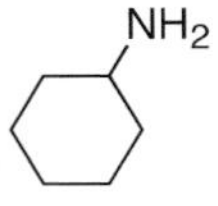   15

Table 8.3 Effect of Amines on Tacticity of Poly(methacrylic acid) at 60 °C[90]

Solvent[a]	Amine	mm:mr:rr[a]
MeOH	none	4.0:34.6:61.4
MeOH	14[b]	3.8:29.1:67.0
CHCl$_3$	none	8.1:41.0:50.9
CHCl$_3$	15	12.3:47.0:40.7
CHCl$_3$	14[b]	16.3:48.8:34.9

a Polymerization of MAA (0.1 M) in presence of amine (stoichiometric NH$_2$) with AIBN initiator (0.004 M). Tacticity determined for PMMA obtained by esterification of PMAA formed. b (R,R)-configuration.

8.3.1.2 Copolymerization

The effects of solvent on radical copolymerization are mentioned in a number of reviews.[69-72,97,98] For copolymerizations involving monomers that are ionizable or form hydrogen bonds (AM, MAM, HEA, HEMA, MAA, *etc.*) solvent effects on reactivity ratios can be dramatic. Some data for MAA-MMA copolymerization are shown in Table 8.4.[99]

For MMA-MAA copolymerizations carried out in the more hydrophobic solvents (toluene, dioxane), MAA is the more reactive towards both propagating species while in water MMA is the more reactive. In solvents of intermediate polarity (alcohols, dipolar aprotic solvents), there is a tendency towards alternation. For these systems, choice of solvent could offer a means of controlling copolymer structure.

For copolymerizations between non-protic monomers solvent effects are less marked. Indeed, early work concluded that the reactivity ratios in copolymerizations involving only non-protic monomers (*e.g.* S, MMA, AN, VAc, *etc.*) should show no solvent dependence.[100,101] More recent studies on these and other systems (*e.g.* AN-S,[102-105] E-VAc,[106] MAN-S,[107] MMA-S,[108-110] MMA-VAc[111]) indicate small yet significant solvent effects (some recent data for AN-S copolymerization are shown in Table 8.5). However, the origin of the solvent effect in these cases is not clear. There have been various attempts to rationalize solvent effects on copolymerization by establishing correlations between radical reactivity and various solvent and monomer properties.[71,72,97,99] None has been entirely successful.

Table 8.4 Solvent Dependence of Reactivity Ratios for MMA-MAA Copolymerization at 70°C[a,99]

solvent	r_{MMA}	r_{MAA}
toluene	0.10	1.06
dioxane	0.12	1.33
acetonitrile	0.27	0.03
acetone	0.31	0.63
DMSO	0.78	0.23
isopropanol	0.78	0.33
ethanol	0.80	0.60
acetic acid	0.80	0.78
DMF	0.98	0.68
water	2.61	0.43

a Reactivity ratios estimated from composition data.

Table 8.5 Solvent Dependence of Penultimate Model Reactivity Ratios for S-AN Copolymerization at 60°C[103]

Solvent	r_{SS}	r_{AS}	r_{SA}	r_{AA}
bulk	0.232	0.566	0.087	0.036
toluene	0.242	0.566	0.109	0.133
acetonitrile	0.322	0.621	0.105	0.052

The solvent in a bulk copolymerization comprises the monomers. The nature of the solvent will necessarily change with conversion from monomers to a mixture of monomers and polymers, and, in most cases, the ratio of monomers in the feed will also vary with conversion. For S-AN copolymerization, since the reactivity ratios are different in toluene and in acetonitrile, we should anticipate that the reactivity ratios are different in bulk copolymerizations when the monomer mix is either mostly AN or mostly S. This calls into question the usual method of measuring reactivity ratios by examining the copolymer composition for various monomer feed compositions at very low monomer conversion. We can note that reactivity ratios can be estimated for a single monomer feed composition by analyzing the monomer sequence distribution. Analysis of the dependence of reactivity ratios determined in this manner of monomer feed ratio should therefore provide evidence for solvent effects. These considerations should not be ignored in solution polymerization either.

Harwood[112] proposed that the solvent need not directly affect monomer reactivity, rather it may influence the way the polymer chain is solvated. Evidence for the proposal was the finding for certain copolymerizations, while the terminal model reactivity ratios appear solvent dependent, copolymers of the same overall composition had the same monomer sequence distribution. This was explained in terms of preferential monomer sorption such that the polymer composition determined the relative monomer concentration in the vicinity of the reactive chain end. This phenomenon was called "the bootstrap effect".[112,113] A partition coefficient K was defined as eq. 1:

$$K = \frac{[M_A]/[M_B]}{[M_{Ao}]/[M_{Bo}]} \tag{1}$$

where $[M_A]/[M_B]$ is the ratio of monomer concentrations in the vicinity of the reactive chain end and $[M_{Ao}]/[M_{Bo}]$ is the global ratio. The conditional probabilities which determine the triad fractions are dependent on $[M_A]/[M_B]$ rather than $[M_{Ao}]/[M_{Bo}]$. The value of $[M_A]/[M_B]$ is determined by the polymer composition.

The apparent terminal model reactivity ratios are then: $r_{AB}^{app} = r_{AB}K$ and $r_{BA}^{app} = r_{BA}/K$. It follows that $r_{AB}^{app} r_{BA}^{app} = r_{AB} r_{BA} = const.$ The bootstrap effect does not require the terminal model and other models (penultimate, complex participation) in combination with the bootstrap effect have been explored.[103,114,115] Variants on the theory have also appeared where the local monomer concentration is a function of the monomer feed composition.[116]

The effects of solvent on reactivity ratios and polymerization kinetics have been analyzed for many copolymerizations in terms of this theory.[98] These include copolymerizations of S with MAH,[117,118] S with MAA,[112] S with MMA,[116,117,119-121] S with HEMA,[122] S with BA,[123,124] S with AN,[103,115,125] S with MAN,[112] S with AM,[113] BA with MMA[126,127] and tBA with HEMA.[128] It must, however, be pointed out that while the experimental data for many systems are consistent with a bootstrap effect, it is usually not always necessary to invoke the bootstrap effect for data interpretation. Many authors have questioned the bootstrap effect and much effort has been put into finding evidence both for or against the theory.[69,70,98,129,130] If a bootstrap effect applies, then reactivity ratios cannot be determined by analysis of composition or sequence data in the normal manner discussed in Section 7.3.3.

Studies on the reactions of small model radicals with monomers provide indirect support but do not prove the bootstrap effect.[131] Krstina et al.[131] showed that the reactivities of MMA and MAN model radicals towards MMA, S and VAc were independent of solvent. However, small but significant solvent effects on reactivity ratios are reported for MMA/VAc[111] and MMA/S[117,119] copolymerizations. For the model systems, where there is no polymer coil to solvate, there should be no bootstrap effect and reactivities are determined by the global monomer ratio $[M_{Ao}]/[M_{Bo}]$.[131]

Other phenomena attributed to a bootstrap or similar effects include

(a) The dependence of copolymer composition on molecular weight in certain copolymerizations.[132-134] There are other explanations for the molecular weight dependence of copolymer composition that relate to specificity shown in the initiation process (Section 7.5.6). However, these effects only apply to relatively low molecular weights (<20 units).

(b) The observation of significant solvent effects in macromonomer copolymerization.[135] Tsukahara et al.[135] found that when copolymerizing macromonomers, the choice of solvent has a substantial influence on the reactivity ratios, the molecular weight of the polymer, and the particle size distribution of the final product. They interpreted their data in terms of the effects of solvent on the degree of interpenetration between unlike polymer chains.

8.3.2 Supercritical Carbon Dioxide

Polymerization, including radical polymerization, in supercritical CO_2 has been reviewed.[136,137] It should be noted supercritical CO_2 while a good solvent for many monomers is a very poor solvent for polymers such as the (meth)acrylates and S. As a consequence, with the exception of certain fluoropolymers and polymerizations taken to very low conversion, most polymerizations in supercritical CO_2 are of necessity precipitation, dispersion or emulsion polymerizations.

Several studies have been directed towards determining the kinetics of radical polymerization in supercritical CO_2 using PLP (Section 4.5.2). While some early results[138,139] suggested that $k_p(CO_2)$ for MMA was not significantly different to k_p(bulk), more recent work has shown that $k_p(CO_2)$ for MMA[140] and various acrylate esters (MA,[74] BA,[140,141] DA[74]) are significantly reduced from values for bulk polymerization. Values of $k_p(CO_2)$ for S[142] and VAc[143] are not significantly different to k_p(bulk).

8.3.3 Ionic liquids

Room temperature ionic liquids are currently receiving considerable attention as environmentally friendly alternatives to conventional organic solvents in a variety of contexts.[144] The ionic liquids have this reputation because of their high stability, inertness and, most importantly, extremely low vapor pressures. Because they are ionic and non-conducting they also possess other unique properties that can influence the yield and outcome of organic transformations. Polymerization in ionic liquids has been reviewed by Kubisa.[145] Commonly used ionic liquids are tetra-alkylammonium, tetra-alkylphosphonium, 3-alkyl-1-methylimidazolium (**16**) or alkyl pyridinium salts (**17**). Counter-ions are typically PF_6^- and BF_4^-, though many others are known.

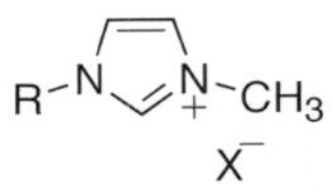

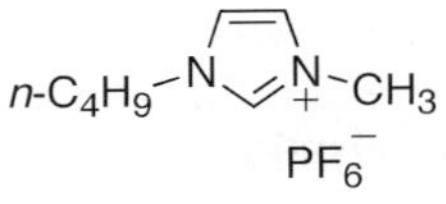

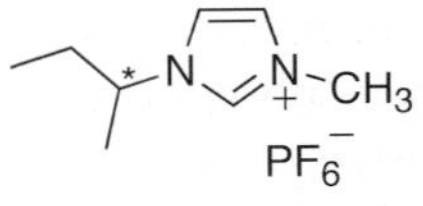

16   **17**   **18**   **19**

Harrison et al.[146,147] have used PLP (Section 4.5.2) to examine the kinetics of MMA polymerization in the ionic liquid **18** (bmimPF$_6$). They report a large (ca 2-fold) enhancement in k_p and a reduction in k_t. This property makes them interesting solvents for use in living radical polymerization (Chapter 9). Ionic liquids have been shown to be compatible with ATRP[148-156] and RAFT.[157,158] However, there are mixed reports on compatibility with NMP.[159,160] Widespread use of ionic liquids in the context of polymerization is limited by the poor solubility of some polymers (including polystyrene) in ionic liquids.

There is also some evidence that the ionic liquid medium affects polymer structure. Biedron and Kubisa[150] reported that the tacticity of PMA prepared in the chiral ionic liquid **19** is different from that prepared in conventional solvent. It is also reported that reactivity ratios for MMA-S copolymerization in the ionic liquid **18**[161] differ from those observed for bulk copolymerization.

8.3.4 Lewis Acids and Inorganics

Lewis acids are known to form complexes both with monomers and with propagating species. Their addition to a polymerization medium, even in catalytic amounts, can bring about dramatic changes in rate constants in homopolymerization (Section 8.3.4.1) and reactivity ratios in copolymerization (Section 8.3.4.2). Early work in this area has been reviewed by Bamford[162] and Barton and Borsig.[71] There is significant current interest in using Lewis Acids in establishing tacticity control in homopolymerization (see 8.3.4.1).

8.3.4.1 Homopolymerization

In 1957, Bamford et al.[163] reported that the addition of small amounts of lithium chloride brought about a significant (up to two-fold) enhancement in the rate of polymerization of AN in DMF and led to a higher molecular weight polymer. Subsequent studies have shown this to be a more general phenomenon for polymerizations involving, in particular, acrylic and vinylheteroaromatic monomers in the presence of a variety of Lewis acids.[71]

For the case of polymerization of AN in DMF, measurements of the absolute rate constants associated with the polymerizations indicated that the rate of initiation (by AIBN) was not significantly affected by added lithium salts. The enhancement in the rate of polymerization was therefore attributed to an increase in k_p. The value of k_t remains essentially unchanged except when very high concentrations of the Lewis acid are employed.

Zubov et al.[164] suggested that during MMA polymerization in the presence of Lewis acids (e.g. AlBr$_3$) complexation occurs preferentially with the propagating radical rather than with monomer. They suggested a mechanism in which the metal ion is transferred to the incoming monomer in the transition state for addition so as to remain with the active chain end. It is known that Lewis acids can bring about significant changes in the appearance of the EPR spectra of MMA propagating radicals and related species.[165]

Although it is clear that added Lewis acids affect the rate of polymerization and the molecular weight of homopolymers formed in their presence,[71] the effect on polymer structure is small. There are reports that Lewis acids affect the tacticity.[67,68,71,90] Otsu and Yamada[166] found a slightly greater proportion of isotactic (*mm*) triads in PMMA formed by bulk polymerization of a 1:1 complex of MMA with zinc chloride than is observed for a similar polymerization of MMA alone. However, for polymerizations carried out in solution or in the presence of lesser amounts of zinc chloride, no effect was observed.[166] For MMA polymerization in solution at 60 °C, a small though significant effect on tacticity (increase in isotactic triads) is seen on addition of 0.2 M scandium triflate[167] and lesser effects with ytterbium triflate and hafnium chloride (Table 8.6).

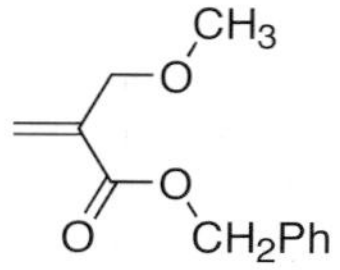

20

Lewis acids have a much greater effect on tacticity in polymerization of α-alkoxymethacrylates such as **20**,[168,169] acrylamides (including AM, NIPAM, DMAM)[170-173] and methacrylamides (including MAM, MMAM) (Table 8.6).[170,171,174] The solvent has a significant effect on the magnitude of the effect observed and little influence is observed for polymerizations carried out in aqueous media. The effect of Lewis acids on tacticity is significantly greater for lower polymerization temperatures. In the polymerizations of acrylamide and methacrylamides a very significant influence on tacticity was seen for 10 mole% Yb(OTf)$_3$ with respect to monomer and the effect was not significantly enhanced for greater concentrations of Lewis acid.

It is also possible that complexation of monomer or propagating species could influence the regiospecificity of addition. However, since the effect is likely to be an enhancement of the usual tendency for head-to-tail addition, perhaps it is not surprising that such effects have not been reported.

Table 8.6. Effect of Lewis Acids on Tacticity of Polymers Formed in High Conversion Radical Polymerizations at 60 °C

Monomer	Solvent	Lewis Acid	Conc. (M)	mm:mr:rr	P(m)[a]
NIPAM[b,172]	CHCl$_3$	none	0	-	0.45
172	CHCl$_3$	Yb(OTf)$_3$	0.20	-	0.58
172	CHCl$_3$	Y(OTf)$_3$	0.20	-	0.62
172	MeOH	Y(OTf)$_3$	0.20	-	0.80
c,172	H$_2$O	Y(OTf)$_3$	0.20	-	0.57
MMAM[d,174]	MeOH	none	0	2:29:69	0.17
174	MeOH	Sc(OTf)$_3$	0.20	28:55:17	0.56
174	MeOH	Y(OTf)$_3$	0.20	46:40:14	0.66
174	MeOH	Yb(OTf)$_3$	0.20	46:44:10	0.68
175	THF	Yb(OTf)$_3$	0.20	32:50:18	0.57
175	H$_2$O	Yb(OTf)$_3$	0.20	2:31:67	0.18
MAM[175]	MeOH	none	0	7:39:54	0.27
175	MeOH	Yb(OTf)$_3$	0.20	36:50:14	0.61
175	MeOH	Y(OTf)$_3$	0.20	33:49:18	0.58
MMA[e,67]	toluene	none	0	3:33:64	0.19
67	toluene	HfCl$_4$	0.20	6:36:58	0.21
67	CHCl$_3$	Yb(OTf)$_3$	0.24	10:36:54	0.23
67	toluene	Sc(OTf)$_3$	0.20	14:46:40	0.30

a P(m) = 0.5 mr + mm. There is evidence of non-Bernoullian statistics for some examples. b NIPAM (2.4 M) with AIBN (0.02 M) polymerized for 24 h at 60°C. c NIPAM (2.4 M) with K$_2$S$_2$O$_8$ (0.02 M) polymerized for 24 h at 60°C. d MMAM (2.0 M) with AIBN (0.02 M) polymerized for 24 h at 60°C. e MAM (2.4 M) with AIBN (0.02 M) polymerized for 2 h at 60°C. f MMA (2.4 M) with AIBN (0.02 M) polymerized for 24 h at 60°C.

8.3.4.2 Copolymerization

The kinetics of copolymerization and the microstructure of copolymers can be markedly influenced by the addition of Lewis acids. In particular, Lewis acids are effective in enhancing the tendency towards alternation in copolymerization of donor-acceptor monomer pairs and can give dramatic enhancements in the rate of copolymerization and much higher molecular weights than are observed for similar conditions without the Lewis acid. Copolymerizations where the electron deficient monomer is an acrylic monomer (e.g. AN, MA, MMA) and the electron rich monomer is S or a diene have been the most widely studied.[164,176-184] Strictly alternating copolymers of MMA and S can be prepared in the presence of, for example, diethylaluminum sesquichloride. In the absence of Lewis acids, there is only a small tendency for alternation in MAA-S copolymerization; terminal model reactivity ratios are ca 0.51 and 0.49 - Section 7.3.1.2.3. Lewis acids used include: EtAlCl$_2$, Et$_2$AlCl, Et$_3$Al$_2$Cl$_3$, ZnCl$_2$, TiCl$_4$, BCl$_3$, LiClO$_4$ and SnCl$_4$.

Various mechanisms (not mutually exclusive) for the influence of Lewis acid on copolymerization have been proposed:

(a) A ternary complex is formed between acceptor, donor, and Lewis acid. An alternating polymer may be formed by homopolymerization of such a complex.[176,177]

(b) The Lewis acid forms a binary complex with the acceptor monomer. The electron deficiency of the double bond is enhanced by complexation with the Lewis acid and thus its reactivity towards nucleophilic radicals is greater.[182]

(c) Spontaneous copolymerization, possibly by a biradical mechanism.[185]

(d) Complexation of the propagating radical to create a species with selectivity different to that of the normal propagating radical.

Most recent work is in accord with mechanism (b). In an effort to distinguish these mechanisms studies on model propagating species have been carried out.[186-189] For S-MMA polymerization initiated by AIBMe-α-^{13}C (Scheme 8.13) it has been established by end group analysis that extremely small amounts of ethyl aluminum sesquichloride (<10^{-3} M with 1.75 M monomers) are sufficient to cause a substantial enhancement in specificity for adding S in the initiation step. This result suggests that complexation of the propagating radical may be sufficient to induce alternating copolymerization but does not rule out other hypotheses.

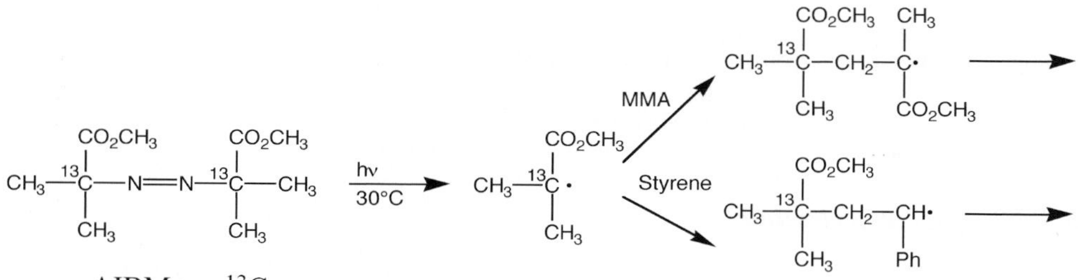

Scheme 8.13

The primary aim of most studies on Lewis acid controlled copolymerization has been the elucidation of mechanism and only low conversion polymerizations are reported. Sherrington et al.[184] studied the high conversion synthesis of alternating MMA-S copolymers in the presence of Lewis acids on a preparative scale. Many Lewis acids were found to give poor control (i.e. deviation from 50:50 composition) and were further complicated by side reactions including cross-linking. They found that the use of catalytic BCl$_3$ as the Lewis acid and photoinitiation gave best results.

Matyjaszewski and coworkers[190,191] have explored living radical copolymerization (ATRP and RAFT) in the presence of Lewis acids.

8.3.5 Template Polymerization

The possibility of using a template polymer to organize the monomer units prior to their being "zipped up" by the attack of a radical species has long attracted interest and the field of template polymerization has been the subject of a number of reviews.[192-195] Template polymerization can also be found under such headings as molecular imprinting, supramolecular chemistry and topological or topochemical polymerization (Section 8.3.7) though some of these terms have additional meaning. Template polymerization, as used here and as its name suggests, involves the formation of a *daughter* polymer on a preformed *parent* polymer.

The interest in this area may be seen to stem from the biological area where the phenomenon is well known and accounts for the regularity in the structure of natural proteins and polynucleotides. Such polymers are efficiently synthesized by enzymes which are capable of organizing monomer units within regularly structured molecular-scale spaces and exploiting weak forces such as hydrogen bonds and Van der Waal forces to control the polymerization process..

The literature distinguishes two limiting forms of template polymerization.[192-194]

(a) Where the monomer is associated with the template and, ideally, initiation, propagation, and termination all occur on the template.
(b) Where only the propagating chain associates with the template. The rate of polymerization is limited by the rate at which monomer is attached from the bulk solution.

The interaction of the template with monomer and/or the propagating radical may involve solely Van der Waals forces or it may involve charge transfer complexation, hydrogen bonding, or ionic forces (Section 8.3.5.1). In other cases, the monomer is attached to the template through formal covalent bonds (Section 8.3.5.2).

8.3.5.1 *Non-covalently bonded templates*

In 1972, Buter et al.[196] reported that polymerization of MMA in the presence of isotactic PMMA leads to a greater than normal predominance of syndiotactic sequences during the early stages of polymerization. Other investigations of this system supporting[197,198] and disputing[199] this finding appeared. The mechanism of the template polymerization is thought to involve initial stereocomplex formation between the oligomeric PMMA propagating radical (predominantly syndiotactic) and the isotactic template polymer with subsequent monomer additions being directed by the environment of the template. Isotactic and syndiotactic PMMA have been shown to form a 1:2 stereocomplex.[197] Recently, Serizawa et al.[200] showed that comparatively pure isotactic PMMA could be prepared within the confines of a matrix comprising a thin porous film of syndiotactic PMMA. The matrix influenced both the molecular weight and the tacticity.

The nature of the interaction between the monomer and the template is more obvious in cases where specific ionic or hydrogen bonding is possible. For example, N-vinylimidazole has been polymerized along a PMAA template[201,202] and acrylic acid has been polymerized on a N-vinylpyrrolidone template.[203] The daughter PAA had a similar degree of polymerization to the template and had a greater fraction of isotactic triads than PAA formed in the absence of the template.

It is well known that rates of polymerizations can increase markedly with the degree of conversion or with the polymer concentration. Some workers have attributed this solely or partly to a template effect. It has been proposed[204] that adventitious template polymerization occurs during polymerizations of AA, MAA and AN, and that the gel or Norrish-Trommsdorff effect observed during polymerizations of these monomers is linked to this phenomenon. However, it is difficult to separate possible template effects from the more generic effects of increasing solution viscosity and chain entanglement at high polymer concentrations on rates of termination and initiator efficiency (Section 5.2.1.4).

There are also reports of template effects on reactivity ratios in copolymerization. For example, Polowinski[205] has reported that both kinetics and reactivity ratios in MMA-MAA copolymerization in benzene are affected by the presence of a PVA template.

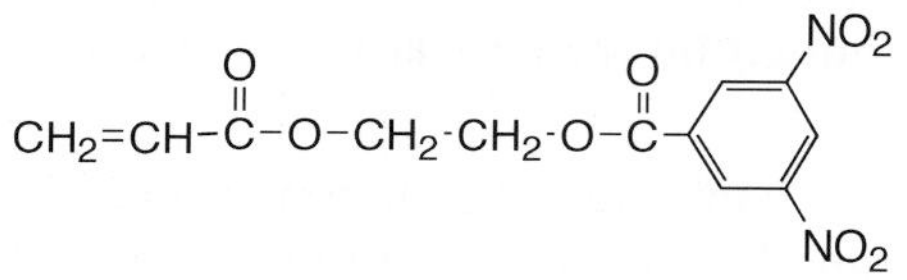

$$CH_2=CH-\overset{O}{\underset{\parallel}{C}}-O-CH_2\cdot CH_2\cdot O-\overset{O}{\underset{\parallel}{C}}-\underset{NO_2}{\overset{NO_2}{\bigcirc}}$$

21

A template polymer may allow the use of monomers that do not otherwise undergo polymerization. An example is the dinitrobenzoate derivative **21**; nitrobenzene derivatives are usually thought of as radical inhibitors (see 5.3.7), thus radical polymerization of monomer with such functionality is unlikely to be successful. Polymerization of **21** on a poly(N-vinylcarbazole) template succeeded in producing a high molecular weight polymer.[206] It was envisaged that the monomer **21** forms a charge transfer complex with the electron donating carbazole group.

8.3.5.2 Covalently bonded templates

Template polymerizations where the monomer is covalently bound to the template clearly have limitations if polymers of high molecular weight or large quantities are required. However, their use offers much greater control over daughter polymer structure. The product in such cases is a ladder polymer and this may be viewed as a special case of cyclopolymerization (Section 4.4.1).

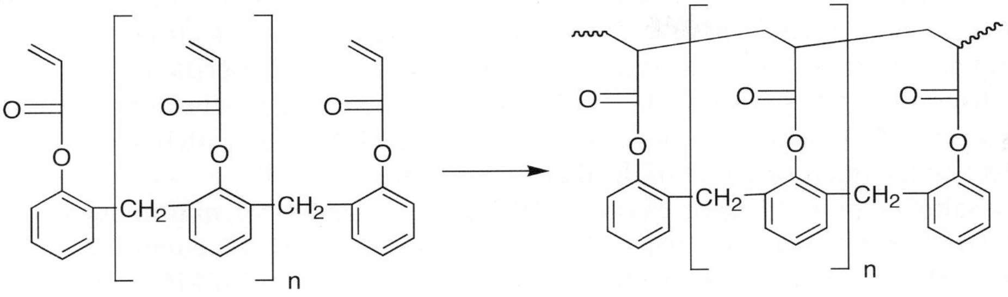

Scheme 8.14

Kämmerer et al.[207-209] have conducted extensive studies on the template polymerization of acrylate or methacrylate derivatives of polyphenolic oligomers **22** with $\bar{X}_n \leq 5$ (Scheme 8.14). Under conditions of low "monomer" and high initiator concentration they found that $\bar{X}_n$ for the daughter polymer was the same as $\bar{X}_n$ for the parent. The possibility of using such templates to control microstructure was considered but not reported.

Feldman et al.[210,211] and Wulff et al.[212] have examined other forms of template controlled oligomerization of acrylic monomers. The template (**23**) has initiator and transfer agent groups attached to a rigid template of precisely defined structure.[210,211] Polymerization of MMA in the presence of **23** gave a 3 unit oligo(MMA) as ca 66% of the polymeric product. The stereochemistry of the oligomer was reported to be "different" from that of atactic PMMA.

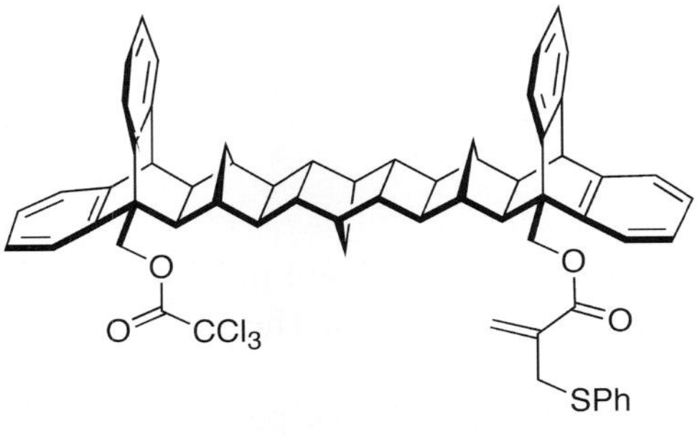

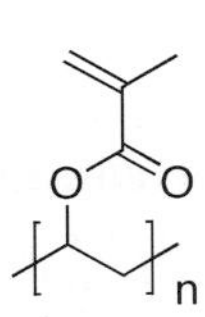

Wulff et al.[212] attached vinyl groups to a large chiral sugar based template molecule and then copolymerized this substrate with various monomers. With MMA and MAN they achieved some optical induction. This approach has been extended in studies of higher molecular weight systems.[213-216] Thus, PVA was esterified with methacryloyl chloride to give a "multimethacrylate" (**24**) and polymerized to give a ladder polymer. "Multimethacrylates" based on PHEMA were also described. The daughter polymer was hydrolyzed to PMMA but only characterized in terms of molecular weight. The value of $\bar{X}_n$ for the daughter polymer was greater than $\bar{X}_n$ for the parent template indicating some inter-template reaction. These workers also examined the copolymerization of partially methacrylated PVA with MMA. It has not been established whether the tacticity of parent PVA or the presence of head-to-head and tail-to-tail linkages has an effect on the microstructure of the daughter polymer.[213]

Saito et al.[217-219] have examined the polymerization of multimethacrylates prepared from β-cyclodextrin. Polymerization using ATRP conditions gave a bimodal molecular weight distribution for the derived PMMA composed predominantly of oligomers of 7 or 14 units indicating that there was little intermolecular reaction

A new form of template polymerization based on ring-opening polymerization of 4-methylenedioxalane has been reported by Endo and coworkers (Scheme 8.15).[220,221] For this system, the monomer is covalently bound and the daughter polymer is released from the template as a consequence of the polymerization process.

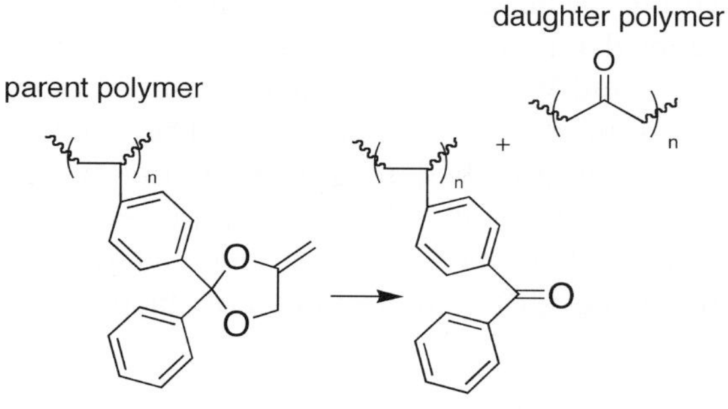

Scheme 8.15

8.3.6 Enzyme Mediated Polymerization

A number of recent papers have explored enzyme-mediated polymerization. Monomers polymerized include MMA, S, AM and derivatives. The area has been reviewed by Singh and Kaplan[222] and Gross et al.[223]

One of the most used systems involves use of horseradish peroxidase, a β-diketone (most commonly 2,4-pentandione), and hydrogen peroxide.[222] Since these enzymes contain iron(II), initiation may involve decomposition of hydrogen peroxide by a redox reaction with formation of hydroxy radicals. However, the proposed initiation mechanism[222] involves a catalytic cycle with enzyme activation by hydrogen peroxide and oxidation of the β-diketone to give a species which initiates polymerization. Some influence of the enzyme on tacticity and molecular weight has been reported. However, further study is required to define the origin of the effects observed.[222,223]

8.3.7 Topological Radical Polymerization

In this section we consider topological or topochemical polymerizations where monomers are constrained by being part of an organic crystal,[68,224] a Langmuir-Blodgett film, a liquid crystal, a lipid bilayer, a micellar aggregate,[225] or a supramolecular assembly.[226] Unlike template polymerization there is no parent polymer to organize the monomer. Rather polymerization occurs in a crystalline or otherwise organized phase that may comprise only the monomer.

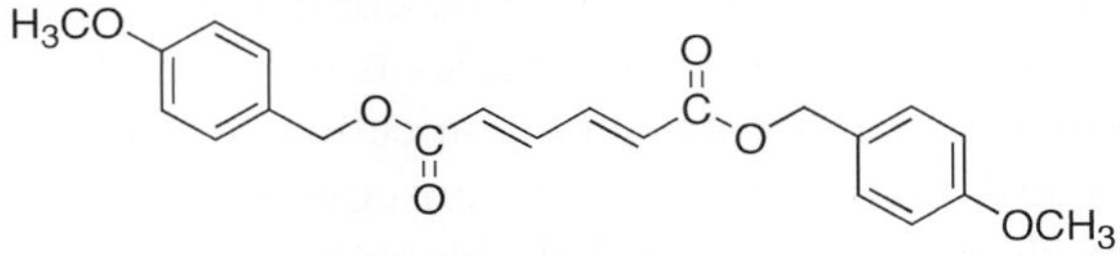

25 (*E,E*)-

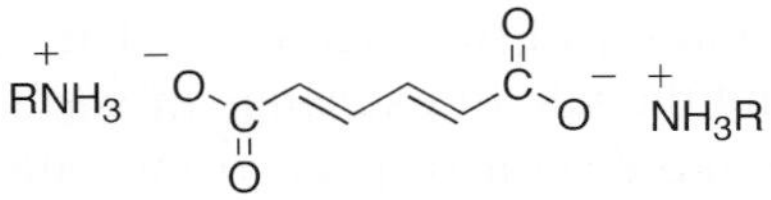

26 (*E,E*)-

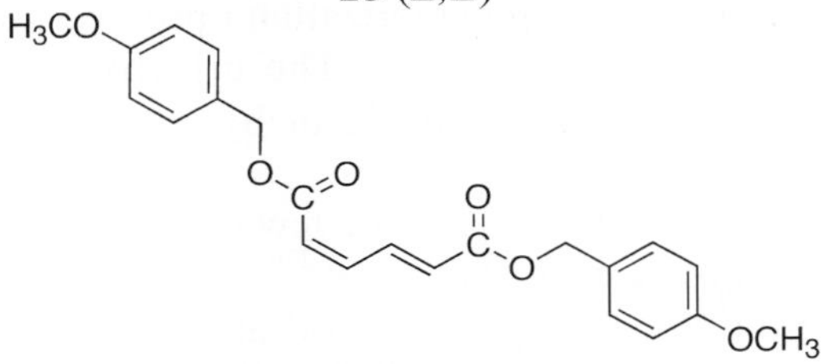

27 (*E,Z*)-

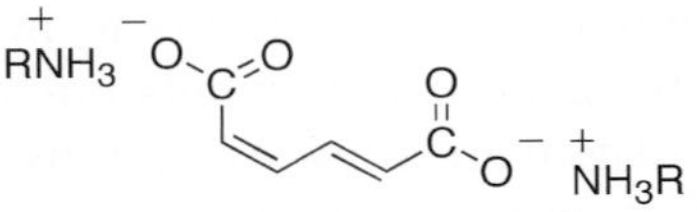

28 (*E,Z*)-

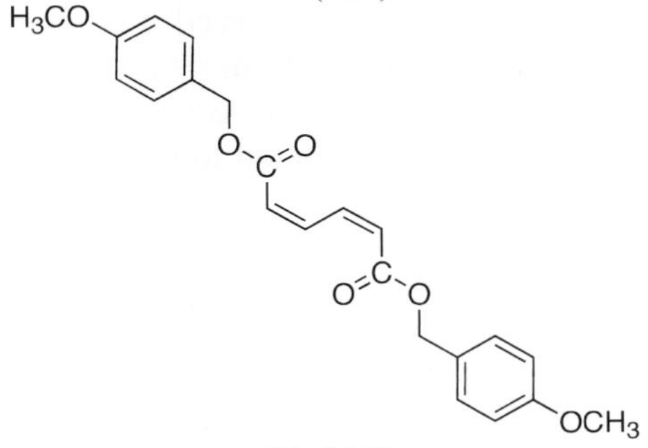

29 (*Z,Z*)-

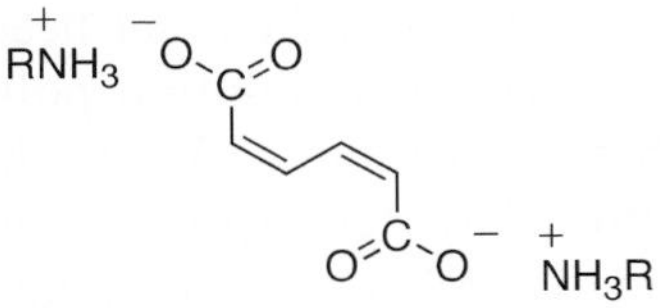

30 (*Z,Z*)-

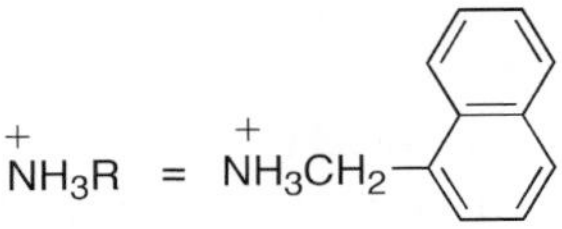

$\overset{+}{N}H_3R = \overset{+}{N}H_3CH_2-$ (naphthyl)

Certain monomers crystallize in a conformation such that they can be zipped together without changing the symmetry of the crystal lattice. In the crystalline state, the arrangement of monomers is strictly determined by crystal packing. Polymerization is usually initiated by irradiation with UV, X- or γ-rays and is assumed to proceed by a radical mechanism. For example, muconic acid esters (**25**, **27**, **29**) and ammonium salts (**26**, **28**, **30**) can be stereospecifically polymerized in the crystalline state to high conversion.[224,227,228] This form of polymerization requires engineering a crystal containing monomer units in appropriate juxtaposition.

Amphiphilic molecules and macromolecules form micelles in aqueous media where the more hydrophobic segments are aggregated to form a core while the more hydrophilic head groups are exposed to the aqueous medium. The molecules can contain monomer functionality within either the more hydrophilic or more hydrophobic segment. The polymerization of surface active monomers has been reviewed by Nagai.[225] Micelles are not static but are dynamic structures and there is rapid exchange of the surfactant monomers between the micellar phase and the aqueous phase and between individual micelles. It has been stated that radical polymerization of linear polymerizable surfactants (surfmers) formed into micelles above the critical micelle concentration (cmc) is unlikely to be controlled by the topology of the micelle.[229] The mobility of the surfactant species is generally high and the rate of exchange of surfactant molecules between the micelle and solution is rapid with respect to the rate of polymerization. As a consequence, neither molecular weight nor polymer stereochemistry is controlled. Nonetheless, rates of polymerization can be high with respect to rates of polymerization observed for a similar monomer concentration in non-organized media. The effective local monomer concentrations in micellar systems can approach, or by organization surpass, those seen in bulk polymerization.

There are a few exceptions to this general rule. One of the few examples of an effect on polymer stereochemistry was provided by Dais et al.[230] who found that polymerization of **31** above the cmc initiated by γ-irradiation at 25 °C yields polymer composed entirely of syndiotactic dyads $P(m) = 0$. When the double bond was distant from the polar head group in **32**, the tacticity observed was similar to that observed in solution polymerization $P(m) \sim 0.18$. Polymerization of **31** at higher temperatures (50 °C) initiated by AIBN also showed no sign of tacticity control. The stereospecific polymerization of **31** was attributed to organization of the methacrylate moiety on the surface of the micelle.

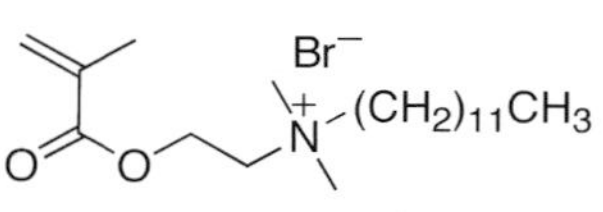

31

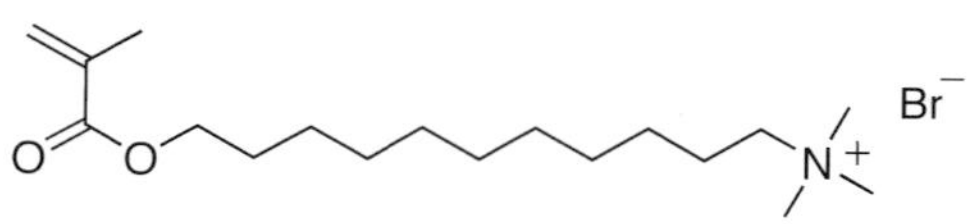

32

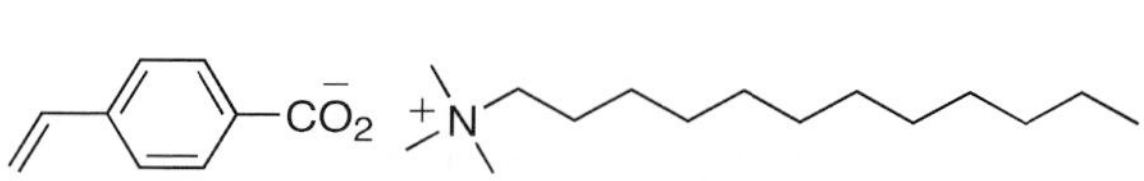

33

Cetyltrimethylammonium 4-vinylbenzoate (**33**) forms rod-like micelles that can be stabilized by radical polymerization. The resulting structure, was observed by small-angle neutron scattering to retain its original rod-like architecture and showed enhanced thermal stability and did not dissociate upon dilution.

Some of the more remarkable examples of this form of topologically controlled radical polymerization were reported by Percec et al.[231-234] Dendron macromonomers were observed to self-assemble at a concentration above 0.20 mol/L in benzene to form spherical micellar aggregates where the polymerizable double bonds are concentrated inside. The polymerization of the aggregates initiated by AIBN showed some living characteristics. Dispersities were narrow and molecular weights were dictated by the size of the aggregate. The shape of the resultant macromolecules, as observed by atomic force microscopy (AFM), was found to depend on $\bar{X}_n$. With $\bar{X}_n<20$, the polymer remained spherical. On the other hand, with $\bar{X}_n>20$, the polymer became cylindrical.[231,232]

Further examples of micellar stabilization when micelles are composed of block copolymers formed by living radical polymerization are mentioned in Section 9.9.2.

8.4 References

1. Moad, G.; Solomon, D.H. *Aust. J. Chem.* **1990**, *43*, 215.
2. Lehrle, R.S.; Peakman, R.E.; Robb, J.C. *Eur. Polym. J.* **1982**, *18*, 517.
3. Rudin, A.; Samanta, M.D.; Reilly, P.M. *J. Appl. Polym. Sci.* **1979**, *24*, 171.
4. Cameron, G.G.; Meyer, J.M.; McWalter, I.T. *Macromolecules* **1978**, *11*, 696.
5. Wall, L.A.; Straus, S.; Florin, R.E.; Fetters, L.J. *J. Res. Nat. Bur. Stand.* **1973**, *77*, 157.
6. Cascaval, C.N.; Straus, S.; Brown, D.W.; Florin, R.E. *J. Polym. Sci., Polym. Symp.* **1976**, *57*, 81.
7. Singh, M.; Nandi, U.S. *J. Polym. Sci., Polym. Lett. Ed.* **1979**, *17*, 121.
8. Howell, B.A.; Cui, Y.M.; Priddy, D.B. *Thermochim. Acta* **2003**, *396*, 167.
9. Guaita, M. *Br. Polym. J.* **1986**, *18*, 226.
10. Grassie, N.; Melville, H.W. *Proc. R. Soc., London* **1949**, *A199*, 1.
11. Grassie, N.; Kerr, W.W. *Trans. Faraday Soc.* **1959**, *55*, 1050.
12. Peterson, J.D.; Vyazovkin, S.; Wight, C.A. *Macromol. Chem. Phys.* **2001**, *202*, 775.
13. Schildknecht, C.E. In *Polymerization Processes*; Schildknecht, C.E.; Skeist, I., Eds.; Wiley: New York, 1977; p 88.
14. Boundy, R.H.; Boyer, R.F., Eds. *Styrene, Its Polymers, Copolymers and Derivatives*; Reinhold: New York, 1952.
15. Krstina, J.; Moad, G.; Solomon, D.H. *Eur. Polym. J.* **1989**, *25*, 767.
16. Moad, G.; Solomon, D.H.; Willing, R.I. *Macromolecules* **1988**, *21*, 855.
17. Cameron, G.G.; MacCallum, J.R. *J. Macromol. Sci., Rev. Macromol. Chem.* **1967**, *C1*, 327.
18. Moad, G.; Solomon, D.H.; Johns, S.R.; Willing, R.I. *Macromolecules* **1984**, *17*, 1094.
19. Krstina, J.; Moad, G.; Willing, R.I.; Danek, S.K.; Kelly, D.P.; Jones, S.L.; Solomon, D.H. *Eur. Polym. J.* **1993**, *29*, 379.
20. Roland, A.I.; Stenzel, M.; Schmidt-Naake, G. *Angew. Makromol. Chem* **1998**, *254*, 69.
21. Roland, A.I.; Schmidt-Naake, G. *J. Anal. Appl. Pyrol.* **2001**, *58*, 143.
22. Postma, A.; Davis, T.P.; Moad, G.; O'Shea, M. *Macromolecules* **2005**, *38*, 5371.
23. Holland, B.J.; Hay, J.N. *Polym. Degr. Stab.* **2002**, *77*, 435.
24. Peterson, J.D.; Vyazovkin, S.; Wight, C.A. *J. Phys. Chem. B* **1999**, *103*, 8087.

25. Stoliarov, S.I.; Westmoreland, P.R.; Nyden, M.R.; Forney, G.P. *Polymer* **2003**, *44*, 883.
26. Manring, L.E. *Macromolecules* **1988**, *21*, 528.
27. Manring, L.E. *Macromolecules* **1991**, *24*, 3304.
28. Hodder, A.N.; Holland, K.A.; Rae, I.D. *J. Polym. Sci., Polym. Lett. Ed.* **1983**, *21*, 403.
29. Manring, L.E.; Sogah, D.Y.; Cohen, G.M. *Macromolecules* **1989**, *22*, 4652.
30. Manring, L.E. *Macromolecules* **1989**, *22*, 2673.
31. Cacioli, P.; Moad, G.; Rizzardo, E.; Serelis, A.K.; Solomon, D.H. *Polym. Bull.* **1984**, *11*, 325.
32. Kashiwagi, T.; Kirata, T.; Brown, J.E. *Macromolecules* **1985**, *18*, 131.
33. Meisters, A.; Moad, G.; Rizzardo, E.; Solomon, D.H. *Polym. Bull.* **1988**, *20*, 499.
34. Bagby, G.; Lehrle, R.S.; Robb, J.C. *Polymer* **1969**, *10*, 683.
35. Kashiwagi, T.; Inaba, A.; Brown, J.E.; Hatada, K.; Kitayama, T.; Masuda, E. *Macromolecules* **1986**, *19*, 2160.
36. Peterson, J.D.; Vyazovkin, S.; Wight, C.A. *Macromol. Rapid Commun.* **1999**, *20*, 480.
37. Rizzardo, E.; Solomon, D.H. *J. Macromol. Sci., Chem.* **1979**, *A13*, 1005.
38. Rizzardo, E.; Solomon, D.H. *Polym. Bull.* **1979**, *1*, 529.
39. Bednarek, D.; Moad, G.; Rizzardo, E.; Solomon, D.H. *Macromolecules* **1988**, *21*, 1522.
40. Aliwi, S.M.; Bamford, C.H. *J. Chem. Soc., Faraday Trans. 1* **1977**, *73*, 776.
41. Bamford, C.H.; White, E.F.T. *Trans. Faraday Soc.* **1958**, *54*, 268.
42. Moad, G.; Ercole, F.; Johnson, C.H.; Krstina, J.; Moad, C.L.; Rizzardo, E.; Spurling, T.H.; Thang, S.H.; Anderson, A.G. *ACS Symp. Ser.* **1998**, *685*, 332.
43. Solomon, D.H.; Rizzardo, E.; Cacioli, P. US 4581429, 1986 (*Chem. Abstr.* **1985**, *102*, 221335q).
44. Starnes, W.H. *Prog. Polym. Sci.* **2002**, *27*, 2133.
45. Endo, K. *Prog. Polym. Sci.* **2002**, *27*, 2021.
46. Ivan, B. *Adv. Chem. Ser.* **1996**, *249*, 19.
47. Vidotto, G.; Crosato-Arnaldi, A.; Talamini, G. *Makromol. Chem.* **1968**, *114*, 217.
48. Puts, R.D.; Sogah, D.Y. *Macromolecules* **1997**, *30*, 3323.
49. Ananchenko, G.; Matyjaszewski, K. *Macromolecules* **2002**, *35*, 8323.
50. Haddleton, D.M.; Duncalf, D.J.; Kukulj, D.; Heming, A.M.; Shooter, A.J.; Clark, A.J. *J. Mater. Chem.* **1998**, *8*, 1525.
51. Johnson, R.M.; Ng, C.; Samson, C.C.M.; Fraser, C.L. *Macromolecules* **2000**, *33*, 8618.
52. Ozerhwskii, B.V.; Reshchupkin, V.P. *Dokl. Phys. Chem. (Engl. Transl.)* **1981**, *254*, 731.
53. Giese, B.; Ghosez, A.; Gobel, T.; Hartung, J.; Huter, O.; Koch, A.; Kroder, K.; Springer, R. In *Free Radicals in Chemistry and Biology*; Minisci, F., Ed.; Kluwer: Dordrecht, 1989; p 97.
54. Clark, A.J.; Jones, K. *Tetrahedron Lett.* **1989**, *30*, 5485.
55. Porter, N.A.; Rosenstein, I.J.; Breyer, R.A.; Bruhnke, J.D.; Wu, W.-X.; McPhail, A.T. *J. Am. Chem. Soc.* **1992**, *114*, 7664.
56. Porter, N.A.; Allen, T.; Breyer, R.A. *J. Am. Chem. Soc.* **1992**, *114*, 7676.
57. Nakano, T.; Tamada, D.; Miyazaki, J.; Kakiuchi, K.; Okamoto, Y. *Macromolecules* **2000**, *33*, 1489.
58. Nakano, T.; Mori, M.; Okamoto, Y. *Macromolecules* **1993**, *26*, 867.

59. Nakano, T.; Okamoto, Y. *Macromolecules* **1999**, *32*, 2391.
60. Nakano, T.; Shikisai, Y.; Okamoto, Y. *Polym. J.* **1996**, *28*, 51.
61. Puzin, Y.I.; Prokudina, E.M.; Yumagulova, R.K.; Muslukhov, R.R.; Kolesov, S.V. *Dokl. Phys. Chem. (Engl. Transl.)* **2002**, *386*, 211.
62. Nakano, T.; Okamoto, Y.; Sogah, D.Y.; Zheng, S. *Macromolecules* **1995**, *28*, 8705.
63. Nakano, T.; Sogah, D.Y. *J. Am. Chem. Soc.* **1997**, *117*, 534.
64. Wulff, G.; Dhal, P.K. *Angew. Chem. Int. Ed. Engl.* **1989**, *28*, 196.
65. Tsuji, M.; Sakai, R.; Satoh, T.; Kaga, H.; Kakuchi, T. *Macromolecules* **2002**, *35*, 8255.
66. Kakuchi, T.; Tsuji, M.; Satoh, T. *ACS Symp. Ser.* **2003**, *854*, 206.
67. Okamoto, Y.; Habaue, S.; Isobe, Y.; Nakano, T. *Macromol. Symp.* **2002**, *183*, 83.
68. Matsumoto, A. In *Handbook of Radical Polymerization*; Davis, T.P.; Matyjaszewski, K., Eds.; John Wiley & Sons: Hoboken, 2002; p 691.
69. Coote, M.L.; Davis, T.P. In *Handbook of Solvents*; Wypych, G., Ed.; William Andrew Publishing: Norwich, New York, 2001; p 777.
70. Coote, M.L.; Davis, T.P.; Klumperman, B.; Monteiro, M.J. *J. Macromol. Sci. Rev. Macromol. Chem. Phys.* **1998**, *C38*, 567.
71. Barton, J.; Borsig, E. *Complexes in Free Radical Polymerization*; Elsevier: Amsterdam, 1988.
72. Gromov, V.F.; Khomiskovskii, P.M. *Russ. Chem. Rev. (Engl. Transl.)* **1979**, *48*, 1040.
73. Kamachi, M. *Adv. Polym. Sci.* **1981**, *38*, 55.
74. Beuermann, S.; Buback, M. *Prog. Polym. Sci.* **2002**, *27*, 191.
75. Reichardt, C. *Solvent Effects in Organic Chemistry*; Verlag Chemie: Weinheim, 1978.
76. Brumby, S. *Polym. Commun.* **1989**, *30*, 13.
77. Paleos, C.M.; Malliaris, A. *J. Macromol. Sci., Rev. Macromol. Chem. Phys.* **1988**, *C28*, 403.
78. Kamachi, M.; Liaw, D.J.; Nozakura, S.-I. *Polym. J.* **1979**, *11*, 921.
79. Kamachi, M.; Satoh, J.; Nozakura, S. *J. Polym. Sci., Polym. Chem. Ed.* **1978**, *16*, 1789.
80. Hatada, K.; Terawaki, Y.; Kitayama, T.; Kamachi, M.; Tamaki, M. *Polym. Bull.* **1981**, *4*, 451.
81. Morrison, B.R.; Piton, M.C.; Winnik, M.A.; Gilbert, R.G.; Napper, D.H. *Macromolecules* **1993**, *26*, 4368.
82. Kamachi, M.; Liaw, D.J.; Nozakura, S. *Polym. J.* **1981**, *13*, 41.
83. Bamford, C.H.; Brumby, S. *Makromol. Chem.* **1967**, *105*, 122.
84. O'Driscoll, K.F.; Monteiro, M.J.; Klumpermann, B. *J. Polym. Sci., Part A: Polym. Chem.* **1997**, *35*, 515.
85. Zammit, M.D.; Davis, T.P.; Willet, G.D.; O'Driscoll, K.F. *J. Polym. Sci. A: Polym. Chem.* **1997**, *35*, 2311.
86. Olaj, O.F.; Schnoll-Bitai, I. *Monatsh. Chem.* **1999**, *130*, 731.
87. Coote, M.L.; Davis, T.P. *Eur. Polym. J.* **2000**, *36*, 2423.
88. Burnett, G.M.; Cameron, G.G.; Joiner, S.N. *Trans. Faraday Soc.* **1973**, *69*, 322.
89. Tsutsumi, K.; Tsukahara, Y.; Okamota, Y. *Polym. J.* **1994**, *26*, 13.
90. Nakano, T.; Okamoto, Y. *ACS Symp. Ser.* **1997**, *685*, 451.
91. Nagara, Y.; Yamada, K.; Nakano, T.; Okamoto, Y. *Polymer J.* **2001**, *33*, 534.
92. Yamada, K.; Nakano, T.; Okamoto, Y. *Macromolecules* **1998**, *31*, 7598.

93. Isobe, Y.; Yamada, K.; Nakano, T.; Okamoto, Y. *J. Polym. Sci., Part A: Polym. Chem.* **2000**, *38*, 4693.
94. Isobe, Y.; Yamada, K.; Nakano, T.; Okamoto, Y. *Macromolecules* **1999**, *32*, 5979.
95. Krakovyak, M.G.; Anufrieva, E.V.; Sycheva, E.A.; Sheveleva, T.V. *Macromolecules* **1993**, *26*, 7375.
96. Ishigaki, Y.; Takahashi, K.; Fukuda, H. *Macromol. Rapid Commun.* **2000**, *21*, 1024.
97. Plochocka, K. *J. Macromol. Sci., Rev. Macromol. Chem.* **1981**, *C20*, 67.
98. Madruga, E.L. *Prog. Polym. Sci.* **2002**, *27*, 1879.
99. Georgiev, G.S.; Dakova, I.G. *Macromol. Chem. Phys.* **1994**, *195*, 1695.
100. Lewis, F.M.; Walling, C.; Cummings, W.; Briggs, E.R.; Mayo, F.R. *J. Am. Chem. Soc.* **1948**, 1519.
101. Price, C.C.; Walsh, J.G. *J. Polym. Sci.* **1951**, *6*, 239.
102. Pichot, C.; Zaganiaris, E.; Guyot, A. *J. Polym. Sci., Polym. Symp.* **1975**, *52*, 55.
103. Hill, D.J.T.; Lang, A.P.; Munro, P.D.; O'Donnell, J.H. *Eur. Polym. J.* **1992**, *28*, 391.
104. Hill, D.J.T.; Lang, A.P.; Munro, P.D.; O'Donnell, J.H. *Eur. Polym. J.* **1989**, *28*, 391.
105. Asakura, J.; Yoshihara, M.; Matsubara, Y.; Maeshima, T. *J. Macromol. Sci., Chem.* **1981**, *A15*, 1473.
106. Van Der Meer, R.; Aarts, M.W.A.M.; German, A.L. *J. Polym. Sci., Polym. Chem. Ed.* **1980**, *18*, 1347.
107. Cameron, G.G.; Esslemont, G.F. *Polymer* **1972**, *13*, 435.
108. Ito, T.; Otsu, T. *J. Macromol. Sci. Chem* **1969**, *A3*, 197.
109. San Roman, J.; Madruga, E.L. *Angew. Makromol. Chem.* **1980**, *86*, 1.
110. Bonta, G.; Gallo, B.M.; Russo, S. *Polymer* **1975**, *16*, 429.
111. Busfield, W.K.; Low, R.B. *Eur. Polym. J.* **1975**, *11*, 309.
112. Harwood, H.J. *Makromol. Chem., Macromol. Symp.* **1987**, *10/11*, 331.
113. Park, K.Y.; Santee, E.R.; Harwood, H.J. *Eur. Polym. J.* **1989**, *25*, 651.
114. Christov, L.K.; Georgiev, G.S. *Macromol. Theory Simul.* **2000**, *9*, 715.
115. Klumperman, B.; Kraeger, I.R. *Macromolecules* **1994**, *27*, 1529.
116. Maxwell, I.A.; Aerdts, A.M.; German, A.L. *Macromolecules* **1993**, *26*, 1956.
117. Klumperman, B.; O'Driscoll, K.F. *Polymer* **1993**, *34*, 1032.
118. Klumperman, B.; Vonk, G. *Eur. Polym. J.* **1994**, *30*, 955.
119. Davis, T.P. *Polym. Commun.* **1990**, *31*, 442.
120. Coote, M.L.; Johnston, L.P.M.; Davis, T.P. *Macromolecules* **1997**, *30*, 8191.
121. Kaim, A. *Macromol. Theory Simul.* **1997**, *6*, 907.
122. Fernandez-Monreal, C.; Sanchez-Chaves, M.; Martinez, G.; Madruga, E.L. *Acta Polym.* **1999**, *50*, 408.
123. Fernandez-Garcia, M.; Fernandez-Sanz, M.; Madruga, E.L.; Cuervo-Rodriguez, R.; Hernandez-Gordo, V.; Fernandez-Monreal, M.C. *J. Polym. Sci., Part A: Polym. Chem.* **2000**, *38*, 60.
124. Chambard, G.; Klumperman, B.; German, A.L. *Polymer* **1999**, *40*, 4459.
125. Kaim, A. *J. Polym. Sci., Part A: Polym. Chem.* **2000**, *38*, 846.
126. Madruga, E.L.; Fernandez-Garcia, M. *Macromol. Chem. Phys.* **1996**, *197*, 3743.
127. de la Fuente, J.L.; Madruga, E.L. *Macromol. Chem. Phys.* **1999**, *200*, 1639.
128. Fernandez-Monreal, C.; Martinez, G.; Sanchez-Chaves, M.; Madruga, E.L. *J. Polym. Sci., Part A: Polym. Chem.* **2001**, *39*, 2043.
129. Coote, M.L.; Davis, T.P. *Prog. Polym. Sci.* **1999**, *24*, 1217.
130. Coote, M.L.; Davis, T.P. In *Handbook of Radical Polymerization*; Davis, T.P.; Matyjaszewski, K., Eds.; John Wiley & Sons: Hoboken, 2002; p 263.
131. Krstina, J.; Moad, G.; Solomon, D.H. *Eur. Polym. J.* **1992**, *28*, 275.

132. Semchikov, Y.D. *Polym. Sci. USSR (Engl. Transl.)* **1990**, *32*, 177.
133. Semchikov, Y.D. *Macromol. Symp.* **1996**, *111*, 317.
134. Kuchanov, S.I.; Russo, S. *Macromolecules* **1997**, *30*, 4511.
135. Tsukahara, Y.; Hayashi, N.; Jiang, X.L.; Yamashita, Y. *Polym. J.* **1989**, *21*, 377.
136. Cooper, A.I. *J. Mater. Chem.* **2000**, *10*, 207.
137. Kendall, J.L.; Canelas, D.A.; Young, J.L.; DeSimone, J.M. *Chem. Rev.* **1999**, *99*, 543.
138. van Herk, A.M.; Manders, B.G.; Canelas, D.A.; Quadir, M.A.; DeSimone, J.M. *Macromolecules* **1997**, *30*, 4780.
139. Quadir, M.A.; DeSimone, J.M.; van Herk, A.M.; German, A.L. *Macromolecules* **1998**, *31*, 6481.
140. Beuermann, S.; Buback, M.; Schmaltz, C.; Kuchta, F.D. *Macromol. Chem. Phys.* **1998**, *199*, 1209.
141. Beuermann, S.; Buback, M.; Schmaltz, C. *Macromolecules* **1998**, *31*, 8069.
142. Beuermann, S.; Buback, M.; Isemer, C.; Lacik, I.; Wahl, A. *Macromolecules* **2002**, *35*, 3866.
143. Beuermann, S.; Buback, M.; Nelke, D. *Macromolecules* **2001**, *34*, 6637.
144. Welton, T. *Chem. Rev.* **1999**, *99*, 2071.
145. Kubisa, P. *Prog. Polym. Sci.* **2004**, *29*, 3.
146. Harrisson, S.; Mackenzie, S.R.; Haddleton, D.M. *Chem. Commun.* **2002**, 2850.
147. Harrisson, S.; Mackenzie, S.R.; Haddleton, D.M. *Macromolecules* **2003**, *36*, 5072.
148. Biedron, T.; Kubisa, P. *Macromol. Rapid Commun.* **2001**, *22*, 1237.
149. Biedron, T.; Kubisa, P. *J. Polym. Sci., Part A: Polym. Chem.* **2002**, *40*, 2799.
150. Biedron, T.; Kubisa, P. *Polym. Int.* **2003**, *52*, 1584.
151. Sarbu, T.; Matyjaszewski, K. *Macromol. Chem. Phys.* **2001**, *202*, 3379.
152. Ma, H.Y.; Wan, X.H.; Chen, X.F.; Zhou, Q.F. *J. Polym. Sci., Part A: Polym. Chem.* **2003**, *41*, 143.
153. Zhao, Y.L.; Zhang, J.M.; Jiang, J.; Chen, C.F.; Xi, F. *J. Polym. Sci., Part A: Polym. Chem.* **2002**, *40*, 3360.
154. Sarbu, T.; Pintauer, T.; McKenzie, B.; Matyjaszewski, K. *J. Polym. Sci., Part A: Polym. Chem.* **2002**, *40*, 3153.
155. Carmichael, A.J.; Haddleton, D.M.; Bon, S.A.F.; Seddon, K.R. *Chem. Commun.* **2000**, 1237.
156. Ma, H.; Wan, X.; Chen, X.; Zhou, Q.-F. *Polymer* **2003**, *44*, 5311.
157. Perrier, S.; Davis, T.P.; Carmichael, A.J.; Haddleton, D.M. *Eur. Polym. J.* **2003**, *39*, 417.
158. Perrier, S.; Davis, T.P.; Carmichael, A.J.; Haddleton, D.M. *Chem. Commun.* **2002**, 2226.
159. Ryan, J.; Aldabbagh, F.; Zetterlund, P.B.; Yamada, B. **2004**, *25*, 930.
160. Zhang, H.W.; Hong, K.; Mays, J.W. *Polym. Bull.* **2004**, *52*, 9.
161. Zhang, H.W.; Hong, K.L.; Jablonsky, M.; Mays, J.W. *Chem. Commun.* **2003**, 1356.
162. Bamford, C.H. In *Alternating Copolymers*; Cowie, J.M.G., Ed.; Plenum: New York, 1985; p 75.
163. Bamford, C.H.; Jenkins, A.D.; Johnston, R. *Proc. R. Soc., London* **1957**, *A241*, 364.
164. Zubov, V.P.; Valuev, L.I.; Kabanov, V.A.; Kargin, V.A. *J. Polym. Sci., Part A-1* **1971**, *9*, 833.
165. Tanaka, T.; Kato, H.; Sakai, I.; Sato, T.; Ota, T. *Makromol. Chem., Rapid Commun.* **1987**, *8*, 223.
166. Otsu, T.; Yamada, B. *J. Macromol. Sci. Chem* **1966**, *A1*, 61.

167. Isobe, Y.; Nakano, T.; Okamoto, Y. *J. Polym. Sci., Part A: Polym. Chem.* **2001**, *39*, 1463.
168. Habaue, S.; Yamada, H.; Uno, T.; Okamoto, Y. *J. Polym. Sci., Part A: Polym. Chem.* **1997**, *35*, 721.
169. Baraki, H.; Habaue, S.; Okamoto, Y. *Macromolecules* **2001**, *34*, 4724.
170. Okamoto, Y.; Habaue, S.; Isobe, Y. *ACS Symp. Ser.* **2003**, *854*, 59.
171. Habaue, S.; Isobe, Y.; Okamoto, Y. *Tetrahedron* **2002**, *58*, 8205.
172. Isobe, Y.; Fujioka, D.; Habaue, S.; Okamoto, Y. *J. Am. Chem. Soc.* **2001**, *123*, 7180.
173. Ray, B.; Isobe, Y.; Matsumoto, K.; Habaue, S.; Okamoto, Y.; Kamigaito, M.; Sawamoto, M. *Macromolecules* **2004**, *37*, 1702.
174. Suito, Y.; Isobe, Y.; Habaue, S.; Okamoto, Y. *J. Polym. Sci., Part A: Polym. Chem.* **2002**, *40*, 2496.
175. Isobe, Y.; Suito, Y.; Habaue, S.; Okamoto, Y. *J. Polym. Sci., Part A: Polym. Chem.* **2003**, *41*, 1027.
176. Hirooka, M.; Yabuuchi, H.; Morita, S.; Kawasumi, S.; Nakaguchi, K. *J. Polym. Sci., Polym. Lett.* **1967**, *5*, 47.
177. Hirooka, M.; Yabuuchi, H.; Morita, S.; Kawasumi, S.; Nakaguchi, K. *J. Polym. Sci., Part A-1: Polym. Chem.* **1968**, *6*, 1381.
178. Momtaz-Afchar, J.; Polton, A.; Tardi, M.; Sigwalt, P. *Eur. Polym. J.* **1985**, *21*, 1067.
179. Rogueda, C.; Polton, A.; Tardi, M.; Sigwalt, P. *Eur. Polym. J.* **1989**, *25*, 1259.
180. Rogueda, C.; Polton, A.; Tardi, M.; Sigwalt, P. *Eur. Polym. J.* **1989**, *25*, 1251.
181. Rogueda, C.; Tardi, M.; Polton, A.; Sigwalt, P. *Eur. Polym. J.* **1989**, *25*, 885.
182. Golubev, V.B.; Zubov, V.P.; Georgiev, G.S.; Stoyachenko, I.L.; Kabanov, V.A. *J. Polym. Sci., Polym. Chem. Ed.* **1973**, *11*, 2463.
183. Seno, M.; Matsumura, N.; Nakamura, H.; Sato, T. *J. Appl. Polym. Sci.* **1997**, *63*, 1361.
184. Sherrington, D.C.; Slark, A.T.; Taskinen, K.A. *Macromol. Chem. Phys.* **2002**, *203*, 1427.
185. Wang, H.; Chu, G.; Srisiri, W.; Padias, A.B.; Hall, H.K. *Acta Polym.* **1994**, *45*, 26.
186. Lyons, R.A.; Moad, G.; Senogles, E. *Eur. Polym. J.* **1993**, *29*, 389.
187. Krstina, J.; Moad, G.; Solomon, D.H. *Polym. Bull.* **1992**, *27*, 425.
188. Fellows, C.M.; Senogles, E. *Eur. Polym. J.* **2001**, *37*, 1091.
189. Fellows, C.M.; Senogles, E. *Eur. Polym. J.* **1999**, *35*, 9.
190. Kirci, B.; Lutz, J.F.; Matyjaszewski, K. *Macromolecules* **2002**, *35*, 2448.
191. Lutz, J.F.; Kirci, B.; Matyjaszewski, K. *Macromolecules* **2003**, *36*, 3136.
192. Bamford, C.H. *Chem. Aust.* **1982**, *49*, 341.
193. Tan, Y.Y. In *Recent Advances in Mechanistic and Synthetic Aspects of Polymerization*; Dordrecht: Reidel, 1987; p 281.
194. Tan, Y.Y. In *Comprehensive Polymer Science*; Eastmond, G.C.; Ledwith, A.; Russo, S.; Sigwalt, P., Eds.; Pergamon: Oxford, 1989; Vol. 3, p 245.
195. Polowinski, S. *Prog. Polym. Sci.* **2002**, *27*, 537.
196. Buter, R.; Tan, Y.Y.; Challa, G. *J. Polym. Sci., Part A-1* **1972**, *10*, 1031.
197. Schomaker, E.; Challa, G. *Macromolecules* **1988**, *21*, 3506.
198. Nodono, M.; Makino, T.; Nishida, K. *React. Funct. Polym.* **2003**, *57*, 157.
199. Matsuzaki, K.; Kanai, T.; Ichijo, C.; Yuzawa, M. *Makromol. Chem.* **1984**, *185*, 2291.
200. Serizawa, T.; Hamada, K.; Akashi, M. *Nature* **2004**, *429*, 52.

201. van de Grampel, H.T.; Tan, Y.Y.; Challa, G. *Macromolecules* **1991**, *24*, 3767.
202. van de Grampel, H.T.; Tan, Y.Y.; Challa, G. *Macromolecules* **1991**, *24*, 3773.
203. Ferguson, J.; Al-Alawi, S.; Grannmayeth, R. *Eur. Polym. J.* **1985**, *19*, 475.
204. Chapiro, A. *Pure Appl. Chem.* **1981**, *53*, 643.
205. Polowinski, S. *Eur. Polym. J.* **1983**, *19*, 679.
206. Natansohn, A. *Polym. Prepr. (Am. Chem. Soc., Div. Polym. Chem)* **1984**, *25(2)*, 65.
207. Kammerer, H. *Angew. Chem., Int. Ed. Engl.* **1965**, *4*, 952.
208. Kern, W.; Kammerer, H. *Pure Appl. Chem.* **1967**, *15*, 421.
209. Kammerer, H.; Onder, N. *Makromol. Chem.* **1968**, *111*, 67.
210. Feldman, K.S.; Bobo, J.S.; Ensel, S.M.; Lee, Y.B.; Weinreb, P.H. *J. Org. Chem.* **1990**, *55*, 474.
211. Feldman, K.S.; Lee, Y.B. *J. Am. Chem. Soc.* **1987**, *109*, 5850.
212. Wulff, G.; Kemmerer, R.; Vogt, B. *J. Am. Chem. Soc.* **1987**, *109*, 7449.
213. Jantas, R. *J. Polym. Sci., Part A: Polym. Chem.* **1990**, *28*, 1973.
214. Jantas, R.; Polowinski, S. *J. Polym. Sci., Part A: Polym. Chem.* **1986**, *24*, 1819.
215. Jantas, R. *J. Polym. Sci., Part A: Polym. Chem.* **1994**, *32*, 295.
216. Bamford, C.H. In *Developments in Polymerization 2*; Howard, R.N., Ed.; Applied Science: London, 1979; Vol. 49, p 215.
217. Saito, R.; Kobayashi, H. *Macromolecules* **2002**, *35*, 7207.
218. Saito, R.; Okuno, Y.; Kobayashi, H. *J. Polym. Sci., Part A: Polym. Chem.* **2001**, *39*, 3539.
219. Saito, R.; Yamaguchi, K. *Macromolecules* **2003**, *36*, 9005.
220. Sugiyama, J.-I.; Yokozawa, T.; Endo, T. *Macromolecules* **1994**, *27*, 1987.
221. Sugiyama, J.-I.; Yokozawa, T.; Endo, T. *J. Am. Chem. Soc.* **1993**, *115*, 2041.
222. Singh, A.; Kaplan, D.L. *J. Polym. Envir.* **2002**, *10*, 85.
223. Gross, R.A.; Kumar, A.; Kalra, B. *Chem. Rev.* **2001**, *101*, 2097.
224. Matsumoto, A. *ACS Symp. Ser.* **2000**, *768*, 93.
225. Nagai, K. *Trends Polym. Sci.* **1996**, *4*, 122.
226. Tajima, K.; Aida, T. *Chem. Commun.* **2000**, 2399.
227. Nagahama, S.; Matsumoto, A. *J. Polym. Sci., Part A: Polym. Chem.* **2004**, *42*, 3922.
228. Nagahama, S.; Tanaka, T.; Matsumoto, A. *Angew. Chem. Int. Ed. Engl.* **2004**, *43*, 3811.
229. Cochin, D.; Zana, R.; Candau, F. *Macromolecules* **1993**, *26*, 5765.
230. Dais, P.; Paleos, C.M.; Nika, G.; Malliaris, A. *Macromol. Chem. Phys.* **1993**, *194*, 445.
231. Percec, V.; Ahn, C.H.; Ungar, G.; Yeardley, D.J.P.; Moller, M.; Sheiko, S.S. *Nature* **1998**, *391*, 161.
232. Percec, V.; Ahn, C.H.; Cho, W.D.; Jamieson, A.M.; Kim, J.; Leman, T.; Schmidt, M.; Gerle, M.; Moller, M.; Prokhorova, S.A.; Sheiko, S.S.; Cheng, S.Z.D.; Zhang, A.; Ungar, G.; Yeardley, D.J.P. *J. Am. Chem. Soc.* **1998**, *120*, 8619.
233. Percec, V.; Mitchell, C.M.; Cho, W.D.; Uchida, S.; Glodde, M.; Ungar, G.; Zeng, X.B.; Liu, Y.S.; Balagurusamy, V.S.K.; Heiney, P.A. *J. Am. Chem. Soc* **2004**, *126*, 6078.
234. Percec, V.; Bera, T.K.; Glodde, M.; Fu, Q.Y.; Balagurusamy, V.S.K.; Heiney, P.A. *Chem. Eur. J.* **2003**, *9*, 921.

9
Living Radical Polymerization

9.1 Introduction

The first demonstration of living polymerization and the current definition of the process can be attributed to Swarc.[1,2] Living polymerization mechanisms offer polymers of controlled composition, architecture and molecular weight distribution. They provide routes to narrow dispersity end-functional polymers, to high purity block copolymers, and to stars and other more complex architectures. Traditional methods of living polymerization are based on ionic, coordination or group transfer mechanisms. Ideally, the mechanism of living polymerization involves only initiation and propagation steps. All chains are initiated at the commencement of polymerization and propagation continues until all monomer is consumed. The combination of a living mechanism with the scope and versatility of the radical process should allow a wider selection of monomers and monomer combinations and more freedom in choosing reaction conditions. This potential and the applications that follow have provided the impetus for the very significant research efforts that have been devoted to this area over the last decade. In this chapter, we discuss the various approaches that have been developed in moving towards a living radical polymerization paying particular attention to the mechanism and the scope of each method.

At the time of the first edition of this book (1995),[3] this field was still very much in its infancy. NMP was described, though little had been published in the open literature, and methods such as ATRP and RAFT had not been reported. Since 1995, the area has expanded dramatically and by themselves living/controlled/mediated processes now account for a very substantial fraction of all research on radical polymerization (Chapter 1). The development of this field over this period can be followed in the publications following successful ACS symposia held in 1997,[4] 2000[5] and 2002[6] and SML meetings held in 1996[7] and 2001.[8] Publications continue to appear at a rapid rate. Matyjaszewski[9] has provided an overview of the history and development of living radical polymerization through 2001 in the *Handbook of Radical Polymerization*.[10]

9.1.1 Living? Controlled? Mediated?

The terminology used in this chapter deserves some mention. Currently there is controversy over the use of the terms "living" and "controlled" in the context of describing a radical polymerization.[11-15] The current IUPAC recommendation, that a living polymerization is "a chain polymerization from which irreversible chain transfer and irreversible chain termination (deactivation) are absent", would preclude use of the term "living" in the context of a radical process.[16] The use of the adjective "controlled" by itself to designate these polymerizations is also contrary to IUPAC recommendations.[16] The adjective "controlled" should only be used when the particular aspect of polymerization that is being controlled is specified. It is not recommended that "controlled" be used in an exclusive sense to mean a particular form of polymerization since the word has an established, much wider, usage. The construct "controlled living polymerization" would seem acceptable when used to refer to those living polymerizations whose outcomes are defined by controlling the reaction conditions or other features. The word "controlled" should not be used to indicate that systems have a lower degree of livingness. Other terms such as "pseudo-living" and "quasi-living" are also discouraged.[16] It has been stated that the definition of living polymerization "tolerates no restrictive adjectives implying something close to but not strictly living".[11]

For this book, we have decided to entitle this chapter "Living Radical Polymerization" and use the term throughout. It is a chapter describing various approaches to living radical polymerization. We do not intend to imply that termination is absent from all or, indeed, any of the polymerizations described, only that the polymerizations display at least some of the observable characteristics normally associated with living polymerization.

9.1.2 Tests for Living (Radical) Polymerization

Following on from the above, various methods have been described to test and/or rank the "livingness" of polymerization processes.[11,12,17-20] All of these tests have limitations.. The following list paraphrases a set of criteria for living polymerization set out by Quirk and Lee[11] who also critically assessed their applicability primarily in the context of living anionic polymerization.

(a) "Living polymerizations proceed until all monomer is consumed and may continue growth if further monomer is added." This criterion paraphrases one of Szwarc's definitions of living polymerization.[1,2] It becomes a rigorous criterion if we add "and the number of living chains remains constant".

(b) "In a living polymerization the molecular weight increases linearly with conversion." This contrasts with observations for conventional radical polymerizations where molecular weights are initially high and decrease with conversion due to monomer depletion (Figure 9.1). However, molecular weights obtained in radical polymerizations with conventional transfer agents with $C_{tr}>1$ will increase with conversion and may meet this test. Expressions for the dependence of molecular weight on conversion for NMP (and similar polymerizations), ATRP and RAFT appear in Sections 9.3.1.2, 9.4.1 and 9.5.1

respectively. A plot of $\overline{M}_n$ vs conversion will remain linear even in circumstances where there is a loss of a substantial fraction of the living chains, although in that case there will be a broadening of the molecular weight distribution.

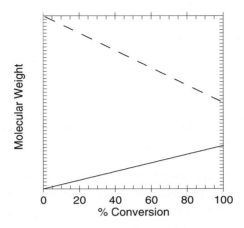

Figure 9.1 Predicted evolution of molecular weight (arbitrary units) with monomer conversion for a conventional radical polymerization with a constant rate of initiation (– – –) and a living polymerization (———).

(c) "In a living polymerization the concentration of active species remains constant." A plot of $\ln([M]_0/[M]_t)$ vs time should be linear. In many conventional radical polymerizations a steady state is established such that, over a wide conversion range, the concentration of active chains remains approximately constant. Thus, these polymerizations will meet this test. Conversely, some living polymerizations with reversible deactivation will not meet this test (Section 9.3.1.3). A rigorous criterion that also covers these cases is that the total concentration of active and dormant chains should remain constant. However, this is more difficult to establish from kinetic measurements alone.

(d) "Living polymerizations provide narrow molecular weight distributions." This is a more qualitative test. What constitutes low dispersity? Theoretically, a dispersity ($\overline{X}_w/\overline{X}_n$) of 1.5 is the narrowest achievable in a conventional radical polymerization with termination by combination for long chains (Section 5.2.1.3). An ideal living polymerization can provide a Poisson molecular weight distribution and $\overline{X}_w/\overline{X}_n = 1+1/\overline{X}_n$; $\overline{X}_w/\overline{X}_n = 1.01$ for $\overline{X}_n = 100$ (Figure 9.2). The better living radical systems produce $\overline{X}_w/\overline{X}_n$ in the range 1.05-1.2. Errors associated with measuring the dispersity can be significant and most cause an underestimate of the actual value. A low dispersity alone does not imply the absence of side reactions.

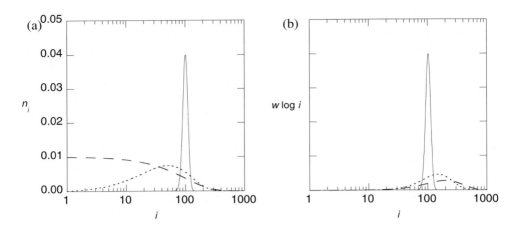

Figure 9.2 Calculated (a) number and (b) GPC distributions for three polymers each with $\bar{X}_n=100$. The number distributions of chains formed by conventional radical polymerization with termination by disproportionation or chain transfer ($----$, $\sum n_i = 1.0$, $\bar{X}_w/\bar{X}_n=2.0$) or termination by combination ($------$, $\sum n_i = 1.0$, $\bar{X}_w/\bar{X}_n=1.5$) were calculated as discussed in Section 5.2.1.3. The number distribution of chains formed in an ideal living polymerization (———, $\sum n_i = 1.0$, $\bar{X}_w/\bar{X}_n=1.01$) was calculated using a Poisson distribution function.

(e) "Block copolymers can be prepared by sequential addition of monomers." This is a special case of (a) above.

(f) "End groups are retained allowing end-functional polymers to be obtained in quantitative yield." Assessment of the fraction of living chains can provide a quantitative measure of the quality of a living polymerization. Currently, the most used methods for end group determination are NMR and mass spectrometry. Some discussion on these techniques is provided in Sections 3.5.3.2 and 3.5.3.4 respectively.

Quirk and Lee concluded "there is no single criterion which is satisfactory for determination of whether a given polymerization is living or not."[11] Most of the radical polymerizations discussed in this chapter meet one or more of these criteria.

9.2 Agents Providing Reversible Deactivation

The kinetics and mechanism of living radical polymerization have been reviewed by Fischer,[21] Fukuda et al.,[22] and Goto and Fukuda.[23] In conventional radical polymerization, new chains are continually formed through initiation while existing chains are destroyed by radical-radical termination. The steady state concentration of propagating radicals is $\sim 10^{-7}$ M and an individual chain will have a lifetime of only 1-10 s before termination within a total reaction time that is typically greater than 10000 s. A consequence is that long chains are formed early in the process and (in the absence of other influences) molecular weights decrease with monomer conversion due to the depletion of monomer (Figure 9.1). In conventional (classical anionic[1,2]) living polymerization all chains are initiated at the beginning of the reaction and grow until all monomer is consumed. As a consequence, molecular weight increases linearly with conversion and the molecular weight distribution is narrow.

The propensity of radicals to undergo self-reaction thus precludes the use of the simple strategy applied in anionic polymerization in developing a living radical polymerization. Radical polymerizations can display the characteristics normally associated with living polymerization in the presence of species that reversibly deactivate or terminate chains. These reagents control the concentration of active propagating species by maintaining a majority of chains in a dormant form. In homogeneous radical polymerization the rate of radical-radical termination is proportional to the square of the radical concentration ($R_t \propto [P_n\bullet]^2$). Thus, the incidence of termination can be reduced relative to propagation ($R_p \propto [P_n\bullet]$) by reducing the radical concentration.

In living radical polymerization, the concentration of propagating radicals is usually similar to or lower than that in conventional radical polymerization (*i.e.* $\leq 10^{-7}$ M). For control, and to retain a high fraction of living chains, the lifetime of chains in their active state must be significantly less than in the conventional process (<<1-10 s). A rapid equilibration between active and dormant forms then ensures that all propagating species have equal opportunity for chain growth. All chains grow intermittently.

It is not necessary that living radical polymerizations be slow. However, it follows from the above discussion that, for a high fraction of living chains, either the final degree of polymerization must be significantly lower than that in an otherwise similar conventional process or that conditions must be chosen such that the rate of polymerization is substantially lower.

Heterogeneous polymerization processes (emulsion, miniemulsion, non-aqueous dispersion) offer another possibility for reducing the rate of termination through what are known as compartmentalization effects. In emulsion polymerization, it is believed that the mechanism for chain stoppage within the particles is not radical-radical termination but transfer to monomer (Section 5.2.1.5). These possibilities have provided impetus for the development of living heterogeneous polymerization (Sections 9.3.6.6, 9.4.3.2, 9.5.3.6).

We can distinguish several sub-classes of activation-deactivation processes according to their mechanism. These are shown in Scheme 9.1-Scheme 9.3.

(a) Those giving deactivation by reversible coupling and involving a unimolecular activation process as shown in Scheme 9.1. $P_n\bullet$ is a propagating radical (an active chain). The deactivator (X) is usually, though not always, a stable radical. However, X may also be an even electron (diamagnetic) species, for example, diphenylethylene (Section 9.3.5). In this case P_n-X would be a persistent radical, or a transition metal complex, for example, a low spin cobalt(II) complex (Section 9.3.9). These systems are discussed in Section 9.3. Possibly the best known process is nitroxide-mediated polymerization (NMP) (Section 9.3.6).

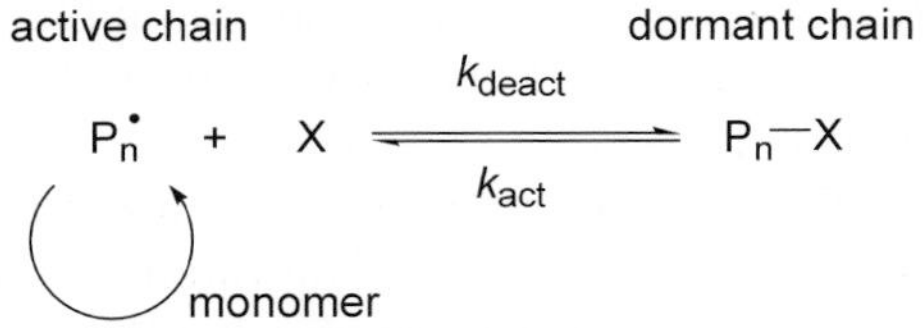

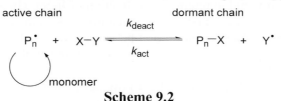

Scheme 9.1

(b) Those giving deactivation by reversible atom or group transfer and involving a bimolecular activation process (Scheme 9.2). For the systems described, the deactivator (X-Y) is a transition metal complex where Y is the metal in a higher oxidation state. Y• is then the metal in a lower oxidation state. Y• is inert with respect to monomer. Y• can be considered as a catalyst for the process shown in Scheme 9.1 and many aspects of the kinetics are similar. The best known example is atom transfer radical polymerization (ATRP - Section 9.4) where the deactivator X-Y is, for example, a copper(II) halide.

active chain   dormant chain

$$P_n^\bullet + X\text{-}Y \underset{k_{act}}{\overset{k_{deact}}{\rightleftharpoons}} P_n\text{-}X + Y^\bullet$$

Scheme 9.2

(c) Those giving simultaneous deactivation and activation by reversible (degenerate) chain transfer (Scheme 9.3). These systems are discussed in Section 9.5. The best known of this class is RAFT (Reversible Addition-Fragmentation chain Transfer) with thiocarbonylthio compounds (Section 9.5.3). In this case, the chain transfer step involves formation of an intermediate adduct. Other examples thought to involve a transfer by homolytic substitution are iodine transfer polymerization (Section 9.5.4) and TERP (telluride-mediated polymerization, Section 9.5.5).

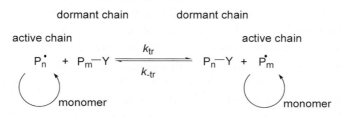

Scheme 9.3

The polymerizations (a) and (b) owe their success to what has become known as the persistent radical effect.[21] Simply stated: when a transient radical and a persistent radical are simultaneously generated, the cross reaction between the transient and persistent radicals will be favored over self-reaction of the transient radical. Self-reaction of the transient radicals leads to a build up in the concentration of the persistent species which favors cross termination with the persistent radical over homotermination. The homotermination reaction is thus self-suppressing. The effect can be generalized to a persistent species effect to embrace ATRP and other mechanisms mentioned in Sections 9.3 and 9.4. Many aspects of the kinetics of the processes discussed under (a) and (b) are similar,[21] the difference being that (b) involves a bimolecular activation process.

The reversible chain transfer process (c) is different in that ideally radicals are neither destroyed nor formed in the activation-deactivation equilibrium. This is simply a process for equilibrating living and dormant species. Radicals to maintain the process must be generated by an added initiator.

Though there is still debate about detailed mechanism, in each of the processes (a-c) the propagating species is believed to be a conventional propagating radical. Thus, termination by radical-radical reaction is not eliminated, though, as we shall see, with appropriate choice of reaction conditions, the significance of this process can be markedly reduced.

9.3 Deactivation by Reversible Coupling and Unimolecular Activation

Most polymerizations in this section can be categorized as stable (free) radical-mediated polymerizations (sometimes abbreviated as SFRMP). In the following discussion systems have been classed according to the type of stable radical involved, which usually correlates with the type of bond homolyzed in the activation process. Those described include systems where the stable radical is a sulfur-centered radical (Section 9.3.2), a selenium-centered radical (Section 9.3.3), a carbon-centered radical (Sections 9.3.4 and 9.3.5), an oxygen-centered radical (Sections 9.3.6, 9.3.7), or a nitrogen-centered radical (Section 9.3.8). We also consider polymerization mediated by cobalt complexes (Section 9.3.9) and certain 'monomers' (Section 9.3.5).

9.3.1 Kinetics and Mechanism

9.3.1.1 *Initiators, iniferters, initers*

In each of the sections below, we will consider the initiation process separately. For each system, various initiation methods have been applied. In some cases the initiator is a low molecular weight analog of the propagating species, in other cases it is a method of generating such a species. The initiators first used in this form of living radical polymerization were called iniferters (*ini*tiator - trans*fer* agent - chain *ter*minator) or initers (*ini*tiator - chain *ter*minator).

These terms were coined by Otsu and Yoshida[24] based on the similar terminology introduced by Kennedy[25] to cover analogous cationic systems. Except for the case of the dithiuram disulfides and related species (Section 9.3.2.1), these expressions have now fallen from favor and are no longer used as a generic terminology. In this chapter, we use the term initiator to denote alkoxyamines in NMP and halo-compounds in ATRP despite the confusion this can create, especially when the process also involves added conventional initiators.

In order for the characteristics of living polymerization to be displayed, initiators should possess the following attributes:

(a) One (in some cases, both) of the radicals formed on initiator decomposition is persistent or long-lived and unable (or slow) to initiate polymerization.
(b) Primary radical termination (or transfer to initiator) should be the only significant mechanism for the interruption of chain growth. Primary radical termination should occur exclusively by combination. Transfer to initiator, when involved, should occur exclusively by group transfer to give a product analogous to that formed by termination by combination.
(c) The bond to the end group (X) formed by these mechanisms must be thermally or photochemically labile under the reaction conditions such that reversible homolysis regenerates the propagating radical.
(d) The initiator must be consumed rapidly with respect to the rate of polymerization.

9.3.1.2 *Molecular weights and distributions*

The initiator or iniferter determines the number of growing chains. Several methods of initiation are used. Only three will be considered here. The first involves direct use of a species I-X (*e.g.* a dithiocarbamate ester - Section 9.3.2 or an alkoxyamine - Section 9.3.6) as shown in Scheme 9.4. Ideally, the degree of polymerization is given by eq. 1 and the molecular weight by eq. 2.

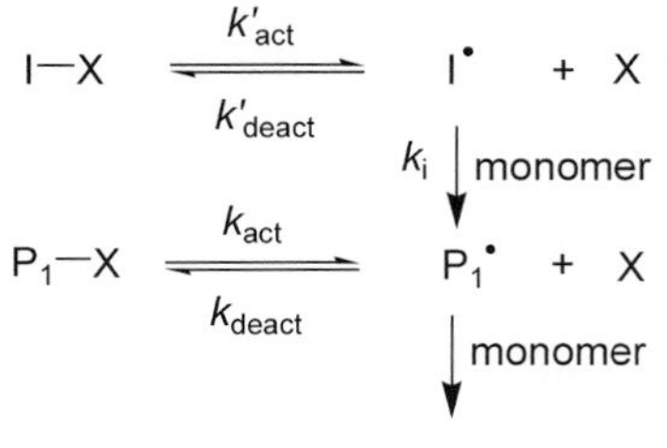

$$\text{I-X} \underset{k'_{deact}}{\overset{k'_{act}}{\rightleftharpoons}} \text{I}^{\bullet} + \text{X}$$

$$k_i \downarrow \text{monomer}$$

$$\text{P}_1\text{-X} \underset{k_{deact}}{\overset{k_{act}}{\rightleftharpoons}} \text{P}_1^{\bullet} + \text{X}$$

$$\downarrow \text{monomer}$$

Scheme 9.4

$$\overline{X}_n = \frac{([M]_o - [M]_t)}{[IX]_o} = \frac{[M]_o}{[IX]_o} c \tag{1}$$

$$\overline{M}_n = \frac{([M]_o - [M]_t)}{[IX]_o} m_{IM} + m_{IX} \qquad (2)$$

where $([M]_o-[M]_t)$ is the amount of monomer consumed, m_M and m_{IX} are the molecular weights of the monomer and the initiator (IX) respectively, and c is the monomer conversion. For a slow decomposing initiator, the term in the denominator should be $([IX]_o-[IX]_t) = [IX](1-\exp(-k_{act}t))$; *i.e.* the amount of initiator consumed. An efficiency term f' that has the usual definition (eq 3) can be introduced which allows for side reactions during the decomposition of IX or in the formation of $P_1\bullet$. The species I• often has different reactivity and specificity for reaction with monomer than the propagating species ($P_n\bullet$). Side reactions involving I• cause the molecular weight to be higher than expected.

$$f' = \frac{[\text{chains initiated}]}{[IX]_o} \qquad (3)$$

For a polymerization with initiation by the process shown in Scheme 9.4 with $k'_{act}=k_{act}$ and $k'_{deact}=k_{deact}$, the dispersity is given by eq. 4

$$\frac{\overline{X}_w}{\overline{X}_n} = 1 + \frac{1}{\overline{X}_n} + \left(\frac{2-c}{c}\right)\frac{k_p[IX]}{k_{deact}} \qquad (4)$$

where c is the monomer conversion. The dispersity depends on the molecular weight, the monomer conversion, and the ratio k_p/k_{deact}. This ratio governs the number of propagation steps per activation cycle and should be large for a narrow molecular weight distribution.

A second process involves use of a conventional initiator (I_2; *e.g.* AIBN, BPO) in the presence of X (*e.g.* a nitroxide) to generate a species IX *in situ* as shown in Scheme 9.5.

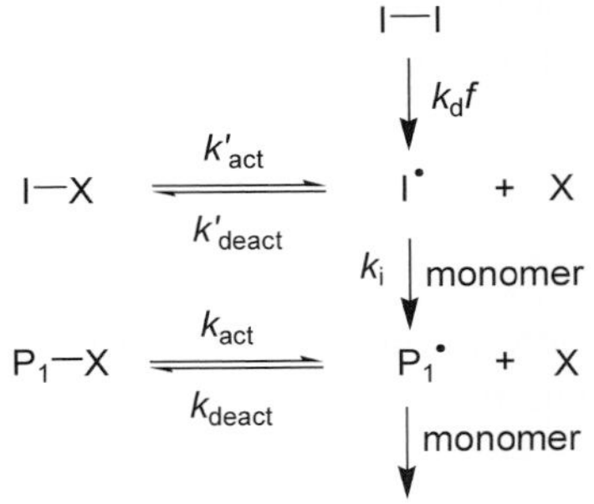

$$
\begin{array}{ccc}
 & & I-I \\
 & & \downarrow k_d f \\
I-X & \underset{k'_{deact}}{\overset{k'_{act}}{\rightleftarrows}} & I\bullet \ + \ X \\
 & & \downarrow k_i \ \text{monomer} \\
P_1-X & \underset{k_{deact}}{\overset{k_{act}}{\rightleftarrows}} & P_1\bullet \ + \ X \\
 & & \downarrow \text{monomer}
\end{array}
$$

Scheme 9.5

The degree of polymerization will usually be determined by the concentration of X. Some X may be lost in side reactions during the formation of IX. In some cases, I• must undergo at least one propagation step before combination with X is likely (*e.g.* in NMP with BPO as initiator). Any processes that irreversibly consume X will raise the molecular weight. Any process that provides additional chains will lower the molecular weight (*e.g.* thermal initiation in S polymerizations or an additional thermal initiator).

A third process involves use of the species (X-X) to generate the 'stable radical' in pairs and relies on the stable radical being able to react with monomer, albeit slowly, to generate P_1X (Scheme 9.6). Polymerizations with dithiuram and other disulfides (Section 9.3.2.1) and hexasubstituted ethanes (Section 9.3.4) belong to this class.

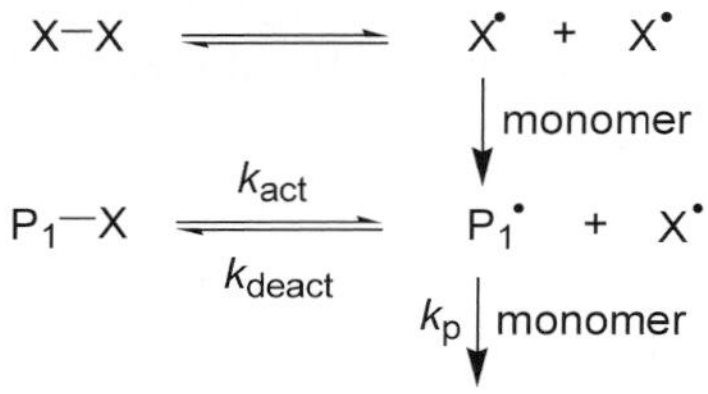

$$\begin{array}{c} X-X \rightleftharpoons X^\bullet + X^\bullet \\ \downarrow \text{monomer} \\ P_1-X \underset{k_{deact}}{\overset{k_{act}}{\rightleftharpoons}} P_1^\bullet + X^\bullet \\ k_p \downarrow \text{monomer} \end{array}$$

Scheme 9.6

Other variations and combinations of these processes are also possible and are described in the following sections.

9.3.1.3 *Polymerization kinetics*

General features of the polymerization kinetics for polymerizations with deactivation by reversible coupling have already been mentioned. Detailed treatments appear in reviews by Fischer,[21] Fukuda *et al.*,[22] and Goto and Fukuda[23] and will not be repeated here.

In conventional radical polymerization the rate of polymerization is described by eq. 5 (Section 5.2.1). As long as the rate of initiation remains constant, a plot of $\ln([M]_o/[M]_t)$ vs time should provide a straight line.

$$\ln \frac{[M]_o}{[M]_t} = k_p \left(\frac{R_i}{k_t} \right)^{1/2} t \qquad (5)$$

For polymerizations where initiation is described by Scheme 9.4, the rate of polymerization is given by eq. 6.[21]

$$\ln \frac{[M]_o}{[M]_t} = \frac{3}{2} k_p \left(\frac{K[IX]_o}{3k_t} \right)^{1/3} t^{2/3} \qquad (6)$$

where $K = k_{act}/k_{deact}$. The derivation of this equation requires that $[X]_0$ is zero and that there is no initiation source other than IX. Note that the relationship between

Living Radical Polymerization

ln($[M]_o/[M]_t$) and time is *not* anticipated to be linear. Under these circumstances, the rate of polymerization is controlled by the value of the activation-deactivation equilibrium constant K.

If there is an external source of free radicals (*e.g.* from thermal initiation in S polymerization or from an added conventional initiator) eq. 5 may again apply. The rate of polymerization becomes independent of the concentration of IX and, as long as the number of radicals generated remains small with respect to $[IX]_o$, a high fraction of living chains and low dispersities is still possible. The validity of these equations has been confirmed for NMP and with appropriate modification has also been shown to apply in the case of ATRP.[23]

9.3.2 Sulfur-Centered Radical-Mediated Polymerization

The carbon sulfur bond of suitably constructed *N,N*-dialkyldithiocarbamates and related compounds undergoes reversible homolysis under irradiation with UV light of appropriate wavelength (Scheme 9.7) allowing monomer insertion into the C-S bond. The *N,N*-dialkyldithiocarbamyl radical is persistent and reacts with monomers only slowly. This form of polymerization has been comprehensively reviewed by Ameduri,[26] Sebenik[27] and Otsu and Matsumoto.[28] The process should be distinguished from RAFT which can involve similar thiocarbonylthio compounds but does not usually involve sulfur-centered radicals as intermediates (Section 9.5.3).

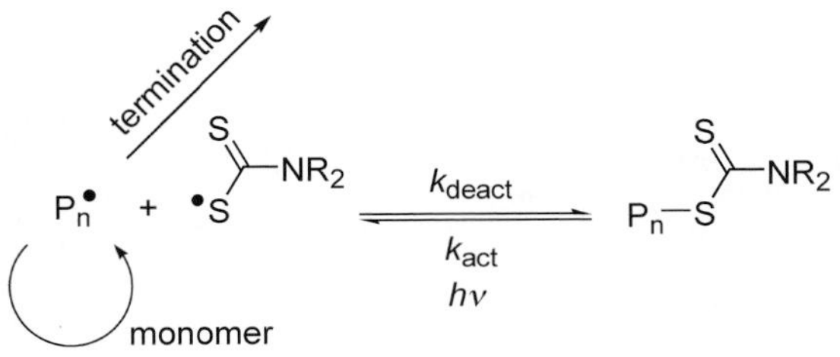

Scheme 9.7

9.3.2.1 Disulfide initiators

The first detailed study of dithiuram disulfides as initiators in polymerizations of MMA and S was reported by Werrington and Tobolsky in 1955.[29] They observed that the transfer constant to the disulfide was relatively high and also found significant retardation. The potential of this and other disulfides as initiators of living radical polymerization was recognized by Otsu and Yoshida in 1982.[24] A wide range of disulfides has now been investigated in this context with varying degrees of success. These include diaryl disulfides *e.g.* diphenyl disulfide (**1**)],[30,31] dibenzoyl disulfide (**2**),[24] dithiuram disulfides [*e.g.* tetraethyldithiuram disulfide (**4**)],[24,32,33] and xanthogen disulfides [*e.g.* bis(isopropylxanthogen) disulfide (**5**)];[34] with the dithiuram disulfides being the most studied in this context.

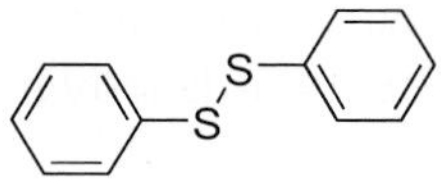

The proposed mechanism of initiation with the dithiuram disulfide **4** is shown in Scheme 9.8. The dithiuram disulfide decomposes thermally or photochemically to give dithiocarbamyl radicals **6**. These radicals **6** add monomer only slowly and relatively high reaction temperatures (typically >80°C) appear necessary even when the initiator is decomposed photochemically.

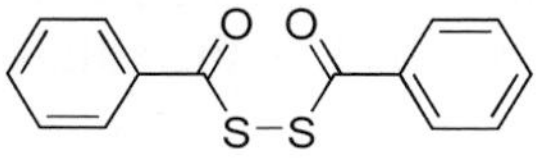

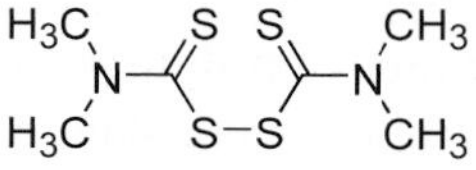

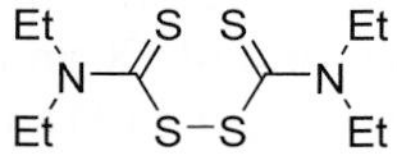

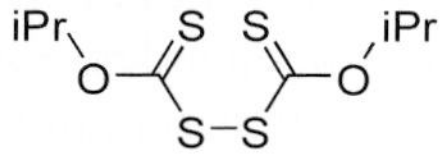

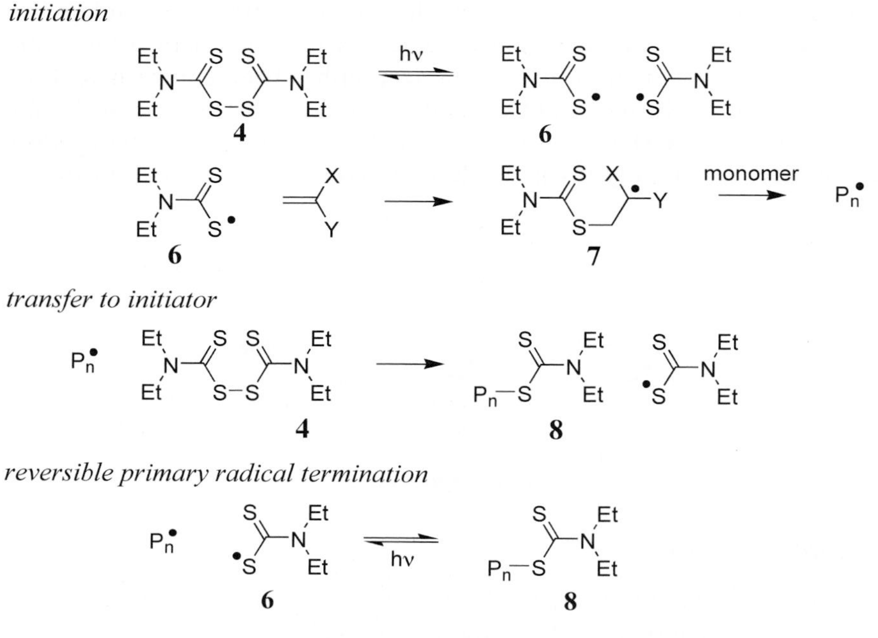

Scheme 9.8

Transfer to the dithiuram disulfide by transfer of the dithiocarbamyl group, probably by addition-fragmentation, is an important mechanism for the termination of polymer chains during the early stages of polymerization. The transfer constant of **3** is reported to be *ca* 0.5 in both S and MMA polymerizations.[35,36] The end groups **8** formed by transfer to the dithiuram disulfide are indistinguishable from those **8** formed by primary radical termination with dithiocarbamyl radicals (**6**, refer Scheme 9.8). While the formation of the end groups **8** is reversible under the photopolymerization conditions (Section 9.3.2.2), the primary dithiocarbamate end groups **7** formed by addition of **6** to monomer are relatively stable to photolysis.

Since the dithiocarbamyl end groups **8** are thermally stable but photochemically labile at usual polymerization temperatures, only photo-initiated polymerizations have the potential to show living characteristics. However, various disulfides, for example, **9** and **10**, have been used to prepare end-functional polymers[37] and block copolymers[38] by irreversible chain transfer in non-living thermally-initiated polymerization (Section 7.5.1).

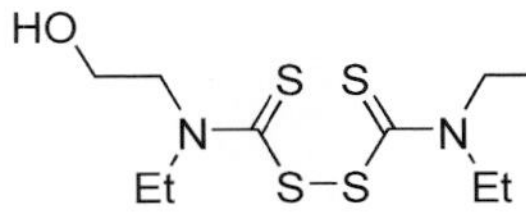

9

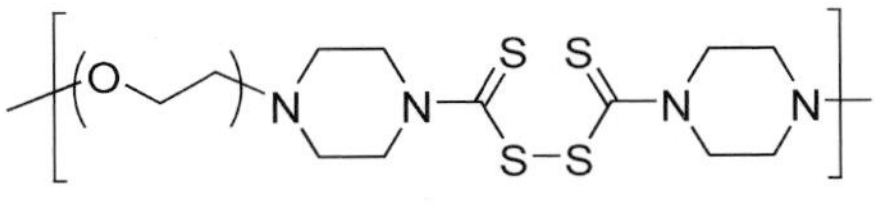

10

Aliphatic disulfides are not thought to be effective as initiators in this context. However, Endo et al.[39] have described the use of the cyclic 1,2-disulfides **11** and **12** as initiators in a controlled radical polymerization. Polymerization of S at 120 °C gave a linear increase in molecular weight with conversion and the PS formed was used as a macroinitiator to form PS-*block*-PMMA. The precise mechanism of the process has not been elucidated.

11

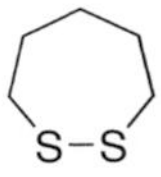

12

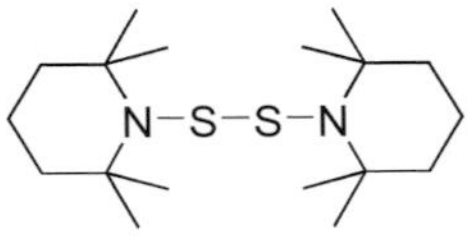

13

The use of the disulfide (**13**), which can dissociate thermally to give a sulfur analog of TEMPO (Section 9.3.6.1), has also been explored for controlling S polymerization though poor results were obtained.[40]

9.3.2.2 Monosulfide initiators

Certain *N,N*-dialkyl dithiocarbamates [*e.g.* benzyl *N,N*-diethyl dithiocarbamate (**14**)] and xanthates have been used as photoinitiators. Photodissociation of the C–S bond of these compounds yields a reactive alkyl radical (to initiate polymerization) and a less reactive sulfur-centered radical (to undergo primary radical termination) as shown in Scheme 9.9.[30,41,42]

Since the experiment is no longer reliant on the dithiocarbamyl radical to both initiate and terminate chains (*cf.* Section 9.3.2.1), lower reaction temperatures may be used (where the dithiocarbamyl radical is slower or unable to add monomer) and better control over the polymerization process can be obtained. The transfer constants for the benzyl dithiocarbamates in polymerization of acrylic and styrenic monomers are very low, thus primary radical termination is the predominant chain termination mechanism.

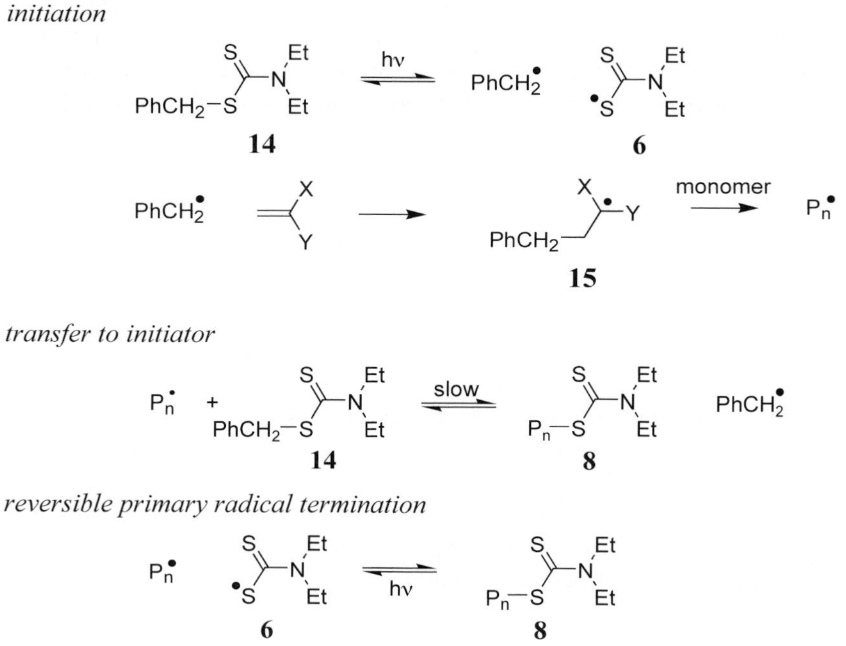

initiation

transfer to initiator

reversible primary radical termination

Scheme 9.9

The processes described in this section should be contrasted with RAFT polymerization (Section 9.5.3), which can involve the use of similar thiocarbonylthio compounds. N,N-dialkyl dithiocarbamates have very low transfer constants in polymerizations of S and (meth)acrylates and are not effective in RAFT polymerization of these monomers. However, N,N-dialkyl dithiocarbamates have been successfully used in RAFT polymerization of VAc. Certain O-alkyl xanthates have been successfully used to control RAFT polymerizations of VAc, acrylates and S. The failure of the earlier experiments using these reagents and monomers to provide narrow molecular weight distributions by a RAFT mechanism can be attributed to the use of non-ideal reaction conditions and reagent choice. A two part photo-initiator system comprising a mixture of a benzyl dithiocarbamate and a dithiuram disulfide has also been described and provides better control (narrower molecular weight distributions).[43]

The use of mono-, di- and multifunctional initiators provides scope for designing polymer architectures. The use of **14**, **18** and **19** in the production of block or star polymers has been demonstrated.[41,44,45] Homopolymers of **20** or copolymers of **20** with S or MMA have been successfully used in photoinitiated graft polymerization of S or MMA.[46-48] The analogous xanthate has also been used in this context. Compounds **20**[49] and **22**[50] have also been used to make hyperbranched polymers. The monomer **20** was reported to have reactivity ratios similar to those of S. It is reported[51] that the xanthate **23** does not copolymerize with MMA, it acts only as a photoiniferter in MMA polymerization and provides a polymer with a relatively narrow molecular weight distribution. In S polymerization **23** also acts as a comonomer.

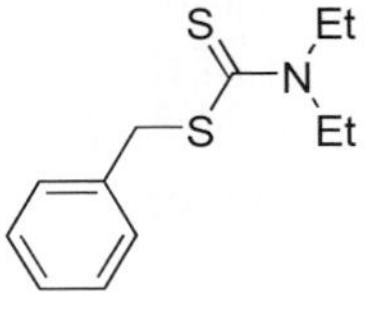

14

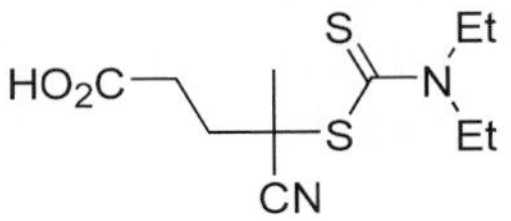

16[52,53]

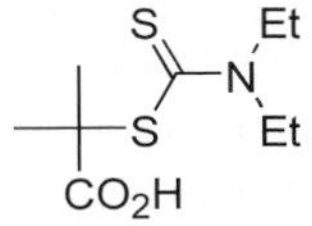

17[53-55]

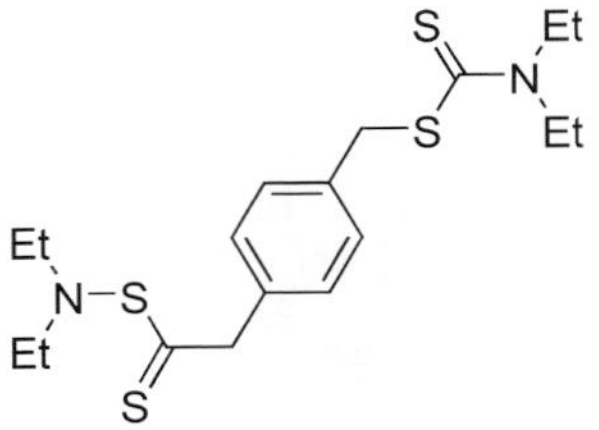

18[41,44]

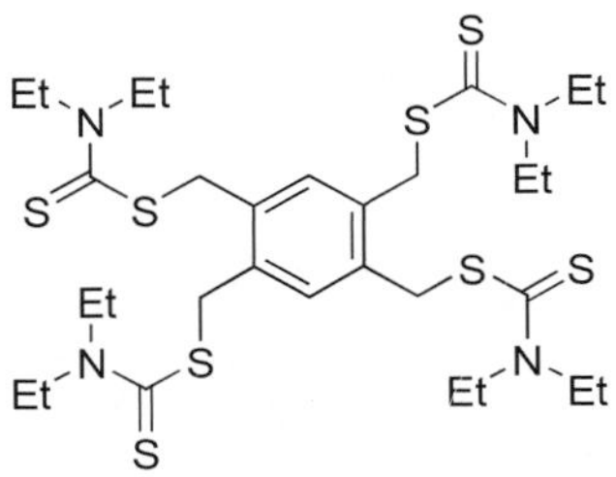

19[45]

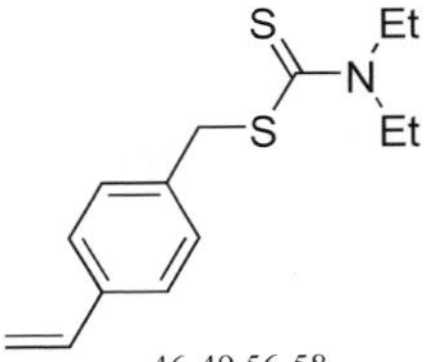

20[46-49,56-58]

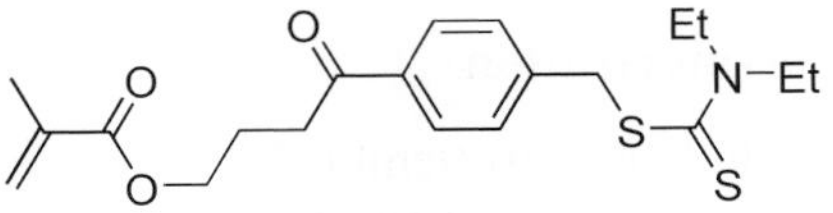

21[59]

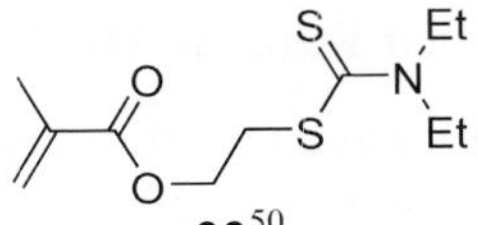

22[50]

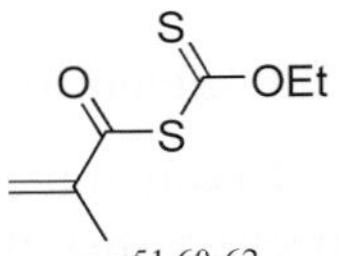

23[51,60-62]

9.3.2.3 Monomers, mechanism, side reactions

The outcome of the polymerization depends strongly on the particular monomer. Polymerizations of S, MMA, MA, VAc and some derivatives have been reported. Studies on model compounds indicate that the primary or secondary dithiocarbamate end groups are much less susceptible to photodissociation than benzyl or tertiary derivatives.

Dithiocarbamate **16** has been used to prepare low dispersity PMAA ($M_w/M_n \sim 1.2$).[52] Photopolymerization of S in the presence of dithiocarbamate **16** also displays some living characteristics (molecular weights that increase with conversion, ability to make block copolymer). However, **17** appears to behave as a conventional initiator in S polymerization.[53] The difference in behavior was attributed to the relatively poor leaving group ability of the 2-carboxyprop-2-yl radical. This hypothesis is supported by MO calculations. Dithiocarbamate **17** was used to control polymerizations of MMA,[54] HEMA[54] and NIPAM.[55]

Chain ends formed with monosubstituted monomers, other than S, appear resistant to photolysis and polymerizations of MA and VAc do not show living characteristics. Most polymerizations involve methacrylate esters or S.

Various side reactions that are likely to lead to a slow loss of "living" ends have been described. With disulfide initiators, one (initiation by the dithiocarbamyl radical) is unavoidable since the experiment relies on the same radical species to both initiate polymerization and terminate chains.

Other side reactions that have been reported are cleavage of the carbon-nitrogen bond to form **24** and an aminyl radical **25** or scission of the thiocarbonyl-sulfur bond to form a thiyl radical **26** and **27** (Scheme 9.10).[33,63,64] Thiocarbonyl-sulfur bond cleavage may be a preferred pathway in the case of primary dithiocarbamates.

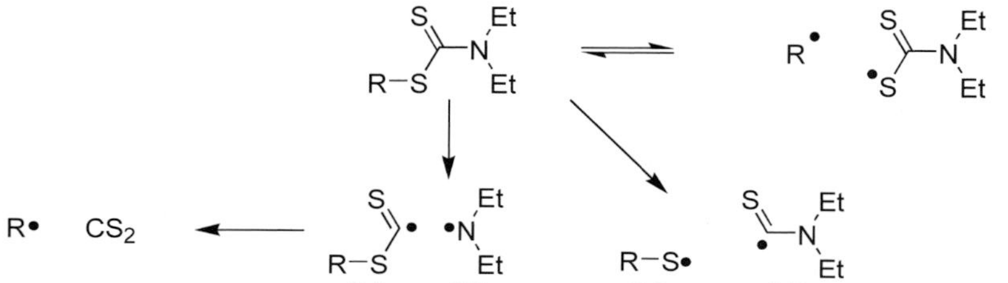

Scheme 9.10

9.3.3 Selenium-Centered Radical-Mediated Polymerization

Kwon and coworkers have reported the use of diphenyl diselenide **28**[65,66] and a variety of benzylic selenides (e.g. **29**,[67,68] **30**,[69] **31**[70,71] and **32**[72]) as photoiniferters for polymerization of S, MMA and some derivatives. Very narrow dispersities were not obtained ($\overline{M}_w / \overline{M}_n$ typically 2-2.5). However, it was possible to prepare block copolymers.[69,71,73] A related visible light photoinitiation system has recently been reported comprising 1-(phenylseleno)ethylbenzene and *t*-butyl(diphenyl)(phenylseleno)silane.[74,75]

The polymerization mechanisms proposed are similar to those discussed for the sulfur compounds described in Sections 9.3.2.1 and 9.3.2.2 and the results obtained are also generically similar. The transfer constant of benzyl selenide (**29**) (C_{tr} is 1.04 in S polymerization at 60 °C) is substantially higher than that of sulfide photoiniferters (Section 9.3.2.2). The value suggests that the incidence of reversible chain transfer should be of significance and that development of a thermal process involving reversible chain transfer may be possible. The transfer constants of diphenyl diselenide **28** are also high (C_{tr} is 1.43 in MMA[66] and 28 in S polymerization[76] at 60 °C). Various methods have been explored for end group transformation and to remove the selenide end group from the final product. These include reduction with tri-*n*-butylstannane and oxidative elimination *via* reaction with hydrogen peroxide.[76]

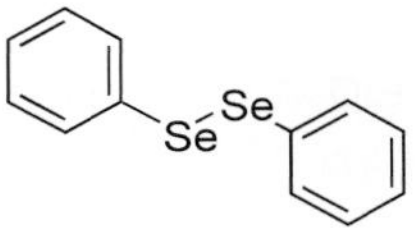

28[65,66] 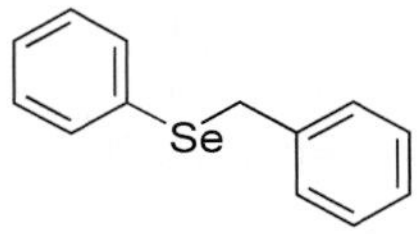   **29**[67,68] 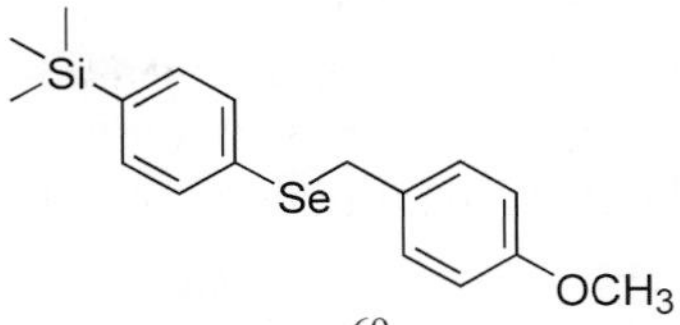   **30**[69]

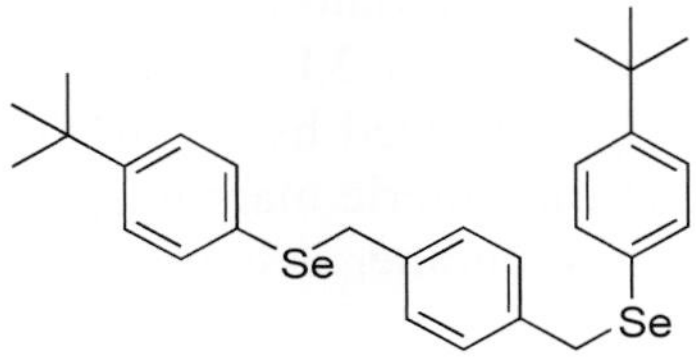

31[70,71] 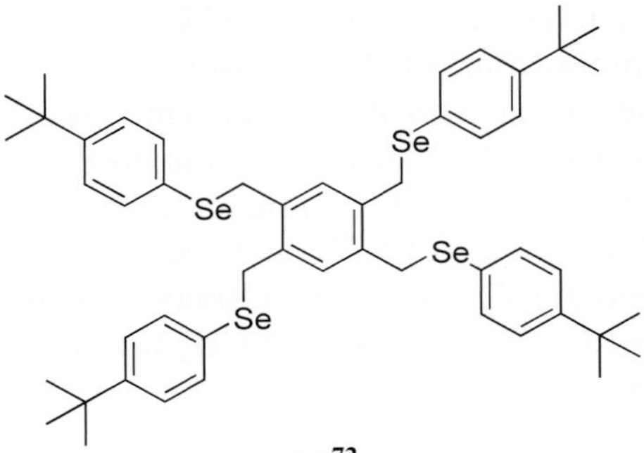   **32**[72]

9.3.4 Carbon-Centered Radical-Mediated Polymerization

Stable carbon-centered radicals, in particular, substituted diphenylmethyl and triphenylmethyl radicals, couple reversibly with propagating radicals (Scheme 9.11). With the carbon-centered radical-mediated polymerization systems described to date, the propagating radical should be tertiary (*e.g.* methacrylate ester) to give reasonable rates of activation.

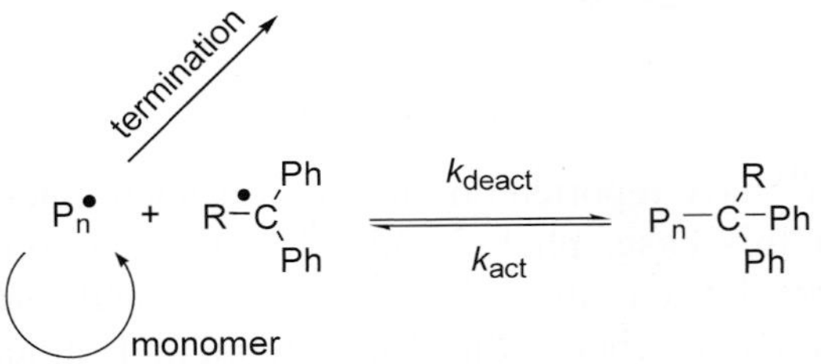

Scheme 9.11

The first use of sterically hindered hexasubstituted ethanes [*e.g.* **33**] as initiators of polymerization was reported by Bledzki *et al.*[77,78] The use of related initiators based on silylated pinacols [*e.g.* **34**, **35**] has been reported by Crivello *et al.*,[79-82] Santos *et al.*,[83] and Roussel and Boutevin.[84,85] Other initiators of this class include **36**[86,87] and **37**.[88] The rates of decomposition of hexasubstituted ethanes and the derived macroinitiators are known to vary according to the degree of steric crowding about the C–C bond undergoing homolysis,[89] though few rate constants have been reported.

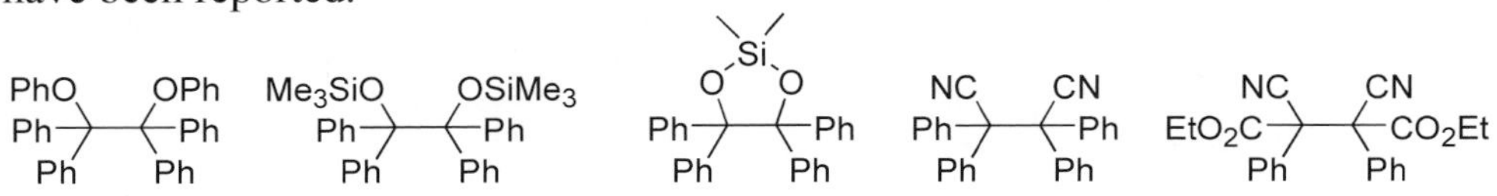

33: PhO, OPh, Ph, Ph, Ph, Ph (hexaphenyl diphenoxy ethane)
34: Me$_3$SiO, OSiMe$_3$, Ph, Ph, Ph, Ph
35: Si-bridged diphenyl compound
36: NC, CN, Ph, Ph, Ph, Ph
37: NC, CN, EtO$_2$C, CO$_2$Et, Ph, Ph

The proposed polymerization mechanism is shown in Scheme 9.12. Thermal decomposition of the hexasubstituted ethane derivative yields hindered tertiary radicals that can initiate polymerization or combine with propagating species (primary radical termination) to form an oligomeric macroinitiator. The addition of the diphenylalkyl radicals to monomer is slow (e.g. k_i for **34** is reported as 10^{-4} M^{-1} s^{-1} at 80 °C[84]) and the polymerization is characterized by an inhibition period during which the initiator is consumed and an oligomeric macroinitiator is formed. The bond to the CH$_2$ formed by addition to monomer is comparatively thermally stable.

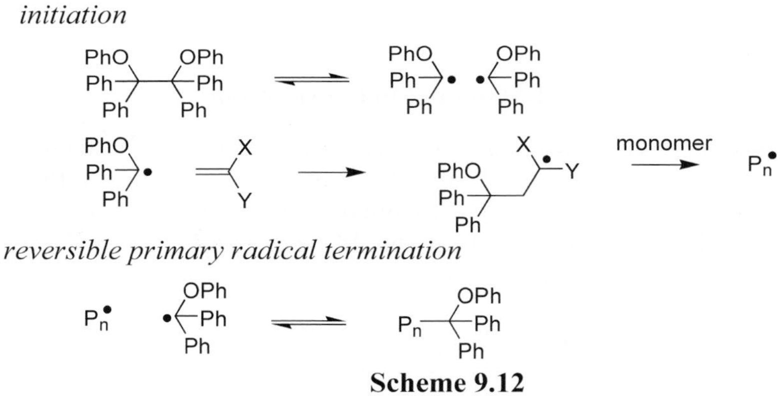

initiation

reversible primary radical termination

Scheme 9.12

Otsu and Tazaki[90] have reported on the use of triphenylmethylazobenzene (**39**) as an initiator. In this case, phenyl radical initiates polymerization and the triphenylmethyl radical reacts mainly by primary radical termination to form a macroinitiator. The early report[91] that triphenylmethyl radical does not initiate MMA polymerization may only indicate a very low rate of polymerization. The addition of triphenylmethyl radical to MMA has been demonstrated in radical trapping experiments.[92]

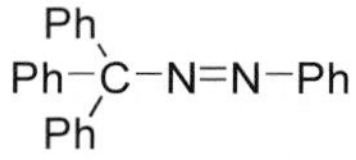

39: Ph$_3$C–N=N–Ph

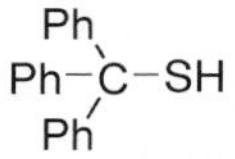

40: Ph$_3$C–SH

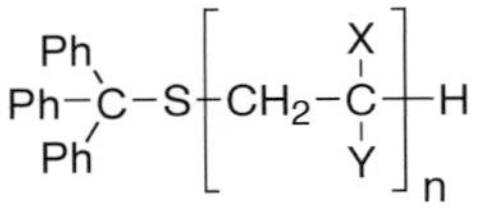

[Structures 41 and 42 shown at top]

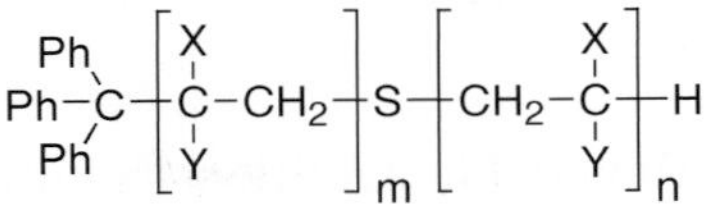

Triphenylmethyl terminated polymers (**41**) are formed in polymerizations conducted in the presence of triphenylmethyl thiol (**40**).[93] Transfer constants for **40** are similar to other thiols (17.8 for S, 0.7 for MMA, compare Section 6.2.2.1). When the polymers (**41**) are heated in the presence of added monomer it is presumed that the S-CPh$_3$ bond is cleaved and triphenylmethyl-mediated polymerization according to Scheme 9.11 can then ensue to yield chain extended or block polymers (**42**).

It is of interest to speculate on the precise structure of the macroinitiator species in these polymerizations. The work of Engel et al.[94] suggests the likelihood of a quinonoid intermediate (e.g. **45**, Scheme 9.13), at least for the polymerizations involving triphenylmethyl radical (**44**).

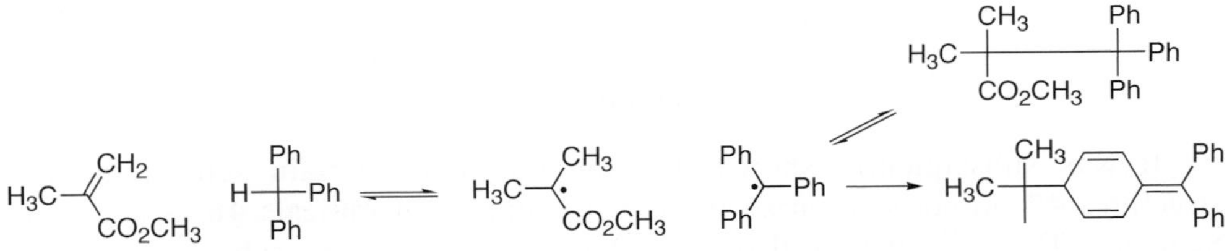

[Structures 43, 44, 45 shown]

Scheme 9.13

9.3.4.1 *Monomers, mechanism, side reactions*

The hindered carbon-centered radicals are most suited as mediators in the polymerization of 1,1-disubstituted monomers (e.g. MMA,[78,95] other methacrylates and MAA,[96] and AMS[97]). Polymerizations of monosubstituted monomers are not thought to be living. Dead end polymerization is observed with S at polymerization temperatures <100°C.[98] Monosubstituted monomers may be used in the second stage of AB block copolymer synthesis (formation of the B block).[95] However the non-living nature of the polymerization limits the length of the B block that can be formed. Low dispersities are generally not achieved.

There will be a gradual loss of stable radical with these systems as the di- or triarylmethyl radicals produced from the macroinitiator can add monomer, albeit slowly.[99,100] This side reaction provides a mechanism for mopping up the excess stable radical formed as a consequence of termination between propagating radicals and may be essential to maintaining polymerization rates.

A further problem with these iniferters is loss of "living" ends through primary radical termination by disproportionation. The ratio of k_{td}/k_{tc} reported for the cross reaction between **43** and triphenylmethyl radicals (**44**) and at 110°C is 0.61 (Scheme 9.13).[94]

9.3.5 Reversible Addition-Fragmentation

Certain monomers may be able to act as reversible deactivators by a reversible addition-fragmentation mechanism. The monomers are 1,1-disubstituted and generate radicals that are unable or extremely slow to propagate or undergo combination or disproportionation. For these polymerizations the dormant species is a radical and the persistent species is the 1,1-disubstituted monomer.

Thus propagating radicals were initially proposed to add reversibly to diphenylethylene as shown in Scheme 9.14.[101]

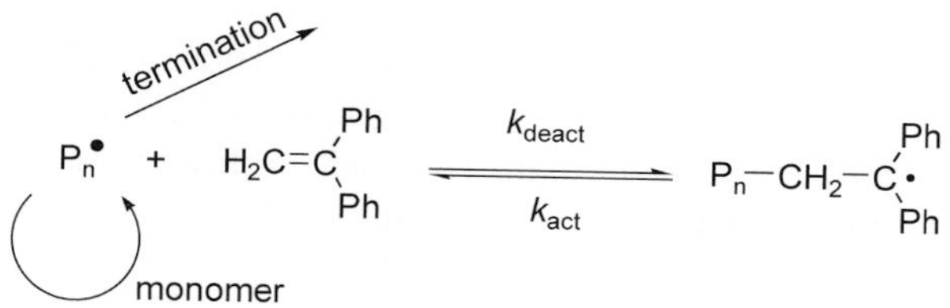

Scheme 9.14

It was subsequently shown that the polymers contain semi-quinonoid structures **47** proposed to arise from α-p coupling of radicals **46** as shown in Scheme 9.15.[102-104] It was also suggested that **47** could be subject to radical-induced decomposition by an addition-fragmentation process.

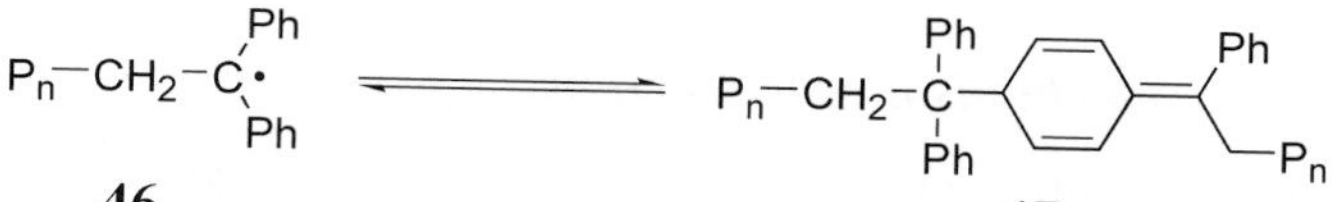

Scheme 9.15

Polymerization in the presence of capto-dative substituted monomers has been proposed[105] to follow a related mechanism (Scheme 9.16) in which the concentration of the radical adduct **48** is additionally controlled by a reversible coupling reaction.

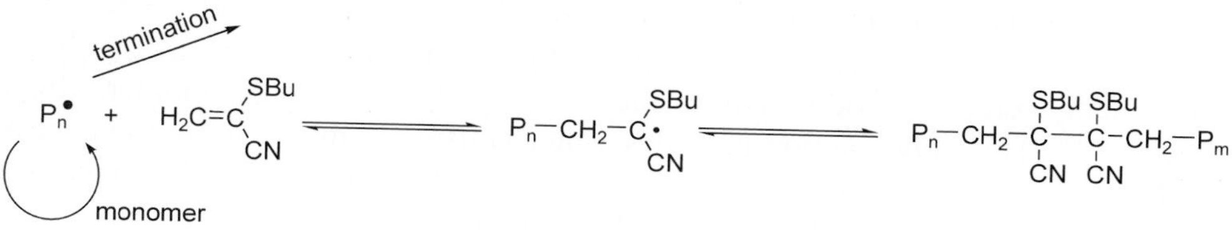

Scheme 9.16

Living Radical Polymerization

To date, the degree of control realized with these methods is poor with respect to those achieved with NMP, ATRP or RAFT.

9.3.6 Nitroxide-Mediated Polymerization

The literature on Nitroxide-Mediated Polymerization (NMP) through 2001 was reviewed by Hawker et al.[106,107] More recently the subject has been reviewed by Studer and Schulte[108] and Solomon.[109] NMP is also discussed by Fischer[110] and Goto and Fukuda[23] in their reviews of the kinetics of living radical polymerization and is mentioned in most reviews on living radical polymerization. A simplified mechanism of NMP is shown in Scheme 9.17.

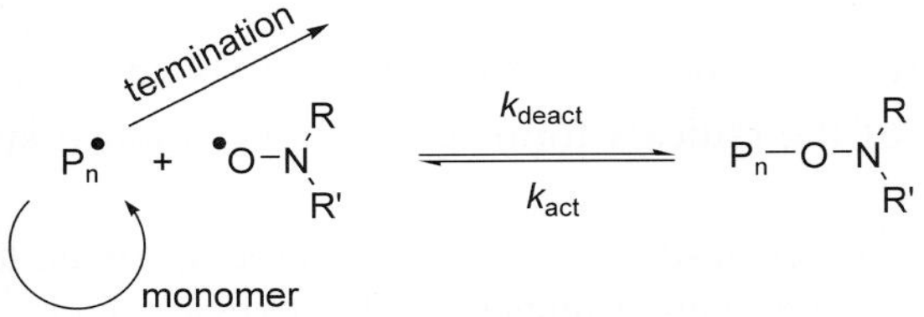

$$P_n^\bullet + {}^\bullet O-N\begin{smallmatrix}R\\R'\end{smallmatrix} \underset{k_{act}}{\overset{k_{deact}}{\rightleftharpoons}} P_n-O-N\begin{smallmatrix}R\\R'\end{smallmatrix}$$

(termination, monomer)

Scheme 9.17

Prior to the development of NMP, nitroxides were well known as inhibitors of polymerization (Section 5.3.1). They and various derivatives were (and still are) widely used in polymer stabilization. Both applications are based on the property of nitroxides to efficiently scavenge carbon-centered radicals by combining with them at near diffusion-controlled rates to form alkoxyamines. This property also saw nitroxides exploited as trapping agents to define initiation mechanisms (Section 3.5.2.4).

The exploitation of alkoxyamines as polymerization initiators and the use of NMP for producing block and end-functional polymers was first described in a patent application by Solomon et al. in 1985.[111] In this work NMP was described as a method of living radical polymerization. This work was mentioned in a communication[112] in 1987 and a conference paper[113] in 1991. In 1990, Johnson et al.[114] described what is now known as the persistent radical effect[115] and showed that NMP, with appropriate selection of alkoxyamine and control of reaction conditions could, in principle, provide narrow dispersity polymers. These early papers focused on NMP of acrylates. However, the method only received significant attention in the wider literature following the demonstration by Georges et al.[116] in 1993 that NMP could be used to prepare PS with a narrow molecular weight distribution. Since that time the literature on NMP has greatly expanded and, along with ATRP and RAFT, NMP is now one of the most cited methods for living radical polymerization.

9.3.6.1 Nitroxides

A wide range of nitroxides and derived alkoxyamines has now been explored for application in NMP. Experimental work and theoretical studies have been carried out to establish structure-property correlations and provide further understanding of the kinetics and mechanism. Important parameters are the value of the activation-deactivation equilibrium constant K and the values of k_{act} and k_{deact} (Scheme 9.17), the combination:disproportionation ratio for the reaction of the nitroxide with the propagating radical (Section 9.3.6.3) and the intrinsic stability of the nitroxide and the alkoxyamine under the polymerization conditions (Section 9.3.6.4). The values of K, k_{act} and k_{deact} are influenced by several factors.[113,117-119]

(a) The degree of steric compression around the C-O bond.[118]
(b) The stabilities of the radicals formed.[118] Higher radical stability lowers k_{act} and raises k_{deact}.
(c) Polar factors.[118] Electron-donating groups on the nitroxide lower k_{act} and raise k_{deact}. Electron-withdrawing groups have the inverse effect.
(d) Hydrogen bonding.[120,121] Hydroxyl substituents on the alkoxyamine (or on the monomer/solvent) lower k_{act}.

The rates thus depend on the structure of both the reactive radical (initiating radical, propagating radical) and the nitroxide fragment. The structures of some nitroxides used in NMP are shown in Table 9.1-Table 9.4. For structurally related nitroxides K and k_{act} are found to increase in the series five-membered ring (*e.g.* **49**, Table 9.1)< six-membered ring (*e.g.* **67**, Table 9.2) < open chain (*e.g.* **83**, Table 9.3) < seven-membered ring (*e.g.* **92**, Table 9.4).[118] Within each series, the incorporation of bulky substituents adjacent to the nitroxide nitrogen increases k_{act}. Thus k_{act} for **58** is less than that for **59**; the value of k_{act} increases in the series **60** < **62** < **64**. In general, factors which increase k_{act} cause k_{deact} to decrease.

These major trends in k_{act} can be qualitatively predicted using semi-empirical molecular orbital calculations.[118,122,123] However, the methods fail to adequately predict some electronic effects, remote substituent effects and the influence of hydrogen bonding. Higher level *ab initio* or DFT calculations provide a better indication of trends in these circumstances.

Another important factor is the stability of the nitroxide. Some degree of instability appears beneficial. This can compensate for the buildup of nitroxide that would occur as a consequence of radical-radical termination and which might otherwise inhibit polymerization.

A number of NMP processes have been reported where the nitroxide is formed *in situ*. Nitrones[124-127] and nitroso-compounds[128] have been used as nitroxide precursors. Control of methacrylate polymerization by mixtures of nitric oxide and nitrogen dioxide has also been attributed to *in situ* formation of a nitroxide.[129,130]

Table 9.1 Five-Membered Ring Nitroxides for NMP

Nitroxide	Structure	Nitroxide	Structure	Nitroxide	Structure
49[118,131]	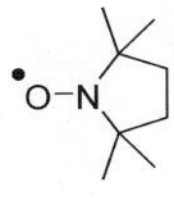	**50**[131-134]	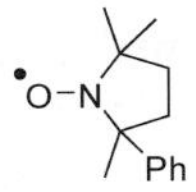	**51**[131]	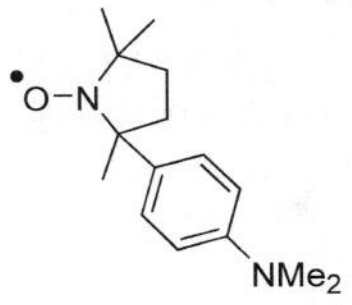
52[a,131]	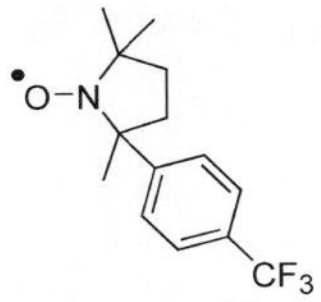	**53**[134]	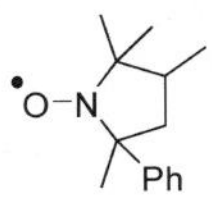	**54**[134]	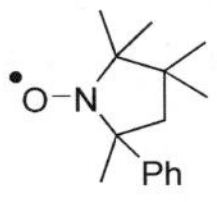
55[135]	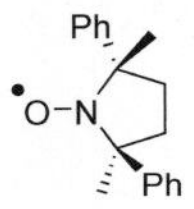	**56**[134]	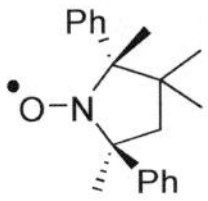		
57[136]	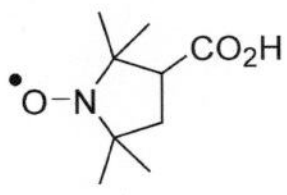	**58**[a,118,120]	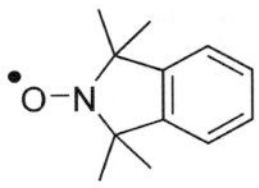	**59**[111,118,120,122,137]	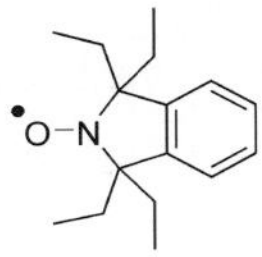
60[138]	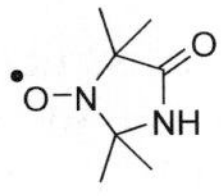	**61**[138]	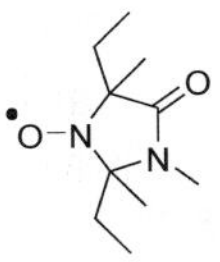		
62[138]	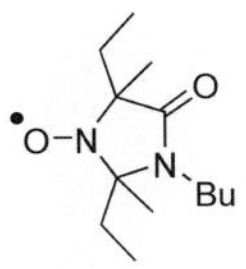	**63**[138]	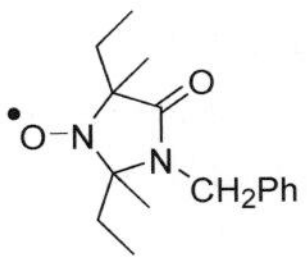	**64**[138,139]	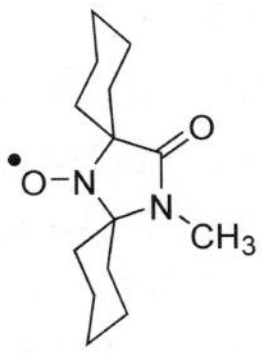
65[a,134]	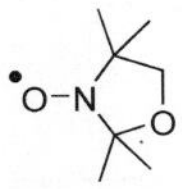	**66**[a,123]	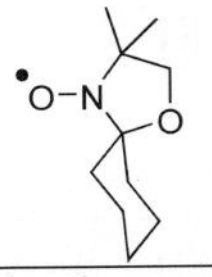		

a These nitroxides were ineffective in NMP under the conditions reported.

Table 9.2 Six-Membered Ring Nitroxides for NMP)

Nitroxide	Structure	Nitroxide	Structure	Nitroxide	Structure
67 TEMPO [111,116,118,120,140-142]	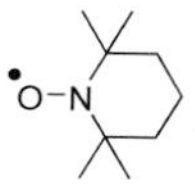	**68**	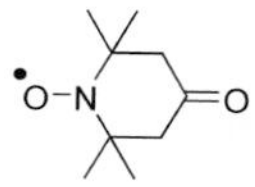	**69**[141]	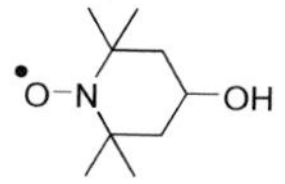 —OH
70[143]	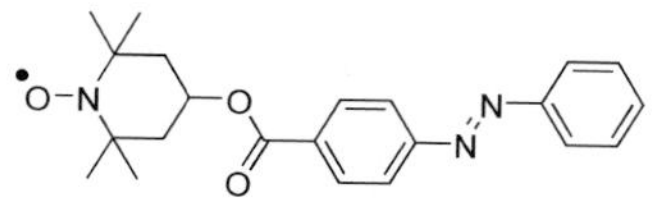			**71**[144-146]	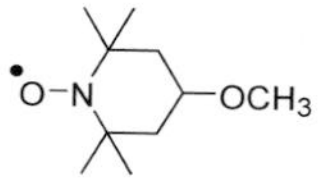 —OCH₃
72	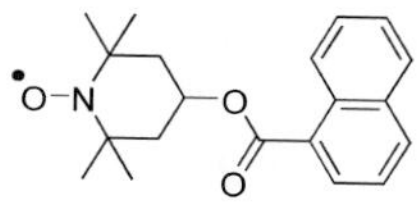			**73**	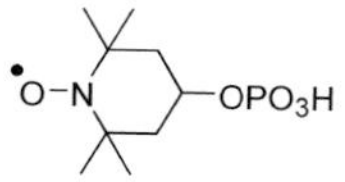 —OPO₃H
74[147]	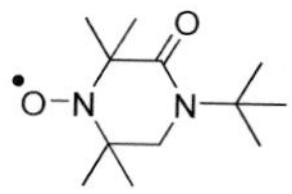	**75**[147]	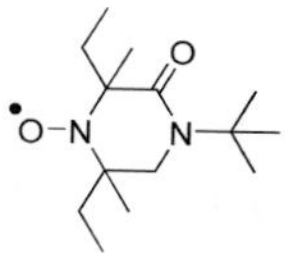	**76**	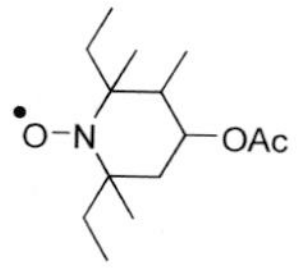 —OAc
77[111]	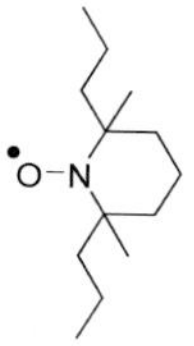	**78**[142,148]	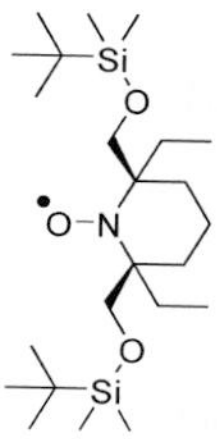	**79**[149]	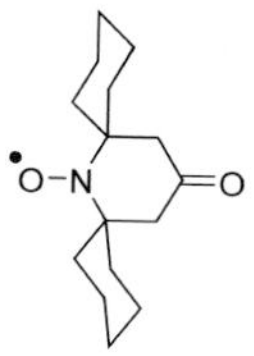
80[149]	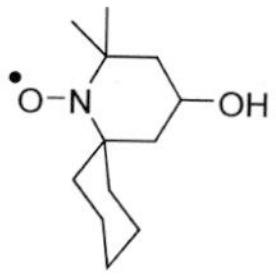 —OH	**81**[149]	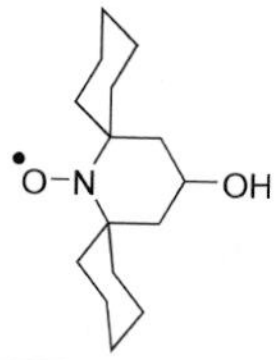 —OH	**82**[150]	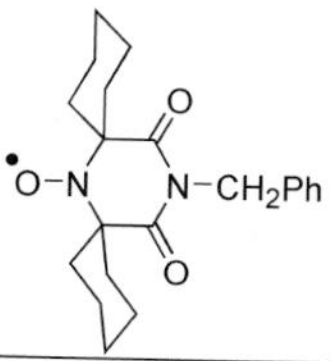 N—CH₂Ph

Table 9.3 Open-Chain Nitroxides for NMP

Nitroxide	Structure	Nitroxide	Structure	Nitroxide	Structure
83 DTBN [111,118,120,151]	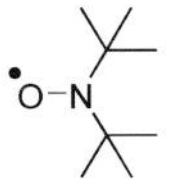	**84**[122]	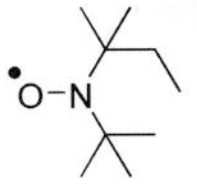	**85**[152]	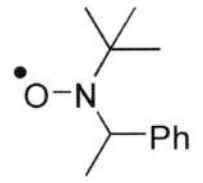
86 [120,142,153,154]	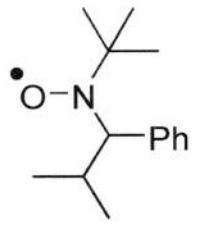	**87**[155]	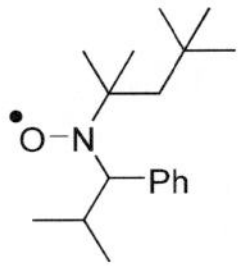	**88**[121]	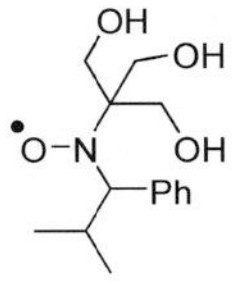
89 SG1 [120,156-158]	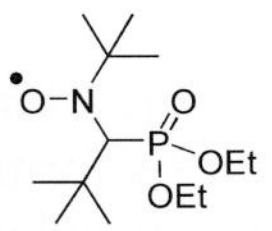	**90**[159,160]	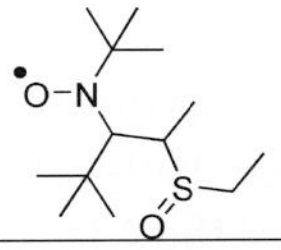	**91**[161]	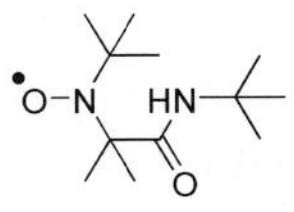

Table 9.4 Seven- and Eight-Membered Ring Nitroxides for NMP

Nitroxide	Structure	Nitroxide	Structure	Nitroxide	Structure
92[118]	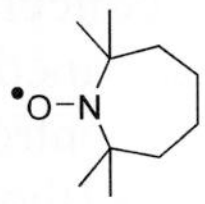	**93**[142,162]	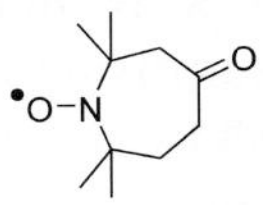	**94**[162]	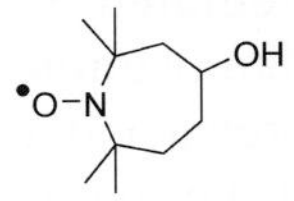
95[163]	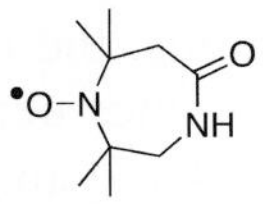	**96**[163]	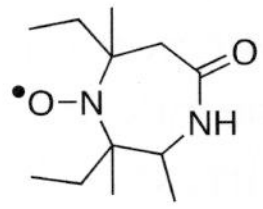	**97**[162]	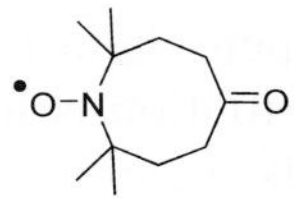
98[164]	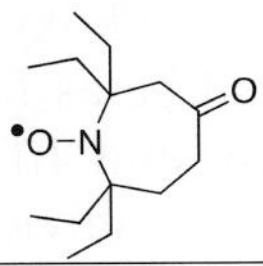	**99**[164]	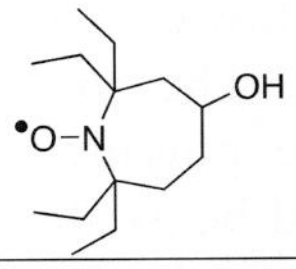		

9.3.6.2 Initiation

Two basic strategies have been applied to initiate NMP. In the first method, the initiator is a low molecular weight alkoxyamine (Scheme 9.4). This approach was used in the original work of Solomon and coworkers.[111-113] Later, Hawker and coworkers[140,165] also exploited this method and coined the term 'unimer' to describe these initiators.

In the second approach, the alkoxyamine is formed *in situ* typically from the nitroxide and radicals generated using a conventional initiator (Scheme 9.5). The initiator used in the early work of Georges et al,[116] was BPO (Scheme 9.18). The yield of alkoxyamine based on BPO is not quantitative and various side reactions are known to accompany alkoxyamine formation (Section 3.5.2.4). When the alkoxyamine is formed *in situ* the initiator efficiency must be known in order to predict molecular weights or rates of polymerization.

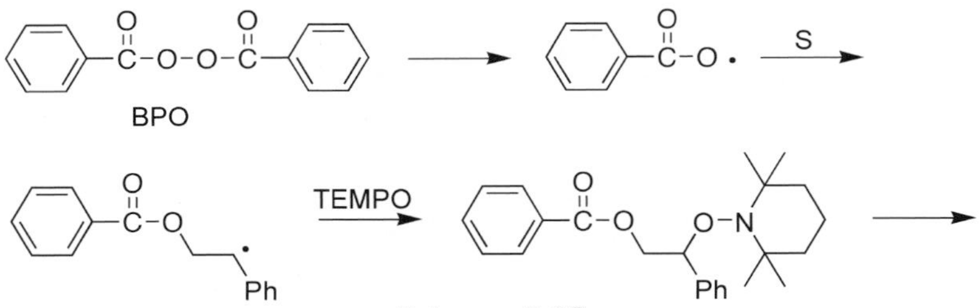

Scheme 9.18

In principle, no added conventional initiator is required for S polymerization within the temperature range 100-130 °C,[166] since radicals formed from monomer through thermal initiation by the Mayo mechanism generate alkoxyamine initiators (Section 3.3.6.1). However, this method is seldom used in practice because the alkoxyamine generation step constitutes a very long inhibition period (~24 hours depending on reaction temperature and nitroxide concentration).

Catala and coworkers[167,168] made the discovery that the rate of TEMPO-mediated polymerization of S is independent of the concentration of the alkoxyamine. This initially surprising result was soon confirmed by others.[23,169] Gretza and Matyjaszewski[169] showed that the rate of NMP is controlled by the rate of thermal initiation. With faster decomposing alkoxyamines (those based on the open-chain nitroxides) at lower polymerization temperatures, the rate of thermal initiation is lower such that the rate of polymerization becomes dependent on the alkoxyamine concentration. Irrespective of whether the alkoxyamine initiator is preformed or formed *in situ*, low dispersities require that the alkoxyamine initiator should have a short lifetime. The rate of initiation should be as fast as or faster than propagation under the polymerization conditions and lifetimes of the alkoxyamine initiators should be as short as or shorter than individual polymeric alkoxyamines.

Various methods have been used to form low molecular weight alkoxyamine initiators for NMP. Most involve forming an appropriate carbon-centered radical in the presence of a nitroxide. Initiators that generate carbon-centered radicals may be thermally decomposed in the presence of a nitroxide. For example, alkoxyamine **100** is formed by decomposition of AIBN in the presence of TEMPO (Scheme 9.19).[111] Carbon-centered radicals may also be generated photochemically.[170]

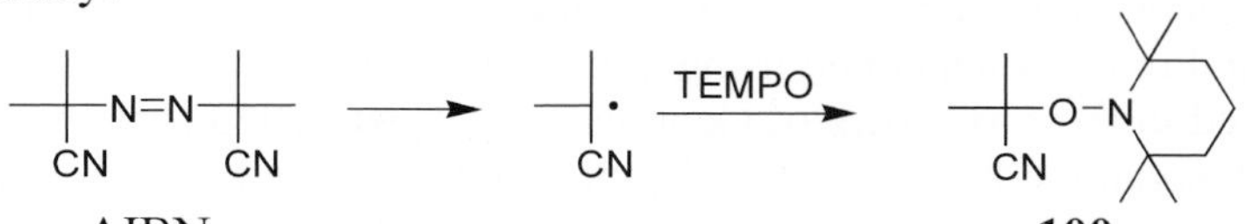

Scheme 9.19

Living Radical Polymerization 477

Another strategy involves decomposition of a peroxide or other initiator in the presence of a monomer. Conditions can be chosen such that only one unit of monomer is consumed. Thus, decomposition of DBPOX in S in the presence of DTBN provides **101** (Scheme 9.20).[111] The monomer initiator and/or combination should be chosen with care to obtain high yield of effective alkoxyamines. Many oxygen-centered radicals react with monomer by multiple pathways. Specificities shown by oxygen-centered radicals in their reaction with monomers have been studied extensively and are discussed in Section 3.4.2. Hydrogen abstraction, often by a source of *t*-butoxy radicals at low temperature [*e.g.* $(tBuO)_2$/hν,[170] DBPOX,[111,171,172] $tBuOOH$/Co(II)[173]], in the presence of a nitroxide is another common method for generating benzylic and other alkoxyamines.

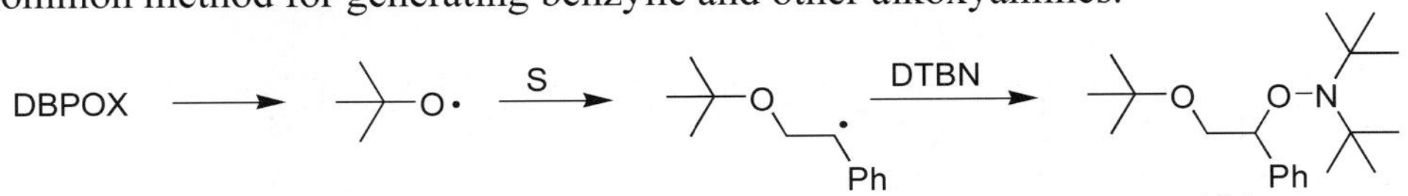

Scheme 9.20

ATRP catalysts may be used to generate radicals and thus alkoxyamines can be produced from alkyl halides in high yield (Scheme 9.21).[174] The alkoxyamine **102** was obtained in 92% yield [174] whereas reaction of TEMPO with PMMA• under ATRP conditions is reported to provide a macromonomer (Section 9.7.2.1).

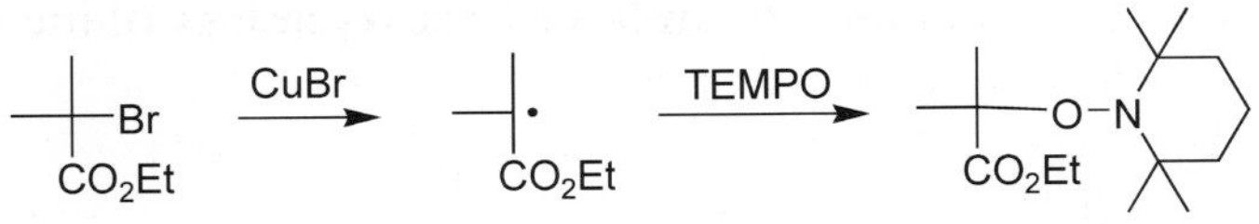

Scheme 9.21

The Manganese(V) catalyzed oxidation of S derivatives in the presence of a nitroxide provides excellent yields of phenylethyl alkoxyamines (Scheme 9.22).[175,176] Alkoxyamines can also be prepared from acrylates by oxymercuration.[177]

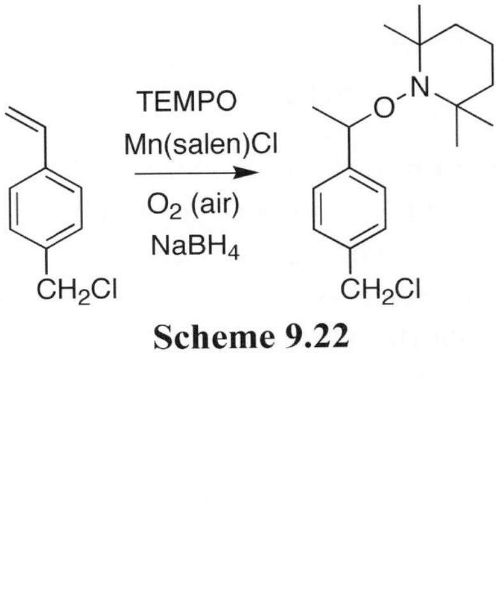

Scheme 9.22

9.3.6.3 Side reactions

The nitroxides appear completely inert towards most monomers under normal polymerization conditions. Nitroxides, in general, do not directly initiate radical polymerization. Alkoxyamines are susceptible to induced decomposition under typical reaction conditions. DTBN can undergo β-scission with formation of a nitroso compound at high temperatures (Scheme 9.23). TEMPO and other cyclic nitroxides appear intrinsically stable under polymerization conditions because of the much higher likelihood of ring closure to reform the nitroxide. The open chain nitroxides (**85-90**) are thought to show greater instability because of the presence of an α-hydrogen.

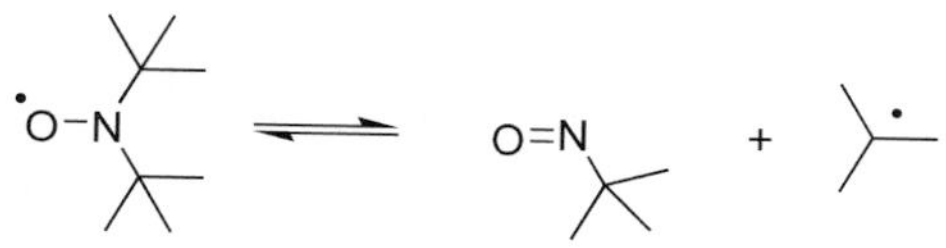

Scheme 9.23

While nitroxides give overwhelmingly combination in their reaction with carbon-centered radicals, the amount of disproportionation is finite (Scheme 9.24). Disproportionation cannot always be rigorously distinguished from elimination and it is possible that both reactions occur. The combination:disproportionation ratio (or extent of elimination) depends on the nitroxide and radical structure and within a series of structurally related systems appears to increase as k_{act} increases.[122,178]

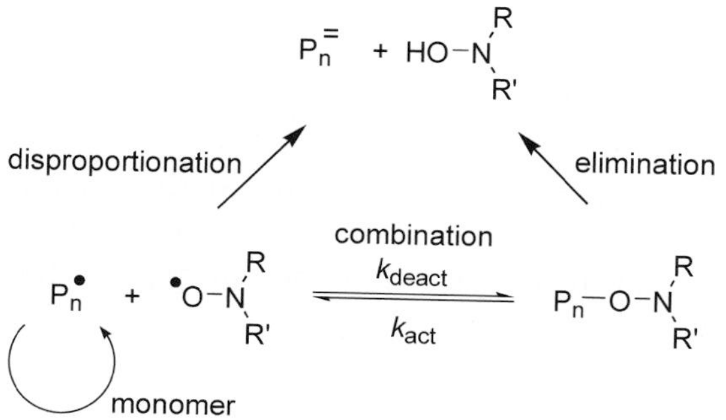

Scheme 9.24

The thermal decomposition of the phenylethyl alkoxyamine with TEMPO and the fraction of living ends in TEMPO-mediated S polymerization has been studied by Priddy and coworkers.[143,179] They concluded that to achieve >90% living ends conversions and/or nitroxide concentrations should be chosen to give M_n less than 10000.[143] However, disproportionation or elimination is most important during polymerizations of methacrylates and accounts for NMP being less successful with these monomers (Scheme 9.25).[122] The process also provides a method of macromonomer synthesis (Sections 9.7.1.1 and 9.7.2.1).

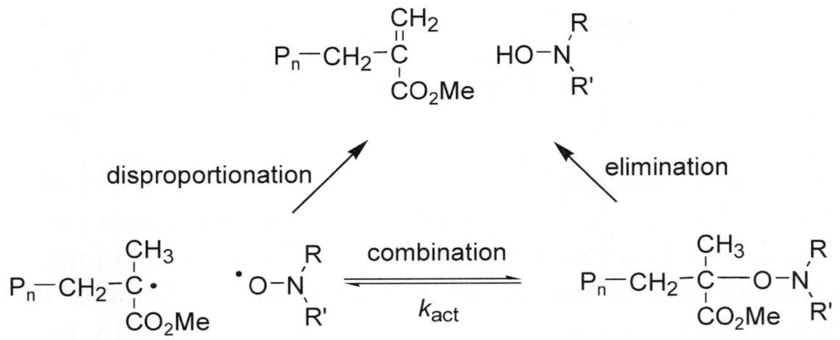

Scheme 9.25

9.3.6.4 Rate enhancement

Various strategies have been used to enhance the rate of NMP and, in particular, that mediated by TEMPO. The effects of some of these strategies on polymerization kinetics have been considered by Souaille and Fischer.[180] Most are based on the use of reagents that directly or indirectly consume and regulate the excess nitroxide that is formed continuously during polymerization as a consequence of radical-radical termination between propagating radicals. The amount of free nitroxide required to significantly retard polymerization is very small ($\sim 10^{-4}$ M). Reagents used include the following.

(a) Anhydrides.[153,181-183]

(b) Sulfonic acids (*e.g.* camphorsulfonic acid,[184,185] sulfoethyl methacrylate[185]), and their salts.[186,187] The sulfonic acid accelerants also inhibit thermal initiation of S polymerization by consuming the intermediate Diels-Alder dimer (Section 3.3.6.1).[185] It has been established that k_p for S is unaffected by sulfonic acid.

(c) Reducing agents (including ascorbic acid).[188] Added ascorbic acid is used to facilitate miniemulsion NMP of acrylates with TEMPO.[189,190] Reduction provides a hydroxylamine which can react as a transfer agent to reform nitroxide.

(d) Additional (conventional) initiators.[191-195] This initiator is chosen to decompose slowly so as to generate a low concentration of additional radicals continuously throughout the experiment. The initiator-derived radicals consume the excess nitroxide but also generate additional polymer chains. The initiator concentration used is thus critical.

Another strategy is to use a nitroxide that is intrinsically unstable. Part of the success of the open chain nitroxides that have an α-hydrogen (**86-90**) has been attributed to this factor.

9.3.6.5 Monomers

Alkoxyamine C-O bond homolysis rates have been shown to increase where propagating radical is:[23,118]

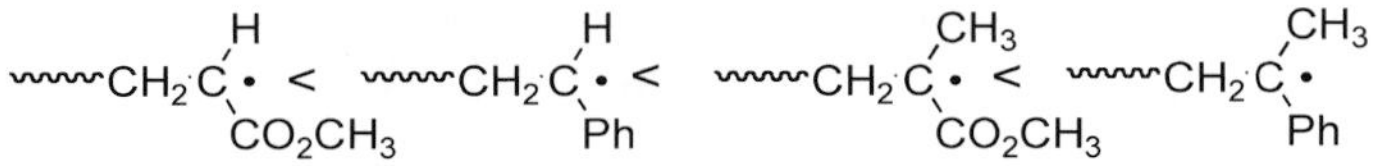

$$\text{\textasciitilde\textasciitilde CH}_2\overset{H}{\underset{CO_2CH_3}{\text{C}\cdot}} < \text{\textasciitilde\textasciitilde CH}_2\overset{H}{\underset{Ph}{\text{C}\cdot}} < \text{\textasciitilde\textasciitilde CH}_2\overset{CH_3}{\underset{CO_2CH_3}{\text{C}\cdot}} < \text{\textasciitilde\textasciitilde CH}_2\overset{CH_3}{\underset{Ph}{\text{C}\cdot}}$$

NMP has mainly been used for S polymerization (9.3.6.5.1) and, to a lesser extent, acrylate (9.3.6.5.2) polymerization. The early and much current work has focused on the use of TEMPO and derivatives. The open chain nitroxides **86-91** (Table 9.3) provide broader though still restricted utility. Some of the previously 'difficult' monomers that have recently been tackled successfully include HEA,[196] DMAM[197] and AA[198,199] with nitroxide **89**.

9.3.6.5.1 Styrene, vinyl aromatics

NMP is most commonly used for S polymerization. For S polymerizations carried out at temperatures greater than 100 °C, thermal initiation provides some rate enhancement and a mechanism for controlling the excess of nitroxide that is formed as a consequence of radical-radical termination and the persistent radical effect.[23,169]

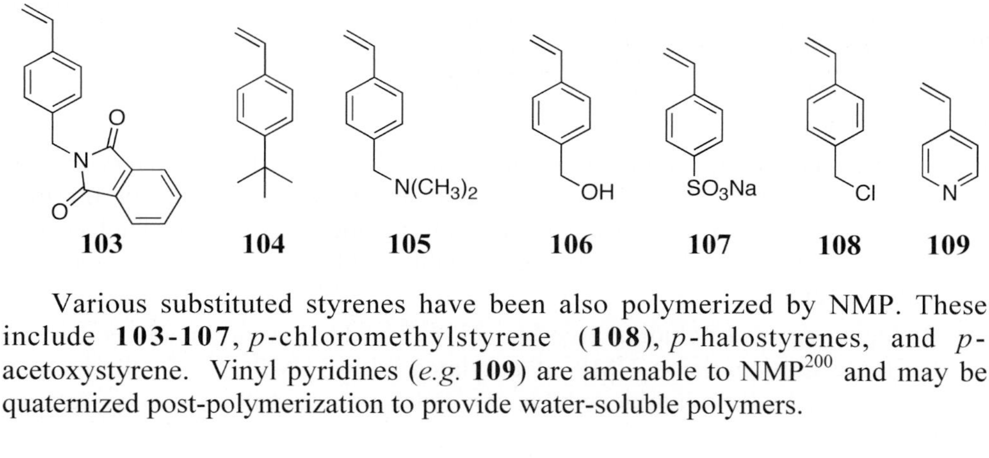

103 **104** **105** **106** **107** **108** **109**

Various substituted styrenes have been also polymerized by NMP. These include **103-107**, *p*-chloromethylstyrene (**108**), *p*-halostyrenes, and *p*-acetoxystyrene. Vinyl pyridines (*e.g.* **109**) are amenable to NMP[200] and may be quaternized post-polymerization to provide water-soluble polymers.

9.3.6.5.2 Acrylates

NMP with acrylates and acrylamides with TEMPO provides only very low conversions. Very low limiting conversions and broad dispersities were reported.[201] Better results were obtained with DTBN (**83**),[111,151] imidazoline (**61-64**)[138] and isoindoline (**59**) nitroxides.[111] However, limiting conversions were still observed. The self-regulation provided in S polymerization by thermal initiation is absent and, as a consequence, polymerization proceeds until inhibited by the build-up of nitroxide. The final product is an alkoxyamine and NMP can be continued following polymer isolation and purification. The use of additives and reaction conditions to control excess nitroxide concentrations also allows higher conversions to be obtained.[151,188-190]

Much better control is obtained with the open chain nitroxides, in particular **86**[153] and **89**,[156,158] where much lower reaction temperatures can be used and high conversions are achieved.

Molecular weights may also be limited by the occurrence of backbiting and fragmentation when high reaction temperatures are used. Backbiting without fragmentation was observed for BA polymerization at 112 °C with **89** (no unsaturated end groups observed by ^{1}H NMR).[158] However, macromonomer chain ends are clearly evident in the ^{1}H NMR of PtBA prepared with DTBN (**83**) at 120 °C.[151] For a system showing limiting conversion behavior side reactions of the propagating radical, such as backbiting-fragmentation or disproportionation, have much greater significance as their rate is not slowed as propagation is slowed by nitroxide build-up through the persistent radical effect.

9.3.6.5.3 *Methacrylates*

NMP with methacrylates is generally recognized as being difficult. It is possible to make PMMA by NMP[122] and examples of PMMA and PMMA block copolymers are provided in the first NMP patent.[111] However, in attempts to obtain high molecular weight polymers, limiting conversion behavior is observed and the product is a macromonomer.[111,122,202] Even though these high conversion polymerizations yield 'dead' polymer, a very close correspondence of found and calculated molecular weights is observed.[122] This demonstrates that the polymer that is produced is formed as a consequence of NMP and that there is little chain transfer or other mechanisms for initiation.

9.3.6.5.4 *Diene monomers*

Of the major methods for living radical polymerization, NMP appears the most successful for polymerization of the diene monomers. There are a number of reports on the use of NMP of diene monomers (B, I) with TEMPO,[188,203] **86**[154,204] and other nitroxides.[127] High reaction temperatures (120-135 °C) were employed in all cases. The ratio of 1,2-:1,4-cis:1,4-trans structures obtained is similar to that observed in conventional radical polymerization (Section 4.3.2).

9.3.6.6 *Heterogeneous polymerization*

NMP of S in heterogeneous media is discussed in reviews by Qiu *et al.*,[205] Cunningham,[206,207] and Schork *et al.*[208] There have been several theoretical studies dealing with NMP and other living radical procedures in emulsion and miniemulsion.[209-213] Butte *et al.*[210,214] concluded that NMP (and ATRP) should be subject to marked retardation as a consequence of the persistent radical effect. Charleux[209] predicted enhanced polymerization rates for miniemulsion with small

(50–100 nm) particles when the persistent radical can be desorbed from the particle phase. Ma et al.[211-213] also concluded that the distribution of nitroxide between the aqueous and organic phases was critical to maintaining livingness and achieving acceptable polymerization rates.

The early attempts at NMP of S in emulsion used TEMPO and related nitroxides and needed to be carried out at high temperatures (100-130 °C) necessitating a pressure reactor. Problems with colloidal stability and molecular weight control and limiting conversions were reported.[215-217]

Successful NMP in emulsion requires use of conditions where there is no discrete monomer droplet phase and a mechanism to remove any excess nitroxide formed in the particle phase as a consequence of the persistent radical effect. Szkurhan and Georges[218] precipitated an acetone solution of a low molecular weight TEMPO-terminated PS into an aqueous solution of PVA to form emulsion particles. These were swollen with monomer and polymerized at 135 °C to yield very low dispersity PS and a stable latex. Nicolas et al.[219] performed emulsion NMP of BA at 90 °C making use of the water-soluble alkoxyamine **110** or the corresponding sodium salt both of which are based on the open-chain nitroxide **89**. They obtained PBA with narrow molecular weight distribution as a stable latex at a relatively high solids level (26%). A low dispersity PBA-*block*-PS was also prepared.

110   **111**

NMP in miniemulsion has been more successful. In miniemulsion polymerization nucleation takes place directly in the monomer droplets that become the polymer particles. Particle sizes are small (<100 nm). Most work has used TEMPO and high reaction temperatures (120-140 °C) with S or BA as monomer.

Various initiation strategies and surfactant/cosurfactant systems have been used. Early work involved *in situ* alkoxyamine formation with either oil soluble (BPO)[220,221] or water soluble initiators (persulfate) and traditional surfactant and hydrophobic cosurfactants. Later work established that preformed polymer could perform the role of the cosurfactant and surfactant-free systems with persulfate initiation were also developed.[190,222,223] Oil soluble (PS capped with TEMPO,[221] **111**,[224] PBA capped with **89**) and water soluble alkoxyamines (**110**, sodium salt[224]) have also been used as initiators. Addition of ascorbic acid, which reduces the nitroxide which exits the particles to the corresponding hydroxylamine, gave enhanced rates and improved conversions in miniemulsion polymerization with TEMPO.[225] Ascorbic acid is localized in the aqueous phase by solubility.

9.3.7 Other Oxygen-Centered Radical-Mediated Polymerizations

A number of other chemistries which involve C-O bond cleavage have been reported.[226,227] Druliner[226] has reported on systems where NCO•, **112**, **113** or related species is the persistent radical. Homolysis rates for these systems were stated to be suitable for MMA polymerization at ambient temperature. The use of NCO• has also been studied by Grande et al.,[228-230] most recently for AA polymerization.[230] Although control during AA homopolymerization was poor the process yielded NCO- terminated PAA that could be used to make PAA-block-PMMA.[230]

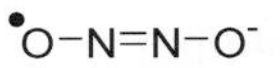

•O−N=N−O⁻
112[226]

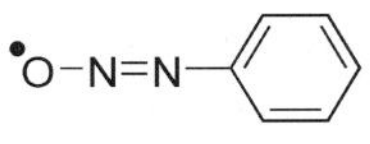

•O−N=N−⟨⟩
113[226]

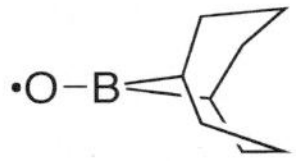

•O−B⟨⟩
114[231,232]

Chung and coworkers have reported on the use of stable borinate or boroxyl radicals (e.g. **114**) to mediate radical polymerization.[231,232] Methacrylates (MMA) and acrylates (trifluoroethyl acrylate) have been polymerized at ambient temperature to yield polymers with relatively narrow molecular weight distributions.[231-233] The method has been used to prepare block copolymers and polyolefin graft copolymers.[234-237]

A living radical polymerization mechanism was proposed for the polymerization of MMA[238-240] and VAc[241] initiated by certain aluminum complexes in the presence of nitroxides. It was originally thought that a carbon-aluminum bond was formed in a reversible termination step. However, a more recent study found the results difficult to reproduce and the mechanism to be complex.[242]

9.3.8 Nitrogen-Centered Radical-Mediated Polymerization

A few studies have appeared on systems based on persistent nitrogen-centered radicals. Yamada et al.[227] examined the synthesis of block polymers of S and MMA initiated by derivatives of the triphenylverdazyl radical **115**. Klapper and coworkers[243] have reported on the use of triazolinyl radicals (e.g. **116** and **117**). The triazolinyl radicals have been used to control S, methacrylate and acrylate polymerization and for the synthesis of block copolymers based on these monomers [S,[243-245] tBA,[243] MMA,[243-245] BMA,[245] DMAEMA,[246] TMSEMA,[247] (DMAEMA-block-MMA),[246] (DMAEMA-block-S)[246] and (TMSEMA-block-S)[247]]. Reaction conditions in these experiments were similar to those used for NMP. The triazolinyl radicals show no tendency to give disproportionation with methacrylate propagating radicals. Dispersities reported are typically in the range 1.4-1.8.[243,246]

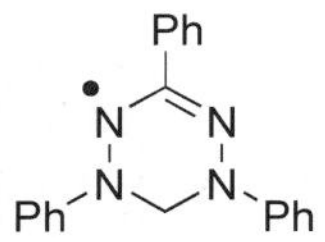

115 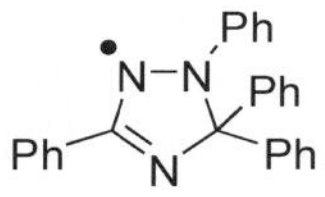 **116** 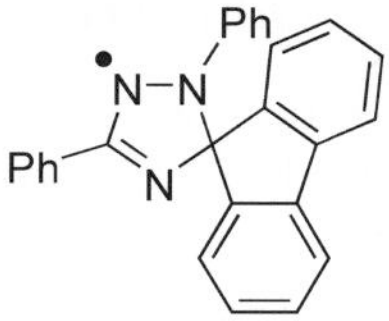 **117**

The triazolinyl radical **116** is thermally unstable with a half-life of ~20 min at 95 °C. The compound **117** is stable under similar conditions. The decomposition mechanism involves loss of a phenyl radical and formation of a stable aromatic triazene (Scheme 9.26).[243] This provides a mechanism for self regulation of the stable radical concentration during polymerization and a supplemental source of initiating radicals.

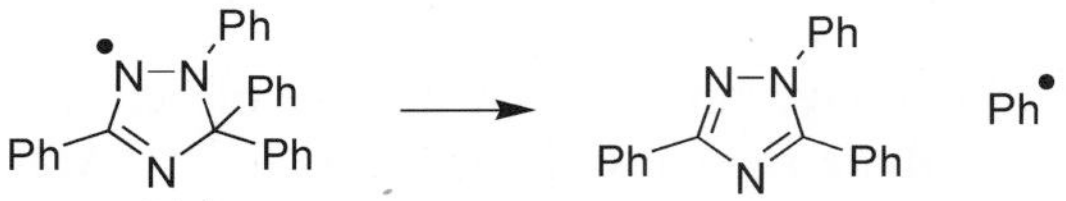

Scheme 9.26

9.3.9 Metal Complex-Mediated Radical Polymerization

Metal complexes may also act as initiators in stable radical-mediated polymerization with the metal complex performing the role of the stable radical. There are reports of titanocene,[248,249] cobalt,[250-253] chromium, iron and molybdenum[254] complexes in this context.

Oganova et al.[255-258] observed that certain cobalt (II) porphyrin complexes reversibly inhibit BA polymerization presumably with formation of a cobalt (III) intermediate as shown in Scheme 9.27. Thus, it seemed reasonable to propose these species may function as initiators in living radical polymerization.[250,259]

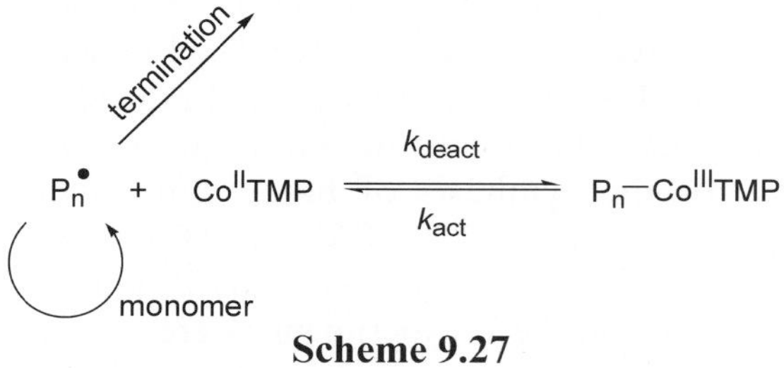

$$P_n^\bullet + Co^{II}TMP \underset{k_{act}}{\overset{k_{deact}}{\rightleftharpoons}} P_n-Co^{III}TMP$$

Scheme 9.27

Wayland et al. reported the use of tetramesitylporphyrin complexes (CoTMP), including **118**[250] and **119**[251] in the synthesis of high molecular weight PMA with very low dispersities (1.1-1.3). Arvanitopoulos et al.[260] have reported similar chemistry with alkylcobaloximes (**120**) as photoinitiators at low temperatures.

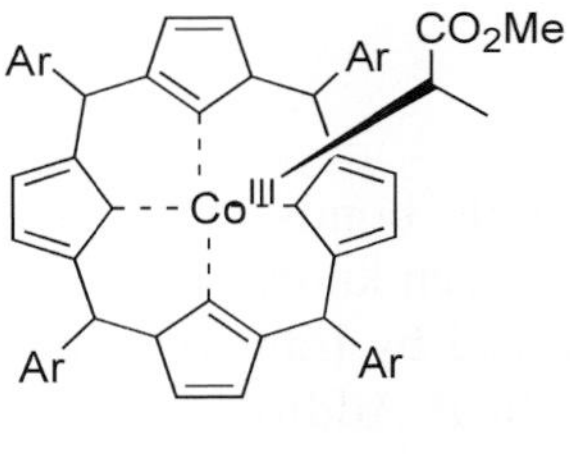

118

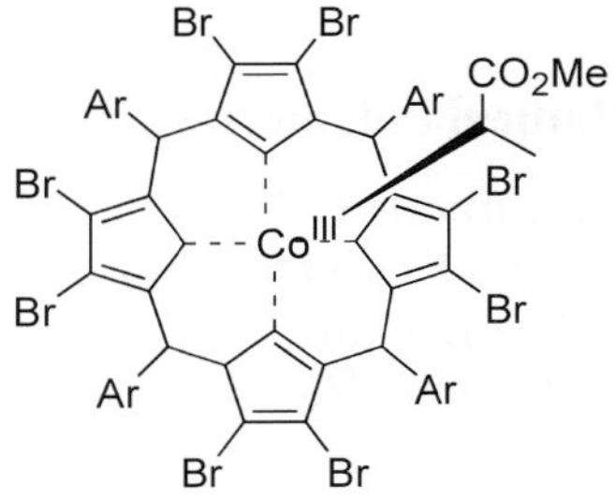

119

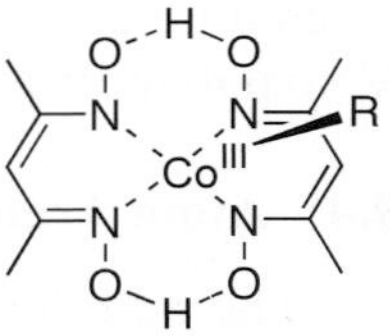

120

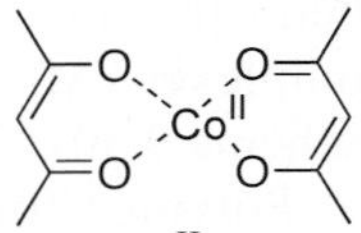

121 Co^{II}(acac)$_2$

The most important side reactions are disproportionation between the cobalt(II) complex and the propagating species and/or β-elimination of an alkene from the cobalt(III) intermediate. Both pathways appear unimportant in the case of acrylate ester polymerizations mediated by Co^{II}TMP but are of major importance with methacrylate esters and S. This chemistry, while precluding living polymerization, has led to the development of cobalt complexes for use in catalytic chain transfer (Section 6.2.5).

It is also known that alkyl cobaloximes are subject to radical-induced decomposition.[257] This suggests an alternative to the mechanism shown in Scheme 9.28 involving reversible chain transfer (Section 9.5).

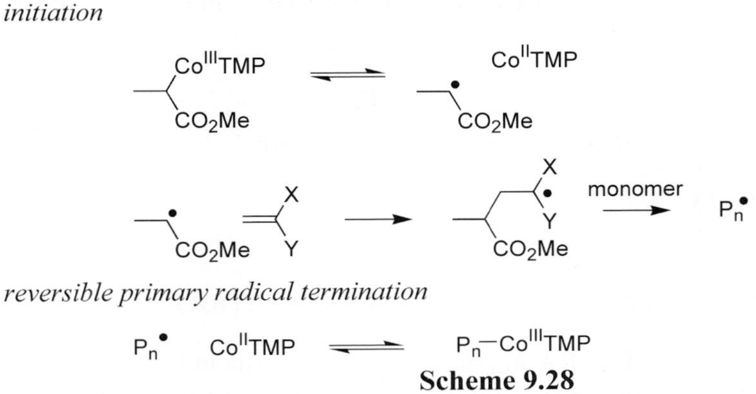

initiation

reversible primary radical termination

$P_n^\bullet$ Co^{II}TMP ⇌ P_n–Co^{III}TMP

Scheme 9.28

It has also been shown that the alkyl cobalt (III) initiator can be generated *in situ*[252] by adding a fast-decomposing azo-initiator [2,2'-azo-bis(4-methoxy-2,4-dimethyl valeronitrile] to a solution of the cobalt (II) complex in monomer. Very narrow dispersity PMA and PMA-*block*-PBA were prepared.

In a very recent development, Debuigne *et al.*[253] have reported polymerization of vinyl acetate at 30 °C mediated by Co^{II}(acac)$_2$ (**121**). They obtained predictable molecular weights up to $\overline{M}_n$=100000 and dispersities < 1.3 and proposed a polymerization mechanism analogous to that shown in Scheme 9.27. The complex offered no control over BA polymerization and the porphyrin complexes inhibited VAc polymerization.

9.4 Atom Transfer Radical Polymerization

The addition of halocarbons (RX) across alkene double bonds in a radical chain process, the Kharasch reaction (Scheme 9.29),[261] has been known to organic chemistry since 1932. The overall process can be catalyzed by transition metal complexes (Mt^n-X); it is then called Atom Transfer Radical Addition (ATRA) (Scheme 9.30).[262]

Polymer formation during the Kharasch reaction or ATRA can occur if trapping of the radical (**123**), by halocarbon or metal complex respectively, is sufficiently slow such that multiple monomer additions can occur. Efficient polymer synthesis additionally requires that the trapping reaction is reversible and that both the activation and deactivation steps are facile.

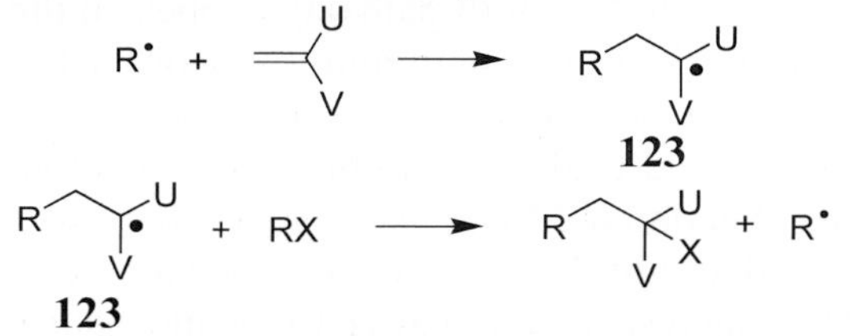

Scheme 9.29 Kharasch Reaction

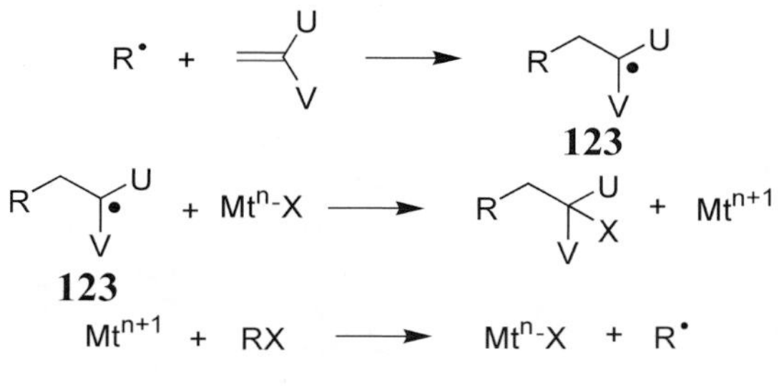

Scheme 9.30 Atom Transfer Radical Addition (ATRA)

The first purposeful use of ATRA in polymer synthesis was in the production of telomers.[263] In this early work, comparatively poor control over the polymerization was achieved and little attempt was made to explore the wider utility of the process. Some analogies may also be drawn with the work of Bamford *et al.* and others on transition metal/organic halide redox initiation (Sections 3.3.5.1 and 7.6.2).[264]

The first reports of ATRP (Atom Transfer Radical Polymerization), which clearly displayed the characteristics of living polymerization, appeared in 1995 from the laboratories of Sawamoto,[265] Matyjaszewski[266] and Percec.[267] The literature on ATRP is now so vast that a comprehensive review cannot be presented here. A number of reviews on ATRP have appeared. Most informative on the scope of the process are those by Matyjaszewski and Xia[268,269] and Kajimoto et al.[270,271] The kinetics of ATRP are considered in reviews by Fischer[110] and Goto and Fukuda.[23] ATRP is sometimes also called transition metal-mediated radical polymerization. We use this latter term for radical polymerizations where control is achieved by a reversible coupling mechanism (Section 9.3.9).

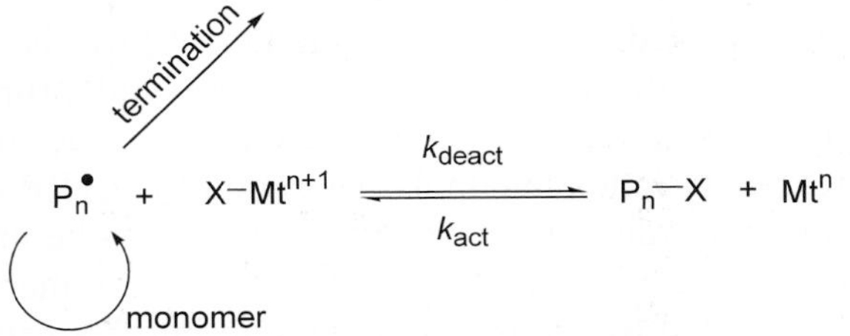

$$P_n^\bullet + X-Mt^{n+1} \underset{k_{act}}{\overset{k_{deact}}{\rightleftharpoons}} P_n-X + Mt^n$$

(with termination and monomer cycle)

Scheme 9.31

A much-simplified mechanism for reversible activation-deactivation of polymer chains during ATRP is shown in Scheme 9.31. In the deactivation process, propagating radicals are trapped by atom or group transfer [most commonly a halogen (Cl, Br, I) although other groups (*e.g.* SCN) are known] from a metal complex in its higher oxidation state. The activation process involves a redox reaction between the polymer end group and the metal complex in its reduced form.

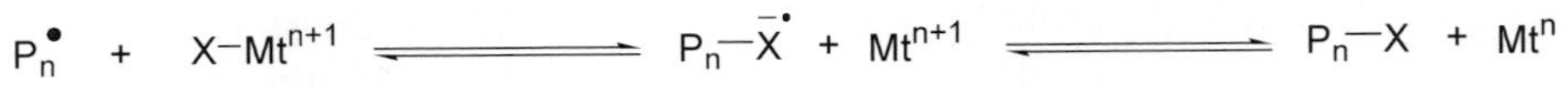

$$P_n^\bullet + X-Mt^{n+1} \rightleftharpoons P_n-\bar{X}^\bullet + Mt^{n+1} \rightleftharpoons P_n-X + Mt^n$$

Scheme 9.32

The atom transfer reaction is generally thought to involve inner sphere electron transfer (ISET) with concerted transfer of the halogen from initiator to the metal complex and various kinetic and other data support this view for most of the common initiator/catalyst/monomer combinations. However, it is possible to write the process as two steps, the first being an outer sphere electron transfer (OSET) process to provide an intermediate radical anion (Scheme 9.32).[268] The living polymerization of vinyl chloride with alkyl iodide initiators and nascent Cu(0) catalyst is considered to involve an OSET process.[272,273] OSET does not require a transition metal catalyst and can involve other single electron reducing agents such as dithionite.[274] For this case it is also possible that the chain equilibration step is, in part, similar to that discussed under iodine transfer polymerization (Section 9.5.4).[274]

Ideally, the metal complex is a catalyst and, in principle, is only required in very small quantities. However, the kinetics of initiation for the systems described to date dictate that relatively large amounts are used and catalyst:initiator ratios are typically in the range 1:1 to 1:10. The most commonly used catalysts are metal complexes based on Cu and Ru. However, a wide range of metals and ligands has been used (Section 9.4.2). Conditions and catalysts have been found such that most monomers polymerizable by a radical mechanism can be used in ATRP. Difficult monomers are vinyl acetate and simple olefins (in homopolymerization) and monomers that coordinate strongly with metal centers. It is extremely important to select the initiator, catalyst and reaction conditions for the particular monomer.

There has been some discussion on whether ATRP is a 'free' radical polymerization.[275,276] Are the reactions of initiating and propagating species produced in ATRP influenced by the presence of the metal complex? Reports[275,276] that reactivity ratios in copolymerization by ATRP differ from those observed in conventional radical polymerization appear to be an effect of chain length (Section 9.6). There is no doubt that the rate of polymerization in ATRP can be dramatically affected by the reaction medium but this can in large part be attributed to changes in the activation/deactivation equilibrium. The current general consensus is that the common forms of ATRP are radical processes and the propagating radicals behave as 'free' propagating radicals under the reaction conditions. The polymerization kinetics can be interpreted on this basis and radical-radical termination occurs to the extent expected given the radical concentration,

Notwithstanding the occurrence of any side reactions, a successful ATRP experiment will generally yield a polymer with halogen end groups. These end-groups are potentially labile and may impair polymer stability. Moreover, corrosive by-products (hydrohalic acids) can be formed by thermal elimination. However, the end groups are also precursors to a wide range of other functionality. It is possible to transform them into groups that are chemically inert or to useful functionalities (Section 9.7.2.1). They also render the polymers useful as precursors to block, star, comb and more complex architectures (Sections 9.8-9.9.3.2).

9.4.1 Initiators

The initiator in ATRP is usually a low molecular weight activated organic halide (RX, R=activated alkyl, X=chlorine, bromine, iodine). However, organic pseudohalides (*e,g,* X=thiocyanate, azide) and compounds with weak N-X (*e.g.* N-bromosuccinimide[277]) or S-X (*e.g.* sulfonyl halides - see below) have been used.

The first reported initiators were polyhalogeno-compounds (*e.g.* CCl_4, $CHCl_3$, CCl_3CH_2OH, CCl_3Br). Trichloromethane derivatives and tetrachloromethane appear effective initiators. Mono- and dichloromethane derivatives are inefficient initiators. Tetrachloromethane may act as a difunctional initiator.

In choosing an initiator the strength of the R-X bond in both the initiator and the dormant propagating species formed should be considered. It is common practice to use a compound such that the radical generated is a monomeric or low molecular weight species structurally analogous to the propagating radical. Thus,

Living Radical Polymerization

α-bromoisopropionates (**124**) are used for acrylates, α-bromoisobutyrates (**125**) are used to initiate polymerization of MMA and other methacrylates (Scheme 9.33), and benzyl bromide (**126**) or phenylethyl bromide (**127**) is used to initiate polymerization of S and derivatives. Initiator activity is discussed further in Section 9.4.1.3.

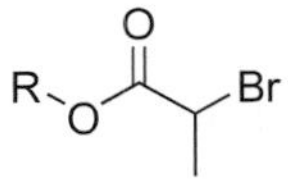

124
a R=C$_2$H$_5$

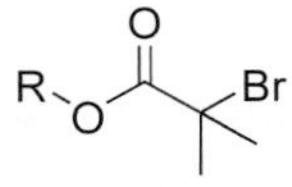

125
a R=C$_2$H$_5$

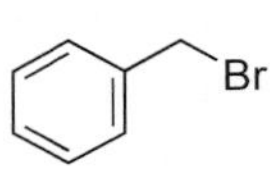

126

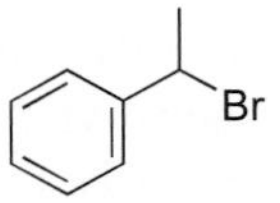

127

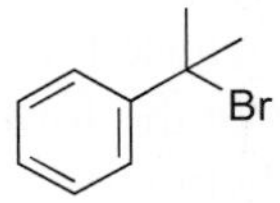

128

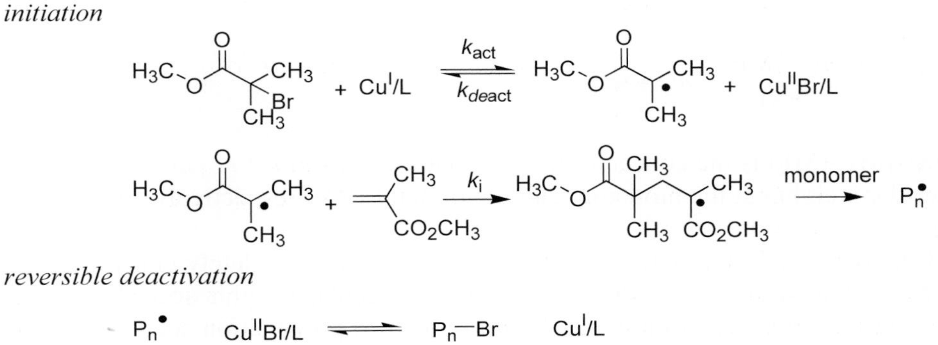

initiation

$$H_3C-O-C(=O)-C(CH_3)(Br)(CH_3) + Cu^I/L \underset{k_{deact}}{\overset{k_{act}}{\rightleftharpoons}} H_3C-O-C(=O)-C^{\bullet}(CH_3)(CH_3) + Cu^{II}Br/L$$

$$H_3C-O-C(=O)-C^{\bullet}(CH_3)(CH_3) + CH_2=CH(CO_2CH_3) \xrightarrow{k_i} H_3C-O-C(=O)-C(CH_3)(CH_3)-CH_2-C^{\bullet}H(CO_2CH_3) \xrightarrow{monomer} P_n^{\bullet}$$

reversible deactivation

$$P_n^{\bullet} + Cu^{II}Br/L \rightleftharpoons P_n-Br + Cu^I/L$$

Scheme 9.33

Other important classes of initiator are the organic sulfonyl chlorides (*e.g.* **129**, **130**)[267,278-280] and bromides (*e.g.* **131**).[281] These are very effective when used in conjunction with copper catalysts with bpy or dNbpy ligands. Functional sulfonyl chloride initiators have also been reported (Section 9.7.2.2). Rates of radical generation are high with respect to propagation such that they can be used with methacrylates, styrenes and acrylates. In some circumstances, initiator efficiencies observed with sulfonyl halide initiators may be lowered by side reactions involving the sulfonyl radicals.[282,283] These side reaction include reaction of sulfonyl radical with the ligand (PMDETA) by hydrogen abstraction.[282] This pathway is not important with bipyridyl ligands (bpy, dNbpy). With ruthenium catalysts that use a Al(OiPr)$_3$ cocatalyst, the cocatalyst may react with the sulfonyl chloride to cause a decrease in the initiator efficiency.[284]

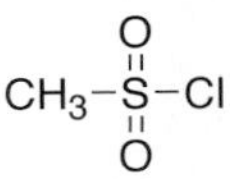

129 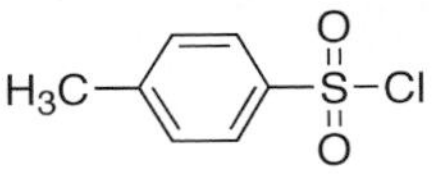 **130** 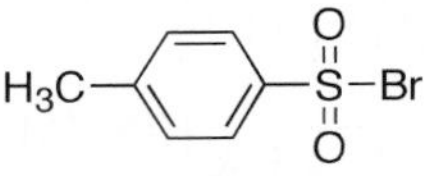 **131**

9.4.1.1 Molecular weights and distributions

In ATRP, the initiator (RX) determines the number of growing chains. Ideally, the degree of polymerization is given by eq. 7 and the molecular weight by eq. 8. Note the appearance of the initiator efficiency (f') in the numerator of these expressions. In practice, the molecular weight is often higher than anticipated because the initiator efficiency is decreased by side reactions. In some cases, these take the form of heterolytic decomposition or elimination reactions. Further redox chemistry of the initially formed radicals is also known. The initiator efficiencies are dependent on the particular catalyst employed.

$$\overline{X}_n = \frac{([M]_o - [M]_t)f'}{[RX]_o} = \frac{[M]_o f'}{[RX]_o} c \tag{7}$$

$$\overline{M}_n = \frac{([M]_o - [M]_t)f'}{[RX]_o} m_M + m_{RX} \tag{8}$$

where $([M]_o-[M]_t)$ is the concentration of monomer consumed m_M and m_{RX} are the molecular weights of the monomer and the initiator (RX) respectively, and c is the monomer conversion.

It is assumed in the derivation of eq. 7 that RX is completely consumed. In order to obtain good control (low dispersities, molecular weights according to eq. 7) it is critical that initiation is rapid with respect to propagation such that RX is consumed before there is any substantial conversion of monomer. Slow usage of RX will give a post-tailing or bimodal molecular weight distribution.

In S polymerization, thermal initiation will be a source of extra chains. Additional chain formation processes will cause the molecular weight to be lower than anticipated by eq. 7. Sometimes conventional thermal initiators are added with similar effect (see also eq. 12). A pre-tailing molecular weight distribution may result.

In ideal circumstances, with polymerization described by Scheme 9.31 and rate of activation of RX equal to that of P_nX, the dispersity is given by eq. 9.[23]

$$\frac{\overline{X}_w}{\overline{X}_n} = 1 + \frac{1}{\overline{X}_n} + \left(\frac{2-c}{c}\right) \frac{k_p[RX]}{k_{deact}[Mt^{n+1}X]} \tag{9}$$

where c is the monomer conversion.

The rate of polymerization is given by eq. 10.

$$R_p = k_p K \frac{[RX][Mt^n]}{[Mt^{n+1}X]}[M] \tag{10}$$

The ATRP experiment is usually commenced with all of the catalyst in its lower oxidation state. The number of propagation events per activation cycle is dependent on the concentration of catalyst in its higher oxidation state. For low dispersities it is important that this number is small. As indicated by eq. 9 dispersity is inversely proportional to the concentration of the deactivator ($Mt^{n+1}X$). Thus, just as in NMP, where it is desirable to have a very low concentration of free nitroxide in the polymerization medium, in ATRP it can be important to have a proportion of the catalyst in its higher oxidation state. However, as implied by eq. 10, a concentration of deactivator that is too high can cause retardation or even inhibition of polymerization.

9.4.1.2 Reverse ATRP

So-called reverse ATRP has been described where a conventional radical initiator (*e.g.* AIBN) and a transition metal complex in its higher oxidation state are used.[285-288] One of the first systems explored was $CuBr_2$/**133**/AIBN/MMA. It is important that the initiator is completely consumed early in the polymerization. The use of peroxide initiators in reverse ATRP can be problematical depending on the catalyst used and the reaction temperature.[286,289] The system $CuBr_2$/**133**/BPO/MMA at 60°C was found to provide no control.[286] In ATRP at lower temperatures (40 °C), the system CuCl/**133**/BPO/MMA was successful though dispersities obtained were relatively broad,[289] Radicals are produced from the redox reaction between the catalyst in its reduced form and BPO.

The molecular weight in reverse ATRP will depend on the concentration of the initiator (I_2) and the initiator efficiency (f) and ideally is given by eq. 11. Side reactions between the catalyst and the initiator and the radicals formed from the initiator may lead to efficiencies being lower than those observed in conventional radical polymerization.

$$\bar{X}_n = \frac{[M]_o}{[I_2]_o f} c \qquad (11)$$

Experiments have been described where a combination of direct and reverse ATRP is used.[290] In this case eq. 12 should apply.

$$\bar{X}_n = \frac{[M]_o}{[I_2]_o f + [RX]_o} c \qquad (12)$$

In combination ATRP, the catalyst is again present in its more stable oxidized form. A slow decomposing conventional initiator (*e.g.* AIBN) is used together with a normal ATRP initiator. Initiator concentrations and rate of radical generation are chosen such that most chains are initiated by the ATRP initiator so dispersities can be very narrow.[290] The conventional initiator is responsible for generating the activator *in situ* and prevents build up of deactivator due to the persistent radical effect. Reverse or combination ATRP are the preferred modes of initiation for ATRP in emulsion or miniemulsion (Section 9.4.3.2).[290,291]

9.4.1.3 Initiator activity

The activity of initiators in ATRP is often judged qualitatively from the dispersity of the polymer product, the precision of molecular weight control and the observed rates of polymerization. Rates of initiator consumption are dependent on the value of the activation-deactivation equilibrium constant (K) and not simply on the activation rate constant (k_{act}). Rate constants and activation parameters are becoming available and some valuable trends for the dependence of these on initiator structure have been established.[292-297]

(a) For compounds with a similar activating group, tertiary halides are substantially more active than secondary halides, which, in turn, are more active than primary halides. Thus activity increases in the series: **126 < 127 < 128**; and **124 < 125**.

(b) In the case of alkyl halide initiators >C(R)-X, activity is reported to decrease in the series where the activating group R is CN>C(O)R>C(O)OR>Ph>>Cl>OCOCH3>Me.[268] Note, this order does not reflect the carbon-halogen bond dissociation energies or the product radical stability. This parallels the trend in activation rate constants for propagating radicals.[268]

$$\text{~~CH}_2\overset{CH_3}{\underset{CO_2CH_3}{C\cdot}} > \text{~~CH}_2\overset{H}{\underset{Ph}{C\cdot}} > \text{~~CH}_2\overset{H}{\underset{CO_2CH_3}{C\cdot}} > \text{~~CH}_2\overset{H}{\underset{OCOCH_3}{C\cdot}}$$

(c) As in other radical processes (and activation NMP and RAFT), penultimate unit effects are important in determining the rate constant for activation.[296] Dimeric (and higher) species are more active than monomeric species particularly in the case of tertiary radicals.

While the above trends appear generic, initiator activity is strongly dependent on the specific catalyst used (Section 9.4.2).

9.4.2 Catalysts

Transition metal catalysts are characterized by their redox chemistry (catalysts can be considered as one electron oxidants/reductants). They may also be categorized by their halogen affinity. While in the initial reports on ATRP (and in most subsequent work) copper[266,267] or ruthenium complexes[265] were used, a wide range of transition metal complexes have been used as catalysts in ATRP.

(a) Group 6: molybdenum (Mo^{IV}-Mo^V).[254,298,299]
(b) Group 7: manganese (Mn^{II}-Mn^{III}),[300,301] rhenium (Re^V-Re^{VI}).[302,303]
(c) Group 8: iron (Fe^{II}-Fe^{III}), (Fe^{I}-Fe^{II}) (Section 9.4.2.3), ruthenium (Ru^{II}-Ru^{III}) (Section 9.4.2.2)
(d) Group 9: cobalt (Co^0-Co^I),[304] rhodium (Rh^I-Rh^{II}),[305-307]
(e) Group 10: nickel (Ni^{II}-Ni^{III}) (Section 9.4.2.4), and palladium (Pd^{II}-Pd^{III}).[308]
(f) Group 11: copper (Cu^I-Cu^{II}) (Section 9.4.2.1).

Most are proposed to involve the general ATRP mechanism. However, it should also be noted that the detailed mechanism has not been elucidated in all cases and not all need be radical processes in the conventional sense. Moreover, in many polymerizations, the active catalyst is formed *in situ* and its exact nature is not rigorously established.

An issue with ATRP is the residual metal catalyst and its removal from the polymer post-polymerization. Many papers have been written on catalyst removal and recycling.[309]

9.4.2.1 Copper complexes

The most common catalysts for ATRP are complexes based on a copper(I) halide and nitrogen based ligand(s). Various ligands have been employed and those most frequently encountered are summarized in Table 9.5. Typically, four nitrogens coordinate to copper. The bidentate bipyridyl (bpy) ligands **132-133** are known to form a 2:1 complex. The tetradentate ligands are expected to form a 1:1 complex.

The first ATRP experiments were conducted with a complex presumed to be of the form $[Cu^I(bpy)_2]^+X^-$ as catalyst and either alkyl halide[266] or sulfonyl chloride initiators.[267] The complexes were formed *in situ* and the experimental process involved mixing Cu^{II} halide and the ligand in the reaction medium. The reactions with 2,2'-bipyridine (bpy, **132**) are generally heterogeneous and the precise structure of the active catalyst in solution was not known. The bpy derivatives with long chain alkyl groups (**134, 133**) were introduced to provide greater solubility for the copper complex and allow a more homogeneous polymerization and therefore improved control over polymerization. Many studies probing the solution and solid-state structures of bipyridine and other complexes have now been carried out.[310]

Certain multidentate ligands also provide for better solubility. Cu^I complexes formed with tetramethylethylenediamine (TMEDA), N,N,N',N',N''-pentamethyldiethylenetriamine (PMDETA, **140**) and 1,1,4,7,10,10-hexamethyltriethylenetetramine (HMTETA, **144**) and Me$_6$TREN (**145**) have been found effective.[311] Transfer to ligand during MMA polymerization has been reported as a side reaction when PMDETA is used.[312,313]

Haddleton and coworkers[314] reported the use of Cu^I complexes based on the methanimine ligands (*e.g.* **136-138**) and have demonstrated their efficacy in the polymerization of methacrylates. The ligands can be prepared *in situ* from the appropriate amine and 2-pyridine carboxaldehyde (Scheme 9.34).

Guidelines for predicting the activity of complexes formed with various ligands have been formulated.[268,269] The activity goes up according to the number of nitrogens coordinated to copper and with the electron donating ability of the nitrogens. Tetradentate ligands appear more effective than tri- or bidentate ligands. Some correlation between k_{act} and k_{deact} and the redox potential of the complex has been observed.[315,316] A lower redox potential results in a higher k_{act}

and a lower k_{deact}. However, there appears no direct correlation with structural features of the complex such as Cu-Br bond lengths.[310]

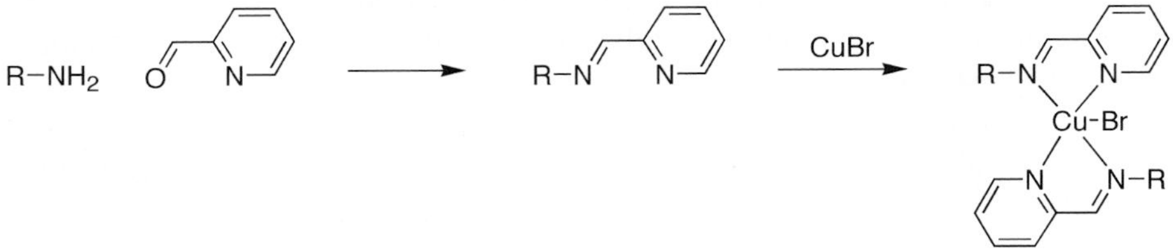

Scheme 9.34

Table 9.5 Structures of Ligands for Copper Based ATRP Catalysts

Ligand	Structure	Ligand	Structure
132 bpy[266,267]	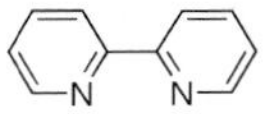	133 dNbpy[317]	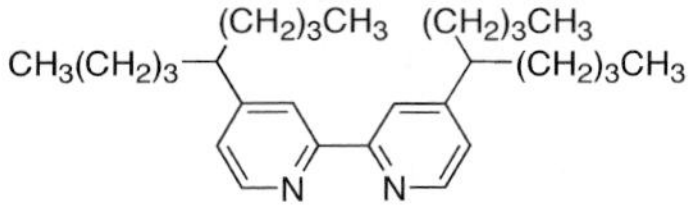
134 dHbpy[318]	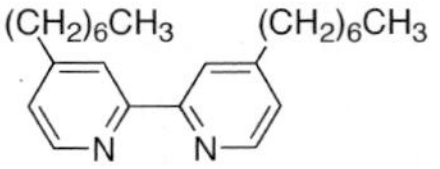	135 phen[319]	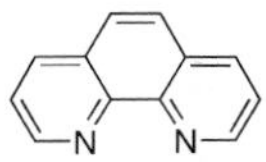
136[314]	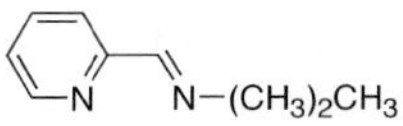	137[320]	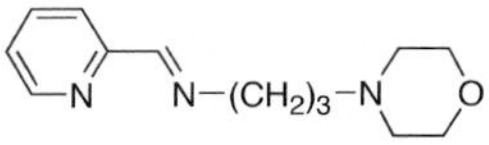
138[321]	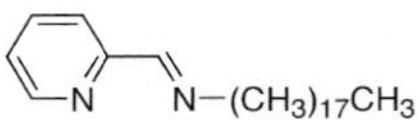	139[311,322]	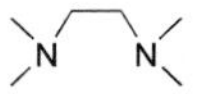
140 PMDETA [311]	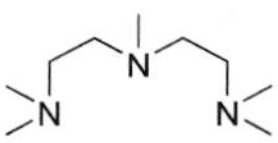	141[323]	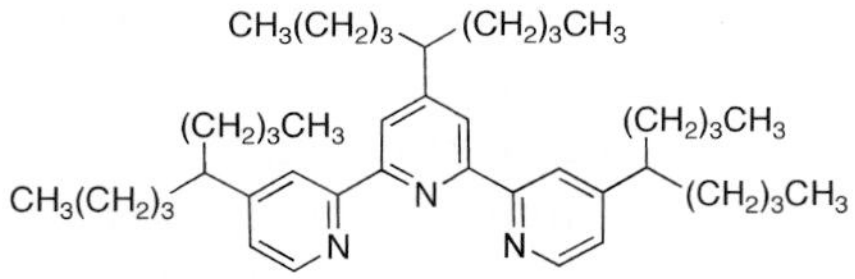
142 BPMODA [324]	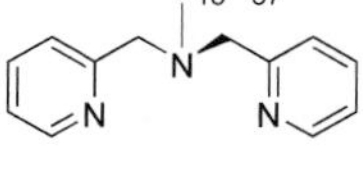	143[325]	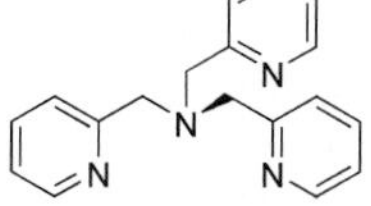
144 HMTETA [311,324]	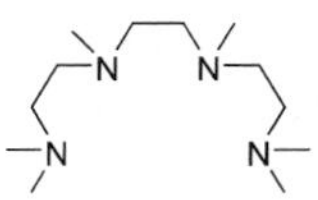	145 Me$_6$TREN [326]	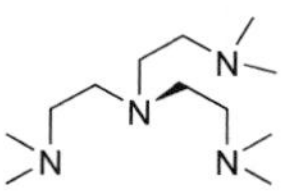

Percec and coworkers[327,328] reported *in situ* formation of active CuCl/CuCl$_2$ catalyst from the initiator, Cu$_2$O, Cu(0) and combinations of these in conjunction with ligand (bpy) and various polyethers or ethylene glycol and suggested that improved control was obtained under these conditions.

Supported copper catalysts have also been described.[329-340] The main impetus for the development of supported ATRP catalysts has been to facilitate catalyst removal and, in some cases, to allow for catalyst recycling.

9.4.2.2 Ruthenium complexes

In contrast to the situation with copper-based catalysts, most studies on ruthenium-based catalysts have made use of preformed metal complexes. The first reports of ruthenium-mediated polymerization by Sawamoto and coworkers appeared in 1995.[265] In the early work, the square pyramidal ruthenium (II) halide **146** was used in combination with a cocatalyst (usually aluminum isopropoxide).

Table 9.6 Ruthenium Complexes Used as ATRP Catalysts

Structure	Monomer	Structure	Monomer
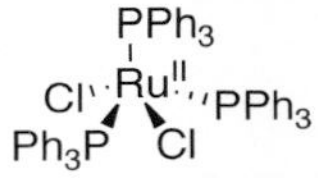 **146** RuCl$_2$(PPh$_3$)$_3$	MMA[265,341,342] EMA[343] BMA[343] DMAM[344] S[345]	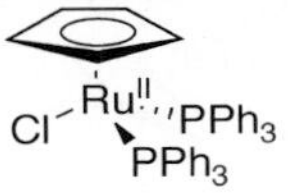 **147** (η^6-arene)RuCl(PPh$_3$)$_2$	MMA[342]
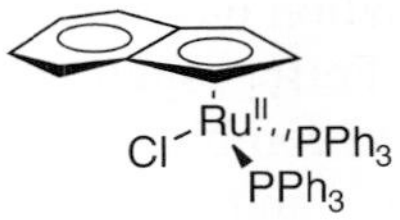 **148** (η^5-indenyl)RuCl(PPh$_3$)$_2$	MMA[342] S[342]	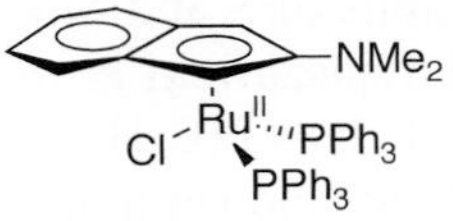 **149** (η^6-C$_6$H$_5$NMe$_2$)RuCl(PPh$_3$)$_2$	MMA[346] MA[346] S[346]
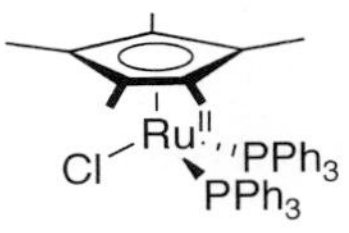 **150** (Cp*)RuCl(PPh$_3$)$_2$	MMA[347] MA[347] S[347]	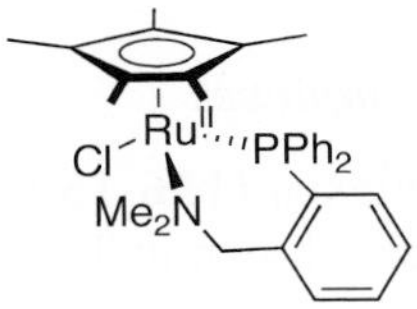 **151** (Cp*)RuCl(PPh$_2$-C$_6$H$_4$-CH$_2$NMe$_2$)	MMA[348]
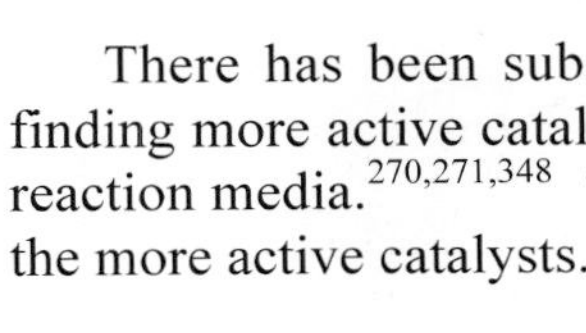 **152** RuCl$_2$(PPh$_2$-C$_6$H$_4$-SO$_3$Na)$_2$	HEMA[349]		

There has been substantial work on catalyst development with the aim of finding more active catalysts and catalysts appropriate for different monomers and reaction media.[270,271,348] The complexes **149-151** (Table 9.6) appear to be some of the more active catalysts.

9.4.2.3 Iron complexes

The catalysts **153-155** shown in Table 9.7 have been used for polymerizations of acrylates and methacrylates and S. The catalyst **155** used in conjunction with an iodo compound initiator has also been employed for VAc polymerization.[350] Catalytic chain transfer (Section 6.2.5) occurs in competition with halogen atom transfer with some catalysts.

Table 9.7 Iron Complexes Used as ATRP Catalysts

Catalyst	Structure	Monomer	Catalyst	Structure	Monomer
153	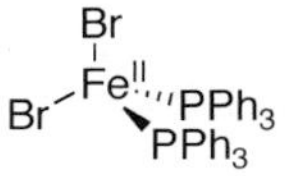 Br–Fe(II)(Br)(PPh$_3$)(PPh$_3$)	MMA[351]	154	 Cp-Fe(II)(I)(CO)(CO)	S[352]
155	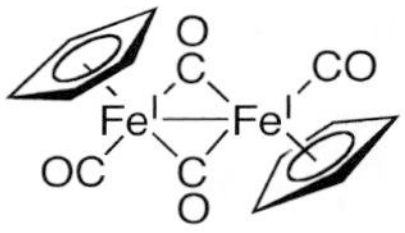 [Cp-Fe(I)(CO)]$_2$ dimer	S VAc[350,353,354]			

Polymerizations of S and MMA with *in situ* catalyst formation have also been carried out. Matyjaszewski *et al.*[355] reported on the use of FeBr$_2$ together with various ligands such as P(C$_4$H$_9$)$_3$, N(C$_4$H$_9$)$_3$ and **133** alone or in combination. The use of dicarboxylic acid (iminodiacetic acid, isophthalic acid)[356] and methanimine ligands[357,358] for MMA polymerization has also been reported.

9.4.2.4 Nickel complexes

Nickel complexes (**156-159**) used as ATRP catalysts for polymerization of (meth)acrylates are shown in Table 9.8.

Table 9.8 Nickel Complexes Used as ATRP Catalysts

Catalyst	Structure	Monomer	Catalyst	Structure	Monomer
156	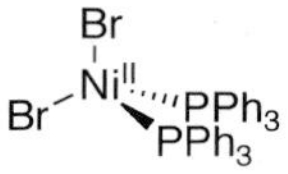 Br–Ni(II)(Br)(PPh$_3$)(PPh$_3$)	MMA[359]	157	 (C$_4$H$_9$)$_3$P–Ni(II)(Br)(Br)(P(C$_4$H$_9$)$_3$)	MMA MA BA[360]
158	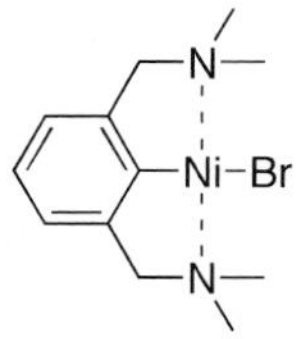 NCN-pincer Ni-Br	MMA BMA[361]	159	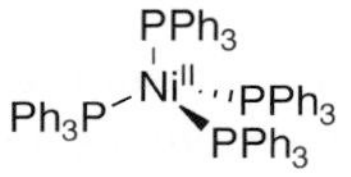 Ph$_3$P–Ni(II)(PPh$_3$)(PPh$_3$)(PPh$_3$)	MMA[362]

The complex **157** is more soluble than **156** in organic solvents; it is more thermally stable and can be used at higher temperatures. Moreover, it can be used without the Al((OiPr)$_3$ cocatalyst that is required with **156**.[360]

9.4.3 Monomers and Reaction Conditions

ATRP has been widely used for the polymerization of methacrylates. However, a very wide range of monomers, including most of those amenable to conventional radical polymerization, has been used in ATRP. ATRP has also been used in cyclopolymerization (*e.g.* of **160**[363,364]) and ring opening polymerization or copolymerization (*e.g.* of **161**[365,366] and **162**[367]).

The selection of reaction conditions for ATRP is dependent on many factors including the particular monomer, initiator and catalyst.

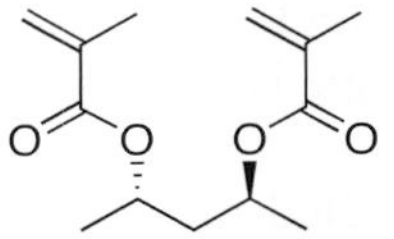

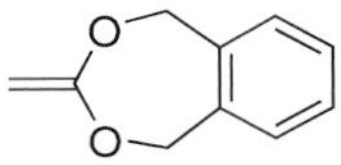

9.4.3.1 Solution polymerization

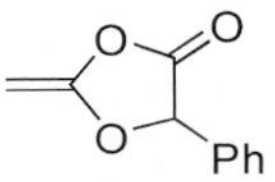

ATRP is usually performed in solution. Many solvents can be used with the proviso that they do not interact adversely with the catalyst. Common solvents include ketones (butanone, acetone) and alcohols (2-propanol). Solvents such as anisole and diphenyl ether are frequently used for polymerizations of S and other less polar monomers to provide greater catalyst solubility.

ATRP of various monomers including HEMA,[368] MAA,[369] α-methoxypoly(ethylene oxide) methacrylate,[370,371] DMAEMA[368,372] and 2-(trimethylammonium)ethyl methacrylate salts[368,373] has been carried out in aqueous media. Rates of ATRP in water can be substantially higher than in organic solvents such that polymerization can be carried out at ambient temperature. This has been attributed to competitive complexation of water and ligand providing a more active catalyst,[374] to a higher equilibrium concentration of propagating radicals, to solvent effects on k_p[371] and to removal of the deactivator by precipitation or hydrolysis. Use of higher reaction temperatures (>60 °C) can lead to loss of catalyst activity.[371,372,374]

9.4.3.2 Heterogeneous polymerization

ATRP in heterogeneous media has been reviewed by Qiu *et al.*[205] Cunningham[206] and Schork *et al.*[208] and is also mentioned in general reviews on ATRP.[268]

Many suspension polymerization recipes have been reported.[375] Some of the more successful that yield polymers of low dispersity are for MMA with **146**,[376] S, BA, MA, tBA and copolymers with **154**,[377] and BMA with **138**.[321] Important considerations are a catalyst that is both hydrophobic (to limit partitioning into the aqueous phase) and hydrolytically stable.

Emulsion polymerization has proved more difficult.[287,288,378] Many of the issues discussed under NMP (Section 9.3.6.6) also apply to ATRP in emulsion. The system is made more complex by both activation and deactivation steps being bimolecular. There is both an activator (Mt^n) and a deactivator (Mt^{n+1}) that may partition into the aqueous phase, although the deactivator is generally more water-soluble than the activator because of its higher oxidation state. Like NMP, successful emulsion ATRP requires conditions where there is no discrete monomer droplet phase and a mechanism to remove excess deactivator built up in the particle phase as a consequence of the persistent radical effect.[210,214] Reverse ATRP (Section 9.4.1.2) with water soluble dialkyl diazenes is the preferred initiation method.[287,288]

ATRP polymerization in miniemulsion has recently attracted more attention and met with greater success. Some difficulties with conventional initiation were attributed to catalyst oxidation during the homogenization/sonication step particularly when more active, less oxidatively stable, catalysts are used. This problem was solved using reverse ATRP or combinations of reverse and normal ATRP[290,291] that meant the catalyst could be added in its oxidized form (Section 9.4.1.2). Better results again were obtained using a conventional ATRP initiation and *in situ* catalyst ($CuBr_2$/BPMODA) reduction by AGET (Activator Generated by Electron Transfer).[379] In this case water soluble ascorbic acid was used as the reducing agent and it was presumed that catalyst reduction occurs in the aqueous phase.

9.5 Reversible Chain Transfer

Radical polymerizations which involve a reversible chain transfer step for chain equilibration and which displayed the characteristics of living polymerizations were first reported in 1995.[380,381] The mechanism of the reversible chain transfer step may involve homolytic substitution (Scheme 9.35) or addition-fragmentation (RAFT) (Scheme 9.36). An essential feature is that the product of chain transfer is also a chain transfer agent with similar activity to the precursor transfer agent. The process has also been termed degenerate or degenerative chain transfer since the polymeric starting materials and products have equivalent properties and differ only in molecular weight.

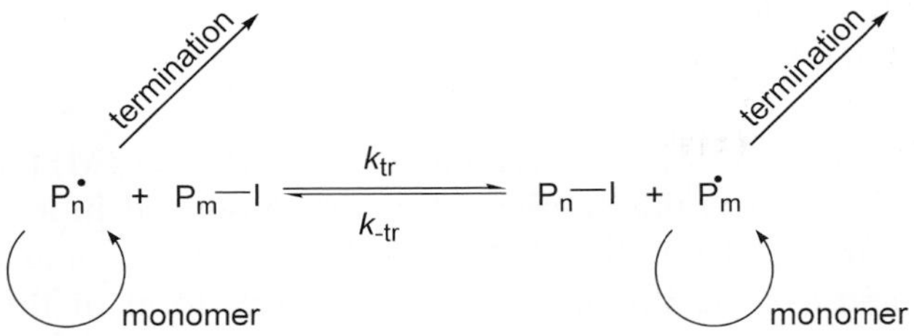

Scheme 9.35

Polymerization of S and certain fluoro-monomers in the presence of alkyl iodides provided the first example of the reversible homolytic substitution process (Scheme 9.35). This process is also known as iodine transfer polymerization (Section 9.5.4).[381] Other examples of reversible homolytic substitution are polymerizations conducted in the presence of certain alkyl tellurides or stibines (Sections 9.5.5 and 9.5.6 respectively).

Polymerizations of methacrylic monomers in the presence of methacrylic macromonomers under monomer-starved conditions display many of the characteristics of living polymerization (Scheme 9.36). These systems involve RAFT (Section 9.5.2). However, RAFT with appropriate thiocarbonylthio compounds is the most well known process of this class (Section 9.5.3). It is also the most versatile having been shown to be compatible with most monomer types and a very wide range of reaction conditions.[382]

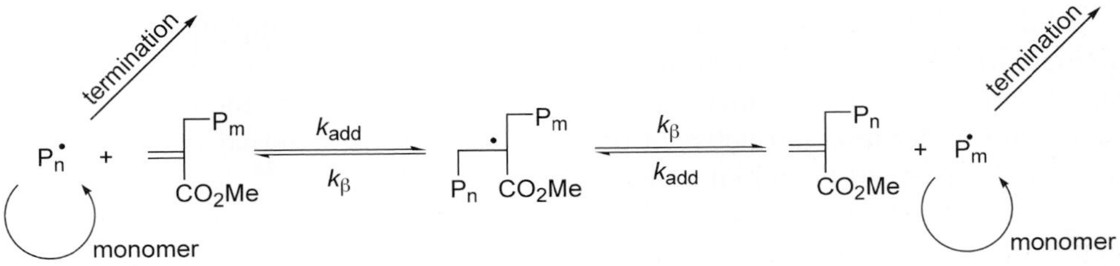

Scheme 9.36

9.5.1 Molecular weights and distributions

As with other forms of living radical polymerization, the degree of polymerization and the molecular weight can be estimated from the concentration of monomer and reagents as shown in eqs. 13 and 14 respectively.[383]

$$\overline{X}_n = \frac{[M]_o - [M]_t}{[T]_o + df([I_2]_o - [I_2])_t} \qquad (13)$$

$$\overline{M}_n = \frac{[M]_o - [M]_t}{[T]_o + df([I_2]_o - [I_2])_t} m_M + m_T \qquad (14)$$

where m_M and m_T are the molecular weights of the monomer (M) and the transfer agent (T) respectively, d is the number of chains produced in a radical-radical termination event (d~1.67 for MMA polymerization and ~1.0 for S polymerization) and f is the initiator efficiency. The form of this term in the denominator is suitable for initiators such as AIBN that produce radicals in pairs but will change for other types of initiator.

Reaction conditions should usually be chosen such that the fraction of initiator-derived chains (should be greater than or equal to the number of chains formed by radical-radical termination) is negligible. The expressions for number average degree of polymerization and molecular weight (eqs. 13 and 14) then simplify to eqs. 15 and 16:

$$\overline{X}_n = \frac{[M]_o - [M]_t}{[T]_o} \qquad (15)$$

$$\overline{M}_n = \frac{[M]_o - [M]_t}{[T]_o} m_M + m_T \qquad (16)$$

These equations suggest that a plot of $\overline{M}_n$ vs conversion should be linear. A positive deviation from the line predicted by eq. 16 indicates incomplete usage of transfer agent (T) while a negative deviation indicates that other sources of polymer chains are significant (*e.g.* the initiator).

Analytical expressions have been derived for calculating dispersities of polymers formed by polymerization with reversible chain transfer. The expression (eq. 17) applies in circumstances where the contributions to the molecular weight distribution by termination between propagating radicals, external initiation, and differential activity of the initial transfer agent are negligible.[23,384]

$$\frac{\overline{X}_w}{\overline{X}_n} = 1 + \frac{1}{\overline{X}_n} + \left(\frac{2-c}{c}\right)\frac{1}{C_{tr}} \qquad (17)$$

where c is the fractional conversion of monomer.

The transfer constant governs the number of propagation steps per activation cycle and should be small for a narrow molecular weight distribution. Rearrangement of eq. 17 to eq. 18 suggests a method of estimating transfer constants on the basis of measurements of the conversion, molecular weight and dispersity.[23]

$$\left(\frac{\overline{X}_w}{\overline{X}_n} - 1 - \frac{1}{\overline{X}_n}\right)^{-1} = C_{tr}\left(\frac{c}{2-c}\right) \qquad (18)$$

Living Radical Polymerization

In more complex cases, kinetic simulation has been used to predict the time/conversion dependence of the polydispersity. Moad et al.[122] first published on kinetic simulation of the RAFT process in 1998. Many papers have now been written on this subject. Zhang and Ray[385] and also Wang and Zhu[386,387] applied a method of moments to obtain molecular weights and dispersities. Peklak et al.[388] used a coarse-graining approach while Shipp and Matyjaszewski,[389] and Barner-Kowollik and coworkers[390-393] used a commercial software package (Predici™) to evaluate complete molecular weight distributions. Moad et al.[122,384,394] applied a hybrid scheme in which the differential equations are solved directly to give the complete molecular weight distribution to a finite limit ($\overline{X}_n$<500) and a method of moments is then used to provide closure to the equations, accurate molecular weights and polydispersities. Much of the research in this area has been carried out with a view to understanding the factors that influence retardation. The main difficulty in modeling RAFT lies in choosing values for the various rate constants.

9.5.2 Macromonomer RAFT

Chain transfer to methacrylate and similar macromonomers has been discussed in Section 6.2.3.4. The first papers on the use of this process to achieve some of the characteristics of living polymerization appeared in 1995.[380] The structure of macromonomer RAFT agents (**163**) is shown in Figure 9.3. An idealized reaction scheme for the case of a MMA terminated macromonomer is shown in Scheme 9.36.

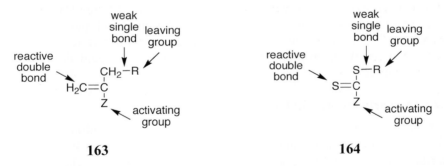

Figure 9.3 General description of macromonomer and thiocarbonylthio RAFT agents.

Macromonomer RAFT polymerization is most effective with methacrylate monomers (Table 9.9).[380,395] With monosubstituted monomers (*e.g.* S, acrylates) graft copolymerization is a significant side reaction which can be mitigated but not eliminated by the use of higher reaction temperatures.

Table 9.9 Block Copolymers Prepared by Macromonomer RAFT Polymerization under Starved-Feed Conditions.[380,395]

Macro[a,b]	$\overline{M}_n$	$\overline{M}_w/\overline{M}_n$	Monomer[a]	Solvent[a]	Temp. °C	$\overline{M}_n$	$\overline{M}_w/\overline{M}_n$
MAA	950	-	MMA	emulsion	80	3000	1.4
MMA	3500	1.6	BMA	emulsion	80	28000	1.4
MMA	2050	1.7	EHMA	emulsion	80	11800	1.3
tBMA	2400	2.1	BMA	emulsion	80	5800	1.3
PhMA	1100	2.2	BMA	emulsion	80	14500	2.3
HEMA	1550	-	MMA	H_2O/iPrOH	80	3600	1.8
BMA	1050	2.0	S	BuAc	125	4700	2.4[c]
MMA-MAA	1030	1.5	BA	BuAc	125	2700	1.8[d]

a Abbreviations: MeS 4-methylstyrene, PhMA phenyl methacrylate, BuAc butyl acetate, iPrOH 2-propanol. Other abbreviations can be found in the Glossary. b Macromonomer **163** made from monomer shown by catalytic chain transfer process. c After subtraction of residual macromonomer. d Contains graft copolymer impurity.

Transfer constants of the macromonomers are typically low (~0.5, Section 6.2.3.4) and it is necessary to use starved feed conditions to achieve low dispersities and to make block copolymers. Best results have been achieved using emulsion polymerization[380,395] where rates of termination are lowered by compartmentalization effects. A 'one-pot' process where macromonomers were made by catalytic chain transfer was developed.[380,395] Molecular weights up to 28000 that increase linearly with conversion as predicted by eq. 16, dispersities that decrease with conversion down to $\overline{M}_w/\overline{M}_n<1.3$ and block purities >90% can be achieved.[380,395] Surfactant-free emulsion polymerizations were made possible by use of a MAA macromonomer as the initial RAFT agent to create 'self-stabilizing lattices'.

9.5.3 Thiocarbonylthio RAFT

Although the term RAFT (an acronym for Reversible Addition-Fragmentation chain Transfer)[382] is sometimes used in a more general sense, it was coined to describe, and is most closely associated with, the reaction when it involves thiocarbonylthio compounds. RAFT polymerization, involving the use of xanthates, is also sometimes called MADIX (Macromolecular Design by Interchange of Xanthate).[396] The process has been reviewed by Rizzardo et al.,[397] Chiefari and Rizzardo,[398] Barner-Kowollik et al.,[399] McCormick et al.,[400] and Moad et al.[401]

Organic chemists have been aware of reversible addition-fragmentation involving xanthate esters in organic chemistry for some time. It is the basis of the Barton-McCombie process for deoxygenation of alcohols (Scheme 9.37).[402-404]

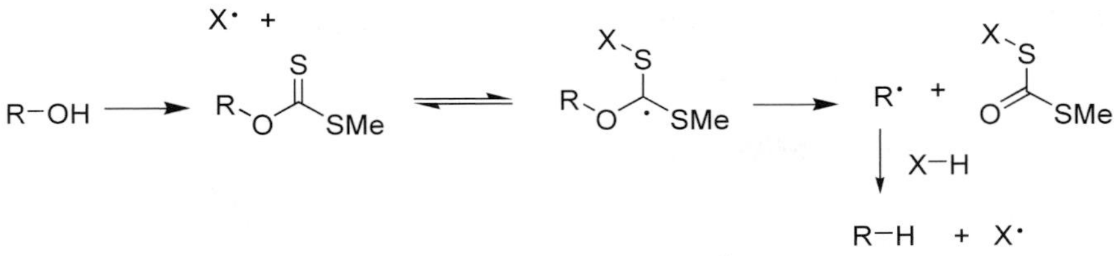

Scheme 9.37 Barton-McCombie deoxygenation reaction

In 1988 a paper by Zard and coworkers[405] reported that xanthates were a convenient source of alkyl radicals by reversible addition-fragmentation and used the chemistry for the synthesis of a monoadduct to monomer (a maleimide). Many applications of the chemistry in organic synthesis have now been described in papers and reviews by the Zard group.[406,407]

Living radical polymerization using thiocarbonylthio RAFT agents (including dithioesters, trithiocarbonates and xanthates) was first described in a patent published in 1998.[408] The first paper describing the process also appeared in 1998.[382] Other patents and papers soon followed. Papers on this method, along with NMP and ATRP, now dominate the literature on radical polymerization.

9.5.3.1 Mechanism

A key feature of the mechanism of RAFT polymerization is the sequence of addition-fragmentation equilibria shown in Scheme 9.38.[382] Initiation and radical-radical termination occur as in conventional radical polymerization. In the early stages of the polymerization, addition of a propagating radical ($P_n^{\bullet}$) to the thiocarbonylthio compound **164** followed by fragmentation of the intermediate radical **165** gives rise to a polymeric thiocarbonylthio compound (**166**) and a new radical ($R^{\bullet}$). Reaction of the radical ($R^{\bullet}$) with monomer forms a new propagating radical ($P_m^{\bullet}$). A rapid equilibrium between the active propagating radicals ($P_n^{\bullet}$ and $P_m^{\bullet}$) and the dormant polymeric thiocarbonylthio compounds (**166**) provides equal probability for all chains to grow and allows for the production of narrow dispersity polymers. With appropriate attention to the reaction conditions, the vast majority of chains will retain the thiocarbonylthio end group when the polymerization is complete (or stopped). Radicals are neither formed nor destroyed in the chain equilibration process. Thus once the equilibria are established, rates of polymerization should be similar to those in conventional radical polymerization. This is borne out by experimental data, which show that, with some RAFT agents, RAFT polymerization is half order in initiator and zero order in the RAFT agent over a wide range of initiator and RAFT agent concentrations.

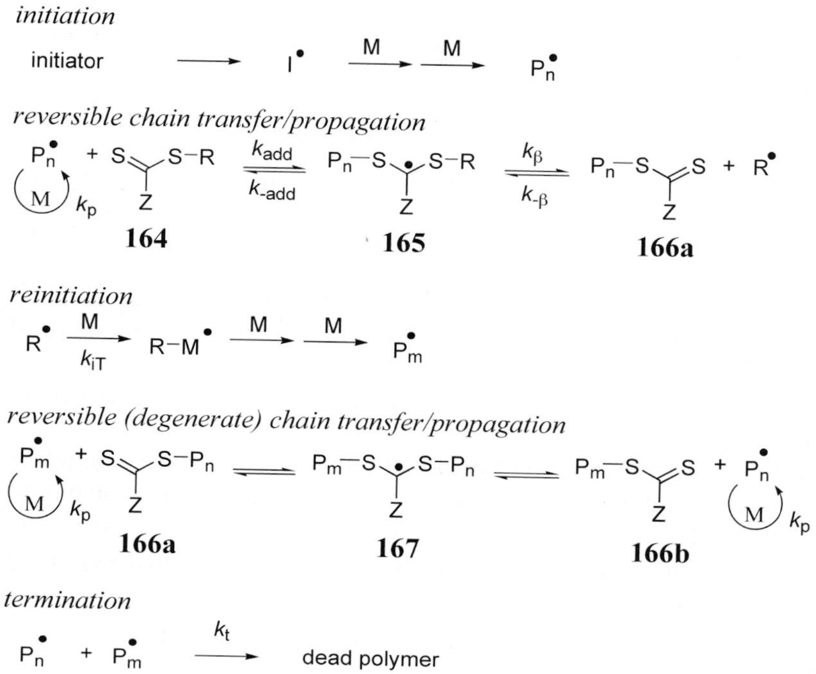

initiation

initiator ⟶ I• —M→ —M→ $P_n^\bullet$

reversible chain transfer/propagation

$P_n^\bullet$ (M, k_p) + S=C(Z)S–R ⇌(k_{add}/k_{-add}) P_n–S–C(Z)•–S–R (**165**) ⇌(k_β/$k_{-\beta}$) P_n–S–C(Z)=S + R• (**166a**)

(**164**)

reinitiation

R• —M, k_{iT}→ R–M• —M→ —M→ $P_m^\bullet$

reversible (degenerate) chain transfer/propagation

$P_m^\bullet$ (M, k_p) + S=C(Z)S–P_n (**166a**) ⇌ P_m–S–C(Z)•–S–P_n (**167**) ⇌ P_m–S–C(Z)=S + $P_n^\bullet$ (M, k_p) (**166b**)

termination

$P_n^\bullet$ + $P_m^\bullet$ —k_t→ dead polymer

Scheme 9.38

For very active RAFT agents, the RAFT agent derived radical (R•) may partition between adding to monomer and reacting with the transfer agent (polymeric or initial). In these circumstances, the transfer constant measured according to the Mayo or related methods will appear to be dependent on the transfer agent concentration and on the monomer conversion. A reverse transfer constant can be defined as follows (eq. 19)

$$C_{-tr} = \frac{k_{-tr}}{k_{iT}} \tag{19}$$

and the rate of RAFT agent consumption is then given by eq. 20.[394]

$$-\frac{d[150]}{d[M]} \approx C_{tr}\frac{[150]}{[M] + C_{tr}[150] + C_{-tr}[152]} \tag{20}$$

For addition-fragmentation chain transfer, the rate constants for the forward and reverse reactions are defined as shown in eqs. 21 and 22 respectively.

$$k_{tr} = k_{add}\frac{k_\beta}{k_{-add} + k_\beta} \tag{21}$$

$$k_{-tr} = k_{-\beta} \frac{k_{-add}}{k_{-add} + k_{\beta}} \tag{22}$$

RAFT polymerization provides the characteristics usually associated with living polymerization. The overall process results in monomer units being inserted into the RAFT agent structure as shown in Scheme 9.38. Expressions (eqs. 13-16) for estimating number average degree of polymerization and molecular weight in RAFT polymerization are provided in section 9.5.1. Dispersities will depend on the chain transfer constants associated with both the initial and the polymeric RAFT agent. The reaction conditions should be chosen such that the initial RAFT agent is rapidly consumed during the initial stages of the polymerization.

9.5.3.2 RAFT agents

Many thiocarbonylthio RAFT agents (**164**) have now been described. Transfer constants are strongly dependent on the Z and R substituents. For an efficient RAFT polymerization (refer Scheme 9.38 and Figure 9.3):

(a) Both the initial (**164**) and polymeric RAFT agents (**166**) should have a reactive C=S double bond (high k_{add}).
(b) The intermediate radicals **165** and **167** should fragment rapidly (high k_β, weak S-R bond) and give no side reactions.
(c) The intermediate **165** should partition in favor of products ($k_\beta \geq k_{-add}$).
(d) The expelled radicals (R•) should efficiently re-initiate polymerization.

The dependence of the transfer constant on the Z substituent, summarized in Figure 9.4, is largely based on studies of the apparent transfer constants of benzyl and cyanoisopropyl RAFT agents in S polymerization[384,409] and qualitative observations of other polymerizations.[397]

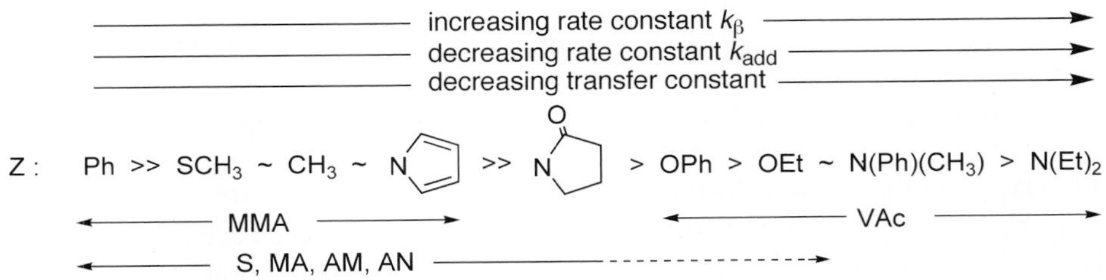

```
                    increasing rate constant k_β     →
                    decreasing rate constant k_add   →
                    decreasing transfer constant     →

Z:   Ph >> SCH₃ ~ CH₃ ~ N(pyrrole)  >>  N(pyrrolidone/O) > OPh > OEt ~ N(Ph)(CH₃) > N(Et)₂
          ←——— MMA ———→                       ←——— VAc ———→
     ←——— S, MA, AM, AN ———→       ----------→
```

Figure 9.4 Effect of Z substituent on effectiveness of RAFT agents **164** in various polymerizations. Dashed line implies limited effectiveness with a particular monomer (broad molecular weight distribution).[401]

Early reports focused on the dithiobenzoate RAFT agents (Z=Ph; *e.g.* **171-180**, Table 9.10).[382,410] Cumyl dithiobenzoate (**175**) shows utility with S and (meth)acrylic monomers.[382] However, retardation is an issue with the acrylates and when high concentrations of RAFT agent are used. For MMA and S, cyanoisopropyl dithiobenzoate (**176**) gives less retardation than **175**.[409] The trithiocarbonates (Z=S-alkyl; *e.g.* **219-232**, Table 9.15 and Table 9.16) are also effective with S and (meth)acrylic monomers and give substantially less retardation than the corresponding dithiobenzoates under similar conditions. Dithioacetate and other RAFT agents with Z=alkyl or aralkyl (*e.g.* **212-218**, Table 9.14) also give less retardation but have lower transfer constants and do not give narrow dispersities with methacrylates.

The trend in relative effectiveness of RAFT agents with varying Z is rationalized in terms of interaction of Z with the C=S double bond to activate or deactivate that group towards free radical addition. Substituents that facilitate addition generally retard fragmentation. *O*-Alkyl xanthates (Z=O-alkyl, Table 9.17) are generally not effective with methacrylates and give relatively broad dispersities with S and acrylates. *N,N*-dialkyl dithiocarbamates (Z=N-alkyl$_2$, Table 9.18) are not effective with S and (meth)acrylic monomers. This is rationalized in terms of the importance of zwitterionic canonical forms as shown in Figure 9.5. Substituents which make the lone pair less available for delocalization with the thiocarbonyl group (C=S) activate the RAFT agent.[384,411-413] Thus, xanthates and dithiocarbamates where the oxygen or nitrogen lone pair is part of an aromatic ring (*e.g.* where Z is pyrole or imidazole) or possesses an adjacent electron-withdrawing (*e.g.* C=O) or conjugating (*e.g.* Ph) substituent are substantially more effective. For examples see Table 9.17 (xanthates) or Table 9.18 (dithiocarbamates). Electron withdrawing substituents also improve the effectiveness (*i.e.* give polymers with lower dispersity) of dithiobenzoate RAFT agents in MMA polymerization.[414]

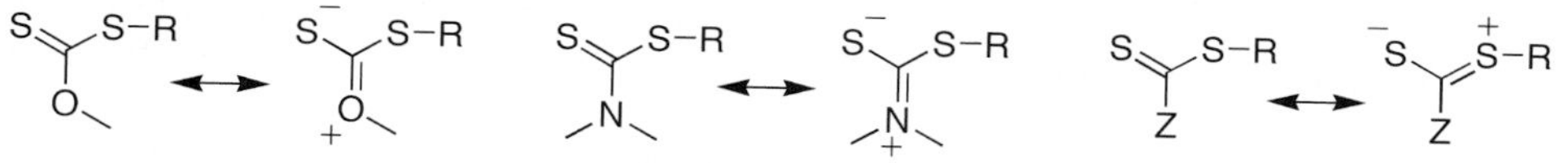

168   **169**   **170**

Figure 9.5 Canonical forms of thiocarbonylthio compounds.

O-Alkyl xanthates and *N*-aryl-*N*-alkyl dithiocarbamates are effective with vinyl acetate.[397] Dithioesters and trithiocarbonates give severe retardation or even inhibition which is attributed to slow fragmentation of the adduct radical.

The choice of R substituent is also extremely important in determining the activity of RAFT agents. The radical R• needs to be a good free radical leaving group with respect to the propagating radical. The order of relative effectiveness shown in Figure 9.6 is largely based on studies of the apparent transfer constants of dithiobenzoate RAFT agents in polymerizations of S and MMA.[394,409] However, the trends appear to be general.

R is made a better leaving group by electrophilic susbstituents (*e.g.* CN), by groups which stabilize the incipient radical, and by bulky subsituents. Penultimate unit effects are important.[394,409] Thus, the 2-carboalkoxy-2-propyl radical [$(CH_3)_2(CO_2R)C\bullet$] is a poor leaving group with respect to PMMA•. The *t*-butyl radical is a poor leaving group with respect to isooctyl radical.

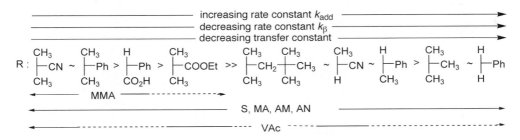

Figure 9.6 Effect of R substituent on effectiveness of RAFT agents **164**. Dashed line implies limited effectiveness with a particular monomer (broad molecular weight distribution or severe retardation).[401]

The dependence of RAFT agent activity on the substituents R and Z can be qualitatively predicted using low level molecular orbital calculations and these also provide a guide to the relative importance of the various factors.[384,394,415] There also appear to be good prospects for more quantitative predictions using higher level *ab initio* and density functional theory (DFT) calculations.[384,414,416,417] The molecular orbital calculations provide insight into the origin of substituent effects and should prove extremely useful in RAFT agent design. However, this work is still in its infancy and the use of these methods to predict absolute values of rate constants or equilibrium constants associated with RAFT must still be treated with caution.

A non-exhaustive tabulation of RAFT agents and the monomers they have been examined with is provided in Table 9.10-Table 9.18. Listing of a monomer or RAFT agent does not mean that that combination provides good results. Combinations shown in parentheses give less than ideal results (dispersity > 1.4 and/or poor molecular weight control) for the reaction conditions used. Even though many RAFT agents have been described, most polymerizations can be performed with just two RAFT agents: one for styrenic and (meth)acrylic monomers (S, AA, MA, MAA, MMA, NIPAM, DMAM, *etc.*) and another for vinyl monomers (VAc, NVP *etc.*). Specific requirements for end group functionality, architecture, ease of RAFT agent synthesis and other considerations may dictate other choices.[401]

Table 9.10 Tertiary Dithiobenzoate RAFT Agents

RAFT Agent	Monomers[a]	RAFT Agent	Monomers[a]
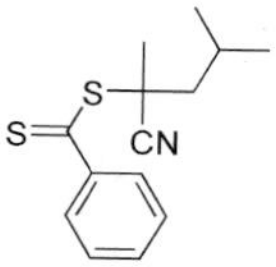 171	MMA[414]	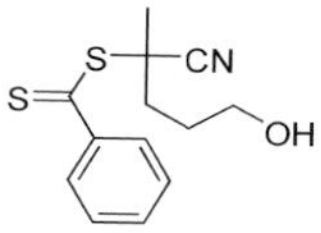 172	MMA[394,397,408]
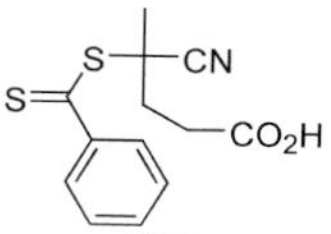 173	S[397,408] DMAEMA[382] AM[418]	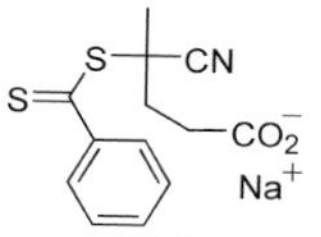 174	SSO$_3$Na[397,408,419] AMPS[419,420]
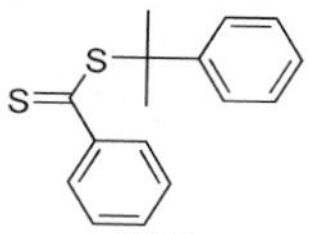 175	S[394,397,409] MA[394,421-423] BA[394,409] (AN)[424] MMA[382,383,394,408,425] BzMA[425] DMAEMA[397] XMA[426,427] AM[418] DMAM[397,428] NIPAM[397,429] 2VP[430] 4VP[430] (S)[409,423] MMA[408] BMA[382,408]	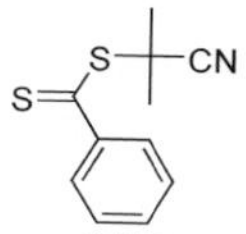 176	S[384,394,408,409] AA[397] MA[394,431] AN[432] MMA[394,397,408,414] XMA[426,427,433] VBz[382,408] MMA[434] BMA[434] EHMA[434]
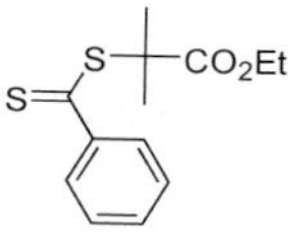 177	S[394,408] (MMA)[394,408]	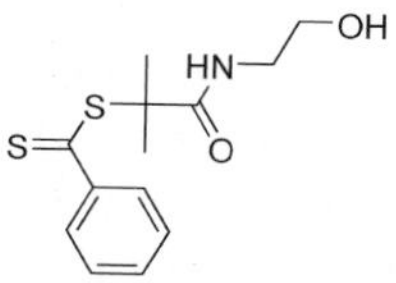 178	(MMA)[397]
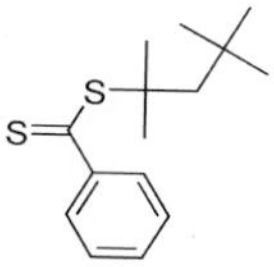 179	S[394,408] (MMA)[394,397]	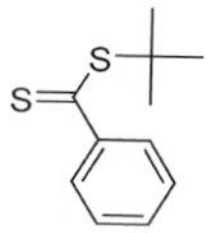 180	(MMA)[394] AA[435]

[a] Abbreviations: AMPS sodium 2-acrylamido-2-methylpropane-1-sulfonate, BMDO 5,6-dibenzo-2-methylene-1,3-dioxepan, SSO$_3$Na sodium styrene-4-sulfonate, 2VP 2-vinylpyridine, 4VP 4-vinylpyridine, XMA functional methacrylate: 2-(acetoacetoxy)ethyl methacrylate;[427] 3-[tris(trimethylsilyloxy)silyl]propyl methacrylate;[426] 6[4-(4'-methoxyphenyl)phenoxy]hexyl methacrylate.[433] For other monomer abbreviations see Glossary. Monomers shown in parentheses give less than ideal results (dispersity > 1.4 and/or poor molecular weight control and/or marked retardation) for the reaction conditions reported. Monomers in italics were polymerized by emulsion or miniemulsion polymerization.

Table 9.11 Other Aromatic Dithioester RAFT Agents

RAFT Agent	M[a]	RAFT Agent	M[a]	RAFT Agent	M[a]
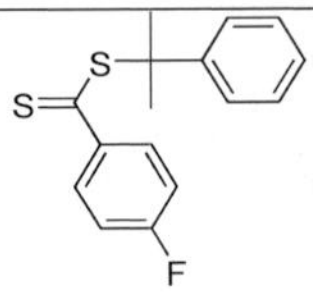 **181** (4-F-C₆H₄-C(=S)S-C(CH₃)₂-Ph)	MA[421]	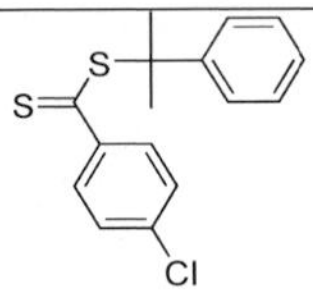 **182** (4-Cl-C₆H₄-C(=S)S-C(CH₃)₂-Ph)	MMA[408]	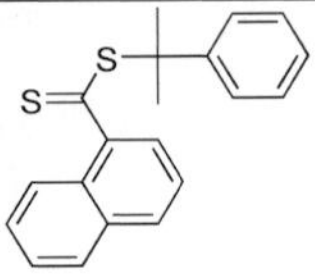 **183** (naphthyl-C(=S)S-C(CH₃)₂-Ph)	MMA[408]
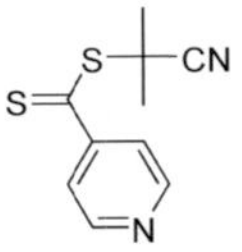 **184** (4-pyridyl-C(=S)S-C(CH₃)(CN)-)	MMA[414]	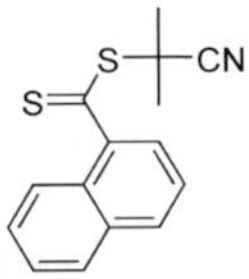 **185** (naphthyl-C(=S)S-C(CH₃)(CN)-)	MMA[414,437], GMA[438]	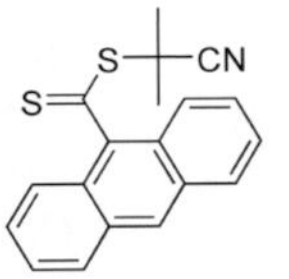 **186** (anthracenyl-C(=S)S-C(CH₃)(CN)-)	MMA[439]
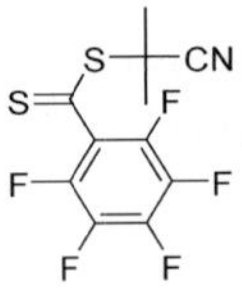 **187** (C₆F₅-C(=S)S-C(CH₃)(CN)F)	MMA[414]	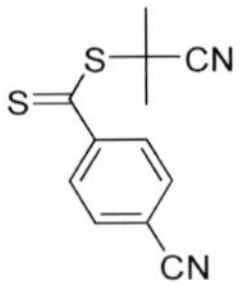 **188** (4-CN-C₆H₄-C(=S)S-C(CH₃)(CN)-)	MMA[414]	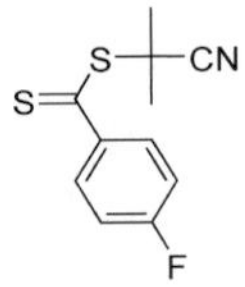 **189** (4-F-C₆H₄-C(=S)S-C(CH₃)(CN)-)	MMA[414]
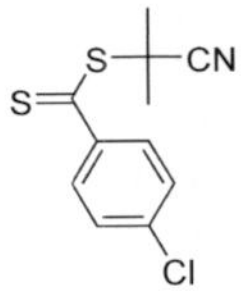 **190** (4-Cl-C₆H₄-C(=S)S-C(CH₃)(CN)-)	MMA[414]	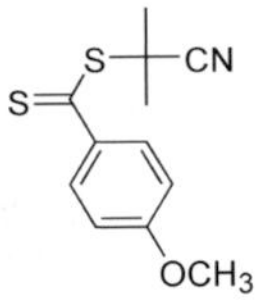 **191** (4-OCH₃-C₆H₄-C(=S)S-C(CH₃)(CN)-)	MMA[414]		

a Monomer. See footnote a of Table 9.10.

Table 9.12 Primary and Secondary Dithiobenzoate RAFT Agents

RAFT Agent	Monomers[a]	RAFT Agent	Monomers[a]
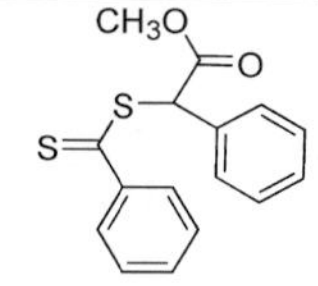 **192** (CH$_3$O-substituted)	S,[440] MA,[440] MMA,[440] DMAM[440]	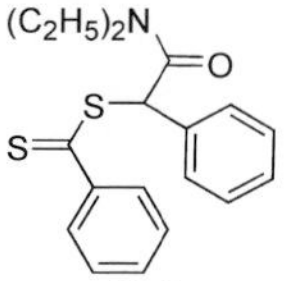 **193** ((C$_2$H$_5$)$_2$N-substituted)	S,[441] MA,[441] DMAM[441]
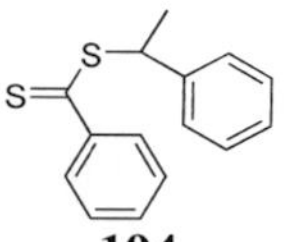 **194**	S[394,442] AA[382,408] MA[408,421,425,431] BA[394,408,409,425] (MMA)[394] S[409,423]	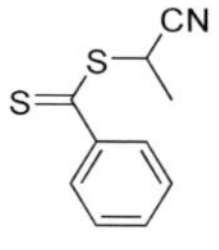 **195** (CN-substituted)	AN[424]
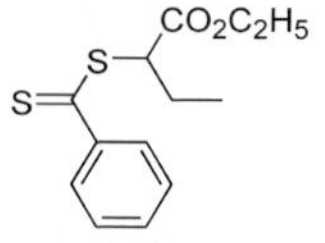 **196** (CO$_2$C$_2$H$_5$)	BMDO[443]	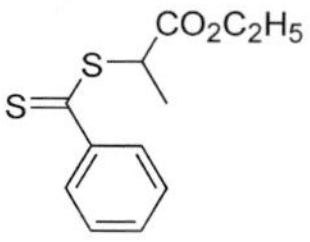 **197** (CO$_2$C$_2$H$_5$)	S[415] BA[415] (MMA)[415]
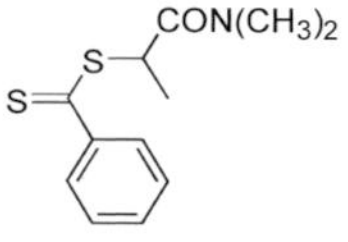 **198** (CON(CH$_3$)$_2$)	DMAM[428]	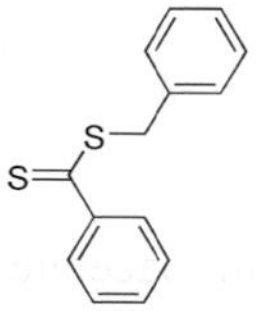 **199**	S[384,394,408,425] MA[394] BA[382,394] (MMA)[394] DMAM[408,428] NIPAM S[408,409] MMA[409]
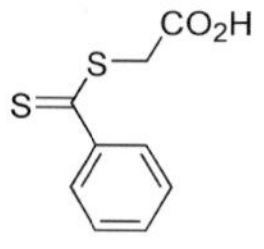 **200** (CO$_2$H)	(S)[408,415] SAc[444] (MA)[408] BA[415] (MMA)[415]	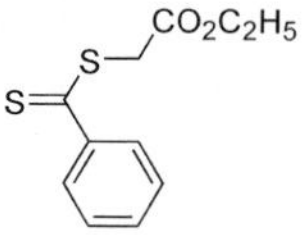 **201** (CO$_2$C$_2$H$_5$)	(S)[415] BA[415] (MMA)[415]
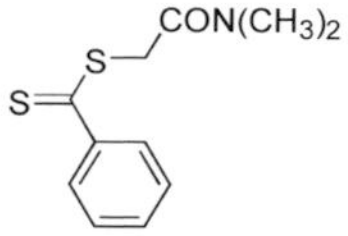 **202** (CON(CH$_3$)$_2$)	DMAM[428]	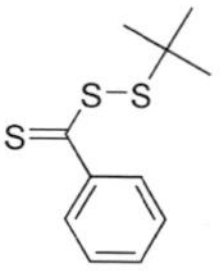 **203**	MMA[394,408] BA[394,408]

a See footnote a of Table 9.10.

Living Radical Polymerization 511

Table 9.13 Bis-RAFT Agents

RAFT Agent	M[a]	RAFT Agent	M[a]
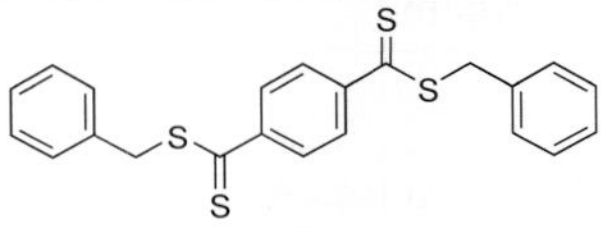 204	S[408]	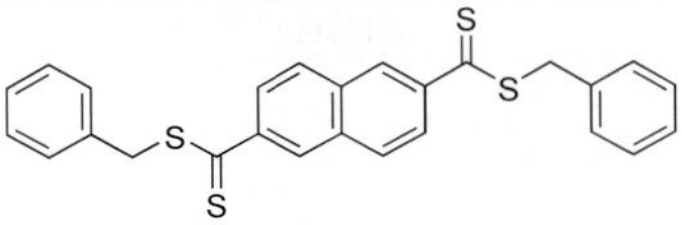 205	S[445] tBA[445]
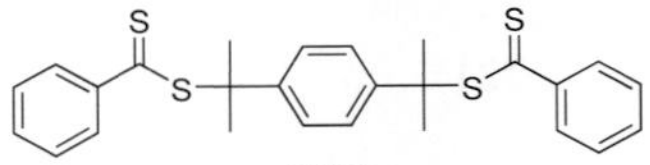 206	MMA[425]	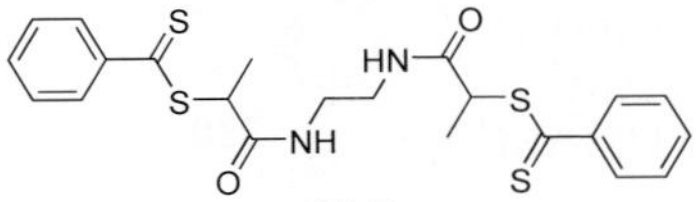 207	DMAM[446]
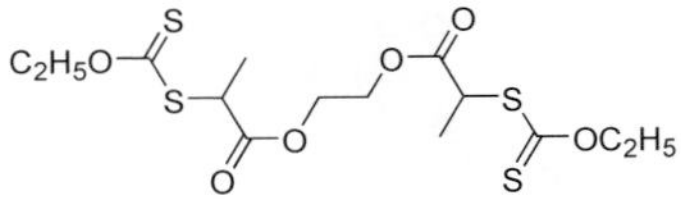 208	AM[447]	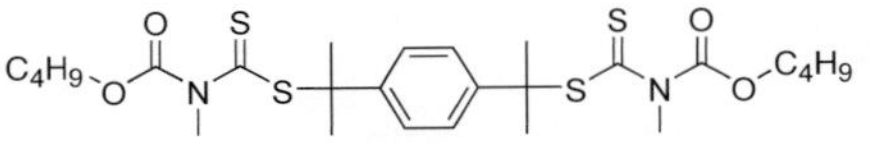 209	(BA)[448]
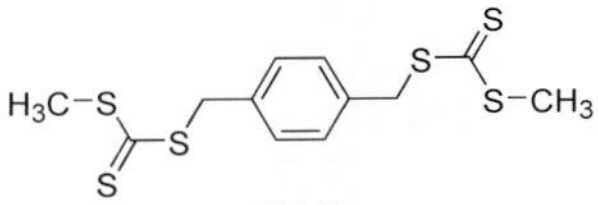 210	BA[449]	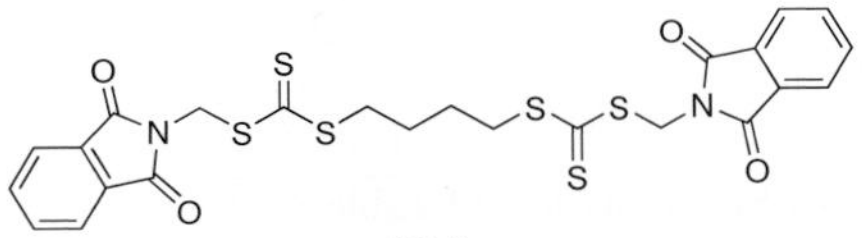 211	S[450]

a Monomer. See footnote a of Table 9.10.

Table 9.14 Dithioacetate and Dithiophenylacetate RAFT Agents

RAFT Agent	Monomers[a]	RAFT Agent	Monomers[a]	RAFT Agent	Monomers[a]
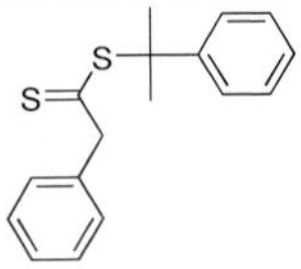 212	S[391] (MMA)[391] AM[418] NIPAM[451]	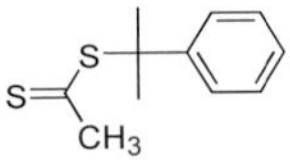 213	MMA[383,397]	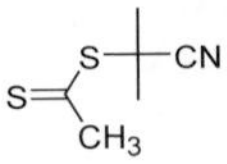 214	S[384] BA[382]
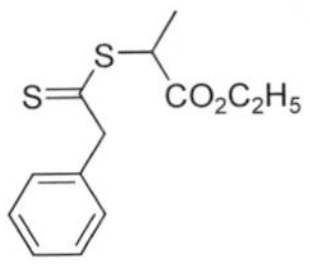 215	MA[452]	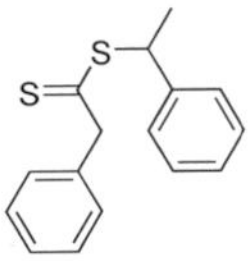 216	S[423] MA[453] NIPAM[451] S[423]	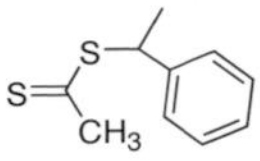 217	S[442]
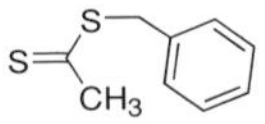 218	S[384] BA[409] S[409]				

a See footnote a of Table 9.10.

Table 9.15 Symmetrical Trithiocarbonate RAFT Agents

RAFT Agent	M[a]	RAFT Agent	M[a]	RAFT Agent	M[a]
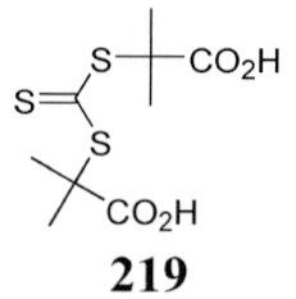 **219**	S,[454] AA,[454] HEA,[454] EA,[454] BA,[454,455] (MMA),[454] AM,[418,456] DMAM[456]	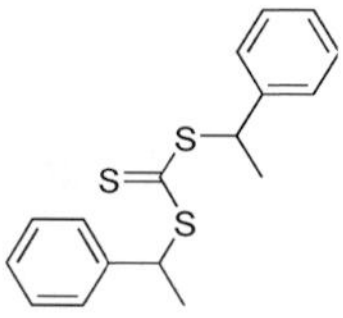 **220**	S[457]	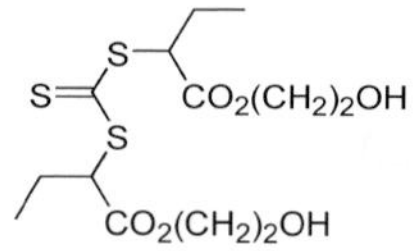 **221**	NIPAM[458]
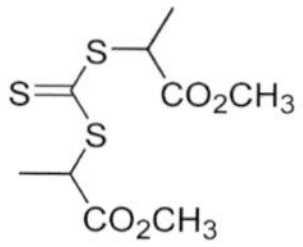 **222**	MA[452]	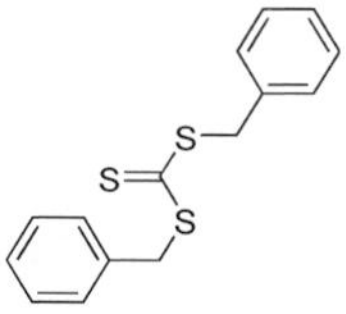 **223**	S,[384,408,457] AA,[435,459] MA[457]	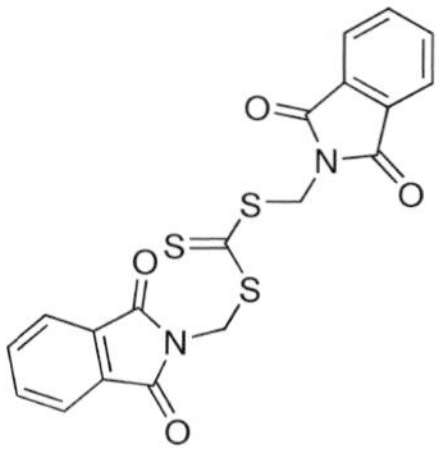 **224**	S[450]

a Monomers. See footnote a of Table 9.10.

Table 9.16 Non-Symmetrical Trithiocarbonate RAFT Agents

RAFT Agent	Monomers[a]	RAFT Agent	Monomers[a]	RAFT Agent	Monomers[a]
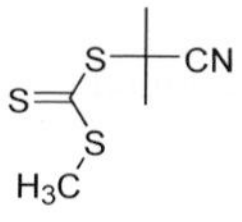 **225**	S,[384,457] MA,[457] MMA[457]	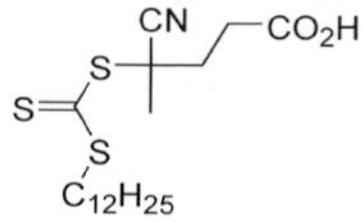 **226**	MMA[450]	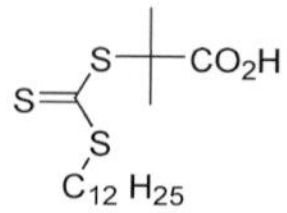 **227**	AA,[454] EA,[454] BA,[455] BAM,[454] NIPAM[460]
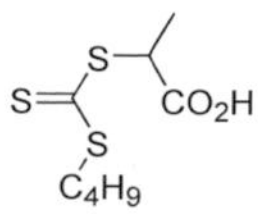 **228**	AA[461,462]	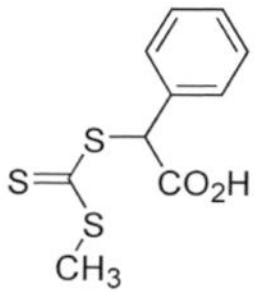 **229**	S[457]	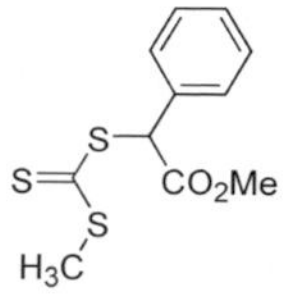 **230**	S,[440] MA,[440] (MMA),[440] DMAM[440]
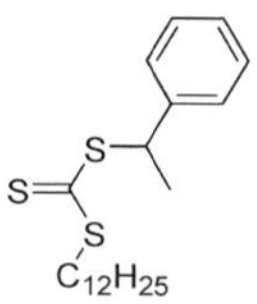 **231**	S,[463] ODA[463]	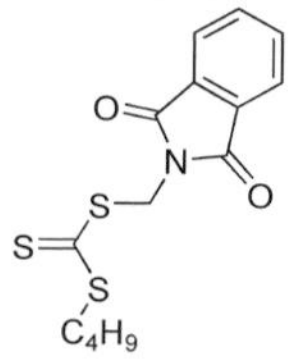 **232**	S,[450,464] BA[464]		

a See footnote a of Table 9.10.

Table 9.17 Xanthate RAFT Agents

RAFT Agent	M[a]	RAFT Agent	M[a]	RAFT Agent	M[a]
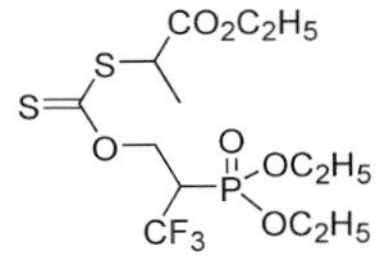 **233**	S[413] EA[413]	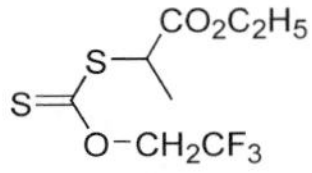 **234**	(S)[413] EA[413] (S)[465]	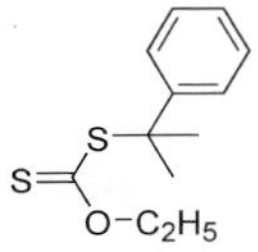 **235**	(S)[466]
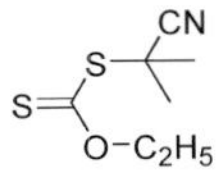 **236**	(S)[466] (tBA)[397] (MMA)[410]	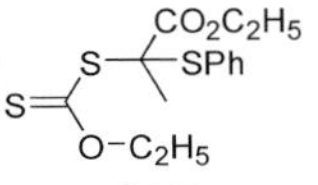 **237**	(S)[396,466] (EA)[396]	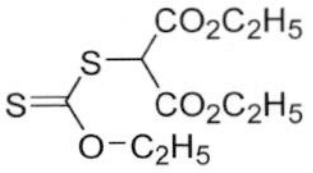 **238**	(S)[396,466]
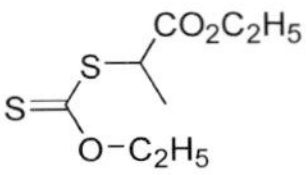 **239**[b]	(S)[396,413,466,467] AA (MA)[396] (EA)[396] AM[447] VAc[396,397] (S)[396] (BA)[396]	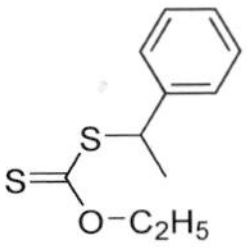 **240**	(S)[396,466-468] (BA)[469] (S)[409]	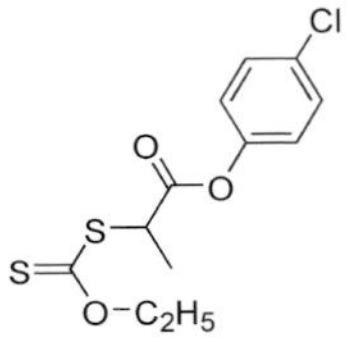 **241**	(S)[466]
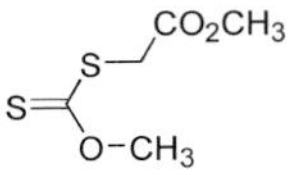 **242**	VAc[470] VAc[471,472]	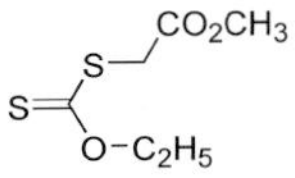 **243**	VAc[470]	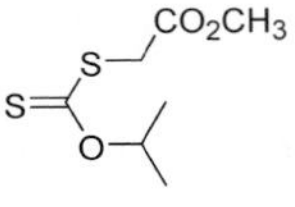 **244**	VAc[470,473]
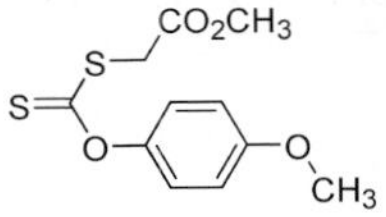 **245**	VAc[470]	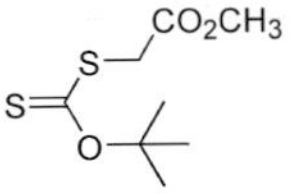 **246**	(VAc)[470]	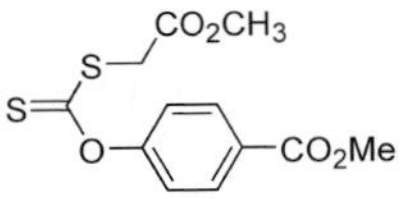 **247**	(VAc)[470]
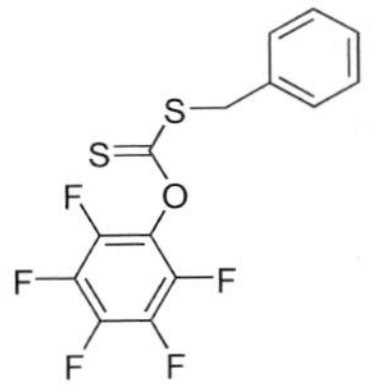 **248**	(S)[384] (tBA)[469]	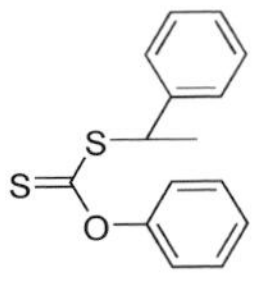 **249**	AA[435]	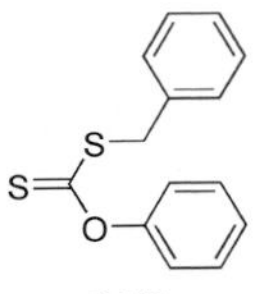 **250**	(S)[384] (AA)[435]
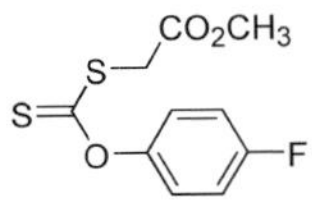 **251**	(VAc)[470]	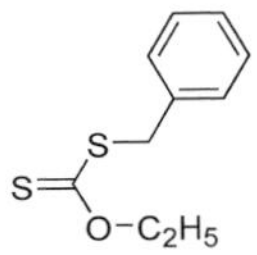 **252**	(S)[384] (S)[409]	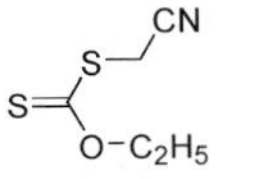 **253**	VAc[397]

a Monomers. See footnote a of Table 9.10. b Some reports relate to the corresponding methyl xanthate.

Table 9.18 Dithiocarbamate RAFT Agents

RAFT Agent	M[a]	RAFT Agent	M[a]	RAFT Agent	M[a]
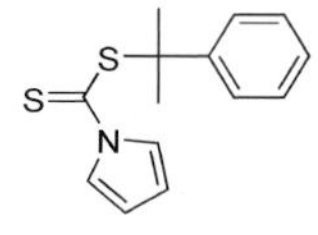 **254**	NIPAM[474]	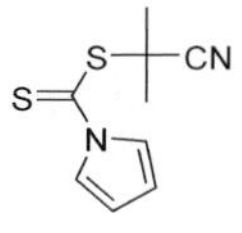 **255**	S MA[469] MMA[384]	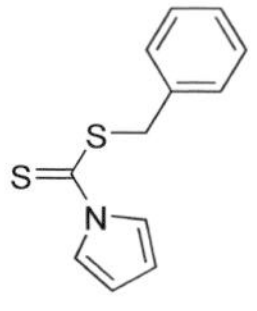 **256**	S[384,397,469] MA[397,411,469] NIPAM[474]
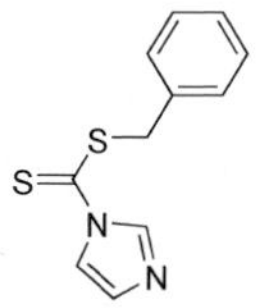 **257**	S[411] MA[411,469]	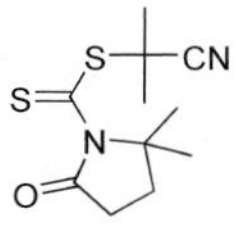 **258**	S[412] (MMA)[412] VAc[412]	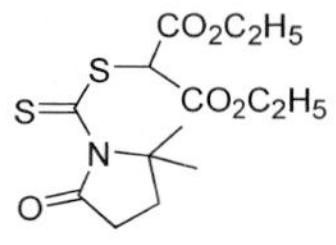 **259**	EA[412]
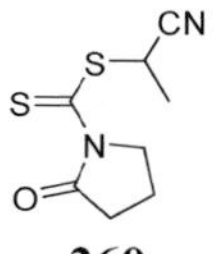 **260**	AN[401] MA[401] AA[475,476]	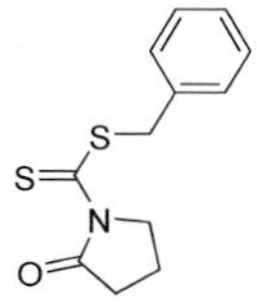 **261**	(S)[401,469] MA[401,469]	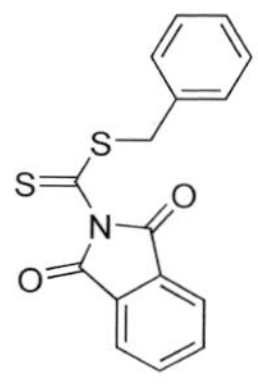 **262**	S[401,469] (MA)[401,469]
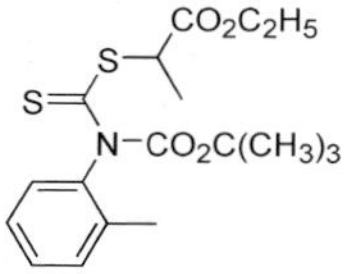 **263**	EA[412]	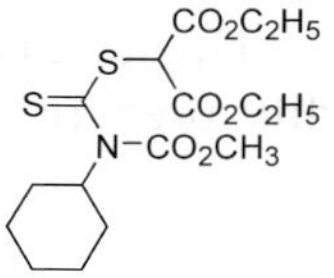 **264**	S[412] EA[412] VAc[412]	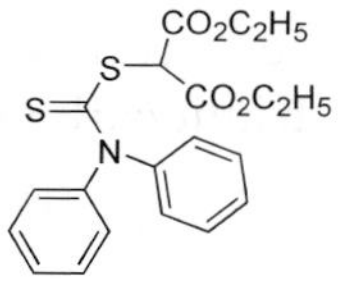 **265**	EA[412] (VAc)[412]
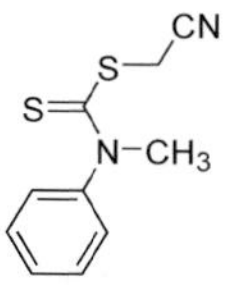 **266**	VAc[397]	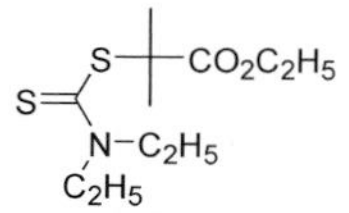 **267**	VAc[397]	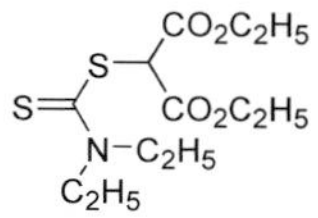 **268**	(EA)[412] (VAc)[412]
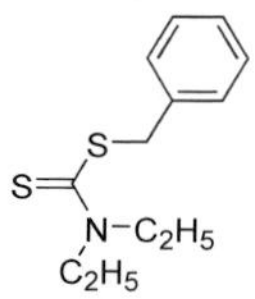 **269**	(S)[384,397]				

a Monomers. See footnote a of Table 9.10.

9.5.3.3 RAFT agent synthesis

Currently, few RAFT agents are commercially available. However, RAFT agents are available in moderate to excellent yields by a variety of methods and syntheses are generally straightforward.

Some of the methods exploited in recent work include:

(a) The reaction of a carbodithioate salt with an alkylating agent.[384,394,409,415,440,450,477,478] Often this will involve sequential treatment of an anionic species with carbon disulfide and an alkylating agent in a one-pot reaction. For example, the process was used to prepare benzyl dithiobenzoate (**199**) from phenyl Grignard reagent (Scheme 9.39),[384] Yields are lower when this method is used to prepare RAFT agents such as 2-(ethoxycarbonyl)prop-2-yl dithiobenzoate (**177**)[384] and 2-cyanoprop-2-yl dithiobenzoate (**176**)[409] from the corresponding tertiary halides.

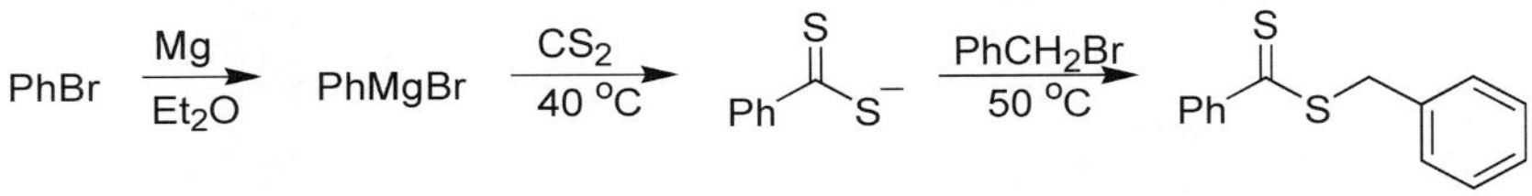

$$PhBr \xrightarrow[Et_2O]{Mg} PhMgBr \xrightarrow[40\ °C]{CS_2} Ph-C(=S)-S^- \xrightarrow[50\ °C]{PhCH_2Br} Ph-C(=S)-S-CH_2Ph$$

199

Scheme 9.39

A similar approach has been used to prepare dithiocarbamates, xanthates and unsymmetrical trithiocarbonates.[478] Thus, unsymmetrical primary and secondary trithiocarbonates are readily prepared in a 'one pot' reaction by treating a thiol with carbon disulfide in the presence of triethylamine to form a carbotrithioate salt and then adding the appropriate alkylating agent.[457,478] The process is shown in Scheme 9.40 for **231**.[463]

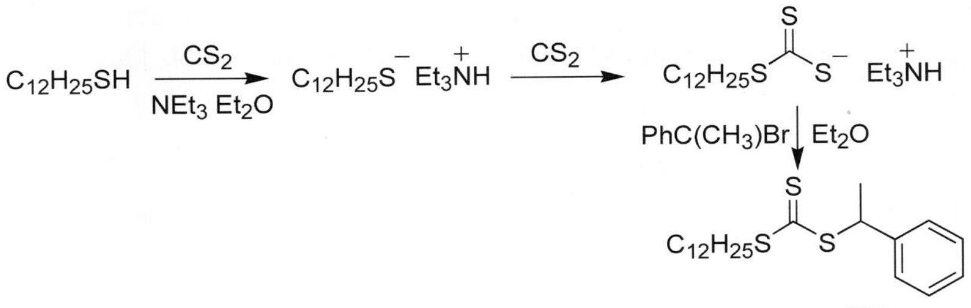

$$C_{12}H_{25}SH \xrightarrow[NEt_3\ Et_2O]{CS_2} C_{12}H_{25}S^-\ Et_3NH^+ \xrightarrow{CS_2} C_{12}H_{25}S-C(=S)-S^-\ Et_3NH^+ \xrightarrow[Et_2O]{PhC(CH_3)Br} C_{12}H_{25}S-C(=S)-S-CH(CH_3)Ph$$

231

Scheme 9.40

(b) Addition of a dithioacid across the double bond of an electron-rich olefin (S, AMS, isooctene and VAc).[394,409,479,480] This procedure has been used to prepare cumyl dithiobenzoate (**175**) from AMS (Scheme 9.41)[409] and isooctyl dithiobenzoate (**179**) from 2,2,4-trimethylpentene.[394]

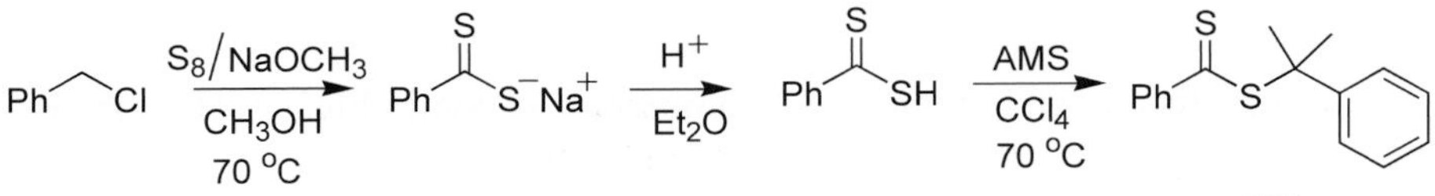

Scheme 9.41

Addition of dithioacids to electron-deficient monomers (MA, MMA, AN) proceeds by Michael addition to put sulfur at the unsubstituted end of the double bond.[479]

(c) Radical-induced decomposition of a bis(thioacyl) disulfide.[384,450,481-483] This is probably the most used method for the synthesis of RAFT agents requiring tertiary R groups. The method was used in preparation of the unsymmetrical trithiocarbonate **226** (Scheme 9.42).[450] It is also possible to use this chemistry to generate a RAFT agent *in situ* during polymerization.

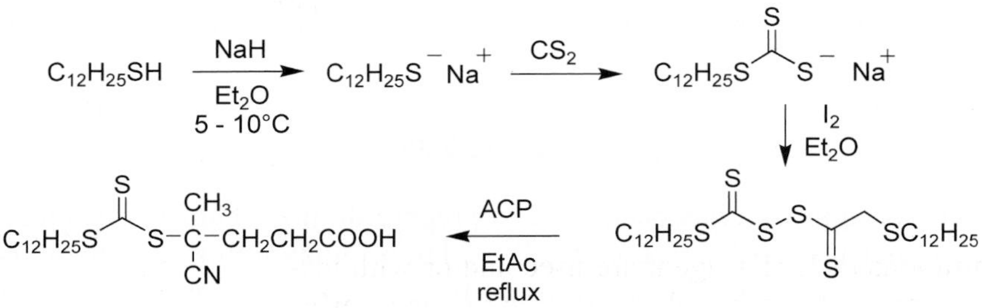

Scheme 9.42

(d) Sulfuration of a thioloester, or a mixture of a carboxylic acid with a halide, olefin, or alcohol, with Lawesson reagent (Scheme 9.43), Davey reagent or P_4S_{10}.[394,484]

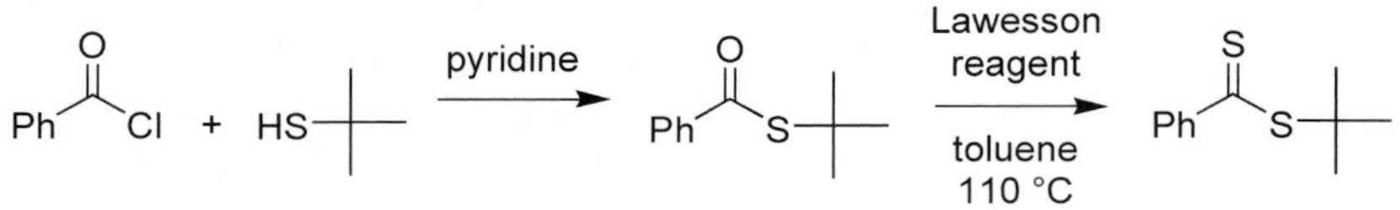

Scheme 9.43

(e) Radical-induced ester exchange.[384,394,442,485] For example, the cyanoisopropyl radical generated from AIBN can replace the cumyl group of cumyl dithiobenzoate (Scheme 9.44). For this method to be most effective the R group of the precursor RAFT agent should be a good free radical leaving group with respect to that of the product RAFT agent.

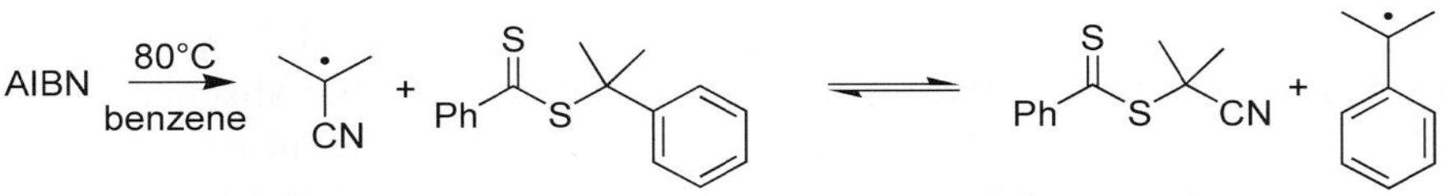

Scheme 9.44

9.5.3.4 Side reactions

Various side reactions may complicate RAFT polymerization. Transfer to solvents, monomer and initiator occur as in conventional radical polymerization. Other potential side reactions involve the intermediate radicals **165** and **167**. These radicals may couple with another radical (Q•) to form **271** or disproportionate with Q• to form **270**. They may also react with oxygen. The intermediate radicals **165** and **167** are not known to add monomer.

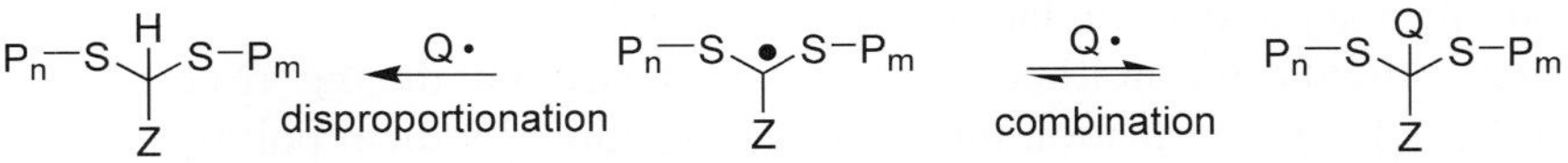

Scheme 9.45 (Q• is an initiator-derived radical or a propagating radical)

Retardation is sometimes observed in RAFT polymerizations when high concentrations of RAFT agent are used and/or with inappropriate choice of RAFT agent. Some decrease in polymerization rate is clearly attributable to a mitigation of the gel (or Norrish-Trommsdorf) effect.[384,394] However, it is also clear that other effects are important.

For example, there is significant retardation in the polymerization of acrylate esters in the presence of dithiobenzoate esters.[392,394,409,431,486-488] With benzyl dithiobenzoate and cyanoisopropyl dithiobenzoate retardation is observed from the onset of polymerization and is not directly related to consumption of the initial RAFT agent which appears to be extremely rapid.[394,409,486] The aliphatic dithioesters (*e.g.* dithioacetate, dithiophenylacetate) and trithiocarbonates give substantially less retardation.[394,409,431,486] Quinn *et al.*[453] observed that dithiophenylacetate RAFT agents enable polymerization of acrylates at ambient temperature whereas cumyl dithiobenzoate (**175**) gives inhibition under these conditions. McCleary *et al.*[488] used cumyl dithiophenylacetate (**212**) and cumyl dithiobenzoate and found an inhibition period corresponding to the time taken to consume the RAFT agent. They called this the initialization step and assigned this to slow reinitiation by cumyl radicals. Moad *et al.* attributed the inhibition period seen with cumyl dithioesters not to slow reinitiation by itself, but to the importance of the back reaction of cumyl radicals with the polymeric RAFT agent.[401]

Retardation has also been observed in polymerizations of S and methacrylates and is pronounced when high concentrations of dithiobenzoate RAFT agent are used.[383,391,409,442,486,489-492] With lower concentrations of RAFT agent, rates of polymerization are little different from those expected in the absence of RAFT agent.[383,409,486] The extent of retardation is markedly dependent on which initial RAFT agent is used and may be manifested as an inhibition period corresponding to the time taken to convert that RAFT agent to the polymeric RAFT agent.[383,409,493] Inconsistencies in reported rates of polymerization suggests that, in some cases, lower rates may in part be attributed to extraneous factors such as impurities in the RAFT[473,494] agent or incomplete degassing.[401,486]

9.5.3.5 Reaction conditions

RAFT polymerization can be performed simply by adding a chosen quantity of an appropriate RAFT agent to an otherwise conventional radical polymerization. Generally, the same monomers, initiators, solvents and temperatures are used. The only commonly encountered functionalities that appear incompatible with RAFT agents are primary and secondary amines and thiols.

Since radicals are neither formed nor destroyed during reversible chain transfer, RAFT polymerization must, like conventional radical polymerization, be initiated by a source of free radicals as shown in Scheme 9.38. RAFT polymerization is usually carried out with conventional radical initiators. Most often thermal initiators (e.g. AIBN, ACP, BPO, $K_2S_2O_8$) are used. S polymerization may be initiated thermally between 100-130°C. Polymerizations initiated with UV irradiation,[495,496] a gamma source[497-503] or a plasma field[504] have been reported. In these polymerizations, radicals generated directly from the RAFT agent may be responsible for initiation. It was initially suggested by Pan and coworkers that the mechanism for molecular weight control in UV[496] and γ-initiated[502] processes might involve reversible coupling and be similar to that seen with dithiocarbamate photoiniferters (Section 9.3.2). However, Quinn et al.[495,497,498] demonstrated that the living behavior observed in these polymerizations could be attributed to the standard RAFT mechanism (Scheme 9.38).

The RAFT process is compatible with a wide range of reaction media including protic solvents such as alcohols and water[382,400,419,505-507] and less conventional solvents such as ionic liquids[508] and supercritical carbon dioxide.[509,510] Even though RAFT polymerization has been successfully carried out in aqueous media, care should be taken because certain RAFT agents show some hydrolytic sensitivity particularly in alkaline media.[400,507,511] Rates of hydrolysis depend on R and Z and roughly correlate with RAFT agent activity (e.g. dithiobenzoates>trithiocarbonates~aliphatic dithioesters). RAFT agents used in aqueous media include **174**, **219** and **228**.

There have been no comprehensive studies of the effect of temperature on the course of RAFT polymerization. Temperatures reported for RAFT polymerization range from ambient to 140 °C. There is evidence with dithiobenzoates that at higher temperatures there is less retardation and also data that suggest narrower molecular weight distributions can be achieved.[398,453] For MMA polymerization with trithiocarbonate **226** there appears to be no dramatic effect of temperature on the molecular weight distribution achieved at a given conversion (Figure 9.7).[401] It should be noted, however, that higher temperatures do offer higher rates of polymerization and allow a given conversion to be achieved in a shorter reaction time.

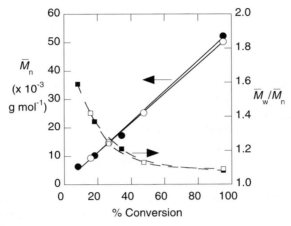

Figure 9.7 Evolution of molecular weight (———) and dispersity (– – –) with conversion for MMA polymerizations in the presence of RAFT agent **226** (0.0112 M) and (a) MMA (7.0 M) with AIBN (0.0061 M) at 60 °C (filled symbols) (b) MMA (6.55 M) with 1,1'-azobis(1-cyclohexanenitrile) (0.0018 M) at 90 °C (open symbols).[401]

RAFT polymerizations under very high pressure (5 kbar) have been reported.[509,512,513] At high pressures, radical-radical termination is slowed and this allows the formation of much higher molecular weight polymers and higher rates of polymerization than are achievable at ambient pressure.

RAFT polymerization can be conducted in the presence of Lewis acids. There are reports of attempts to control the tacticity of homopolymers[451,514-516] (to enable the synthesis of stereoblock copolymers[517]) and the alternating tendency for copolymerizations[518,519] through the use of Lewis acids as additives. For MMA polymerization, the addition of scandium triflate $Sc(OTf)_3$ increases the fraction of isotactic triads and enhances the rate of polymerization in conventional radical (Chapter 8) and RAFT processes.[451,514,516,517] Polymerizations with dithiobenzoate in the presence of $Sc(OTf)_3$ and with dithiobenzoate RAFT agents **175**[451,514,516,517] or **176**[514] gave comparatively poor control over molecular weight and dispersity. NMR studies show[514] that the poor results can be attributed to the Lewis acid causing degradation of the dithiobenzoate group. Polymerizations with the trithiocarbonate RAFT agent **225** provided polymer with narrow molecular weight distributions and molecular weights as anticipated for the RAFT process, as well as the expected effect on tacticity.[514]

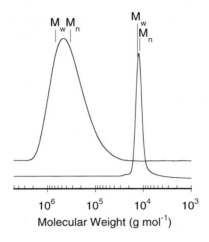

Figure 9.8 Comparison of molecular weight distributions for a conventional and RAFT polymerization. Data shown are GPC distributions (upper trace) for PS prepared by thermal polymerization of S at 110°C for 16 h ($\overline{M}_n$ 324000, $\overline{M}_w / \overline{M}_n$ 1.74, 72% conversion) and (lower trace) with cumyl dithiobenzoate (**175**) (0.0029 M) ($\overline{M}_n$ 14400, $\overline{M}_w / \overline{M}_n$ 1.04, 55% conversion).[401,409]

9.5.3.6 *Heterogeneous polymerization*

Much has been written on RAFT polymerization under emulsion and miniemulsion conditions. Most work has focused on S polymerization,[409,520,521] although polymerizations of BA,[461,522] methacrylates[382,409] and VAc[471,472] have also been reported. The first communication on RAFT polymerization briefly mentioned the successful semi-batch emulsion polymerization of BMA with cumyl dithiobenzoate (**175**) to provide a polymer with a narrow molecular weight distribution.[382] Additional examples and discussion of some of the important factors for successful use of RAFT polymerization in emulsion and miniemulsion were provided in a subsequent paper.[409] Much research has shown that the success in RAFT emulsion polymerization depends strongly on the choice of RAFT agent and polymerization conditions.[214,409,520-527]

The early emulsion recipes[382,409] were feed processes in which conversion of monomer to polymer was maintained at a very high level (often > 90%). In a first step a low molecular weight polymeric RAFT agent was prepared *ab initio*. Control during this stage was not always good. However, poor dispersity obtained in this step need not substantially affect control exerted during the later stages of polymerization.

The use of cumyl dithiobenzoate (**175**) and other dithiobenzoates as RAFT agent either in *ab initio* or in semi-batch emulsion polymerization of S is not recommended.[409] Much has been written on failings of these systems and they will not be detailed here. Xanthates have also been recommended[465] over dithioester RAFT agents for *ab initio* batch emulsion polymerization of S because the kinetics more closely approximate those of conventional emulsion polymerization. Substantially better control over S polymerization is also observed with RAFT agents such as trithiocarbonates, dithioacetates and these reagents also offer narrow molecular weight distributions. Dithiobenzoates have been successfully used in RAFT emulsion polymerization of methacrylates to produce low dispersity polymers where again transfer constants are lower.[382]

Some of the issues associated with RAFT emulsion polymerization have been attributed to an effect of chain length-dependent termination.[528] In conventional emulsion polymerization, most termination is between a long radical and a short radical. For RAFT polymerization at low conversion most chains are short thus the rate of termination is enhanced. Conversely, at high conversion most chains are long and the rate of termination is reduced.

A novel approach to RAFT emulsion polymerization has recently been reported.[461,529] In a first step, a water-soluble monomer (AA) was polymerized in the aqueous phase to a low degree of polymerization to form a macro RAFT agent. A hydrophobic monomer (BA) was then added under controlled feed to give amphiphilic oligomers that form micelles. These constitute a RAFT-containing seed. Continued controlled feed of hydrophobic monomer may be used to continue the emulsion polymerization. The process appears directly analogous to the 'self-stabilizing lattices' approach previously used in macromonomer RAFT polymerization (Section 9.5.2). Both processes allow emulsion polymerization without added surfactant.

RAFT in miniemulsion has also been reported[210,409,423,462,530-532] and is more readily used to produce polymers with a narrow molecular weight distribution. Moad *et al*.[409] used RAFT in miniemulsion to provide narrow dispersity PS in a batch process. Significant retardation was observed with the dithiobenzoate RAFT agent used. However, this is markedly reduced when aliphatic dithioesters[423] or trithiocarbonate RAFT agents are used.[462] One of the issues with traditional miniemulsion polymerization is the high level of surfactant and co-stabilizer that is typically employed. Pham *et al*.[462] have recently described surfactant-free miniemulsion polymerization. Amphipathic macro RAFT agents synthesized *in situ* by polymerization of AA were used as the sole stabilizers. This process eliminated secondary nucleation of new particles and lead to a latex with no labile surfactant and good particle size control.

9.5.4 Iodine-Transfer Polymerization

The history of iodine transfer polymerization may be traced back to telomerization experiments carried out in the 1940's.[26,533] Iodine-transfer polymerization as a method of living radical polymerization was reported by Tatemoto in 1992.[534] The process involves conducting a polymerization with a conventional initiator (AIBN, BPO) in the presence of an activated alkyl iodide. Iodine-transfer polymerization has been used for S,[381,535,536] acrylates,[535] VAc[537,538] and various fluoro-olefins.[534,539] Narrow dispersity PS was not obtained and this can be attributed to the transfer constant (C_{tr}~3.6 at 80 °C).

Side reactions observed in VAc polymerization include head addition during propagation (Scheme 9.46) (Section 4.3.1.1).[538] The primary alkyl iodide (**273**) is much less effective as a transfer agent than the secondary iodide (**272**) derived from the normal propagating radical. Thus, formation of **273** constitutes a chain termination reaction. Another side reaction is the formation of an aldehyde end group by acid catalyzed decomposition of end group **272**.[538] Despite these side reactions relatively narrow dispersities <1.4 are observed for molecular weights less than 20000. Use of higher transfer agent concentrations gives slower rates but better control over polymer dispersity.

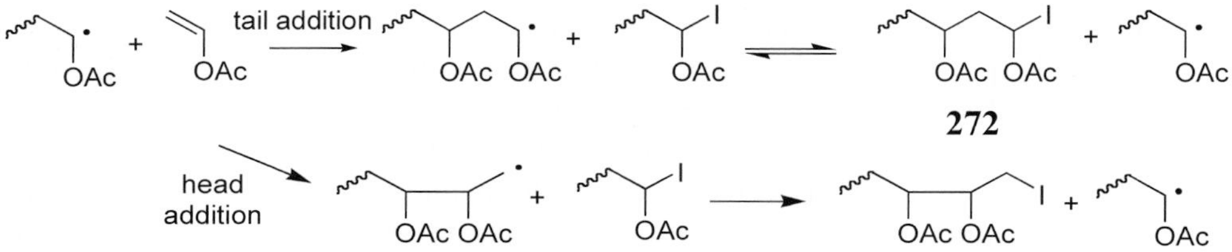

Scheme 9.46

9.5.5 Telluride-Mediated Polymerization

Telluride-mediated polymerization (TERP) has been described.[23,540-542] The importance of chain transfer to the organic chalcogenides Z-X-R where R is a free radical leaving group and Z is an activating group (Figure 9.9) increases in the series where X is O<S<Se<Te. In this series, only the alkyl tellurides appear effective in lending living characteristics to a thermally initiated polymerization. The application of alkyl sulfides and selenides in photoinitiated polymerizations has already been discussed in Sections 9.3.2 and 9.3.3 respectively. It is believed that these agents control polymerization by a reversible coupling mechanism with the sulfur or selenium-centered radical as the mediating agent. When alkyl tellurides are used as control agents, it is possible that reversible activation/deactivation by reversible coupling and reversible chain transfer mechanisms are simultaneously operative (Scheme 9.47). However, the reversible chain transfer by homolytic substitution appears to be the dominant mechanism. The kinetics and mechanism of radical polymerizations in the presence of the tellurides has been studied by Goto et al.[23,540]

reversible activation/deactivation

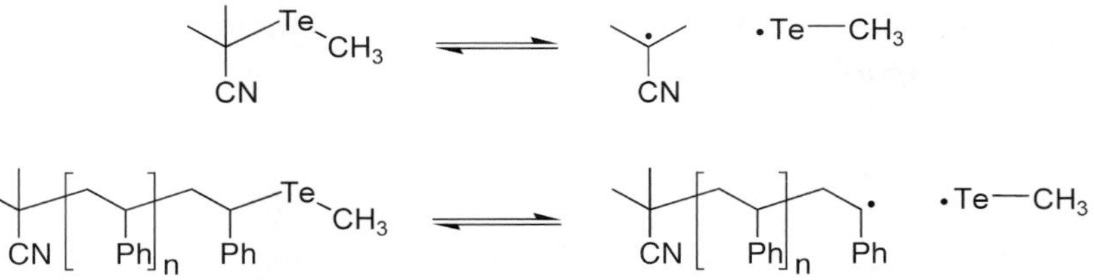

reversible chain transfer

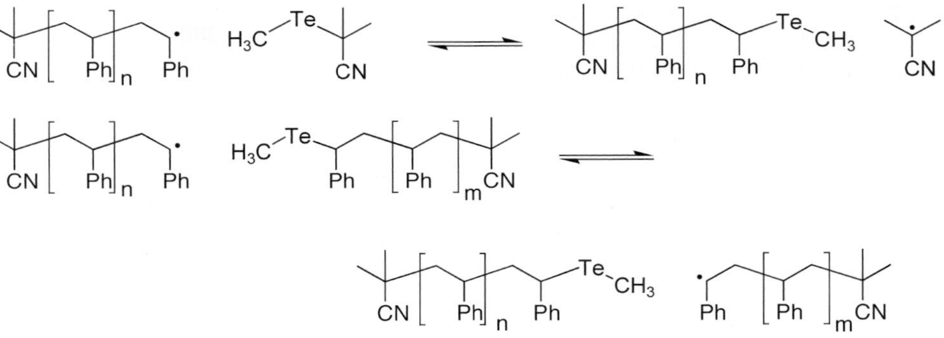

Scheme 9.47

Alkyl tellurides appear very effective in controlling thermally initiated polymerization of a very wide range of monomers (Table 9.19).[540-542] In the first experiments the telluride was both a thermal initiator and a reversible chain transfer agent. This required reaction temperatures of 80-100 °C.[541,542] In later work AIBN was used as coinitiator to enable the use of lower reaction temperatures (60 °C).[540] Narrowest molecular weight distributions are obtained with methyl tellurides **276-278**. The phenyl telluride **280** and the methyl benzyl telluride **279** give poorer control. In polymerization of methacrylates, narrow dispersities are only obtained in the presence of added ditelluride (**274** or **275**).[540,542] This may reflect the monomeric radical being a much poorer leaving group than the propagating radical as has been seen in RAFT polymerization. Polymerizations can also be carried out with AIBN as initiator in the presence of dimethyl ditelluride (**274**) to form the dormant species *in situ*.[543]

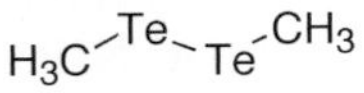

274

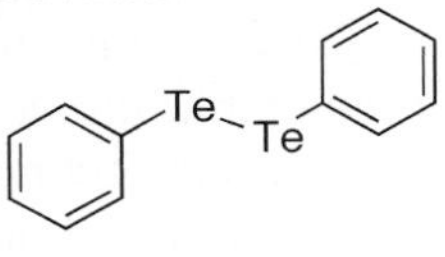

275

Table 9.19 Initiators for Telluride-Mediated Polymerization[a]

Telluride	Monomers	Telluride	Monomers	Telluride	Monomers
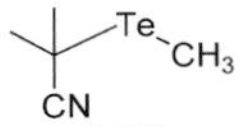 **276** (Te-CH3, CN, CH3, CH3)	S, BA, (MMA), AN, NIPAM[540]	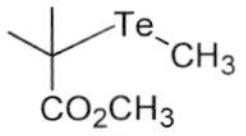 **277** (Te-CH3, CO2CH3, CH3, CH3)	MA, tBA, (MMA)[542]		
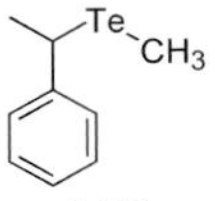 **278** (Te-CH3, phenyl, CH3)	S[541] MA, BA, DMAEA, DMAM, AN[542]	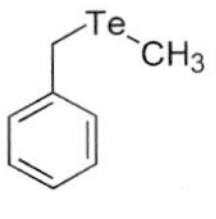 **279** (Te-CH3, benzyl)	S[541]	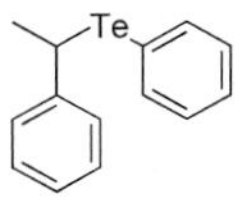 **280** (Te-phenyl, CH-phenyl-CH3)	S[541]

a Dispersities <1.2 except for systems shown in parentheses.

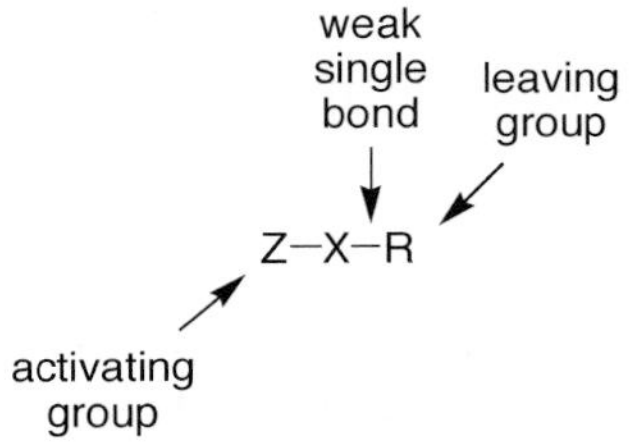

Figure 9.9 General description of organochalcogenide transfer agents

Z—X—R with weak single bond, leaving group (R), and activating group (Z)

9.5.6 Stibine-Mediated Polymerization

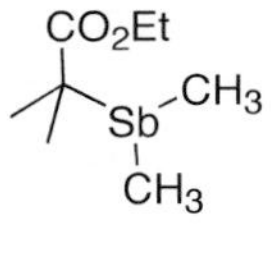

281 (CO2Et, Sb(CH3)2, CH3, CH3)

Very recently stibine-mediated polymerization has been reported by Yamago and coworkers[544,545] The living characteristics are thought to be imparted by a reversible chain transfer mechanism similar to that involved with the tellurides (Section 9.5.5). Thus far only one organostibine transfer agent (**281**) has been reported.[544] However, a class of reagents as shown in Figure 9.10 can be envisaged. Narrow molecular weight distributions (dispersity<1.3, with most <1.2) and predictable molecular weights were obtained with a remarkably wide range of monomers including S and (meth)acrylics (BA, MMA, NIPAM and AN) and vinyl monomers (NVP and VAc). Polymerizations were carried out at 60 °C with unusually large concentrations of AIBN (up to 0.5 molar equivalents with respect to **281**).

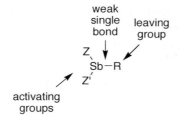

Figure 9.10 General description of organostibine transfer agents

9.6 Living Radical Copolymerization

One of the major advantages of radical polymerization over most other forms of polymerization, (anionic, cationic, coordination) is that statistical copolymers can be prepared from a very wide range of monomer types that can contain various unprotected functionalities. Radical copolymerization and the factors that influence copolymer structure have been discussed in Chapter 7. Copolymerization of macromonomers by NMP, ATRP and RAFT is discussed in Section 9.10.1.

An issue in living radical copolymerization is that the conditions for dormant chain activation can vary substantially according to the particular propagating radical. The problem may be mitigated by two factors.

(a) In copolymerization the steady state concentration of the propagating radical from the slower propagating monomer at the chain end will be higher than that of the faster propagating monomer. Deactivation events, which proceed at close to diffusion-controlled rates, should preferentially involve the species that is present in highest concentration.

(b) For many monomer pairs the reactivity ratios are both less than unity and cross propagation is substantially faster than homopropagation.

9.6.1 Reactivity Ratios

Although, there are reports on differences in reactivity ratios observed for conventional radical copolymerization *vs* living radical copolymerization (ATRP[275,276,546-548] or RAFT[548]), most research suggests that reactivity ratios are identical[398,549] and any discrepancies in composition should be attributed to other factors.

In comparing observed reactivity ratios between various polymerization systems, it is important to take into account the possible effect of molecular weight on copolymer composition.[547,549] In conventional radical copolymerization, the specificity shown in the initiation and termination steps can have a significant effect on the composition of low molecular weight copolymers (usually <10 units). These effects are discussed in Section 7.4.5. In a living polymerization molecular weights are low at low conversion and increase with conversion. In these circumstances, the overall copolymer composition will also depend on conversion. The usual methods of determining reactivity ratios, which involve the evaluation of copolymer composition or sequence distribution for low conversion samples, are not directly applicable in these circumstances. Either, molecular weights must be sufficiently high for statistical averaging of the composition to take place, or the effects of specificity in initiation and termination steps must be explicitly included in any calculations.

One might also anticipate that the influence of 'bootstrap effects' (Section 8.3.1.2) would be quite different in living and non-living processes.[268] A comprehensive study of reactivity ratios in living and conventional radical polymerization may provide a test of the various hypotheses for the origin of this effect.

9.6.2 Gradient Copolymers

Copolymers produced by living polymerization processes differ from those produced by conventional polymerization in one important aspect. Living polymerization processes produce gradient or tapered copolymers. Such copolymers are known from anionic living polymerization.

Disparate reactivity ratios cause unequal rates of monomer consumption and a drift in the composition of the monomer feed with conversion. In conventional radical copolymerization this means that the copolymer macromolecules formed at the beginning of the experiment will be different from those formed at higher monomer conversion; the high conversion product will be a polymer blend. In a living polymerization process, any compositional drift is captured within each chain. Such copolymers will have a blocky character with the degree of blockiness depending on the values of the reactivity ratios and the monomer feed ratio. For example, copolymerization of a 1:0.91 (mole ratio) mixture of MMA and BA (r_{MMA} 1.7 and r_{BA} 0.2) in the presence of cumyl dithiobenzoate (**175**) provides a narrow polydispersity copolymer with a gradient in composition of [MMA]:[BA] from *ca* 1:0.45 at the initiated end to *ca* 2:1 at the RAFT agent end.[398,425] The overall composition of the copolymer was the same as that of a copolymer prepared in the absence of **175**.

If reactivity ratios are particularly disparate then it is possible to form a block copolymer from a batch polymerization. Thus the copolymerization of MAH with S by NMP[550] or RAFT[551,552] with excess S provides P(MAH-*alt*-S)-*block*-PS. There is a similar outcome in other copolymerizations which show a strong alternating tendency such as S with maleimides (*e.g.* NPMI[204,401]) or AN. The copolymerization of tBA with VAc by RAFT provides P(tBA-*co*-VAc)-*block*-VAc.[449] Similarly, that of MA with VAc provides P(MA-*co*-VAc)-*block*-VAc.[401] The copolymerization of S with VAc or NVP by NMP is also reported to give a blocky copolymer but the process becomes non-living once the S is exhausted.[553]

9.6.3 NMP

A generic scheme for nitroxide-mediated copolymerization is shown in Scheme 9.48. The literature through 2001 has been summarized by Davis and Matyjaszewski.[554] A non-exhaustive summary of nitroxide-mediated copolymerizations is provided in Table 9.20; most involve S or isoprene (I).

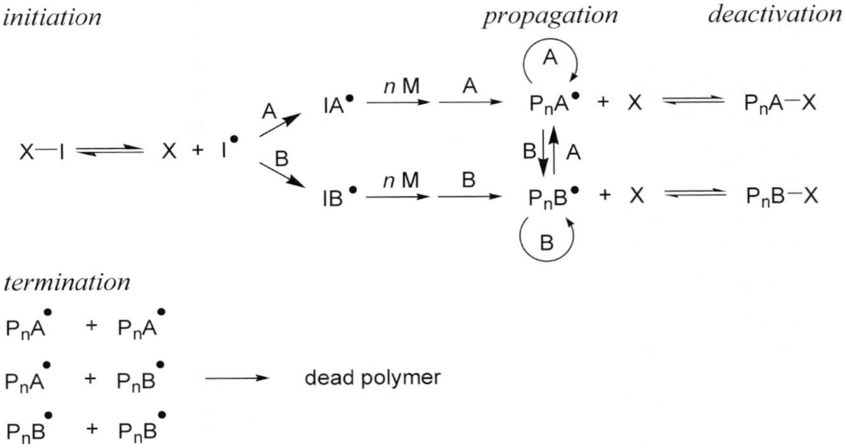

initiation *propagation* *deactivation*

$$X-I \rightleftharpoons X + I^\bullet \xrightarrow{A} IA^\bullet \xrightarrow{nM\ A} P_nA^\bullet + X \rightleftharpoons P_nA-X$$

$$\xrightarrow{B} IB^\bullet \xrightarrow{nM\ B} P_nB^\bullet + X \rightleftharpoons P_nB-X$$

with cross-propagation $B \updownarrow A$ between $P_nA^\bullet$ and $P_nB^\bullet$, and self-propagation loops A and B.

termination

$$P_nA^\bullet + P_nA^\bullet$$
$$P_nA^\bullet + P_nB^\bullet \longrightarrow \text{dead polymer}$$
$$P_nB^\bullet + P_nB^\bullet$$

Scheme 9.48 (For NMP: X—I is an alkoxyamine and X is a nitroxide, A and B are specific monomers, M is any monomer, P_n is a copolymer chain; note that $P_nB\bullet$, $P_nAB\bullet$, $P_nBB\bullet$ and $P_nA\bullet$, $P_nBA\bullet$, $P_nAA\bullet$ are not distinguished)

Monomers not amenable to direct homopolymerization using a particular reagent can sometimes be copolymerized. For example, NMP often fails with methacrylates (*e.g.* MMA, BMA), yet copolymerizations of these monomers with S are possible even when the monomer mix is predominantly composed of the methacrylate monomer.[153] This is attributed to the facility of cross propagation and the relatively low steady state concentration of propagating radicals with a terminal MMA (Section 7.4.3.1). MMA can also be copolymerized with S or acrylates at low temperature (60 °C).[111] Under these conditions, only deactivation of propagating radicals with a terminal MMA unit is reversible, deactivation of chains with a terminal S or acrylate unit is irreversible. Molecular weights should then be controlled by the reactivity ratios and the comonomer concentration rather than by the nitroxide/alkoxyamine concentration.

Table 9.20 Statistical/Gradient Copolymers Synthesized by NMP

Monomers[a]	Nitroxide[b]	Monomers[a]	Nitroxide[b]
S-SMe	83[555]	I-S	86[154]
S-SMeCl	TEMPO[556]	I-SMeCl	86[154]
S-SAc	83[557]	I-SAc	86[154]
S-SOMe	69[558]	I-BA	86[154]
S-SOBu	69[194,558]	I-AA	86[154]
S-SOCOBu	86[153]	I-NVP	86[154]
S-MMA	TEMPO,[556] 86[153]	I-MMA	86[154]
S-BMA	TEMPO[194,559]	I-HEMA	86[154]
S-MA	TEMPO[560]		
S-EA	TEMPO[560]		
S-BA	TEMPO,[556] 86,[153] 89[561]		
S-AN	TEMPO[132,138,560,562] 61, 63, 64[138]		
S-4VP	TEMPO[563]		
S-VCz	TEMPO[193,560,564]		
S-MAH	TEMPO[c,550] 86[550]		

a Abbreviations: SAc 4-acetoxystyrene, SMe 4-methylstyrene, SMeCl 4-chloromethystyrene, SOMe 4-methoxystyrene, SOBu 4-*t*-butoxystyrene, SOCOBu 4-(*t*-butoxycarbonyloxy)styrene, VCz *N*-vinylcarbazole, 4VP 4-vinylpyridine. Other abbreviations can be found in the Glossary. b Nitroxide structures in Table 9.1-Table 9.4. c Poor control/non living behavior observed.

9.6.4 ATRP

Atom transfer radical copolymerization can be described by a scheme similar to that shown in Scheme 9.48 except that bimolecular activation steps must be added (Section 9.4). Copolymerization by ATRP through 2001 has been reviewed by Kelly and Matyjaszewski.[554] A summary of ATRP copolymerizations appears in Table 9.21.

Lewis acids (diethylaluminum chloride, ethyl aluminum sesquichloride) have been used in conjunction with ATRP to provide greater alternating tendency in S-MMA copolymerization.[519] However, poor control was obtained because of interaction between the catalyst (CuCl/dNbpy) and the Lewis acid. Better results were obtained by RAFT polymerization.[519] Copper catalysts, in particular Cu(II)Br/PMDETA, have been shown to coordinate monomer but this has negligible influence on the outcome of copolymerization.[565]

As with NMP there are examples of copolymerizations providing good control where homopolymerization is unsuccessful. Copolymerization of MA with small amounts of 1-octene is thought to provide control[283,566] because the propagating radical with a terminal 1-octene unit undergoes rapid cross propagation. It has been established that the ATRP catalyst is unable to efficiently activate the polymeric bromo-compound with a terminal 1-octene unit.[566]

Table 9.21 Statistical/Gradient Copolymers Synthesized by ATRP

Monomers[a]	Catalyst/Ligand	Monomers[a]	Catalyst/Ligand
S-BA	133[b,567]	tBA-ODMA	133[b,568]
S-MMA	146[569]	tBA-ODA	133[b,568]
BA-MMA	145,[c,570] 144[571] 140,[547]	MMA-BMDO[d,]	140[366]
MA-O	133,[283] 140[566]	MA-NFH	140[572]
S-SAc	132[573]	MMA-BMA	138[546]
MMA-282	140[367]	MMA-TBAEMA	138[275]
BA-iB	133,140[574]	MMA-DEAEMA	138[275]
S-AN	132,[575] 140[575]	MMA-DMAEMA	138[275]

a Abbreviations: iB isobutylene, BMDO 5,6-benzo-2-methylene-1,3-dioxepane, DEAEMA N,N-diethylaminoethyl methacrylate, NFH 3,3,4,4,5,5,6,6,6-nonafluoro-1-hexene, O 1-octene, TBAEMA t-butylaminoethyl methacrylate. Other abbreviations can be found in the Glossary. b Catalyst formed in situ with CuBr and ligand indicated. c Hybrid catalyst system. d Ring-opening copolymerization.

9.6.5 RAFT

The reaction scheme for RAFT copolymerization is relatively complex (Scheme 9.49) when considered alongside that for NMP or ATRP (Scheme 9.48). A summary of RAFT copolymerizations is provided in Table 9.22. An advantage of RAFT over other methods is its greater compatibility with monomers containing protic functionality though as yet few have taken advantage of this in the synthesis of functional copolymers.

RAFT of MMA with benzyl dithiobenzoate provides very poor control[394] yet copolymerization of S with MMA with this RAFT agent provides low dispersities with as little as 5% S in the monomer feed.

Table 9.22 Statistical/Gradient Copolymers Synthesized by RAFT Polymerization

Monomers[a]	RAFT Agent[b]	Monomers	RAFT Agent[b]
S-MMA	175[408] 216[c,576]	tBA-VAc	248[449]
S-AN	175[382,408]	NIPAM-XMA	176[577]
S-MAH	176[551] 199[552,578-580]	AMBS-AMPS	174[420]
AMS-MAH	199[d,579]	MMA-HEMA	175[382,408]
		MMA-BA	175[398]

a Abbreviations: AMBS sodium 2-acrylamido-3-methylbutanoate, AMPS sodium 2-acrylamido-2-methylpropane-1-sulfonate, XMA N-hydroxysuccinimide methacrylate.[577] Other abbreviations can be found in the Glossary. b Structures in Table 9.10-Table 9.18 c Miniemulsion copolymerization. d Poor control.

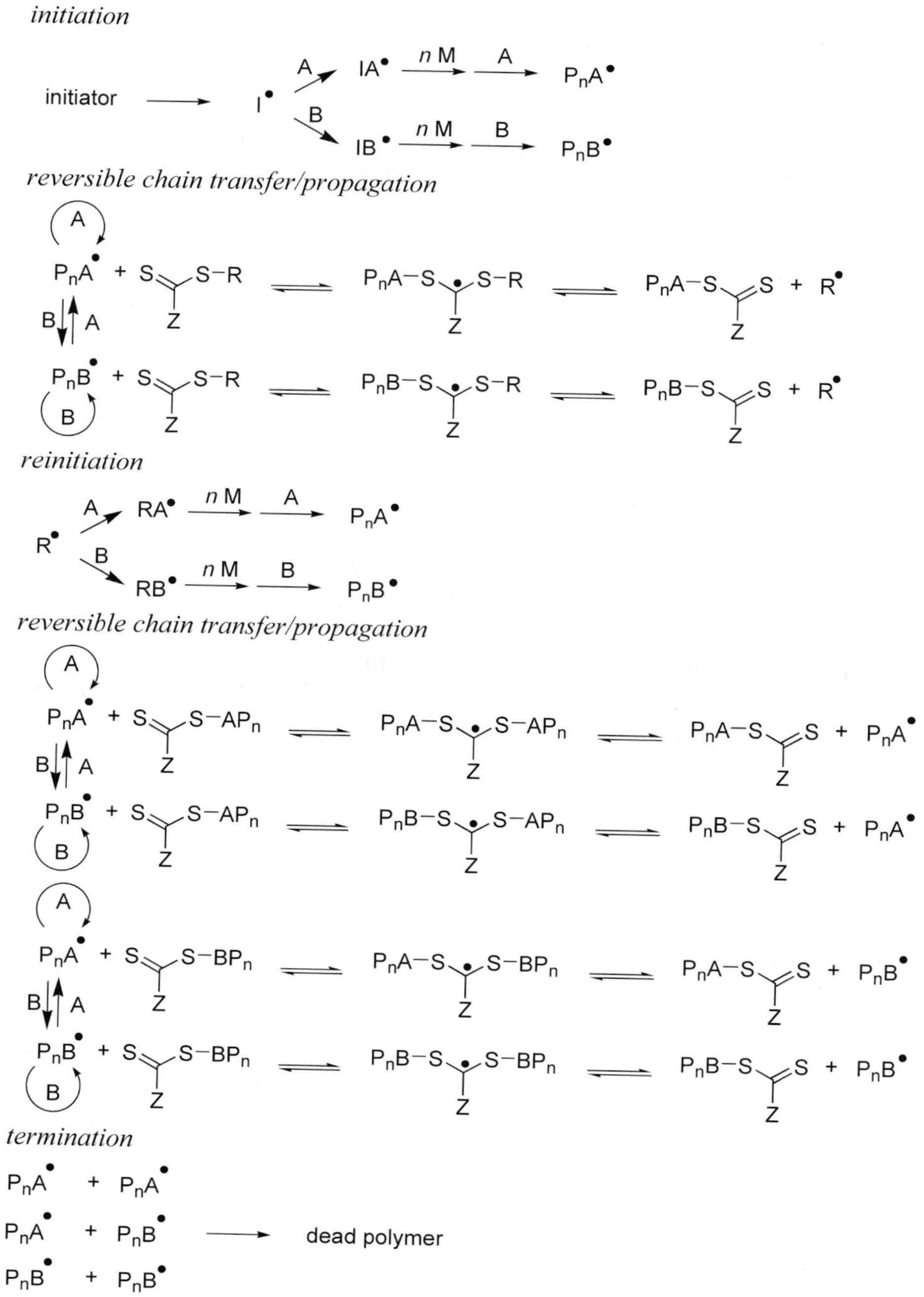

initiation

$$\text{initiator} \longrightarrow I^\bullet \underset{B}{\overset{A}{\nearrow}} \begin{matrix} IA^\bullet \xrightarrow{nM} P_nA^\bullet \\ IB^\bullet \xrightarrow{nM} P_nB^\bullet \end{matrix}$$

reversible chain transfer/propagation

reinitiation

$$R^\bullet \underset{B}{\overset{A}{\nearrow}} \begin{matrix} RA^\bullet \xrightarrow{nM} P_nA^\bullet \\ RB^\bullet \xrightarrow{nM} P_nB^\bullet \end{matrix}$$

reversible chain transfer/propagation

termination

$$\begin{matrix} P_nA^\bullet + P_nA^\bullet \\ P_nA^\bullet + P_nB^\bullet \\ P_nB^\bullet + P_nB^\bullet \end{matrix} \longrightarrow \text{dead polymer}$$

Scheme 9.49 (A and B are specific monomers, M is any monomer (A or B), P_n is a copolymer chain; note that $P_nB^\bullet$, $P_nAB^\bullet$, $P_nBB^\bullet$ and $P_nA^\bullet$, $P_nBA^\bullet$, $P_nAA^\bullet$ are not distinguished)

9.7 End-Functional Polymers

Most reviews on living radical polymerization mention the application of these methods in the synthesis of end-functional polymers. In that ideally all chain ends are retained, and no new chains are formed (Section 9.1.2), living polymerization processes are particularly suited to the synthesis of end-functional polymers. Living radical processes are no exception in this regard. We distinguish two main processes for the synthesis of end-functional polymers.

(a) The α-functionalization approach makes use of a functional initiator (alkoxyamine, halo-compound) or transfer agent (RAFT agent) to generate a functional initiating radical. All chains should then possess this functionality. The level of functionality will be reduced if there are other processes for initiation (*e.g.* thermal initiation in the case of S polymerization at high temperatures) and by reinitiation after chain transfer to monomer, solvent or other species present in the polymerization medium. It may be increased by the incidence of chain termination by combination.

(b) The ω-functionalization route involves chemical transformation of the dormant chain end in a post-polymerization reaction. It is also possible to introduce ω-functionality by building it in to the nitroxide fragment of an alkoxyamine NMP initiator or the 'Z' activating group of a RAFT agent (**164**). The level of functionality will generally equate to the fraction of living (dormant) chain ends and will be reduced by chain termination by radical-radical reaction and further reduced by any chain transfer to monomer, solvent or other species present in the polymerization medium.

There are additional factors that may reduce functionality which are specific to the various polymerization processes and the particular chemistries used for end group transformation. These are mentioned in the following sections. This section also details methods for removing dormant chain ends from polymers formed by NMP, ATRP and RAFT. This is sometimes necessary since the dormant chain-end often constitutes a "weak link" that can lead to impaired thermal or photochemical stability (Sections 8.2.1 and 8.2.2). Block copolymers, which may be considered as a form of end-functional polymer, and the use of end-functional polymers in the synthesis of block copolymers are considered in Section 9.8. The use of end functional polymers in forming star and graft polymers is dealt with in Sections 9.9.2 and 9.10.3 respectively.

9.7.1 NMP

9.7.1.1 ω-Functionalization

Two methods for cleaving the nitroxide functionality from polymers made by NMP are summarized in Table 9.23. Transfer agents such as thiols[111] or dithiuram disulfides (Scheme 9.50)[581] can be used for end group replacement and lead to the nitroxide moiety being substituted by a transfer agent-derived group (hydrogen-atom or dithiocarbamyl respectively). The reaction shown in Scheme 9.50 is a method for preparing functional dithiocarbamates[581] and might reasonably be applied to synthesize functional RAFT agents allowing conversion between NMP and RAFT polymerization (Section 9.8.2).

Table 9.23 Methods for End Group Transformation of Polymers Formed by NMP

Reaction	Monomer/Nitroxide
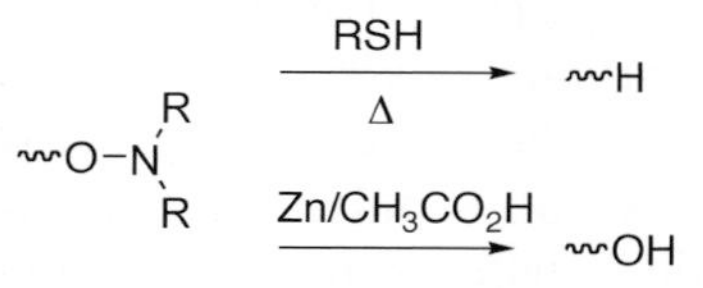 RSH, Δ → ~H	S/83[111]
Zn/CH$_3$CO$_2$H → ~OH	MA/59[111]

A method for ω-functionalization involves polymerization in the presence of a comonomer that does not propagate under the reaction conditions. Monomers that have been used include MAH and maleimide derivatives such as NPMI (Scheme 9.51).[582] In these cases, elimination of hydroxylamine under the reaction conditions provides an unsaturated end group.

When these methodologies involving the use of a non-propagating monomer or a transfer agent are applied *in situ* during polymerization, the comonomer/transfer agent concentration and the respective reactivity ratios or transfer constants control molecular weights.

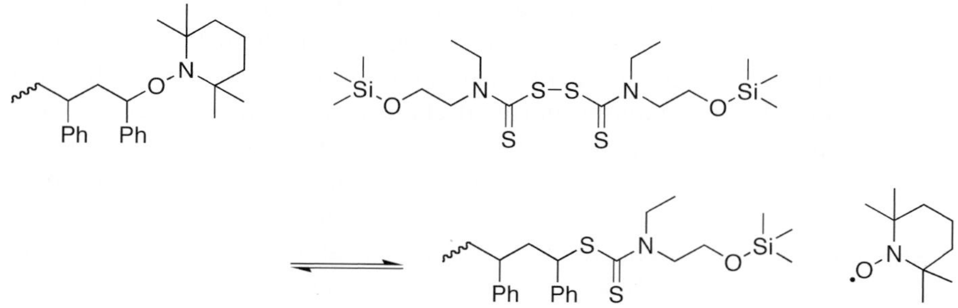

Scheme 9.50

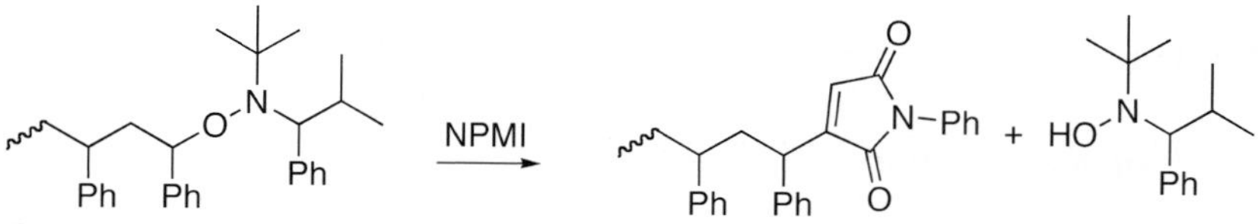

Scheme 9.51

Living Radical Polymerization 533

A side reaction in NMP is loss of nitroxide functionality by thermal elimination. This may occur by disproportionation of the propagating radical with nitroxide or direct elimination of hydroxylamine as discussed in Section 9.3.6.3. In the case of methacrylate polymerization this leaves an unsaturated end group.[111] The chemistry has also been used to prepare macromonomers from PMMA prepared by ATRP (Section 9.7.2.1).

Heating an alkoxyamine in the presence of another nitroxide provides nitroxide exchange[111,118,583] and a process for ω-functionalization.[584] The product distribution will be determined by the relative stability of the alkoxyamines and the excess of nitroxide. Exchange is also observed when two alkoxyamines are heated together.[585,586]

9.7.1.2 α-Functionalization

Functional alkoxyamines used as initiators for NMP include **283-287**. The functional alkoxyamines can be formed *in situ* by use of a functional azo compound or peroxide. NMP has been shown to be compatible with hydroxy, epoxy, amide and tertiary amine groups in the initiator. Carboxylic acid groups can cause problems but may be tolerated in some circumstances.[106]

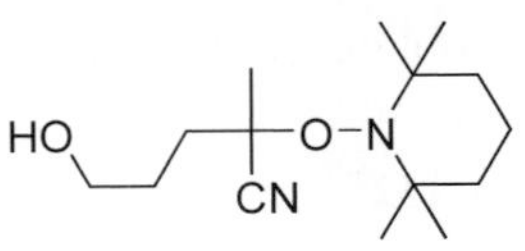

283[111] 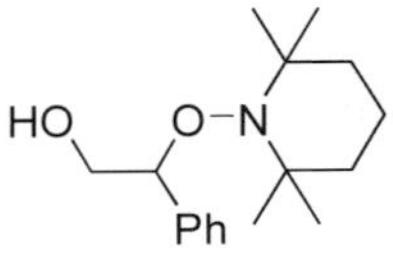   **284**[587,588] 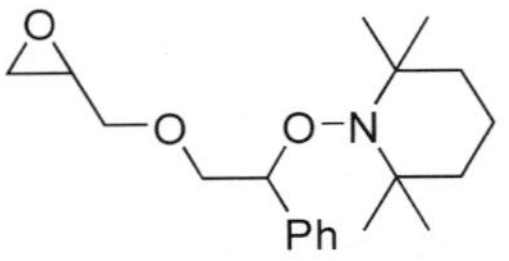   **285**[589]

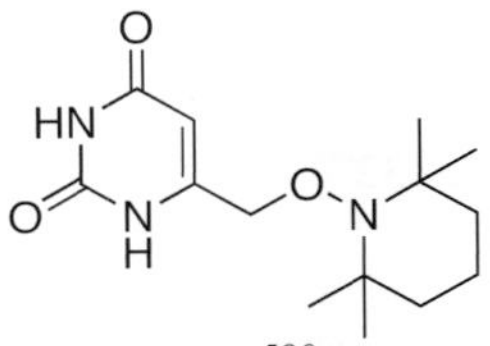

286[590] 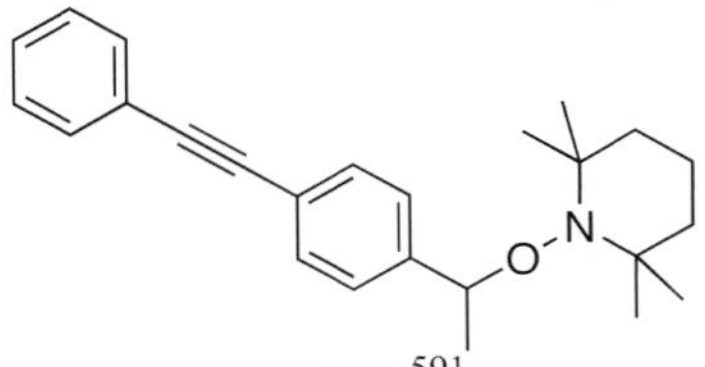   **287**[591]

9.7.2 ATRP

The literature on synthesis of end-functional polymers by ATRP through 2000 is discussed in a review by Coessens and Matyjaszewski.[592] The topic also has coverage in more general reviews on ATRP.[268,269]

9.7.2.1 ω-Functionalization

Polymers formed by ATRP should retain a halogen (typically bromine) on the dormant chain end and this is confirmed by analysis for many polymerizations.

Transformation of the end group may be required to confer greater stability or to introduce new functionality. The various methods include reactions with addition-fragmentation transfer agents or non-propagating monomers (Table 9.24) added at the end of the polymerization. The extent of functionalization will depend on the efficiency of the particular reaction. Those with addition-fragmentation chain transfer agents and MAH appear highly effective with yields >95%. Processes involving the less active non-propagating monomers are prone to side reactions.[593] An unusual non-propagating 'monomer' is buckminsterfullerene (C_{60}).[594,595] For example, P(MAA-*block*-DMAEMA)-C_{60} was prepared from P(MAA-*block*-DMAEMA)-Cl with CuCl/**144** in the presence of C_{60}.[594]

Table 9.24 Methods for End Group Transformation of Polymers Formed by ATRP by Addition or Addition-Fragmentation.

Reaction		Monomer/Catalyst(Ligand)
CO₂R–Br	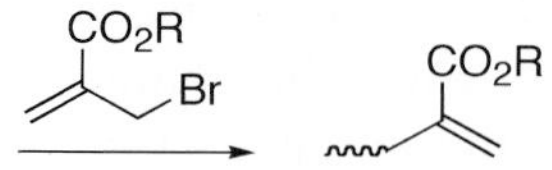 CO₂R	MMA/**138**,[596,597] MA/**132**[598]
Ph-O-SiMe₃	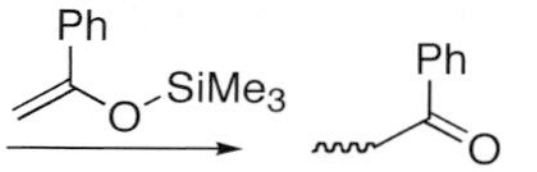 Ph (C=O)	MMA/**146**[599] MMA/**138**[a,597] PBA/**140**[a,593]
SnBu₃	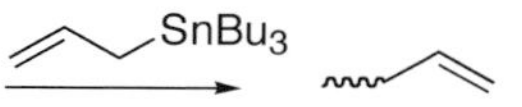	MA/**132**[600]
OH	 Br, OH	MA/**140**[600] PBA/**140**[593]
epoxide	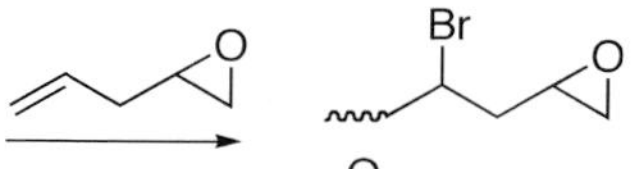 Br, epoxide	MA/**132**[600]
MAH	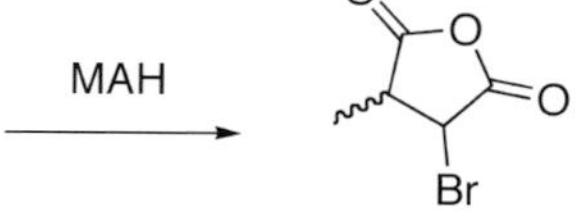 Br-lactone	MMA/**138**[597,601]

(~Br ⇌ ~ •)

[a] 4-trimethylsilyloxy derivative used to give phenoxy functional polymer after deprotection.

Addition of TEMPO post-polymerization to a methacrylate polymerization provides an unsaturated chain end (Scheme 9.52)[597,599] presumably by disproportionation of the PMMA propagating radical with the nitroxide. For polymers based on monosubstituted monomers (PS,[602] PBA[593,602]) the alkoxyamine is formed in high yield. A functional nitroxide (*e.g.* **69**[593]) can be used to yield an end-functional polymer.

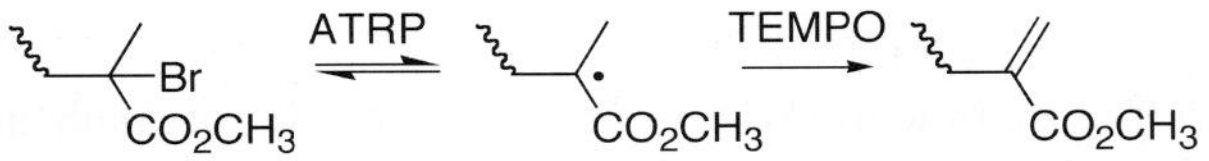

Scheme 9.52 (MMA/146,[599] MMA/138[597])

The chain end functionality may be reduced by the incidence of various side reactions. In that ATRP is a radical process, we should expect an amount of radical-radical termination consistent with the concentration of propagating radicals and the reaction time. Radical-radical termination cannot be eliminated, however, it can be minimized through choice of polymerization conditions. The incidence of other side reactions depends on the particular initiator, monomer(s) and catalyst used. During the (co)polymerization of S[603,604] a slow elimination of HBr from the initiator or dormant species occurs to yield an unsaturated end group. The reaction is catalyzed by Cu(II) and limits the molecular weight of PS that can be prepared with high end group functionality to ~10000.[603,604] For ATRP with Cu(I) and aliphatic amine ligands (e.g. 140), chain transfer to the ligand occurs to yield a saturated chain end.[312]

Table 9.25 End Group Transformations for Polymers Formed by ATRP

Reaction[c]	Polymer
 ∼∼X →(Bu$_3$SnH) ∼∼H	PMA-Br, PMA-Cl, PMMA-Br, PS-Br[a,605]
∼∼Br →(NaN$_3$) ∼∼N$_3$ →(PPh$_3$) ∼∼N=PPh$_3$; →(LiAlH$_4$) ∼∼NH$_2$	PMA-Br[606]
∼∼Br →(K-phthalimide) ∼∼N(phthalimide) →(NH$_2$NH$_2$) ∼∼NH$_2$	PS-Br[607]
	PS-Br[608]
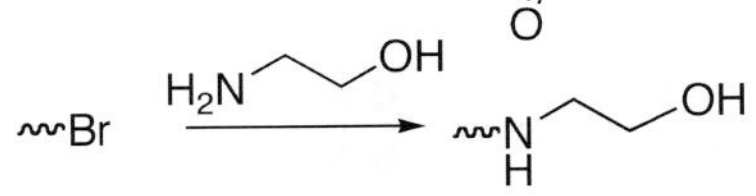 ∼∼Br →(H$_2$N∼OH) ∼∼N(H)∼OH	PMA-Br, PBA-Br, PS-Br,[b,593,609-611]
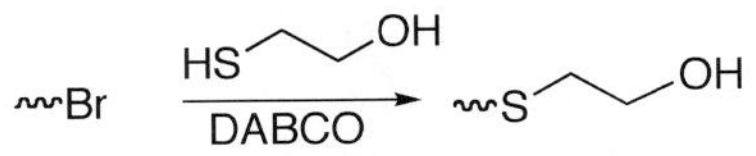 ∼∼Br →(HS∼OH, DABCO) ∼∼S∼OH	PBA-Br[593]
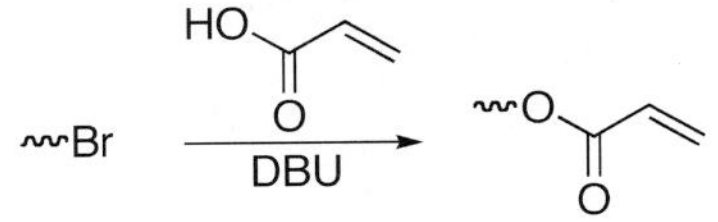 ∼∼Br →(HO-acrylate, DBU) ∼∼O-acrylate	PBA-Br[612]

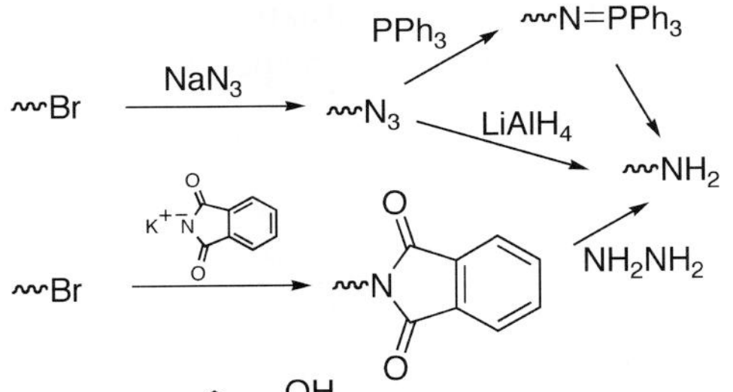

a May be carried out as a 'one-pot' reaction. b Various amino-alcohols have been used. c Abbreviations: DABCO, 1,4-diazabicyclo[2.2.2]octane, DBU 1,8-Diazabicyclo[5.4.0]undec-7-ene.

The nucleophilic displacement reactions with azide, primary amines, thiols and carboxylate salts are reported to be highly efficient giving high (>95%) yields of the displacement product (Table 9.25). The latter two reactions are carried out in the presence of a base (DBU, DABCO). Radical-induced reduction with tin hydrides is quantitative. The displacement reaction with phenolates,[613] phosphines,[614] and potassium phthalimide[608] gives elimination of HBr as a side reaction.

9.7.2.2 α-Functionalization

Initiators containing a wide range of functional groups have been applied in ATRP (*e.g.* **288-313**).[268] These include olefin (**293**,[615] **296**,[616] **297**[615]) hydroxy (**298**,[610] **302**,[611,617,618] **311**[613]), tertiary amine (**303**,[617] **308**[617]), epoxy (**299**[615]), oxazoline (**312**[613]), *t*-butyl ester (**300**,[619] precursor to carboxylic acid), amide (**304**,[617] **309**,[617] **310**[617]) and lactone (**295**[615]). Unprotected acid functionality and primary and secondary amine groups are an issue as these groups may interfere with the stability of the metal complex[615] though, with appropriate catalyst/initiator design, even these groups may be tolerated.[268,619,620] The ATRP process is tolerant of aromatic amine and carboxylic acid groups in initiators **288c**, **288f**[620] **291** and **313**.[613]

N-Bromomethylphthalimide (**294**)[608] and 2-bromopropanenitrile (**301**)[621] and the tBOC derivative (**305**)[312,313] have been used as initiators in the synthesis of PS with primary amine functionality (deprotection involves hydrazinolysis, $LiAlH_4$ reduction or treatment with CF_3COOH at room temperature respectively). The initiator **306** contains a protected thiol functionality.[622]

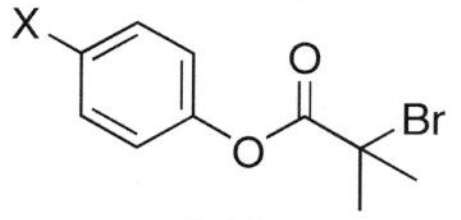

288

a X=Ph_2CH_2O **b** X=CH_3O **c** X=NH_2 **d** X=H
e X=CHO **f** X=CO_2H **g** X=$PhCO_2$ **h** X=NO_2

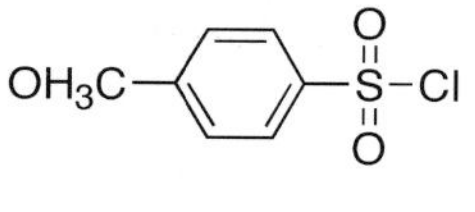

289

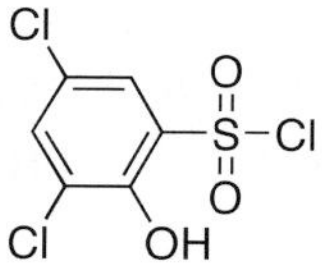

290

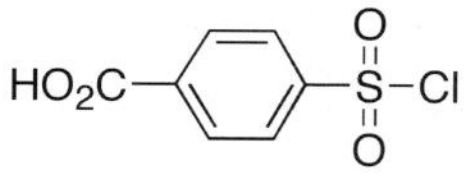

291

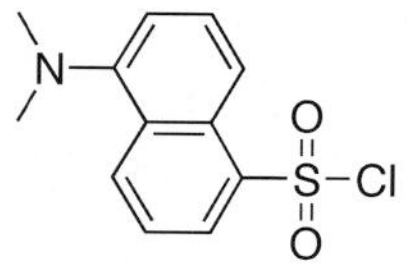

292

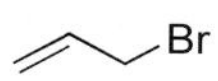

293

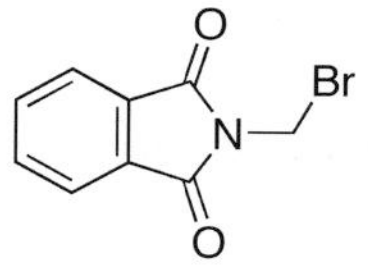

294

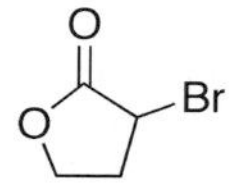

295

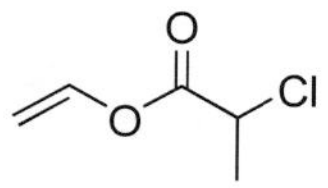

296

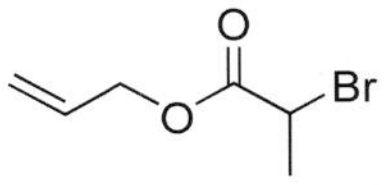

Living Radical Polymerization

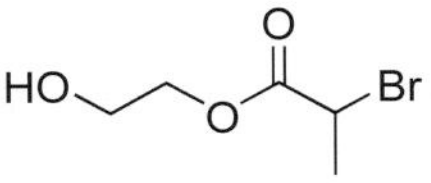

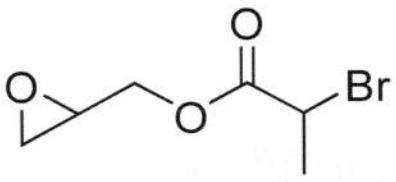

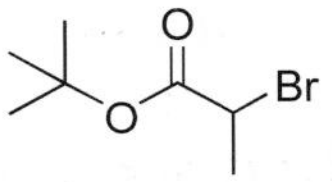

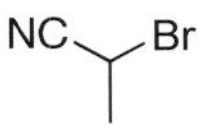

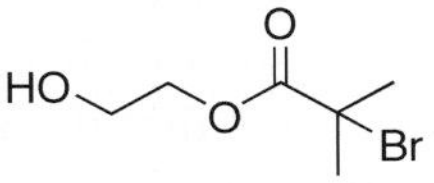

297 — allyl 2-bromopropanoate

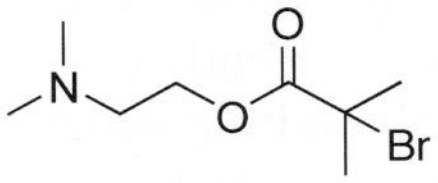

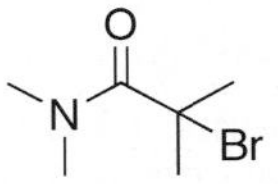

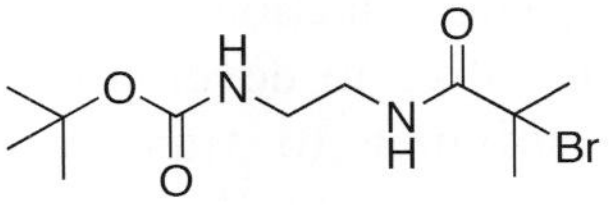

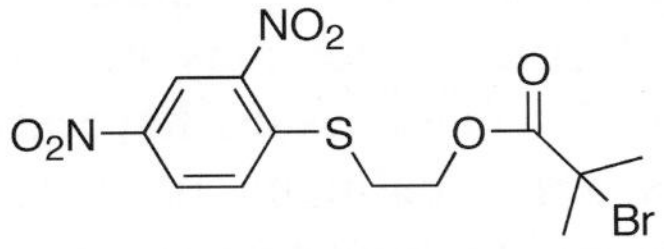

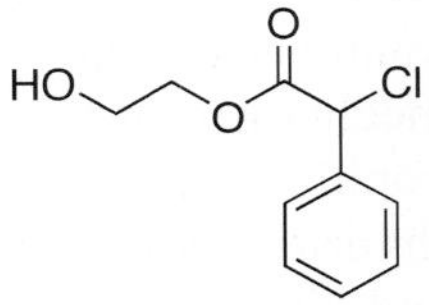

298 — 2-hydroxyethyl 2-bromopropanoate

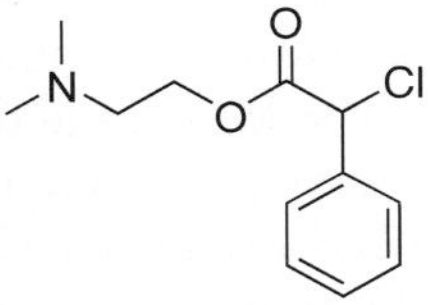

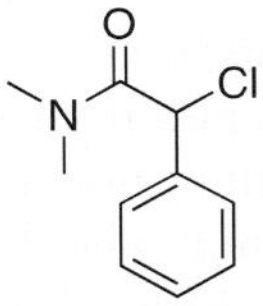

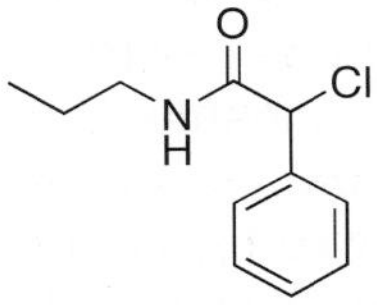

299 — glycidyl 2-bromopropanoate

300 — tert-butyl 2-bromopropanoate

301 — 2-bromopropanenitrile

302 — 2-hydroxyethyl 2-bromoisobutyrate

303 — 2-(dimethylamino)ethyl 2-bromoisobutyrate

304 — N,N-dimethyl 2-bromoisobutyramide

305

306 — 2,4-dinitrophenylthioethyl 2-bromoisobutyrate

307 — 2-hydroxyethyl 2-chloro-2-phenylacetate

308 — 2-(dimethylamino)ethyl 2-chloro-2-phenylacetate

309 — N,N-dimethyl 2-chloro-2-phenylacetamide

310 — N-propyl 2-chloro-2-phenylacetamide

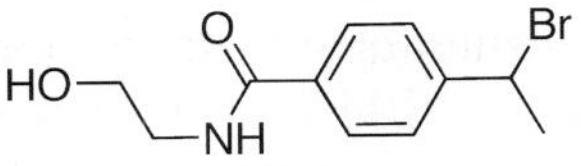

311

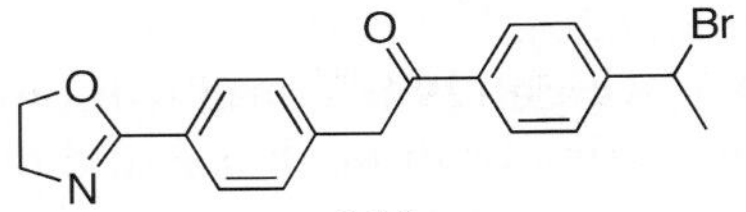

312

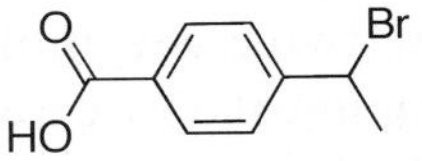

313

Telechelic polymers can be produced by a combination of α- and ω-functionalization[611] or by ω-functionalization of a polymer produced using a bis-functional initiator. Another method is to couple α-functionalized chains. Atom transfer radical coupling (ATRC) has been used to couple α-functional PS-Br made by ATRP and produce telechelic PS (Scheme 9.53).[618] This approach requires an appropriate rate of radical generation and should only be applied to systems where the propagating radicals undergo termination predominantly by combination. The telechelic purity is limited by the ratio of combination to disproportionation (greater than 85:15 in the case of PS• - Section 5.2.2). The technique can be applied to other polymers with the addition of small amounts of S to form propagating radicals with a terminal S in situ.[618]

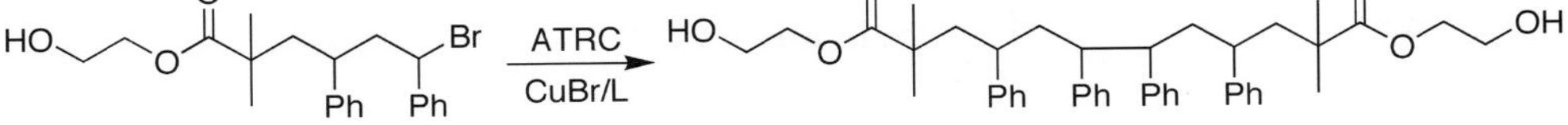

Scheme 9.53

9.7.3 RAFT

9.7.3.1 ω-Functionalization

The thiocarbonylthio group can be transformed post-polymerization in a variety of ways to produce end-functional polymers or it can be removed. The presence of the thiocarbonylthio groups also means that the polymers synthesized by RAFT polymerization are usually colored and they possess a labile end group that may decompose to produce sometimes odorous byproducts. Even though the color and other issues may be modified by appropriate selection of the initial RAFT agent, these issues have provided further incentive to develop effective methods for treatment of RAFT-synthesized polymer to transform the thiocarbonylthio groups post-polymerization.

It is well known that thiocarbonylthio groups can be transformed into thiols by reaction with nucleophiles that include pyridines, primary and secondary amines, ammonia, other thiols and hydroxide. The kinetics and mechanism of the reaction of compounds containing thiocarbonyl groups with nucleophiles has been reviewed by Castro.[623] They may also be reduced to thiols with hydride reducing agents such as sodium borohydride, lithium aluminum hydride and zinc in acetic acid. The thiocarbonylthio groups in RAFT-synthesized polymers are subject to the same reactions (Table 9.26).[382] Oxidation to the disulfide to form an impurity of twice the molecular weight is a complication in aminolysis that can be minimized by careful degassing or through use of dithionite.[455] RAFT-synthesized thiols have been used to make protein conjugates.[475,624]

Radical-induced reduction with, for example, tri-*n*-butylstannane can be used to replace the thiocarbonylthio group with hydrogen. Other transfer agents offer the possibility of introducing different functionality by group transfer. The RAFT end group is also light sensitive and can be removed under UV irradiation and it may be oxidized with reagents such as peroxides or sodium hypochlorite.[382,551]

RAFT end groups are known to be unstable at very high temperatures (>200 °C). Thermal elimination has been used as a means of trithiocarbonate end group removal. For PS[450,464] direct elimination is observed (Scheme 9.54). For poly(butyl acrylate)[464] the major product suggests a homolysis/backbiting/β-scission reaction is involved (Scheme 9.55).

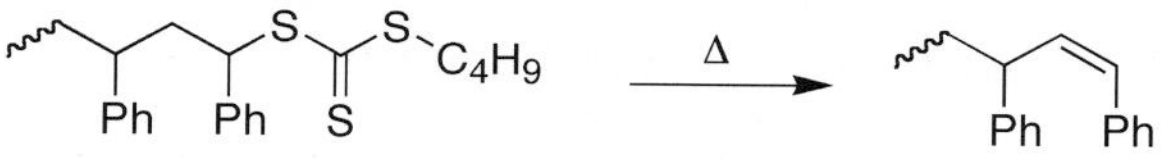

Scheme 9.54

Table 9.26 Methods for End Group Removal from Polymers Formed by RAFT Polymerization

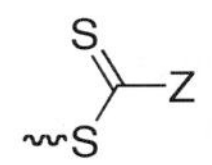

Reaction	Polymer[a]
$\xrightarrow{\text{AIBN}}$ ∽S-C(CN)	PMA, PMMA, PS[485]
$\xrightarrow{\text{Bu}_3\text{SnH}}$ ∽H	PS,[450] PAc[625]
$\xrightarrow{\text{R}_2\text{NH}}$ ∽SH	PS,[450,626,627] PMA[457,626] PMMA[455]
$\xrightarrow{\text{OH}^-}$ ∽SH	PNIPAM[474]
$\xrightarrow{\text{NaBH}_4}$ ∽SH	PDMAM, NaPSS, others[400,628]
$\xrightarrow{\text{Zn/CH}_3\text{CO}_2\text{H}}$ ∽SH	PS[450,627]

(Starting material: ∽S-C(=S)-Z)

a PAc = polyacenaphthalene, NaPSS = poly(sodium 4-styrenesulfonate). For other abbreviations see Glossary.

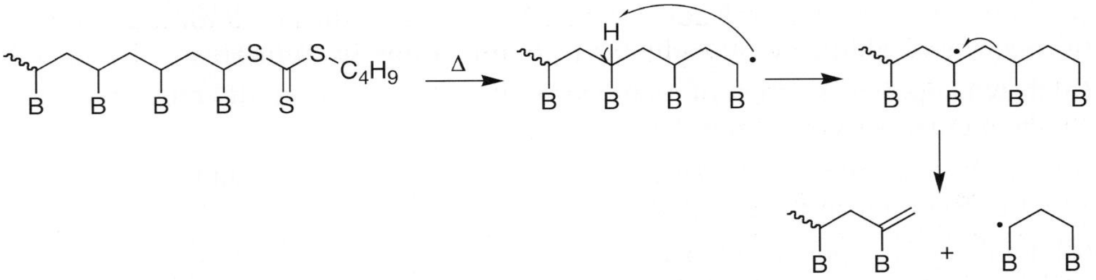

Scheme 9.55 ($B=CO_2C_4H_9$)

9.7.3.2 α-Functionalization

One significant advantage of the RAFT process is its compatibility with a wide range of functionality present in the monomer or the RAFT agent. This makes the technique eminently suitable for the synthesis of end functional polymers by incorporating the functionality into the Z or R groups of the RAFT agent. RAFT agents with unprotected functionality that have been used successfully include: **172, 178, 221** (-OH); **173, 219, 226, 227, 229** (-CO$_2$H); **174** (CO$_2$Na).

Polymers with primary or secondary amine functionality cannot be prepared directly by RAFT polymerization; these groups undergo facile reaction with thiocarbonylthio compounds. Such polymers can be prepared indirectly using RAFT agents with latent amine functionality, such as the phthalimido group in

RAFT agents (**211, 224, 232**), which can be subsequently deprotected by hydrazinolysis.[450]

9.8 Block Copolymers

Block copolymers are composed of two or more covalently connected segments of differing composition. The simplest case is an AB diblock, which consists of two segments. These may be extended to form ABA or BAB triblocks and further extended to form higher-order $(AB)_n$ multi blocks. Introduction of a third block type creates ABC triblocks. A wide range of block copolymer architectures is possible including radial or star-blocks and graft copolymers with block copolymer arms. These structures are mentioned in the sections devoted to the synthesis of star and graft copolymers (Sections 9.9 and 9.9.3.2 respectively).

Living polymerization processes immediately lend themselves to block copolymer synthesis and the advent of techniques for living radical polymerization has lead to a massive upsurge in the availability of block copolymers. Block copolymer synthesis forms a significant part of most reviews on living polymerization processes. This section focuses on NMP,[106] ATRP,[268,270] and RAFT.[397] Each of these methods has been adapted to block copolymer synthesis and a substantial part of the literature on each technique relates to block synthesis.

Four processes for block copolymer synthesis can be distinguished.

(a) sequential addition of monomers to a living chain end (9.8.1).
(b) batch copolymerization of monomers with disparate reactivity ratios to form a gradient block copolymer (9.6.2).
(c) use of a functional polymer prepared by another process as an initiator (NMP, ATRP) or transfer agent (RAFT) (9.8.2).
(d) joining of pre-prepared blocks in a post-polymerization coupling reaction.

Block copolymers have a wide range of applications from surfactants and dispersants to compatibilizers and thermoplastic elastomers and are found in areas as diverse as biomaterials, drug delivery, nanocomposites and electronics. Many applications depend on the propensity of block copolymers to self assemble into micelles and more complex supramolecular structures.[629] Any detailed discussion of applications is, however, beyond the scope of this book.

Some comment should be made on block copolymer purities. The usual and often the only method of assessment is GPC. For the usual case, where the molecular weight of the block is 2-5 times higher than that of the precursor, baseline resolution between block and precursor will seldom be obtained unless dispersities are very low. A complicating factor is that in a GPC trace signal intensity is proportional to molecular weight squared. This has the effect of emphasizing the block copolymer with respect to any first block impurity. Tailing to low molecular weight is deemphasized. There are additional issues that relate to the composition dependence of the refractive index and the elution behavior. A consequence is that simple inspection of a GPC trace may not be a particularly good indicator of block purity and quantitative assessment is problematical.

9.8.1 Direct Diblock Synthesis

The most direct method for synthesizing block copolymers involves the sequential addition of two monomers in a polymerization reaction. Isolation and purification of the first block may sometimes be desirable. An advantage of living radical methods over classical (anionic) polymerization is that the product of polymerization is a dormant polymer that is usually sufficiently stable that it can be isolated and purified before being used in another polymerization process. This is important since a disadvantage of living radical methods is that it is seldom desirable to operate at very high conversion because, irrespective of method, the likelihood of side reactions is high under these conditions.

9.8.1.1 NMP

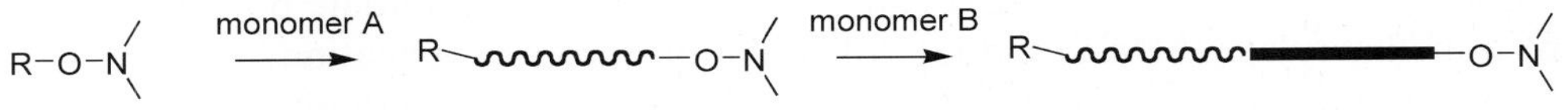

Scheme 9.56

The process for block synthesis by NMP with sequential monomer addition is shown in Scheme 9.56. Block synthesis is generally subject to the same limitations as polymer synthesis. Optimal conditions for NMP depend strongly on the particular monomer(s) and this should be taken into account when designing syntheses of block copolymers. TEMPO and similar nitroxides are most suited to controlling polymerizations of styrenic monomers and a majority of reported block copolymers prepared by NMP with TEMPO and TEMPO derivatives have a first block and often a second block based on S or a S derivative [*e.g.* PS-*block*-PBA[201], PS-*block*-P(S-*co*-BMA),[559] PS-*block*-PBMA,[202] PS-*block*-PB,[203] and PS-*block*-PI[203]]. S derivatives include **103** (protected 4-aminostyrene),[630,631] **104**,[632] **108**,[633] **105**,[634] **106**[634] and **107**.[634] Polymers containing 4-chloromethylstyrene (**108**) and 4-vinylpyridine (**109**) often serve as precursors to other structures.[635] However, with the use of other nitroxides, and lower reaction temperatures, a much wider range of block copolymers is possible including PS-*block*-PtBA with nitroxide **89**.[636]

An issue when making the second (and subsequent) blocks from styrenic monomers is that thermal initiation or an added initiator will provide a homopolymer impurity.

9.8.1.2 ATRP

Although, ATRP appears most suited to polymerization of methacrylate monomers, a very wide range of monomers can and have been used as is illustrated by Table 9.27. A survey of block syntheses by ATRP is provided in the review by Davis and Matyjaszewski.[554] A general reaction scheme for block synthesis by ATRP is shown in Scheme 9.57.

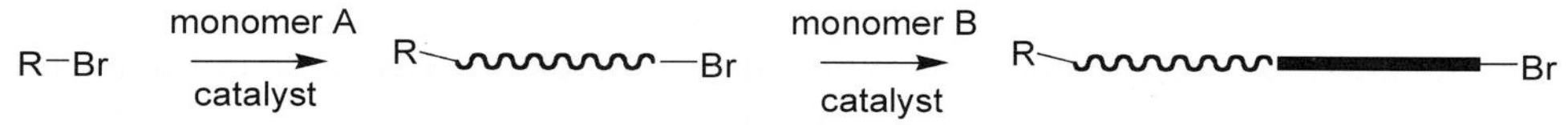

Scheme 9.57

Optimal conditions for ATRP depend strongly on the particular monomer(s) to be polymerized. This is mainly due to the strong dependence of the activation-deactivation equilibrium constant (K), and hence the rate of initiation, on the type of propagating radical (Section 9.4.1.3). When using monomers of different types, polymer isolation and changes in the catalyst are frequently necessary before making the second block

For example, when using an macroinitiator based on a monosubstituted monomer (e.g. PMA-Br, PS-Br) and Cu(I)Br/L catalyst to initiate polymerization of a methacrylate (MMA) the rate of initiation (cross-propagation) is slow with respect to the rate of propagation of the second monomer and reinitiation from the new macroinitiator (PMMA-Br). The result can be a broad or bimodal molecular weight distribution. The process known as halogen exchange can be used to adjust the rate of initiation.[571] This involves use of a Cu(I)Cl/L catalyst such that a less active macroinitiator is formed following propagation (PMMA-Cl) (Scheme 9.58). Several examples of where halogen exchange has been used to prepare low dispersity block copolymers are provided in Table 9.27.

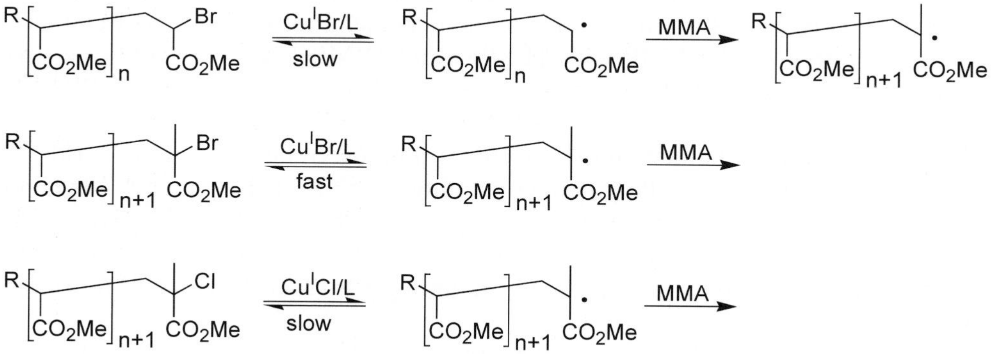

Scheme 9.58

Table 9.27 Diblock Copolymers Prepared by ATRP

Macroinitiator[a]	Monomer 2[a]	Catalyst /Ligand	Solvent	Temp. °C
PMMA-Cl	DMAEMA[637]	CuCl/**144**	o-$C_6H_4Cl_2$	90
PMA-Br	DMAEMA[637]	CuCl/**144**[b]	o-$C_6H_4Cl_2$	90
PMMA-Cl	4VP	CuCl/**145**	2-C_3H_7OH	40
P(SAN)-Br	MMA[575]	CuCl/**132**[b]	butanone	80
P(SAN)-Br	GA[575]	CuBr/**132**	anisole	80
P(SAN)-Br	tBA, BA[575]	CuBr/**140**	acetone	60

a Abbreviations: 4VP 4-vinylpyridine, PSAN P(S-*co*-AN). b Halogen exchange process used.

9.8.1.3 RAFT

The synthesis of block copolymers by macromonomer RAFT polymerization has been discussed in Section 9.5.2 and examples are provide in Table 9.9. RAFT polymerization with thiocarbonylthio compounds has been used to make a wide variety of block copolymers and examples are provided below in Table 9.28. The process of block formation is shown in Scheme 9.59. Of considerable interest is the ability to make hydrophilic-hydrophobic block copolymers directly with monomers such as AA, DMA, NIPAM and DMAEMA. Doubly hydrophilic blocks have also been prepared.[476,638] The big advantage of RAFT polymerization is its tolerance of unprotected functionality.

Table 9.28 Diblock Copolymers Prepared by RAFT Polymerization[a]

Macro-RAFT[b,c]	$\overline{M}_n$	$\overline{M}_w/\overline{M}_n$	Monomer[c]	Solvent	T °C	$\overline{M}_n$	$\overline{M}_w/\overline{M}_n$
S-**199**[408,425]	20300	1.15	SMe	benzene	60	25400	1.19
S-**199**[408,425]	20300	1.15	DMAM	benzene	60	43000	1.24
S-**199**[397]	13200	1.22	MA	bulk	60	53300	1.19
SNHMe$_2$Cl-**174**[639]	6700	1.12	DMAM	water	80	11300	1.12
MMA-**175**[408,425]	17400	1.20	S	bulk	60	35000	1.24
MMA-**255**[411]	6700	1.27	S	bulk	60	25600	1.15
MMA-**175**[408,425]	3200	1.17	MAA	DMF	60	4700	1.18
BzMA-**175**[425]	1800	1.13	DMAEMA	EtAc	60	3500	1.06
MA-**194**[408,425]	24100	1.07	BA	benzene	60	30900	1.10
BA-**194**[408,425]	33600	1.13	AA	DMF	60	52000	1.19
AA-**260**[476]	7900	1.19	NIPAM	CH_3OH	60	13600	-
AMPS-**174**[420]	16100	1.17	AMBA	water	70	24200	1.10
DMAM-**174**[639]	4900	1.17	SNHMe$_2$Cl	water	70	14900	1.17

a For other examples and further details of reaction conditions see references cited. b Monomer - initial RAFT agent used in synthesis of Macro-RAFT agent. c Abbreviations: AMBA 3-acrylamido-3-methylbutanoate, AMPS 2-acrylamido-3-methylbutanoate, SNHMe$_2$Cl N,N-dimethylvinylbenzylammonium chloride, SCO$_2$Na sodium 4-styrenecarboxylate, SSO$_3$Na sodium 4-styrenesulfonate, EtAc ethyl acetate

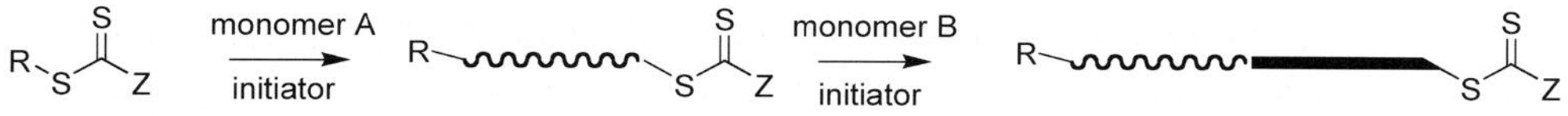

544   The Chemistry of Radical Polymerization

[Scheme 9.59: R-S-C(=S)-Z + monomer A/initiator → R-[~~]-S-C(=S)-Z + monomer B/initiator → R-[~~]-[■■■]-S-C(=S)-Z]

Scheme 9.59

In RAFT polymerization, the order of constructing the blocks of a block copolymer can be very important.[394,425] The propagating radical for the first formed block must be a good homolytic leaving group with respect to that of the second block. For example, in the synthesis of a methacrylate-acrylate or methacrylate-S diblock, the methacrylate block should be prepared first.[425,442] The S or acrylate propagating radicals are poor leaving groups with respect to methacrylate propagating radicals.

The problem of macro-RAFT agents with low transfer constants is mitigated by use of a starved-feed polymerization protocol to maximize the concentration of [RAFT agent]:monomer. It is then important to use a RAFT agent that gives minimal retardation *(e.g.* a dithioacetate or trithiocarbonate rather than a dithiobenzoate).[409] Use of emulsion polymerization conditions is also beneficial. This strategy is also used in block copolymer synthesis when using macromonomer RAFT agents (Section 9.5.2)

9.8.2 Transformation Reactions

Many block and graft copolymer syntheses involving 'transformation reactions' have been described. These involve preparation of polymeric species by a mechanism that leaves a terminal functionality that allows polymerization to be continued by another mechanism. Such processes are discussed in Section 7.6.2 for cases where one of the steps involves conventional radical polymerization. In this section, we consider cases where at least one of the steps involves living radical polymerization. Numerous examples of converting a preformed end-functional polymer to a macroinitiator for NMP or ATRP or a macro-RAFT agent have been reported.[554] The overall process, when it involves RAFT polymerization, is shown in Scheme 9.60.

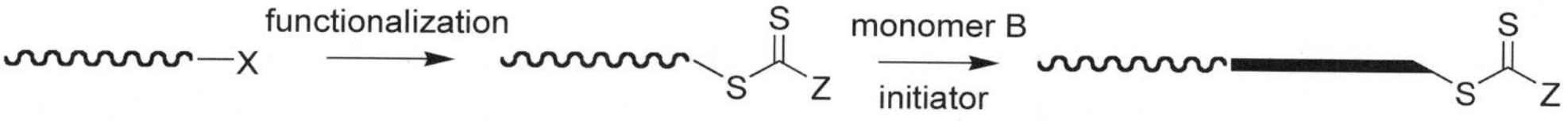

[Scheme 9.60: ~~~-X → (functionalization) → ~~~-S-C(=S)-Z → (monomer B/initiator) → ~~~-■■■-S-C(=S)-Z]

Scheme 9.60

The alternative strategy of using a polymer prepared by one of the living radical methods as a precursor to using another (non-radical) polymerization technique is also frequently encountered. Methods for synthesizing the end-functional polymers used in such experiments are described in Section 9.7. Techniques for the interconversion of halo end-groups (formed by ATRP) alkoxyamine end-groups (formed by NMP) and thiocarbonylthio groups (formed by RAFT polymerization) have also been devised and are also mentioned in Section 9.7. This enables one living radical method to be followed by another so that use can be made of the beneficial features of each method. In all cases triblocks by use of a bis-functional precursor and stars and grafts can be prepared from precursors with a greater number of functional groups (Sections 9.9 and 9.10).

9.8.2.1 Second step NMP

Commercial end functional polymers have been converted to alkoxyamines and used to prepare PEO-*block*-PS.[640] The hydroxyl group of alkoxyamine **284** was used to initiate ring-opening polymerization of caprolactone catalyzed by aluminum tris(isopropoxide) and the product subsequently was used to initiate S polymerization by NMP thus forming polycaprolactone-*block*-PS.[641] The alternate strategy of forming PS by NMP and using the hydroxyl chain end of the product to initiate polymerization of caprolactone was also used.

Kobetaki et al.[589,642] have examined the combination of conventional free radical and NMP to prepare PBMA-*block*-PS and the combination of anionic and NMP to prepare PB-*block*-PS.

Other block copolymers prepared using similar strategies include PEO (anionic) with second block poly(4-vinylpyridine).[643]

4-hydroxyTEMPO (**69**) has been used to initiate polymerization of caprolactone *via* the hydroxy group and the polymeric nitroxide formed used in NMP to give polycaprolactone-*block*-poly(4-vinylpyridine).[644] The polymerization process can be described by Scheme 9.61.

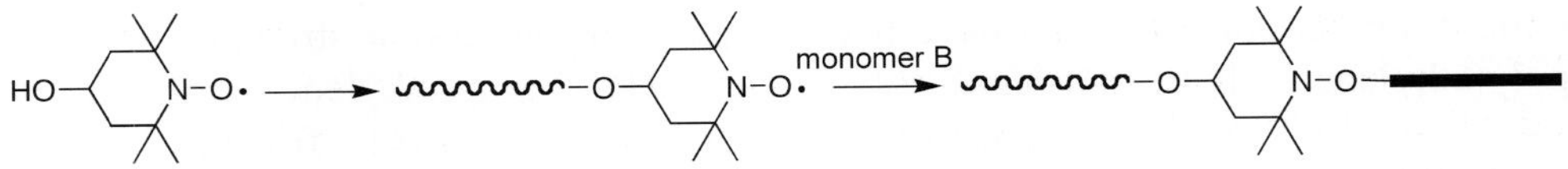

Scheme 9.61

9.8.2.2 Second step ATRP

Many examples exist where a polymerization has been continued by ATRP.[554] Often the procedure involves functionalization of a hydroxy-terminated polymer with bromoisobutyroyl (BriBBr, **314**) or bromoisopropionoyl (BriPBr, **315**) bromide. Examples include poly(ethylene oxide)[645,646] and poly(propylene oxide).[646]

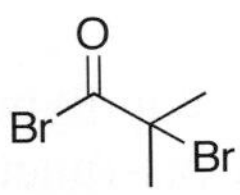

314 BriBBr

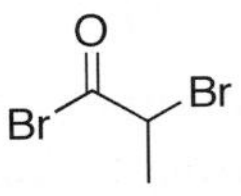

315 BriPBr

Poly(dimethyl siloxane) with vinyl or hydrosilane (Si-H) chain ends have been converted to ATRP initiator ends (*e.g.* Scheme 9.62) by hydrosilylation. Bis-functional dimethyl siloxane polymers prepared in this way were used in polymerizations of S, MA, isobornyl acrylate and BA to form ABA triblock copolymers.

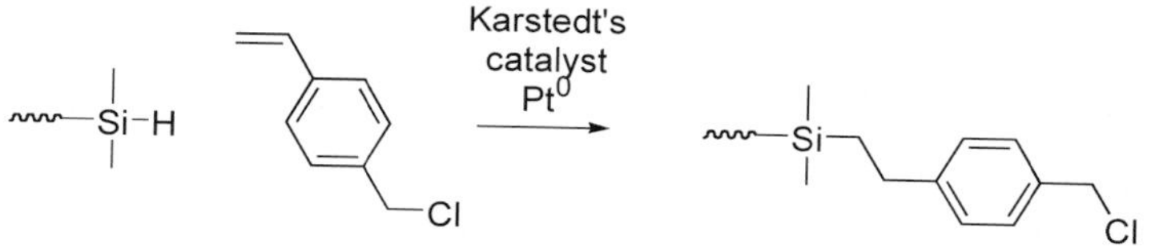

Scheme 9.62

Ring-opening metathesis polymerization (ROMP) of 1,4-cyclooctadiene was used to prepare poly(1,4-B) terminated with halo end groups.[647] This was then used as a macroinitiator of ATRP with heterogeneous Cu bpy catalysts to form PS-*block*-poly(1,4-B)-*block*-PS and PMMA-*block*-poly(1,4-B)-*block*-PMMA.

Polymers prepared with the trichloromethyl-functional initiators[648] or with chloroform or carbon tetrachloride as a transfer agent[649] have been used as macroinitiators for ATRP. The method has been used to make PVAc-*block*-PS.[649,650]

9.8.2.3 Second step RAFT

RAFT polymerization has been used to prepare poly(ethylene oxide)-*block*-PS from commercially available hydroxy end-functional poly(ethylene oxide).[425,449] Other block copolymers that have been prepared using similar strategies include poly(ethylene-*co*-butylene)-*block*-poly(S-*co*-MAH),[551] poly(ethylene oxide)-*block*-poly(MMA),[440] poly(ethylene oxide)-*block*-poly(N-vinyl formamide),[651] poly(ethylene oxide)-*block*-poly(NIPAM),[652] poly(ethylene oxide)-*block*-poly(1,1,2,2-tetrahydroperfluorodecyl acrylate),[653] poly(lactic acid)-*block*-poly(MMA)[440] and poly(lactic acid)-*block*-poly(NIPAM).[458,654]

Low molecular weight or polymeric ATRP initiators have been converted to dithiobenzoate RAFT agents by reaction with phenylethyl dithiobenzoate RAFT agent[442,655] or by reaction with bis(thiobenzoyl) disulfides under ATRP conditions.[483] It is likely that ATRP initiators can be transformed to other forms of RAFT agent by similar methods.

9.8.3 Triblock Copolymers

Triblock copolymers can be prepared from diblock copolymers by a third monomer addition. They can also be prepared using a bis-functional NMP or ATRP initiator or a bis-RAFT agent (for examples, see Table 9.13). Symmetrical trithiocarbonates (Table 9.15) should also be considered as bis-RAFT agents in this context. For NMP and RAFT there are two limiting strategies for triblock synthesis that lead to the dormant group being in the center or at the ends of the triblock copolymer (shown in Scheme 9.63 and Scheme 9.64 respectively for the case of RAFT polymerization). The two methods are then subject to the same limitations as star polymer synthesis (triblocks may be considered as two arm stars) and these are discussed in Section 9.9.

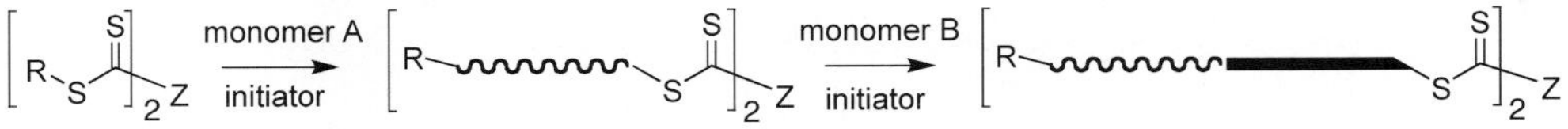

Scheme 9.63

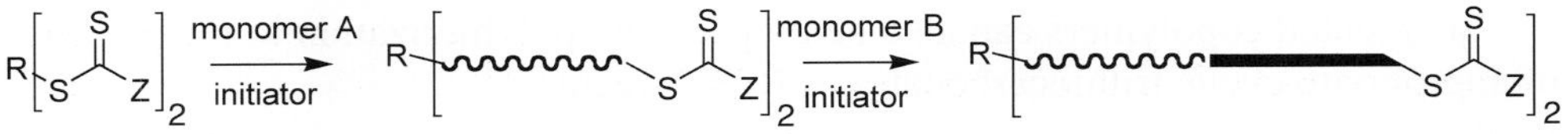

Scheme 9.64

9.8.4 Segmented Block Copolymers

NMP and RAFT polymerization can be used to prepare segmented or multi-block copolymers directly. Polymer with in-chain alkoxyamine functionality such as **316** or **317** can be heated in S to form segmented block copolymers containing PS blocks by NMP.[656,657] Heating a mixture of the polyester (**316**) and polyurethane (**317**) provides a polymer containing novel polyester-urethane units (**318**) by a chain reorganization involving alkoxyamine exchange.[585,586] The exchange process can be followed by NMR.

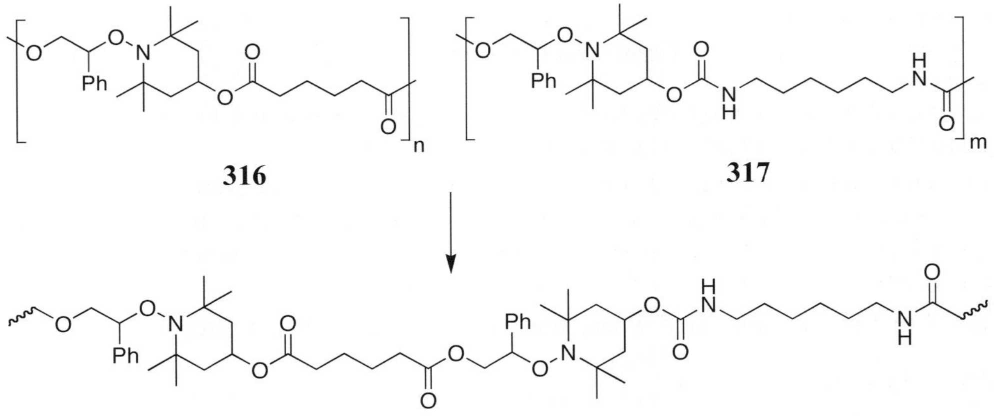

Scheme 9.65

548 The Chemistry of Radical Polymerization

Multi-RAFT agents have also been used to prepare segmented block copolymers.[448,658-660] The molecular weight distributions obtained in these experiments are broad when compared to those obtained using analogous mono- or bis-RAFT agents.

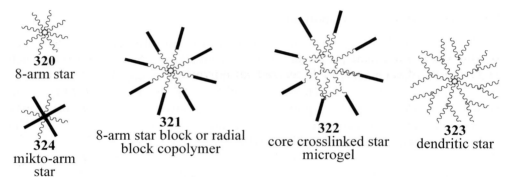

319

Segmented copolymers can also be prepared by polymerization in the presence of appropriate cyclic trithiocarbonates as RAFT agents.[661]

9.9 Star Polymers

320 8-arm star

321 8-arm star block or radial block copolymer

322 core crosslinked star microgel

323 dendritic star

324 mikto-arm star

Figure 9.11 Star Architectures

Several basic approaches to star polymer synthesis leading to architectures as shown in Figure 9.11 will be considered in this section.

(a) The core first approach to star copolymers requires a core containing the appropriate functionality such that the arms can be grown from the core (Section 9.9.1). The number of arms is dictated by the functionality of the core.

(b) The arm-first approach where the arms are grown then aggregated to form a star (*e.g.* **322**, Section 9.9.2).

(c) Self-condensing vinyl polymerization to provide a hyperbranched polymer (Section 9.9.3.1).

(d) The synthesis of dendritic polymers (*e.g.* **323**) by an iterative approach (Section 9.9.3.2).

Living Radical Polymerization

The generic features of these approaches are known from experience in anionic polymerization. However, radical polymerization brings some issues and some advantages. Combinations of strategies (a-d) are also known. Following star formation and with appropriate experimental design to ensure dormant chain end functionality is retained, the arms may be chain extended to give star block copolymers (**321**). In other cases the dormant functionality can be retained in the core in a manner that allows synthesis of mikto-arm stars (**324**).

A comment should be made on the dispersity of star polymers. If the arms each have a 'most probable' distribution ($\bar{M}_w/\bar{M}_n=2$), dispersity of the star polymers is expected to be $\sim 1+1/a$, where a is the number of arms of the star polymer, simply as a consequence of statistical averaging.[662] This explains why polymers formed by conventional radical polymerization with termination by combination (*i.e.* 2 arms) have $\bar{M}_w/\bar{M}_n=1.5$. When we additionally take into account the fact that living polymerizations are capable of producing arms of much lower dispersity, we should anticipate that low dispersities are the norm for multi-armed star polymers.

9.9.1 Core-first Star Synthesis

The possibility of attaching appropriate functionality to a multifunctional core to grow star and dendritic polymers was recognized and evaluated early during the development of each form of living radical polymerization. Thus, Ostu *et al.*[45] used the tetrakis(dithiocarbamate) photoiniferter **19** to form a four armed star (Section 9.3.2.2). A partially soluble product indicating some crosslinking was observed.

Precursors for stars by NMP (Table 9.29), ATRP (Table 9.30) and RAFT (Table 9.31) are shown below. Hawker *et al.*[663] used NMP with **325** to form a three-armed star. Matyjaszewski *et al.*[664] used ATRP with **328** to form a six-arm star. Chen *et al.*[665] formed a six-arm star based on the organometallic RAFT agent **337**. Barner *et al.*[666] prepared crosslinked poly(divinylbenzene) microspheres by precipitation polymerization in the presence of phenylethyl dithiobenzoate (**194**) and used these to form particles with dithiobenzoate terminated PS chains. There exist numerous other examples which make use of NMP, ATRP,[667-672] RAFT and other techniques using a wide range of cores. The core may be organic, inorganic or organometallic, it may be a dendrimer(ATRP,[673,674] RAFT[675-677]), a hyperbranched polymer (RAFT[678]), a (poly)saccharide (ATRP,[670,679,680] RAFT[681,682], a polymer particle (NMP,[683,684] ATRP,[685,686] RAFT[666]), a macromolecular species, or indeed, any moiety possessing multiple thiocarbonylthio groups (though here the distinction between star and graft copolymers becomes blurred; graft copolymers are discussed further in Section 9.10).

Table 9.29 Star Precursors for NMP

Precursor	Structure	Precursor	Structure
325[663]	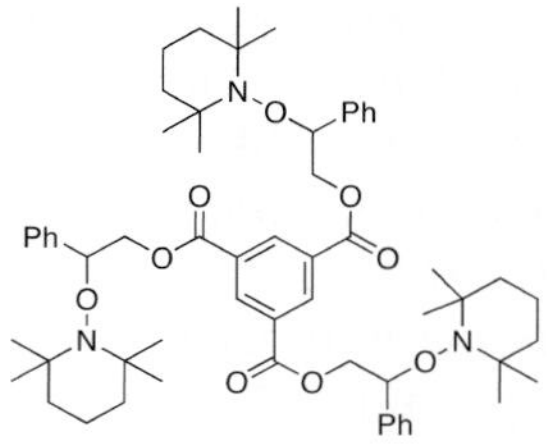	326[591]	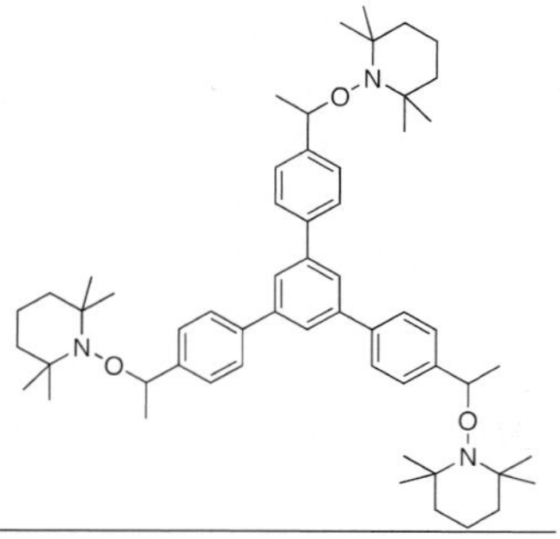

Table 9.30 Star Precursors for ATRP

Precursor	Structure	Precursor	Structure
327[672]		328 [664,668]	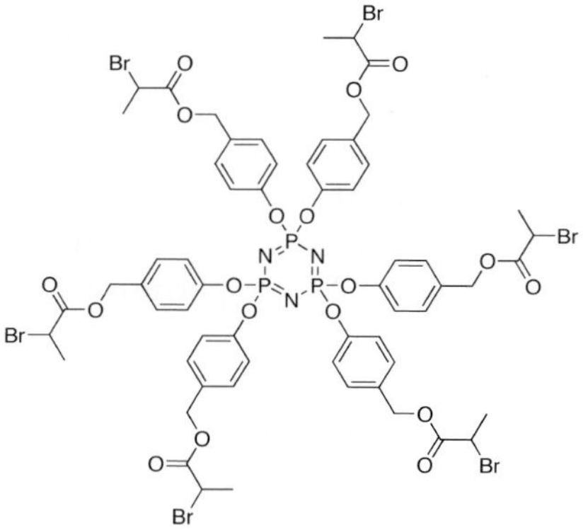
329[687] R=H			
330[600] R=CH$_3$			
331[280]	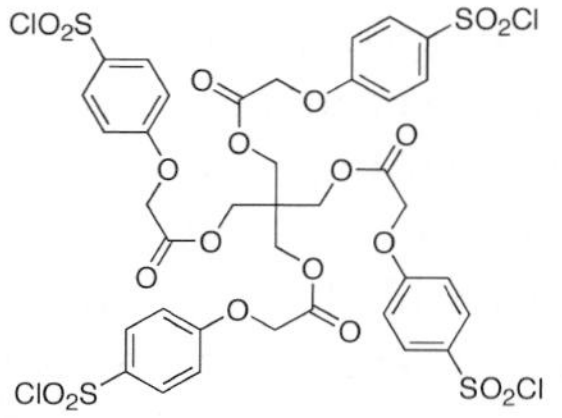	332[688]	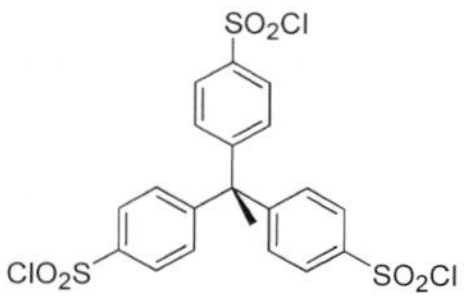

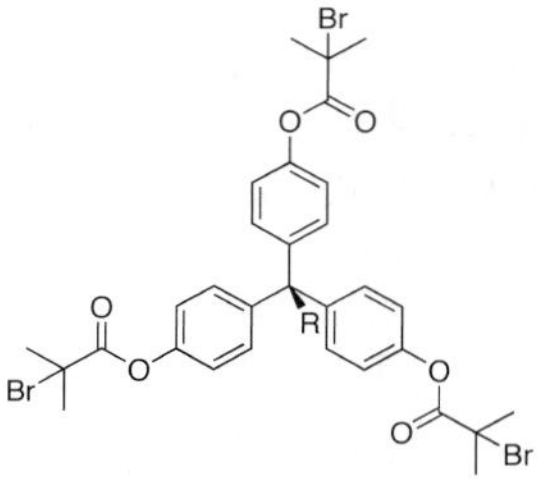

The method of polymerization needs to be chosen for compatibility with functionality in the cores and the monomers to be used. Star block copolymers have also been reported. Multi(bromo-compounds) may be used directly as ATRP initiators or they can be converted to RAFT agents. One of the most common methods of core synthesis involves functionalization of an appropriate polyhydroxy compound.[478,669]

For the case of NMP and RAFT, there exist two basic ways of growing star copolymers (this discussion also applies to block and graft copolymer synthesis).

(a) In the first approach, the polymer chains remain directly attached to the core and chain grow occurs at the periphery (Scheme 9.66). Examples of precursors are **325** (NMP) and **333-338** (RAFT). ATRP star syntheses with halo-compound initiators will always involve this approach.

Table 9.31 Star Precursors for RAFT Polymerization

Agent	Structure	Agent	Structure
333[445]	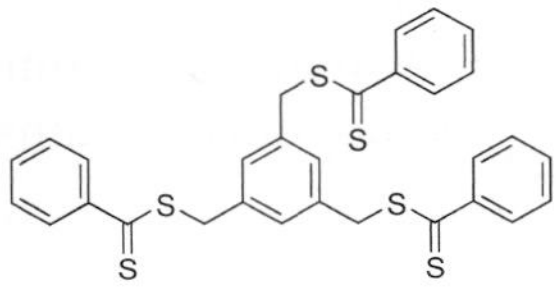	**334**[425]	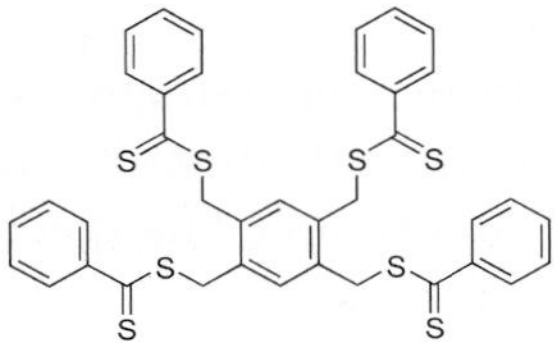
335[425,689]	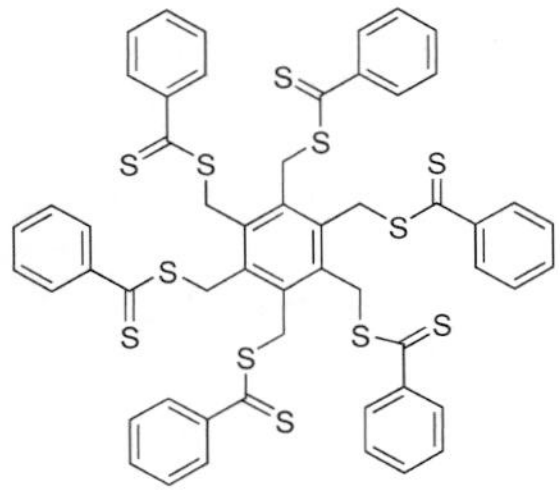	**336**[690]	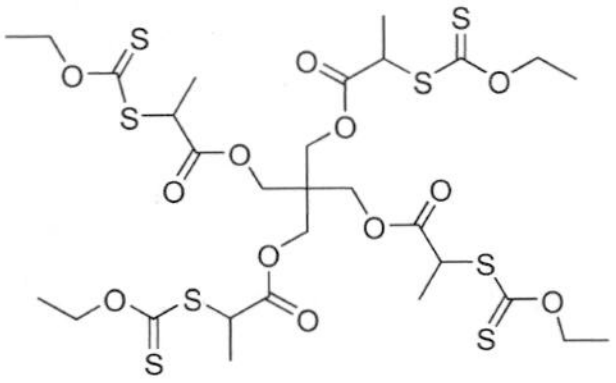
337[665]	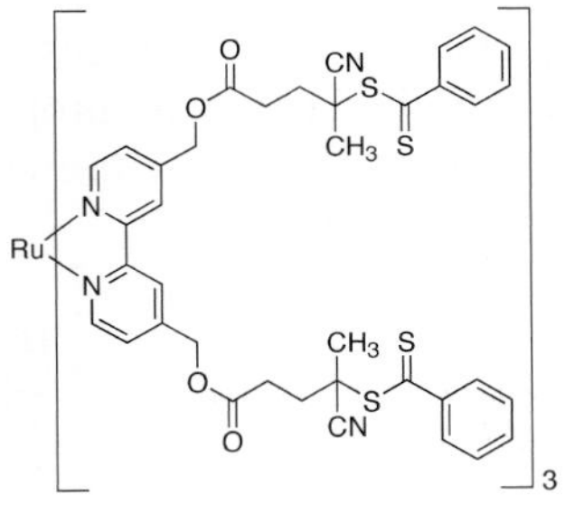	**338**[626]	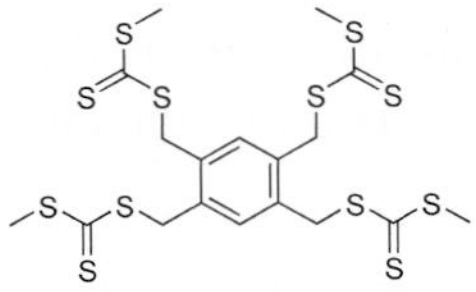
339[626]	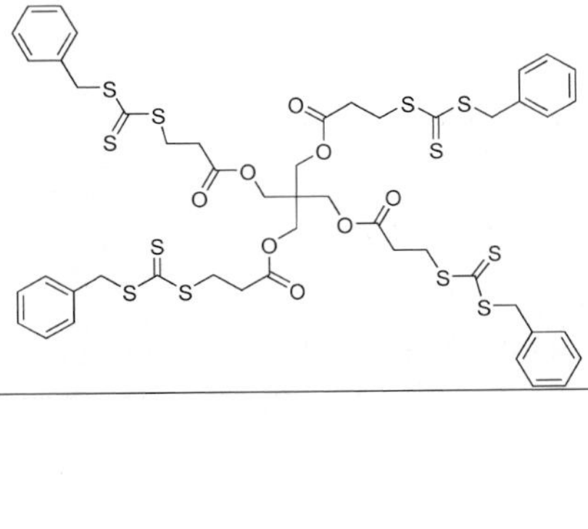		

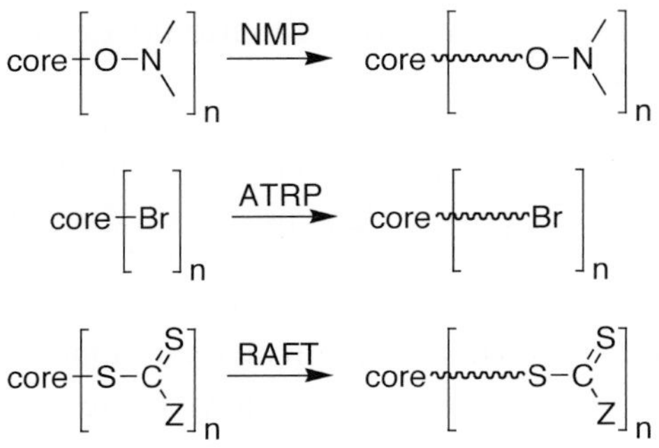

Scheme 9.66

(b) In the second approach, the polymer chains dissociate from the core during each activation-deactivation cycle and grow as linear chains (Scheme 9.67). An example of a precursor is **339** (RAFT).

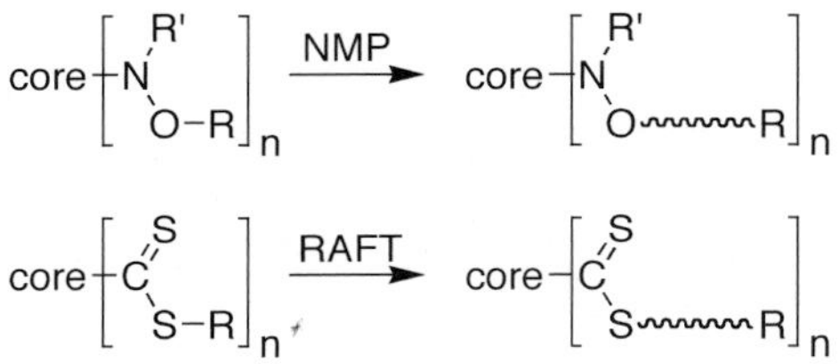

Scheme 9.67

The two strategies for star synthesis each have advantages and limitations. Star-star coupling only occurs with strategy method (a). The propagating radicals remain attached to the core as shown in Scheme 9.68 for the case of a RAFT polymerization and an example is shown in Figure 9.12a.[626]

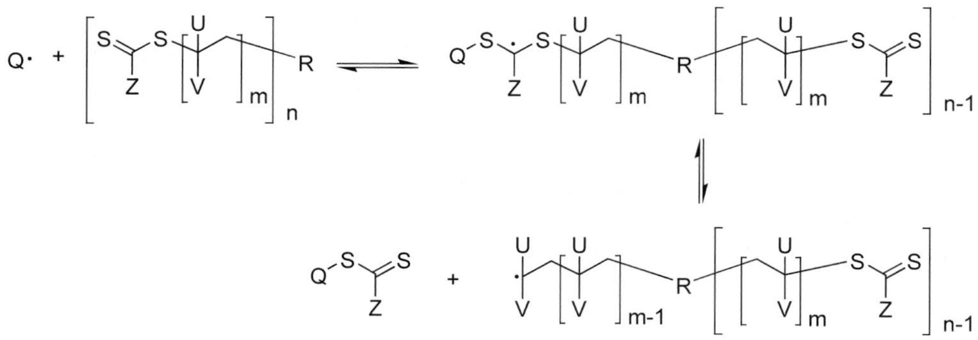

Scheme 9.68 (Q• is an initiator-derived or a propagating radical)

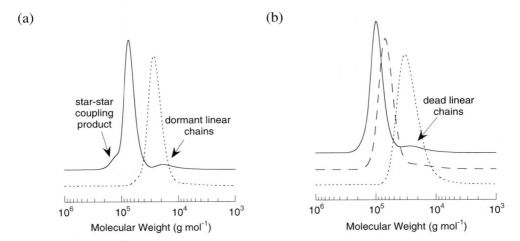

Figure 9.12. GPC distributions obtained during bulk thermal polymerization of S at 110 °C (a) with tetrafunctional RAFT agent **339** (0.0074M) at 6 h, 25% conversion (------), $\overline{M}_n$ =25550, $\overline{M}_w/\overline{M}_n$ =1.2; at 20h, 63% conversion (– – –), $\overline{M}_n$ =63850, $\overline{M}_w/\overline{M}_n$ =1.1; at 64 h, 96% conversion (———), $\overline{M}_n$=92100, $\overline{M}_w/\overline{M}_n$ =1.2) and (b) tetrafunctional RAFT agent **338** (0.0074M) at 6 h, 24% conversion (------), $\overline{M}_n$ =24300, $\overline{M}_w/\overline{M}_n$ =1.1; at 48 h, 96% conversion(———), $\overline{M}_n$ =70700, $\overline{M}_w/\overline{M}_n$ =1.2).[626]

In method (b) the propagating radicals are never attached to the core. Star-star coupling by combination of propagating radicals is not possible. The process is illustrated in Scheme 9.69 for RAFT polymerization and an example is shown in Figure 9.12b.[626] Termination products are from arm-arm reaction and are always of lower molecular weight than the star. A potential disadvantage of strategy (b) is that the products are intrinsically unstable because there is a weak C-ON (NMP) or C-S (RAFT) bond attaching the polymer chains to the core. This may be used to advantage in some applications including polymer-supported synthesis.

It has been suggested that because the RAFT functionality remains at the core with method (b) at higher conversions as the arms grow longer they may shield the RAFT functionality from the propagating radicals and chain growth may be limited (PVAc).[691] Studies by Mayadunne et al.[626] for the case of a 4-armed star based on precursor **339** found excellent agreement between found and calculated arm lengths to high conversion and suggest that this limit is not reached with an arm molecular weight $\overline{M}_n$ 30000 (PMA) or $\overline{M}_n$ 18750 (PS).

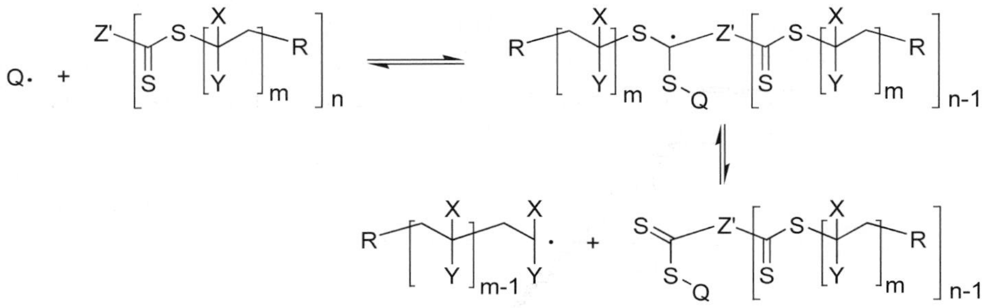

554   The Chemistry of Radical Polymerization

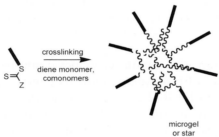

Scheme 9.69 (Q• is an initiator-derived or a propagating radical)

9.9.2 Arm-first Star Synthesis

In the arm-first approach the arms are prepared and then self-assembled to form the core. There are two main variants that will be considered.

(a) *In-situ* microgel formation by polymerization or copolymerization of a non-conjugated diene or a divinyl benzene initiated by an ATRP or a NMP macroinitiator, or carried out in the presence of a macroRAFT agent (Scheme 9.70).

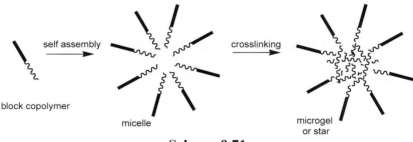

Scheme 9.70

(b) Self-assembly of diblock copolymers to form a micelle that is then crosslinked to form a stable structure. Core-crosslinked micelles (Scheme 9.71) and shell cross-linked micelles and other variants have been reported.

Scheme 9.71

The arm-first synthesis of star microgels by initiating polymerization or copolymerization of a divinyl monomer such as divinylbenzene or a bis-maleimide with a polystyryl alkoxyamine was pioneered by Solomon and coworkers.[692,693] The general approach had previously been used in anionic polymerization. The method has now been exploited in conjunction with NMP,[692-696] ATRP[697-700] and RAFT.[449,701,702] The product contains dormant functionality in the core. This can be used as a core for subsequent polymerization of a monoene monomer to yield a mikto-arm star (NMP,[703] ATRP[704]).

The shell-crosslinking of self assembled micelles based on block copolymers made by NMP or ATRP has been exploited extensively by Wooley and coworkers[705-707] and others[708-710] to make a variety of structures. RAFT has also been used both in this context[449,654] and to make core crosslinked structures.[420,449] A difficulty with this route to microgels is that the crosslinking step must typically be carried out in very dilute solution to avoid network formation and gelation. Armes and coworkers[709,710] have found that this problem is mitigated when crosslinking the central block of micelles formed from ABC triblocks.

9.9.3 Hyperbranched Polymers

Hyperbranched or dendritic polymers have recently attracted significant attention.[711-714] The possibility of generating highly branched soluble polymers by polymerization of AB_x monomers was first brought to the attention of the scientific community by Flory[715] in a theoretical paper. Early work in this field focused on the synthesis of dendrimers by iterative approaches. These methods suffer practical disadvantages in that the syntheses are both labor and purification intensive. As a consequence, other more viable routes to the generation of branched polymers have been sought, leading to the formation of hyperbranched polymers that are polydisperse systems both in terms of molecular weight and branching distribution. These include the self-condensing vinyl polymerization of so-called AB* monomers such as **340-343** which contain monomer and initiator functionality in the one molecule.

9.9.3.1 Self-condensing vinyl polymerization

With appropriate choice of reaction conditions, hyperbranched polymers can be formed by self-condensing vinyl polymerization of monomers that additionally contain the appropriate initiator (NMP, ATRP), when the compounds are called inimers, or RAFT agent functionality. Monomers used in this process include **340**,[716] **341**[717] and **342**[640] (for NMP), **108**[718,719] and **344** and related monomers[720-723] (for ATRP) and **343**[408] (for RAFT). Careful control of reaction conditions is required to avoid network formation.

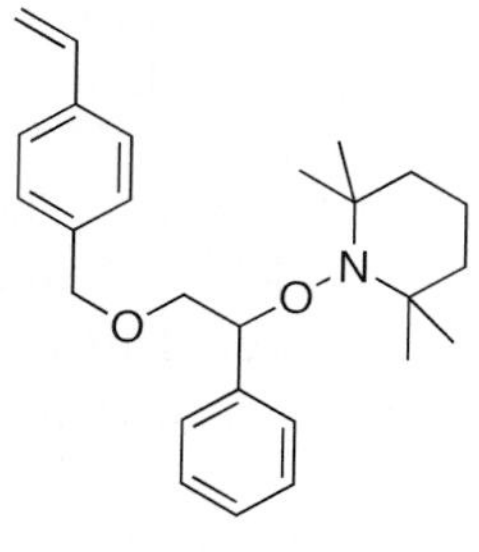

340[716] 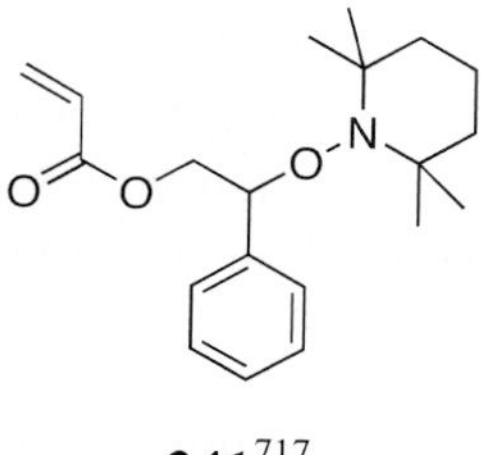 **341**[717] 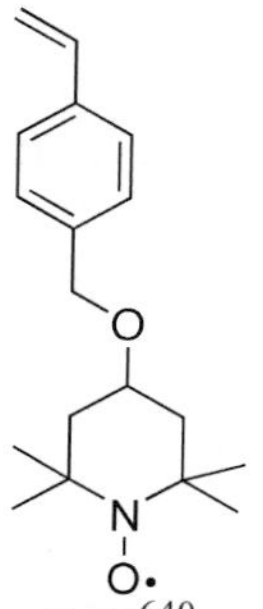 **342**[640] 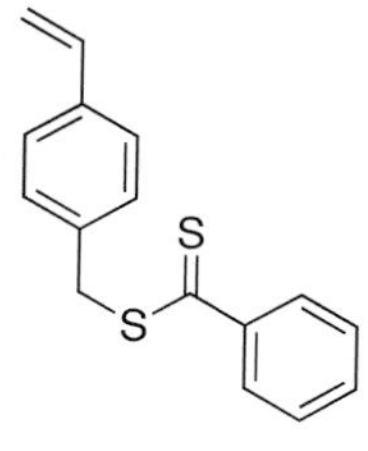 **343**[408]

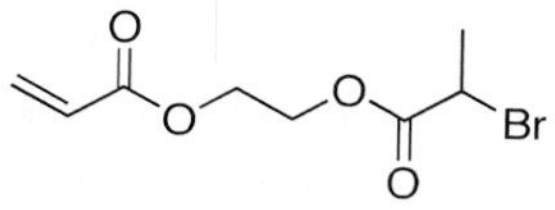

344[720-723]

9.9.3.2 Dendritic polymers

The use of dendritic cores in star polymer synthesis by NMP, ATRP and RAFT polymerization was mentioned in Section 9.9.1. In this section we describe the synthesis of multi-generation dendritic polymers by an iterative approach.

Percec et al.[688,724] developed what they termed the TERMINI approach to dendritic polymers. TERMINI is an acronym for TERminator Multifunctional INItiator. Polymerization of MMA by ATRP initiated by **332** with CuO/bpy catalyst provided a three armed star in the first generation which was multiplied by the TERMINI process with **345** to provide six and twelve arms in the second and third generation respectively. The first few steps of the process are shown in Scheme 9.72. The TERMINI agent **345** is an addition-fragmentation chain transfer agent. The thiocarbamate groups are converted to sulfonyl chloride groups to initiate further ATRP by treatment with chlorine in formic acid.

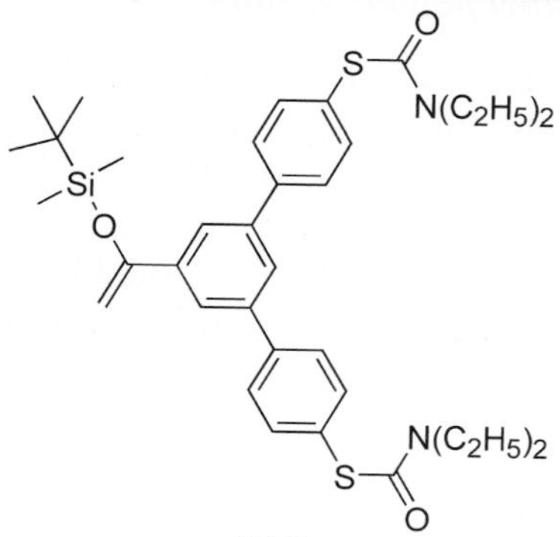

345

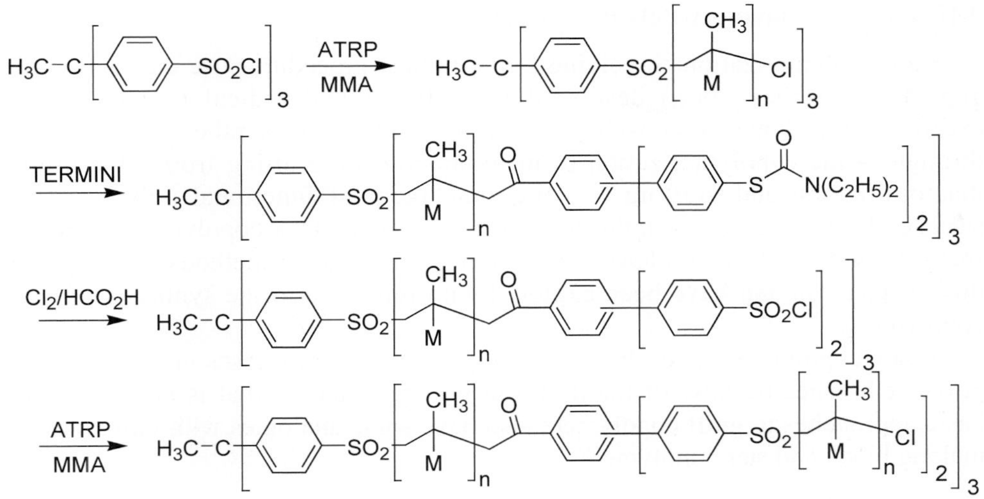

Scheme 9.72 (M = CO$_2$CH$_3$, TERMINI agent is **345**)

The first steps of a second process for divergent synthesis of dendritic polymers by ATRP are shown in Scheme 9.73.[725] In this case, a calixarene core was used.

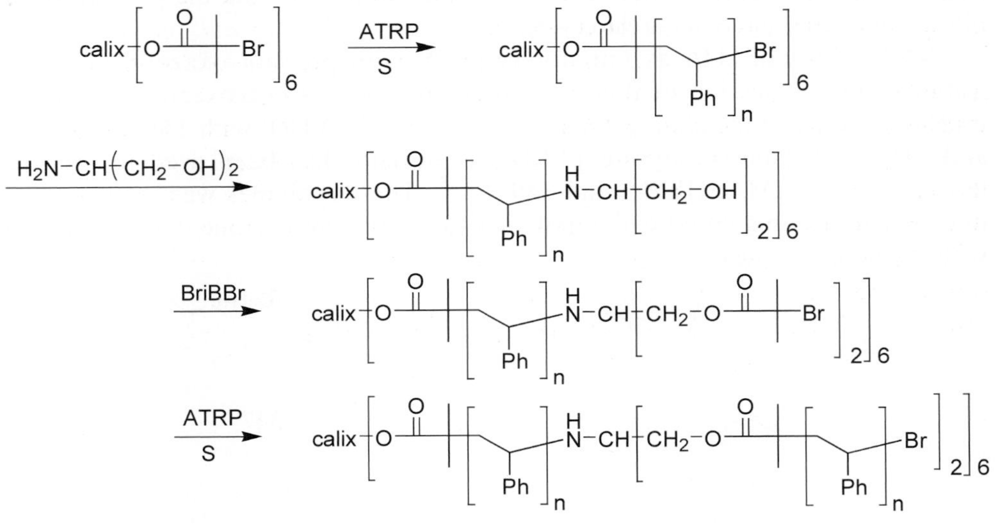

Scheme 9.73 (calix = 4-*t*-butylcalix[6]arene, BriBBr = bromoisobutyroyl bromide (**314**))

9.10 Graft Copolymers/Polymer Brushes

Graft polymerizations involving living radical procedures use the same basic approaches as have been described for conventional radical polymerization (Section 7.6). Thus we consider in turn graft copolymer synthesis by "grafting through" - the copolymerization of macromonomers, grafting from - the use of macroinitiators, and grafting to - the attachment of functional polymers to a surface. In this section, as in the preceding section on block copolymers, there is a focus on NMP, ATRP and RAFT though most of the other methods mentioned in this chapter can and have been explored with reference to the synthesis of graft copolymers.

Graft copolymers made by living polymerization processes are often called polymer brushes because of the uniformity in graft length that is possible. The basic approaches to graft copolymers also have some analogies with those used in making block and star copolymers.

9.10.1 Grafting Through - Copolymerization of Macromonomers

The 'grafting through' approach involves copolymerization of macromonomers. NMP, ATRP and RAFT have each been used in this context. The polymerizations are subject to the same constraints as conventional radical polymerizations that involve macromonomers (Section 7.6.5). However, living radical copolymerization offers greater product uniformity and the possibility of blocks, gradients and other architectures.

NMP of S with **346** as initiator gave PS with pendant oxazoline groups. Cationic polymerization of this macromonomer gave a polyoxazoline with PS grafts that retained the alkoxyamine functionality.[726] ATRP with **156** as catalyst and ethyl 2-bromoisobutyrate (**125a**) as initiator has been used to prepare terpolymers of MMA, HEMA and **349**.[727,728] The terpolymer was then used to initiate ring-opening coordination polymerization of caprolactone or (L,L)-lactide with stannous octoate catalyst.

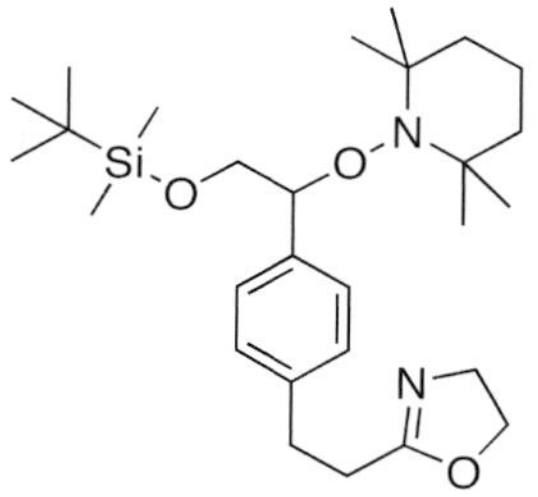

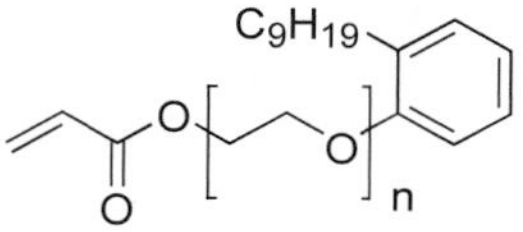

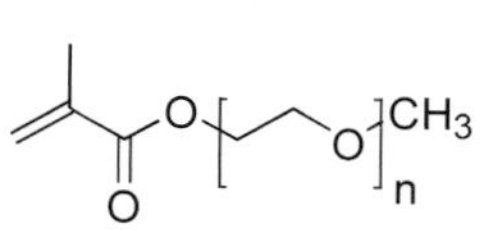

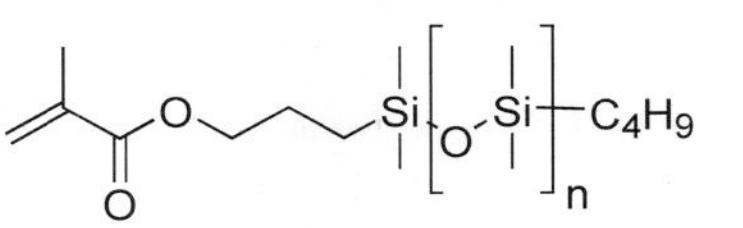

349 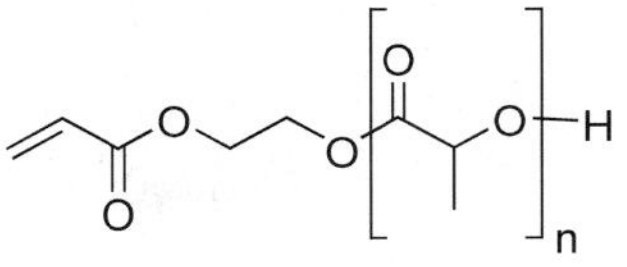 **350**

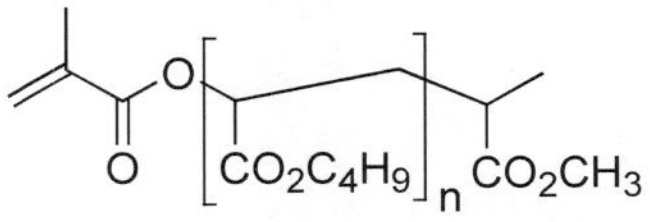

351 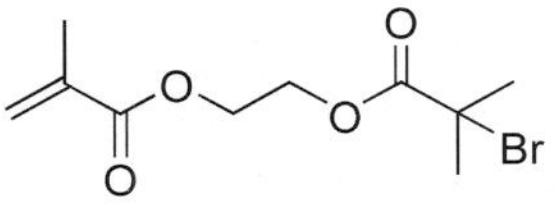 **352**

ATRP has also been used to synthesize macromonomers subsequently used to make graft copolymers by conventional radical polymerization. Thus, low molecular weight PBA formed by ATRP was converted in near quantitative yield to the methacrylate ester (**351**) or the corresponding acrylate ester.[612]

There have been several studies on the use of RAFT to form polymer brushes by polymerization or copolymerization of macromonomers **348-350**.[548,729-735] Systems examined include copolymerizations of **349** with MMA using RAFT agent **175**,[548,731] **348** with AA using **220**[730] and **348** with **352** using **176**.[734] The latter copolymerization created a precursor for a grafting from reaction by ATRP using the bromoisobutyrate group as an initiator.[734]

A number of reviews on dendronized polymers and their synthesis by various methods including radical polymerization of dendron macromonomers have appeared.[736-738] This macromonomer strategy for the synthesis of dendronized polymers is seen to have an advantage over other strategies that involve a post-polymerization reaction to attach dendrons to a polymer chain. The polymerization of dendron macromonomers when designed with a propensity to self assemble can show some living characteristics even when polymerized with a conventional initiator such as AIBN (Section 8.3.7). The steric demand of the dendrons has a large effect on the polymerization kinetics and the rate of polymerization is extremely sensitive to the monomer concentration. The polymerization of dendron macromonomers by ATRP[739,740] (*e.g.* **355-357** with CuBr/**145** catalyst and **124a** as initiator)[741] and RAFT polymerization (**353** and **354** with RAFT agent **176**)[729] has been described.

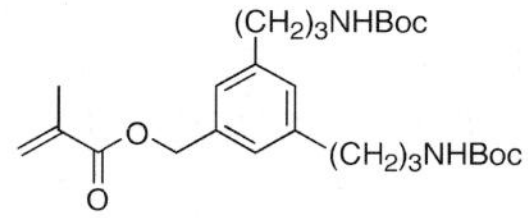

353 (G1)

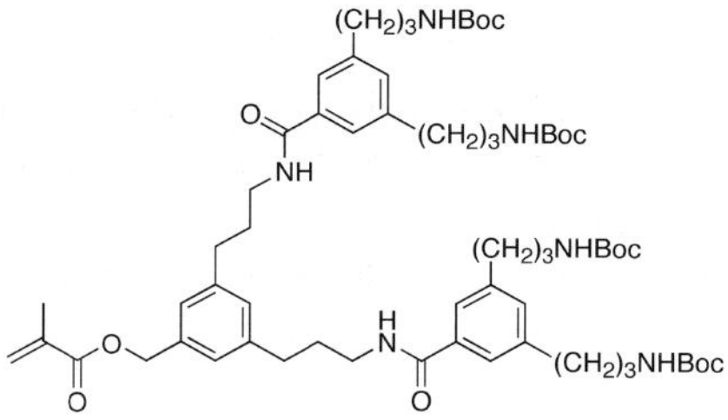

354 (G2)

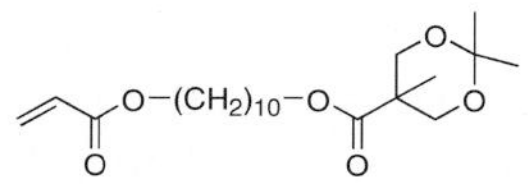

355 (G1)

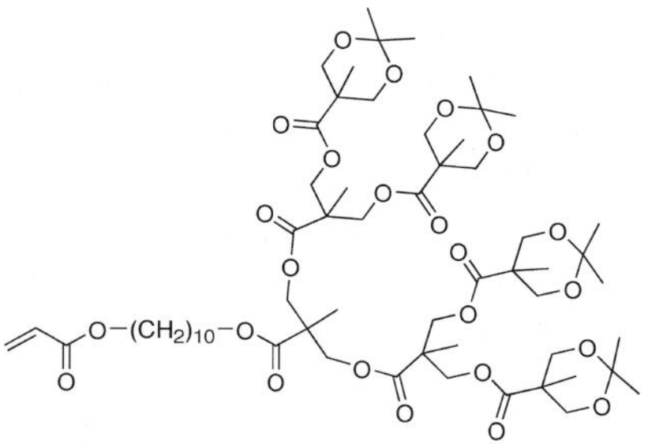

356 (G3)

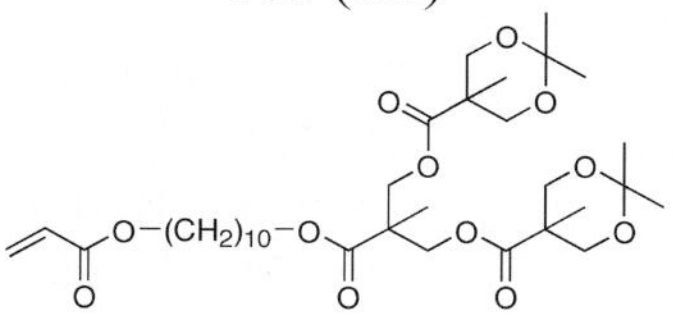

357 (G2)

9.10.2 Grafting From - Surface Initiated Polymerization

The preparation of polymer brushes by controlled radical polymerization from appropriately functionalized polymer chains, surfaces or particles by a grafting from approach has recently attracted a lot of attention.[742,743] The advantages of growing a polymer brush directly on a surface include well-defined grafts, when the polymerization kinetics exhibit living character, and stability due to covalent attachment of the polymer chains to the surface. Most work has used ATRP or NMP, though papers on the use of RAFT polymerization in this context also have begun to appear.

Several routes have been reported for preparation of the required functional polymer/surface. Most methods are analogous to techniques used to form grafts by conventional radical polymerization (Section 7.6). However the living processes allow control over graft length, architecture and composition and a means of avoiding or limiting the concomitant formation of non-grafted polymer.

9.10.2.1 Grafting from polymer surfaces

Several techniques have been applied in attaching the appropriate functionality to the polymer surface. For example, copolymerization of a monomer containing functionality (alkoxyamine *e.g.* **358** or **359**,[744] ATRP initiator, *e.g.* **352**,[734] RAFT agent, *e.g.* **360**) appropriate for producing a copolymer with pendant groups that subsequently can be used to initiate graft copolymerization. The copolymerization is carried out under conditions where said functionality is inert, otherwise the likely product is a hyperbranched polymer or a network as discussed in Section 9.9.3.1. These conditions correspond to the use of low reaction temperatures in the case of alkoxyamines **358** or **359**, the absence of catalyst in the case of ATRP initiator **352,** or a copolymerization with MMA where transfer is negligible in the case of RAFT agent **360**.

The monomer **359** has been formed *in situ* by decomposing the initiator (AIBN) in the presence of the corresponding nitroxide in a solution of S or 2-ethoxyethyl acrylate.[744] The kinetics dictate that alkoxyamine formation, by coupling of the nitroxide with cyanoisopropyl radicals, will take place before copolymerization.

358[111]   **359**[744]   **360**

A second approach is to graft functionality onto a pre-existing polymer by generating radicals on the polymer surface in the presence of a nitroxide or a RAFT agent. Radicals may be formed on the surface by a number of methods including abstraction by radicals generated from a peroxide, decomposition of initiator groups on the surface, γ-irradiation or exposure to a plasma field. Alkoxyamine functionality can be attached to pre-existing polymers by generating *t*-butoxy radicals in the presence of a polymer and a nitroxide. This procedure was applied to PB and poly(isobutyl methacrylate) and the resultant polymeric alkoxyamines formed were used to initiate MA and EA polymerization respectively.[111] Recent papers describe RAFT polymerization from plasma-treated Teflon surfaces[504] and ozonolyzed polyimide films.[745]

A third technique involves reaction of a functional alkoxyamine, ATRP initiator or RAFT agent with a functional surface. An early example involved the reaction of the sodium salt of **284** with poly(S-*co*-chloromethylstyrene) to provide multi-alkoxyamines which were used for the synthesis of a variety of graft copolymers.[663] Interchain coupling reactions, evidenced by broadening of the molecular weight distribution, became significant when there were greater than six alkoxyamine functions per chain.[663] The hydroxyl functionalities of ethylene-vinyl alcohol films[746] were esterified with BriBBr (**314**). Acid functionality of ethylene-acrylic acid copolymer films was transformed to hydroxyl functionality and then esterified with BriBBr.[747] Perrier and coworkers[440,748] attached RAFT moieties to cellulose (cotton) in order to form PS, PMA or PMMA grafts. This involved derivatization of the cellulosic OH groups with thiocarbonylthio functionality.

The very small number of growing polymer chains, when compared to the monomer concentration results in a very low overall concentration of free control agent and leads to inefficient capping of chain ends. One solution to this problem is the addition of a free or unbound control agent to the polymerization medium. This can take the form of a low molecular weight alkoxyamine, ATRP initiator, RAFT agent or, alternatively, free deactivator such as nitroxide or Cu(II). This species is often called a sacrificial agent. This solution also leads to the formation of free polymer that must ultimately be removed from the brush.

9.10.2.2 Grafting from inorganic surfaces

Grafting from silica particles, silicon wafers, and related surfaces usually involves attaching a chlorosilane or alkoxysilane derivative. Thus alkoxyamines (*e.g*, **361**,[744,749] **362**[750]) and a wide variety of ATRP initiators (*e.g.* **363**[751]) have been attached directly to surfaces and used to initiate "grafting from" processes.

Wu *et al*.[752] developed a technique called MAPA (acronym for Mechanically Assisted Polymer Assembly) to produce polymer brushes on a cross-linked polydimethylsiloxane (PDMS) surface. The technique involves stretching the PDMS substrate, then generating surface silanol groups by ozonolysis. The functional surface was then treated with a trichlorosilane-based ATRP initiator. PAM brushes were then grown from the surface by ATRP. The strain was then released, allowing the PDMS substrate to return to its former size thereby producing densely grafted polymer brushes. By altering the amount by which the PDMS substrate was stretched the grafting density could be controlled.

The application of RAFT polymerization in grafting to surfaces was first investigated by Tsujii *et al*.[655] and Brittain and coworkers.[753,754] The approach used in these and other more recent studies[755,756] was to immobilize the initiator functionality on the surface (*e.g.* an ATRP initiator **364**[655] or a conventional initiator **365**[753,754]) and use this to initiate polymerization in the presence of a dithioester RAFT agent. Tsujii *et al*.[655] reported that some difficulties arise in using RAFT for grafting from particles which they attributed to an abnormally high rate of radical-radical termination caused by the locally high concentration of the RAFT functionality.

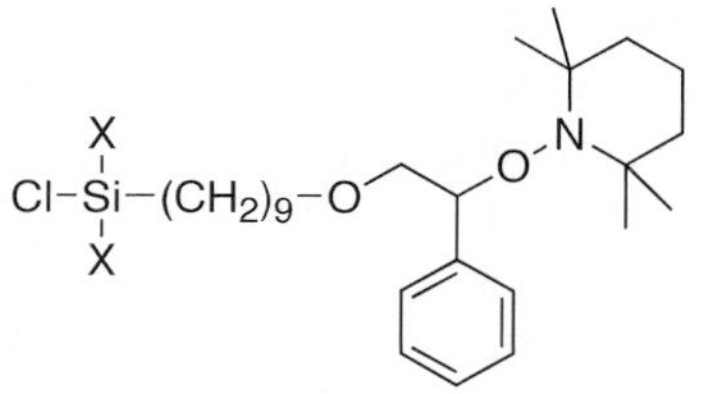

361a X=CH$_3$
b X=Cl

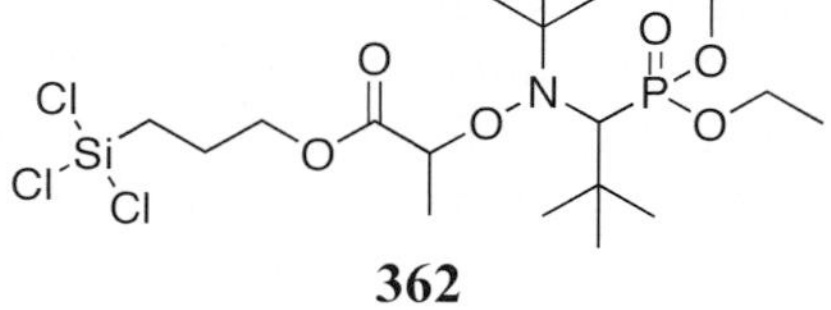

362

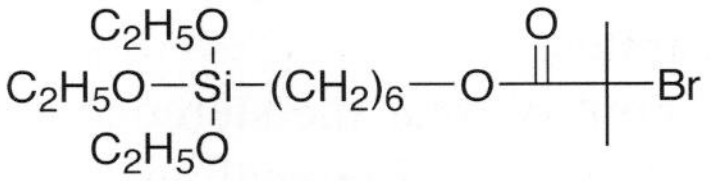

363 — C₂H₅O–Si(OC₂H₅)₂–(CH₂)₆–O–C(=O)–C(CH₃)₂–Br

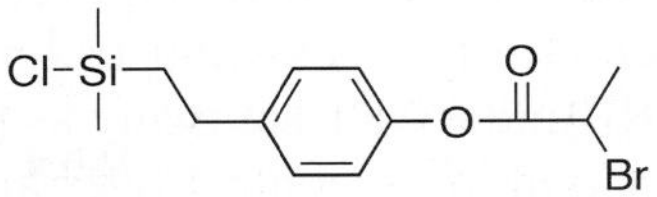

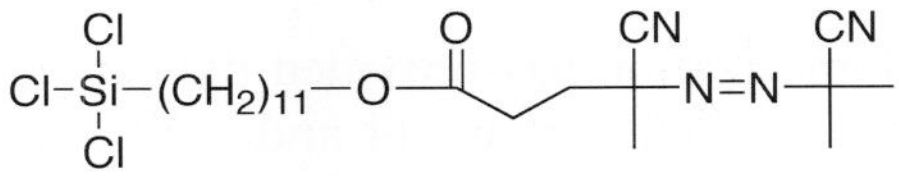

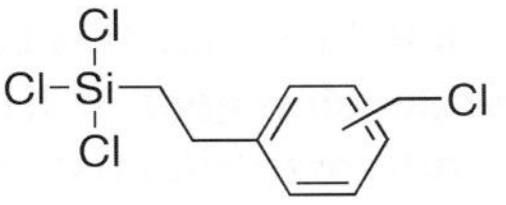

364 — (Cl)₃Si–CH₂CH₂–C₆H₄–O–C(=O)–CH(CH₃)–Br

365 — Cl₂(Cl)Si–(CH₂)₁₁–O–C(=O)–C(CH₃)₂–N=N–C(CH₃)₂–CN (with two CN groups)

366 — Cl₃Si–CH₂CH₂–C₆H₄–CH₂–Cl

Treatment of a gold surface with disulfide **367** left bromoisobutyrate groups on the surface to initiate ATRP of various methacrylate esters[757] including HEMA.[758,759] Skaff and Emrick[760] bound RAFT agent functionality to cadmium selenide nanoparticles by a ligand exchange process and grew various narrow MWD polymers (PS, PMA, PBA, PS-*co*-MA, PS-*co*-AA, PS-*co*-IP, PS-*block*-PMA, PS-*block*-PBA) from these particles.

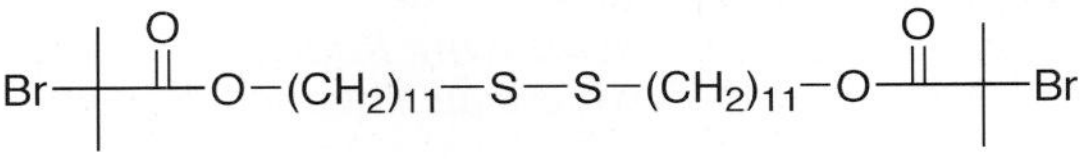

367 — Br–C(CH₃)₂–C(=O)–O–(CH₂)₁₁–S–S–(CH₂)₁₁–O–C(=O)–C(CH₃)₂–Br

9.10.3 Grafting To - Use of End-Functional Polymers

The synthesis of end functional polymers by NMP, ATRP and RAFT has already been discussed in Section 9.7. The "grafting to" approach involves the covalent attachment of an end-functionalized polymer with reactive surface groups on the substrate. The approach is inherently limited by the crowding of chains at the surface and the limit this places on the final graft density.

RAFT polymerization lends itself to the synthesis of polymers with thiol end groups. Several groups have utilized the property of thiols and dithioesters to bind heavy metals such as gold or cadmium in preparing brushes based on gold film or nanoparticles[628,761,762] and cadmium selenide nanoparticles.[763,764]

9.11 Outlook for Living Radical Polymerization

Living radical polymerization currently dominates patents, publications and conferences on radical polymerization. The most popular systems, NMP, ATRP and RAFT, while offering unprecedented versatility are not without drawbacks and still have some limitations. Thus, while the progress in this field since the first edition of this book is substantial by any standard, there remains significant scope for new and improved processes. Further studies of the detailed kinetics and mechanism are also required to enable better understanding so that the full potential of the existing techniques can be realized. The complexities of NMP, ATRP and RAFT are many as this chapter illustrates.

Combining control over architecture with control over the stereochemistry of the propagation process remains a *holy grail* in the field of radical polymerization. Approaches to this end based on conventional polymerization were described in Chapter 8. The development of living polymerization processes has yet to substantially advance this cause.

The development of living radical polymerization has provided the capability for the polymer chemist to synthesize a wide range of novel and well-defined structures. The transformation of this capability into commercial outcomes and novel products has only just commenced.

9.12 References

1. Szwarc, M. *Nature* **1956**, *178*, 1168.
2. Szwarc, M. *J. Polym. Sci., Part A: Polym. Chem.* **1998**, *36*, ix.
3. Moad, G.; Solomon, D.H. *The Chemistry of Free Radical Polymerization*; Pergamon: Oxford, 1995.
4. Matyjaszewski, K., Ed. *ACS Symposium Series, Controlled Radical Polymerization*; American Chemical Society: Washington, 1998; Vol. 685.
5. Matyjaszewski, K., Ed. *ACS Symposium Series, Controlled/Living Radical Polymerization: Progress in ATRP, NMP and RAFT*; American Chemical Society: Washington DC, 2000; Vol. 768.
6. Matyjaszewski, K., Ed. *ACS Symposium Series, Advances in Controlled/Living Radical Polymerization*; American Chemical Society: Washington DC, 2003; Vol. 854.
7. German, A.L., Ed. *Macromol. Symp., Free Radical Polymerization: Kinetics and Mechanism*; Wiley-VCH: Weinheim, 1996; Vol. 111.
8. Buback, M.; German, A.L., Eds. *Macromol. Symp., Free Radical Polymerization: Kinetics and Mechanism*; Wiley-VCH: Weinheim, 2002; Vol. 182.
9. Matyjaszewski, K. In *Handbook of Radical Polymerization*; Davis, T.P.; Matyjaszewski, K., Eds.; John Wiley & Sons: Hoboken, 2002; p 361.
10. Davis, T.P.; Matyjaszewski, K., Eds. *Handbook of Radical Polymerization*; John Wiley & Sons: Hoboken, 2002.
11. Quirk, R.P.; Lee, B. *Polym. Int.* **1992**, *27*, 359.
12. Matyjaszewski, K.; Mueller, A.H.E. *Polym. Prepr. (Am. Chem. Soc., Div. Polym. Chem.)* **1997**, *38(1)*, 6.
13. Ivan, B. *Polym. Prepr. (Am. Chem. Soc., Div. Polym. Chem.)* **2000**, *41(2)*, 6a.
14. Darling, T.R.; Davis, T.P.; Fryd, M.; Gridnev, A.A.; Haddleton, D.M.; Ittel, S.D.; Matheson, R.R., Jr.; Moad, G.; Rizzardo, E. *J. Polym. Sci., Part A: Polym. Chem.* **2000**, *38*, 1706.
15. Darling, T.R.; Davis, T.P.; Fryd, M.; Gridnev, A.A.; Haddleton, D.M.; Ittel, S.D.; Matheson, R.R., Jr.; Moad, G.; Rizzardo, E. *J. Polym. Sci., Part A: Polym. Chem.* **2000**, *38*, 1709.
16. Penczek, S. *J. Polym. Sci., Part A: Polym. Chem.* **2002**, *40*, 1665.
17. Matyjaszewski, K. *Macromolecules* **1993**, *26*, 1787.
18. Matyjaszewski, K. *J. Phys. Org. Chem.* **1995**, *8*, 197.
19. Penczek, S. *Polimery* **1995**, *40*, 384.
20. Russell, G.T. *Aust. J. Chem.* **2002**, *55*, 399.
21. Fischer, H. *Chem. Rev.* **2001**, *101*, 3581.

22. Fukuda, T.; Goto, A.; Tsujii, M. In *Handbook of Radical Polymerization*; Davis, T.P.; Matyjaszewski, K., Eds.; John Wiley & Sons: Hoboken, 2002; p 407.
23. Goto, A.; Fukuda, T. *Prog. Polym. Sci.* **2004**, *29*, 329.
24. Otsu, T.; Yoshida, M. *Makromol. Chem., Rapid Commun.* **1982**, *3*, 127.
25. Kennedy, J.P. *J. Macromol. Sci., Chem.* **1979**, *A13*, 695.
26. Ameduri, B.; Boutevin, B.; Gramain, P. *Adv. Polym. Sci.* **1997**, *127*, 87.
27. Sebenik, A. *Prog. Polym. Sci.* **1998**, *23*, 875.
28. Otsu, T.; Matsumoto, A. *Adv. Polym. Sci.* **1998**, *136*, 75.
29. Ferington, T.E.; Tobolsky, A.V. *J. Am. Chem. Soc.* **1955**, *77*, 4510.
30. Otsu, T.; Kuriyama, A. *J. Macromol. Sci., Chem.* **1984**, *A21*, 961.
31. Shefer, A.; Grodzinsky, A.J.; Prime, K.L.; Busnel, J.P. *Macromolecules* **1993**, *26*, 2240.
32. Otsu, T.; Yoshida, M. *Polym. Bull.* **1982**, *7*, 197.
33. Turner, S.R.; Blevina, R.W. *Macromolecules* **1990**, *23*, 1856.
34. Niwa, M.; Matsumoto, T.; Izumi, H. *J. Macromol. Sci., Chem.* **1987**, *A24*, 567.
35. Nair, C.P.R.; Clouet, G.; Chaumont, P. *J. Polym. Sci., Polym. Chem. Ed.* **1982**, *27*, 1795.
36. Staudner, E.; Kysela, G.; Beniska, J.; Mikolaj, D. *Eur. Polym. J.* **1978**, *14*, 1067.
37. Nair, C.P.R.; Clouet, G. *J. Macromol. Sci., Rev. Macromol. Chem. Phys.* **1991**, *C31*, 311.
38. Nair, C.P.R.; Chaumont, P.; Clouet, G. *J. Macromol. Sci., Chem* **1990**, *A27*, 791.
39. Endo, T.; Shiroi, T.; Murata, K. *J. Polym. Sci., Part A: Polym. Chem.* **2001**, *39*, 145.
40. Bricklebank, N.; Pryke, A. *J. Chem. Soc., Perkin Trans. 1* **2002**, 2048.
41. Otsu, T.; Kuriyama, A. *Polym. J.* **1985**, *17*, 97.
42. Otsu, T.; Matsunaga, T.; Kuriyama, A.; Yoshioka, M. *Eur. Polym. J.* **1989**, *25*, 643.
43. Doi, T.; Matsumoto, A.; Otsu, T. *J. Polym. Sci., Part A: Polym. Chem.* **1994**, *32*, 2911.
44. Otsu, T.; Kuriyama, A. *Polym. Bull.* **1984**, *11*, 135.
45. Kuriyama, A.; Otsu, T. *Polym. J.* **1984**, *16*, 511.
46. Otsu, T.; Yamashita, K.; Tsuda, K. *Macromolecules* **1986**, *19*, 287.
47. Yamashita, K.; Ito, K.; Tsuboi, H.; Takahama, S.; Tsuda, K.; Otsu, T. *J. Appl. Polym. Sci.* **1990**, *40*, 1445.
48. Yamashita, K.; Kanamori, T.; Tsuda, K. *J. Macromol. Sci., Chem.* **1990**, *A27*, 897.
49. Ishizu, K.; Ohta, Y.; Kawauchi, S. *Macromolecules* **2002**, *35*, 3781.
50. Ishizu, K.; Shibuya, T.; Kawauchi, S. *Macromolecules* **2003**, *36*, 3505.
51. Ajayaghosh, A.; Francis, R. *J. Am. Chem. Soc.* **1999**, *121*, 6599.
52. Ishizu, K.; Katsuhara, H.; Itoya, K. *J. Polym. Sci., Part A: Polym. Chem.* **2005**, *43*, 230.
53. Ishizu, K.; Katsuhara, H.; Kawauchi, S.; Furo, M. *J. Appl. Polym. Sci.* **2005**, *95*, 413.
54. Ishizu, K.; Khan, R.A.; Ohta, Y.; Furo, M. *J. Polym. Sci., Part A: Polym. Chem.* **2004**, *42*, 76.
55. Ishizu, K.; Khan, R.A.; Furukawa, T.; Furo, M. *J. Appl. Polym. Sci.* **2004**, *91*, 3233.
56. Nakayama, Y.; Matsuda, T. *Macromolecules* **1996**, *29*, 8622.
57. Nakayama, Y.; Matsuda, T. *Langmuir* **1999**, *15*, 5560.
58. Higashi, J.; Nakayama, Y.; Marchant, R.E.; Matsuda, T. *Langmuir* **1999**, *15*, 2080.
59. Luo, N.; Hutchison, J.B.; Anseth, K.S.; Bowman, C.N. *J. Polym. Sci., Part A: Polym. Chem.* **2002**, *40*, 1885.
60. Francis, R.; Ajayaghosh, A. *Polymer* **1995**, *36*, 1091.
61. Ajayaghosh, A.; Francis, R. *Macromolecules* **1998**, *31*, 1436.
62. Francis, R.; Ajayaghosh, A. *Macromolecules* **2000**, *33*, 4699.
63. Lambrinos, P.; Tardi, M.; Polton, A.; Sigwalt, P. *Eur. Polym. J.* **1990**, *26*, 1125.

64. Doi, T.; Matsumoto, A.; Otsu, T. *J. Polym. Sci., Part A: Polym. Chem.* **1994**, *32*, 2241.
65. Kwon, T.S.; Kumazawa, S.; Yokoi, T.; Kondo, S.; Kunisada, H.; Yuki, Y. *J. Macromol. Sci., Chem.* **1997**, *A34*, 1553.
66. Kwon, T.S.; Suzuki, K.; Takagi, K.; Kunisada, H.; Yuki, Y. *J. Macromol. Sci., Chem.* **2001**, *38*, 591.
67. Kwon, T.S.; Ochiai, H.; Kondo, S.; Takagi, K.; Kunisada, H.; Yuki, Y. *Polym. J.* **1999**, *31*, 411.
68. Kwon, T.S.; Kondo, S.; Kunisada, H.; Yuki, Y. *Polym. J.* **1998**, *30*, 559.
69. Kwon, T.S.; Takagi, K.; Kunisada, H.; Yuki, Y. *J. Macromol. Sci., Chem.* **2000**, *37*, 1461.
70. Kwon, T.S.; Kumazawa, S.; Kondo, S.; Takagi, K.; Kunisada, H.; Yuki, Y. *J. Macromol. Sci., Chem.* **1998**, *A35*, 1895.
71. Kwon, T.S.; Takagi, K.; Kunisada, H.; Yuki, Y. *J. Macromol. Sci., Chem.* **2001**, *A38*, 605.
72. Kwon, T.S.; Takagi, K.; Kunisada, H.; Yuki, Y. *Eur. Polym. J.* **2003**, *39*, 1437.
73. Kwon, T.S.; Takagi, K.; Kunisada, H.; Yuki, Y. *J. Macromol. Sci., Chem.* **2002**, *A39*, 991.
74. Patwa, A.N.; Tomer, N.S.; Singh, R.P. *J. Mater. Sci.* **2004**, *39*, 1047.
75. Rathore, K.; Reddy, K.R.; Tomer, N.S.; Desai, S.M.; Singh, R.P. *J. Appl. Polym. Sci.* **2004**, *93*, 348.
76. Kwon, T.S.; Kondo, S.; Kunisada, H.; Yuki, Y. *Eur. Polym. J.* **1999**, *35*, 727.
77. Bledzki, A.; Braun, D.; Titzschkau, K. *Makromol. Chem.* **1983**, *184*, 745.
78. Bledzki, A.; Braun, D. *Makromol. Chem.* **1981**, *182*, 1047.
79. Crivello, J.V.; Lee, J.L.; Conlon, D.A. In *Advances in Elastomers and Rubber Elasticity*; Lal, J., Ed.; Plenum: New York, 1986; p 157.
80. Crivello, J.V.; Lee, J.L.; Conlon, D.A. *J. Polym. Sci., Part A: Polym. Chem.* **1986**, *24*, 1251.
81. Crivello, J.V.; Lee, J.L.; Conlon, D.A. *Polym. Bull.* **1986**, *16*, 95.
82. Crivello, J.V.; Lee, J.L.; Conlon, D.A. *J. Polym. Sci., Part A: Polym. Chem.* **1986**, *24*, 1197.
83. Santos, R.G.; Chaumont, P.R.; Herz, J.E.; Beinert, G.J. *Eur. Polym. J.* **1994**, *30*, 851.
84. Roussel, J.; Boutevin, B. *Polym. Int.* **2001**, *50*, 1029.
85. Roussel, J.; Boutevin, B. *J. Fluor. Chem.* **2001**, *108*, 37.
86. Braun, D.; SteinhauerBeisser, S. *Eur. Polym. J.* **1997**, *33*, 7.
87. Braun, D.; SteinhauerBeisser, S. *Angew. Makromol. Chem.* **1996**, *239*, 43.
88. Qin, S.H.; Qiu, K.Y.; Swift, G.; Westmoreland, D.G.; Wu, S.G. *J. Polym. Sci., Part A: Polym. Chem.* **1999**, *37*, 4610.
89. Rüchardt, C. *Top. Curr. Chem.* **1980**, *88*, 1.
90. Otsu, T.; Tazaki, T. *Polym. Bull.* **1986**, *16*, 277.
91. McElvain, S.M.; Aldridge, C.L. *J. Am. Chem. Soc.* **1953**, *75*, 3987.
92. Moad, G.; Rizzardo, E.; Solomon, D.H. *Aust. J. Chem.* **1983**, *36*, 1573.
93. Demircioglu, P.; Acar, M.H.; Yagci, Y. *J. Appl. Polym. Sci.* **1992**, *46*, 1639.
94. Engel, P.S.; Chen, Y.; Wang, C. *J. Org. Chem.* **1991**, *56*, 3073.
95. Otsu, T.; Matsumoto, A.; Tazaki, T. *Polym. Bull.* **1987**, *17*, 323.
96. Bledzki, A.; Braun, D.; Menzel, W.; Titzschkau, K. *Makromol. Chem.* **1983**, *184*, 287.
97. Bledzki, A.; Braun, D. *Polym. Bull.* **1986**, *16*, 19.
98. Bledzki, A.; Braun, D. *Makromol. Chem.* **1986**, *187*, 2599.
99. Moad, G.; Rizzardo, E.; Solomon, D.H. *Macromolecules* **1982**, *15*, 909.
100. Marvel, C.S.; Dec, J.; Corner, J.O. *J. Am. Chem. Soc.* **1945**, *67*, 1855.
101. Wieland, P.C.; Raether, B.; Nuyken, O. *Macromol. Rapid Commun.* **2001**, *22*, 700.

102. Wieland, P.C.; Nuyken, O.; Heischkel, Y.; Raether, B.; Strissel, C. *ACS Symp. Ser.* **2003**, *854*, 619.
103. Viala, S.; Antonietti, M.; Tauer, K.; Bremser, W. *Polymer* **2003**, *44*, 1339.
104. Viala, S.; Tauer, K.; Antonietti, M.; Kruger, R.P.; Bremser, W. *Polymer* **2002**, *43*, 7231.
105. Tanaka, H. *Prog. Polym. Sci.* **2003**, *28*, 1171.
106. Hawker, C.J.; Bosman, A.W.; Harth, E. *Chem. Rev.* **2001**, *101*, 3661.
107. Hawker, C.J. In *Handbook of Radical Polymerization*; Davis, T.P.; Matyjaszewski, K., Eds.; John Wiley & Sons: Hoboken, 2002; p 463.
108. Studer, A.; Schulte, T. *Chem. Rec.* **2005**, *5*, 27.
109. Solomon, D.H. *J. Polym. Sci., Part A: Polym. Chem.* **2005**, in press.
110. Fischer, H.; Radom, L. *Angew. Chem., Int. Ed. Engl.* **2001**, *40*, 1340.
111. Solomon, D.H.; Rizzardo, E.; Cacioli, P. US 4581429, 1986 (*Chem. Abstr.* **1985**, *102*, 221335q).
112. Rizzardo, E. *Chem. Aust.* **1987**, *54*, 32.
113. Rizzardo, E.; Chong, Y.K. In *2nd Pacific Polymer Conference, Preprints*; Pacific Polymer Federation: Tokyo, 1991; p 26.
114. Johnson, C.H.J.; Moad, G.; Solomon, D.H.; Spurling, T.H.; Vearing, D.J. *Aust. J. Chem.* **1990**, *43*, 1215.
115. Fischer, H. *Macromolecules* **1997**, *30*, 5666.
116. Georges, M.K.; Veregin, R.P.N.; Kazmaier, P.M.; Hamer, G.K. *Macromolecules* **1993**, *26*, 2987.
117. Moad, G.; Rizzardo, E. In *Pacific Polymer Conference Preprints*; Polymer Division, Royal Australian Chemical Institute: Brisbane, 1993; Vol. 3, p 651.
118. Moad, G.; Rizzardo, E. *Macromolecules* **1995**, *28*, 8722.
119. Georges, M.K.; Veregin, R.P.N.; Kazmaier, P.M.; Hamer, G.K. *Trends Polym. Sci.* **1994**, *2*, 66.
120. Marque, S.; Le Mercier, C.; Tordo, P.; Fischer, H. *Macromolecules* **2000**, *33*, 4403.
121. Marque, S.; Fischer, H.; Baier, E.; Studer, A. *J. Org. Chem.* **2001**, *66*, 1146.
122. Moad, G.; Ercole, F.; Johnson, C.H.; Krstina, J.; Moad, C.L.; Rizzardo, E.; Spurling, T.H.; Thang, S.H.; Anderson, A.G. *ACS Symp. Ser.* **1998**, *685*, 332.
123. Kazmaier, P.M.; Moffat, K.A.; Georges, M.K.; Veregin, R.P.N.; Hamer, G.K. *Macromolecules* **1995**, *28*, 1841.
124. Grishin, D.F.; Semenycheva, L.L.; Kolyakina, E.V. *Russ. J. Appl. Chem.* **2001**, *74*, 494.
125. Golubev, V.B.; Zaremski, M.Y.; Orlova, A.P.; Olenin, A.V. *Polym. Sci.* **2004**, *46*, 295.
126. Zaremski, M.Y.; Orlova, A.P.; Garina, E.S.; Olenin, A.V.; Lachinov, M.B.; Golubev, V.B. *Polym. Sci.* **2003**, *45*, 502.
127. Detrembleur, C.; Sciannamea, V.; Koulic, C.; Claes, M.; Hoebeke, M.; Jerome, R. *Macromolecules* **2002**, *35*, 7214.
128. Catala, J.M.; Jousset, S.; Lamps, J.P. *Macromolecules* **2001**, *34*, 8654.
129. Detrembleur, C.; Claes, M.; Jerome, R. *ACS Symp. Ser.* **2003**, *854*, 496.
130. Detrembleur, C.; Teyssie, P.; Jerome, R. *Macromolecules* **2002**, *35*, 1611.
131. Cameron, N.R.; Reid, A.J.; Span, P.; Bon, S.A.F.; van Es, J.; German, A.L. *Macromol. Chem. Phys.* **2000**, *201*, 2510.
132. Brinkman-Rengel, S.; Niessner, N. *ACS Symp. Ser.* **2000**, *768*, 394.
133. Cameron, N.R.; Reid, A.J. *Macromolecules* **2002**, *35*, 9890.
134. Yamada, B.; Miura, Y.; Nobukane, Y.; Aota, M. *ACS Symp. Ser.* **1998**, *685*, 200.
135. Puts, R.D.; Sogah, D.Y. *Macromolecules* **1997**, *30*, 3323.
136. Veregin, R.P.N.; Georges, M.K.; Hamer, G.K.; Kazmaier, P.M. *Macromolecules* **1995**, *28*, 4391.

137. Cresidio, S.P.; Aldabbagh, F.; Busfield, W.K.; Jenkins, I.D.; Thang, S.H.; Zayas-Holdsworth, C.; Zetterlund, P.B. *J. Polym. Sci., Part A: Polym. Chem.* **2001**, *39*, 1232.
138. Chong, Y.K.; Ercole, F.; Moad, G.; Rizzardo, E.; Thang, S.H.; Anderson, A.G. *Macromolecules* **1999**, *32*, 6895.
139. Dervan, P.; Aldabbagh, F.; Zetterlund, P.B.; Yamada, B. *J. Polym. Sci., Part A: Polym. Chem.* **2003**, *41*, 327.
140. Hawker, C.J. *J. Am. Chem. Soc.* **1994**, *116*, 11185.
141. Skene, W.G.; Belt, S.T.; Connolly, T.J.; Hahn, P.; Scaiano, J.C. *Macromolecules* **1998**, *31*, 9103.
142. Schulte, T.; Knoop, C.A.; Studer, A. *J. Polym. Sci., Part A: Polym. Chem.* **2004**, *42*, 3342.
143. Zhu, Y.; Li, I.O.; Howell, B.; Priddy, D.B. *ACS Symp. Ser.* **1997**, *685*, 214.
144. Yoshida, E. *J. Polym. Sci., Part A: Polym. Chem.* **1996**, *34*, 2937.
145. Yoshida, E.; Fuji, T. *J. Polym. Sci., Part A: Polym. Chem.* **1997**, *35*, 2371.
146. Yoshida, E.; Fujii, T. *J. Polym. Sci., Part A: Polym. Chem.* **1998**, *36*, 269.
147. Marque, S.; Sobek, J.; Fischer, H.; Kramer, A.; Nesvadba, P.; Wunderlich, W. *Macromolecules* **2003**, *36*, 3440.
148. Knoop, C.A.; Studer, A. *J. Am. Chem. Soc.* **2003**, *125*, 16327.
149. Miura, Y.; Nakamura, N.; Taniguchi, I. *Macromolecules* **2001**, *34*, 447.
150. Aldabbagh, F.; Dervan, P.; Phelan, M.; Gilligan, K.; Cunningham, D.; McArdle, P.; Zetterlund, P.B.; Yamada, B. *J. Polym. Sci., Part A: Polym. Chem.* **2003**, *41*, 3892.
151. Goto, A.; Fukuda, T. *Macromolecules* **1999**, *32*, 618.
152. Grubbs, R.B.; Wegrzyn, J.K.; Xia, Q. *Chem. Commun.* **2005**, 80.
153. Benoit, D.; Chaplinski, V.; Braslau, R.; Hawker, C.J. *J. Am. Chem. Soc* **1999**, *121*, 3904.
154. Benoit, D.; Harth, E.; Fox, P.; Waymouth, R.M.; Hawker, C.J. *Macromolecules* **2000**, *33*, 363.
155. Studer, A.; Harms, K.; Knoop, C.; Muller, C.; Schulte, T. *Macromolecules* **2004**, *37*, 27.
156. Benoit, D.; Grimaldi, S.; Robin, S.; Finet, J.P.; Tordo, P.; Gnanou, Y. *J. Am. Chem. Soc* **2000**, *122*, 5929.
157. Benoit, D.; Grimaldi, S.; Finet, J.P.; Tordo, P.; Fontanille, M.; Gnanou, Y. *ACS Symp. Ser.* **1998**, *685*, 225.
158. Farcet, C.; Belleney, J.; Charleux, B.; Pirri, R. *Macromolecules* **2002**, *35*, 4912.
159. Drockenmuller, E.; Lamps, J.P.; Catala, J.M. *Macromolecules* **2004**, *37*, 2076.
160. Drockenmuller, E.; Catala, J.M. *Macromolecules* **2002**, *35*, 2461.
161. Nesvadba, P.; Bugnon, L.; Sift, R. *Polym. Int.* **2004**, *53*, 1066.
162. Schulte, T.; Studer, A. *Macromolecules* **2003**, *36*, 3078.
163. Nesvadba, P.; Bugnon, L.; Sift, R. *J. Polym. Sci., Part A: Polym. Chem.* **2004**, *42*, 3332.
164. Wetter, C.; Gierlich, J.; Knoop, C.A.; Muller, C.; Schulte, T.; Studer, A. *Chem. Eur. J.* **2004**, *10*, 1156.
165. Hawker, C.J.; Barclay, G.G.; Orellana, A.; Dao, J.; Devonport, W. *Macromolecules* **1996**, *29*, 5245.
166. Georges, M.K.; Kee, R.A.; Veregin, R.P.N.; Hamer, G.K.; Kazmaier, P.M. *J. Phys. Org. Chem.* **1995**, *8*, 301.
167. Catala, J.M.; Bubel, F.; Hammouch, S.O. *Macromolecules* **1995**, *28*, 8441.
168. Hammouch, S.O.; Catala, J.M. *Macromol. Rapid Commun.* **1996**, *17*, 683.
169. Greszta, D.; Matyjaszewski, K. *Macromolecules* **1996**, *29*, 5239.
170. Connolly, T.J.; Baldovi, M.V.; Mohtat, N.; Scaiano, J.C. *Tetrahedron Lett.* **1996**, *37*, 4919.
171. Miura, Y.; Hirota, K.; Moto, H.; Kobetake, S. *Macromolecules* **1998**, *37*, 4051.

172. Li, I.Q.; Knauss, D.M.; Priddy, D.B.; Howell, B.A. *Polym. Int.* **2003**, *52*, 805.
173. Sugimoto, N.; Narumi, A.; Satoh, T.; Kaga, H.; Kakuchi, T. *Polym. Bull.* **2003**, *49*, 337.
174. Matyjaszewski, K.; Woodworth, B.E.; Zhang, X.; Gaynor, S.G.; Metzner, Z. *Macromolecules* **1998**, *31*, 5955.
175. Dao, J.; Benoit, D.; Hawker, C.J. *J. Polym. Sci., Part A: Polym. Chem.* **1998**, *36*, 2161.
176. Bothe, M.; Schmidt-Naake, G. *Macromol. Rapid Commun.* **2003**, *24*, 609.
177. Lukkarila, J.L.; Hamer, G.K.; Georges, M.K. *Tetrahedron Lett.* **2004**, *45*, 5317.
178. Gridnev, A.A. *Macromolecules* **1997**, *30*, 7651.
179. Li, I.; Howell, B.; Matyjasewski, K.; Shigemoto, T.; Smith, P.B.; Priddy, D.B. *Macromolecules* **1995**, *28*, 6692.
180. Souaille, M.; Fischer, H. *Macromolecules* **2002**, *35*, 248.
181. Malmstrom, E.; Miller, R.D.; Hawker, C.J. *Tetrahedron* **1997**, *53*, 15225.
182. Goto, A.; Tsujii, Y.; Fukuda, T. *Chem, Lett.* **2000**, 788.
183. Baumann, M.; Schmidt-Naake, G. *Macromol. Chem. Phys.* **2001**, *202*, 2727.
184. Georges, M.K.; Veregin, R.P.N.; Kazmaier, P.M.; Hamer, G.K.; Saban, M. *Macromolecules* **1994**, *27*, 7228.
185. Matthews, B.R.; Pike, W.; Rego, J.M.; Kuch, P.D.; Priddy, D.B. *J. Appl. Polym. Sci.* **2003**, *87*, 869.
186. Odell, P.G.; Veregin, R.P.N.; Michalak, L.M.; Brousmiche, D.; Georges, M.K. *Macromolecules* **1995**, *28*, 8453.
187. Odell, P.G.; Veregin, R.P.N.; Michalak, L.M.; Georges, M.K. *Macromolecules* **1997**, *30*, 2232.
188. Keoshkerian, B.; Georges, M.; Quinlan, M.; Veregin, R.; Goodbrand, B. *Macromolecules* **1998**, *31*, 7559.
189. Georges, M.K.; Lukkarila, J.L.; Szkurhan, A.R. *Macromolecules* **2004**, *37*, 1297.
190. Keoshkerian, B.; Szkurhan, A.R.; Georges, M.K. *Macromolecules* **2001**, *34*, 6531.
191. Goto, A.; Fukuda, T. *Macromolecules* **1997**, *30*, 4272.
192. Greszta, D.; Matyjaszewski, K. *J. Polym. Sci., Part A: Polym. Chem.* **1997**, *35*, 1857.
193. Baethge, H.; Butz, S.; Han, C.-H.; Schmidt-Naake, G. *Angew. Makromol. Chem.* **1999**, *267*, 52.
194. Butz, S.; Baethge, H.; Schmidt-Naake, G. *Angew. Makromol. Chem.* **1999**, *270*, 42.
195. He, J.; Chen, J.; Li, L.; Pan, J.; Li, C.; Cao, J.; Tao, Y.; Hua, F.; Yang, Y.; McKee, G.E.; Brinkmann, S. *Polymer* **2000**, *41*, 4573.
196. Bian, K.; Cunningham, M.F. *Macromolecules* **2005**, *38*, 695.
197. Diaz, T.; Fischer, A.; Jonquieres, A.; Brembilla, A.; Lochon, P. *Macromolecules* **2003**, *36*, 2235.
198. Couvreur, L.; Lefay, C.; Belleney, J.; Charleux, B.; Guerret, O.; Magnet, S. *Macromolecules* **2003**, *36*, 8260.
199. Lefay, C.; Belleney, J.; Charleux, B.; Guerret, O.; Magnet, S. *Macromol. Rapid Commun.* **2004**, *25*, 1215.
200. Bohrisch, J.; Wendler, U.; Jaeger, W. *Macromol. Rapid Commun.* **1997**, *18*, 975.
201. Listigovers, N.A.; Georges, M.K.; Odell, P.G.; Keoshkerian, B. *Macromolecules* **1996**, *29*, 8992.
202. Burguiere, C.; Dourges, M.A.; Charleux, B.; Vairon, J.P. *Macromolecules* **1999**, *32*, 3883.
203. Georges, M.K.; Hamer, G.K.; Listigovers, N.A. *Macromolecules* **1998**, *31*, 9087.
204. Benoit, D.; Harth, E.; Helms, B.; Rees, I.; Vestberg, R.; Rodlert, M.; Hawker, C.J. *ACS Symp. Ser.* **2000**, *768*, 123.
205. Qiu, J.; Charleux, B.; Matyjaszewski, K. *Prog. Polym. Sci.* **2001**, *26*, 2083.
206. Cunningham, M.F. *Prog. Polym. Sci.* **2002**, *27*, 1039.
207. Cunningham, M.F. *Compt. Rend. Chim.* **2003**, *6*, 1351.

208. Schork, F.J.; Y., L.; Smulders, W.; Russum, J.P.; Butte, A.; Fontenot, K. *Adv. Polym. Sci.* **2005**, *175*, 129.
209. Charleux, B. *Macromolecules* **2000**, *33*, 5358.
210. Butte, A.; Storti, G.; Morbidelli, M. *Macromolecules* **2001**, *34*, 5885.
211. Ma, J.W.; Cunningham, M.F.; McAuley, K.B.; Keoshkerian, B.; Georges, M. *Chem. Eng. Sci.* **2003**, *58*, 1177.
212. Ma, J.W.; Cunningham, M.F.; McAuley, K.B.; Keoshkerian, B.; Georges, M.K. *Macromol. Theory Simul.* **2003**, *12*, 72.
213. Ma, J.W.; Smith, J.A.; McAuley, K.B.; Cunningham, M.F.; Keoshkerian, B.; Georges, M.K. *Chem. Eng. Sci.* **2003**, *58*, 1163.
214. Butte, A.; Storti, G.; Morbidelli, M. In *DECHEMA Monograph.*, 2001; Vol. 137, p 273.
215. Bon, S.A.F.; Bosveld, M.; Klumperman, B.; German, A.L. *Macromolecules* **1997**, *30*, 324.
216. Cao, J.Z.; He, J.P.; Li, C.M.; Yang, Y.L. *Polym. J.* **2001**, *33*, 75.
217. Marestin, C.; Noel, C.; Guyot, A.; Claverie, J. *Macromolecules* **1998**, *31*, 4041.
218. Szkurhan, A.R.; Georges, M.K. *Macromolecules* **2004**, *37*, 4776.
219. Nicolas, J.; Charleux, B.; Guerret, O.; Magnet, S.P. *Angew. Chem. Int. Ed. Engl.* **2004**, *43*, 6186.
220. Prodpran, T.; Dimonie, V.L.; Sudol, E.D.; El-Aasser, M.S. *Macromol. Symp.* **2000**, *155*, 1.
221. Pan, G.F.; Sudol, E.D.; Dimonie, V.L.; El-Aasser, M.S. *Macromolecules* **2001**, *34*, 481.
222. MacLeod, P.J.; Barber, R.; Odell, P.G.; Keoshkerian, B.; Georges, M.K. *Macromol. Symp.* **2000**, *155*, 31.
223. Cunningham, M.F.; Tortosa, K.; Ma, J.W.; McAuley, K.B.; Keoshkerian, B.; Georges, M.K. *Macromol. Symp.* **2002**, *182*, 273.
224. Nicolas, J.; Charleux, B.; Guerret, O.; Magnet, S. *Macromolecules* **2004**, *37*, 4453.
225. Cunningham, M.; Lin, M.; Buragina, C.; Milton, S.; Ng, D.; Hsu, C.C.; Keoshkerian, B. *Polymer* **2005**, *46*, 1025.
226. Druliner, J.D. *Macromolecules* **1991**, *24*, 6079.
227. Yamada, B.; Tanaka, H.; Konishi, K.; Otsu, T. *J. Macromol. Sci., Chem.* **1994**, *A31*, 351.
228. Grande, D.; Baskaran, S.; Baskaran, C.; Gnanou, Y.; Chaikof, E.L. *Macromolecules* **2000**, *33*, 1123.
229. Grande, D.; Baskaran, S.; Chaikof, E.L. *Macromolecules* **2001**, *34*, 1640.
230. Grande, D.; Guerrero, R.; Gnanou, Y. *J. Polym. Sci., Part A: Polym. Chem.* **2005**, *43*, 519.
231. Chung, T.C.; Janvikul, W.; Lu, H.L. *J. Am. Chem. Soc.* **1996**, *118*, 705.
232. Chung, T.C.; Hong, H. *ACS Symp. Ser.* **2003**, *854*, 481.
233. Hong, H.; Chung, T.C. *Macromolecules* **2004**, *37*, 6260.
234. Chung, T.C.; Lu, H.L.; Janvikul, W. *Polymer* **1997**, *38*, 1495.
235. Lu, B.; Chung, T.C. *Macromolecules* **1998**, *31*, 5943.
236. Chung, T.C.; Xu, G.; Lu, Y.Y.; Hu, Y.L. *Macromolecules* **2001**, *34*, 8040.
237. Chung, T.C. *Israel J. Chem.* **2002**, *42*, 307.
238. Dimonie, M.; Mardare, D.; Matyjaszewski, K.; Coca, S.; Dragutan, V.; Ghiviriga, J. *Macromol. Rapid Commun.* **1992**, *13*, 283.
239. Mardare, D.; Matyjaszewski, K.; Coca, S. *Macromol. Rapid Commun.* **1994**, *15*, 37.
240. Mardare, D.; Matyjaszewski, K. *Polym. Prepr. (Am. Chem. Soc., Div. Polym. Chem)* **1993**, *34(2)*, 566.
241. Mardare, D.; Matyjaszewski, K. *Macromolecules* **1994**, *27*, 645.
242. Granel, C.; Jerome, R.; Teyssie, P.; Jasieczek, C.B.; Shooter, A.J.; Haddleton, D.M.; Hastings, J.J.; Gigmes, D.; Grimaldi, S.; Tordo, P.; Greszta, D.; Matyjaszewski, K. *Macromolecules* **1998**, *31*, 7133.

243. Klapper, M.; Brand, T.; Steenbock, M.; Müllen, K. *ACS Symp. Ser.* **2000**, *768*, 152.
244. Steenbock, M.; Klapper, M.; Muellen, K.; Bauer, C.; Hubrich, M. *Macromolecules* **1998**, *31*, 5223.
245. Steenbock, M.; Klapper, M.; Muellen, K. *Macromol. Chem. Phys.* **1998**, *199*, 763.
246. Khelfallah, N.S.; Peretolchin, M.; Klapper, M.; Mullen, K. *Polym. Bull.* **2005**, *53*, 295.
247. Dasgupta, A.; Brand, T.; Klapper, M.; Mullen, K.R. *Polym. Bull.* **2001**, *46*, 131.
248. Grishin, D.F.; Ignatov, S.K.; Shchepalov, A.A.; Razuvaev, A.G. *Appl. Organometal. Chem.* **2004**, *18*, 271.
249. Asandei, A.D.; Moran, I.W. *J. Am. Chem. Soc.* **2004**, *126*, 15932.
250. Wayland, B.B.; Poszmik, G.; Mukerjee, S.L.; Fryd, M. *J. Am. Chem. Soc.* **1994**, *116*, 7943.
251. Wayland, B.B.; Basickes, L.; Mukerjee, S.; Wei, M.; Fryd, M. *Macromolecules* **1997**, *30*, 8109.
252. Lu, Z.; Fryd, M.; Wayland, B.B. *Macromolecules* **2004**, *37*, 2686.
253. Debuigne, A.; Caille, J.-R.; Jerome, R. *Angew. Chem. Int. Ed. Engl.* **2005**, *44*, 1101–1104.
254. Le Grognec, E.; Claverie, R.; Poli, R. *J. Am. Chem. Soc.* **2001**, *123*, 9513.
255. Oganova, A.G.; Smirnov, B.R.; Ioffe, N.T.; Enikolopyan, N.S. *Bull. Acad. Sci. USSR* **1983**, 1837.
256. Oganova, A.G.; Smirnov, B.R.; Ioffe, N.T.; Kim, I.P. *Bull. Acad. Sci. USSR* **1984**, 1154.
257. Oganova, A.G.; Smirnov, B.R.; Ioffe, N.T.; Enikopyan, N.S. *Doklady Akad. Nauk SSR (Engl. Transl.)* **1983**, *268*, 66.
258. Morozova, I.S.; Oganova, A.G.; Nosova, V.S.; Novikov, D.D.; Smirnov, B.R. *Bull. Acad. Sci. USSR* **1987**, 2628.
259. Smirnov, V.R. *Polym. Sci. USSR (Engl. Transl.)* **1990**, *32*, 524.
260. Arvanitopoulos, L.D.; Greuel, M.P.; Harwood, H.J. *Polym. Prepr. (Am. Chem. Soc., Div. Polym. Chem.)* **1994**, *35(2)*, 549.
261. Kharasch, M.S.; Jensen, E.V.; Urry, W.H. *Science* **1945**, *102*, 128.
262. Minisci, F. *Acc. Chem. Res.* **1975**, *8*, 165.
263. Ameduri, B.; Boutevin, B. *Macromolecules* **1990**, *23*, 2433.
264. Bamford, C.H. In *Comprehensive Polymer Science*; Eastmond, G.C.; Ledwith, A.; Russo, S.; Sigwalt, P., Eds.; Pergamon: Oxford, 1989; Vol. 3, p 123.
265. Kato, M.; Kamigaito, M.; Sawamoto, M.; Higashamura, T. *Macromolecules* **1995**, *28*, 1721.
266. Wang, J.-S.; Matyjaszewski, K. *Macromolecules* **1995**, *28*, 7901.
267. Percec, V.; Barboiu, B. *Macromolecules* **1995**, *28*, 7970.
268. Matyjaszewski, K.; Xia, J. *Chem. Rev.* **2001**, *101*, 2921.
269. Matyjaszewski, K.; Xia, J. In *Handbook of Radical Polymerization*; Davis, T.P.; Matyjaszewski, K., Eds.; John Wiley & Sons: Hoboken, 2002; p 523.
270. Kamigaito, M.; Ando, T.; Sawamoto, M. *Chem. Rev.* **2001**, *101*, 3689.
271. Kamigaito, M.; Ando, T.; Sawamoto, M. *Chem. Rec.* **2004**, *4*, 159.
272. Percec, V.; Popov, A.V.; Ramirez-Castillo, E.; Monteiro, M.; Barboiu, B.; Weichold, O.; Asandei, A.D.; Mitchell, C.M. *J. Am. Chem. Soc.* **2002**, *124*, 4940.
273. Percec, V.; Popov, A.V.; Ramirez-Castillo, E.; Weichold, O. *J. Polym. Sci., Part A: Polym. Chem.* **2003**, *41*, 3283.
274. Percec, V.; Popov, A.V.; Ramirez-Castillo, E.; Coelho, J.F.J. *J. Polym. Sci., Part A: Polym. Chem.* **2005**, *43*, 773.
275. Lad, J.; Harrisson, S.; Haddleton, D.M. *ACS Symp. Ser.* **2003**, *854*, 148.
276. Lad, J.; Harrisson, S.; Mantovani, G.; Haddleton, D.M. *Dalton Trans.* **2003**, 4175.
277. Jiang, J.G.; Zhang, K.D.; Zhou, H. *J. Polym. Sci., Part A: Polym. Chem.* **2004**, *42*, 5811.
278. Percec, V.; Barboiu, B.; Kim, H.J. *J. Am. Chem. Soc* **1998**, *120*, 305.

279. Percec, V.; Kim, H.J.; Barboiu, B. *Macromolecules* **1997**, *30*, 8526.
280. Percec, V.; Barboiu, B.; Bera, T.K.; van der Sluis, M.; Grubbs, R.B.; Frechet, J.M.J. *J. Polym. Sci., Part A: Polym. Chem.* **2000**, *38*, 4776.
281. Grigoras, C.; Percec, V. *J. Polym. Sci., Part A: Polym. Chem.* **2005**, *43*, 319.
282. Gurr, P.A.; Mills, M.F.; Qiao, G.G.; Solomon, D.H. *Polymer* **2005**, *46*, 2097.
283. Venkatesh, R.; Harrisson, S.; Haddleton, D.M.; Klumperman, B. *Macromolecules* **2004**, *37*, 4406.
284. Matsuyama, M.; Kamigaito, M.; Sawamoto, M. *J. Polym. Sci., Part A: Polym. Chem.* **1996**, *34*, 3585.
285. Wang, J.-S.; Gaynor, S.; Matyjaszewski, K. *Macromolecules* **1995**, *28*, 7572.
286. Xia, J.H.; Matyjaszewski, K. *Macromolecules* **1999**, *32*, 5199.
287. Qiu, J.; Gaynor, S.G.; Matyjaszewski, K. *Macromolecules* **1999**, *32*, 2872.
288. Qiu, J.; Pintauer, T.; Gaynor, S.G.; Matyjaszewski, K.; Charleux, B.; Vairon, J.P. *Macromolecules* **2000**, *33*, 7310.
289. Wang, W.; Yan, D. *ACS Symp. Ser.* **2000**, *768*, 263.
290. Gromada, J.; Matyjaszewski, K. *Macromolecules* **2001**, *34*, 7664.
291. Li, M.; Min, K.; Matyjaszewski, K. *Macromolecules* **2004**, *37*, 2106.
292. Matyjaszewski, K.; Paik, H.J.; Zhou, P.; Diamanti, S.J. *Macromolecules* **2001**, *34*, 5125.
293. Pintauer, T.; Zhou, P.; Matyjaszewski, K. *J. Am. Chem. Soc.* **2002**, *124*, 8196.
294. Nanda, A.K.; Matyjaszewski, K. *Macromolecules* **2003**, *36*, 599.
295. Nanda, A.K.; Matyjaszewski, K. *Macromolecules* **2003**, *36*, 1487.
296. Nanda, A.K.; Matyjaszewski, K. *Macromolecules* **2003**, *36*, 8222.
297. Pintauer, T.; Braunecker, W.; Collange, E.; Poli, R.; Matyjaszewski, K. *Macromolecules* **2004**, *37*, 2679.
298. Brandts, J.A.M.; van de Geijn, P.; van Faassen, E.E.; Boersma, J.; van Koten, G. *J. Organometal. Chem.* **1999**, *584*, 246.
299. Stoffelbach, F.; Haddleton, D.M.; Poli, R. *Eur. Polym. J.* **2003**, *39*, 2099.
300. Endo, K.; Yachi, A. *Polym. Bull.* **2001**, *46*, 363.
301. Endo, K.; Yachi, A. *Polym. J.* **2002**, *34*, 320.
302. Kotani, Y.; Kamigaito, M.; Sawamoto, M. *Macromolecules* **2000**, *33*, 6746.
303. Kotani, Y.; Kamigaito, M.; Sawamoto, M. *Macromolecules* **1999**, *32*, 2420.
304. Wang, B.Q.; Zhuang, Y.; Luo, X.X.; Xu, S.S.; Zhou, X.Z. *Macromolecules* **2003**, *36*, 9684.
305. Moineau, G.; Granel, C.; Dubois, P.; Jerome, R.; Teyssie, P. *Macromolecules* **1998**, *31*, 542.
306. Opstal, T.; Zednik, J.; Sedlacek, J.; Svoboda, J.; Vohlidal, J.; Verpoort, F. *Collect. Czech. Chem. Commun.* **2002**, *67*, 1858.
307. Percec, V.; Barboiu, B.; A., N. *Macromolecules* **1996**, *29*, 3665.
308. Lecomte, P.; Drapier, I.; Dubois, P.; Teyssie, P.; Jerome, R. *Macromolecules* **1997**, *30*, 7631.
309. Shen, Y.Q.; Tang, H.D.; Ding, S.J. *Prog. Polym. Sci.* **2004**, *29*, 1053.
310. Pintauer, T.; Matyjasewski, K. *Coord. Chem. Rev.* **2005**, *249*, 1155.
311. Xia, J.; Matyjaszewski, K. *Macromolecules* **1997**, *30*, 7697.
312. Sadhu, V.B.; Pionteck, J.; Voigt, D.; Komber, H.; Fischer, D.; Voit, B. *Macromol. Chem. Phys.* **2004**, *205*, 2356.
313. Sadhu, V.B.; Pionteck, J.; Voigt, D.; Komber, H.; Voit, B. *Macromol. Symp.* **2004**, *210*, 147.
314. Haddleton, D.M.; Jasieczek, C.B.; Hannon, M.J.; Shooter, A., J. *Macromolecules* **1997**, *30*, 2190.
315. Qiu, J.; Matyjaszewski, K.; Thouin, L.; Amatore, C. *Macromol. Chem. Phys.* **2000**, *201*, 1625.
316. Xia, J.H.; Matyjaszewski, K. *Macromolecules* **1997**, *30*, 7697.
317. Matyjaszewski, K.; Patten, T.E.; Xia, J. *J. Am. Chem. Soc.* **1997**, *119*, 674.

318. Ohno, K.; Goto, A.; Fukuda, T.; Xia, J.H.; Matyjaszewski, K. *Macromolecules* **1998**, *31*, 2699.
319. Destarac, M.; Bessiere, J.M.; Boutevin, B. *Macromo. Rapid Commun.* **1997**, *18*, 967.
320. Wang, X.S.; Malet, F.L.G.; Armes, S.P.; Haddleton, D.M.; Perrier, S. *Macromolecules* **2001**, *34*, 162.
321. Limer, A.; Heming, A.; Shirley, I.; Haddleton, D. *Eur. Polym. J.* **2005**, *41*, 805.
322. Zhang, X.; Xia, J.H.; Matyjaszewski, K. *Macromolecules* **1998**, *31*, 5167.
323. Kickelbick, G.; Matyjaszewski, K. *Macromol. Rapid Commun.* **1999**, *20*, 341.
324. Xia, J.H.; Matyjaszewski, K. *Macromolecules* **1999**, *32*, 2434.
325. Xia, J.; Zhang, K.; Matyjasewski, K. *ACS Symp. Ser.* **2000**, *760*, 207.
326. Xia, J.H.; Gaynor, S.G.; Matyjaszewski, K. *Macromolecules* **1998**, *31*, 5958.
327. Percec, V.; Barboiu, B.; van der Sluis, M. *Macromolecules* **1998**, *31*, 4053.
328. van der Sluis, M.; Barboiu, B.; Pesa, N.; Percec, V. *Macromolecules* **1998**, *31*, 9409.
329. Haddleton, D.M.; Kukulj, D.; Radigue, A.P. *Chem. Commun.* **1999**, 99.
330. Kotre, T.; Nuyken, O.; Weberskirch, R. *Macromol. Chem. Phys.* **2004**, *205*, 1187.
331. Faucher, S.; Zhu, S.P. *Macromol. Rapid Commun.* **2004**, *25*, 991.
332. Kumar, K.R.; Kizhakkedathu, J.N.; Brooks, D.E. *Macromol. Chem. Phys.* **2004**, *205*, 567.
333. Nguyen, J.V.; Jones, C.W. *J. Polym. Sci., Part A: Polym. Chem.* **2004**, *42*, 1367.
334. Honigfort, M.E.; Brittain, W.J. *Macromolecules* **2003**, *36*, 3111.
335. Hong, S.C.; Matyjaszewski, K. *Macromolecules* **2002**, *35*, 7592.
336. Shen, Y.Q.; Zhu, S.P. *Macromolecules* **2001**, *34*, 8603.
337. Hong, S.C.; Paik, H.J.; Matyjaszewski, K. *Macromolecules* **2001**, *34*, 5099.
338. Shen, Y.Q.; Zhu, S.P.; Pelton, R. *Macromolecules* **2001**, *34*, 3182.
339. Liou, S.; Rademacher, J.T.; Malaba, D.; Pallack, M.E.; Brittain, W.J. *Macromolecules* **2000**, *33*, 4295.
340. Honigfort, M.E.; Liou, S.; Rademacher, J.; Malaba, D.; Bosanac, T.; Wilcox, C.S.; Brittain, W.J. *ACS Symp. Ser.* **2003**, *854*, 250.
341. Nishikawa, T.; Ando, T.; Kamigaito, M.; Sawamoto, M. *Macromolecules* **1997**, *30*, 2244.
342. Takahashi, H.; Ando, T.; Kamigaito, M.; Sawamoto, M. *Macromolecules* **1999**, *32*, 3820.
343. Kotani, Y.; Kato, M.; Kamigaito, M.; Sawamoto, M. *Macromolecules* **1996**, *29*, 6979.
344. Senoo, M.; Kotani, Y.; Kamigaito, M.; Sawamoto, M. *Macromolecules* **1999**, *32*, 8005.
345. Kotani, Y.; Kamigaito, M.; Sawamoto, M. *ACS Symp. Ser.* **2000**, *768*, 168.
346. Kamigaito, M.; Watanabe, Y.; Ando, T.; Sawamoto, M. *J. Am. Chem. Soc.* **2002**, *124*, 9994.
347. Watanabe, Y.; Ando, T.; Kamigaito, M.; Sawamoto, M. *Macromolecules* **2001**, *34*, 4370.
348. Kamigaito, M.; Ando, T.; Sawamoto, M. *ACS Symp. Ser.* **2003**, *854*, 102.
349. Fuji, Y.; Watanabe, K.; Baek, K.Y.; Ando, T.; Kamigaito, M.; Sawamoto, M. *J. Polym. Sci., Part A: Polym. Chem.* **2002**, *40*, 2055.
350. Wakioka, M.; Baek, K.Y.; Ando, T.; Kamigaito, M.; Sawamoto, M. *Macromolecules* **2002**, *35*, 330.
351. Ando, T.; Kamigaito, M.; Sawamoto, M. *Macromolecules* **1997**, *30*, 4507.
352. Kotani, Y.; Kamigaito, M.; Sawamoto, M. *Macromolecules* **1999**, *32*, 6877.
353. Kotani, Y.; Kamigaito, M.; Sawamoto, M. *Macromolecules* **2000**, *33*, 3543.
354. Kamigaito, M.; Onishi, I.; Kimura, S.; Kotani, Y.; Sawamoto, M. *Chem. Commun.* **2002**, 2694.

355. Matyjaszewski, K.; Wei, M.L.; Xia, J.H.; McDermott, N.E. *Macromolecules* **1997**, *30*, 8161.
356. Zhu, S.; Yan, D.; Van Beylen, M. *ACS Symp. Ser.* **2003**, *854*, 221.
357. Zhang, H.Q.; Schubert, U.S. *J. Polym. Sci., Part A: Polym. Chem.* **2004**, *42*, 4882.
358. Zhang, H.Q.; Schubert, U.S. *Chem. Commun.* **2004**, 858.
359. Uegaki, H.; Kotani, Y.; Kamigaito, M.; Sawamoto, M. *Macromolecules* **1997**, *30*, 2249.
360. Uegaki, H.; Kotani, Y.; Kamigaito, M.; Sawamoto, M. *Macromolecules* **1998**, *31*, 6756.
361. Granel, C.; Dubois, P.; Jerome, R.; Teyssie, P. *Macromolecules* **1996**, *29*, 8576.
362. Uegaki, H.; Kamigaito, M.; Sawamoto, M. *J. Polym. Sci., Part A: Polym. Chem.* **1999**, *37*, 3003.
363. Tsuji, M.; Sakai, R.; Satoh, T.; Kaga, H.; Kakuchi, T. *Macromolecules* **2002**, *35*, 8255.
364. Kakuchi, T.; Tsuji, M.; Satoh, T. *ACS Symp. Ser.* **2003**, *854*, 206.
365. Pan, C.Y.; Lou, X.D. *Macromol. Chem. Phys.* **2000**, *201*, 1115.
366. Wickel, H.; Agarwal, S.; Greiner, A. *Macromolecules* **2003**, *36*, 2397.
367. Chung, I.S.; Matyjaszewski, K. *Macromolecules* **2003**, *36*, 2995.
368. Tsarevsky, N.V.; Pintauer, T.; Matyjaszewski, K. *Macromolecules* **2004**, *37*, 9768.
369. Ashford, E.J.; Naldi, V.; O'Dell, R.; Billingham, N.C.; Armes, S.P. *Chem. Commun.* **1999**, 1285.
370. Perrier, S.; Armes, S.P.; Wang, X.S.; Malet, F.; Haddleton, D.M. *J. Polym. Sci., Part A: Polym. Chem.* **2001**, *39*, 1696.
371. Coullerez, G.; Carlmark, A.; Malmstrom, E.; Jonsson, M. *J. Phys. Chem. A* **2004**, *108*, 7129.
372. Lee, S.B.; Russell, A.J.; Matyjaszewski, K. *Biomacromolecules* **2003**, *4*, 1386.
373. Li, Y.T.; Armes, S.P.; Jin, X.P.; Zhu, S.P. *Macromolecules* **2003**, *36*, 8268.
374. Perrier, S.; Haddleton, D.M. *Macromol. Symp.* **2002**, *182*, 261.
375. Zhu, C.Y.; Sun, F.; Zhang, M.; Jin, R. *Polymer* **2004**, *45*, 1141.
376. Nishikawa, T.; Kamigaito, M.; Sawamoto, M. *Macromolecules* **1999**, *32*, 2204.
377. Fuji, Y.; Ando, T.; Kamigaito, M.; Sawamoto, M. *Macromolecules* **2002**, *35*, 2949.
378. Gaynor, S.G.; Qiu, J.; Matyjaszewski, K. *Macromolecules* **1998**, *31*, 5951.
379. Min, K.; Gao, H.F.; Matyjaszewski, K. *J. Am. Chem. Soc* **2005**, *127*, 3825.
380. Krstina, J.; Moad, G.; Rizzardo, E.; Winzor, C.L.; Berge, C.T.; Fryd, M. *Macromolecules* **1995**, *28*, 5381.
381. Matyjaszewski, K.; Gaynor, S.; Wang, J.-S. *Macromolecules* **1995**, *28*, 2093.
382. Chiefari, J.; Chong, Y.K.; Ercole, F.; Krstina, J.; Jeffery, J.; Le, T.P.T.; Mayadunne, R.T.A.; Meijs, G.F.; Moad, C.L.; Moad, G.; Rizzardo, E.; Thang, S.H. *Macromolecules* **1998**, *31*, 5559.
383. Moad, G.; Chiefari, J.; Moad, C.L.; Postma, A.; Mayadunne, R.T.A.; Rizzardo, E.; Thang, S.H. *Macromol. Symp.* **2002**, *182*, 65.
384. Chiefari, J.; Mayadunne, R.T.A.; Moad, C.L.; Moad, G.; Rizzardo, E.; Postma, A.; Skidmore, M.A.; Thang, S.H. *Macromolecules* **2003**, *36*, 2273.
385. Zhang, M.; Ray, W.H. *Ind. Eng. Chem. Res.* **2001**, *40*, 4336.
386. Wang, A.R.; Zhu, S.P. *J. Polym. Sci., Part A: Polym. Chem.* **2003**, *41*, 1553.
387. Wang, A.R.; Zhu, S. *Macromol. Theory Simul.* **2003**, *12*, 196.
388. Peklak, A.D.; Butte, A.; Storti, G.; Morbidelli, M. *Macromol. Symp.* **2004**, *206*, 481.
389. Shipp, D.A.; Matyjaszewski, K. *Macromolecules* **1999**, *32*, 2948.
390. Barner-Kowollik, C. *Aust. J. Chem.* **2001**, *54*, 343.
391. Barner-Kowollik, C.; Quinn, J.F.; Nguyen, T.L.U.; Heuts, J.P.A.; Davis, T.P. *Macromolecules* **2001**, *34*, 7849.
392. Vana, P.; Davis, T.P.; Barner-Kowollik, C. *Macromol. Theory Simul.* **2002**, *11*, 823.

393. Wulkow, M.; Busch, M.; Davis, T.P.; Barner-Kowollik, C. *J. Polym. Sci., Part A: Polym. Chem.* **2004**, *42*, 1441.
394. Chong, Y.K.; Krstina, J.; Le, T.P.T.; Moad, G.; Postma, A.; Rizzardo, E.; Thang, S.H. *Macromolecules* **2003**, *36*, 2256.
395. Krstina, J.; Moad, C.L.; Moad, G.; Rizzardo, E.; Berge, C.T.; Fryd, M. *Macromol. Symp.* **1996**, *111*, 13.
396. Charmot, D.; Corpart, P.; Adam, H.; Zard, S.Z.; Biadatti, T.; Bouhadir, G. *Macromol. Symp.* **2000**, *150*, 23.
397. Rizzardo, E.; Chiefari, J.; Mayadunne, R.T.A.; Moad, G.; Thang, S.H. *ACS Symp. Ser.* **2000**, *768*, 278.
398. Chiefari, J.; Rizzardo, E. In *Handbook of Radical Polymerization*; Davis, T.P.; Matyjaszewski, K., Eds.; John Wiley & Sons: Hoboken, NY, 2002; p 263.
399. Barner-Kowollik, C.; Davis, T.P.; Heuts, J.P.A.; Stenzel, M.H.; Vana, P.; Whittaker, M. *J. Polym. Sci., Part A: Polym. Chem.* **2003**, *41*, 365.
400. McCormick, C.L.; Lowe, A.B. *Acc. Chem. Res.* **2004**, *37*, 312.
401. Moad, G.; Rizzardo, E.; Thang, S. *Aust. J. Chem.* **2005**, *58*, 379.
402. Barton, D.H.R.; McCombie, S.W. *J. Chem. Soc., Perkin Trans. 1* **1975**, 1574.
403. Barton, D.H.R.; Parekh, S.I.; Tse, C.L. *Tetrahedron Lett.* **1993**, *34*, 2733.
404. Forbes, J.E.; Zard, S.Z. *Tetrahedron Lett.* **1989**, *30*, 4367.
405. Delduc, P.; Tailhan, C.; Zard, S.Z. *J. Chem. Soc., Chem. Commun.* **1988**, 308.
406. Quiclet-Sire, B.; Zard, S.Z. *Pure Appl. Chem.* **1997**, *69*, 645.
407. Zard, S.Z. *Angew. Chem. Int. Ed. Engl.* **1997**, *36*, 672.
408. Le, T.P.; Moad, G.; Rizzardo, E.; Thang, S.H. Int. Patent Appl. WO 9801478, 1998 (*Chem. Abstr.* (1997) 128: 115390).
409. Moad, G.; Chiefari, J.; Chong, Y.K.; Krstina, J.; Postma, A.; Mayadunne, R.T.A.; Rizzardo, E.; Thang, S.H. *Polym. Int.* **2000**, *49*, 933.
410. Rizzardo, E.; Chiefari, J.; Chong, Y.K.; Ercole, F.; Krstina, J.; Jeffery, J.; Le, T.P.T.; Mayadunne, R.T.A.; Meijs, G.F.; Moad, C.L.; Moad, G.; Thang, S.H. *Macromol. Symp.* **1999**, *143*, 291.
411. Mayadunne, R.T.A.; Rizzardo, E.; Chiefari, J.; Chong, Y.K.; Moad, G.; Thang, S.H. *Macromolecules* **1999**, *32*, 6977.
412. Destarac, M.; Charmot, D.; Franck, X.; Zard, S.Z. *Macromol. Rapid Commun.* **2000**, *21*, 1035.
413. Destarac, M.; Bzducha, W.; Taton, D.; Gauthier-Gillaizeau, I.; Zard, S.Z. *Macromol. Rapid Commun.* **2002**, *23*, 1049.
414. Benaglia, M.; Rizzardo, E.; Alberti, A.; Guerra, M. *Macromolecules* **2005**, *38*, 3129.
415. Farmer, S.C.; Patten, T.E. *J. Polym. Sci., Part A: Polym. Chem.* **2002**, *40*, 555.
416. Coote, M.L. *Macromolecules* **2004**, *37*, 5023.
417. Coote, M.L.; Radom, L. *J. Am. Chem. Soc.* **2003**, *125*, 1490.
418. Thomas, D.B.; Convertine, A.J.; Myrick, L.J.; Scales, C.W.; Smith, A.E.; Lowe, A.B.; Vasilieva, Y.A.; Ayres, N.; McCormick, C.L. *Macromolecules* **2004**, *37*, 8941.
419. Mertoglu, M.; Laschewsky, A.; Skrabania, K.; Wieland, C. *Macromolecules* **2005**, *38*, 3601.
420. Sumerlin, B.S.; Lowe, A.B.; Thomas, D.B.; McCormick, C.L. *Macromolecules* **2003**, *36*, 5982.
421. Vana, P.; Albertin, L.; Barner, L.; Davis, T.P.; Barner-Kowollik, C. *J. Polym. Sci., Part A: Polym. Chem.* **2002**, *40*, 4032.
422. Ah Toy, A.; Vana, P.; Davis, T.P.; Barner-Kowollik, C. *Macromolecules* **2004**, *37*, 744.
423. Lansalot, M.; Davis, T.P.; Heuts, J.P.A. *Macromolecules* **2002**, *35*, 7582.
424. Tang, C.B.; Kowalewski, T.; Matyjaszewski, K. *Macromolecules* **2003**, *36*, 8587.

425. Chong, Y.K.; Le, T.P.T.; Moad, G.; Rizzardo, E.; Thang, S.H. *Macromolecules* **1999**, *32*, 2071.
426. Saricilar, S.; Knott, R.; Barner-Kowollik, C.; Davis, T.P.; Heuts, J.P.A. *Polymer* **2003**, *44*, 5169.
427. Krasia, T.; Soula, R.; Boerner, H.G.; Schlaad, H. *Chem. Commun.* **2003**, 538.
428. Donovan, M.S.; Lowe, A.B.; Sumerlin, B.S.; McCormick, C.L. *Macromolecules* **2002**, *35*, 4123.
429. Ganachaud, F.; Monteiro, M.J.; Gilbert, R.G.; Dourges, M.A.; Thang, S.H.; Rizzardo, E. *Macromolecules* **2000**, *33*, 6738.
430. Convertine, A.J.; Sumerlin, B.S.; Thomas, D.B.; Lowe, A.B.; McCormick, C.L. *Macromolecules* **2003**, *36*, 4679.
431. Perrier, S.; Barner-Kowollik, C.; Quinn, J.F.; Vana, P.; Davis, T.P. *Macromolecules* **2002**, *35*, 8300.
432. An, Q.F.; Qian, J.W.; Yu, L.Y.; Luo, Y.W.; Liu, X.Z. *J. Polym. Sci., Part A: Polym. Chem.* **2005**, *43*, 1973.
433. Hao, X.J.; Heuts, J.P.A.; Barner-Kowollik, C.; Davis, T.P.; Evans, E. *J. Polym. Sci., Part A: Polym. Chem.* **2003**, *41*, 2949.
434. de Brouwer, H.; Tsavalas, J.G.; Schork, F.J.; Monteiro, M.J. *Macromolecules* **2000**, *33*, 9239.
435. Ladaviere, C.; Doerr, N.; Claverie, J.P. *Macromolecules* **2001**, *34*, 5370.
436. Zhu, J.; Zhu, X.; Zhou, D.; Chen, J. *e-Polymers* **2003**, [43].
437. Zhu, J.; Zhu, X.L.; Cheng, Z.P.; Liu, F.; Lu, J.M. *Polymer* **2002**, *43*, 7037.
438. Zhu, J.; Zhou, D.; Zhu, X.L.; Chen, G.J. *J. Polym. Sci., Part A: Polym Chem* **2004**, *42*, 2558.
439. Zhu, J.; Zhu, M.L.; Zhou, D.; Chen, J.Y.; Wang, X.Y. *Eur. Polym. J.* **2004**, *40*, 743.
440. Perrier, S.; Takolpuckdee, P.; Westwood, J.; Lewis, D.M. *Macromolecules* **2004**, *37*, 2709.
441. Takolpuckdee, P.; Mars, C.A.; Perrier, S.; Archibald, S.J. *Macromolecules* **2005**, *38*, 1057.
442. Goto, A.; Sato, K.; Tsujii, Y.; Fukuda, T.; Moad, G.; Rizzardo, E.; Thang, S.H. *Macromolecules* **2001**, *34*, 402.
443. He, T.; Zou, Y.F.; Pan, C.Y. *Polym. J.* **2002**, *34*, 138.
444. Kanagasabapathy, S.; Sudalai, A.; Benicewicz, B.C. *Macromol. Rapid Commun.* **2001**, *22*, 1076.
445. Dureault, A.; Taton, D.; Destarac, M.; Leising, F.; Gnanou, Y. *Macromolecules* **2004**, *37*, 5513.
446. Donovan, M.S.; Lowe, A.B.; Sanford, T.A.; McCormick, C.L. *J. Polym. Sci., Part A: Polym. Chem.* **2003**, *41*, 1262.
447. Taton, D.; Wilczewska, A.Z.; Destarac, M. *Macromol. Rapid Commun.* **2001**, *22*, 1497.
448. Bussels, R.; Bergman-Gottgens, C.; Meuldijk, J.; Koning, C. *Macromolecules* **2004**, *37*, 9299.
449. Moad, G.; Mayadunne, R.T.A.; Rizzardo, E.; Skidmore, M.; Thang, S. *Macromol. Symp.* **2003**, *192*, 1.
450. Moad, G.; Chong, Y.K.; Rizzardo, E.; Postma, A.; Thang, S.H. *Polymer* **2005**, *46*, 8458–8468.
451. Ray, B.; Isobe, Y.; Matsumoto, K.; Habaue, S.; Okamoto, Y.; Kamigaito, M.; Sawamoto, M. *Macromolecules* **2004**, *37*, 1702.
452. Theis, A.; Feldermann, A.; Charton, N.; Stenzel, M.H.; Davis, T.P.; Barner-Kowollik, C. *Macromolecules* **2005**, *38*, 2595.
453. Quinn, J.F.; Rizzardo, E.; Davis, T.P. *Chem. Commun.* **2001**, 1044.
454. Lai, J.T.; Filla, D.; Shea, R. *Macromolecules* **2002**, *35*, 6754.
455. Lima, V.; Jiang, X.L.; Brokken-Zijp, J.; Schoenmakers, P.J.; Klumperman, B.; Van Der Linde, R. *J. Polym. Sci., Part A: Polym. Chem.* **2005**, *43*, 959.

456. Convertine, A.J.; Lokitz, B.S.; Lowe, A.B.; Scales, C.W.; Myrick, L.J.; McCormick, C.L. *Macromol. Rapid Commun.* **2005**, *26*, 791.
457. Mayadunne, R.T.A.; Rizzardo, E.; Chiefari, J.; Krstina, J.; Moad, G.; Postma, A.; Thang, S.H. *Macromolecules* **2000**, *33*, 243.
458. You, Y.Z.; Hong, C.Y.; Wang, W.P.; Lu, W.Q.; Pan, C.Y. *Macromolecules* **2004**, *37*, 9761.
459. Loiseau, J.; Doeerr, N.; Suau, J.M.; Egraz, J.B.; Llauro, M.F.; Ladaviere, C.; Claverie, J. *Macromolecules* **2003**, *36*, 3066.
460. Convertine, A.J.; Ayres, N.; Scales, C.W.; Lowe, A.B.; McCormick, C.L. *Biomacromolecules* **2004**, *5*, 1177.
461. Ferguson, C.J.; Hughes, R.J.; Pham, B.T.T.; Hawkett, B.S.; Gilbert, R.G.; Serelis, A.K.; Such, C.H. *Macromolecules* **2002**, *35*, 9243.
462. Pham, B.T.T.; Nguyen, D.; Ferguson, C.J.; Hawkett, B.S.; Serelis, A.K.; Such, C.H. *Macromolecules* **2003**, *36*, 8907.
463. Moad, G.; Li, G.; Rizzardo, E.; S.H., T.; Pfaendner, R.; Wertmer, H. *Polym. Prepr. (Am. Chem. Soc., Div. Polym. Chem.)* **2005**, *46(2)*, in press.
464. Postma, A.; Davis, T.P.; Moad, G.; O'Shea, M. *Macromolecules* **2005**, *38*, 5371.
465. Monteiro, M.J.; Adamy, M.M.; Leeuwen, B.J.; van Herk, A.M.; Destarac, M. *Macromolecules* **2005**, *38*, 1538.
466. Adamy, M.; van Herk, A.M.; Destarac, M.; Monteiro, M.J. *Macromolecules* **2003**, *36*, 2293.
467. Destarac, M.; Brochon, C.; Catala, J.-M.; Wilczewska, A.; Zard, S.Z. *Macromol. Chem. Phys.* **2002**, *203*, 2281.
468. Monteiro, M.J.; de Barbeyrac, J. *Macromol. Rapid Commun.* **2002**, *23*, 370.
469. Chiefari, J.; Mayadunne, R.T.; Moad, G.; Rizzardo, E.; Thang, S.H. PCT Int. Appl. WO 9931144, 1999 (*Chem. Abstr.* 131:45250).
470. Stenzel, M.H.; Cummins, L.; Roberts, G.E.; Davis, T.P.; Vana, P.; Barner-Kowollik, C. *Macromol. Chem. Phys.* **2003**, *204*, 1160.
471. Simms, R.W.; Davis, T.P.; Cunningham, M.F. *Macromol. Rapid Commun.* **2005**, *26*, 592.
472. Russum, J.P.; Barbre, N.D.; Jones, C.W.; Schork, F.J. *J. Polym. Sci., Part A: Polym. Chem.* **2005**, *43*, 2188.
473. Favier, A.; Barner-Kowollik, C.; Davis, T.P.; Stenzel, M.H. *Macromol. Chem. Phys.* **2004**, *205*, 925.
474. Schilli, C.; Lanzendoerfer, M.G.; Mueller, A.H.E. *Macromolecules* **2002**, *35*, 6819.
475. Schilli, C.M.; Muller, A.H.E.; Rizzardo, E.; Thang, S.H.; Chong, Y.K. *ACS Symp. Ser.* **2003**, *854*, 603.
476. Schilli, C.M.; Zhang, M.F.; Rizzardo, E.; Thang, S.H.; Chong, Y.K.; Edwards, K.; Karlsson, G.; Muller, A.H.E. *Macromolecules* **2004**, *37*, 7861.
477. Meijer, J.; Vermeer, P.; Brandsma, L. *Recueil* **1973**, *92*, 601.
478. Mayadunne, R.A.; Moad, G.; Rizzardo, E. *Tetrahedron Lett.* **2002**, *43*, 6811.
479. Oae, S.; Yagihara, T.; Okabe, T. *Tetrahedron* **1972**, *28*, 3203.
480. Kanagasabapathy, S.; Sudalai, A.; Benicewicz, B.C. *Tetrahedron Lett.* **2001**, *42*, 3791.
481. Thang, S.H.; Chong, Y.K.; Mayadunne, R.T.A.; Moad, G.; Rizzardo, E. *Tetrahedron Lett.* **1999**, *40*, 2435.
482. Bouhadir, G.; Legrand, N.; Quiclet-Sire, B.; Zard, S.Z. *Tetrahedron Lett.* **1999**, *40*, 277.
483. Wager, C.M.; Haddleton, D.M.; Bon, S.A.F. *Eur. Polym. J.* **2004**, *40*, 641.
484. Sudalai, A.; Kanagasabapathy, S.; Benicewicz, B.C. *Org. Lett.* **2000**, *2*, 3213.
485. Perrier, S.; Takolpuckdee, P.; Mars, C.A. *Macromolecules* **2005**, *38*, 2033.
486. Moad, G.; Mayadunne, R.T.A.; Rizzardo, E.; Skidmore, M.; Thang, S. *ACS Symp. Ser.* **2003**, *854*, 520.

487. Chernikova, E.; Morozov, A.; Leonova, E.; Garina, E.; Golubev, V.; Bui, C.O.; Charleux, B. *Macromolecules* **2004**, *37*, 6329.
488. McLeary, J.B.; Calitz, F.M.; McKenzie, J.M.; Tonge, M.P.; Sanderson, R.D.; Klumperman, B. *Macromolecules* **2005**, *38*, 3151.
489. Monteiro, M.J.; de Brouwer, H. *Macromolecules* **2001**, *34*, 349.
490. Barner-Kowollik, C.; Davis, T.P. *Macromol. Theory Simul.* **2001**, *10*, 255.
491. Kwak, Y.; Goto, A.; Fukuda, T. *Macromolecules* **2004**, *37*, 1219.
492. Kwak, Y.; Goto, A.; Tsujii, Y.; Murata, Y.; Komatsu, K.; Fukuda, T. *Macromolecules* **2002**, *38*, 3026.
493. McLeary, J.B.; Calitz, F.M.; McKenzie, J.M.; Tonge, M.P.; Sanderson, R.D.; Klumperman, B. *Macromolecules* **2004**, *37*, 2383.
494. Plummer, R.; Goh, Y.-K.; Whittaker, A.K.; Monteiro, M.J. *Macromolecules* **2005**, *38*, 5352.
495. Quinn, J.F.; Barner, L.; Barner-Kowollik, C.; Rizzardo, E.; Davis, T.P. *Macromolecules* **2002**, *35*, 7620.
496. You, Y.Z.; Hong, C.Y.; Bai, R.K.; Pan, C.Y.; Wang, J. *Macromol. Chem. Phys.* **2002**, *203*, 477.
497. Quinn, J.F.; Barner, L.; Davis, T.P.; Thang, S.H.; Rizzardo, E. *Macromol. Rapid Commun.* **2002**, *23*, 717.
498. Quinn, J.F.; Barner, L.; Rizzardo, E.; Davis, T.P. *J. Polym. Sci., Part A: Polym. Chem.* **2002**, *40*, 19.
499. Barner, L.; Quinn, J.F.; Barner-Kowollik, C.; Vana, P.; Davis, T.P. *Eur. Polym. J.* **2003**, *39*, 449.
500. Bai, R.K.; You, Y.Z.; Zhong, P.; Pan, C.Y. *Macromol. Chem. Phys.* **2001**, *202*, 1970.
501. Hong, C.Y.; You, Y.Z.; Bai, R.K.; Pan, C.Y.; Borjihan, G. *J. Polym. Sci., Part A: Polym. Chem.* **2001**, *39*, 3934.
502. Bai, R.K.; You, Y.Z.; Pan, C.Y. *Macromol. Rapid Commun.* **2001**, *22*, 315.
503. You, Y.Z.; Bai, R.K.; Pan, C.Y. *Macromol. Chem. Phys.* **2001**, *202*, 1980.
504. Chen, G.; Zhu, X.; Zhu, J.; Cheng, Z. *Macromol. Rapid Commun.* **2004**, *25*, 818.
505. Lowe, A.B.; Sumerlin, B.S.; Donovan, M.S.; Thomas, D.B.; Hennauxz, P.; McCormick, C.L. *ACS Symp. Ser.* **2003**, *854*, 586.
506. Lowe, A.B.; McCormick, C.L. *Aust. J. Chem.* **2002**, *55*, 367.
507. Baussard, J.F.; Habib-Jiwan, J.L.; Laschewsky, A.; Mertoglu, M.; Storsberg, J. *Polymer* **2004**, *45*, 3615.
508. Perrier, S.; Davis, T.P.; Carmichael, A.J.; Haddleton, D.M. *Chem. Commun.* **2002**, 2226.
509. Arita, T.; Buback, M.; Janssen, O.; Vana, P. *Macromol. Rapid Commun.* **2004**, *25*, 1376.
510. Arita, T.; Beuermann, S.; Buback, M.; Vana, P. *Macromol. Mater. Eng.* **2005**, *290*, 283.
511. Thomas, D.B.; Convertine, A.J.; Hester, R.D.; Lowe, A.B.; McCormick, C.L. *Macromolecules* **2004**, *37*, 1735.
512. Monteiro, M.J.; Bussels, R.; Beuermann, S.; Buback, M. *Aust. J. Chem.* **2002**, *55*, 433.
513. Rzayev, J.; Penelle, J. *Angew. Chem. Int. Ed. Engl.* **2004**, *43*, 1691.
514. Chong, Y.K.; Moad, G.; Rizzardo, M.; Skidmore, M.A.; Thang, S. In *27th Australian Polymer Symposium*; RACI Polymer Division: Adelaide, SA, 2004; p C1/3.
515. Ray, B.; Isobe, Y.; Morioka, K.; Habaue, S.; Okamoto, Y.; Kamigaito, M.; Sawamoto, M. *Macromolecules* **2003**, *36*, 543.
516. Lutz, J.F.; Jakubowski, W.; Matyjaszewski, K. *Macromol. Rapid Commun.* **2004**, *25*, 486.
517. Lutz, J.F.; Neugebauer, D.; Matyjaszewski, K. *J. Am. Chem. Soc.* **2003**, *125*, 6986.

518. Kirci, B.; Lutz, J.F.; Matyjaszewski, K. *Macromolecules* **2002**, *35*, 2448.
519. Lutz, J.F.; Kirci, B.; Matyjaszewski, K. *Macromolecules* **2003**, *36*, 3136.
520. Monteiro, M.J.; Hodgson, M.; De Brouwer, H. *J. Polym. Sci., Part A: Polym. Chem.* **2000**, *38*, 3864.
521. Monteiro, M.J.; de Barbeyrac, J. *Macromolecules* **2001**, *34*, 4416.
522. Monteiro, M.J.; Sjoberg, M.; van der Vlist, J.; Gottgens, C.M. *J. Polym. Sci., Part A: Polym. Chem.* **2000**, *38*, 4206.
523. Uzulina, I.; Kanagasabapathy, S.; Claverie, J. *Macromol. Symp.* **2000**, *150*, 33.
524. Smulders, W.; Gilbert, R.G.; Monteiro, M.J. *Macromolecules* **2003**, *36*, 4309.
525. Prescott, S.W.; Ballard, M.J.; Rizzardo, E.; Gilbert, R.G. *Aust. J. Chem.* **2002**, *55*, 415.
526. Prescott, S.W.; Ballard, M.J.; Rizzardo, E.; Gilbert, R.G. *Macromolecules* **2002**, *35*, 5417.
527. Nozari, S.; Tauer, K. *Polymer* **2005**, *46*, 1033.
528. Prescott, S.W. *Macromolecules* **2003**, *36*, 9608.
529. Such, C.H.; Rizzardo, E.; Serelis, A.K.; Hawkett, B.S.; Gilbert, R.G.; Ferguson, C.J.; Hughes, R.J. WO 03055919, 2003 (*Chem. Abstr.* 139, 101540v).
530. Lansalot, M.; Farcet, C.; Charleux, B.; Vairon, J.P.; Pirri, R. *Macromolecules* **1999**, *32*, 7354.
531. Russum, J.P.; Jones, C.W.; Schork, F.J. *Macromol. Rapid Commun.* **2004**, *25*, 1064.
532. Tsavalas, J.G.; Schork, F.J.; de Brouwer, H.; Monteiro, M.J. *Macromolecules* **2001**, *34*, 3938.
533. Haszeldine, R.N. *J. Chem. Soc.* **1949**, 2859.
534. Tatemoto, M. *Kobunshi Ronbunshu* **1992**, *49*, 765.
535. Gaynor, S.; Wang, J.-S.; Matyjaszewski, K. *Macromolecules* **1995**, *28*, 8051.
536. Goto, A.; Ohno, K.; Fukuda, T. *Macromolecules* **1998**, *31*, 2809.
537. Ueda, N.; Kamigaito, M.; Sawamoto, M. *Polym. Prepr. Japan* **1996**, *45*, E622.
538. Iovu, M.C.; Matyjaszewski, K. *Macromolecules* **2003**, *36*, 9346.
539. Ameduri, B.; Boutevin, B. *J. Fluorine Chem.* **1999**, *100*, 97.
540. Goto, A.; Kwak, Y.; Fukuda, T.; Yamago, S.; Iida, K.; Nakajima, M.; Yoshida, J. *J. Am. Chem. Soc.* **2003**, *125*, 8720.
541. Yamago, S.; Iida, K.; Yoshida, J. *J. Am. Chem. Soc.* **2002**, *124*, 2874.
542. Yamago, S.; Iida, K.; Yoshida, J.-i. *J. Am. Chem. Soc.* **2002**, *124*, 13666.
543. Yamago, S.; Iida, K.; Nakajima, M.; Yoshida, J. *Macromolecules* **2003**, *36*, 3793.
544. Yamago, S.; Ray, B.; Iida, K.; Yoshida, J.; Tada, T.; Yoshizawa, K.; Kwak, Y.; Goto, A.; Fukuda, T. *J. Am. Chem. Soc.* **2004**, *126*, 13908.
545. Kwak, Y.W.; Goto, A.; Fukuda, T.; Yamago, S.; Ray, B. *Z. Phys. Chem.* **2005**, *219*, 283.
546. Haddleton, D.M.; Crossman, M.C.; Hunt, K.H.; Topping, C.; Waterson, C.; Suddaby, K.G. *Macromolecules* **1997**, *30*, 3992.
547. Klumperman, B.; Chambard, G.; Brinkhuis, R.H.G. *ACS Symp. Ser.* **2003**, *854*, 180.
548. Shinoda, H.; Matyjaszewski, K.; Okrasa, L.; Mierzwa, M.; Pakula, T. *Macromolecules* **2003**, *36*, 4772.
549. Feldermann, A.; Toy, A.A.; Phan, H.; Stenzel, M.H.; Davis, T.P.; Barner-Kowollik, C. *Polymer* **2004**, *45*, 3997.
550. Benoit, D.; Hawker, C.J.; Huang, E.E.; Lin, Z.Q.; Russell, T.P. *Macromolecules* **2000**, *33*, 1505.
551. De Brouwer, H.; Schellekens, M.A.J.; Klumperman, B.; Monteiro, M.J.; German, A.L. *J. Polym. Sci., Part A: Polym. Chem.* **2000**, *38*, 3596.
552. Du, F.S.; Zhu, M.Q.; Guo, H.Q.; Li, Z.C.; Li, F.M.; Kamachi, M.; Kajiwara, A. *Macromolecules* **2002**, *35*, 6739.
553. Zaremski, M.Y.; Plutalova, A.V.; Lachinov, M.B.; Golubev, V.B. *Macromolecules* **2000**, *33*, 4365.

554. Davis, K.A.; Matyjasewski, K. *Adv. Polym. Sci.* **2002**, *159*, 2.
555. Gray, M.K.; Zhou, H.Y.; Nguyen, S.T.; Torkelson, J.M. *Polymer* **2004**, *45*, 4777.
556. Hawker, C.J.; Elce, E.; Dao, J.; Volksen, W.; Russell, T.P.; Barclay, G.G. *Macromolecules* **1996**, *29*, 2686.
557. Gray, M.K.; Zhou, H.Y.; Nguyen, S.T.; Torkelson, J.M. *Macromolecules* **2004**, *37*, 5586.
558. Yoshida, E.; Takiguchi, Y. *Polym. J.* **1999**, *31*, 429.
559. Butz, S.; Baethge, H.; Schmidt-Naake, G. *Macromol. Rapid Commun.* **1997**, *18*, 1049.
560. Fukuda, T.; Terauchi, T.; Goto, A.; Tsujii, Y.; Miyamoto, T. *Macromolecules* **1996**, *29*, 3050.
561. Cuervo-Rodriguez, R.; Bordege, V.; Fernandez-Monreal, M.C.; Fernandez-Garcia, M.; Madruga, E.L. *J. Polym. Sci., Part A: Polym. Chem.* **2004**, *42*, 4168.
562. Baumert, M.; Mülhaupt, R. *Macromol. Rapid Commun.* **1997**, *18*, 787.
563. Taube, C.; Garcia, M.F.; Schmidt-Naake, G.; Fischer, H. *Macromol. Chem. Phys.* **2002**, *203*, 2665.
564. Baethge, H.; Butz, S.; Schmidt-Naake, G. *Macromol. Rapid Commun.* **1997**, *18*, 911.
565. Braunecker, W.A.; Tsarevsky, N.V.; Pintauer, T.; Gil, R.R.; Matyjaszewski, K. *Macromolecules* **2005**, *38*, 4081.
566. Venkatesh, R.; Vergouwen, F.; Klumperman, B. *Macromol. Chem. Phys.* **2005**, *206*, 547.
567. Arehart, S.V.; Matyjaszewski, K. *Macromolecules* **1999**, *32*, 2221.
568. Qin, S.H.; Saget, J.; Pyun, J.R.; Jia, S.J.; Kowalewski, T.; Matyjaszewski, K. *Macromolecules* **2003**, *36*, 8969.
569. Kotani, Y.; Kamigaito, M.; Sawamoto, M. *Macromolecules* **1998**, *31*, 5582.
570. Hong, S.C.; Lutz, J.F.; Inoue, Y.; Strissel, C.; Nuyken, O.; Matyjaszewski, K. *Macromolecules* **2003**, *36*, 1075.
571. Matyjaszewski, K.; Shipp, D.A.; McMurtry, G.P.; Gaynor, S.G.; Pakula, T. *J. Polym. Sci., Part A: Polym. Chem.* **2000**, *38*, 2023.
572. Borkar, S.; Sen, A. *Macromolecules* **2005**, *38*, 3029.
573. Gao, B.; Chen, X.; Ivan, B.; Kops, J.; Batsberg, W. *Polym. Bull.* **1997**, *39(5)*, 559.
574. Lutz, J.F.; Pakula, T.; Matyjaszewski, K. *ACS Symp. Ser.* **2003**, *854*, 268.
575. Tsarevsky, N.V.; Sarbu, T.; Gobelt, B.; Matyjaszewski, K. *Macromolecules* **2002**, *35*, 6142.
576. Luo, Y.W.; Liu, X.Z. *J. Polym. Sci., Part A: Polym Chem* **2004**, *42*, 6248.
577. Savariar, E.N.; Thayumanavan, S. *J. Polym. Sci., Part A: Polym Chem* **2004**, *42*, 6340.
578. Chernikova, E.; Terpugova, P.; Bui, C.; Charleux, B. *Polymer* **2003**, *44*, 4101.
579. Davies, M.C.; Dawkins, J.V.; Hourston, D.J. *Polymer* **2005**, *46*, 1739.
580. Zhu, M.Q.; Wei, L.H.; Li, M.; Jiang, L.; Du, F.S.; Li, Z.C.; Li, F.M. *Chem. Commun.* **2001**, 365.
581. Beyou, E.; Chaumont, P.; Chauvin, F.; Devaux, C.; Zydowicz, N. *Macromolecules* **1998**, *31*, 6828.
582. Harth, E.; Hawker, C.J.; Fan, W.; Waymouth, R.M. *Macromolecules* **2001**, *34*, 3856.
583. Bon, S.A.F.; Chambard, G.; German, A.L. *Macromolecules* **1999**, *32*, 8269.
584. Turro, N.J.; Lem, G.; Zavarine, I.S. *Macromolecules* **2000**, *33*, 9782.
585. Otsuka, H.; Aotani, K.; Higaki, Y.; Takahara, A. *J. Am. Chem. Soc.* **2003**, *125*, 4064.
586. Higaki, Y.; Otsuka, H.; Takahara, A. *Macromolecules* **2004**, *37*, 1696.
587. Chen, X.; Gao, B.; Kops, J.; Batsberg, W. *Polymer* **1998**, *39*, 911.
588. Hawker, C.J.; Malmstroem, E.E.; Frechet, J.M.J.; Leduc, M.R.; Grubbs, R.B.; Barclay, G.G. *ACS Symp. Ser.* **1998**, *685*, 433.

589. Kobatake, S.; Harwood, H.J.; Quirk, R.P.; Priddy, D.B. *Macromolecules* **1997**, *30*, 4238.
590. Mather, B.D.; Lizotte, J.R.; Long, T.E. *Macromolecules* **2004**, *37*, 9331.
591. Miura, Y.; Yoshida, Y. *Macromol. Chem. Phys.* **2002**, *203*, 879.
592. Coessens, V.; Pintauer, T.; Matyjaszewski, K. *Prog. Polym. Sci.* **2001**, *26*, 337.
593. Snijder, A.; Klumperman, B.; Van der Linde, R. *J. Polym. Sci., Part A: Polym. Chem.* **2002**, *40*, 2350.
594. Teoh, S.K.; Ravi, P.; Dai, S.; Tam, K.C. *J. Phys. Chem. B* **2005**, *109*, 4431.
595. Zhou, P.; Chen, G.Q.; Hong, H.; Du, F.S.; Li, Z.C.; Li, F.M. *Macromolecules* **2000**, *33*, 1948.
596. Bon, S.A.F.; Morsley, S.R.; Waterson, C.; Haddleton, D.M. *Macromolecules* **2000**, *33*, 5819.
597. Bon, S.A.F.; Steward, A.G.; Haddleton, D.M. *J. Polym. Sci., Part A: Polym. Chem.* **2000**, *38*, 2678.
598. Bielawski, C.W.; Jethmalani, J.M.; Grubbs, R.H. *Polymer* **2003**, *44*, 3721.
599. Ando, T.; Kamigaito, M.; Sawamoto, M. *Macromolecules* **1998**, *31*, 6708.
600. Coessens, V.; Pyun, J.; Miller, P.J.; Gaynor, S.G.; Matyjaszewski, K. *Macromol. Rapid Commun.* **2000**, *21*, 103.
601. Koulouri, E.G.; Kallitsis, J.K.; Hadziioannou, G. *Macromolecules* **1999**, *32*, 6242.
602. Chambard, G.; Klumperman, B.; German, A.L. *Macromolecules* **2000**, *33*, 4417.
603. Lutz, J.F.; Matyjaszewski, K. *J. Polym. Sci., Part A: Polym. Chem.* **2005**, *43*, 897.
604. Lutz, J.F.; Matyjaszewski, K. *Macromol. Chem. Phys.* **2002**, *203*, 1385.
605. Coessens, V.; Matyjaszewski, K. *Macromol. Rapid Commun.* **1999**, *20*, 66.
606. Coessens, V.; Nakagawa, Y.; Matyjaszewski, K. *Polym. Bull.* **1998**, *40*, 135.
607. Matyjaszewski, K.; Nakagawa, Y.; Gaynor, S.G. *Macromol. Rapid Commun.* **1997**, *18*, 1057.
608. Postma, A.; Moad, G.; Davis, T.P.; O'Shea, M. *React. Funct. Polym.* **2005**, in press.
609. Coessens, V.; Matyjaszewski, K. *J. Macromol. Sci. Chem.* **1999**, *A36*, 811.
610. Coessens, V.; Matyjaszewski, K. *Macromol. Rapid Commun.* **1999**, *20*, 127.
611. Zhang, H.; Jiang, X.; van der Linde, R. *Polymer* **2004**, *45*, 1455.
612. Muehlebach, A.; Rime, F. *J. Polym. Sci., Part A: Polym. Chem.* **2003**, *41*, 3425.
613. Malz, H.; Komber, H.; Voigt, D.; Hopfe, I.; Pionteck, J. *Macromol. Chem. Phys.* **1999**, *200*, 642.
614. Coessens, V.; Matyjaszewski, K. *J. Macromol. Sci. Chem.* **1999**, *A36*, 653.
615. Matyjaszewski, K. *ACS Symp. Ser.* **2000**, *768*, 2.
616. Matyjaszewski, K.; Beers, K.L.; Kern, A.; Gaynor, S.G. *J. Polym. Sci., Part A: Polym. Chem.* **1998**, *36*, 823.
617. Baek, K.Y.; Kamigaito, M.; Sawamoto, M. *J. Polym. Sci., Part A: Polym. Chem.* **2002**, *40*, 1937.
618. Sarbu, T.; Lin, K.Y.; Spanswick, J.; Gil, R.R.; Siegwart, D.J.; Matyjaszewski, K. *Macromolecules* **2004**, *37*, 9694.
619. Zhang, X.; Matyjaszewski, K. *Macromolecules* **1999**, *32*, 7349.
620. Haddleton, D.M.; Waterson, C. *Macromolecules* **1999**, *32*, 8732.
621. Ji, S.; Hoye, T.R.; Macosko, C.W. *Macromolecules* **2005**, *38*, 4679
622. Carrot, G.; Hilborn, J.; Hedrick, J.L.; Trollsas, M. *Macromolecules* **1999**, *32*, 5171.
623. Castro, E.A. *Chem. Rev.* **1999**, *99*, 3505.
624. Kulkarni, S.; Schilli, C.; Muller, A.H.E.; Hoffman, A.S.; Stayton, P.S. *Bioconjugate Chemistry* **2004**, *15*, 747.
625. Chen, M.; Ghiggino, K.P.; Thang, S.H.; White, J.; Wilson, G.J. *J. Org. Chem* **2005**, *70*, 1844.
626. Mayadunne, R.T.A.; Jeffery, J.; Moad, G.; Rizzardo, E. *Macromolecules* **2003**, *36*, 1505.
627. Wang, Z.M.; He, J.P.; Tao, Y.F.; Yang, L.; Jiang, H.J.; Yang, Y.L. *Macromolecules* **2003**, *36*, 7446.

628. Sumerlin, B.S.; Lowe, A.B.; Stroud, P.A.; Zhang, P.; Urban, M.W.; McCormick, C.L. *Langmuir* **2003**, *19*, 5559.
629. Riess, G. *Prog. Polym. Sci.* **2003**, *28*, 1107.
630. Mariani, M.; Lelli, M.; Sparnacci, K.; Laus, M. *J. Polym. Sci., Part A: Polym. Chem.* **1999**, *37*, 1237.
631. Ohno, K.; Izu, Y.; Tsujii, Y.; Fukuda, T.; Kitano, H. *Eur. Polym. J.* **2004**, *40*, 81.
632. Jousset, S.; Hammouch, S.O.; Catala, J.M. *Macromolecules* **1997**, *30*, 6685.
633. Lacroix-Desmazes, P.; Delair, T.; Pichot, C.; Boutevin, B. *J. Polym. Sci. Pol. Chem.* **2000**, *38*, 3845.
634. Gabaston, L.I.; Furlong, S.A.; Jackson, R.A.; Armes, S.P. *Polymer* **1999**, *40*, 4505.
635. Stancik, C.M.; Lavoie, A.R.; Schutz, J.; Achurra, P.A.; Lindner, P.; Gast, A.P.; Waymouth, R.M. *Langmuir* **2004**, *20*, 596.
636. Burguiere, C.; Pascual, S.; Bui, C.; Vairon, J.P.; Charleux, B.; Davis, K.A.; Matyjaszewski, K.; Betremieux, I. *Macromolecules* **2001**, *34*, 4439.
637. Zhang, X.; Matyjaszewski, K. *Macromolecules* **1999**, *32*, 1763.
638. Mitsukami, Y.; Donovan, M.S.; Lowe, A.B.; McCormick, C.L. *Macromolecules* **2001**, *34*, 2248.
639. Sumerlin, B.S.; Lowe, A.B.; Thomas, D.B.; Convertine, A.J.; Donovan, M.S.; McCormick, C.L. *J. Polym. Sci., Part A: Polym. Chem.* **2004**, *42*, 1724.
640. Chen, X.Y.; Gao, B.; Kops, J.; Batsberg, W. *Polymer* **1998**, *39*, 911.
641. Hawker, C.J.; Hedrick, J.L.; Malmstroem, E.E.; Trollss, M.; Mecerreyes, D.; Moineau, G.; Dubois, P.; Jerome, R. *Macromolecules* **1998**, *31*, 213.
642. Kobatake, S.; Harwood, H.J.; Quirk, R.P.; Priddy, D.B. *Macromolecules* **1999**, *32*, 10.
643. Lu, G.Q.; Jia, Z.F.; Yi, W.; Huang, J.L. *J. Polym. Sci., Part A: Polym. Chem.* **2002**, *40*, 4404.
644. Li, Z.Y.; Lu, G.Q.; Huang, J.L. *J. Appl. Polym. Sci.* **2004**, *94*, 2280.
645. Jankova, K.; Chen, X.; J., k.; Batsberg, W. *Macromolecules* **1998**, *31*, 538.
646. Save, M.; Weaver, J.V.M.; Armes, S.P.; McKenna, P. *Macromolecules* **2002**, *35*, 1152.
647. Bielawski, C.W.; Morita, T.; Grubbs, R.H. *Macromolecules* **2000**, *33*, 678.
648. Destarac, M.; Boutevin, B. *Macromol. Rapid Commun.* **1999**, *20*, 641.
649. Destarac, M.; Pees, B.; Boutevin, B. *Macromol. Chem. Phys.* **2000**, *201*, 1189.
650. Semsarzadeh, M.A.; Mirzaei, A.; Vasheghani-Farahani, E.; Haghighi, M.N. *Eur. Polym. J.* **2003**, *39*, 2193.
651. Shi, L.J.; Chapman, T.M.; Beckman, E.J. *Macromolecules* **2003**, *36*, 2563.
652. Hong, C.Y.; You, Y.Z.; Pan, C.Y. *J. Polym. Sci., Part A: Polym Chem* **2004**, *42*, 4873.
653. Ma, Z.; Lacroix-Desmazes, P. *Polymer* **2004**, *45*, 6789.
654. Hales, M.; Barner-Kowollik, C.; Davis, T.P.; Stenzel, M.H. *Langmuir* **2004**, *20*, 10809.
655. Tsujii, Y.; Ejaz, M.; Sato, K.; Goto, A.; Fukuda, T. *Macromolecules* **2001**, *34*, 8872.
656. Higaki, Y.; Otsuka, H.; Endo, T.; Takahara, A. *Macromolecules* **2003**, *36*, 1494.
657. Higaki, Y.; Otsuka, H.; Takahara, A. *Polymer* **2003**, *44*, 7095.
658. Motokucho, S.; Sudo, A.; Sanda, F.; Endo, T. *Chem. Commun.* **2002**, 1946.
659. You, Y.Z.; Hong, C.Y.; Pan, C.Y. *Chem. Commun.* **2002**, 2800.
660. You, Y.Z.; Hong, C.Y.; Wang, P.H.; Wang, W.P.; Lu, W.Q.; Pan, C.Y. *Polymer* **2004**, *45*, 4647.
661. Hong, J.; Wang, Q.; Lin, Y.Z.; Fan, Z.Q. *Macromolecules* **2005**, *38*, 2691.
662. Schaefgen, J.R.; Flory, J. *J. Am. Chem. Soc.* **1948**, *70*, 2709.
663. Hawker, C.J. *Angew. Chem., Int. Ed. Engl.* **1995**, *34*, 1456.
664. Matyjaszewski, K.; Miller, P.J.; Pyun, J.; Kickelbick, G.; Diamanti, S. *Macromolecules* **1999**, *32*, 6526.

665. Chen, M.; Ghiggino, K.P.; Launikonis, A.; Mau, A.W.H.; Rizzardo, E.; Sasse, W.H.F.; Thang, S.H.; Wilson, G.J. *J. Mater. Chem.* **2003**, *13*, 2696.
666. Barner, L.; Li, C.; Hao, X.J.; Stenzel, M.H.; Barner-Kowollik, C.; Davis, T.P. *J. Polym. Sci., Part A: Polym. Chem.* **2004**, *42*, 5067.
667. Ueda, J.; Matsuyama, M.; Kamigaito, M.; Sawamoto, M. *Macromolecules* **1998**, *31*, 557.
668. Matyjaszewski, K.; Miller, P.J.; Fossum, E.; Nakagawa, Y. *Appl. Organomet. Chem.* **1998**, *12*, 667.
669. Lecolley, F.; Waterson, C.; Carmichael, A.J.; Mantovani, G.; Harrisson, S.; Chappell, H.; Limer, A.; Williams, P.; Ohno, K.; Haddleton, D.M. *J. Mater. Chem.* **2003**, *13*, 2689.
670. Haddleton, D.M.; Edmonds, R.; Heming, A.M.; Kelly, E.J.; Kukulj, D. *New J. Chem.* **1999**, *23*, 477.
671. Angot, S.; Murthy, K.S.; Taton, D.; Gnanou, Y. *Macromolecules* **1998**, *31*, 7218.
672. Wang, J.-S.; Greszta, D.; Matyjaszewski, K. *Polym. Mater. Sci. Eng.* **1995**, *73*, 416.
673. Hedrick, J.L.; Trollsas, M.; Hawker, C.J.; Atthoff, B.; Claesson, H.; Heise, A.; Miller, R.D.; Mecerreyes, D.; Jerome, R.; Dubois, P. *Macromolecules* **1998**, *31*, 8691.
674. Zhao, Y.L.; Jiang, J.; Liu, H.W.; Chen, C.F.; Xi, F. *J. Polym. Sci., Part A: Polym. Chem.* **2001**, *39*, 3960.
675. Darcos, V.; Dureault, A.; Taton, D.; Gnanou, Y.; Marchand, P.; Caminade, A.M.; Majoral, J.P.; Destarac, M.; Leising, F. *Chem. Commun.* **2004**, 2110.
676. Hao, X.J.; Nilsson, C.; Jesberger, M.; Stenzel, M.H.; Malmstrom, E.; Davis, T.P.; Ostmark, E.; Barner-Kowollik, C. *J. Polym. Sci., Part A, Polym Chem* **2004**, *42*, 5877.
677. You, Y.Z.; Hong, C.Y.; Pan, C.Y.; Wang, P.H. *Adv. Mater.* **2004**, *16*, 1953.
678. Jesberger, M.; Barner, L.; Stenzel, M.H.; Malmstrom, E.; Davis, T.P.; Barner-Kowollik, C. *J. Polym. Sci., Part A: Polym. Chem.* **2003**, *41*, 3847.
679. Ohno, K.; Wong, B.; Haddleton, D.M. *J. Polym. Sci., Part A: Polym. Chem.* **2001**, *39*, 2206.
680. Haddleton, D.M.; Ohno, K. *Biomacromolecules* **2000**, *1*, 152.
681. Stenzel-Rosenbaum, M.H.; Davis, T.P.; Chen, V.K.; Fane, A.G. *Macromolecules* **2001**, *34*, 5433.
682. Stenzel, M.H.; Davis, T.P. *J. Polym. Sci., Part A: Polym. Chem.* **2002**, *40*, 4498.
683. Bian, K.J.; Cunningham, M.F. *J. Polym. Sci., Part A: Polym. Chem.* **2005**, *43*, 2145.
684. Hodges, J.C.; Harikrishnan, L.S.; Ault-Justus, S. *J. Comb. Chem.* **2000**, *2*, 80.
685. Ayres, N.; Haddleton, D.M.; Shooter, A.J.; Pears, D.A. *Macromolecules* **2002**, *35*, 3849.
686. Angot, S.; Ayres, N.; Bon, S.A.F.; Haddleton, D.M. *Macromolecules* **2001**, *34*, 768.
687. Narrainen, A.P.; Pascual, S.; Haddleton, D.M. *J. Polym. Sci., Part A: Polym. Chem.* **2002**, *40*, 439.
688. Percec, V.; Barboiu, B.; Grigoras, C.; Bera, T.K. *J. Am. Chem. Soc.* **2003**, *125*, 6503.
689. Stenzel-Rosenbaum, M.; Davis, T.P.; Chen, V.; Fane, A.G. *J. Polym. Sci., Part A: Polym. Chem.* **2001**, *39*, 2777.
690. Stenzel, M.H.; Davis, T.P.; Barner-Kowollik, C. *Chem. Commun.* **2004**, 1546.
691. Barner, L.; Barner-Kowollik, C.; Davis, T.P.; Stenzel, M.H. *Aust. J. Chem.* **2004**, *57*, 19.
692. Abrol, S.; Kambouris, P.A.; Looney, M.G.; Solomon, D.H. *Macromol. Rapid Commun.* **1997**, *18*, 755.
693. Abrol, S.; Caulfield, M.J.; Qiao, G.G.; Solomon, D.H. *Polymer* **2001**, *42*, 5987.
694. Bosman, A.W.; Heumann, A.; Klaerner, G.; Benoit, D.; Frechet, J.M.J.; Hawker, C.J. *J. Am. Chem. Soc.* **2001**, *123*, 6461.

695. Bosman, A.W.; Vestberg, R.; Heumann, A.; Frechet, J.M.J.; Hawker, C.J. *J. Am. Chem. Soc.* **2003**, *125*, 715.
696. Pasquale, A.J.; Long, T.E. *J. Polym. Sci., Part A: Polym. Chem.* **2001**, *39*, 216.
697. Xia, J.H.; Zhang, X.; Matyjaszewski, K. *Macromolecules* **1999**, *32*, 4482.
698. Zhang, X.; Xia, J.H.; Matyjaszewski, K. *Macromolecules* **2000**, *33*, 2340.
699. Gurr, P.A.; Qiao, G.G.; Solomon, D.H.; Harton, S.E.; Spontak, R.J. *Macromolecules* **2003**, *36*, 5650.
700. Connal, L.A.; Gurr, P.A.; Qiao, G.G.; Solomon, D.H. *J. Mater. Chem.* **2005**, *15*, 1286.
701. Lord, H.T.; Quinn, J.F.; Angus, S.D.; Whittaker, M.R.; Stenzel, M.H.; Davis, T.P. *J. Mater. Chem.* **2003**, *13*, 2819.
702. Zheng, G.H.; Pan, C.Y. *Polymer* **2005**, *46*, 2802.
703. Tsoukatos, T.; Pispas, S.; Hadjichristidis, N. *J. Polym. Sci., Part A: Polym. Chem.* **2001**, *39*, 320.
704. Matyjaszewski, K. *Polym. Int.* **2003**, *52*, 1559.
705. Wooley, K.L. *J. Polym. Sci., Part A: Polym. Chem.* **2000**, *38*, 1397.
706. Becker, M.L.; Liu, J.Q.; Wooley, K.L. *Biomacromolecules* **2005**, *6*, 220.
707. Becker, M.L.; Liu, J.Q.; Wooley, K.L. *Chem. Commun.* **2003**, 180.
708. Butun, V.; Lowe, A.B.; Billingham, N.C.; Armes, S.P. *J. Am. Chem. Soc* **1999**, *121*, 4288.
709. Fujii, S.; Cai, Y.L.; Weaver, J.V.M.; Armes, S.P. *J. Am. Chem. Soc* **2005**, *127*, 7304.
710. Liu, S.Y.; Weaver, J.V.M.; Save, M.; Armes, S.P. *Langmuir* **2002**, *18*, 8350.
711. Yates, C.R.; Hayes, W. *Eur. Polym. J.* **2004**, *40*, 1257.
712. Gao, C.; Yan, D. *Prog. Polym. Sci.* **2004**, *29*, 183.
713. Inoue, K. *Prog. Polym. Sci.* **2000**, *25*, 453.
714. Voit, B. *J. Polym. Sci., Part A: Polym. Chem.* **2000**, *38*, 2505.
715. Flory, P.J. *J. Am. Chem. Soc* **1952**, *74*, 2718.
716. Hawker, C.J.; Frechet, J.M.J.; Grubbs, R.B.; Dao, J. *J. Am. Chem. Soc.* **1995**, *117*, 10763.
717. Ignatova, M.; Voccia, S.; Gilbert, B.; Markova, N.; Mercuri, P.S.; Galleni, M.; Sciannamea, V.; Lenoir, S.; Cossement, D.; Gouttebaron, R.; Jerome, R.; Jerome, C. *Langmuir* **2004**, *20*, 10718.
718. Gaynor, S.G.; Edelman, S.; Matyjaszewski, K. *Macromolecules* **1996**, *29*, 1079.
719. Weimer, M.W.; Frechet, J.M.J.; Gitsov, I. *J. Polym. Sci., Part A: Polym. Chem.* **1998**, *36*, 955.
720. Matyjaszewski, K.; Gaynor, S.G.; Kulfan, A.; Podwika, M. *Macromolecules* **1997**, *30*, 5192.
721. Matyjaszewski, K.; Gaynor, S.G.; Muller, A.H.E. *Macromolecules* **1997**, *30*, 7034.
722. Matyjaszewski, K.; Gaynor, S.G. *Macromolecules* **1997**, *30*, 7042.
723. Matyjaszewski, K.; Pyun, J.; Gaynor, S.G. *Macromol. Rapid Commun.* **1998**, *19*, 665.
724. Percec, V.; Grigoras, C.; Kim, H.J. *J. Polym. Sci., Part A: Polym. Chem.* **2004**, *42*, 505.
725. Lepoittevin, N.; Matmour, R.; Francis, R.; Taton, D.; Gnanou, Y. *Macromolecules* **2005**, *38*, 3120.
726. Puts, R.D.; Sogah, D.Y. *Macromolecules* **1997**, *30*, 7050.
727. Ydens, I.; Degee, P.; Libiszowski, J.; Duda, A.; Penczek, S.; Dubois, P. *ACS Symp. Ser.* **2003**, *854*, 283.
728. Ydens, I.; Degee, P.; Dubois, P.; Libiszowski, J.; Duda, A.; Penczek, S. *Macromol. Chem. Phys.* **2003**, *204*, 171.
729. Zhang, A.; Wei, L.H.; Schluter, A.D. *Macromol. Rapid Commun.* **2004**, *25*, 799.
730. Khousakoun, E.; Gohy, J.F.; Jerome, R. *Polymer* **2004**, *45*, 8303.
731. Shinoda, H.; Matyjaszewski, K. *Macromol. Rapid Commun.* **2001**, *22*, 1176.

732. Lutz, J.F.; Jahed, N.; Matyjaszewski, K. *J. Polym. Sci., Part A: Polym. Chem.* **2004**, *42*, 1939.
733. Sprong, E.; De Wet-Roos, D.; Tonge, M.P.; Sanderson, R.D. *J. Polym. Sci., Part A: Polym. Chem.* **2003**, *41*, 223.
734. Venkatesh, R.; Yajjou, L.; Koning, C.E.; Klumperman, B. *Macromol. Chem. Phys.* **2004**, *205*, 2161.
735. Li, Y.G.; Shi, P.J.; Zhou, Y.S.; Pan, C.Y. *Polym. Int.* **2004**, *53*, 349.
736. Frauenrath, H. *Prog. Polym. Sci.* **2005**, *30*, 325.
737. Frey, H. *Angew. Chem. Int. Ed. Engl.* **1998**, *37*, 2193.
738. Schluter, A.D.; Rabe, J.P. *Angew. Chem. Int. Ed. Engl.* **2000**, *39*, 864.
739. Malkoch, M.; Carlmark, A.; Wodegiorgis, A.; Hult, A.; Malmstrom, E.E. *Macromolecules* **2004**, *37*, 322.
740. Cheng, C.X.; Tang, R.P.; Zhao, Y.L.; Xi, F. *J. Appl. Polym. Sci.* **2004**, *91*, 2733.
741. Carlmark, A.; Malmstrom, E.E. *Macromolecules* **2004**, *37*, 7491.
742. Edmondson, S.; Osborne, V.L.; Huck, W.T.S. *Chem. Soc. Rev.* **2004**, *33*, 14.
743. Pyun, J.; Kowalewski, T.; Matyjaszewski, K. *Macromol. Rapid Commun.* **2003**, *24*, 1043.
744. Appelt, M.; Schmidt-Naake, G. *Macromol. Mater. Eng.* **2004**, *289*, 245.
745. Fu, G.D.; Zong, B.Y.; Kang, E.T.; Neoh, K.G. *Ind. Eng. Chem. Res.* **2004**, *43*, 6723.
746. Luo, N.; Husson, S.M.; Hirt, D.E.; Schwark, D.W. *ACS Symp. Ser.* **2003**, *854*, 352.
747. Luo, N.; Husson, S.M.; Hirt, D.E.; Schwark, D.W. *J. Appl. Polym. Sci.* **2004**, *92*, 1589.
748. Takolpuckdee, P.; Westwood, J.; Lewis, D.M.; Perrier, S. *Macromol. Symp.* **2004**, *216*, 23.
749. Husseman, M.; Malmstrom, E.E.; McNamara, M.; Mate, M.; Mecerreyes, D.; Benoit, D.G.; Hedrick, J.L.; Mansky, P.; Huang, E.; Russell, T.P.; Hawker, C.J. *Macromolecules* **1999**, *32*, 1424.
750. Parvole, J.; Laruelle, G.; Khoukh, A.; Billon, L. *Macromol. Chem. Phys.* **2005**, *206*, 372.
751. Ohno, K.; Morinaga, T.; Koh, K.; Tsujii, Y.; Fukuda, T. *Macromolecules* **2005**, *38*, 2137.
752. Wu, T.; Efimenko, K.; Genzer, J. *Macromolecules* **2001**, *34*, 684.
753. Baum, M.; Brittain, W.J. *Macromolecules* **2002**, *35*, 610.
754. Boyes, S.G.; Granville, A.M.; Baum, M.; Akgun, B.; Mirous, B.K.; Brittain, W.J. *Surface Science* **2004**, *570*, 1.
755. Yu, W.H.; Kang, E.T.; Neoh, K.G. *Ind. Eng. Chem. Res.* **2004**, *43*, 5194.
756. Zhai, G.Q.; Yu, W.H.; Kang, E.T.; Neoh, K.G.; Huang, C.C.; Liaw, D.J. *Ind. Eng. Chem. Res.* **2004**, *43*, 1673.
757. Shah, R.R.; Merreceyes, D.; Husemann, M.; Rees, I.; Abbott, N.L.; Hawker, C.J.; Hedrick, J.L. *Macromolecules* **2000**, *33*, 597.
758. Huang, W.X.; Kim, J.B.; Bruening, M.L.; Baker, G.L. *Macromolecules* **2002**, *35*, 1175.
759. Brantley, E.L.; Jennings, G.K. *Macromolecules* **2004**, *37*, 1476.
760. Skaff, H.; Emrick, T. *Angew. Chem. Int. Ed. Engl.* **2004**, *43*, 5383.
761. Lowe, A.B.; Sumerlin, B.S.; Donovan, M.S.; McCormick, C.L. *J. Am. Chem. Soc* **2002**, *124*, 11562.
762. Shan, J.; Nuopponen, M.; Jiang, H.; Kauppinen, E.; Tenhu, H. *Macromolecules* **2003**, *36*, 4526.
763. Matsumoto, K.; Tsuji, R.; Yonemushi, Y.; Yoshida, T. *Chem. Lett.* **2004**, *33*, 1256.
764. Matsumoto, K.; Tsuji, R.; Yonemushi, Y.; Yoshida, T. *J, Nanoparticle Res.* **2004**, *6*, 649.

Abbreviations

AA	acrylic acid	C_M	transfer constant to monomer
AM	acrylamide	C_P	transfer constant to polymer
ACP	azocyanovaleric acid, 4,4'-azobis(4-cyanopentanoic acid)	C_S	transfer constant to solvent or added transfer agent
AFM	atomic force microscopy	C_T	transfer constant to transfer agent
AIBN	azobisisobutyronitrile, 2,2'-azobis(2-cyanopropane)	C_{tr}	transfer constant ($=k_{tr}/k_p$)
AIBMe	azobis(methyl isobutyrate), 2,2'-azobis(methyl 2-methylpropionate)	C_{-tr}	reverse transfer constant ($=k_{-tr}/k_i$)
		DFT	density functional theory
AMS	α-methylstyrene	EA	ethyl acrylate
AN	acrylonitrile	EMA	ethyl methacrylate
ATRP	atom transfer radical polymerization	EPR	electron paramagnetic resonance (spectroscopy), also ESR
α-	initial position, attached to	D	dispersity/polydispersity of a molecular weight distribution ($\overline{M}_w / \overline{M}_n$)
B	butadiene		
BA	n-butyl acrylate		
tBA	t-butyl acrylate	DBPOX	di-t-butyl peroxyoxalate
BMA	n-butyl methacrylate	DMAEMA	2-(dimethylamino)ethyl methacrylate
tBMA	t-butyl methacrylate	DMAM	N,N-dimethylacrylamide
BPB	t-butyl perbenzoate	DMF	N,N-dimethylformamide
BPO	benzoyl peroxide	DMSO	dimethylsulfoxide
BriBBr	bromoisobutyroyl bromide	DPPH	diphenylpicrylhydrazyl
Bu	n-butyl	DTBP	di-t-butyl peroxide
tBu	t-butyl	E	ethylene
β-	adjacent position, next to α	EHMA	2-ethylhexyl methacrylate
c	conversion	EP	poly(ethylene-co-propylene)
C_I	transfer constant to initiator	Et	ethyl

EtAc	ethyl acetate	k_i	rate constant for initiator-derived radical adding to monomer
f	initiator efficiency		
f_X	instantaneous mole fraction of monomer X in monomer feed during copolymerization	k_p	rate constant for propagation
		k_{prt}	rate constant for primary radical termination
F_X	instantaneous mole fraction of monomer X in a copolymer	k_T	rate constant for tail addition to monomer
γ-	next to β	k_t	rate constant for radical-radical termination
GPC	gel permeation chromatography	k_{tc}	rate constant for radical-radical termination by combination
h	hour(s)		
HEA	2-hydroxyethyl acrylate	k_{td}	rate constant for radical-radical termination by disproportionation
HDPE	high density polyethylene		
HEMA	2-hydroxyethyl methacrylate		
HPMA	2-hydroxypropyl methacrylate	k_{tr}	rate constant for reaction with chain transfer agent
ΔH_p	enthalpy of polymerization	k_{trI}	rate constant for chain transfer to initiator
I	isoprene		
I_2	symmetrical initiator	k_{trM}	rate constant for chain transfer to monomer
I•	initiator-derived radical	k_{trP}	rate constant for chain transfer to polymer
IR	infra-red		
K	degrees Kelvin	k_{trS}	rate constant for chain transfer to polymer
k_{act}	rate constant for activation		
k_{add}	rate constant for addition	k_{trT}	rate constant for chain transfer to chain transfer agent T
k_β	rate constant for β-scission (fragmentation)		
		k_z	rate constant for reaction with inhibitor
K_{eq}	propagation/ depropagation equilibrium constant		
		LDPE	low density polyethylene
k_d	rate constant for initiator decomposition	LLDPE	linear low density polyethylene
		LPO	lauroyl (dodecanoyl) peroxide
k_{deact}	rate constant for deactivation		
k_H	rate constant for head addition to monomer	m	minutes
		M	monomer

Abbreviations

m-	meta-	$P_i\bullet$	propagating radical of length i (i is an integer)
$[M]_{eq}$	equilibrium monomer concentration	P_i^H	saturated disproportionation product of length i (i is an integer)
MA	methyl acrylate		
MAA	methacrylic acid	$P_i^=$	unsaturated disproportionation product of length i (i is an integer)
MALDI	matrix assisted laser desorption ionization		
MAM	methacrylamide	P_i^T	product from chain transfer of length i (i is an integer)
MAH	maleic anhydride		
MAN	methacrylonitrile	Ph	phenyl
Me	methyl	PP	polypropylene
MMA	methyl methacrylate	Pr	propyl
MMAM	N-methyl methacrylamide	PX	poly(X)
MPK	methyl isopropenyl ketone	PX•	poly(X) propagating radical
MVK	methyl vinyl ketone	$P_X\bullet$	propagating radical ending in monomer X
$\overline{M}_n$	number average molecular weight	p-	para-
		r_{IJ}	terminal model reactivity ratio
$\overline{M}_w$	weight average molecular weight	r_{IJK}	penultimate model monomer reactivity ratio
$\overline{M}_v$	viscosity average molecular weight	RAFT	reversible addition-fragmentation chain transfer
		s	second(s)
$\overline{M}_z$	Z average molecular weight	s-	secondary-
n-	normal-	S	styrene
NIPAM	N-isopropyl acrylamide		
NMP	nitroxide-mediated polymerization	s_I	penultimate model radical reactivity ratio
NMR	nuclear magnetic resonance (spectroscopy)	ΔS_p	entropy of polymerization
		SFRMP	stable free radical mediated polymerization
NVP	N-vinylpyrrolidone		
o-	ortho-	Σ	summation
OTf	triflate, trifluoromethanesulfonate	T	transfer agent
		T	temperature
P_i	polymer chain of length i (i is an integer)	t-	tertiary-

T•	transfer agent-derived radical
T_c	ceiling temperature
tBA	*t*-butyl acrylate
TBAEMA	2-(*t*-butylamino)ethyl methacrylate
tBMA	*t*-butyl methacrylate
tBu	*t*-butyl
TEMPO	2,2,6,6-tetramethylpiperidin-*N*-oxyl
THF	tetrahydrofuran
TMSEMA	trimethylsilyloxyethyl methacrylate
UV	ultraviolet
VA	vinyl alcohol
VAc	vinyl acetate
VC	vinyl chloride
VF	vinyl fluoride
VF2	vinylidene fluoride
VF3	trifluoroethylene
ω-	terminal (remote chain end) position
x	ratio of monomers in feed (f_A/f_B)
$\overline{X}_n$	number average degree of polymerization
$\overline{X}_w$	weight average degree of polymerization
$\overline{X}_z$	z average degree of polymerization
y	ratio of monomer units in copolymer (F_A/F_B)

Index

AA *see* acrylic acid
abstraction *see* hydrogen atom transfer
abstraction *vs* addition
   and nucleophilicity 35
   by alkoxy radicals 34–5, 124–5, 392
   by alkoxycarbonyloxy radicals 103, 127–8
   by alkyl radicals 34–5, 113, 116
   by *t*-amyloxy radicals 124
   by arenethiyl radicals 132
   by aryl radicals 35, 118
   by benzoyloxy radicals 35, 53, 120, 126
      with MMA 53, 120
   by *t*-butoxy radicals 35, 53, 55, 124
      solvent effects 54, 55, 123
      with alkenes 122–3
      with allyl acrylates 122
      with AMS 120, 123
      with BMA 53, 123
      with isopropenyl acetate 121
      with MA 120
      with MAN 121
      with MMA 53, 55, 120, 419
      with VAc 121
      with vinyl ethers 123
   by carbon-centered radicals 34–5, 113, 116
   by cumyloxy radicals 120
   by cyanoisopropyl radicals 116
   by heteroatom-centered radicals 35, 131–133
   by hydroxy radicals 35, 103
      with AMS 120, 128
      with MA 120
      with MMA 120, 128
   by isopropoxycarbonyloxy radicals 103, 127–8
      with MMA 103
   by oxygen-centered radicals 34, 35, 118–131
   by sulfate radical anion 129–30
   prediction
      from bond dissociation energies 34–5
      from FMO theory 35
      from radical polarity 35
   solvent effects 55, 123
abstraction-fragmentation chain transfer 309
acceptor monomers
   copolymerization with donor monomers 351
   interaction with Lewis acids 435–6
   list 351

thermal initiation 110–1
acetone, chain transfer to 295
p-acetoxystyrene, NMP 480
acetyl radicals, structure 13
acrylamide (AM) polymerization
   enzyme-mediated polymerization 440
   head *vs* tail addition 182
   NMP 480
   tacticity 174–5
      effect of Lewis acids 434
acrylate esters *see also* allyl acrylate, butyl acrylate; methyl acrylate
   reaction with radicals 120, 122–3
      relative rate constants, substituent effects 18–19
acrylate ester polymerization
   see also butyl acrylate polymerization; dodecyl acrylate polymerization; ethyl acrylate polymerization, methyl acrylate polymerization; trifluoroethyl acrylate polymerization; vinyl acrylate polymerization
   backbiting 211–2, 481
   chain transfer to polymer 322
   combination *vs* disproportionation 262
   head *vs* tail addition 182
   in supercritical CO_2 432
   iodine transfer polymerization 522
   NMP 480
   template polymerization 439
   thermal initiation 109–10
acrylic acid (AA), reaction with carbon-centered radicals, rate constants 114
acrylic acid (AA) copolymerization
   Q-e values 365
   with I 528
   with NMP 528
   with S, thermal initiation 110
acrylic acid (AA) polymerization
   amphipathic macro RAFT agent 521
   block copolymers 521, 543
   k_p solvent effects 426
   polymer brushes 559
   acticity 173
   template polymerization 438
   thermodynamics 215
      with NMP 480

with RAFT 507–508, 510, 512–514, 521, 543, 559
   with SFRMP 483
acrylic monomers, polymerization, head *vs* tail addition 182
acrylic polymers, head-to-head linkages 182
acrylonitrile (AN)
  induced decomposition of BPO 86
  reaction with alkyl radicals, penultimate unit effects 345
  reaction with dithioacids 516
  reaction with carbon-centered radicals, rate constants 115
  reaction with heteroatom-centered radicals, rate constants 131
  reaction with oxygen-centered radicals
    rate constants 119
    specificity 121
acrylonitrile (AN) copolymerization
  Q-e values 365
  reactivity ratios 339
  with AMS 353
  with E 209
  with S
    ATRP 529, 543
    bootstrap effect 431
    gradient copolymers 526
    NMP 528
    RAFT 529
    solvent effects 429–30
    thermal initiation 110
acrylonitrile (AN) polymerization
  chain transfer
    to halocarbons 293
    to polymer 320
    to solvent 295
    to thiols 290
  combination *vs* disproportionation 262
  head *vs* tail addition 182
  inhibition constants 265
  k_p 221
    effect of Lewis acid 433
    solvent effects 426
  tacticity 175
  template polymerization 438
  thermodynamics 215
  with RAFT 508, 510, 514
  with TERP 524
  with stibine-mediated polymerization 524
activation energy
  for hydrogen-atom transfer 30
  for initiator decomposition
    azo-compounds 70, 71
    peroxides 81
  for propagation 218
    in copolymerization 349, 350
  for radical addition 20, 26
    calculation of 26
  for termination 234, 254
  for radical-radical reaction 36
activation entropy *see* Arrhenius A factor
activation-deactivation processes
  equilibrium constant 461
  in ATRP 492
  in living radical polymerization 455–7
active species, in living radical polymerization 6, 453, 455–457
acyl peroxides *see* diacyl peroxides; dibenzoyl peroxide; didecanoyl peroxide; dilauroyl peroxide
acyl phosphine oxides
  as photoinitiators 98, 101–2, 132
  radicals from 101, 132
acyl radicals 117–8
  decarbonylation rate constants 118
  from hydrogen abstraction from aldehydes 118
  nucleophilicity 118
acyloxy radicals 125–6
  see also benzoyloxy radicals
  aliphatic
    fragmentation to alkyl radicals 83, 112, 126
    from diacyl peroxides 83, 86
  decarboxylation 83
  initiator efficiency 83
  from α-acylperoxydiazenes 97
  from diacyl peroxides 82
  from hyponitrites 78
  from diaroyl peroxides 82, 215
  from hyponitrites 78
  from peroxyesters 88, 125
acylphosphine oxides, radical generation 117
addition *see* radical addition
addition-abstraction polymerization 208, 212
addition-fragmentation chain transfer 296–309
  see also reversible addition-fragmentation chain transfer (RAFT)
  mechanisms 287, 296–7
  reverse transfer constant 288–9
  reviews 296
  to allyl halides 299–300, 302–303
  to allyl peroxides 303–5
  to allyl phosphonates 299, 303
  to allyl silanes 299, 303
  to allyl sulfides 299–300, 377–8
  to allyl sulfonates 299–302
  to allyl sulfones 299
  to allyl sulfoxides 302
  to benzyl vinyl ethers 298–9, 377
  to macromonomers 305–8
  to methacrylate macromonomer 252, 305–8, 322, 400–1, 419, 420, 501–502

Index

to TERMINI agent 556–7
to thiohydroxamic esters 308, 309
to thionoesters 308, 309
to VC 180, 296
to vinyl ethers 298–9
transfer constants 299–300, 302–4, 307, 309
AIBMe *see* azobis(methyl isobutyrate)
AIBN *see* azobisisobutyronitrile
aldehydes, acyl radicals from 118
alkanethiyl radicals
  from allyl sulfides 300
  from disulfides 291–2
  from thiols 290, 291
  polarity 290
  reaction with monomers 132
alkenyl radicals
  cyclization 23
    3-butenyl radicals 23, 197
    6-heptenyl radicals 23, 193
    5-hexenyl radicals 5, 54, 187, 192
alkoxy radicals 118–25
  see also t-amyloxy radicals; t-butoxy radicals; cumyloxy radicals
  abstraction *vs* addition 34, 35, 124–5
  combination *vs* disproportionation 41
  fragmentation to alkyl radicals 51–7, 66–7, 88–9, 91–3, 105, 112, 123–5
    rate constant 124
    solvent effects 123–4
    substituent effects 124
  from alkyl hydroperoxides 92
  from dialkyl hyponitrites 78
  from dialkyl peroxides 90
  from dialkyl peroxydicarbonates 87
  from peroxides 391–2
  from peroxyesters 88
  from peroxyketals 91
  isopropoxy radicals 79, 87, 125, 127-128, 138
  methoxy radicals 35, 125
  polarity 27, 30–1, 35, 122
  primary 35, 41, 125
  reviews 119
  secondary 35, 41, 125
  solvent effects, hydrogen atom transfer 34
  temperature dependence on reactivity 56
  tertiary 35 119–125
alkoxyamines
  as initiators of NMP 458, 471, 475–477, 544, 560–2
  functional 533, 558, 561
  water-soluble 482
  formation/synthesis 138–9, 471, 475–6
  induced decomposition 478
  radical trapping 138–9
  thermal stability 140
alkoxycarbonyloxy radicals 125, 127–8

abstraction *vs* addition 127–8
fragmentation to alkoxy radicals 127
from dialkyl peroxydicarbonates 87, 125
α-alkoxymethacrylate polymerization
  ceiling temperature 216
  tacticity, effect of Lewis acids 434
alkoxysilane derivatives, grafting on to silicon particles and wafers 561
α-alkyl benzoin derivatives, radical generation 100
alkyl cobaloximes
  as chain transfer catalysts 485
  radical-induced decomposition 485
alkyl halides
  formation of alkoxyamines 477
  reduction by stannyl or germyl radicals 137
alkyl hydroperoxides
  see also t-amyl hydroperoxide ; t-butyl hydroperoxide
  as initiators 56–7, 80, 88, 92–3
  as source of alkoxy and hydroxy radicals 92
  as transfer agents 93
  induced decomposition 93
  kinetic data for decomposition 80–1
  non-radical decomposition 93
  reaction with transition metals 93
alkyl iodides, as control agents 499, 522
alkyl pyridinium salts, as ionic liquids 432
alkyl radicals 112–13
  see also α-aminoalkyl radicals; benzyl radicals; t-butyl radicals; 2-carboalkoxy-2-propyl radicals; α-cyanoalkyl radicals; cyanoisopropyl radicals; cyclohexyl radicals; ethyl radicals; methyl radicals; undecyl radicals
  abstraction *vs* addition 34–5
  combination *vs* disproportionation 40, 42
  disproportionation pathways 38
  fluorine substitution 14
  from diacyl peroxides 83
  from dialkyldiazenes 68
  from fragmentation
    of acyl radicals 117–8
    of acyloxy radicals 51–4, 66–7, 82–5, 88, 112, 117, 126–7, 146
      rate constant 126–7
    of alkoxy radicals 51–7, 66–7, 88–9, 91–3, 105, 112, 123–5
      rate constant 124
    of N,N-dimethylaniline radical cation
  from reduction of alkyl halides 137
  from xanthates 502–3
  head *vs* tail addition 112–3
  hydrogen abstraction by 116
  isopropyl radicals 35
  nucleophilicity 13, 21, 31–2

polarity 35, 113
rate constants for addition 26, 114–5
    table of 114–5
reaction with fumarodinitrile 25
reaction with inhibitors 268–73
    nitroxides 138, 266
    oxygen 56–7, 130, 266
    rate constants 266
reaction with methyl α-chloroacrylate 25
reaction with monomers 112–3
    rate constants 112–3
    penultimate unit effects 345
reaction with oxygen 50, 116, 130
reviews 112
stability 14
structure 12
alkyl stibines, as control agents 499
alkyl tellurides, as control agents 499, 523
O-alkyl xanthates *see* xanthates
alkylcobaloximes
    as initiators of SFRMP 484
    as chain transfer catalysts 314–5
    radical-induced decomposition 314–5, 485
3-alkyl-1-methylimidazolium salts, as ionic liquids 432
alkylperoxy radicals 130–1
    epoxidation by 130–1
    from alkyl hydroperoxides 93, 130
    from alkyl radicals and oxygen 56, 130
    in autoxidation processes 130
    reaction with monomers 130
allyl acetate polymerization, chain transfer to monomer 319
allyl acrylamide, cyclopolymerization 189
allyl acrylate
    cyclopolymerization 189
    reaction with t-butoxy radicals 122
allyl amines, polymerization, chain transfer to monomer 191, 319
allyl esters
    polymerization
        chain transfer to monomer 319
        head *vs* tail addition 181–2
allyl ethers, polymerization, chain transfer to monomer 319
allyl halides
    chain transfer constants 303
    chain transfer to 299–300, 302, 317
allyl methacrylamide, cyclopolymerization 189
allyl methacrylate
    cyclopolymerization 189
    reaction with t-butoxy radicals 122
allyl monomers *see also* allyl acetate; allyl amines; allyl esters;
    polymerization
        chain transfer to monomer 191, 319–20

    head *vs* tail addition 181–2, 188
allyl peroxides
    chain transfer constants 304
    chain transfer to 303–5
allyl phosphonates
    chain transfer constants 303
    chain transfer to 299
allyl polymers, head-to-head linkages 181–2, 188
allyl radical, structure 13
allyl silanes
    as comonomers 301
    chain transfer constants 303
    chain transfer to 299, 301
allyl sulfides
    chain transfer constants 300
    chain transfer to 292, 299–300
    synthesis of end-functional polymers 377–8
allyl sulfonates
    chain transfer constants 302
    chain transfer to 299–301
allyl sulfones, chain transfer to 299
allyl sulfoxides, chain transfer constants 302
alternating copolymerization 333
    cyclocopolymerization
        of divinyl ether with MAH 170
    Lewis acid induced 435
        of MMA with S 435, 436
    of donor-acceptor monomer pairs 350–351
        of MAH with S 340, 350, 395
    thiol-ene polymerization 378–9
alternating copolymers, definition 333
aluminum isopropoxide cocatalysts, in ATRP 489, 495
AM *see* acrylamide
α-aminoalkyl radicals 86, 102
    from amine-peroxide redox couples
AMS *see* α-methylstyrene
t-amyl hydroperoxide, as initiator 92
t-amyloxy radicals
    abstraction vs addition 124
    fragmentation to ethyl radicals 124
t-amyl peroxides, peroxyesters
    source of ethyl radicals 124
anhydrides, rate enhancement of NMP 479
anilinomethyl radical, formation 86
anionic polymerization
    transformation to radical 387–8, 545
    tests for living 452–4
    comparison with radical 455, 525, 541
arenethiyl radicals
    dimethylamino substituted, zwitterionic quinonoid form 26
    from disulfides 291–2
    from thiols 290
    polarity 132
    reaction with monomers 132

Index

MMA 132
S 132
arm-first star synthesis 548, 554–5
   cross-linked micelles 554–5
   in-situ microgel formation 554–5
   self-assembly of diblock copolymers to form micelles 554–5
   through ATRP 554–5
   through NMP 554–5
   through RAFT 554–5
aromatic substitution
   by benzoyloxy radicals
      of benzene 127
      of PS 127
      of S 52–54, 127
      reversibility 126
   by chlorine atoms, of benzene 34
   by isopropoxycarbonyloxy radicals, of S 128
   by hydroxy radicals
      of S 128
      of AMS 128
   by phenyl radicals, of S 52, 117
aroyloxy radicals 125–7
   fragmentation into aryl radicals 117
Arrhenius *A* factor
   see also activation energy
   for addition *vs* abstraction 56
   for backbiting 208
   for hydrogen atom transfer 32
   for initiator decomposition
      azo-compounds 70–1
      peroxides 81
   for propagation 218, 221
      in copolymerization 349
   for radical addition 20–1, 24, 56, 221
      calculation of 26
aryl ethers, unsuitability as photoinitiators 100
aryl peroxyesters, unsuitability as initiators 88
aryl radicals 117
   see also phenyl radicals
   abstraction *vs* addition 35
   p-bromophenyl radicals 35
   from fragmentation of aroyloxy radicals 117
   head *vs* tail addition 117
   p-nitrophenyl radicals 35
   polarity 35, 117
   reaction with monomers
      rate constants 117
      specificity 117
ascorbic acid, rate enhancement of NMP 479, 482
atom transfer radical coupling (ATRC), telechelic polymer synthesis 537
atom transfer radical polymerization (ATRP) 7, 250, 456–7, 486–98

activation-deactivation equilibrium constant 492
alkoxyamine formation 477
block copolymer synthesis 541–3
   diblock copolymers prepared by 543
   halogen exchange 542
   mechanisms 542
   optimal conditions 542
   transformation reactions 544–6
   triblock copolymers 546
catalysts 487–9, 492–7, 528, 535
   copper complexes 489, 492, 492–5
   iron complexes 492, 496
   nickel complexes 492, 496
   ruthenium complexes 492, 495
cocatalysts 481, 492
combination ATRP 491
copolymerization 528–9
   effect of Lewis acids 528
   mechanism 528
   gradient copolymers 529
end-functional polymer synthesis 544
   α-functionalization 536–7
   dormant chain end removal 531
   ω-functionalization 533–6
graft copolymer synthesis
   by grafting from inorganic surfaces 562
   by grafting through 558–9
   grafting from polymer surfaces 560–1
heterogeneous polymerization 497–8
   activator generated by electron transfer (AGET) 498
   emulsion 491, 498
   miniemulsion 491, 498
   reverse ATRP 491, 498
   reviews 497
   suspension 498
initiators 458, 488–92, 536–7
   bromoisobutyrates 489, 536–7, 545, 557–9, 563
   initiator efficiency 489–91
   initiator activity 492
   functional 536–7
   sulfonyl halides 488–9, 493, 550, 556–7
molecular weights and distributions 490–1
   degree of polymerization 490
   dispersity 490–1
mechanisms 486–7
   inner sphere electron transfer 487
   outer sphere electron transfer 487
monomers
   BA 495–6, 498
   BMA 495–6, 498
   DMAEMA 497
   DMAM 495
   HEMA 495, 497

MA 495–6, 498
MAA 497
MMA 489, 491, 495–6, 498
α-methoxypoly(ethylene oxide)
  methacrylate
S 495–6
VAc 496
reaction conditions 497–8
polymerization kinetics 461, 487–8, 490
reactivity ratios, copolymerization 488
reverse ATRP 491, 498
  AIBN 491
  peroxide initiators 491
reviews 487, 497
side reactions 488–9, 534–5
star synthesis
  mechanisms 551–2
  precursors 549–50, 554
  microgel formation 555
  self-condensing vinyl polymerization 555–6
  shell-crosslinking of micelles 555
  dendritic cores 556–7
azeotropic copolymer composition
  in terpolymerization 359
  penultimate model 343
  terminal model 340–1
azo-compounds
  see also azobisisobutyronitrile; azobis(methyl isobutyrate); azo-*t*-butane; azoisooctane; azonitriles; azoperoxides; dialkyl diazenes; dialkyl hyponitrites; dialkyl diazenes; triphenylmethylazobenzene
  as initiators 64–79
azobisisobutyronitrile (AIBN)
  ^{13}C labeled 146–8, 260–1, 263
  and transition metal complex, in reverse ATRP 491
  as initiator 68–9, 442–3, 459, 518, 522, 524
  cage reaction 60–1, 76
  cage return 68, 73
  chain transfer to 77
  decomposition 68, 113, 476
  decomposition mechanism 60–1, 68, 76, 258
  decomposition products 60–1, 76
  diazenyl radicals from 68
  effect of viscosity on k_d 73
  ketenimine from 76, 257, 258
  kinetic data for decomposition 70–1
  MAN from 60–1, 77
  solvent effect on decomposition 68
  synthesis of end-functional polymers 376
  toxicity 77
azobis(methyl isobutyrate) (AIBMe)
  ^{13}C labeled 146, 148–9, 436
  advantages 77
  as initiator 68–9, 73, 77, 148–9, 436

chain transfer to 77
decomposition rate 77
kinetic data for decomposition 70–1, 73
solvent effect on k_d 73
synthesis of end-functional polymers 376
zero-conversion initiator efficiency 76
azo-*t*-butane
  high temperature initiator 69
  kinetic data for decomposition 70–1
azoisoctane 69
  high temperature initiator 69
  kinetic data for decomposition 70–1
azonitriles *see also* azobisisobutyronitrile
  as initiators 68–9
  4,4'-azobis(4-cyanopentanoic acid) 72, 518
  1,1'-azobis(1-cyclohexanenitrile) 69
  2,2'-azobis(2-methylbutanenitrile) 69
  decomposition 113
  decomposition kinetics 77
azoperoxides
  decomposition mechanism 97
  synthesis of block and graft copolymers 386

backbiting 32, 208–12
  addition-abstraction polymerization 208, 212
  chain transfer to polymer 208, 210, 320
  effect of monomer concentration 209
  effect of temperature 209, 211–2
  in acrylate ester polymerization 211–2
  in E polymerization 208–10
  in monosubstituted monomers 211–2
  in NMP with acrylates 481
  in polymerization of vinyl polymers 211
  in S polymerization 211
  in VAc polymerization 208–10
  in VC polymerization 208
  mechanism 168
  transition state 209
backbiting-fragmentation
  in acrylate polymerization 212
  macromonomer synthesis 212
Barton-McCombie reaction 296, 502–3
Beer-Lambert law 59
benzene
  chain transfer to 295
  solvent effect on VAc polymerization 324–5
benzil, as photoredox initiator 102
benzil monooxime, fragmentation 99
benzoin derivatives
  acyl phosphine oxides 101–2, 132
  as photoinitiators 99–102
  benzil monooxime 99
  benzoin esters 100
  benzoin ethers
    α-ether 100
    α-sulfonyl 101

Index

as photoinitiators 61–2, 98, 100
   shelf life 100
  polymerizable 102
  radical generation 99–100, 117
p-benzoquinone
  copolymerization 271
  inhibition by 266
  radical addition, polar effects 271
benzoyl radical, from benzoin derivatives 100
benzoyloxy radicals 126–7
  abstraction *vs* addition 35
  aromatic substitution
    of benzene 127
    of PS 127
    of S 52, 127
  fragmentation to phenyl radicals 52, 54
    rate constants 127
  from BPO 51, 82, 84, 86
  head *vs* tail addition 126
  polarity 35, 126
  reaction with monomers
    AN 126
    MA 18, 19
    MC 18
    MMA 18–9, 21, 52, 126
    rate constants 119, 127
    S 5, 52, 126, 415
    specificity 52, 120–1, 126
    VAc 126
α-benzoyloxystyrene, chain transfer to 298
benzoylthiyl radicals 132
benzyl bromide, as ATRP initiator 489
benzyl dithiobenzoate, retardation in RAFT 517
benzyl dithiocarbamates
  as RAFT agents 514
  in SFRMP 463–4
  transfer constants 463–4
benzyl radicals
  combination pathways 37, 254
  from benzoin derivatives 100
  from toluene 55
  radical–radical reactions 254–5
  rate constants for addition 114–5
  reaction with nitroxides 138
benzyl selenide, as initiator 466
benzyl thionobenzoate, chain transfer to 308
benzyl vinyl ethers
  chain transfer constants 299
  chain transfer to 298–9, 377
  synthesis of end-functional polymers 377
β-scission *see* fragmentation
bicyclo[n,1,0]alkanes, reaction with *t*-butoxy radicals 33
bicyclobutanes, ring-opening polymerization 195
bicyclo[2,2,1]heptadiene derivatives, cyclopolymerization 193

binary copolymerization *see* copolymerization
biomolecular termination 234
2,2'-bipyridine ligands (bpy, dNbpy)
  ATRP catalysts 493–4
  in situ formation 494
bis-diazenes
  as multifunctional initiators 97
  decomposition 97
bis-methacrylates, cyclopolymerization 424
bis-RAFT agents 511, 546
block copolymer synthesis 384–5
  by coupling reaction 540
  by sequential monomer addition 540–3
    using ATRP 541–3
    using NMP 541
    using RAFT 502, 543–4
  by living radical polymerization 291, 454, 463, 465, 483, 526, 540–8
  by transformation reactions 387–9, 544–6
    to ATRP 545–6
    to NMP 545
    to RAFT 546
  direct diblock synthesis 541–4
  from end-functional polymers 374, 377, 540
  segmented block copolymers 547–8
  triblock copolymer synthesis 546–7
  use of multifunctional initiators 98, 252, 386, 387
block copolymers 540–1
  applications 540
  composition 540
  definition 334
  analysis of by GPC 540–1
BMA *see* *n*-butyl methacrylate
bond dissociation energies
  abstraction *vs* addition 34–5
  and fragmentation of *t*-butoxy radicals 124
  and hydrogen atom transfer 30, 34–5
  and radical addition 17, 23
  and radical stability 14, 17
  C–H bond 15
  fluorine substitution effects 14, 23
  substituent effects
    C–C bond 23, 34
    C–H bond 30, 34
    C–O bond 23, 34
    O–H bond 34
bootstrap effect
  in copolymerization 357, 431–2, 526
  definition 431
  in living radical copolymerization 526
boroxyl radicals, in SFRMP 483
BPB *see* *t*-butyl perbenzoate
BPO *see* dibenzoyl peroxide
branched copolymers, definition 334–5

α-bromoisobutyrates, as ATRP initiators 489, 536–7, 545, 557-9, 563
α-bromoisoproprionates, as ATRP initiators 489
N-bromomethylphthalimide, as ATRP initiator 536
2-bromopropanenitrile, as ATRP initiator 536
butadiene (B) polymerization
   1,2- vs 1,4-addition 183–4
   kinetic parameters 219
   k_p 217
   NMP with TEMPO 481
butadiene (B) copolymerization, Q-e values 365
t-butoxy radicals
   abstraction vs addition 35, 53, 55, 120–3
   fragmentation 51–7, 66–7, 88–9, 91–3, 105, 112, 123–5
      rate constant 124
      radical clock 54
   from DBPOX 51, 89, 119
   from dialkyl peroxides 90, 119
   from di-t-butyl hyponitrites 78, 119
   Hammett parameters
      for abstraction from substituted toluenes 22
      for addition to substituted styrenes 22
   head vs tail addition
      fluoro-olefins 16,17, 28, 122
      MA 120
      olefins 122
      VAc 121
      VF 17, 23
      VF2 16–8, 23, 28, 122
      VF3 17
   hydrogen atom transfer 32–3
   polarity 22, 35, 122
   reaction with bicyclo[n,1,0]alkanes 33
   reaction with ethers 32–3
   reaction with monomers
      alkenes 122–3
      allyl acrylate 122
      allyl methacrylate 122
      AMS 120
      AN 121
      BMA 123
      E 17
      fluoro-olefins 16–8, 22–23, 28, 122
      isopropenyl acetate 121
      MA 18–9, 120
      MAN 121
      MC 18
      MMA 18–9, 52–3, 55–6, 118, 120
      rate constants 114–5, 119
      S 52–3, 120, 134
      specificity 120–1
      VAc 121
      VF 17, 23
      VF2 16–8, 23, 122
      VF3 17
   vinyl ethers 123
   reaction with spiro[2,n]alkanes 33
   reaction with toluene 55
   specificity of hydrogen transfer 33
n-butyl acrylate (BA) copolymerization
   Q-e values 365
   with AMS 353
   with MMA 526
   with VAc 526
n-butyl acrylate (BA) polymerization
   backbiting 481
   chain transfer
      to allyl sulfonates 302
      to halocarbons 293
   inhibition by cobalt (II) porphyrin complexes 484
   kinetic parameters 219
   k_p 247
   k_t, chain length dependence 247
   emulsion polymerization
      NMP 482
      RAFT 520
   stibine-mediated polymerization 524
   with ATRP 495–6, 498
t-butyl hydroperoxide
   as initiator 92
   radicals from 113–14
n-butyl methacrylate (BMA), reaction with t-butoxy radicals 53, 123
n-butyl methacrylate (BMA) copolymerization
   Q-e values 365
   with MAN, combination vs disproportionation 374
   with MMA, combination vs disproportionation 374
   with S, using NMP 527
i-butyl methacrylate (iBMA) polymerization, kinetic parameters 219
n-butyl methacrylate (BMA) polymerization
   combination vs disproportionation 255, 262
   thermodynamics 215
   kinetic parameters 219
   with ATRP 495–6, 498
t-butyl methacrylate (tBMA) polymerization, tacticity, solvent effects 428
t-butyl perbenzoate (BPB)
   as initiator 80
   kinetic data for decomposition 80–1
t-butyl peroxybenzoate, as initiator 88
t-butyl peroxyesters, variation in efficiency 88
t-butyl peroxypentanoate, ethylene polymerization 88
2-butyl radicals, disproportionation pathways 38
t-butyl radicals
   abstraction vs addition 35

Index

combination *vs* disproportionation 41
   effect of temperature and solvent 42, 43
   Hammett parameters
      for abstraction from substituted toluenes 22
      for addition to substituted styrenes 22
   polarity 35
   rate constants for addition 26, 114–5
      table of 114-5
   reaction with fumarodinitrile 25
   reaction with methyl α-chloroacrylate 25
   reaction with oxygen 130
t-butylperoxy radicals 130
   Hammett parameters
      for abstraction from substituted toluenes 22
      for addition to substituted styrenes 22

cage reaction
   by-products from 60–1
      AIBN 60, 76
      BPO 83
      dialkyl hyponitrites 78
   definition 60
   effect of initiator structure 65
   effect of magnetic field 61
   effect of solvents 73, 89
   effect of viscosity 60–1, 75, 78, 84, 89
      AIBN 75
      DBPOX 89
      diacyl peroxides 84
      dialkyl hyponitrites 78
      dialkyldiazenes 74, 75
   effect on initiator efficiency 57–8, 60, 76, 84
   effect on k_d 60, 84
   effect on product distribution 42–3
   initiator decomposition
      AIBMe 75
      AIBN 60–1, 68, 76
      alkyl hydroperoxides 93
      BPO 60, 83, 84
      DBPOX 89
      diacyl peroxides 63, 83–5
      dialkyl hyponitrites 78
      dialkyl peroxydicarbonates 87
      dialkyldiazenes 72–5
      di-*t*-butyl peroxides 92–3
      dicumyl peroxides 92
      DTBP 60, 91–2
      hyponitrites 78–9
      LPO 82–3
      peroxyesters 88
   photodecomposition, BPO 83–4
   vs encounter reactions 42–3
cage return
   AIBN decomposition 68, 73
   BPO decomposition 60, 82
   diacyl peroxides 82, 84

dialkyldiazenes 68
dilauroyl peroxide 83
DTBP decomposition 60
   effect on initiator efficiency 60
camphoroquinone, as photoredox initiator 102
camphorsulfonic acid, rate enhancement of NMP 479
caprolactone block copolymers 545, 558
captodative olefins
   copolymerization 270
   inhibition by 269–70
   reversible addition-fragmentation 470
2-carboalkoxy-2-propyl radicals
   as leaving group in RAFT agent 507
2-carbomethoxy-2-propyl radicals
   from AIBMe 436
   reaction with monomers 436
   radical-radical reactions 255
   self-reaction 256
carbon-centered radical-mediated polymerization 467–70
   monomer effects 469
carbon-centered radicals 12–13, 112–18
   see also acyl radicals; alkyl radicals; aryl radicals
carbon monoxide, copolymerization with depropagation 353
carbonyl compound-tertiary amine systems
   as photoinitiators 102–3
   mechanisms 102–3
catalytic chain transfer 310
   advantages over conventional agents 310
   catalysts
      chromium complexes 315
      cobalt (II) cobaloximes 313–5
      cobalt (III) cobaloximes 314–5
      cobalt porphyrins 313
      iron complexes 315
      molybdenum complexes 315
      phthalocyanine complexes 313
   mechanism 310–12
   reaction condition limitations 315–6
   reviews 310
   to cobalt complexes 310–6
   to PMMA macromonomer 322
cationic polymerization
   of oxazoline groups 558
   transformation to radical polymerization 389
ceiling temperature 213–4
   in copolymerization 353–4
   in homopolymerization 214, 216
   in ring-opening polymerization 196
   table 215
ceric ions
   as trap for carbon-centered radicals 106
   reaction with 1,2-diols 105–6

in redox initiation 104–6, 386
cetyltrimethylammonium 4-vinylbenzoate, micelle formation 442
chain length, and copolymer composition 381
chain length dependence
  and diffusion mechanisms, living radical polymerization 251
  combination *vs* disproportionation
    for PE 258
    for PMAN 256
    for PS 253
  of chain transfer constant 282–4, 294
  of chain transfer to polymer 320
    PE 321
  of copolymer composition 431–2
  of k_p 213, 218–21
  of k_t 234, 245–9
    common monomers 247
chain length distribution 240–1
chain transfer
  see also abstraction-fragmentation chain transfer; addition-fragmentation chain transfer; backbiting; catalytic chain transfer; hydrogen atom transfer
  by atom or group transfer 280, 289–95
  control of, defect groups 418–19
  definition 234
  homolytic substitution 289–95
  in thiol-ene polymerization 378
  in VAc polymerization 178–9
  mechanisms 279–80
  penultimate unit effects 282–3, 291, 294
  rate constant for reinitiation 280
  reversible 288–9
  reviews 280
  synthesis of end-functional polymers 211–12, 279, 291, 295, 298–9, 377–8
  termination by 234, 279
    combination *vs* disproportionation 283–4
    rate constant 236
  to disulfides 291–2
    polar effects 292
    substituent effects 292
  to functional transfer agents 377–8
  to halocarbons 283, 289, 293–4
    catalysis 294
    penultimate unit effects 294
  to initiator *see* chain transfer to initiator
  to monomer *see* chain transfer to monomer
  to polymer *see* chain transfer to polymer
  to solvents 294–5
  to sulfides 292–3
  to thiols 289–91, 382–3
    penultimate unit effects 291
    polar effects 290
    substituent effects 290

  uses 279
chain transfer constants
  chain length dependence 282–4
  definition 281
  effect of conversion 281–2
  for allyl halides 302
  for allyl peroxides 304
  for allyl phosphonates 302
  for allyl silanes 302
  for allyl sulfides 302
  for allyl sulfonates 302
  for allyl sulfoxides 302
  for cobalt complexes 316
  for disulfides 292
  for macromonomers 307
  for monomers 317
  for polymers 320
  for thiohydroxamic esters 309
  for thiols 290
  for thionoesters 309
  for vinyl ethers 299
  ideal 282
  Mayo equation 282–3
  measurement 283–7
    integrated forms of Mayo equation 286
    log CLD method 283–4
    Mayo equation 283, 285
    through evaluation of usage of transfer agent and monomer conversion 285–6
  prediction 282, 283
  reversible 288
  temperature effects 282
chain transfer to initiator 58, 62–3, 325
  AIBN 77
  aliphatic diacyl peroxides 82–3, 126
  alkyl hydroperoxides 56–7, 93
  BPO
    in S polymerization 63, 85, 127, 415
    in VAc polymerization 85
    in VC polymerization 85
    mechanisms 63
  diacyl peroxides 63, 85
  dialkyl peroxides 92
  dialkyl peroxydicarbonates 87
  dialkyldiazenes 75, 77
  disulfides 103, 291–2
  dithiuram disulfides 103
  effect of conversion 85
  effect on initiator efficiency 63
  effect on k_d 63, 82–3
  α-hydroperoxy diazenes 97
  ketenimine from AIBN 76
  peroxyesters 89
  persulfate 94
chain transfer to monomer 316–17
  allyl acetate 317

allyl amines 191, 319
allyl chloride 317
allyl esters 319
allyl ethers 319
allyl monomers 191, 319–20
  AN 317
  MA 317
  mechanisms 316–17
  MMA 316–7
  PMMA macromonomer 252, 305–8
  S 316–7
  transfer constants 317
  VAc 316–8
  VC 179–80, 296, 316–9
chain transfer to polymer 320–1
  AN 320
  backbiting 208, 210, 320
  E 320, 321
  MA 320
  mechanisms 320
  MMA 320–2
  molecular weight dependence 321
  PE 320–1
  PMMA 321–2
  PMMA macromonomer 321–2
  poly(alkyl acrylate) 322
  poly(alkyl methacrylate) 321–2
  PVA 323–4
  PVAc 323–5
  PVC 325
  PVF 325
  S 320
  transfer constants 320
  VAc 320
  VC 320, 325
chemical control model, of termination in copolymerization 366–8
chemical methods, for measurement of end groups 144–5
chlorine atoms
  1,2-chlorine shifts 179–80
  aromatic substitution 34
  hydrogen atom transfer
    solvent effects 34
    specificity 32
  in VC polymerization 179–80
  polarity 29, 31
chlorocyclohexadienyl radical 34
p-chloromethylstyrene, polymerized by NMP 480, 541
chloroprene polymerization, 1,2- *vs* 1,4-addition 183, 184–5
chlorosilane derivatives, grafting on to silicon particles and wafers 562
chromium complexes
  as chain transfer catalysts 315
  in SFMRP 484
cobaloximes
  as chain transfer catalysts 310–6
    chain transfer constants 316
  in situ generation 314–5
cobalt complexes
  see also alkyl cobaloximes; catalytic chain transfer; cobaloximes; cobalt porphyrin complexes
  as chain transfer catalysts 310–6
  as initiators in SFMRP 484
  as catalysts in ATRP 492
  catalytic chain transfer 310–6
    chain transfer constants 316
  control of propagation 423
  in SFMRP
    of acrylates 484–5
    of VAc 485
cobalt porphyrin complexes
  as chain transfer catalysts 313
  cobalt tetramesitylporphyrin (CoTMP) 484
    in SFMRP of acrylates 484–5
combination 36, 37–8 , 233–4
  definition 251
  pathways
    for benzyl radicals 37, 254
    for cyanoalkyl radicals 37–8, 116, 257–8, 262
    for PAN 262
    for PMAN models 256–7
    for PS models 253
    for triphenylmethyl radicals 37
combination *vs* disproportionation
  for alkoxy radicals 41, 78
  for alkyl radicals 40, 42
  for 2-carboalkoxy-2-propyl radicals 255
  for cumyl radicals 40, 253, 255
  for cyanoisopropyl radicals 116–17, 256–7
  for ethyl radicals 41–2
  for fluoromethyl radicals 41
  for nitroxides 478
    with carbon-centered radicals 138
  for PAMS, model studies 253–4
  for PAN 262
  for PBMA 262
    model studies 255
    with PMAN 257
    with PMMA 255
  for PE
    chain length dependence 258
    model studies 258
    with PMAN 257
  for PEMA 262
    model studies 255
  for 1-phenylalkyl radicals 253, 255
  for PMA 262

for PMAN 262
  chain length dependence 256
  model studies 256–8
  with PS 257
for PMMA 252, 261–2
  effect of temperature 255, 261–2
  model studies 255–6
for PS 260–1
  effect of temperature 254, 260–1
  model studies 253–5
for PVAc 263
for PVC 263
for *t*-butyl radicals 41, 42
guidelines 43–4, 264
in copolymerization 252, 255–7
in primary radical termination 61
in radical-radical reactions 39–43
in radical-radical terminations 251–64
measurement
  end group determination 259
  Gelation technique 258–9
  mass spectrometry 259
  model studies 252–3
  molecular weight distribution 259
  molecular weight measurement 259
molecular weight distributions 241–2
oligostyryl radicals 253–4
polar effects 41
reaction condition effects 42–3
statistical factors 39–40
stereoelectronic factors 41–2
steric effects 40–1
substituent effects 40, 42
summary 263
synthesis of end-functional polymers 376–7
temperature dependence 254, 260–2
complex dissociation model, copolymerization 352–3
complex participation model, copolymerization 352
compositional heterogeneity
  HEA:BA:S copolymer 382–3
  in functional copolymers 336, 381–4
computer modeling
  of chain end functionality 377
  of molecular weight distribution 217
concentration
  initiator, effect on k_d 82, 87, 94
  monomer
    effect on backbiting 209, 211
    effect on ceiling temperature 215
    effect on ring-opening polymerization 196
conformationally constrained compounds, hydrogen atom abstraction 33
controlled living polymerization 452

controlled radical polymerization *see* living radical polymerization
conversion
  effect on branching in PVAc 324
  effect on chain transfer constants 281–2
  effect on initiation 55
  effect on initiator efficiency 237, 244
    AIBN in S polymerization 75, 76
    BPO in MMA polymerization 84
  effect on polymer composition 336
  effect on radical-radical termination
    low conversion 244–8
    medium to high conversion 248–9
    very high conversion 244
copolymerization 333–5
  binary
    kinetic models 354–7
    penultimate model 355–6
    terminal model 354–5
  by ATRP, reactivity ratios 488
  chemical control model 366–8
  combination and disproportionation 370–1
  composition drift 336
  compositional heterogeneity 336, 381–4
  cross-termination radical precursors 72
  depropagation model 353–4
    AMS 353
    carbon monoxide 353
    ceiling temperature 353–4
    sulfur dioxide 353
  diffusion control models 368–70
  donor-acceptor monomer pairs, model studies 353
  effect of conversion 360–1
  effect of Lewis acids 435–6
  effect of molecular weight 431–2
  effect of solvent 336, 357, 361, 429–32
  models for
    complex dissociation model 352–3
    complex participation model 352
    discrimination 348, 350
      by composition data 361
    monomer complex models 350–3
    penultimate model 342–51, 355–6, 429–30
    terminal model 337–42, 366, 431–2
  monomer sequence distribution 336
    effect of solvent 357, 430–1
    effect of tacticity 356–7
    other models 364
    penultimate model 355–6
    terminal model 354–5
  of AN with S 345
  of BA with AMS 380
  of BMA with MAN 374
  of BMA with MMA 374
  of E 209–10

with MAN 374
of macromonomers 400–1, 558–60
of MAH with S 351
of MAN with S 373
of MAN from AIBN 77
of MMA with S
  chemical control model 368
  combination *vs* disproportionation 371–3
  penultimate unit effects 346–9
of quinones 271
of VAc with S 269
penultimate model 342
  binary copolymerization 355–6
  instantaneous composition equation 343
  model description 348–51
  model discrimination 348
  reactivity ratios 347–8
  remote substituent effects 344–7
  solvent effects 429–30
polar effects 21
prediction 287
reactivity ratios 363
  estimation 359–63
  Patterns of Reactivity scheme 365–6
  Q-e scheme 363–5
radical-radical termination
  combination *vs* disproportionation 252, 255, 257, 263
  model studies 255, 257
ring-opening 195, 199, 206
thermal initiation 351
template effects 438
terminal model 337–42, 366
  assumptions 337–9
  azeotropic composition 340–1
  binary copolymerization 354–5
  bootstrap effect 357, 431–2
  implications 340–1
  Mayo-Lewis equation 338
  overall rate constant for propagation 341–2
  reactivity ratios for common monomers 339
termination 366
  chemical control model 366–8
  combination and disproportionation 370–4
  diffusion control models 368–70
terpolymerization 356–7
thermal initiation 110–11
vs living radical copolymerization 525
copolymers
see also block copolymers; graft copolymers
alternating 333
branched 334–5
chain statistics 354–9
compositional heterogeneity 336, 381–4
depiction 335
functional 381–4

gradient 334, 527
monomer sequence distribution 336, 354–7
segmented 334
statistical 333
terminology 333–5
copper complexes as catalysts for ATRP 488, 492–5, 535
  activity guidelines 493–4
  halogen exchange mechanism 542
  in copolymerization 528
  ligands 493–5
    structures 494
core crosslinked microgels 548
core-first star synthesis 548–54
  importance of compatibility and method of polymerization 550
  mechanisms 551–2
    advantages and limitations 552–4
    ATRP 551, 552
    NMP 551, 552
    RAFT 551, 552, 553–4
  multi(bromo-compounds) as ATRP initiators 550
  precursors
    for ATRP 549, 550
    for NMP 549, 550
    for RAFT 549, 551
  through functionalization of polyhydroxy compounds 551
cross termination
  ethyl and fluoromethyl radicals 41
  in copolymerization
    PBMA• with PMMA• 255, 257
    PE• with PMAN• 257
    PMAN• with PS• 257
cumene hydroperoxide, as initiator 92
cumyl dithiobenzoate *see* dithiobenzoate RAFT agents
cumyl dithiophenylacetate *see* dithioacetate RAFT agents
cumyl radicals
  from α-methylstyrene dimer 140
  combination *vs* disproportionation 40, 253, 255
cumyloxy radicals
  abstraction from isopropylbenzene 56
  addition to styrene 56
  β-scission to methyl radicals 56
  fragmentation
    rate constants 125
    solvent effects 125
  from dialkyl peroxides 90
  from dicumyl hyponitrites 78
  reaction with monomers
    rate constants 119
    specificity 120, 125

temperature dependence of rate constants 56
cupric salts, radical trapping 134, 136
α-cyanoalkyl radicals *see also* cyanoisopropyl radicals
   adding monomer via nitrogen 195
   from dialkyldiazenes 68, 113
   pathways for combination 37–8, 116, 257
   polarity 116
   reaction with monomers 116
      rate constants 113–15
cyanoisopropyl dithiobenzoate, retardation in RAFT 517
cyanoisopropyl radicals 13
   abstraction *vs* addition 116
   combination pathways 37
   combination *vs* disproportionation 116–7, 256–7
   from AIBN 51
   ketenimine formation from 76, 116, 257
   mechanism of formation from AIBN 60–1
   primary radical termination, combination and disproportionation pathways 61–2
   radical-radical reactions 256
   reaction with monomers 116
      MMA 45, 52, 116
      rate constants 113–16
      S 52, 116
      specificity 116
      styrene 52
      VAc 116
   reaction with oxygen 56–7, 116
   structure 13
cyclic 1,2-disulfides, as initiators 463
cyclization
   see also cyclopolymerization
   3-butenyl radicals 23, 197
   6-heptenyl radicals 23, 193
   5-hexenyl radicals 5–6, 23, 54, 192
      radical clock 54
      substituent effects 24, 187–8, 192
   ω-alkenyl radicals 23
   4-pentenyl radicals 23
   stereoelectronic effects 23–4
cyclobutylmethyl radicals, ring-opening 198
cyclocopolymerization, of divinyl ether with maleic anhydride 194
cyclohexadienyl radicals
   structure 13
   trapped by disproportionation 140
cyclohexyl radicals
   disproportionation pathways 38–9
   Hammett parameters
      for abstraction from substituted toluenes 22
      for addition to substituted styrenes 22
   reaction with fumarodinitrile 25
   reaction with MA 18, 19
   reaction with MC 18
   reaction with methyl α-chloroacrylate 25
   reaction with MMA 17–9, 21
   reaction with monomers 114–15
cyclohexylmethyl radicals, ring-opening 198
1,4-cyclooctadiene, ring-opening metathesis polymerization 546
cyclopentadiene, reaction with *t*-butoxy radicals 123
cyclopentylmethyl radicals, ring-opening 198
cyclopolymerization 185–94
   1,4-dienes 192
      divinyl ether 192–3
   1,5-dienes 192–3
      bicyclo[2,2,1]heptadiene derivatives 193
      o-divinylbenzene 192–3
      vinyl acrylate 192–3
      vinyl methacrylate 192–3
   1,6-dienes 5, 6, 16, 186–2
      allyl acrylate 189
      allyl methacrylate 189
      bis-acrylamides 189
      bis-methacrylamides 189
      diacrylic anhydride 188
      diallyl monomers 5, 187, 188
      diallylammonium salts 186
      dimethacrylic anhydride 188
      dimethacrylic imides 188
      dimethallyl monomers 188
      o-isopropenylstyrene 189
      kinetic *vs* thermodynamic control 187
      propagation kinetics 189
      ring size 5, 186–7, 190–1
      stereochemistry of ring closure 187–8
      stereoelectronic effects 186
      substituent effects 187–8
      symmetrical 190–1
      unsymmetrical 189
   1,7-dienes 193
      bis-acryloylhydrazine 193–4
      ethylene glycol divinyl ether 193–4
   1,8-dienes, methylene-bis-acrylamide 193
   1,9-dienes, *o*-dimethacryloylbenzene 194
   1,10-dienes, 2,4-pentanediol dimethacrylate 194
   1,11-dienes, diallyl phthalate 193–4
   ATRP in 497
   bis-methacrylates 424
   bis-styrene derivatives 424
   diene monomers, review 185, 191
   triene monomers 191–2
      double ring closure 192
      triallylamine 191–2
cyclopropyl radical, structure 12
cyclopropylmethyl radicals

Index

ring-opening 196–7
    rate constants 196
    reversibility 196
cyclopropylstyrene, ring-opening polymerization 196

DBPOX *see* di-*t*-butyl peroxyoxalate
deactivation by reversible chain transfer and biomolecular activation 456
    atom transfer radical polymerization 7, 250, 456, 457, 458, 461, 486–98
deactivation by reversible coupling and unimolecular activation 455–6, 457–86
    carbon-centered radical-mediated polymerization 467–70
    initiators, inferters and initers 457–8
    metal complex-mediated radical polymerization 484
    molecular weights and distributions 458–60
    nitrogen-centered radical-mediated polymerization 483–4
    nitroxide-mediated polymerization 471–82
    oxygen-centered radical-mediated polymerizations 483
    polymerization kinetics 460–1
    reversible addition-fragmentation 470–1
    selenium-centered radical-mediated polymerization 466–7
    sulfur-centered radical-mediated polymerization 461–6
dead-end polymerization
    definition 375
    synthesis of end-function polymers 375
    with disubstituted monomers 469
defect groups 3
    see also structural irregularities
    control 413
    effect on polymer properties 3, 50, 414
    from radical-radical termination 252
    in PMMA 417–20
        anionic *vs* radical initiation 2, 417–8
        from addition-fragmentation chain transfer 419–20
        from *t*-butoxy radical initiation 53, 419
        from radical-radical termination 417–18
        head-to-head linkages 417, 418
        unsaturated chain ends 53, 417–20
    in PS 53, 414–17
        anionic *vs* radical initiation 414
        benzoate end groups 415–6
        peroxide linkages 414–5
        prepared with AIBN initiator 416
        prepared with BPO initiator 415
        unsaturated end groups 415
    in PVA/PVAc 323–4
    in PVC 2, 420–1
        anionic *vs* radical initiation 421
        formation 179
degradative chain transfer
    definition 234
    to allyl monomers 190, 319
dendrimer synthesis 555
dendritic polymer synthesis 548, 556–7
    by ATRP 556–7
    by NMP 556
    by RAFT 556
    TERMINI approach 556–7
    use of calixarene core 557
dendritic star 548
dendron macromonomers 443, 559
depropagation
    ceiling temperature 213–4
    in copolymerization 353–4
    temperature dependence 213–4
diacetylenes, thiol-ene polymerization 379
diacyl peroxides
    see also diaroyl peroxide; dibenzoyl peroxide; dilauroyl peroxide
    acyloxy radicals from 82–3
    alkyl radicals from 83
    as initiators 65, 66–7, 79–80, 86, 421
    cage decomposition products 83
    cage return 82, 84
    chain transfer to 63, 85
    concerted decomposition 82
    decomposition
    diacetyl peroxide 82
    didecanoyl peroxide 82
    explosive decomposition 83
    induced decomposition 63, 83, 85, 139
    initiator efficiency 84–5
    kinetic data for decomposition 80–1
    non-radical decomposition 85
    photochemical decomposition 83–4
    reaction with nitroxides 85, 139
    redox reactions 85–7
    substituent effects on k_d 82
    synthesis of end-functional polymers 377
    thermal decomposition 82–3
        rate constants (k_d) 80-81
        Arrhenius parameters 80-81
        effect of solvent on k_d 83
        effect of viscosity on k_d 83
    transfer to initiator 63, 85
N,N-dialkyl dithiocarbamates *see* dithiocarbamates
dialkyl hyponitrites
    as initiators 66–7, 78–9
    cage reactions 78
    induced decomposition 79
    kinetic data for decomposition 70–1
    triplet sensitized decomposition 78

viscosity effects 78
dialkyl ketone peroxides 78
   as initiators 66–7
dialkyl peroxides
   alkoxy radicals from 90
   as initiators 66–7, 79–80, 90–2
   decomposition mechanisms 91
   formation 78
   induced decomposition 91
   kinetic data for decomposition 80–1
   solvent dependence 91
dialkyl peroxydicarbonates
   abstraction of α-hydrogen 87
   alkoxy radicals from 87
   alkoxycarbonyloxy radicals from 87, 125
   as initiators 66–7, 79–80, 87–8, 421
   effect of concentration 87
   effect of solvent on k_d 87
   induced decomposition 87
   kinetic data for decomposition 80–1
   synthesis of end-functional polymers 376
dialkyl peroxyketals
   as initiators 90–91
   decomposition mechanism 91, 97
   initiator efficiency 91
dialkyl peroxyoxalates
   see also di-t-butyl peroxyoxalate
   as initiators 66–7
   as multifunctional initiators 97
dialkylamino radicals, polarity 31
dialkyldiazenes
   see also azobisisobutyronitrile; azobis(methyl isobutyrate); azonitriles
   as initiators 65, 66–78, 376–7
   cage return 68
   cis-trans isomerization 74
   commercially available 68–9, 72
   decomposition by-products 76–7
   decomposition mechanism 68
   initiator efficiency 74–7
      initial hydrogen abstraction 75
      solvent effects 75
      tautomerization 74
      temperature effects 75–6
      viscosity effects 75
   kinetic data for decomposition 70–1
   photodecomposition 74
   photolability 74
   rate constants for decomposition 68
   selection 68
   solvent effects 73
   substituent effects 72–3
   symmetrical, decomposition 68
   synthesis of end-functional polymers 376–7
   thermal decomposition 72–4
      by Lewis acids 73

      by transition metal salts 73
      steric effects 73
      thermolysis rates (k_d) 72–3
   transfer to initiator 77
   unsymmetrical 68
      as high temperature initiators 72
      as hydroxy radical sources 72
      as initiators for quasi-living polymerization 72
      for enhanced solubility in organic solvents 72
      water soluble 72
diallyl phthalate, cyclopolymerization 193, 194
diallyl trimethylolmethane, thiol-ene polymerization 379
diallylammonium salts, cyclopolymerization 186
diaroyl peroxides
   see also dibenzoyl peroxide
   as initiators 66–7, 82
   thermal decomposition 82–3, 125
diaryl disulfides, as initiators 461
diazenyl radicals 68
dibenzoyl disulfide
   as initiator 461
   radicals from 132
dibenzoyl peroxide (BPO)
   ^{13}C labeled 146–7, 260, 415
   ^{19}F labeled 261
   as initiator 80–2, 459–60, 518, 522
   cage reaction 82
   cage return 83
   chain transfer to 63, 85, 127
   decomposition mechanism
      photochemical 83–4
      redox 85–6
      thermal 83
   in alkoxyamine formation 475–6
   in living radical polymerization 459–60
   induced decomposition 85
      by nitroxides 85, 139
   initiator efficiency 83–4
   initiator of MMA polymerization 51, 84, 141
   initiator of S polymerization 51, 53, 63, 85, 141, 415–16
   initiator of VAc polymerization 85
   initiator of VC polymerization 85
   kinetic data for decomposition 80–1
   radiolabeled 145–6
   redox initiation systems
      with N,N-dimethylaniline 86
      with transition metals 85–6
diblock copolymers
   definition 334
   synthesis 541
      using ATRP 541–3
      using NMP 451

using RAFT 543–4
3,5-di-*t*-butyl catechol, inhibition by 270
di-*t*-butyl hyponitrites
   as initiators 78
   as source of *t*-butoxy radicals 78
   kinetic data for decomposition 70–1
di-*t*-butyl methyl radical, persistent radical 40
di-*t*-butyl nitroxide (DTBN)
   alkoxyamines from 477
   in NMP 475, 480
   decomposition 478
di-*t*-butyl peroxide (DTBP)
   as initiator 80, 90
   from DBPOX decomposition 89
   induced decomposition 91–2
   initiator efficiency 91
      S polymerization 92
   kinetic data for decomposition 80–1
   radical yield 92
   solvent dependence 91
1,1-di-*t*-butyl peroxycyclohexane, decomposition rate 90
di-*t*-butyl peroxyoxalate (DBPOX)
   decomposition 89
      in presence of DTBN 477
   initiator of MMA polymerization 51, 55
   initiator of S polymerization 51
   kinetic data for decomposition 80–1
dicumyl hyponitrite
   as initiator 78
   as source of cumyloxy radicals 78
   cage reaction 78
   kinetic data for decomposition 70–1
dicumyl peroxide
   as initiator 90
   radical yield 92
   stability 78
didodecanoyl peroxide *see* dilauroyl peroxide (LPO)
Diels-Alder reaction of S 108–11, 317
diene monomers *see also* butadiene; chloroprene; cyclopolymerization; isoprene
diene monomer polymerization
   head *vs* tail addition 182–5
   1,2- vs 1,4-addition 182–5
diethylaluminum chloride, in ATRP copolymerization 528
diethylene glycol diacrylate, thiol-ene polymerization 379
diffusion control of radical-radical termination
   in copolymerization 368–70
   in hompolymerization 234, 242–8
      conversion dependence 244–9
      at very high conversion 244
   mechanisms for
      reaction diffusion 243, 248

reptation 243, 248
segmental motion 243, 248, 251
translational diffusion 251
diffusion controlled limit, for radical-radical reaction 36
1,1-difluoroethylene *see* vinylidene fluoride
diisopropyl peroxydicarbonate
   t-amine couple 87–8
   radicals from 87, 125
   reaction with tertiary amines 87–8
α-diketones
   as photoredox initiator 102
   visible light systems 103
β-diketones, in enzyme-mediated polymerization 440
dilauroyl peroxide (LPO)
   as initiator 80, 82
   cage return 83
   induced decomposition 82–3
   initiator efficiency 83
   kinetic data for decomposition 80–1
dimethacrylate polymerization
   formation of head-to-head linkages 193–4
   kinetic parameters 219
   k_p 217, 247
   k_t, chain length dependence 247
o-dimethacryloylbenzene, cyclopolymerization 194
dimethoxybenzyl radical, from benzoin ether photoinitiation 62
N,N-dimethylacrylamide (DMAM)
   polymerization
      ATRP 495
      block copolymers 543
      NMP 480
      RAFT 507–8, 510–12, 539, 543
      tacticity 174–5
      TERP 524
N,N-dimethylamino radicals
   Hammett parameters
      for abstraction from substituted toluenes 22
      for addition to substituted styrenes 22
2-(dimethylamino)ethyl methacrylate (DMAEMA) polymerization
   ATRP 497, 534, 543
   block copolymers 483, 534, 543
   RAFT 508, 543
   triazolinyl radical mediated 483
2-(dimethylamino)ethyl methacrylate (DMAEMA) copolymerization
   ATRP 529
N,N-dimethylaniline, as photoredox initiator 102
N,N-dimethylaniline/BPO couple, radical production 86
dimethylglyoxime, catalytic chain transfer 314–5
2-dioxolanyl radicals, ring-opening 201

diphenyl diselenide, as initiator 466
diphenyl disulfide, as initiator 461
diphenyl ether, for S polymerizaation 497
diphenylcyclobutanes, from S polymerization 107, 109
diphenylethylene
  as deactivator 455, 470
  inhibition by 269
diphenylmethyl radicals 467
diphenylpicrylhydrazyl (DPPH)
  inhibition by 76, 268
  stability 14
direct detection of end groups 141–5
  chemical methods 144–5
  EPR spectroscopy 143
  IR spectroscopy 141
  mass spectrometry 143–4
  NMR spectroscopy 142–3
  UV spectroscopy 141
diselenides, as initiators 466–7
dispersity
  and molecular weight distribution 239–42, 461
  in ATRP 490–1
  in living radical polymerization 459
  in RAFT 519
  in reversible chain transfer 500–1
  in telluride-mediated polymerization 523
  star polymers 549
disproportionation 36, 38–9
  see also combination vs disproportionation
  between cobalt complexes and propagating species 485
  cross termination, copolymerization 370
  definition 251
  end groups
    from PMAN 256–7
    from PMMA 256, 262, 418
  homotermination, copolymerization 370
  of alkoxy radicals 78
  of alkyl radicals 38
  of but-2-yl radicals 38
  of cyclohexyl radicals 38–9
  of triazolinyl radicals 483
  specificity of hydrogen atom transfer 38, 256–7, 262
  stereoelectronic control 39
  termination by 233–4, 469
    rate constant 236
  transition state 41
dissociation energies see bond dissociation energies
disulfides
  aliphatic, as initiators 463
  as initiators 66–7, 291, 460, 461–3, 465
  as photoinitiators 103–4
  chain transfer constants 292

chain transfer to 291–2
  in living radical polymerization 103, 291, 460–3, 465
  in synthesis of end-functional polymers 377
  in synthesis of telechelics 103, 291
  radicals from 132
  transfer to initiator 103
dithioacetate RAFT agents 505–506, 511, 517, 521
  cumyl dithiophenylacetate 511, 517
  benzyl dithioacetate 511
  dithiophenylacetates 505–506, 511, 517
  in emulsion polymerization 521
dithioacids, for RAFT agent synthesis 515–16
dithiobenzoate RAFT agents 505–510, 517, 519, 546
  benzyl 510, 515, 529, 543
  cumyl 505–508, 515-517, 519–521, 526, 529, 543, 559
  cyanoisopropyl 506–508, 515, 517, 519, 529, 559
  in emulsion polymerization 521
  primary and secondary alkyl 510
  tertiary alkyl 508
  retardation 517
  synthesis 515–517
dithiocarbamates see also benzyl dithiocarbamate
  as RAFT agents 506, 514
  as photoinitiators 103–4, 458, 461–3, 465–6, 549
dithiocarbamyl radicals 462–3
dithiocarbonates see xanthates
dithioester RAFT agents 505-511
  see also dithiobenzoate RAFT agents; dithioacetate RAFT agents
dithiols, in thiol-ene polymerization 379
dithiophenylacetate RAFT agents see dithioacetate RAFT agents
dithiuram disulfides
  as photoinitiators 103–4, 461-3
  as transfer agents 103, 292, 377
    in NMP 531–2
  in living radical polymerization 460
  initiation mechanism 462–3
  synthesis of end-functional polymers 377
divinylbenzene
  cyclopolymerization 192–3
  thiol-ene polymerization 378–9
divinyl ether
  cyclo-copolymerization with MAH 194
  cyclopolymerization 192–3
DMAEMA see 2-(dimethylamino)ethyl methacrylate
dodecyl acrylate polymerization
  kinetic parameters 219

k_p 247
k_t, chain length dependence 247
dodecyl methacrylate, polymerization, k_t 247
donor monomers
   copolymerization with acceptor monomers
      interaction with Lewis acids 435–6
      list 351
      thermal initiation 110–1
dormant species, in living radical polymerization 6, 455–457
double-ring opening polymerization *see* ring-opening polymerization
DPPH *see* diphenylpicrylhydrazyl
DTBN *see* di-*t*-butyl nitroxide
DTBP *see* di-*t*-butyl peroxide
dye-partition method, to determine end groups 144

electron paramagnetic resonance spectroscopy *see* EPR spectroscopy
electron spin resonance spectroscopy *see* EPR spectroscopy
electron transfer pathway, for radical-radical reactions 36–7
electrophilicity
   see also polarity
   haloalkyl radicals 21
   Hammett correlation 21
   oxygen-centered radicals 21
electrospray ionization (ESI) *see* mass spectrometry
EMA *see* ethyl methacrylate
emulsion polymerization *see also* miniemulsion polymerization
   catalytic chain transfer 316
   compartmentalization effects 455, 502
   distinction from mini- and microemulsion 64
   entry 62–63
   initiation 51
   initiation kinetics 64
   inorganic peroxides as initiators 94
   of BA
      with ATRP 498
      with NMP 482
      with RAFT 520–521
   of BMA
      with ATRP 498
      with RAFT 520
   of methacrylates
      with macromonomer RAFT 502
      with thiocarbonylthio RAFT 520
   of S
      with ATRP 498
      with RAFT 520–521
   of VAc, with RAFT 520
   particle size 63–64

particle phase termination rates 249
reverse or combination ATRP 491
surfactant-free RAFT
   with MAA macromonomer RAFT agent 502
   with PAA amphipathic RAFT agent 521
termination kinetics 249–50
with ATRP 491, 497–498
with NMP 481–2
with RAFT 502, 520–1, 544
zero-one kinetics 250
z-mer 63
end-functional polymers 384
   synthesis
      α-functionalization 531, 533, 536–7, 539–40
      methods 384
      ω-functionalization 531–6, 538–9
      with functional inhibitors 381
      with functional initiators 375–7, 533–4, 536–7
      with functional monomers 379–80
      with functional transfer agents 289, 377–8
      with halogen transfer agents 293
   synthesis by ATRP 544
      α-functionalization 536–7
      dormant chain end removal 531
      ω-functionalization 533–6
   synthesis by living radical polymerization 454, 463, 531–40
   synthesis by NMP 544
      α-functionalization 533
      dormant chain end removal 531
      ω-functionalization 531–3
   synthesis by RAFT 545, 563
      dormant chain end removal 531
      ω-functionalization 538–9
end groups
   determination, combination *vs* disproportionation 259
   direct detection 141–5
      by chemical methods 144–5
      by EPR spectroscopy 143
      by IR spectroscopy 141
      by mass spectrometry 143–4
      by NMR spectroscopy 142–3
      by UV-visible spectroscopy 141
   ester from aliphatic diacyl peroxides 126
   formation 4
      by initiation 49
   in persulfate initiated PS 129
   in polymers initiated with dialkyl peroxides 91
   in PVAc 178, 324
   initiator-derived 53
   peroxy groups 91
   solvent-derived 55

entropy of activation *see* Arrhenius A factor
entropy of polymerization 213, 216
    steric effects 215
    table 215
enzyme-mediated polymerization 437, 440–1
EPR spectroscopy 15, 84
    initiation mechanisms 134–5, 143
    kinetic studies 133
    measurement
        of k_p 217
        of k_t 238
    spin trapping 134–5
ESI mass spectrometry *see* mass spectrometry
ESR spectroscopy *see* EPR spectroscopy
ethyl acetate, chain transfer to 295
ethyl acrylate (EA) polymerization, transfer constants, to macromonomers 307
ethyl methacrylate (EMA) polymerization
    combination *vs* disproportionation 255, 262
    kinetic parameters 219
    tacticity, solvent effects 428
    thermodynamics 215
ethyl radicals
    Arrhenius A factors 24
    combination *vs* disproportionation 41–2
    disproportionation with fluoromethyl radicals 41
ethylene (E)
    reaction with *t*-butoxy radicals 17
    reaction with methyl radicals 17
    reaction with trichloromethyl radicals 17
    reaction with trifluoromethyl radicals 17
ethylene-acrylic acid copolymer films, acid functionality 561
ethylene (E) copolymerization
    backbiting 209–10
    with AN 209
    with MAN, combination *vs* disproportionation 374
    with (meth)acrylate esters 209
    with VAc 209
        solvent effects 429
ethylene glycol divinyl ether, cyclopolymerization 193–4
ethylene (E) polymerization
    backbiting 208–10
    chain transfer, to polymer 320–1
    combination *vs* disproportionation 258
    k_p 221
    k_p solvent effects 25
    t-butyl peroxypentanoate efficiency 88
2-ethylhexyl methacrylate (EHMA), polymerization, kinetic parameters 219
Evans-Polanyi equation 30

Fenton's Reagent, reaction of organic substrates with 96
ferric salts, radical trapping 134, 136
fluorescence spectrophotometry 15–16
fluorine substituents
    effect on alkyl radicals 14
    effect on bond dissociation energies 23
fluoro-olefins
    polar factors role 21
    rate constants and regiospecificities for radical addition 17–18, 22
    reaction with *t*-butoxy radicals 22, 122
    reaction with radicals 22, 122
    synthesis via iodine-transfer polymerization 521
    use in radiation-induced grafting 390
fluoromethyl radicals
    combination *vs* disproportionation 41
    disproportionation with ethyl radicals 41
FMO theory *see* frontier molecular orbital theory
fragmentation (β-scission)
    initiator-derived radicals 54
    of acyloxy to alkyl radicals 82–3, 112
        rate constants 127
    of *t*-alkoxy to alkyl radicals 56, 91
        rate constants 124
        substituent effects 123–5
    of alkoxycarbonyloxy to alkoxy radicals 127
    of benzoyloxy to phenyl radicals 52, 54, 84, 127
        as radical clock 127
        photogenerated 84
        rate constants 127
    of *t*-butoxy to methyl radicals 54, 123–4
        as radical clock 54
        solvent effects 123
    of cumyloxy to methyl radicals 125
    of 6-heptenoyloxy to 5-hexenyl radical 54
    of isopropoxycarbonyloxy to isopropoxy radicals 87
    of photoinitiators 98–9
Fremy's Salt, inhibition by 268
frequency factor *see* Arrhenius *A* factor
Frontier Molecular Orbital (FMO) theory
    prediction of abstraction *vs* addition 35
    prediction of head *vs* tail addition 27
functional copolymers, compositional heterogeneity 381–4
functional inhibitors 381
functional initiators 375–7, 536–7
    alkoxyamines 533, 558, 561–2
    benzoin derivatives 101
    dialkyldiazenes 68, 376–7
    disulfides 103
    nitroxides 534
    peroxides 376–7

Index

functional monomers 379–80
functional transfer agents
  chain transfer 377–8
  synthesis
    end-functional polymers 289, 377–8
    telechelic polymers 289
α-functionalization, end-functional polymer
  synthesis 531
  ATRP
    initiators 536–7
    methods 537
    telechelic polymers 537
  NMP 533
  RAFT 539–40
    RAFT agents used 539
ω-functionalization, end-functional polymer
  synthesis 531
  ATRP 533–6
    end group transformation 535
    nucleophilic displacement reactions 536
    side reactions 534–5
  NMP 531–3
    end group transformation 531–2
  RAFT 538–9
    light sensitive end groups 538
    methods for end group removal 538–9

galvinoxyl radical
  diacyl peroxides decomposition 85
  inhibition by 76, 268
  stability 14
gel effect *see* Norrish-Trommsdorf effect
gelation technique 258–9
germyl hydrides, as radical traps 137
glycidyl methacrylate polymerization
  kinetic parameters 219
  grafting onto polyolefins 397–8
gold surface
  ATRP initiation, graft copolymer synthesis 563
  RAFT polymerization, synthesis of end-functional polymers 563
GPC distributions 241–2
  purities, block copolymers 540–1
  with tetrafunctional RAFT agents 553
gradient copolymers 334, 526
  compositional drift 526
  degree of blockiness 525
  reactivity ratios 526
  synthesized
    by ATRP 529
    by NMP 528
    by RAFT 529
graft copolymer synthesis 90, 384–5
  by living radical polymerization 558–63

grafting from - surface initiated
  polymerization 560
  advantages 560
  ATRP initiator use 560–2
  grafting from inorganic surfaces 562–3
  grafting from polymer surfaces, mechanisms 560–2
  NMP initiator use 560
  RAFT use 560–3
grafting through - copolymerization of macromonomers 558–60
  ATRP use 558–9
  NMP use 558
  RAFT use 558–9
grafting to - use of end-functional polymers 374, 563
  with ceric ion redox initiation 105
  with living radical polymerization 483
  with macromonomers 400–1
  with multifunctional initiators 98, 386–7
  with photoinitiation 98
  with transformation reactions 387–9, 544–6
graft copolymerization
  as side reaction with macromonomer RAFT polymerization 501
  categorization by method of formation 385
graft copolymers
  alternative names 385
  definition 334–5
  termination by combination, Gelation technique 258–9
graft polymerization
  of MMA 464–5
  of S 464–5
graft polyolefins
  maleate ester 396–7
  maleic anhydride 392–6
  maleimide 396–7
  (meth)acrylate 397–9
  styrene 399
  vinyloxazoline 400
  vinylsilane 399–400
grafting
  from inorganic surfaces 562–3
  from polymer surfaces 560–2
  radiation-induced 389–90
  radical-induced 390–400
group transfer polymerization, transformation to radical polymerization 309
Group VI hydrides, radical trapping 134, 137
guidelines
  for initiator selection 66–7
  for predicting, combination *vs* disproportionation ratio 43–4, 264
  for predicting outcomes
    of hydrogen atom abstraction reactions 36

of radical addition 28–9
half-lives
    for initiator decomposition
        azo-compounds 71
        peroxides 81
        typical values 64–5
haloalkyl radicals
    electrophilicity 21
    from halocarbons 293
halocarbons
    as initiators 488–9
    chain transfer to 283, 289, 293–4
    effect of chain length on C_{tr} 294
    in preparation of telomers 293
    polarity 293
α-haloketones, fragmentation 98–9
halo-olefins
    see also fluoro-olefins
    rate constants and regiospecificities for radical addition 17–18
    reaction with t-butoxy radicals 122
    reaction with radicals 122
p-halostyrenes, polymerized by NMP 480
Hammett parameters
    and hydrogen atom transfer reactions 31
    and radical polarity 21
    for abstraction from substituted toluenes 22
    for addition to substituted styrenes 21–2
    for radical reactions 21
    Patterns of reactivity scheme 365
Hammond postulate, radical addition 20
head addition, definition 5, 16
head-to-head linkages
    from captodative olefins 270
    from head addition 167, 176, 179, 183
    from radical-radical reaction 36, 37
    from radical-radical termination 176, 217, 417–18
    in acrylic polymers 182
    in allyl polymers 181–2
    in cyclopolymers 5
    in dimethacrylate polymers 193–4
    in fluoro-olefin polymers 181
    in PAM 182
    in PAN 182
    in PMA 182
    in PMMA 217, 417–8
        effect on thermal stability 418
    in PVAc 324
    in PVF 5, 181
head vs tail addition
    definition 176
    effect of temperature 116
    initiation
        by alkyl radicals 112–13
        by aryl radicals 117
        by benzoyloxy radicals 52, 120–1, 126, 415
        by t-butoxy radicals 120, 122
        by cyanoisopropyl radicals 116
        by hydroxy radicals 120, 128
        with acrylate esters 18–19, 120
        with halo-olefins 18, 122
        with methyl crotonate 18–9
    measurement, NMR spectroscopy 178–9
    mechanism 16
    of t-butoxy radicals with fluoro-olefins 22
    polar effects 21–2, 122, 178, 181
    propagation 4–5, 167, 176–85, 421
        acrylic monomers 182
        allyl monomers 181–2
        diene monomers 182–5
        fluoro-olefins 180–1
        in cyclopolymerization 6, 185–94
        monoene polymers 176–82
        VAc 178–9, 522
        VC 179–80
    steric effects 16, 21, 176–8
heat of polymerization 213, 216
    steric effects 215
    table 215
HEMA see hydroxyethyl methacrylate
6-heptenyl radicals, cyclization 23, 193
heteroatom-centered radicals 131–3
    abstraction vs addition 35, 132
    polar effects in hydrogen atom transfer 31
heterogeneous polymerization
    see also emulsion polymerization; miniemulsion polymerizatiom
    compartmentalization effects 455
    in ATRP 497–8
    in NMP 481–2
    in RAFT 520–1
    initiation in 63–4
    termination in 249–50
hexasubstituted ethanes
    in living radical polymerization 460, 467–8
    in synthesis of end-functional polymers 377
    polymerization mechanism 468
5-hexenyl radicals
    cyclization 5–6, 23, 24, 54, 192
        as radical clock 54
        substituent effects 24, 187–8, 192
    from 6-heptenoyloxy radical fragmentation 54
    reaction with monomers 114–15
n-hexyl radicals
    Hammett parameters
        for abstraction from substituted toluenes 22
        for addition to substituted styrenes 22
    reaction with α–chloroacrylate 25
    reaction with fumarodinitrile 25

history
  initiation 49–50
  living radical polymerization 6
  polymer structure 1–2
  radical polymerization 1–2
  radical reactions 11–12
homolytic substitution chain transfer agents 289–90
  alkyl iodides 499, 522
  alkyl stibines 499
  alkyl tellurides 499, 523
  disulfides 291–2
  monosulfides 292–3
  thiols 290–1
horseradish peroxidase, in enzyme-mediated polymerization 440
hydrogen abstraction *see* abstraction *vs* addition; hydrogen atom transfer
hydrogen atom transfer 11, 29–36, 289
  see also abstraction *vs* addition
  activation energy 29
  bond dissociation energies 30, 34–5
  effect of adjacent lone pair 33
  effect of radical structure on site of attack 32
  Evans-Polanyi equation 30
  from acetone 295
  from benzene 295
  from cobalt hydride 311
  from initiator
    alkyl hydroperoxides 93, 130
    dialkyl hyponitrites 78
    dialkyl peroxydicarbonates 87
    dialkyldiazenes 75
    ketenimines 77
  from metal hydrides 137
  from polymer 320
    poly(alkyl acrylates) 322
    PVAc 323
    PVC 325
  from solvent 294–5
  from toluene 113, 295
  guidelines for prediction of outcome 36
  Hammett relationship 31
  in disproportionation 264
    from PMAN 256–7
    from PMMA 256, 262
  intramolecular 32, 208–12
  mechanism 29
  order of reactivity of X-H compounds 30
  polar effects 30–2
  reaction conditions 33–4
  reactivity-selectivity principle 30
  specificity of
    in disproportionation 38
    intramolecular hydrogen abstraction 32
  stereoelectronic effects 32–3
  steric effects 30–1
  summary 36
  to *t*-alkoxy radicals 124–5
  to alkyl radicals 116
  to alkylperoxy radicals 56–7
  to aryl radicals 117
  to benzoyloxy radicals 52, 120
  to *t*-butoxy radicals 32–3
    from alkenes 123
    from allyl acrylates 122
    from amines 33
    from AMS 120, 123
    from axial *vs* equatorial positions 33
    from BMA 123
    from conformationally constrained compounds 33
    from ethers 32–3
    from MA 120
    from MAN 121
    from MMA 52, 55, 120
    from toluene 55
    from VAc 121
    from vinyl ethers 123
    with BMA 53
  to carbon-centered radicals, polar effects 36
  to chlorine atoms, specificity 32
  to cumyloxy radicals 120
  to cyanoalkyl radicals 116
  to α-cyanoisopropyl radicals 116
  to heteroatom-centered radicals, polar effects 31
  to hydroxy radicals 120, 128, 130
    specificity 32
  to methyl radicals 31
    specificity 32
  to nitroxides 140
  to oxygen-centered radicals, polar effects 36
  to phenyl radicals 117
    specificity 32
  to sulfate radical anion 130
  transition state for 29
hydrogen peroxide
  as initiator 93, 96
  as source of hydroperoxy radicals 96
  in enzyme-mediated polymerization 440
  photodecomposition 96
  reaction with transition metal ions 96
  redox system 96
  synthesis of end-functional polymers 377
  thermal decomposition 96
hydroperoxides, as initiators 66–7, 79
hydroperoxy radicals, from hydrogen peroxide 96
α-hydroperoxydiazenes
  as multifunctional initiators 72, 97
  as source of hydroxy radicals 97

decomposition mechanism 97
induced decomposition 97
hydroperoxyketals, decomposition mechanism 97
hydroquinone, inhibition by 270–1
hydrosilylation, of poly(dimethyl siloxane) 546
hydroxy radicals 128
   abstraction vs addition 35, 128
   from alkyl hydroperoxides 92
   from hydrogen peroxide 96
   from α-hydroperoxy diazenes 72, 97
   from sulfate radical anion 130
   polarity 35, 128
   reaction with aliphatic esters 32
   reaction with monomers
      rate constants 119
      specificity 128
α-hydroxyalkyl radicals, from ceric ion initiation 105
hydroxyethyl methacrylate (HEMA)
  polymerization
    kinetic parameters 219
    by ATRP 497
    with dithiocarbamate photoiniferter 465
hyperbranched polymers 548, 555–7
   dendritic polymers 548, 556–7
   self-condensing vinyl polymerization 548, 555–6
hyperconjugation 13
hyponitrites see dialkyl hyponitrites

'ideal' inhibitors 264
induced decomposition of initiator
   alkyl hydroperoxides 93, 130
   diacyl peroxides 63, 83, 85, 139
   dialkyl hyponitrites 79
   dialkyl peroxides 91–2
   dialkyl peroxydicarbonates 87
   dialkyldiazenes 75
   di-t-butyl peroxide (DTBP) 91–2
   dilauroyl peroxide 82–3
   α-hydroperoxy diazenes 97
   isopropyl hyponitrite 79
   ketenimines 77
   peroxyesters 89
   persulfate 95
inhibition
   definition 233
   kinetics 265–7
   mechanisms 266
inhibitors 233–4, 264–7
   aromatic nitro-compounds 272–3
   captodative olefins 269–70
   definition 264
   1,1-diphenylethylene 269
   DPPH 14, 76, 268

Fremy's Salt 268
functional, synthesis of end-functional polymers 381
galvinoxyl radical 14, 76, 268
'ideal' 264
Koelsch radical 268
nitro compounds 272–3
nitrones and nitroso-compounds 134–5, 272
nitroxides 14, 76, 138–40, 266, 268, 381
oxygen 234, 266, 268–9
phenols 234, 270–1
phenothiazine 272
quinones 271–2
'stable' radicals 267–8
TEMPO 14, 266, 268
transition metal salts 136, 265–6, 273, 381
triphenylmethyl radical 268
triphenylverdazyl radical 76, 268
iniferters
  see also photoiniferters
  living radical polymerization 457, 465, 469
  loss of 'living' ends through primary radical termination by disproportionation 469–70
inimers 555
initers, living radical polymerization 457
initiation
  cage reaction and initiator-derived by-products 60–1
  chain transfer to initiator 62–3
  definition 49
  history 49–50
  in heterogeneous polymerization 63–4
  initiator-derived radicals
    formation 49–51
    fragmentation 54
    reaction with monomer 51–3
    reaction with oxygen 56–7
    solvent effects 55
  photoinitiation 58–60, 74, 78, 90
  primary radical termination 61–2
  reaction with oxygen 56–7
  reaction with solvents, additives or impurities 55
  structural irregularities from 3–4, 49–50
  temperature effects and reaction medium on radical reactivity 55–6
initiator efficiency
  definition 57
  effect of 75–6
  effect of cage reaction 57–8, 60, 76, 84
  effect of cage return 60
  effect of chain transfer to initiator 62–3
  effect of conversion 84, 237, 244
  effect of initiator structure 65
  effect of primary radical termination 61–2
  effect of radicals formed 60

effect of temperature 75–6
effect of viscosity 60–1, 75
in thermal initiation 57–8
measurement 145
of AIBMe 76
of AIBN 76
of BPO 83–4
    photochemical decomposition 84
of BPO/dimethylaniline redox couple 86
of DBPOX 89
of diacyl peroxides 84–5
of dialkyldiazenes 74–7
of di-*t*-butyl peroxide 92
of LPO 83
of persulfate redox system 95–6
solvent effects 75
tautomerization 74
initiators 64–7
  see also functional initiators; multifunctional initiators; photoinitiators; redox initiators
  ^{13}C labeled 146–8
  aqueous phase, particle formation 63–4
  atom transfer mediated polymerizations 348
  azo-compounds 64, 66–8
    dialkyl hyponitrites 66–7, 78–9
    dialkyldiazenes 65–78, 376–7
  chain transfer to 62–3
  disulfides 66–7, 103, 291, 460–3, 465, 488
  half lives
    azo-compounds 71
    peroxides 81
    typical values 64–5
  halo-compounds 458, 488–9
  high temperature 72
  organometallics 104–5, 423
  peroxides 64, 66–7, 79–96, 376
  properties 66–7
  radiolabeled 145–6
  rate constants for decomposition
    azo-compounds 71
    peroxides 81
    typical values 64
  review 65
  selection 53, 65–7
  stable (free) radical-mediated polymerizations 457–8
  thermal decomposition rates 57–8, 65
  thermal initiation 106–11
inorganic peroxides
  see also hydrogen peroxide; persulfate
  as initiators 79– 80, 93–6
  kinetic data for decomposition 80–1
intramolecular atom transfer *see* backbiting
iodine transfer polymerization 456, 499, 521–2
  mechanism 522
  of acrylates 522
  of fluoro-olefins 522
  of S 522
  of VAc 522
  side reactions 522
ionic liquids, in polymerization 432–3
IR spectroscopy
  measurement of
    end groups 141
    tacticity 173
iron complexes
  as catalysts for ATRP 492, 496
  as initiators 484
  catalytic chain transfer 315
  in MMA polymerization 496
  in S polymerization 496
  in VAc polymerization 496
isobutylene, abstraction *vs* addition 123
isoprene (I) polymerization
  1,2- *vs* 1,4-addition 183, 185
  NMP with TEMPO 481
isopropenyl acetate
  reaction with carbon-centered radicals, rate constants 115
  reaction with oxygen-centered radicals, specificity 121
isopropoxycarbonyloxy radicals 110
  aromatic substitution by 128
  fragmentation 87
  from diisopropyl peroxydicarbonate 87, 108
  reaction with monomers
    MMA 118
    S 128
    specificity 120
N-isopropyl acrylamide (NIPAM)
  copolymerization by RAFT 529
N-isopropyl acrylamide (NIPAM) polymerization
  block copolymers 543, 546
  with dithiocarbamate photoinitiator 465
  with RAFT 507–508, 510–512, 514, 539, 543, 546
  with stibine-mediated polymerization 524
  with TERP 524
  tacticity 174, 175
    Lewis acid effects 435
IUPAC recommendations
  copolymer depiction 335
  living polymerization 452
  polymer structure 2

Jablonski diagram 58, 59

ketene acetals
  ring-opening copolymerization 195, 379–80
  ring-opening polymerization 199–203
    ring size effects 199–200

substituent effects 199–201
synthesis of end-functional polymers 195, 379–80
ketenimine
  from AIBN decomposition 76–7
  from combination of cyanoalkyl radicals 37–9
  from cyanoisopropyl radicals 76, 116, 257
  from dimeric PMAN 257–8
  thermal stability 257
α-ketoalkyl radicals, combination pathways 37–8
ketyl radicals 102–3
Kharasch reaction 486
kinetic vs thermodynamic control
  cyclization of hexenyl radicals 5–6
  in 1,6-dienes cyclopolymerization 187
  of radical ring-opening 201
  of ring-opening polymerization 196
  radical addition 4, 12, 16–17, 50
  radical reactions 50
Koelsch radical, inhibition by 268

labeling techniques 145–9
  radiolabeling 145–6
  NMR 146–9
    ^{13}C labeled initiators 146–8
    ^{19}F labeled initiators 146–8
ladder polymers
  from covalently bonded templates 438–40
  from methacryloyl derivatives of poly(hydroxy compounds) 194
Lewis acids
  control of propagation 425, 433–6
    head vs tail addition 434
    k_p 432–3
    tacticity 434–5
  control of copolymerization 435–6
    with ATRP 436, 528
    with RAFT 436, 519, 528
  control of tacticity 174–176, 434–435
    with RAFT 519
  effect on decomposition of azo-compounds 73
  effect on RAFT agents 519
  ethyl aluminum sesquichloride, in ATRP 528
ligands, for copper based ATRP catalysts 493–4
linear low density polyethylene (LLDPE) see polyolefins
living/controlled radical polymerization see living radical polymerization
living radical copolymerization 525–30
  atom transfer radical polymerization 528–9
  bootstrap effects 526
  cross propagation 525, 527
  end-functional polymer synthesis, by RAFT 538–40
  gradient copolymers 526
  major advantages of 525
  molecular weight distribution 525
  nitroxide-mediated polymerization 527–8
  reactivity ratios 525–6
  reversible addition-fragmentation chain transfer (RAFT) 529–30
  steady state composition of propagating radical 525
  vs conventional radical copolymerization 525
living radical polymerization
  see also atom transfer radical polymerization (ATRP); iodine transfer polymerization; nitroxide-mediated polymerization (NMP); metal complex-mediated radical polymerization; reversible addition fragmentation chain transfer (RAFT); tellurium-mediated radical polymerization (TERP)
  activation-deactivation processes
    deactivation by reversible coupling and unimolecular activation 455–86
    deactivation by reversible atom or group transfer and biomolecular activation 456, 486–98
    reversible chain transfer 456, 498–525
  active species 6, 455–6
  agents providing reversible deactivation 454–7
  ATRP 7, 250, 456–7, 486–98
  block copolymer synthesis 291, 454, 463, 465, 483, 526, 540–8
    by ATRP 541–3, 545–6
    by NMP 541, 545–7
    by RAFT 543–4, 546–7
    segmented block copolymers 547–8
    transformation reactions 544–6
    triblock copolymers 546–7
  carbon-centered radical-mediated polymerization 467–70
  chain length
    and conversion 250–1
    and diffusion mechanisms 251
  criteria for 452–4
  definition 451
  dormant species 6, 455–6
    labile functionality 416–17, 420
    chain end removal 531
  end-functional polymer synthesis 454, 463, 531–40
    by ATRP 533–7
    by NMP 531–3
    by RAFT 538–40
    α-functionalization 531, 533, 536–7, 539–40
    ω-functionalization 531–6, 538–9
  graft copolymer synthesis 558–63
    grafting from 560–3
    grafting through 558–60

Index

grafting to 563
history 6
hyperbranched polymers 555–7
telechelic polymers 103, 537
initiators for
  NMP 457–60
    AIBN 459
    alkoxyamines 458
    BPO 459, 460
    characteristics 458
  SFMRP
    characteristics of 458
    diselenides 466–7
    disulfides 103, 291, 460–3
    dithiocarbamates 458, 463–6
    dithiuram disulfides 461–3
    iniferters, initers 457–8
    hexasubstituted ethanes 460, 467–8
    triphenylmethylazobenzene 72, 468
iodine transfer polymerization 456, 499, 521–2
Lewis acids 436
metal complex-mediated radical polymerization 484–6
molecular weight distributions 251, 453–4, 458–60, 490–1, 499–501
molecular weight conversion dependence 452–3, 455
nitrogen-centered radical-mediated polymerization 483–4
NMP 250, 416, 456, 458, 461, 471–82
outlook 563–4
oxygen-centered radical mediated polymerizations 483
persistent radical effect 457
publication rate 7
RAFT 470–1, 501–21
selenium-centered radical-mediated polymerization 466–7
stable (free) radical-mediated polymerizations 457–60
star polymers 547–57
  arm-first approach 548, 554–5
  core-first approach 548, 549–54
  dendritic polymers 548, 556–7
  self-condensing vinyl polymerization 548, 555–6
stibine-mediated polymerization 499, 524–5
sulfur-centered radical-mediated polymerization 461–6
TERP 456, 499, 522–4
termination kinetics 250–1
terminology 451–2
long chain branches
  by transfer to polymer 320
  in PE 231

  in PVA 324
  in PVAc 323–4
  in PVC 325
  in PVF 325
low density polyethylene (LDPE) *see* polyethylene; polyolefins
LPO *see* dilauroyl peroxide

MA *see* methyl acrylate
MAA *see* methacrylic acid
macromonomer RAFT agents 305–8, 501–2
  see also macromonomers; methacrylic
  block copolymers synthesis 502
  emulsion polymerization 502
  transfer constants 307, 502
macromonomers
  acrylate esters 410, 481
  addition-fragmentation chain transfer, mechanism 305
  chain transfer constants 307
  copolymerization 400–1, 558–60
  definition 400
  dendron 443, 559
  methacrylic
    see also macromonomer RAFT agents
    addition-fragmentation chain transfer 252, 305–8, 322, 400–1, 501-2
    copolymerization 400
    polymerization 401
    solvent effects 428
  styrenic 400
  synthesis
    by ATRP 477, 559
    by catalytic chain transfer 311–2
    by NMP 481
    by thermal decomposition of alkoxyamines 477, 479, 481, 559
    with addition fragmentation chain transfer agents 296–7, 299–303, 305–8
MADIX (Macromolecular Design by Interchange of Xanthate) *see* reversible addition-fragmentation chain transfer (RAFT) polymerization
magnetic field, effect on cage reaction 61
MAH *see* maleic anhydride
MALDI-TOF mass spectrometry *see* mass spectrometry
maleate ester graft polyolefins 396–7
maleic anhydride (MAH) copolymerization
  reactivity ratios 339
  with S 351, 526
    bootstrap effect 431
    thermal initiation 110
    ω-functionalization, NMP 532
  cyclocopolymerization with divinyl ether 194
maleic anhydride (MAH) graft polyolefins 392–6

factors affecting grafting yields 394–6
maleimide graft polyolefins 396–7
MAM *see* methacrylamide
MAN *see* methacrylonitrile
manganese complexes
   ATRP catalysts 492
   as initiators 104
mass spectrometry
   end group determination
      initiation 143–4
         combination *vs* disproportionation 259
      molecular weight determination 143–4
      ESI 143–4
      MALDI-TOF 143–4, 259
Mayo equation 282–3
Mayo-Lewis equation, in copolymerization 338
MC *see* methyl crotonate
mechanically assisted polymer assembly (MAPA) 562
mediated radical polymerization *see* living radical polymerization
mercaptoethanol
   as transfer agent 291, 377
   synthesis of end-functional polymers 377
mercuric hydrides, radical trapping 134, 137
metal complex-mediated radical polymerization 484–6
   side reactions 485
metal complex-organic halide redox systems 104–5
metal hydrides, radical trapping 134, 137
methacrylamide polymerization, tacticity, effect of Lewis acids 434–5
methacrylate esters copolymerization
   see also butyl methacrylate copolymerization; methyl methacrylate copolymerization
   with ethylene 209
methacrylate esters polymerization
   see also butyl methacrylate polymerization; ethyl methacrylate polymerization; hydroxyethyl methacrylate polymerization; methyl methacrylate polymerization
   by emulsion RAFT 520
   by NMP 481
   by stibine-mediated polymerization 524
   catalytic chain transfer 311
   combination *vs* disproportionation 255–6, 261–2, 478
   head *vs* tail addition 182
   isotacticity 174
   template 439
   thermal initiation 109–10
methacrylate graft polyolefins 397–9
methacrylate macromonomers
   chain transfer constants 307
   chain transfer to 305–8

methacrylic acid (MAA)
   chain transfer to 305
   reaction with carbon-centered radicals, rate constants 114
methacrylic acid (MAA) copolymerization
   Q-e values 365
   with MMA
      solvent effects 429–30
      template effects 438
   with S, bootstrap effect 431
methacrylic acid (MAA) polymerization
   kinetic parameters 219
   k_p solvent effects 426
   tacticity
      amine effects 428–9
      solvent effects 428
   thermodynamics 215
   with ATRP in solution 497
methacrylonitrile (MAN)
   by-product from AIBN 60–1, 77
   from disproportionation of cyanoisopropyl radicals 62
   reaction with carbon-centered radicals, rate constants 115
   reaction with heteroatom-centered radicals, rate constants 131
   reaction with oxygen-centered radicals
      rate constants 119
      specificity 121
methacrylonitrile (MAN) copolymerization 77
   AIBN by-product 60
   Q-e values 365
   with S
      bootstrap effect 431
      combination *vs* disproportionation 373
      solvent effects 429
methacrylonitrile (MAN) polymerization
   chain transfer to allyl sulfides 300
   combination *vs* disproportionation 256–8, 262
   kinetic parameters 219
   k_p 221
   tacticity 173, 175
   thermodynamics 215
α-methoxypoly(ethylene oxide) methacrylate polymerization, with ATRP in solution 497
methyl acrylate (MA)
   rate of addition to hexenyl radicals 26
   reaction with benzoyloxy radicals 18–9
   reaction with *t*-butoxy radicals 19, 122
   reaction with carbon-centered radicals, rate constants 114
   reaction with cyclohexyl radicals 18–9
   reaction with heteroatom-centered radicals, rate constants 131
   reaction with 5-hexenyl radicals, solvent effects 26

Index

reaction with oxygen-centered radicals
   rate constants 119
   specificity 120
reaction with phenyl radicals 18–9
methyl acrylate (MA) copolymerization
  $Q\text{-}e$ values 365
  reactivity ratios 339
methyl acrylate (MA) polymerization
  catalytic chain transfer 316
  chain transfer
    to allyl halides 303
    to allyl sulfides 300
    to cobalt complexes 316
    to disulfides 292
    to polymer 320
    to solvent 295
    to thiohydroxamic esters 309
    to thiols 283, 290
    to thionoesters 309
    to vinyl ethers 299
  combination *vs* disproportionation 262
  inhibition constants 265
  kinetic parameters 219
  k_p 221, 247
  k_t, chain length dependence 247
  tacticity 175
    ionic liquids effect 433
  thermodynamics 215
  with ATRP 495–6, 498
  with SFRMP
    CoTMP initiator 484
    dithiocarbamate photoiniferters 465
methyl α-chloroacrylate, temperature effect on alkyl radical addition 25
methyl crotonate
  reaction with benzoyloxy radicals 18
  reaction with cyclohexyl radicals 18–9
  reaction with phenyl radicals 18
  reaction with *t*-butoxy radicals 18
methyl ethacrylate (MEA) polymerization, thermodynamics 215–216
N-methyl methacrylamide (MMAM) polymerization, tacticity, effect of Lewis acid 435
methyl methacrylate (MMA)
  reaction with alkyl radicals, penultimate unit effects 346
  reaction with arenethiyl radicals 132
  reaction with benzoyloxy radicals 18–9, 21, 52, 120
  reaction with *t*-butoxy radicals 19, 52–3, 55, 118, 120
    solvent and temperature effects 55–6
  reaction with carbon-centered radicals, rate constants 114
  reaction with cumyloxy radicals 120, 125

reaction with cyanoisopropyl radicals 52
reaction with cyclohexyl radicals 18–9, 21
reaction with heteroatom-centered radicals, rate constants 131
reaction with hydroxy radicals 120, 128
reaction with isopropoxycarbonyloxy radicals 118, 120
reaction with methyl radicals 52
reaction with oxygen-centered radicals
  rate constants 119
  specificity 120
reaction with phenyl radicals 19, 52
  solvent effects 26
reaction with radicals 18
methyl methacrylate (MMA) copolymerization
  $Q\text{-}e$ values 365
  reactivity ratios 339
  with AMS 353
  with BA 526
  with MAA
    effect of solvent 429–30
    template copolymerization 438
  with S
    bootstrap effect 31
    combination vs disproportionation 371–3
    effect of Lewis acid 435–6
    effect of solvent 429
    in ionic liquids 433
    penultimate unit effects 346, 347–9
    using NMP 527
  with VAc, solvent effects 429
methyl methacrylate (MMA) polymerization
  catalytic chain transfer 310–12, 314–6
  catalytic inhibition 311
  chain transfer
    to allyl halides 303
    to allyl phosphonates 303
    to allyl silanes 303
    to allyl sulfides 300
    to allyl sulfonates 302
    to allyl sulfoxides 302
    to cobaloximes 316
    to disulfides 292
    to halocarbons 293
    to macromonomers 307
    to polymer 320–2
    to solvent 295
    to thiohydroxamic esters 309
    to thiols 290
    to thionoesters 309
    to vinyl ethers 299
  combination *vs* disproportionation 255, 258, 261–2
  enzyme-mediated polymerization 440
  head *vs* tail addition 182
  inhibition constants 265

in ionic liquids 433
in supercritical CO_2 432
kinetic parameters 219, 244
k_p 217, 221, 247
    effect of Lewis acids 434
    effect of solvent 427
k_t, chain length dependence 246–7
tacticity 173–4
    effects of Lewis acid 435, 519
    effect of solvent 428
template polymerization 437–8
thermal initiation 109–10
thermodynamics 215
with AIBN initiator 51, 75–6
with ATRP 489, 491, 495–6, 498
with BPO initiator 51, 84, 141
with DBPOX initiator 51, 55
with NMP 481
with RAFT 519
with SFMRP
    disulfide initiators 462–3
    dithiocarbamate photoiniferters 465
with stibine-mediated polymerization 524
methyl radicals
    abstraction vs addition 35
    addition to E 20
    Arrhenius frequency factors 24
    from fragmentation of t-butoxy radicals 54
    Hammett parameters
        for abstraction from substituted toluenes 22
        for addition to substituted styrenes 22
    hydrogen atom transfer 31–2
    polarity 22, 31–2, 35
    reaction with aliphatic esters 32
    reaction with E 17
    reaction with fluoro-olefins 17–8, 22
    reaction with monomers 114–15
        MMA 52, 114
        rate constants 114–15
        S 52, 114
    reaction with propionic acid 32
methylene-bis-acrylamide, cyclopolymerization 193
2-methylene-1,3-dioxolanes, ring-opening polymerization 200–3
4-methylene-1,3-dioxolanes, ring-opening polymerization 202–3, 440
2-methylenetetrahydrofurans, ring-opening polymerization 204
2-methylenetetrahydropyrans, ring-opening polymerization 204
α-methylstyrene (AMS)
    aromatic substitution, by hydroxy radicals 128
    reaction with arenethiyl radicals
        rate constants 131
        solvent effects 26

reaction with t-butoxy radicals
    rate constants 119
    solvent effects 123
    specificity 120
reaction with carbon-centered radicals, rate constants 114
reaction with heteroatom-centered radicals 132
reaction with hydroxy radicals
    rate constants 119
    specificity 120, 128
α-methylstyrene (AMS) copolymerization
    Q-e values 365
    with depropagation 353
α-methylstyrene dimer
    chain transfer to 305
    mechanism 141
    radical trapping 140–1
    reaction with oxygen-centered radicals 140–1
α-methylstyrene (AMS) polymerization
    combination vs disproportionation 253–4
    depropagation 214
    thermodynamics 215–6
α-methylvinyl monomers
    see also methyl methacrylate; methacrylonitrile; α-methylstyrene; methacrylate esters
    abstraction vs addition, solvent effects 55
    ene reaction with nitroso-compounds 134–5
    polymerization
        combination vs disproportionation 264
        thermodynamics 216
Michler's ketone, as photoredox initiator 102
micro-emulsion polymerization 64, 250
    distinction from emulsion and miniemulsion 64, 250
mikto-arm star 548–549
miniemulsion polymerization 64, 250
    compartmentalization effects 455
    distinction from emulsion and microemulsion 64, 250
    reverse or combination ATRP 491
    with ATRP 491, 498
    with NMP 481–482
    with RAFT 520–521
MMA see methyl methacrylate
MMAM see N-methyl methacrylamide
molecular mechanics calculations, on radical addition 26
molecular orbital calculations 16, 472
    on radical addition 27
    RAFT agent activity 507
molecular weight 238–40
    see also chain length dependence
    control 413
        by template polymerization 438–40

Index

with transfer agents 279–2
with living radical polymerization 251, 453–4, 458–60, 490–1, 519–20
effect on copolymerization 431–2
evaluation after ultrasonication 259–60
predicted evolution
conventional *vs* living radical polymerization 452–3
molecular weight averages 238–40, 283
molecular weight moments 239–40
molecular weight distributions 239–42, 282, 458–61, 490–1, 500–1, 519, 523
see also dispersity
and dispersity
chain length distribution 240–1
combination *vs* disproportionation 241–2, 259
in ATRP 490–1
in living radical polymerization 251, 453–4, 458–60, 490–1, 519–20
in RAFT 519–20
in polymerization with reversible chain transfer 499–501
in polymerization with conventional chain transfer 240, 282
in stibine-mediated polymerization 524
in telluride-mediated polymerizations 523
living radical copolymerization 525
most probable distribution 240
Schultz-Flory distribution 240
molybdenum complexes
as catalysts for ATRP 492
as chain transfer catalysts 315
as initiators 484
monoene polymers, propagation, head *vs* tail addition 176–82
monomers, reaction with 51–3
monomethylhydroquinone, inhibition by 270
monosulfides, as initiators 463–5
muconic acid esters, topological polymerization 441
multi-diazenes, as multifunctional initiators 97
multiblock copolymers 334
from NMP and RAFT 547
multifunctional initiators 96–7
applications 98, 386–7
azo-peroxides 386
bis- and multi-diazenes 97
concerted decomposition 97
dialkyl peroxyketals 97
α-hydroperoxy diazenes 97
hydroperoxyketals 97
non-concerted decomposition 97–8
peroxyesters 97
peroxyoxalate esters 97
synthesis of block and graft copolymers 98, 252, 386–7

multimethacrylate, template polymerization 439–40

neoprene *see* polychloroprene
network polymers, from thiol-ene polymerization 379
nickel complexes
as catalysts for ATRP 492, 496–7
in meth(acrylate) polymerization 496
NIPAM *see* N-isopropyl acrylamide
nitro-compounds
diacyl peroxide decomposition 86
inhibition by 272–3
nitrogen-centered radical-mediated polymerization 483–4
nitrones
as nitroxide precursors 472
as spin traps 134–5
inhibition by 272
nitroso-compounds
as nitroxide precursors 472
as spin traps 134–5
diacyl peroxide decomposition 86
inhibition by 234, 272
nitroxide-mediated polymerization (NMP) 6–7. 144, 250, 416, 420, 422–3, 451–7, 471–82
activation-deactivation equilibrium constant 472
k_{act}, prediction, molecular orbital calculations 472
alkoxyamine initiators 458, 471–2, 475, 560–1
block copolymer synthesis 541
diblock copolymers 541
segmented block copolymers 547
transformation reactions 544–5
triblock copolymers 546–7
copolymerization 525–9
of MMA 527–8
of MAH with S 526
mechanism 528
gradient copolymers 526–8
table 528
end-functional polymer synthesis 531–3, 544–5
dormant chain end removal 531
α-functionalization 533
ω-functionalization 531–3
in block copolymer synthesis
graft copolymer synthesis
grafting from 560
grafting through 558–9
heterogeneous NMP 481–2
emulsion 481–2
miniemulsion 481–2
surfactant/cosurfactants 482
initiation 475–7

macromonomer synthesis 478–9
mechanism 471
monomers 480–1
  acrylates 480–1
  diene monomers 481
  methacrylates 481
  S 480–1
  vinyl pyridines 480
nitroxides for 472–5
  eight-membered ring 475
  five-membered ring 473
  open-chain 475, 478, 480
  seven-membered ring 475
  six-membered ring 474
persistent radical effect 471
polymerization kinetics 461
product stability 416, 420, 478–9
rate enhancement 479, 482
reviews 471
side reactions 478–9, 533
  nitroxide decomposition 478
  disproportionation/elimination 478, 533
star synthesis
  mechanisms 52, 551
  precursors 549, 550, 554–7
  microgel formation 555
  self-condensing vinyl polymerization 555–6
  shell-crosslinking of micelles 555
  dendritic cores 556–7
nitroxides
  see also 2,2,6,6-tetramethylpiperidin-N-oxyl (TEMPO)
  alkoxyamine homolysis 472
  combination vs disproportionation 478
  exchange, in NMP 533
  fluorescence quenching 139
  formation
    from nitrones 135
    from nitroso-compounds 135, 472
  in macromonomer synthesis
    by ATRP 534
    by NMP 478–9
  inhibition by 14, 76, 138–40, 233, 266, 471, 472
  radical trapping 134, 138–40
  reaction with diacyl peroxides 85, 139
  reaction with radicals 138–9
    combination vs disproportionation 138, 478–9
    rate constants 138
  stability 472, 478
  structures
    eight-membered ring, for NMP 475
    five-membered ring, for NMP 473
    open-chain, for NMP 475, 478, 480–2
    seven-membered ring, for NMP 475

six-membered ring, for NMP 474
synthesis of end-functional polymers 381
NMP see nitroxide-mediated polymerization
NMR spectroscopy
  and ^{13}C labeled initiators 146–9
  measurement of
    end groups 142–3, 146–9
    head vs tail addition 178–9
    monomer sequence distribution 362–3
    tacticity 173
  monomer reactivity correlation 365
non-aqueous dispersion polymerization, compartmentalization effects 455
non-covalently bonded templates, template polymerization 437–8
Norrish-Trommsdorff effect
  and k_t 244, 248
  and template polymerization 438
  control of with transfer agents 279
  in acrylate polymerization 322
  in RAFT 517
nucleophilicity
  see also polarity
  acyl radicals 118
  alkyl radicals 13, 21, 31
  and abstraction vs addition 35
  Hammett correlation 21
  oxygen-centered radicals 35

oligostyryl radicals, combination vs disproportionation 253, 254
organochalcogenides
  selenium-centered radical-mediated polymerization 466–7
  stibine-mediated polymerization 524–5
  sulfur-centered radical-mediated polymerization 461–6
  TERP 522–4
organostibine transfer agents 524–5
outer sphere electron transfer (OSET), in atom transfer reaction 487
oxygen
  as chain transfer facilitator 116, 269
  as comonomer 269
  effect on initiation 56–7
  inhibition by 234, 266, 268–9
  reaction with alkyl radicals 50, 56, 116, 130
  reaction with cyanoisopropyl radicals 56–7, 116
  reaction with polymeric anions 387–8
oxygen-centered radical mediated polymerizations 483
oxygen-centered radicals 118–31
  see also acyloxy radicals; alkoxy radicals; alkoxycarbonyloxy radicals; benzoyloxy radicals; t-butoxy radicals; hydroxy radicals;

Index

methoxy radicals, isopropoxycarbonyloxy
  radicals; sulfate radical anion
  abstraction vs addition 35
  polarity 21, 35
  rate constants 118–9
  specificity 120–1, 477
oxygen radical anion
  abstraction vs addition 35
  polarity 35

PAA *see* poly(acrylic acid)
PAM *see* polyacrylamide
PAN *see* polyacrylonitrile
palladium complexes, as catalysts for ATRP 492
Patterns of Reactivity scheme 21, 26, 31
  for prediction of reactivity ratios 365–6
  for prediction of transfer constants 287
PB *see* polybutadiene
PE *see* polyethylene
pentafluorostyrene, thermal initiation 109
1,2,3,3,3-pentafluorovinylcyclopropane, ring-
  opening polymerization 197
2,4-pentanediol dimethacrylate,
  cyclopolymerization 194
4-pentenyl radicals, cyclization 23
penultimate model
  copolymerization 348, 355–6
    MMA-S copolymerization 347–9
    model description 342–4
    reactivity ratios 347–8
    remote substituent effects on radical
      addition 344–7
    solvent effect on reactivity ratios 429–30
penultimate unit effects
  see also chain length dependence
  in copolymerization 342–4, 347–8
    origin 349–50
  on chain transfer 282–3
    to halocarbons 294
    to thiols 291
    to RAFT agents 507
  on propagation
    backbiting 211
    k_p 220
    tacticity 171–2, 346–7
  on radical addition 344–7
  on radical-radical reactions 253
  on termination 253
  ATRP initiation 492
  RAFT agent activity 507
peresters *see* peroxyesters
peroxide linkages
  formation 56–7, 269, 387–8
  in PS 414–5
peroxides
  see also alkyl hydroperoxides; diacyl
    peroxides; dialkyl peroxides; dialkyl
    peroxyketals; peroxydicarbonates; hydrogen
    peroxide; peroxyesters; persulfate
  as component in redox system 79
  as initiators 64, 66–7, 79–96, 391–2
  as photoinitiators 58, 79
  kinetic data for decomposition 80–1
  peroxymonosulfate 93
α-peroxyalkoxy radical, from peroxyketals 91
peroxydicarbonates *see* dialkyl
  peroxydicarbonates
peroxydiphosphate, as initiator 93
peroxydisulfate *see* persulfate
peroxyesters
  as alkoxy and acyloxy radical source 88
  as initiators 66–7, 79–80, 88–90
  as multifunctional initiators 97
  chain transfer to 305
  kinetic data for decomposition 80–1
  non-radical decomposition 89–90
  photochemical decomposition 90, 125
  thermal decomposition 88–90
  transfer to initiator 89
peroxyketals *see* dialkyl peroxyketals
peroxyoxalates *see* dialkyl peroxyoxalates
persistent radical effect 471
persistent radicals *see also* stable free radical
  mediated polymerization (SFRMP)
  di-*t*-butylmethyl radical 40
  triisopropylmethyl radical 40
persulfate
  as initiator 66–7, 79, 93–6, 518
    in aqueous and organic media 94–5
  crown ether complexes 94
  decomposition mechanisms 94
  effect of reaction conditions on k_d 94
  non-radical decomposition 94
  photodecomposition 95
  reaction with transition metals ions 95
  redox initiation 95–6
phenacyl radicals
  carbon monoxide loss from 118
  from photodecomposition of initiators 117
9,10-phenanthrene quinone, as photoredox
  initiator 102
phenols, inhibition by 234, 270–1
phenothiazine, inhibition by 272
phenyl radicals
  abstraction vs addition 35
  aromatic substitution of S 52
  from BPO 82
  from fragmentation of benzoyloxy radicals 52,
    54, 127
  from triphenylmethylazobenzene
  hydrogen atom transfer 32

polarity 35, 117
  reaction with monomers 114–5
    MA 18–9
    MC 18
    MMA 18–9, 52
    S 52
  reaction with tolyl chloride 32
  solvent effects on relative reactivity towards S
    and MMA 26
  structure 13
1-phenylethyl radicals
  combination vs disproportionation 255
  reaction with monomers 114–15
  self reaction 253
phenylseleno radicals 132
1-(phenylseleno)ethylbenzene 466
phosphinyl radicals
  from acyl phosphine oxides 101–2, 132
  from transfer to phosphines 132
  polarity 132–3
  reaction with monomers 132
phosphorus-centered radicals 132–3
  rate constants 131
photoiniferters 465–6
  dithiocarbamates 465–6
  dithiuram disulfides 103–4
  xanthates 463–5
photoinitiation 58–60, 98
  Jablonski diagram 58–9
  quantum yield 59–60
  rate and efficiency 59
  rate of radical generation 60
  reviews 98
photoinitiators 98–104
  see also photoiniferters; photoredox initiators
  acyl phosphine oxides 98, 101
  acyl phosphonates 101
  aromatic carbonyl compounds 98–9
  azo-compounds 58, 74, 78
  benzil monooxime 99
  benzoin esters 100
  benzoin ethers 61–2, 98, 100
  BPO 83–4
  characteristics 58
  diacyl peroxides 83–4
  dialkyldiazenes 74
  dialkyl hyponitrites 78
  dialkyldiazenes 74
  diselenides 466–7
  disulfides 102-3, 458, 460–3
  dithiocarbamates 102-3, 461–6
  dithiuram disulfides 102-3, 458, 460–3
  hydrogen peroxide 96
  peroxides 58, 79
  peroxyesters 90
  persulfate 95
  polymerizable 102
  selenides 466–7
  sulfur compounds 102–3
  visible light 103
  xanthates 463–5
  xanthogen disulfides 461–2
photoredox initiators
  carbonyl compound/tertiary amine 102–3
  metal complex/organic halide 104–5, 388–9
    in block/graft copolymer synthesis 388–9
phthalocyanine complexes, chain transfer to 313
PI see polyisoprene
PMA see poly(methyl acrylate)
PMAA see poly(methacrylic acid)
PMAN see poly(methacrylonitrile)
PMMA see poly(methyl methacrylate)
polar effects 21–2
  in hydrogen atom transfer 31–2, 290
  on chain transfer
    to disulfides 292
    to halocarbons 294
    to thiols 290
  on fragmentation of t-alkoxy radicals 124
  on radical addition 117
    copolymerization 364
    head vs tail addition 21–2, 122, 178, 181
    rate 271
    to p-benzoquinone 271
    to fluoro-olefins 122, 181
  on radical reactions 16
  on radical-radical reactions 41
    combination 257
    combination vs disproportionation 255
polarity
  see also electrophilicity; nucleophilicity
  Hammett parameters 21
  in hydrogen atom transfer 30
  of alkanethiyl radicals 290
  of alkoxy radicals 27, 30, 35
  of alkyl radicals 35, 113
  of arenethiyl radicals 132
  of aryl radicals 35, 117
  of benzoyloxy radicals 35, 126
  of t-butoxy radicals 22, 35, 122
  of t-butyl radicals 35
  of cyanoalkyl radicals 116
  of halocarbons 293
  of hydroxy radicals 35, 128
  of methyl radicals 22, 35
  of phenyl radicals 35, 117
  of phosphinyl radicals 132–3
  of radicals, Hammett parameters 21
  of selenium-centered radicals 132
  of silicon-centered radicals 131
  of sulfur-centered radicals 132
  of thiols 290

Index

of trichloromethyl radicals 22
of trifluoromethyl radicals 22
polyacrylamide (PAM), head-to-head linkages 182
poly(acrylic acid), tacticity 173
polyacrylonitrile (PAN)
  combination *vs* disproportionation 262
  head-to-head linkages 182
  tacticity 175
poly(alkyl acrylates), combination *vs* disproportionation 262
poly(alkyl methacrylates)
  chain transfer to polymer 321–2
  combination *vs* disproportionation 255–6, 261–2
poly(allyl esters), head-to-head linkages 182
polyamides, from ring-opening polymerization 199
polybutadiene (PB)
  alkoxyamine functionality attached to 561
  microstructure 183–4
poly(butyl acrylate) (PBA)
  block-PS, prepared by NMP 482
  homolysis/backbiting/β-scission reaction, RAFT end group synthesis 538–9
poly(butyl methacrylate-co-methacrylonitrile), combination *vs* disproportionation 374
poly(butyl methacrylate-co-methyl methacrylate), combination *vs* disproportionation 374
polycaprolactone-*block*-poly(4-vinylpyridine) synthesis 545
polycaprolactone-*block*-PS synthesis 545
polychloroprene, microstructure 183–5
poly(dimethyl siloxane) (PDMS)
  conversion to ATRP initiator ends by hydrosilylation 546
  polymer brush synthesis from 562
polydispersity
  of molecular weight distribution *see* dispersity
polyesters, from ring-opening polymerization 195, 199
polyethers, from ring-opening polymerization 197
polyethylene (PE)
  combination *vs* disproportionation 258
  long chain branches 321
  short chain branches 208–10
poly(ethylene oxide)-*block*-PS synthesis
  from alkoxyamines, using NMP 545
  using RAFT 546
poly(ethylene-co-methacrylonitrile), combination *vs* disproportionation 374
poly(isobutyl methacrylate), attachment of alkoxyamine functionality to 561
polyisoprene (PI), microstructure 183, 185

polyketones, from ring-opening polymerization 195, 202–3
polymer brushes *see* graft copolymer
polymer structure
  see also defect groups; structural irregularities
  control 413, 421–2
  copolymers 333–5
    monomer sequence distribution 354–7, 430–1
  end groups
    from chain transfer 279, 282, 287, 418, 420, 421
    from radical-radical termination 233, 415, 417–8, 421
    initiator-derived 49, 63, 414–5, 420–1
  general formula 1–4, 49
  history 1–2
  in chain groups
    head-to-head linkages 167, 176, 178–9, 184, 252, 434
    rings 186–7, 193
    short-chain branches 208–10, 211
    tacticity 434–5
    tail-to-tail linkages 167, 176, 178–9
    unsaturation 182–4, 195, 206, 251–2
  IUPAC recommendations 1
  repeat units 1, 4
  Staudinger concept 1–2
  tacticity 167–9, 175
polymeric initiators, synthesis of block copolymers 386
polymerization thermodynamics 213–16
  autoacceleration phenomenon 281
  degree of polymerization 281–2
  reviews 213
poly(methacrylic acid) (PMAA), template polymer 438
poly(methacrylonitrile) (PMAN)
  combination vs disproportionation 256–8, 262
  tacticity 173, 175
poly(methacrylonitrile-co-styrene), combination vs disproportionation 373
poly(methyl acrylate) (PMA), tacticity 175
poly(methyl acrylate)-block-PBA synthesis 485
poly(methyl methacrylate) (PMMA)
  defect groups 417–20
    control 418–19
    head-to-head linkages 182, 252, 417–18
    unsaturated chain ends 53, 252, 417–20
  end groups
    determination 142–144, 146–148
    from benzoyloxy radicals 52–3, 120
    from *t*-butoxy radicals 52–3, 120, 144
    from cyanoisopropyl radicals 52–3, 144, 147–148
    from phenyl radicals 120

tacticity 173, 175, 424, 438
thermal stability 2, 252, 417–20
prepared with NMP 420
poly(methyl methacrylate-co-styrene),
combination vs disproportionation 371–3
polyol-redox system, as multifunctional initiators 386
polyolefins (PO)
radiation-induced grafting 390
melt-phase grafting to 392–400
monomers for grafting to
maleate ester 396–7
maleic anhydride 392–6
maleimide 396–7
(meth)acrylate 397–9
styrene 399
vinyl oxazoline 400
vinylsilane 399–400
polypropylene (PP)
see polyolefins
polystyrene (PS)
aromatic substitution, by benzoyloxy radicals 127
combination vs disproportionation 260–1
model studies 253–5
defect groups 414–7
benzoate end groups 53, 415–6
peroxide linkages 414–5
unsaturated end groups 415
end groups
determination 141–144, 146–148
from benzoyloxy radicals 52–3, 120, 139, 415–6
from *t*-butoxy radicals 52–3, 120, 144
from cyanoisopropyl radicals 52–3, 144, 147–148
from phenyl radicals 120
tacticity 175
thermal stability
anionic vs free-radical 414
prepared with AIBN 416
prepared with BPO 53, 85, 415–16
produced by ATRP and RAFT 416, 417
produced by NMP with TEMPO 416
weathering 53
polythioesters, from ring-opening polymerization 199
poly(trifluoroethylene), head-to-head linkages 181
poly(vinyl acetate) (PVAc)
chain transfer to polymer 323–5
combination vs disproportionation 263
emulsion particle formation 482
end groups, FAB-MS analysis 178
head-to-head linkages 178–9
long-chain branches 323–4

effect of reaction conditions 324
measurement 324
short-chain branches 211
tacticity 175
poly(vinyl alcohol) (PVA)
ceric ion initiated graft copolymerization 105–6
chain transfer to polymer 323–4
long-chain branches 324
tacticity 173, 175
poly(vinyl chloride) (PVC)
combination vs disproportionation 263
defect groups 420–1
long-chain branches 325
propagation 179–80
short chain branches 211
structural irregularities 179
tacticity 173, 175
thermal stability 2, 179, 420–1
poly(vinyl fluoride) (PVF)
head-to-head linkages 5, 181
long-chain branches 325
tacticity 173
poly(vinylidene fluoride) (PVF2), head-to-head linkages 181
poly(4-vinylpyridine) block polymer. from PEO 545
potassium persulfate, as initiator 518
pressure effects
backbiting in E polymerization 209
on chain transfer to PVF 325
on radical-radical reactions 43
on RAFT process 519
primary alkoxy radicals
atom abstraction 35
reactions with monomers 125
primary radical termination 58, 61–2
benzoin ethers 61–2, 100
by sulfur-centered radicals 103, 463–4
combination vs disproportionation 61
cyanoisopropyl radicals 61–2
definition 61, 233
effect on initiator efficiency 61
in living radical polymerization 458
in telechelic synthesis 62
of PBMA, with cyanoisopropyl radicals 257
of PE, with cyanoisopropyl radicals 257
of PS
with benzoyloxy radicals 415
with cyanoisopropyl radicals 116, 257, 376
reversible 62
primary radicals
definition 49
from azo-compound decomposition 112
propagation 167–8
control 421–43

with nitroxide 422–3
  with organometallic reagent 422–3
  with titanocene dichloride 424
 effect of solvent 425–32
  copolymerization 429–32
  homopolymerization 426–9
  mechanisms 426
 head vs tail addition 5, 167, 176–85
 kinetics 216–21
 k_p 216–21
  chain length dependence 213, 218–21
  conversion dependence 218
  measurement 216–18
  monomer structure dependence 218–20
  values 216
 polymerization thermodynamics 213–16
 regiosequence isomerism 167, 176–85, 421
  acrylic polymers 182
  allyl polymers 181–2
  diene polymers 182–5
  fluoro-olefin polymers 180–1
  monoene polymers 176–82
  PAM 182
  PAN 182
  PVAc 178–9
  PVC 179–80
  terminology 176
 stereosequence isomerism/tacticity 167–76, 421
 structural irregularities 3, 167
 structural isomerism 167, 185–212, 421
  addition-abstraction polymerization 208
  backbiting 208–12
  cyclopolymerization 185–94
  ring-opening polymerization 194–208
PS see polystyrene
pseudo-living polymerization see living radical polymerization
pulsed laser photolysis (PLP)
 measurement of k_p 217
 measurement of k_t 238
PVA see poly(vinyl alcohol)
PVAc see poly(vinyl acetate)
PVC see poly(vinyl chloride)
PVF see poly(vinyl fluoride)
2-pyridine carboxaldehyde, methanimine ligands from 493

Q-e scheme 21, 26, 31, 287
 for prediction of reactivity ratios 363–5
 Q-e values 364–5
quantum yield, definition 59
quasi-living polymerization see living radical polymerization
quinones
 copolymerization 271

 inhibition by 234, 271–2
quinonoid intermediates
 from combination of benzyl radicals 37
 from reaction of triphenylmethyl radicals 469

radial block copolymer 548
radiation-induced grafting processes 389–90
radical addition
 and product radical stability 17
 application of FMO theory 27
 bond strength effects 22–3
 carbon-carbon double bonds 11, 16–29
 entropic considerations 24
 factors affecting specificity 16–17, 19
 guidelines 28–9
 Hammond postulate 20
 kinetic vs thermodynamic control 4, 12, 16–17, 50
 mechanism 16
 polar effects 21–2
 reaction condition effects 24–6
 regiospecificity 17–19
 solvent effects 25–6
 stability vs resonance factors 19–20
 stereoelectronic effects 23–4
 steric effects 19–21
 substituent effects 20, 22
  acrylate esters 18
  halo-olefins 17–18
 summary 28–9
 temperature effects 24–5
 theoretical studies 20–1, 26–8
 transition state for 17, 20, 23
radical clock
 cyclization of 5-hexenyl radicals 54
 fragmentation of benzoyloxy radicals 127
 fragmentation of t-butoxy radicals 54
 reaction of carbon-centered radicals
  with metal hydrides 137
  with nitroxides 138
radical cyclization see cyclization
radical detection 14–16
 EPR 15
 fluorescence 15–16
 molecular orbital calculations 16
 UV 15–6
radical-induced grafting 390–9
 maleic anhydride graft polyolefins 392–6
 melt phase grafting 390
 side reactions 390–1
radical polymerization
 benefits of 1
 general mechanism 2
 history 1–2
 publication rate 7
radical properties 12–16

radical-radical reactions 11, 36–44
  see also combination pathways;
    disproportionation pathways; primary
    radical termination
  cage *vs* encounter reactions 43–4, 252
  combination pathways 36, 37–8
  combination *vs* disproportionation 39–43, 251–2
  cross termination 255, 257
  diffusion control 36
  disproportionation pathways 36, 38–9
  electron transfer pathway 36–7
  head-to-head coupling 36–8
  polar effects 41
  pressure effects 43
  rate and specificity
    steric effects 254–5
    temperature effects 254
  rate constants 36
  solvent effects 42–3
  statistical factors 39–40
  stereoelectronic effects 41–2
  steric effects 37, 40–1
  summary 43–4
  techniques 252
  temperature effects 42–3
  termination, in living radical polymerization 455, 457
  transition states 39
  viscosity effects 43
radical-radical termination
  activation energy 234
  at low conversion 244–8
  at medium to high conversion 244, 248–9
  at very high conversion 244
  by chain transfer, rate constant 236
  by combination, rate constant 236
  by disproportionation, rate constant 236
  chain length dependence
    of combination *vs* disproportionation 253, 256, 258
    of rate constant 234, 245–8
  classical kinetics 235–8
  combination *vs* disproportionation 251–64
    model studies 252–8
    polymerizations 258–63
    summary 263–4
  diffusion controlled 242–3
    low conversions 244–8
    medium to high conversions 244, 248–9
    very high conversions 244
  in copolymerization, combination *vs* disproportionation 255, 257
  kinetics, reviews 235
  model studies 252–3
  pathways for 233–4
  rate constant 234–8
    definition 235–6
    diffusion mechanisms 242–51
    effect of conversion 244–9
    gel or Norrish-Trommsdorff effect 244, 248
    molecular weight distributions 240–2
    molecular weights and molecular weight averages 238–40
    prediction 244–5
    terminology 234
  structural irregularities from 176, 252
radical reactions
  addition *see* radical addition; abstraction *vs* addition
  historical beliefs 11–12
  hydrogen atom transfer *see* hydrogen atom transfer; abstraction *vs* addition
  kinetic *vs* thermodynamic control 4, 12, 50
  radical-radical reaction *see* radical-radical reactions; combination pathways; disproportionation pathways; primary radical termination
  steps in radical polymerization 11
  stereoselectivity 12
  temperature dependence 55–6
  medium effects 55–6
radical stability 14
  and hyperconjugation 13
  effect on regioselectivity of radical addition 17
  for calculating thermolysis rates 73
  of primary, secondary and tertiary radicals 17
radical structure 12–13
  see also specific radicals, e.g. acetyl radical
  pi-radicals 12
  sigma-radicals 12–13
radical trapping 133–41
  agents for 132–3
  metal hydrides 134, 137
  α-methylstyrene dimer 134, 140–1
  mercuric hydrides 137
  nitrones 134–6
  nitroso-compounds 134–6
  nitroxides 134, 138–40
  spin traps 134–6
  transition metal salts 134, 136
  tri-*n*-butylstannane 137
radicals
  carbon-centered *see* carbon-centered radicals; alkyl radicals; aryl radicals
  heteroatom-centered 131–3
  initiator-derived, classification 53
  oxygen-centered *see* oxygen-centered radicals; acyloxy radicals; alkoxy radicals
  phosphorus-centered 132–3
  polarity 21–2
  properties *see* radical properties

selenium-centered 132
silicon-centered 131
structure *see* radical structure
sulfur-centered 132
terminology 12
trapping *see* radical trapping
radiolabeling 145–6
RAFT *see* reversible addition-fragmentation chain transfer (RAFT) polymerization
RAFT agents *see* dithiocarbamates; dithioesters; macromonomer RAFT agents; thiocarbonylthio RAFT agents; reversible addition-fragmentation chain transfer (RAFT) polymerization; trithiocarbonates; xanthates
random copolymers 335
  definition 333
rate constants
  see also activation-deactivation equilibrium constant
  for abstraction, effect of bond dissociation energies 34–5
  for addition
    effect of bond dissociation energies 34–5
    to acrylate esters 18–9
    to halo-olefins 17–8, 22
  for β-scission, *t*-alkoxy radicals 124
  for decarbonylation, acyl radicals 118
  for fragmentation
    of benzoyloxy radicals 127
    of cumyloxy radicals 125
  for inhibitor-radical reaction (k_z)
    inhibitors and carbon-centered radicals 266
    nitroxides and carbon-centered radicals 138, 266
  for initiation (k_i) 236
    see also for addition
  for initiator decomposition (k_d)
    azo-compounds 71
    effect of cage return 60
    effect of chain transfer to initiator 63
    peroxides 81
    typical values 64
  for propagation (k_p) 236
    chain length dependence 213, 218–21
    conversion dependence 218
    effect of Lewis acids 433–4
    measurement 216–18
    monomer structure dependence 218–20
    solvent effects 25, 426–8
  for radical addition (k_i)
    aryl radicals 117
    carbon-centered radicals 113–5, 221
    cyanoisopropyl radicals 116
    heteroatom-centered radicals 131
    oxygen-centered radicals 118–9
    solvent effects 25–6

    temperature effects 25
  for radical-radical reaction
    typical values 36
  for radical-radical termination (k_t)
    by combination 236
    by disproportionation 236
    chain length dependence 234, 245–8
    definition 235–6
    effect of conversion 244–9
    non-steady state conditions 238
    prediction 244–5
    steady state approximation 236–7
    typical values 238
  for reinitiation (k_{iT}, k_{iM}) 235
  for ring opening
    cyclobutylmethyl radicals 198
    cyclopropylmethyl radicals 196
    in ring-opening polymerization 195–6
  temperature dependence, cumyloxy radicals 56
  use of radical clocks in calibration 54
rate of polymerization, deactivation by reversible coupling 460–1
reaction conditions 24
  see also pressure effects; solvent effects; temperature effects; viscosity effects
  effect on hydrogen abstraction 33
  effect on radical-radical reactions 42–3
  influence on combination *vs* disproportionation 42–3
  solvent effects 25–6
  temperature effects 24–5
reaction diffusion 243, 248–9, 251
reactivity ratios
  definition
    penultimate model 342
    terminal model 338
  estimation
    from composition data 360–1
    from monomer sequence distribution 362–3
  for common monomers 339
  for gradient copolymers 526
  implicit penultimate model 347–8
  in ATRP 488
  prediction 363
    NMR chemical shifts 364
    Q-e scheme 363–5
  solvent effects 361, 364, 429–32
  substituent effects 344–7
  template effects 438
  terpolymerization 357
reactivity-selectivity principle
  and hydrogen atom transfer 30
  and radical addition 24–5
rearrangement of radicals during polymerization 185–212

cyclopolymerization 185–94
  intramolecular atom transfer 208–12
  ring-opening polymerization 194–208
redox initiators 104
  see also photoredox initiators
  metal complex/organic halide 104–5
  transition metal salts effects 85–6, 95–6
  with alkyl hydroperoxides 93
  with ceric ions 104–6
  with diacyl peroxides 85–6
  with hydrogen peroxide 96
  with inorganic peroxides 94–6
  with *N,N*-dimethylaniline/BPO 86–7
  with persulfate 95–6
regiosequence isomerism *see* head *vs* tail addition
reptation 243, 248–9
residual termination 243, 249
retardation
  definition 234
  in RAFT 517–8
  kinetics 266–7
  k_p effects 280
  mechanisms 266
  with transfer agents 279
    addition-fragmentation chain transfer 297
    solvents 294
    thiols 290
    VAc polymerization 294
retarders 264–7
  definition 264
reverse transfer constant
  in RAFT process 288–9, 504–5
reversible addition-fragmentation
  with 1,1-diphenylethylene 455, 470
  with captodative monomers 470
reversible addition-fragmentation chain transfer (RAFT) polymerization 7, 250–1, 288, 297, 456, 470–1, 499, 501–21
  block copolymer synthesis 543–4
    diblock copolymers prepared by 543
    hydrophilic-hydrophobic blocks 543
    macromonomer RAFT 502, 543
    mechanism 544
    order of constructing blocks 544
    segmented block copolymers 547–8
    transformation reactions 544, 546
    triblock copolymers 546–7
  copolymerization 529–30
    mechanisms 530
    statistical/gradient copolymers 529
  dispersities 505, 519
  effect of Lewis acids 519
  effect of pressure 519
  effect of solvents 518
  effect of temperature 518–9

emulsion polymerization 520–1, 544
end-functional polymer synthesis 545
  α-functionalization 539–40
  dormant chain end removal 531
  ω-functionalization 538–9
gel effect 517
graft copolymer synthesis 558–63
heterogeneous polymerization 520–1
kinetic simulation 501
macromonomer RAFT agents 501–2
mechanism 498–9, 503–5
miniemulsion polymerization 520–1
molecular weight distributions 519–21
  of acrylates, by xanthates 464
  of MAH with S 526
  of VAc
    with dithiocarbamates 464
    with xanthates 502–3, 506, 513, 521
RAFT agent synthesis 515–7
rates of polymerization 503
reaction conditions 518–20
retardation 517–8
  and inhibition period 517–8
  methacrylate polymerization 517–8
  S polymerization 517–8
reverse transfer constants 504–5
review 502
side reactions 517–18
star synthesis
  GPC distributions 553
  mechanisms 551–4
  precursors 549, 551
  microgel formation 554–5
  self-condensing vinyl polymerization 555–6
  shell-crosslinking of micelles 555
  dendritic cores 556–7
thiocarbonylthio RAFT agents 464, 501–2, 505–14
  transfer constants 504
xanthates 502–3, 506, 513, 521
reversible chain transfer 288–9, 456, 485, 498–525
  see also reversible addition-fragmentation chain transfer (RAFT) polymerization
  by homolytic substitution 498–9
  by RAFT 498–9
  degenerative chain transfer 498
  effect of reaction conditions 500
  in iodine transfer polymerization 456, 499, 521–2
  in stibine-mediated polymerization 499
  in telluride-mediated polymerization 456, 499
  molecular weights and distributions 499–501
  with methacrylic macromonomers 499
ring-opening
  of cyclobutylmethyl radicals 198

of cyclopropylmethyl radicals 196–7
of dioxolan-2-yl radicals 201
ring-opening copolymerization
    ATRP in 497
    of ketene acetals 195
    of methylenecyclohexadiene spiro compounds 199
    of spiroorthocarbonates 206
    of spiroorthoesters 206
    synthesis of end-functional polymers 195
ring-opening polymerization 194–208
    double-ring opening 205–8
    effect of concentration 196
    effect of temperature 196, 200, 202–3
    of bicyclobutanes 195
    of caprolactone 545
    of cyclic allyl sulfides 204–5
    of cyclopropylstyrene 196
    of ketene acetals 199–203, 379–80
        effect of temperature 199
        polyester synthesis 199
        reversibility 201
        substituent effects 199–201
    of methylene substituted cyclic compounds 199–205
    of 2-methylene-1,3-dioxolanes 200–3
    of 4-methylene-1,3-dioxolanes 202–3
        effect of temperature 202–3
        polyketone synthesis 202–3
        substituent effects 202–3
    of 2-methylenetetrahydrofurans 204
    of 2-methylenetetrahydropyrans 204
    of ring cyclic allyl sulfides 204–5
    of spirodithioacetal rings 108, 206
    of spiroorthocarbonates 206–7
    of spiroorthoesters 206–7
    of vinylcycloalkanes 196–9
        ring size effects 198
    of vinylcyclobutanes 197–8
    of vinylcyclohexane 198
    of vinylcyclopentane 198
    of vinylcyclopropanes 195, 206–7
        rate constant for ring opening 196
        reversibility 196
        substituent effects 196–7
    of vinyloxiranes 197–8
    of vinylsulfones 198–9
    rate constant for ring-opening 195–6
    synthesis of polyamides 199
    synthesis of polyesters 195, 199
    synthesis of polyketones 195, 202–3
    template polymerization 440
    volume expansion 194–5, 205
ruthenium complexes as catalysts for ATRP 488, 489, 492, 495

scandium triflate, effect on RAFT process 519
Schultz-Flory distribution 240
β-scission see fragmentation
secondary radicals
    definition 54
    from peroxides by β-scission 112
segmented block copolymers 547–8
    definition 334
    synthesis
        by NMP 547
        by RAFT 547–8
        mechanisms 547–8
selenides, as initiators 466–7
selenium-centered radical-mediated polymerization 466–7
selenium-centered radicals 132, 522
    phenylseleno radicals 132
    polarity 132
    rate constants 131
self-condensing vinyl polymerization 548, 555–6
    with ATRP initiators 555–6
    with NMP initiators 555–6
    with RAFT agents 555–6
self-initiated initiation see thermal initiation
self-reaction of carbon-centered radicals 11
SFRMP see stable (free) radical-mediated polymerizations
SG1 nitroxide, in NMP 475
short chain branches, by backbiting 320
side reactions
    in ATRP 488–9, 534–5
    in iodine transfer polymerization 522
    in metal complex-mediated radical polymerization 485
    in NMP 478–9, 533
    in radical-induced grafting 390–1
    in RAFT polymerization 517–18
    in sulfur-centered radical-mediated polymerization 466
silicon-centered radicals 131
    polarity 131
    rate constants 131
solvent effects
    see also cage reaction
    on chain transfer, to PVAc 324–5
    on copolymerization 25, 336, 357, 429–32
        bootstrap effect 357, 431–2
        model studies 431
        monomer sequence distribution 357
        of macromonomers 401
        reactivity ratios 361, 429–32
    on fragmentation
        benzoyloxy radicals 127
        cumyloxy radicals 125
        t-butoxy radicals 54, 56, 123
    on hydrogen atom transfer

to alkoxy radicals 34
to *t*-butoxy radicals 55, 123
to chlorine atoms 34
on initiation 55
on initiator decomposition
   alkoxyamines 140
   DBPOX 89
   diacyl peroxides 83
   di-*t*-alkyl peroxides 91
   dialkyl peroxydicarbonates 87
   dialkyldiazenes 73
on initiator efficiency 75
on polymerization 25, 425–9, 497
   t-BMA 428
   EMA 428
   k_p 426–8
   MAA 428–9
   mechanisms 426
   MMA 427–8
   molecular weight 428
   propagation rate constants 25
   S 427
   tacticity 174–175, 428–9
   VAc 179, 324–5, 427–8
on radical addition 25–6, 426–9
   to arenethiyl radicals 26
on radical reactivity
   t-butoxy radicals 56, 123–4
   mechanisms 426
   phosphinyl radicals 132
on radical-radical reactions 42
   combination *vs* disproportionation 43, 255
on radical-radical termination, PMMA 262
on RAFT process 518
SOMO-HOMO orbital interactions 27
SOMO-LUMO orbital interactions 27
spin trapping
   initiation mechanism 134–6
   limitations 135
spiroorthocarbonates, ring-opening polymerization 206–7
spiroorthoesters, ring-opening polymerization 206–7
spontaneous initiation *see* thermal initiation
stable (free) radical-mediated polymerizations (SFRMP) 457–86
   carbon-centered radical-mediated 467–71
   kinetics and mechanism
      initiators, iniferters and initers 457–8
   molecular weight distributions 458–60
      degree of polymerization 458–9
      dispersity 459, 461
   nitrogen-centered radical-mediated 483–4
   oxygen-centered radical-mediated 471–84
      nitroxide-mediated (NMP) 471–83
   polymerization kinetics 460–1
   selenium-centered radical-mediated 466–7
   sulfur-centered radical-mediated 461–6
stable radicals
   see also nitroxides; stable (free) radical-mediated polymerizations (SFRMP)
   as radical traps, for initiation mechanism 138–40
   inhibition by 233–4, 267–8
star polymers 547–57
   star block or radial block copolymer 548, 550
   core crosslinked star microgel 548
   dendritic polymers 548, 556–7
   dispersity of 549
   hyperbranched polymers 555–7
   mikto-arm star 548–9
   synthesis 547–57
      arm-first approach 548, 554–5
      core-first approach 548, 549–54
      self-condensing vinyl polymerization 548, 555–6
State Correlation Diagram (SCD) approach 27–8
statistical copolymerization *see* copolymerization
statistical copolymers *see also* copolymers
   definition 333
Staudinger concept of polymer structure 1–2
stereoelectronic effects
   disproportionation pathways 39
   in combination *vs* disproportionation 41–2
   on cyclopolymerization 186
   on hydrogen atom transfer 32–3
      backbiting 209
   on radical addition 23–4
   on radical cyclization 23–4
   on radical-radical reactions 41–2
   on ring-opening polymerization 196, 200
stereosequence isomerism *see* tacticity
steric effects 19–21
   B-strain 19
   in hydrogen atom transfer 30–1
   in radical reactions 16
   on chain transfer 283
      to halocarbons 294
   on fragmentation of *t*-alkoxy radials 124
   on initiator decomposition, dialkyldiazenes 73
   on radical addition
      head *vs* tail addition 16, 21, 117, 176–7, 178
      intramolecular 187, 189
      polymerization thermodynamics 215–16
   on radical cyclization 23
   on radical-radical reactions 37, 40–1
      combination 257
      combination *vs* disproportionation 254–5, 256, 263
   on rate of radical addition 19–21
steric inhibition 31

stibine-mediated polymerization 499, 524–5
  molecular weight distributions 524
structural irregularities
  control 279, 413–22
  from anomalous propagation 167–8
  from chain transfer 3–4, 287
  from initiation 3–4, 49–50
  from radical-radical termination 176
structural isomerism *see* backbiting; cyclopolymerization; ring-opening polymerization
styrene (S)
  aromatic substitution
    by benzoyloxy radicals 5, 52, 127
    by hydroxy radicals 128
    by isopropoxycarbonyloxy radicals 128
    by phenyl radicals 52, 117
  Diels-Alder reaction 107–9, 251, 317
  Hammett parameters 21–2
  inhibition of VAc polymerization 265–6
  reaction with alkyl radicals
    penultimate unit effects 345, 346
    rate constants 114
  reaction with arenethiyl radicals 132
  reaction with benzoyloxy radicals 5, 52, 120, 127
  reaction with *t*-butoxy radicals 52–3, 120, 134
  reaction with cumyloxy radicals 56, 120, 125
  reaction with cyanoisopropyl radicals 52, 114
  reaction with hydroxy radicals 120
  reaction with isopropoxycarbonyloxy radicals 128
  reaction with methyl radicals 52, 114
  reaction with oxygen-centered radicals 120
  reaction with phenyl radicals 52
    rate constants 114
    solvent effects 26
  reaction with sulfate radical anion 129
  ring-opening copolymerization 195, 199, 379
styrene (S) copolymerization
  Q-e values 365
  reactivity ratios 339
  thermal initiation 110–11
  with AMS 353
  with AN
    bootstrap effect 431
    solvent effects 430
  with BMA, using NMP 527
  with MAA, bootstrap effect 431
  with MAH 526
    bootstrap effect 431
  with MAN
    bootstrap effect 431
    solvent effects 429
  with MMA
    bootstrap effect 431

    ionic liquids effect 433
    Lewis acid effects 435, 436
    solvent effects 429
    using NMP 527
styrene graft polyolefins 399
styrene (S) polymerization
  backbiting 211–12
  biradical mechanism 107
  catalytic chain transfer 311, 316
  chain transfer
    to allyl halides 303
    to allyl sulfides 300
    to allyl sulfonates 302
    to cobalt complexes 316
    to disulfides 292
    to halocarbons 283, 293
    to macromonomers 307
    to monomer 317
    to polymer 320
    to silylcyclohexadienes 309
    to solvent 295
    to thiohydroxamic esters 309
    to thiols 283, 290
    to thionoesters 309
    to vinyl ethers 299
  combination *vs* disproportionation 253–5, 258
  effect of oxygen 269
  enzyme-mediated polymerization 440
  inhibition constants 265
  initiator efficiency in 75–6, 92
  kinetic parameters 219
    k_p 217, 221, 247, 427
      solvent effects 427
    k_t, chain length dependence 246–7
  tacticity 175
  thermal initiation 107–9, 251, 317
    Mayo mechanism 107–8
  thermodynamics 215
  with AIBN initiator 51, 75–7, 116, 141, 416
  with BPO initiator 51, 63, 85, 141, 415–16
  with DBPOX initiator 51
  with disulfide initiators 462–3
  with dithiocarbamate initiators 465
  with DTBP initiator 92
  with NMP 108, 476, 480-1
  with RAFT 520
  with iodine transfer polymerization 522
  with ATRP 496–7
  with stibine-mediated polymerization 524
substituent effects
  fluorine, on alkyl radicals 14
  on C-C bond dissociation energies 23, 34
  on C-H bond dissociation energies 34
  on C-O bond dissociation energies 23, 34
  on ceiling temperature 215
  on chain transfer

to disulfides 292
to thiols 290
on combination *vs* disproportionation 40, 42
  for 2-carboalkoxy-2-propyl radicals 255
  for phenylethyl radicals 253–4
on cyclopolymerization of 1,6-dienes 187–8
on fragmentation of alkoxy radicals 123–5
on hydrogen atom transfer from toluene 22
on k_d
  diacyl peroxides 83
  dialkyldiazenes 73
on monomer reactivity ratios 344–7
on O-H bond dissociation energies 34
on radical addition 20
  to acrylate esters 18
  to halo-olefins 17–18
  to styrene 22
on radical cyclization 23–4
on regiospecificity of radical addition 19
on ring-opening polymerization
  of 2-methylene-1,3-dioxolanes 200–1
  of vinylcyclobutanes 198
  of vinylcyclopropanes 196–7
  of vinyloxiranes 197–8
on stereochemistry of radical addition 20
sulfate radical anion 129–30
  abstraction *vs* addition 129–30
  conversion to hydroxyl radical 130
  effect of pH 129
  hydrolysis to hydroxy radical 130
sulfides, chain transfer to 292–3
sulfonic acids, rate enhancement of NMP 479
sulfonyl halides, as ATRP initiators 488–9, 493, 550, 556–7
  and TERMINI agents 556–7
  side reactions 489
  star precursor 550
  sulfonyl bromides 489
  sulfonyl chlorides 489, 550, 556–7
sulfur-centered radical-mediated polymerization 461–6
  initiators *see* disulfides, dithiocarbamates, dithiuram disulfides, xanthates
  side reactions 466
sulfur-centered radicals 132
  alkanethiyl radicals 132
  arenethiyl radicals 132
  benzoylthiyl radicals 132
  N,N-dialkyl dithiocarbamate radicals 461, 463–4
  polarity 132
  rate constants 131
sulfur dioxide, copolymerization with
  depropagation 353
supercritical carbon dioxide
  as solvent in polymerization 432, 518

for RAFT 518
suspension polymerization
  initiation 63
  termination kinetics 249
  with ATRP 498

tacticity 167
  chain statistics 170–5
    1st order Markov 171–2, 175
    Bernoullian 171, 173, 175
    Coleman-Fox 172
    random 170–1
  control
    ATRP 423
    by topological polymerization 441–2
    effect of amines 428–9
    effect of ionic liquids 433
    effect of Lewis acids 425, 434, 519
    effect of solvent 174–5, 425, 428–9, 433
    effect of steric factors 174
    effect of temperature 174–5
    effect of template polymer 437–440
    NMP 423
    RAFT 423, 519–20
    with cobalt complexes 423–4
    with cyclopolymerization 424
  definition 168–9
  diastereoisomers 169–70
  dyad composition 170–2, 421
    meso 169–70
    racemic 169–70
  effect on backbiting 211
  measurement of 173
  of diene polymers 183
  of 1,1-disubstituted monomers 174–5
  of bulky methacrylate polymers 174, 423–4
  in RAFT 519–20
  of PAA 173
  of PAM 174–5
  of PAN 175
  of PtBMA 428
  of PDMAM 174–5
  of PEMA 428
  of PMA 175, 433
  of PMAA 428–9
  of PMAN 173, 175
  of PMMA 173, 175, 424, 427, 434–5
  of PNIPAM 174–5, 434–5
  of polyacrylamides 174–5, 423–4, 434–5
  of poly(α-alkoxymethacrylates) 434
  of PS 175
  of PVAc 173, 175, 428
  of PVC 173, 175
  of PVF 173
  of vinyl polymers 174
  penultimate unit effects 171–2, 346–7

reviews 168
structure
   atactic chain 169
   heterotactic chain 169
   isotactic chain 168
   syndiotactic chain 169
terminology 168–72
tail addition
   see also head *vs* tail addition
   definition 4, 16
tail-to-tail linkages
   in PVAc 324
   with head-to-head linkages 167, 176
tapered copolymers *see* gradient copolymers
techniques for measurement
   of branching 208–10, 320
   of chain transfer constants 283–7, 321
   of combination *vs* disproportionation 259–60
      end group determination 259
      Gelation technique 258–9
      mass spectrometry 259
      model studies 252–3
      molecular weight distribution 259
      molecular weight measurement 259
   of compositional heterogeneity 381
   of initiation mechanism 133
      chemical methods 144–5
      direct detection of end groups 141–5
      kinetic studies 133
      labeling techniques 145–9
      radical trapping 133–41
      spectroscopic methods 133, 141–4
      spin trapping 134–6
   of k_p 216–18
   of k_t 238
   of monomer sequence distribution 354–7, 362–3
   of reactivity ratios 359–63
   of tacticity 173
telechelic polymers
   definition 374
   synthesis 62, 375–6
      by ATRP 537
      by copolymerizing butadiene or acetylene derivatives 380
      by living radical polymerization 103, 537
      with functional transfer agents 289
telluride-mediated polymerization (TERP) 456, 499, 522–4
   dispersities 523
   initiators for 524
   kinetics 522
   mechanisms
      reversible activation/deactivation 522–3
      reversible chain transfer by homolytic substitution 522–3
   molecular weight distributions 523
   substituent effects 522
telomer synthesis 293, 486
temperature effects
   on backbiting in E polymerization 209
   on chain transfer
      backbiting 209
      C_{tr} 282
      to PVAc 324
      to PVF 325
   on combination *vs* disproportionation, for PS 254, 260–2
   on copolymerization
      ceiling temperature 353–4
   on inhibition, by nitroxides 268
   on initiation 56, 123
   on initiator efficiency 75–6
   on propagation
      backbiting 209, 211–2
      ceiling temperature 213–4
      diene monomers 183–4
      k_p 282
      ring-opening polymerization 196, 200, 202–3
      VAc 179
      VF 181
   on radical addition 24–5
   on radical-radical reactions 42
      combination *vs* disproportionation of *t*-butyl radicals 43
   on RAFT process 518–9
   on tacticity 175
template polymerization 437–40
   and gel or Norrish-Trommsdorff effect 438
   covalently bonded templates 438–40
   definition 437
   enzyme-mediated polymerization 437
   ladder polymerization 194, 439
   mechanisms 437
   non-covalently bonded templates 437–8
   of MMA 437–8
   of multimethacrylate 439–40
   optical induction 439
   ring-opening polymerization 440
TEMPO *see* 2,2,6,6-tetramethylpiperidin-*N*-oxyl
terminal model for copolymerization 337–42, 366
   binary copolymerization 354–5
   chemical control model 366–8
   diffusion control models 368–70
termination 4
   chain transfer constants 283–4
   disproportionation *vs* combination 251–64
   during copolymerization 370–4
   in statistical copolymerization 366–74
   pathways for 233–4

structural irregularities 3
termination kinetics 235
  classical kinetics 235–8
  diffusion controlled termination 242–4
    at low conversion 244–8
    at medium to high conversion 244, 248–9
    at very high conversion 244
  heterogeneous polymerization 249–50
  living radical polymerization 250–1
  molecular weight averages 238–40
  molecular weight distributions 239–42
  molecular weights 238–40
TERMINI agents 556–7
ternary copolymerization *see* terpolymerization
TERP *see* telluride-mediated polymerization
terpolymerization 357–9
  azeotropic compositions 359
  composition equation 357–9
  mechanism 357
  reactivity ratios 357
tetraalkylammonium salts, as ionic liquids 432
tetraalkylphosphonium salts, as ionic liquids 432
tetrachloromethane, as ATRP initiator 488
tetraethyldithiuram disulfide, in SFRMP 461
tetrahydronaphthalene, formation of, in S polymerization 107–8
tetramethylenes, unifying mechanism, donor-acceptor polymerization 111
2,2,6,6-tetramethylpiperidin-*N*-oxyl (TEMPO)
  alkoxyamine formation from AIBN decomposition 476
  alkoxyamine concentration, S polymerization 476
  in block copolymer synthesis 541, 545
  in methacrylate polymerization, unsaturated chain end production 534–5
  in S polymerization 476, 480, 482
  inhibition by 14, 266, 268
  rate enhancement of NMP 479
  reaction with alkyl radicals, rate constants 138
  reaction with benzyl radicals, rate constants 138
  stability 14, 478
  sulfur analog 463
  thermal decomposition of phenylethyl alkoxyamine 478
tetramethylsuccinonitrile, by-product from AIBN 60–1, 77
tetrathiol, thiol-ene polymerization 379
theoretical studies
  on 1,2-chlorine shifts 179–80
  on reactivity ratios in copolymerization 365–6
  radical addition 20–1, 26–8
thermal initiator decomposition
  and effective rate of initiation 57–8
  diacyl peroxides 82–3

dialkyldiazenes 72–4
diaroyl peroxides 82–3
thermal initiation 106–11
  of acrylate ester polymerization 109–10
  of copolymerization 110–1, 351
  of MMA polymerization 109–10
  of pentafluorostyrene polymerization 109
  of S polymerization 107–9
  unified mechanism for 111
thermal stability
  of PMMA 2, 417–20
  of PS 53, 85, 414–7
    prepared by ATRP 416–7
    prepared with AIBN initiator 416
    prepared with BPO initiator 53, 85, 415–6
    prepared by NMP 416–7
    prepared by RAFT 416–7
  of PVC 2, 179, 420–1
thiocarbonylthio RAFT agents 501–2, 505–14
  see also dithioacetate RAFT agents; dithiobenzoate RAFT agents; dithiocarbamates; reversible addition-fragmentation chain transfer (RAFT) polymerization; trithiocarbonate RAFT agents; xanthates
  activity prediction, molecular orbital calculations 507
  bis-RAFT agents 511, 546
  canonical forms 506
  characteristics for efficient RAFT polymerization 505
  end-functional polymer synthesis 539
  relative effectiveness 506
  synthesis of 515–6
  transfer constants 505
thioglycolic acid, as transfer agent 291
thiohydroxamic esters
  chain transfer constants 309
  chain transfer to 308–9
thiol-ene polymerization 378–9
thiols
  as transfer agents, in NMP 531
  chain transfer constants 290
  chain transfer to 289, 290–1, 382–3
  in thiol-ene polymerization 378–9
  polarity 290
  polymeric 388
  synthesis of end-functional polymers 377
  protein conjugates 538
thionoesters
  chain transfer constants 309
  chain transfer to 308–9
thiophenol, radicals from 132
titanocene dichloride
  as initiator 484
  control of propagation 424

titanous salts, radical trapping 134–6
toluene
   chain transfer to 295
   Hammett parameters for abstraction 22
   reaction with t-butoxy radicals 55
topological radical polymerization 441–3
transfer to initiator *see* chain transfer to initiator
transformation reactions 387–9, 544–6
   anionic-radical 387–8
   cationic-radical 389
   group transfer-radical 389
   in synthesis of block and graft copolymers 387–9, 544–6
   to ATRP 544–6
   to NMP 544–5
   to RAFT 544, 546
transition metal salts/complexes
   see also catalytic chain transfer; chromium complexes; cobalt complexes; iron complexes; molybdenum complexes
   catalysts for ATRP 487–9, 492–7
      cobalt complexes 492
      copper complexes 488, 492–5
      iron complexes 492, 496
      manganese complexes 492
      molybdenum complexes 492
      nickel complexes 492, 496–7
      palladium complexes 492
      rhenium complexes 492
      rhodium complexes 492
      ruthenium complexes 488–9, 492, 495
   catalysis for chain transfer to halocarbons 294
   catalysts for catalytic chain transfer
      chromium complexes 315
      cobalt complexes 310-316
      molybdenum complexes 315
      iron complexes 315
   inhibition by 136, 234, 265–6, 273
   mediators for SFRMP
      chromium complexes 484
      cobalt complexes 484-486
      molybdenum complexes 484
      titanium complexes 484
   radical trapping
      cupric salts 136
      ferric salts 136
      mercuric hydride 137
      tin hydrides 137
      titanous salts 136
   reaction with carbon-centered radicals 136
   reaction with dialkyldiazenes 73
   redox initiation 85–6, 93, 95–6, 104–5, 381
      with alkyl hydroperoxides 93
      with diacyl peroxides 85–6
      with hydrogen peroxide 96
      with persulfate 95

transition metal-mediated radical polymerization *see* atom transfer radical polymerization
transition state
   for combination 39
   for disproportionation 39, 41
   for hydrogen atom transfer 30
   for radical addition 17, 20
   for radical cyclization 23
   for radical-radical reaction 39
triarylmethyl radicals 469
   see also triphenylmethyl radical
triazolinyl radicals
   decomposition mechanism 484
   disproportionation 483
   in block copolymer synthesis 483
   in SFRMP 483–4
   stability 484
triblock copolymer synthesis 546–7
   by ATRP 546
   by NMP 546
   by RAFT 546
trichloroacetyl isocyanate, facile reaction with protic end groups 144–5
trichloromethane derivatives, as ATRP initiators 488
trichloromethyl-functional initiators, as ATRP macroinitiators 546
trichloromethyl radicals
   Hammett parameters
      for abstraction from substituted toluenes 22
      for addition to substituted styrenes 22
   polarity 22
   reaction with ethylene 17
   reaction with fluoro-olefins 17–8, 22
triethylamine, chain transfer to 295
trifluoroethyl acrylate polymerization, with oxygen-centered radicals 483
trifluoroethylene
   polymerization, head *vs* tail addition 180–1
   reaction with t-butoxy radicals 17
   reaction with methyl radicals 17
   reaction with trichloromethyl radicals 17
   reaction with trifluoromethyl radicals 17
trifluoromethyl radicals
   polarity 22
   reaction with ethylene 17
   reaction with fluoro-olefins 17–18, 22
   structure 12
2-(trimethylammonium)ethyl methacrylate salts, with ATRP in solution 497
triphenylverdazyl radical
   in SFRMP 483
   in block copolymer synthesis 483
triphenylmethyl radicals
   abstraction *vs* addition 35
   diacyl peroxides decomposition 85

from triphenylmethylazobenzene 468
  in SFRMP 467–9
  inhibition by 268
  pathways for combination 37
  quinonoid intermediate 469
  stability 14
triphenylmethylazobenzene
  as initiator 70–2, 468
  in SFRMP 72, 468
  kinetic data for decomposition 70–1
triphenylverdazyl radical, inhibition by 76, 268
trithiocarbonate RAFT agents 506, 512, 515–6, 519, 548
  see also reversible addition-fragmentation chain transfer (RAFT) polymerization; thiocarbonylthio RAFT agents
  cyclic, in segmented block copolymer synthesis 548
  non-symmetrical 512
  symmetrical 512, 546
  synthesis of 515–6
Trommsdorf effect see Norrish-Trommsdorf effect

undecyl radicals
  Hammett parameters
    for abstraction from substituted toluenes 22
    for addition to substituted styrenes 22
unsaturated chain ends
  from addition-fragmentation chain transfer 296–7, 303, 308
  from catalytic chain transfer 310–2
  from transfer to monomer 311–2, 319
  in PMAN, from disproportionation 257
  in PMMA
    effect on thermal stability 418–20
    from catalytic chain transfer 310–12
    from disproportionation 256, 262, 418
    from initiation 53
    from TEMPO post-polymerization 534–5
  in PS
    effect on thermal stability 415–6
    from benzoate chain ends 415–6
    from disproportionation 253–5, 260–1
UV-visible spectrophotometry 15–16
  from radical-radical termination 251–2
  kinetic studies 133
UV-visible spectroscopy, measurement of end groups 141

verdazyl radicals, inhibition by 76, 268
VAc see vinyl acetate
VC see vinyl chloride
VF see vinyl fluoride
VF2 see vinylidene fluoride
vinyl acetate (VAc)

abstraction vs addition 121
radical addition
  rate constants 114–5, 119, 131
  carbon centered radicals s 114–5
  t-butoxy radicals 122
  heteroatom-centered radicals 131
  oxygen-centered radicals 119, 121
  specificity 121
vinyl acetate (VAc) copolymerization
  Q-e values 365
  reactivity ratios 339
  with tBA 526
  with ethylene 209–10
  solvent effects 429
  with MMA, solvent effects 429
vinyl acetate (VAc) polymerization
  backbiting 209–11
  chain transfer
    intramolecular 209–11
    to allyl sulfides 300
    to allyl sulfonates 302
    to BPO 85
    to disulfides 292
    to halocarbons 283, 293
    to initiator 85
    to monomer 318
    to polymer 209–11, 320, 323–5
    to solvent 294–5
    to thiohydroxamic esters 309
    to thiols 290
    to thionoesters 309
    to vinyl ethers 299
  combination vs disproportionation 263
  effect of oxygen 269
  head vs tail addition 178–9, 522
  inhibition
    by styrene 265–6
    inhibition constants 265
  kinetic parameters 219
  k_p 217
  retardation 294
  solvent effects 25 324–5, 427
  tacticity 173, 175
  thermodynamics 215
  with AIBN initiator 77
  with ATRP 488
    iron catalysts 496
  with iodine transfer polymerization 522
    side reactions 522
  with ^{13}C labeled initiator
  with RAFT 520
  with SFMRP 465, 483, 485–6
  with stibine-mediated polymerization 524
vinyl acrylate, cyclopolymerization 192–3
vinyl chloride (VC)
  radical addition

Index

carbon-centered radicals 114–5
  rate constants 114–5
vinyl chloride (VC) copolymerization
  Q-e values 365
  reactivity ratios 339
vinyl chloride (VC) polymerization
  1,2-chlorine shifts 179–80
  backbiting 208, 211
  chain transfer
    intramolecular 208, 211
    to BPO 85
    to halocarbons 283
    to initiator 85
    to monomer 180, 318–19
    to polymer 208, 211, 320, 325
  combination vs disproportionation 263
  head vs tail addition 179–80
  tacticity 173, 175
  thermodynamics 215
  with ^{13}C labeled AIBN 147
vinyl ethers
  chain transfer to 298–9
    chain transfer constants 299
vinyl fluoride (VF)
  radical addition
    t-butoxy radicals 17
    methyl radicals 17
    trichloromethyl radicals 17
    trifluoromethyl radicals 17
vinyl fluoride (VF) polymerization
  chain transfer to polymer 325
  head vs tail addition 180–1
  tacticity 173
vinyl methacrylate, cyclopolymerization 192–3
N-vinylcarbazole, induced decomposition of BPO 86
vinylcycloalkanes
  ring-opening polymerization 196–9
    ring size effects 198
vinylcyclobutanes, ring-opening polymerization 197–8
vinylcyclohexane, ring-opening polymerization 198
vinylcyclopentane, ring-opening polymerization 198
vinylcyclopropanes
  double ring-opening polymerization 206–7
  ring-opening polymerization 195–7
vinylidene cyanide, copolymerization with S, thermal initiation 110
vinylidene fluoride (VF2)
  polymerization, head vs tail addition 180–1
  radical addition 21–2
    t-butoxy radicals 16–7, 23, 122
    methyl radicals 17
    trichloromethyl radicals 17

trifluoromethyl radicals 17
N-vinylimidazole
  induced decomposition of BPO 86
  template polymerization 438
N-vinyloxazoline, grafting to polyolefins 400
vinyloxiranes, ring-opening polymerization 197–8
vinylpyridine polymerization
  graft of 309
  with ATRP 543
  with NMP 480, 541
  with RAFT 508
N-vinylpyrrolidone (NVP) polymerization
  induced decomposition of BPO 86
  RAFT 507
  stibine-mediated 524
N-vinylpyrrolidone (NVP) copolymerization
  with NMP 526, 528
vinylsilane graft polyolefins 399–400
vinylsulfones
  ring-opening polymerization 198–9
viscosity effects
  dialkyldiazenes 74, 75
  on cage reaction 61
    DBPOX 89
    diacyl peroxides 83, 84
    hyponitrites 78
  on copolymerization of macromonomers 401
  on initiator efficiency 60–1, 75
  on k_d 83
  on radical-radical reactions 43
  on radical-radical termination 238, 242, 248
volume expansion
  in double ring-opening polymerization 205
  in ring-opening polymerization 194–5

xanthates see also reversible addition-fragmentation chain transfer (RAFT) polymerization; thiocarbonylthio RAFT agents
  as alkyl radical source 296, 502–3
  Barton-McCombie reaction 296, 502–3
  as RAFT agents 502–3, 506, 513, 521
    table 513
    monomers polymerized 513
    emulsion polymerization 521
  as photoiniferter 463–5
  MADIX 502
  synthesis 515
xanthogen disulfides
  as initiators 461
  chain transfer to 292

z-mer, in emulsion polymerization 63–4